Gauge/Gravity Duality

Foundations and Applications

Gauge/gravity duality creates new links between quantum theory and gravity. It has led to new concepts in mathematics and physics, and provides new tools for solving problems in many areas of theoretical physics. This book is the first comprehensive textbook on this important topic, enabling graduate students and researchers in string theory and particle, nuclear and condensed matter physics to become acquainted with the subject.

Focusing on the fundamental aspects as well as on applications, this textbook guides readers through a thorough explanation of the central concepts of gauge/gravity duality. For the AdS/CFT correspondence, it explains in detail how string theory provides the conjectured map. Generalisations to less symmetric cases of gauge/gravity duality and their applications are then presented, in particular to finite temperature and density, hydrodynamics, QCD-like theories, the quark–gluon plasma and condensed matter systems. The textbook features a large number of exercises, with solutions available online at www.cambridge.org/9781107010345.

Johanna Erdmenger is a Research Group Leader at the Max Planck Institute for Physics (Werner Heisenberg Institute), Munich, Germany, and Honorary Professor at Ludwig Maximilian University, Munich. She is one of the pioneers of applying gauge/gravity duality to elementary particle, nuclear and condensed matter physics.

Martin Ammon is a Junior Professor at Friedrich Schiller University, Jena, Germany, leading a research group on gauge/gravity duality. He was awarded the prestigious Otto Hahn Medal of the Max Planck Society for his Ph.D. thesis on applying gauge/gravity duality to condensed matter physics.

Gauge/Gravity Duality

Foundations and Applications

MARTIN AMMON

Friedrich Schiller University, Jena

JOHANNA ERDMENGER

Max Planck Institute for Physics, Munich

CAMBRIDGE
UNIVERSITY PRESS

CAMBRIDGE
UNIVERSITY PRESS

University Printing House, Cambridge CB2 8BS, United Kingdom

One Liberty Plaza, 20th Floor, New York, NY 10006, USA

477 Williamstown Road, Port Melbourne, VIC 3207, Australia

4843/24, 2nd Floor, Ansari Road, Daryaganj, Delhi - 110002, India

79 Anson Road, #06-04/06, Singapore 079906

Cambridge University Press is part of the University of Cambridge.

It furthers the University's mission by disseminating knowledge in the pursuit of
education, learning and research at the highest international levels of excellence.

www.cambridge.org
Information on this title: www.cambridge.org/9781107010345

First published 2015

A catalogue record for this publication is available from the British Library

Library of Congress Cataloging in Publication data
Ammon, Martin, 1981-
Gauge-gravity duality : foundations and applications / Martin Ammon,
Friedrich Schiller University, Jena, Johanna Erdmenger,
Max Planck Institute for Physics, Munich.
pages cm.
Includes bibliographical references and index.
ISBN 978-1-107-01034-5 (hardback : alk. paper)
1. Gauge fields (Physics). 2. Gravity. 3. Holography.
4. Mathematical physics. I. Erdmenger, Johanna. II. Title.
QC793.3.G38A56 2015
530.14'35–dc23 2015003328

ISBN 978-1-107-01034-5 Hardback

Contents

Preface

Gauge/gravity duality is a major new development within theoretical physics. It brings together string theory, quantum field theory and general relativity, and has applications to elementary particle, nuclear and condensed matter physics. Gauge/gravity duality is of fundamental importance since it provides new links between quantum theory and gravity which are based on string theory. It has led to both new insights about the structure of string theory and quantum gravity, and new methods and applications in many areas of physics. In a particular case, the duality maps strongly coupled quantum field theories, which are generically hard to describe, to more tractable classical gravity theories. In this way, it provides a wealth of applications to strongly coupled systems. Examples include theories similar to low-energy quantum chromodynamics (QCD), the theory of strong interactions in elementary particle physics, and models for quantum phase transitions relevant in condensed matter systems.

Gauge/gravity duality realises the *holographic principle* and is therefore referred to as *holography*. The holographic principle states that the entire information content of a quantum gravity theory in a given volume can be encoded in an effective theory at the boundary surface of this volume. The theory describing the boundary degrees of freedom thus encodes all information about the bulk degrees of freedom and their dynamics, and vice versa. The holographic principle is of very general nature and is expected to be realised in many examples. In many of these cases, however, the precise form of the boundary theory is unknown, so that it cannot be used to describe the bulk dynamics.

String theory, however, gives rise to a precise realisation of the holographic principle, in which both bulk and boundary theory are known: this is gauge/gravity duality. In this case, a quantum field theory at the boundary, which involves a gauge symmetry, is conjectured to be equivalent to a theory involving gravity in the bulk. Moreover, string theory provides many examples of *dualities*: a physical theory may generically have different equivalent formulations which are referred to as being dual to each other. Two formulations are equivalent if there is a one-to-one map between the states in each of them, and the dynamics are the same. Duality is particularly useful if physical processes are hard to calculate in one formulation, but easy to obtain in another. An example of a duality of this type is a map between two equivalent formulations in different coupling constant regimes. For instance, in a particular limit gauge/gravity duality maps a strongly coupled gauge theory, which generally is hard to describe, to a weakly coupled gravity theory, in which it is much more straightforward to perform explicit calculations.

The most prominent and best understood example of gauge/gravity duality is the *AdS/CFT correspondence*, the celebrated proposal by Maldacena. The AdS/CFT correspondence is characterised by a very high degree of symmetry. Here, 'AdS' stands for

Anti-de Sitter space and 'CFT' for conformal field theory. The field and gravity theories involved in the AdS/CFT correspondence display both supersymmetry and conformal symmetry. These symmetries are realised by the isometries of Anti-de Sitter space and further internal spaces on the one hand, and by the covariance of the quantum fields on the other. The high degree of symmetry allows for very non-trivial tests of the duality conjecture. These have led in particular to an increased understanding of the mathematical properties of $\mathcal{N} = 4$ Super Yang–Mills theory, the four-dimensional superconformal quantum field theory which is the most studied example of the AdS/CFT correspondence.

Motivated by the successes of the AdS/CFT correspondence in its original form, many physicists have begun to ask the question whether the AdS/CFT correspondence can be used to shed new light onto open problems in theoretical physics which are linked to strong coupling. There are many important strongly coupled systems in physics. However, although approaches to describing subsets of their properties exist, there is no general method to calculate their observables which as well established and ubiquitous as perturbation theory is for weakly coupled systems. Consequently, new ideas for describing strongly coupled systems are very welcome, and generalisations of the AdS/CFT correspondence to *gauge/gravity dualities* have made useful contributions to new descriptions of at least some aspects of strongly coupled systems. The best established example is given by the combination of gauge/gravity duality methods with linear response theory, for describing transport processes.

There are many interesting phenomena of strong coupling which have been investigated using gauge/gravity duality. These include the description of theories related to QCD at low energies. The most extensively studied examples are applications to the physics of the quark–gluon plasma, a new strongly coupled state of matter at temperatures above the QCD deconfinement temperature. The quark–gluon plasma has been observed experimentally and continues to be under experimental study, in particular at the RHIC accelerator in Brookhaven and at the LHC at CERN, Geneva. In this context, gauge/gravity duality has contributed the celebrated result for η/s, the ratio of shear viscosity to entropy density, of Kovtun, Son and Starinets, which agrees well with experimental observations. This result provides an example of *universality* in gauge/gravity duality, which means that gravity theories with different structure, dimensionality and field content all give the same result for η/s. On the field theory side, this implies that the precise form of the microscopic degrees of freedom is irrelevant for the dynamics.

More recently, gauge/gravity duality has also been applied to strongly coupled systems in condensed matter physics. In this context, the concept of universality is also of central importance, and is realised for instance near quantum phase transitions. These are phase transitions at zero temperature generated by quantum fluctuations.

Given the central importance of the new research area of gauge/gravity duality, this book aims to introduce a wide audience, including beginning graduate students and researchers from neighbouring areas, to its central ideas and concepts. The book is structured in three parts. The first part covers the prerequisites for explaining the duality. In the second part, the duality is established. The third part is devoted to applications.

To explain the subtle relations provided by gauge/gravity duality, in part I we first present the many ingredients which the duality relates. This involves elements of gauge theory, such as the large N expansion, conformal symmetry and supersymmetry on the one hand, and the geometry and gravity of Anti-de Sitter spaces on the other. Moreover, since the duality is firmly rooted within string theory, we also present an overview of relevant string theory topics.

Part II of the book is devoted to establishing the duality. We explain in detail the motivation for the AdS/CFT correspondence. We state the associated conjecture and give a number of examples of the compelling evidence supporting the conjecture. A particularly important approach is based on the use of integrability. Moreover, the correspondence is generalised to non-conformal examples, and we introduce holographic renormalisation group (RG) flows. We also discuss generalisations to finite temperature, which are obtained by considering a black hole in Anti-de Sitter space.

In part III, applications of gauge/gravity duality are presented. As examples, we consider holographic hydrodynamics, as relevant in particular to applications to the quark–gluon plasma. We also consider applications to theories similar to low-energy QCD. Finally, we present applications to systems of relevance in condensed matter physics, such as quantum phase transitions, superfluids and superconductors as well as Fermi surfaces. We conclude with a discussion of holographic entanglement entropy.

Let us give a more detailed guide to these three parts. Part I contains four chapters reviewing the relevant aspects of quantum field theory, general relativity, symmetries such as conformal and supersymmetry, and string theory, respectively. Part I is intended primarily for graduate students. However, experienced readers may use it as a glossary of concepts used in parts II and III. Moreover, researchers interested in applications of gauge/gravity duality may find it useful to read chapter 4, which contains a short summary of string theory and supergravity as relevant for understanding the string theory origin of gauge/gravity duality.

In part II, the AdS/CFT correspondence is stated and non-trivial tests as well as extensions of the AdS/CFT correspondence are presented. The key chapter is chapter 5 in which the AdS/CFT correspondence is motivated within string theory, considering in particular the near-horizon limit of D3-branes. Moreover, the field-operator map is established and the important concept of holographic renormalisation is introduced. Also, an explanation of how to realise Wilson loops in AdS/CFT is given. Chapter 6 contains non-trivial tests of the AdS/CFT correspondence, such as the calculation of correlation functions and of the conformal anomaly. In chapter 7, aspects of integrability and scattering amplitudes are introduced, providing further tests, as well as further elucidating string theory aspects of the correspondence. In chapter 8, further examples of the AdS/CFT correspondence are presented, such as AdS/CFT for branes at singularities and for M2-branes. Moreover, as a first step towards generalising the correspondence, we consider examples of the duality in which conformal symmetry is broken. In chapter 9 we discuss holographic renormalisation group (RG) flows. We consider simple cases of flows linking a UV to an IR fixed point, as well as explicit realisations of RG flows within IIB supergravity. In chapter 10 we describe models with additional branes in supergravity. In particular, we consider flavour branes, which provide descriptions of particles with similar properties to quarks and electrons within gauge/gravity duality. In chapter 11, we formulate the correspondence at finite temperature in Lorentzian signature and explain how to obtain a causal structure which allows us to introduce retarded Green's functions.

Readers interested primarily in applications may omit chapters 7, 8 and the second half of chapter 9 (RG flows within IIB supergravity) at first reading. Readers interested primarily in foundations are encouraged to concentrate on all chapters of part II, including chapter 11 on finite temperature. Within part II, there are a few sections denoted by an asterisk *. These provide material at a more advanced level and are not a prerequisite for reading the subsequent sections and chapters.

Part III, devoted to applications, is organised as follows. In chapter 12 we introduce the linear response formalism and hydrodynamics and explain how both are implemented in gauge/gravity duality. This provides the tools for calculating transport coefficients. As an important example, we consider the shear viscosity over entropy ratio. We also discuss the important concept of quasinormal modes and their relation to the pole structure of Green's functions. In chapters 13 and 14 we introduce aspects of applications of gauge/gravity duality to theories related to QCD. Chapter 13 is devoted in particular to confinement, chiral symmetry breaking and light mesons. Chapter 14 deals with applications to QCD-like theories at finite temperature and density. In chapter 15, we introduce applications of gauge/gravity duality to systems of relevance in condensed matter physics. We review the concept of quantum phase transitions, calculate conductivities, introduce holographic superconductors, review the electron star and hyperscaling models and give an introduction to the gauge/gravity duality approach to entanglement entropy.

There are three appendices, on Grassmann numbers (appendix A), on Lie algebras, superalgebras and their representations (appendix B), and an appendix summarising our conventions (appendix C). Appendix B contains important information on group theory which is essential for establishing the field-operator map for the AdS/CFT correspondence.

We have chosen to list the relevant references at the end of each chapter. Each of these reference lists is preceded by a 'Further reading' section, which briefly describes the references used in preparing the text. Moreover, an outlook on further relevant literature is given.

There are exercises given in the text which are intended to help the reader become acquainted with the standard tools and methods of gauge/gravity duality.

Gauge/gravity duality is a fast growing area of research with a wealth of different aspects. This implies that certain topics had to be selected for inclusion in this book. Our main guiding principle is to provide a textbook style introduction to the subject. This implies that there is an extensive introduction, and examples of generalisations and applications. Our choice of generalisations and applications is influenced by our own research experience and interests. We hope that after studying this book, readers will be able to read and understand original research papers on many other exciting aspects of gauge/gravity duality, and to become involved with research in this fascinating area themselves.

Johanna Erdmenger and Martin Ammon

Acknowledgements

We are indebted to a large number of colleagues for joint work in gauge/gravity duality. In particular, we would like to thank our past and present collaborators for numerous discussions and inspiring thoughts:

Riccardo Apreda, Mario Araújo, Daniel Arean, James Babington, Nicolas Boulanger, Yan-Yan Bu, Alejandra Castro, Neil Constable, Nick Evans, Eric D'Hoker, Viviane Graß, Johannes Große, Michael Gutperle, Daniel Fernández, Veselin Filev, Mario Flory, Dan Freedman, Kazuo Ghoroku, Zachary Guralnik, Sebastian Halter, Michael Haack, Benedikt Herwerth, Carlos Hoyos, Nabil Iqbal, Matthias Kaminski, Patrick Kerner, Ingo Kirsch, Steffen Klug, Per Kraus, Karl Landsteiner, Shu Lin, Dieter Lüst, René Meyer, Thanh Hai Ngo, Carlos Núñez, Andy O'Bannon, Hugh Osborn, Da-Wei Pang, Jeong-Hyuck Park, Manolo Pérez-Victoria, Eric Perlmutter, Felix Rust, Robert Schmidt, Jonathan Shock, Christoph Sieg, Charlotte Sleight, Corneliu Sochichiu, Stephan Steinfurt, Migael Strydom, Gianmassimo Tasinato, Derek Teaney, Timm Wrase, Jackson Wu, Amos Yarom and Hansjörg Zeller.

Moreover we would like to thank the students who attended our lecture and examples class 'Introduction to gauge/gravity duality' at LMU Munich for participation and useful questions, in particular Yegor Korovin, Mario Flory, Alexander Gussmann and Tehseen Rug. Special thanks go to Oliver Schlotterer for typesetting the lecture notes for this course. These provided the starting point for this book.

Many people have contributed to the completion of this book. We are grateful to Frank Dohrmann, Benedikt Herwerth, Patrick Kerner, Felix Rust, Jonathan Shock and Migael Strydom for help with figures. We are particularly grateful to Biagio Lucini for providing figure 13.8. We also would like to thank Nick Evans, Livia Ferro, Mario Flory, Felix Karbstein, Andreas Karch, Karl Landsteiner, Javier Lizana, Julian Leiber, Johanna Mader, Sebastian Möckel, Andy O'Bannon, Hugh Osborn, Da-Wei Pang, Tehseen Rug, Charlotte Sleight, Stephan Steinfurt, Ann-Kathrin Straub, Migael Strydom and Hansjörg Zeller, as well as Matthias Kaminski, Steffen Klug, René Meyer, Birger Böning, Markus Gardemann, Sebastian Grieninger, Stefan Lippold, Attila Lüttmerding and Tim Nitzsche for proofreading the manuscript and very useful comments. Moreover, we thank Charlotte Sleight and Migael Strydom for help with compiling the index.

Finally, we are grateful to our families for moral support while writing this book, and for their interest in our work throughout our lives.

PART I

PREREQUISITES

1 Elements of field theory

In this chapter we review some elements of quantum field theory which are essential in the study of gauge/gravity duality. A full development of quantum field theory is clearly beyond the scope of this book. We refer the reader to the many excellent textbooks on the subject, some of which are listed in the further reading section at the end of this chapter.

For simplicity, we will restrict ourselves to scalar fields in flat spacetime in the first part of the chapter and explain the most important concepts. We begin by introducing symmetries and conserved currents in the classical theory. A particularly important conserved quantity is the energy-momentum tensor which plays a central role in tests of the AdS/CFT correspondence. We discuss its derivation and its properties in detail. We then move on to the quantisation of field theories, beginning with the quantisation of the free scalar field. We review the definitions and concepts of generating functionals, of correlation functions and of the Feynman propagator. We then move on to interacting fields and discuss perturbation theory. Next we consider fermions as well as Abelian and non-Abelian gauge theories, both classically and in the quantised case. We discuss the energy-momentum tensor for classical gauge theories, as well as quantisation involving Faddeev–Popov ghost fields. An approximation of significance for gauge/gravity duality is the large N limit of non-Abelian gauge theories. Moreover, we discuss Ward identities and anomalies, which provide important examples of checks of the AdS/CFT correspondence later on.

1.1 Classical scalar field theory

Let us begin by introducing a real scalar field in flat d-dimensional Minkowski spacetime $\mathbb{R}^{d-1,1}$, with $d-1$ spatial directions. The points of the Minkowski spacetime are denoted by x with components x^{μ}, where μ runs from 0 to $d-1$. While $x^0 = ct$ is the time, x^i with $i = 1, \ldots, d-1$ are the spatial directions. In the following we set the speed of light c to one, $c = 1$, thus using the same units of measure for space and time. Sometimes it is also convenient to collect all the spatial components into a $(d-1)$-dimensional vector $\vec{x}$.

Minkowski spacetime is equipped with a metric. The infinitesimal length ds of a spacetime interval dx is given by

$$(ds)^2 \equiv ds^2 = -\left(dx^0\right)^2 + \sum_{i=1}^{d-1}\left(dx^i\right)^2 \equiv \eta_{\mu\nu}dx^{\mu}dx^{\nu}. \tag{1.1}$$

By definition, $\eta_{\mu\nu}$ is thus a diagonal matrix of the form

$$\eta_{\mu\nu} = \mathrm{diag}(-1, \underbrace{1, \ldots, 1}_{(d-1)\ \mathrm{times}}). \tag{1.2}$$

Using $\eta_{\mu\nu}$ or $\eta^{\mu\nu}$, which is the inverse of $\eta_{\mu\nu}$ satisfying $\eta^{\mu\nu}\eta_{\nu\sigma} = \delta^\mu_\sigma$, we may raise and lower the indices of x^μ, for example $x_\mu = \eta_{\mu\nu}x^\nu$. Equation (1.1) implies that $\mathrm{d}s^2$ can also be negative and therefore we do not have a metric in the strict mathematical sense. If $\mathrm{d}s^2 < 0$, the spacetime interval $\mathrm{d}x$ is *timelike*. For $\mathrm{d}s^2 = 0$ or $\mathrm{d}s^2 > 0$ the spacetime interval $\mathrm{d}x$ is *lightlike* or *spacelike*, respectively.

Let us now consider those transformations Λ of spacetime points $x \overset{\Lambda}{\mapsto} x'$ which leave $\mathrm{d}s^2$ invariant, i.e. for which

$$\eta_{\mu\nu}\mathrm{d}x^\mu\mathrm{d}x^\nu = \eta_{\mu\nu}\mathrm{d}x'^\mu\mathrm{d}x'^\nu. \tag{1.3}$$

It is easy to check that all transformations which satisfy equation (1.3) can be decomposed into translations of x by a constant vector a (with components a^μ), and into *Lorentz transformations* Λ given by the matrix components $\Lambda^\mu{}_\nu$ obeying

$$\Lambda^\mu{}_\rho\Lambda^\nu{}_\sigma\eta_{\mu\nu} = \eta_{\rho\sigma}. \tag{1.4}$$

For example, rotations in the spatial directions and boosts along a spatial direction are examples of Lorentz transformations. The Lorentz transformations Λ form a group, the Lorentz group $SO(d-1, 1)$.

Both transformations, translations by a constant vector a and Lorentz transformations Λ, form a group, the *Poincaré* group $ISO(d-1, 1)$, consisting of pairs (Λ, a) which act on spacetime as

$$x \mapsto x' = \Lambda x + a, \tag{1.5}$$

or in components $x'^\mu = \Lambda^\mu{}_\nu x^\nu + a^\mu$. The group multiplication of two such operations (Λ_1, a_1) and (Λ_2, a_2) is given by

$$(\Lambda_1, a_1) \circ (\Lambda_2, a_2) = (\Lambda_1\Lambda_2, a_1 + \Lambda_1 a_2) \tag{1.6}$$

and is again in $ISO(d-1, 1)$.

As an example of a field theory, we consider real scalar fields in d-dimensional Minkowski space. A real scalar field ϕ is a map which assigns a real number $\phi(x)$ to each spacetime point x. Under a Lorentz transformation $x \mapsto x' = \Lambda x$ the scalar field transforms as $\phi \mapsto \phi'$ where $\phi'(x') = \phi(x)$, or in terms of an active transformation $\phi'(x) = \phi(\Lambda^{-1}x)$. The dynamics of the scalar field is specified by an action functional $\mathcal{S}[\phi]$ which can be written as a spacetime integral of the *Lagrangian density* $\mathcal{L}(\phi, \partial_\mu\phi)$, or *Lagrangian* for short,

$$\mathcal{S}[\phi] = \int \mathrm{d}t\,\mathrm{d}^{d-1}\vec{x}\,\mathcal{L}(\phi, \partial_\mu\phi) \equiv \int \mathrm{d}^d x\,\mathcal{L}(\phi, \partial_\mu\phi). \tag{1.7}$$

The Lagrangian $\mathcal{L}$, and therefore also the action $\mathcal{S}$, depends on ϕ as well as its derivatives $\partial_\mu\phi$. For the partial derivative we use the shorthand notation $\partial_\mu \equiv \partial/\partial x^\mu$. We follow the usual approach to allow only first derivatives in the action functional and not second or higher derivatives of the scalar field. Moreover, we only consider local terms in the

Lagrangian, which means that terms of the form $\phi(x)\phi(x + a)$, where a is a spacetime vector, are not used. In order to formulate a scalar field theory which is invariant under Poincaré transformations, the action functional can only depend on ϕ, as well as on

$$- (\partial_t\phi(t,\vec{x}))^2 + (\nabla\phi(t,\vec{x}))^2 \equiv \eta^{\mu\nu}\partial_\mu\phi(x)\partial_\nu\phi(x). \tag{1.8}$$

The simplest example is the free scalar field theory given by the Lagrangian $\mathcal{L}_{\text{free}}$,

$$S[\phi] = \int d^dx\,\mathcal{L}_{\text{free}} = -\frac{1}{2}\int d^dx\left(-(\partial_t\phi(t,\vec{x}))^2 + (\nabla\phi(t,\vec{x}))^2 + m^2\phi(t,\vec{x})^2\right)$$

$$= -\frac{1}{2}\int d^dx\left(\eta^{\mu\nu}\partial_\mu\phi(x)\partial_\nu\phi(x) + m^2\phi(x)^2\right). \tag{1.9}$$

The parameter m in the Lagrangian $\mathcal{L}_{\text{free}}$ is the mass of the scalar field ϕ. Varying the action S as given by (1.7) with respect to ϕ we obtain

$$\frac{\delta S}{\delta \phi} = \frac{\partial \mathcal{L}}{\partial \phi} - \partial_\mu\left(\frac{\partial \mathcal{L}}{\partial(\partial_\mu\phi)}\right). \tag{1.10}$$

As usual, the classical equation of motion corresponding to the action S is determined by the principle of least action, $\delta S/\delta\phi = 0$, and reads

$$\partial_\mu\left(\frac{\partial \mathcal{L}}{\partial(\partial_\mu\phi)}\right) = \frac{\partial \mathcal{L}}{\partial \phi}. \tag{1.11}$$

For the free field Lagrangian

$$\mathcal{L}_{\text{free}}(\phi, \partial_\mu\phi) = -\frac{1}{2}\eta^{\mu\nu}\partial_\mu\phi(x)\partial_\nu\phi(x) - \frac{1}{2}m^2\phi(x)^2 \tag{1.12}$$

the equation of motion (1.11) simplifies to

$$(\Box - m^2)\phi(x) = 0, \tag{1.13}$$

where $\Box = \partial^\mu\partial_\mu = -\partial_t^2 + \nabla^2$ is the *D'Alembert operator*. Equation (1.13) is known as the *Klein–Gordon equation*.

It is possible to add interaction terms to the free field Lagrangian $\mathcal{L}_{\text{free}}$, which are summarised in the *interaction Lagrangian* $\mathcal{L}_{\text{int}}$. Typically $\mathcal{L}_{\text{int}}$ is a polynomial of the field ϕ, for example

$$\mathcal{L}_{\text{int}}(\phi) = -\frac{g_n}{n!}\phi(x)^n, \tag{1.14}$$

where $n \geq 3, n \in \mathbb{N}$. The constant $g_n \in \mathbb{R}$ controls the strength of the interaction and is therefore referred to as the coupling constant.

Exercise 1.1.1   Show that the equations of motion (1.13) of a free scalar field are satisfied by

$$\phi(x) = \frac{1}{(2\pi)^{d-1}}\int\frac{d^{d-1}\vec{k}}{2\omega_k}\left[a(\vec{k})e^{-ikx} + a^*(\vec{k})e^{ikx}\right]\Big|_{k^0=\omega_k}, \tag{1.15}$$

where $\omega_k = (\vec{k}\cdot\vec{k} + m^2)^{1/2}$ and $kx = -k^0x^0 + \vec{k}\vec{x}$.

Exercise 1.1.2   Derive the equations of motion for a scalar field with mass m and the interaction Lagrangian $\mathcal{L}_{\text{int}} = -\frac{g}{4!}\phi(x)^4$.

Exercise 1.1.3   Consider two non-interacting real scalar fields ϕ_1 and ϕ_2 with common mass m. Show that the Lagrangian can be written in terms of the *complex scalar field* $\phi = 1/\sqrt{2}\,(\phi_1 + i\phi_2)$ and its complex conjugate, $\phi^* = 1/\sqrt{2}\,(\phi_1 - i\phi_2)$ in the form

$$\mathcal{L}_{\text{free}}(\phi, \partial\phi) = -\partial_\mu\phi^*\partial^\mu\phi - m^2\phi^*\phi. \tag{1.16}$$

Derive the equations of motion for ϕ and ϕ^* assuming that ϕ and ϕ^* are independent fields. Are the equations of motion consistent with those for ϕ_1 and ϕ_2?

1.2 Symmetries and conserved currents

Symmetries are essential within field theory, and also play an essential role in the AdS/CFT correspondence. Let us first review the role of symmetries within classical field theory. One of the fundamental ingredients of theoretical physics is the intimate relation between continuous symmetries and conserved charges, as expressed in *Noether's theorem*. According to this theorem, a continuous symmetry gives rise to a conserved current which we now determine.

Let us assume that the action $\mathcal{S}[\phi]$ is invariant under the transformation

$$\phi(x) \;\mapsto\; \tilde{\phi}(x) \;=\; \phi(x) + \alpha\,\delta\phi(x), \tag{1.17}$$

where α denotes an arbitrary infinitesimal parameter associated with some deformation $\delta\phi$. The invariance of the action,

$$\mathcal{S}[\phi] = \mathcal{S}[\tilde{\phi}], \tag{1.18}$$

is ensured if the Lagrangian is also invariant under this deformation, up to a total derivative of some vector field X^μ,

$$\mathcal{L}\big(\tilde{\phi}, \partial_\mu\tilde{\phi}\big) \;=\; \mathcal{L}\big(\phi, \partial_\mu\phi\big) + \alpha\,\partial_\nu X^\nu \tag{1.19}$$

implying

$$
\begin{aligned}
\alpha\,\partial_\nu X^\nu &\stackrel{!}{=} \mathcal{L}\big(\tilde{\phi}, \partial_\mu\tilde{\phi}\big) - \mathcal{L}\big(\phi, \partial_\mu\phi\big) = \mathcal{L}\big(\phi + \alpha\delta\phi, \partial_\mu\phi + \alpha\partial_\mu\delta\phi\big) - \mathcal{L}\big(\phi, \partial_\mu\phi\big) \\
&= \alpha\left\{\frac{\partial\mathcal{L}}{\partial\phi}\,\delta\phi + \frac{\partial\mathcal{L}}{\partial(\partial_\mu\phi)}\,\partial_\mu\delta\phi\right\} + \mathcal{O}(\alpha^2) \\
&= \alpha\left\{\underbrace{\left(\frac{\partial\mathcal{L}}{\partial\phi} - \partial_\mu\left(\frac{\partial\mathcal{L}}{\partial(\partial_\mu\phi)}\right)\right)}_{=0\ \text{by}\ ((1.11))}\,\delta\phi + \partial_\mu\left(\frac{\partial\mathcal{L}}{\partial(\partial_\mu\phi)}\,\delta\phi\right)\right\} + \mathcal{O}(\alpha^2) \tag{1.20}
\end{aligned}
$$

or equivalently

$$0 \stackrel{!}{=} -\alpha\,\partial_\mu\left(\frac{\partial\mathcal{L}}{\partial(\partial_\mu\phi)}\,\delta\phi\right) + \alpha\,\partial_\mu X^\mu \;=\; \alpha\,\partial_\mu\left(-\frac{\partial\mathcal{L}}{\partial(\partial_\mu\phi)}\,\delta\phi + X^\mu\right). \tag{1.21}$$

This identifies a *conserved current* $\mathcal{J}^\mu$ associated with the *symmetry transformation* $\delta\phi$ of the field ϕ,

$$\mathcal{J}^\mu \;=\; -\frac{\partial\mathcal{L}}{\partial(\partial_\mu\phi)}\,\delta\phi + X^\mu, \qquad \partial_\mu\mathcal{J}^\mu \;=\; 0. \tag{1.22}$$

Due to the conserved current $\mathcal{J}$, we may define an associated charge $\mathcal{Q}$, the *Noether charge*, by integration of the temporal component of $\mathcal{J}$, denoted by $\mathcal{J}^t$, over the spatial directions (given by $\mathbb{R}^{d-1}$) for a fixed value of time,

$$Q = \int_{\mathbb{R}^{d-1}} \mathrm{d}^{d-1}\vec{x}\, \mathcal{J}^t. \tag{1.23}$$

Exercise 1.2.1   By using Gauss' law, show that Q is time independent.

Let us discuss a few explicit examples of symmetries and associated Noether charges. Since the action S is invariant under Poincaré transformations by construction, we first construct the conserved current associated with spacetime translations of the form $x^\mu \mapsto x'^\mu = x^\mu + a^\mu$. Such transformations can be described alternatively as transformations of the field configuration

$$\phi(x) \mapsto \tilde{\phi}(x) = \phi(x-a) = \phi(x) - a^\mu \partial_\mu \phi(x) + \mathcal{O}(a^2), \tag{1.24}$$

under which the Lagrangian transforms as

$$\mathcal{L} \mapsto \tilde{\mathcal{L}} = \mathcal{L} - a^\nu \partial_\mu(\delta^\mu_\nu \mathcal{L}) + \mathcal{O}(a^2). \tag{1.25}$$

Let us now apply the Noether theorem with $\delta\phi = -a^\nu \partial_\nu \phi$ and $X^\mu = -\delta^\mu_\nu a^\nu \mathcal{L}$. We obtain a conserved current $\mathcal{J}^\mu = -a^\nu \Theta^\mu_\nu$, where

$$\Theta^\mu_\nu = -\frac{\partial \mathcal{L}}{\partial(\partial_\mu \phi)} \partial_\nu \phi + \mathcal{L}\delta^\mu_\nu. \tag{1.26}$$

Note that $\Theta_{\mu\nu}$ is not manifestly symmetric by construction. However, if the Lagrangian takes the form $\mathcal{L} = \mathcal{L}_{\text{free}} + \mathcal{L}_{\text{int}}$, with $\mathcal{L}_{\text{free}}$ given by (1.12) and $\mathcal{L}_{\text{int}}$ independent of $\partial_\mu \phi$, then $\Theta_{\mu\nu}$ is given by

$$\Theta_{\mu\nu} = \partial_\mu \phi \partial_\nu \phi + \eta_{\mu\nu}\mathcal{L} \tag{1.27}$$

and it turns out that Θ is symmetric, $\Theta_{\mu\nu} = \Theta_{\nu\mu}$. The associated conserved Noether charges are given by

$$H \equiv \int_{\mathbb{R}^{d-1}} \mathrm{d}^{d-1}\vec{x}\, \mathcal{H} = \int_{\mathbb{R}^{d-1}} \mathrm{d}^{d-1}\vec{x}\, \Theta^{tt} = \int_{\mathbb{R}^{d-1}} \mathrm{d}^{d-1}\vec{x}\, (\Pi\, \partial_t \phi - \mathcal{L}) \tag{1.28}$$

for time translations as well as

$$P^i = \int_{\mathbb{R}^{d-1}} \mathrm{d}^{d-1}\vec{x}\, \Theta^{ti} = -\int_{\mathbb{R}^{d-1}} \mathrm{d}^{d-1}\vec{x}\, \Pi\, \partial^i \phi \tag{1.29}$$

for space translations. H is the *Hamiltonian* and $\mathcal{H}$ the *Hamiltonian density*. Moreover, we have introduced the canonical momentum density $\Pi(t,\vec{x})$ conjugate to the field $\phi(t,\vec{x})$

$$\Pi(t,\vec{x}) = \frac{\partial \mathcal{L}}{\partial(\partial_t \phi(t,\vec{x}))}. \tag{1.30}$$

Furthermore, P_i is the *physical momentum* of the field ϕ. Equations (1.28) and (1.29) imply that the conserved current $\Theta_{\mu\nu}$ as given by equation (1.26) is the *energy-momentum tensor*.

Box 1.1   Energy-momentum tensor in general relativity

The energy-momentum tensor $T_{\mu\nu}$ is a key ingredient in general relativity since it determines the curvature of space by entering the Einstein equation. In section 2.2 we will introduce a second way of calculating $T_{\mu\nu}$ which by construction makes sure that $T_{\mu\nu}$ is symmetric in μ and ν.

Exercise 1.2.2   Show that for a free real scalar field ϕ with mass m, the Hamiltonian density is given by

$$\mathcal{H} = \frac{1}{2}\Pi^2 + \frac{1}{2}(\nabla\phi)^2 + \frac{1}{2}m^2\phi^2. \tag{1.31}$$

Instead of translations in space and time we can consider Lorentz transformations which are also a symmetry of the action. Under an infinitesimal Lorentz transformation, $\Lambda^\mu{}_\nu = \delta^\mu{}_\nu + \omega^\mu{}_\nu$ with $\omega_{\mu\nu} = -\omega_{\nu\mu}$, the scalar field $\phi(x)$ transforms as $\phi(x^\mu) \mapsto \tilde{\phi}(x^\mu) = \phi(x^\mu - \omega^\mu{}_\rho x^\rho)$, i.e. with an x-dependent translation parameter $a^\mu = \omega^\mu{}_\rho x^\rho$. Using the same methods as above we conclude that

$$N^{\mu\nu\rho} = x^\nu \Theta^{\mu\rho} - x^\rho \Theta^{\mu\nu} \tag{1.32}$$

is conserved, i.e. $\partial_\mu N^{\mu\nu\rho} = 0$, and that the associated Noether charge is

$$M^{\nu\rho} = \int_{\mathbb{R}^{d-1}} \mathrm{d}^{d-1}\vec{x}\, N^{t\nu\rho}(x). \tag{1.33}$$

Exercise 1.2.3   Use the conservation laws of $N^{\mu\nu\rho}$ and $\Theta^{\mu\nu}$ to show that any Poincaré invariant field theory has to have a symmetric energy-momentum tensor.

Note that the energy-momentum tensor $\Theta_{\mu\nu}$ as defined by (1.26) is not necessarily symmetric by construction. For the Lagrangian $\mathcal{L}_{\text{free}} + \mathcal{L}_{\text{int}}$ given by (1.12) and (1.14), $\Theta_{\mu\nu}$ is a symmetric tensor. Later we will see examples where the energy-momentum tensor as defined by (1.26) is not symmetric but is still conserved. However, note that we may add a term of the form $\partial_\lambda f^{\lambda\mu\nu}$ to $\Theta^{\mu\nu}$, with $f^{\lambda\mu\nu} = -f^{\mu\lambda\nu}$ antisymmetric in its first two indices, without spoiling the conservation laws. Due to the statement of exercise 1.2.3, there has to be a clever choice of $f^{\lambda\mu\nu}$ such that the tensor $T^{\mu\nu} = \Theta^{\mu\nu} + \partial_\lambda f^{\lambda\mu\nu}$ is still conserved but is also symmetric. $T^{\mu\nu}$ is the *Belinfante* or *canonical energy-momentum tensor*. Moreover, if we replace $\Theta^{\mu\nu}$ by $T^{\mu\nu}$ in (1.32) then $N^{\mu\nu\rho}$ is still conserved.

Exercise 1.2.4   For the massless free scalar field, we can refine the energy-momentum tensor even further to impose tracelessness in addition to conservation and index symmetry. In particular, show that the modified energy-momentum tensor given by

$$T_{\mu\nu} = \partial_\mu\phi\partial_\nu\phi - \frac{1}{2}\eta_{\mu\nu}\partial_\rho\phi\partial^\rho\phi - \frac{d-2}{4(d-1)}\left(\partial_\mu\partial_\nu - \eta_{\mu\nu}\Box\right)\phi^2 \tag{1.34}$$

is symmetric, conserved and traceless, i.e. $T^\mu{}_\mu = \eta^{\mu\nu}T_{\mu\nu} = 0$, if we use the equations of motion of ϕ. Equation (1.34) is referred to as an *improved* energy-momentum tensor. In chapter 2 we will see that the last term in (1.34) is generated by coupling the scalar field to the Ricci scalar in a particular way referred to as *conformal*. The consequences of tracelessness of the energy-momentum tensor will be explored in chapter 3 when discussing conformal field theories.

In addition to spacetime symmetries, a further interesting type of symmetries is *internal symmetries*. For example, consider a complex scalar field as discussed in exercise 1.1.3. The Lagrangian (1.16) is invariant under the $U(1)$ transformation

$$\phi(x) \mapsto \phi(x)' = e^{i\alpha}\phi(x), \qquad \phi^*(x) \mapsto \phi^*(x)' = e^{-i\alpha}\phi^*(x). \qquad (1.35)$$

This is an example of an internal symmetry. Since the parameter α is not spacetime dependent, the transformation is *global*.

Exercise 1.2.5 Determine the Noether currents associated with the global $U(1)$ transformation (1.35).

Exercise 1.2.6 Consider n free, real (or complex) fields ϕ^j with $j = 1, \ldots, n$ numbering the different fields. We assume the fields to be of the same mass, i.e. $m_j = m$. Determine the action and show that it is invariant under the transformation $\phi'^j(x) = R^j{}_k \phi^k(x)$ where $R^j{}_k$ are the components of a matrix R. In particular show that in the case of real scalar fields $R \in O(n)$, while for complex scalar fields $R \in O(2n) \supseteq U(n)$.[1]

1.3 Quantisation

Let us now quantise the classical scalar field theory using two different approaches: *canonical quantisation* and *path integral quantisation*. For canonical quantisation, the classical fields are promoted to operator valued quantum fields. On the other hand, the idea of path integral quantisation is to sum over all possible field configurations. Both approaches are discussed for free fields in 1.3.1 and 1.3.2.

In 1.3.3 we discuss interacting field theories. Particle scattering processes may be related to *correlation functions* of quantised fields which can be deduced from a *generating functional*. For quantisation of interacting fields, the approach that is best understood is *perturbation theory* which requires the couplings to be small. This implies that the majority of our current understanding of physical systems described by quantum field theories refers to weak coupling.

1.3.1 Canonical quantisation of free fields

We consider a massive real scalar field with equation of motion

$$(-\Box + m^2)\phi = 0. \qquad (1.36)$$

We already discussed its solution in exercise 1.1.1 in terms of modes $a(\vec{k})$ and $a^*(\vec{k})$. The starting point of quantising the real scalar field is to promote these modes to operators $\hat{a}(\vec{k})$

---

[1] For generic interactions of complex scalar fields, the symmetry $O(2n)$ is typically broken down to $U(n)$ or even further.

and $\hat{a}^\dagger(\vec{k})$. The field $\phi(x)$ is then also operator valued and therefore denoted by $\hat{\phi}(x)$,

$$\hat{\phi}(x) = \frac{1}{(2\pi)^{d-1}} \int \frac{d^{d-1}\vec{k}}{2\omega_k} \left[\hat{a}(\vec{k})e^{-ikx} + \hat{a}^\dagger(\vec{k})e^{ikx} \right]\Big|_{k^0 = \omega_k}, \tag{1.37}$$

with $\omega_k = (\vec{k} \cdot \vec{k} + m^2)^{1/2}$. The operators $a(\vec{k})$ and $a^\dagger(\vec{k})$ satisfy the commutation relations

$$[\hat{a}(\vec{k}), \hat{a}^\dagger(\vec{k}')] = 2\omega_k (2\pi)^{d-1} \delta^{d-1}(\vec{k} - \vec{k}'), \quad [\hat{a}(\vec{k}), \hat{a}(\vec{k}')] = [\hat{a}^\dagger(\vec{k}), \hat{a}^\dagger(\vec{k}')] = 0. \tag{1.38}$$

Exercise 1.3.1   Using the commutation relations (1.38) show that $\hat{\phi}(t,\vec{x})$ and $\hat{\Pi}(t,\vec{x}) = \frac{\partial}{\partial t}\hat{\phi}(t,\vec{x})$ satisfy the equal-time commutation relations

$$\begin{aligned}
\left[\hat{\phi}(t,\vec{x}), \hat{\Pi}(t,\vec{y})\right] &= i\delta^{d-1}(\vec{x} - \vec{y}), \\
\left[\hat{\phi}(t,\vec{x}), \hat{\phi}(t,\vec{y})\right] &= \left[\hat{\Pi}(t,\vec{x}), \hat{\Pi}(t,\vec{y})\right] = 0.
\end{aligned} \tag{1.39}$$

Exercise 1.3.2   Show that the measure $d^{d-1}\vec{k}/(2\omega_k)$ is invariant under Lorentz transformations by rewriting it in the form

$$\int \frac{d^{d-1}\vec{k}}{2\omega_k} = \int d^{d-1}\vec{k} \int dk^0 \delta^d(k^2 + m^2)\Theta(k^0), \tag{1.40}$$

where Θ is the step function defined by $\Theta(k^0) = 1$ for $k^0 > 0$ and $\Theta(k^0) = 0$ for $k^0 < 0$.

The commutation relations (1.38) are similar to those of a quantum harmonic oscillator. Therefore we interpret the operators $\hat{a}^\dagger(\vec{k})$ and $\hat{a}(\vec{k})$ as creation and annihilation operators of particles with momentum $\vec{k}$, respectively. The vacuum state $|0\rangle$ of the theory is then given by

$$\hat{a}(\vec{k})|0\rangle = 0. \tag{1.41}$$

We assume the normalisation $\langle 0|0\rangle = 1$. A single-particle state with momentum $\vec{k}$, denoted by $|\vec{k}\rangle$ can be created by acting on the vacuum state with the creation operator $\hat{a}^\dagger(\vec{k})$,

$$|\vec{k}\rangle = \hat{a}^\dagger(\vec{k})|0\rangle. \tag{1.42}$$

Multi-particle states $|\vec{k}_1, \vec{k}_2, \ldots\rangle$ can be similarly constructed by applying a product of creation operators $\hat{a}^\dagger(\vec{k}_1)\hat{a}^\dagger(\vec{k}_2)\ldots$ to the vacuum state $|0\rangle$.

1.3.2 Path integral quantisation of free fields

Within quantum mechanics, the path integral sums over all possible paths which start at some position q_i at time t_i and end at a position q_f at time t_f. In quantum field theory, this translates into summing over all field configurations ϕ in configuration space. The integration measure becomes formally

$$\mathcal{D}\phi \propto \prod_{t_i \leq t \leq t_f} \prod_{\vec{x} \in \mathbb{R}^{d-1}} d\phi(t,\vec{x}). \tag{1.43}$$

The transition from an initial state $|\phi_i, t_i\rangle$ to a final state $|\phi_f, t_f\rangle$ where

$$\hat{\phi}(t_i,\vec{x})|\phi_i, t_i\rangle = \phi_i(\vec{x})|\phi_i, t_i\rangle, \qquad \hat{\phi}(t_f,\vec{x})|\phi_f, t_f\rangle = \phi_f(\vec{x})|\phi_f, t_f\rangle \tag{1.44}$$

is then given by

$$\langle \phi_{\mathrm{f}}, t_{\mathrm{f}} | \phi_{\mathrm{i}}, t_{\mathrm{i}} \rangle = N \int \mathcal{D}\phi \exp \left[i \int_{t_{\mathrm{i}}}^{t_{\mathrm{f}}} \mathrm{d}t \int_{\mathbb{R}^{d-1}} \mathrm{d}^{d-1}\vec{x}\, \mathcal{L}_{\mathrm{free}}(\phi, \partial\phi) \right], \qquad (1.45)$$

with N a (possibly divergent) normalisation factor which will be determined below. In (1.45) we integrate over those field configurations $\phi(t, \vec{x})$ satisfying $\phi(t_{\mathrm{i}}, \vec{x}) = \phi_{\mathrm{i}}(\vec{x})$ and $\phi(t_{\mathrm{f}}, \vec{x}) = \phi_{\mathrm{f}}(\vec{x})$. It is not clear whether this integral exists in a strict mathematical sense. A common trick to improve the convergence of the path integral is to replace the mass m^2 by $m^2 - i\epsilon$ and to take the limit $\epsilon \to 0$ at the end of the calculation. This $i\epsilon$-prescription will be used in the following implicitly. From now on we restrict ourselves to vacuum–vacuum transitions, i.e. we take the limits $t_{\mathrm{i}} \to -\infty$ and $t_{\mathrm{f}} \to +\infty$ and consider $\phi_{\mathrm{i}}(\vec{x}) = \phi_{\mathrm{f}}(\vec{x}) = 0$. Moreover, we use the abbreviation $\langle 0, -\infty | 0, +\infty \rangle \equiv \langle 0 | 0 \rangle$. This vacuum transition amplitude is given by

$$\langle 0 | 0 \rangle = N \int \mathcal{D}\phi \exp \left[i \int \mathrm{d}^d x\, \mathcal{L}_{\mathrm{free}}(\phi, \partial\phi) \right], \qquad (1.46)$$

where we choose N such that $\langle 0 | 0 \rangle = 1$. We are also interested in *correlation functions* of the form

$$\langle 0 | \mathcal{T}\hat{\phi}(x_1)\hat{\phi}(x_2)\ldots\hat{\phi}(x_n) | 0 \rangle \equiv \langle \phi(x_1)\phi(x_2)\ldots\phi(x_n) \rangle \equiv G^{(n)}(x_1, \ldots, x_n). \qquad (1.47)$$

Here, $\mathcal{T}$ denotes the *time ordering prescription* which states that a product of operators $\hat{\phi}(x_1)\hat{\phi}(x_2)\ldots\hat{\phi}(x_n)$ to the right of the symbol $\mathcal{T}$ has to be ordered such that fields at later times stand to the left of those at earlier times. In particular, for two operators $\hat{\phi}(x)\hat{\phi}(y)$ the time ordering is given by

$$\mathcal{T}\hat{\phi}(x)\hat{\phi}(y) \equiv \Theta(x^0 - y^0)\hat{\phi}(x)\hat{\phi}(y) + \Theta(y^0 - x^0)\hat{\phi}(y)\hat{\phi}(x), \qquad (1.48)$$

where Θ is the step function. The correlation functions $\langle \phi(x_1)\phi(x_2)\ldots\phi(x_n) \rangle$ in the path integral formulation are given by

$$\langle \phi(x_1)\phi(x_2)\ldots\phi(x_n) \rangle = N \int \mathcal{D}\phi\, \phi(x_1)\ldots\phi(x_n) \exp \left[i \int \mathrm{d}^d x\, \mathcal{L}_{\mathrm{free}}(\phi, \partial\phi) \right]. \qquad (1.49)$$

In particular we see that in the path integral formalism, the time ordering appears naturally by definition. In order to calculate correlation functions such as (1.49), it is convenient to introduce the generating functional $Z_0[J]$ defined by

$$Z_0[J] \equiv \left\langle \exp \left[i \int \mathrm{d}^d x\, J(x)\phi(x) \right] \right\rangle, \qquad (1.50)$$

where $J(x)$ stands for the source dual to the operator $\phi(x)$. The subscript 0 indicates that we are considering a free theory. Functionally varying $Z_0[J]$ with respect to the sources $J(x_i)$ and setting them to zero after the variation, we obtain

$$\langle \phi(x_1)\phi(x_2)\ldots\phi(x_n) \rangle = (-i)^n \frac{\delta^n Z_0[J]}{\delta J(x_1)\ldots\delta J(x_n)} \bigg|_{J=0}. \qquad (1.51)$$

In analogy with (1.49), the generating functional for a free field theory reads

$$Z_0[J] = N \int \mathcal{D}\phi \exp \left[i \int \mathrm{d}^d x\, (\mathcal{L}_{\mathrm{free}}(\phi, \partial\phi) + J(x)\phi(x)) \right]. \qquad (1.52)$$

Let us now determine $Z_0[J]$ for a free real scalar field with mass m,

$$Z_0[J] = N \int \mathcal{D}\phi \, \exp\left[i \int d^d x \left(-\tfrac{1}{2}\phi(-\Box + m^2 - i\epsilon)\phi + J\phi \right) \right]. \quad (1.53)$$

Note that the integrals over the fields ϕ are Gaussian since ϕ is at most quadratic in (1.53). Performing the Gaussian integrals, (1.53) may be rewritten as

$$Z_0[J] = \exp\left[\frac{i}{2} \int d^d x \, d^d y \, J(x) \Delta_F(x - y) J(y) \right], \quad (1.54)$$

where Δ_F is the Feynman propagator for a scalar field

$$\Delta_F(x - y) = \int \frac{d^d k}{(2\pi)^d} \frac{e^{ik(x-y)}}{k^2 + m^2 - i\epsilon} \quad (1.55)$$

satisfying the differential equation

$$(-\Box + m^2)\Delta_F(x - y) = \delta^d(x - y). \quad (1.56)$$

In other words, the Feynman propagator is a Green's function for the Klein–Gordon equation. From (1.54) we determine the two-point function to be

$$G^{(2)}(x_1, x_2) \equiv \langle \phi(x_1)\phi(x_2) \rangle = -i\Delta_F(x_1 - x_2). \quad (1.57)$$

Exercise 1.3.3 A further important two-point function is the *retarded Green function* G_R given by

$$G_R(x - y) = \int \frac{d^d k}{(2\pi)^d} \frac{e^{ik(x-y)}}{-(k^0 + i\epsilon)^2 + \vec{k}^2 + m^2}. \quad (1.58)$$

Compare the pole structure of (1.58) and of (1.55) in the complex k^0 plane. Show that $G_R(x - y)$ satisfies $(-\Box + m^2)G_R(x - y) = \delta^d(x - y)$ and that $G_R(x - y)$ vanishes for $x^0 < y^0$. Moreover, check explicitly by using the mode expansion (1.37) that $G_R(x - y)$ may be written as

$$G_R(x - y) = -i\,\Theta(x^0 - y^0)\, \langle 0|[\hat{\phi}(x), \hat{\phi}(y)]|0 \rangle, \quad (1.59)$$

where Θ is the step function.

1.3.3 Beyond free fields: interactions and Feynman rules

Within quantum field theories we are interested not only in free but also in interacting field theories. Interactions are introduced by adding terms such as $\mathcal{L}_{\text{int}}$ given by (1.14) to the Lagrangian. However, note that in this case the integrand of the generating functional $Z[J]$ given by

$$Z[J] = N \int \mathcal{D}\phi \, \exp\left[i \int d^d x \, (\mathcal{L}_{\text{free}} + \mathcal{L}_{\text{int}} + J\phi) \right] \quad (1.60)$$

is no longer Gaussian and we cannot perform the integration explicitly. Whenever the coupling constants in (1.14), such as g_n, are small, we may use *perturbation theory*. The starting point of perturbation theory is to write

$$Z[J] = N\exp\left[i \int d^d x\, \mathcal{L}_{\text{int}}\left(\frac{1}{i}\frac{\delta}{\delta J(x)}\right)\right] \int \mathcal{D}\phi \exp\left[i \int d^d x\, (\mathcal{L}_{\text{free}} + J\phi)\right]$$
$$= \exp\left[i \int d^d x\, \mathcal{L}_{\text{int}}\left(\frac{1}{i}\frac{\delta}{\delta J(x)}\right)\right] Z_0[J]. \tag{1.61}$$

According to (1.61), if we know $Z_0[J]$ as well as $\mathcal{L}_{\text{int}}$, we can determine $Z[J]$. This is done most conveniently in a graphical representation known as a *Feynman diagram*: while $Z_0[J]$ encodes the propagator of the free field theory, $\mathcal{L}_{\text{int}}$ can be represented as interaction vertices. The rules for drawing Feynman diagrams for ϕ^4 theory are derived in exercise 1.3.4. Moreover, in exercise 1.3.5 we discuss the validity of the reformulation of (1.60) to (1.61) using a simple toy model.

Exercise 1.3.4 Show that the Feynman rules for ϕ^4 theory with Lagrangian

$$\mathcal{L} = -\frac{1}{2}\partial^\mu \phi \partial_\mu \phi - \frac{1}{2}m^2\phi^2 - \frac{g}{4!}\phi^4 \tag{1.62}$$

are given by the following set of rules:

- a line from x_i to y_i represents a propagator $-i\Delta_F(x_i - y_i)$,

$$x_i \xrightarrow{\hspace{2cm}} y_i$$

- a vertex at y_i connecting four lines corresponds to a factor ig,

$$y_i$$

- an integration has to be performed over the position coordinates of all vertices, including appropriate symmetry factors.

Transforming all n-point correlation functions $G^{(n)}(x_1, \ldots, x_n)$ into momentum space using

$$G^{(n)}(p_1, \ldots, p_n) = \int d^d x_1 \ldots \int d^d x_n\, G^{(n)}(x_1, \ldots, x_n) e^{-i(p_1 x_1 + \cdots + p_n x_n)},$$
$$G^{(n)}(x_1, \ldots, x_n) = \int \frac{d^d p_1}{(2\pi)^d} \ldots \int \frac{d^d p_n}{(2\pi)^d}\, G^{(n)}(p_1, \ldots, p_n) e^{i(p_1 x_1 + \cdots + p_n x_n)},$$

where all momenta p_i are taken to be ingoing, the Feynman rules are the following.

- Each line with associated momentum k represents a factor $(k^2 + m^2 - i\epsilon)^{-1}$,

$$\xrightarrow{\hspace{2cm}}\\ k$$

- For each vertex, add a factor $ig(2\pi)^d \delta^d(\sum_i p_i)$ where the delta function ensures conservation of momentum,

- Integrate over undetermined loop momenta with integration measure $\int \frac{d^d k}{(2\pi)^d}$ and add appropriate symmetry factors.

Exercise 1.3.5 Convergence of the perturbative expansion In view of the fact that gauge/gravity duality is a non-perturbative approach to quantum field theory, let us examine the convergence properties of the *trick* used to go from (1.60) to (1.61). In particular we consider an ordinary one-dimensional integral which we can perform analytically and compare to the results which we obtain from perturbation theory using (1.61).

For example, let us consider the integral

$$f(\lambda) = \int_{-\infty}^{\infty} dx \, e^{-\frac{1}{2}m^2 x^2 - \frac{\lambda}{4!}x^4 + jx}. \tag{1.63}$$

(i) For $j = 0$ but $\lambda \in \mathbb{R}$, $\lambda > 0$, evaluate $f(\lambda)$ exactly. The result is

$$f(\lambda) = \sqrt{\frac{3m^2}{\lambda}} e^{\frac{3m^4}{4\lambda}} K_{1/4}\left(\frac{3m^4}{4\lambda}\right) \tag{1.64}$$

with $K_\nu(x)$ being the modified Bessel function of the second kind.

(ii) Show that the integral (1.63) may be rewritten as

$$
\begin{aligned}
f(\lambda) &= \int_{-\infty}^{\infty} dx \, e^{-\frac{1}{2}m^2 x^2 + jx} \sum_{k=0}^{\infty} \frac{(-\lambda x^4)^k}{k!(4!)^k} \\
&= \sum_{k=0}^{\infty} \frac{(-\lambda)^k}{k!(4!)^k} \int_{-\infty}^{\infty} dx \, x^{4k} e^{-\frac{1}{2}m^2 x^2 + jx},
\end{aligned} \tag{1.65}
$$

assuming that we can exchange the infinite sum and the integral in the last step.

(iii) Show that

$$\int_{-\infty}^{\infty} dx \, x^{2n} e^{-\frac{1}{2}m^2 x^2} = \sqrt{2\pi} \frac{(2n)!}{n! \, 2^n \, m^{2n+1}}, \tag{1.66}$$

for example by considering $\int_{-\infty}^{\infty} dx \, e^{-1/2m^2 x^2 + jx}$ and taking derivatives with respect to j.

(iv) Argue why the steps (ii) and (iii) to evaluate the integral (1.63) are similar to those involved from (1.60) to (1.61).

(v) For $j = 0$, compare the partial sums

$$f_n(\lambda) = \sqrt{2\pi} \sum_{k=0}^{n} \frac{(-\lambda)^k (4k)!}{k! \, (2k)! \, (4!)^k 2^{2k} m^{4k+1}} \tag{1.67}$$

as a function of n to the exact result obtained from (i). What about $\lim_{n \to \infty} f_n(\lambda)$?

So far we have considered time ordered correlation functions. Now we turn to the scattering of particles. Interaction processes are described by how a set of incoming states |in⟩ evolves into a set of outgoing states |out⟩. While incoming states are prepared at $t_i \to -\infty$ the outgoing states are measured at $t_f \to \infty$. An example of this are n incoming particles which scatter into m outgoing particles. The |in⟩ and |out⟩ states are assumed to be free, i.e. to satisfy the equations of motion (1.36) and commutation relations (1.38). The states describing the ingoing and outgoing particles, respectively, span Fock spaces of the form

$$\mathcal{V}^{\text{in}} = \{|k_1, \ldots, k_n, \text{in}\rangle = \hat{a}^\dagger_{\text{in}}(\vec{k}_1)\hat{a}^\dagger_{\text{in}}(\vec{k}_2)\ldots\hat{a}^\dagger_{\text{in}}(\vec{k}_n)|0\rangle\},$$
$$\mathcal{V}^{\text{out}} = \{|k_1, \ldots, k_m, \text{out}\rangle = \hat{a}^\dagger_{\text{out}}(\vec{k}_1)\hat{a}^\dagger_{\text{out}}(\vec{k}_2)\ldots\hat{a}^\dagger_{\text{out}}(\vec{k}_m)|0\rangle\}. \tag{1.68}$$

In what follows we assume that the Fock spaces $\mathcal{V}^{\text{in}}$ and $\mathcal{V}^{\text{out}}$ are isomorphic, $\mathcal{V}^{\text{in}} \simeq \mathcal{V}^{\text{out}} \simeq \mathcal{V}$. Moreover, we choose a complete and orthonormal basis $|\alpha_{\text{in}}\rangle$ of $\mathcal{V}^{\text{in}}$ and $|\beta_{\text{out}}\rangle$ of $\mathcal{V}^{\text{out}}$, i.e. the basis vectors satisfy

$$\sum_\alpha |\alpha_{\text{in}}\rangle\langle\alpha_{\text{in}}| = \sum_\beta |\beta_{\text{out}}\rangle\langle\beta_{\text{out}}| = 1. \tag{1.69}$$

In scattering processes we are interested in how a set of incoming particles given by $|\alpha_{\text{in}}\rangle$ evolves into a set of outgoing particles $|\beta_{\text{out}}\rangle$. This information is encoded in the *S-matrix* $\hat{S}$ which maps *in* states to *out* states,

$$\hat{S}|\alpha_{\text{in}}\rangle = |\alpha_{\text{out}}\rangle. \tag{1.70}$$

Using the S-matrix we can rewrite

$$\langle\beta_{\text{out}}|\alpha_{\text{in}}\rangle = \langle\beta_{\text{in}}|\hat{S}^\dagger|\alpha_{\text{in}}\rangle. \tag{1.71}$$

The modulus squared of $\langle\beta_{\text{out}}|\alpha_{\text{in}}\rangle$ gives the probability of finding the *out* state $|\beta_{\text{out}}\rangle$ when starting from the *in* state $|\alpha_{\text{in}}\rangle$. Moreover, the completeness relation (1.69) implies that the S-matrix is unitary.

How can we calculate the S-matrix? It turns out that the S-matrix and therefore also $\langle\beta_{\text{out}}|\alpha_{\text{in}}\rangle$ are related to correlation functions using the *Lehmann–Symanzik–Zimmermann (LSZ) reduction formula*. This is given by

$$\langle\vec{k}_1 \cdots \vec{k}_m, \text{out}|\vec{p}_1 \cdots \vec{p}_n, \text{in}\rangle$$
$$= \prod_{i=1}^m \left(i \int \mathrm{d}^d x_i\, e^{ik_i x_i}(-\Box_{x_i} + m^2)\right) \prod_{j=1}^n \left(i \int \mathrm{d}^d y_j\, e^{-ip_j y_j}(-\Box_{y_j} + m^2)\right)$$
$$\times \langle\phi(x_1)\cdots\phi(x_m)\phi(y_1)\cdots\phi(y_n)\rangle. \tag{1.72}$$

We will see in section 1.5 that we have to modify the LSZ reduction formula slightly due to wave function renormalisation.

1.3.4 Further generating functionals

We saw that generating functionals such as $Z[J]$ are very useful since, in a given quantum field theory, $Z[J]$ determines all correlation functions and thus due to the LSZ reduction formula also all scattering amplitudes. In addition to the generating functional $Z[J]$, there are two other important generating functionals, $W[J]$ and $\Gamma[\varphi]$, which encode the same physical information as $Z[J]$ but are easier to obtain. The generating functional $W[J]$ is defined by

$$Z[J] \equiv e^{iW[J]}. \tag{1.73}$$

It can be shown that $W[J]$ is the generating functional for *connected* Green's functions, which are those in which all links are connected in the perturbative expansion in Feynman diagrams. For example, connected two-point functions are obtained by defining

$$\langle \phi(x)\phi(y)\rangle_c = \langle \phi(x)\phi(y)\rangle - \langle \phi(x)\rangle \langle \phi(y)\rangle \tag{1.74}$$

and a similar definition holds for n-point functions $\langle \phi(x_1)\phi(x_2)\dots\phi(x_n)\rangle_c$. We have

$$\langle \phi(x_1)\phi(x_2)\dots\phi(x_n)\rangle_c = (-i)^{n-1} \frac{\delta^n W[J]}{\delta J(x_1)\cdots\delta J(x_n)}\bigg|_{J=0}. \tag{1.75}$$

An important subset of connected diagrams are *one-particle irreducible* (1PI) Feynman diagrams. While one-particle *reducible* diagrams are those which can be made disconnected by cutting a single internal line, one-particle *irreducible* diagrams are the converse; cutting any one of their lines will not render them disconnected. While connected Feynman diagrams are generated by $W[J]$, the one-particle irreducible Feynman diagrams are generated by the *effective action* $\Gamma[\varphi]$ which is defined as the Legendre transform of $W[J]$,

$$\Gamma[\varphi] \equiv W[J] - \int d^d x\, J(x)\varphi(x). \tag{1.76}$$

Here, the field $\varphi(x)$ is defined by

$$\varphi(x) = \langle \phi(x)\rangle_J = \frac{\delta W[J]}{\delta J(x)}. \tag{1.77}$$

Note that we do not set the source J to zero in the above equation. The argument $\varphi(x)$ of the effective action is thus the expectation value of $\phi(x)$ in the presence of the sources J. Let us assume from now on that we do not have any *tadpoles*, i.e. that for $J = 0$ also $\langle \phi(x)\rangle_J = 0$. The functionals $\Gamma^{(n)}$ defined by

$$\Gamma^{(n)}(x_1,\dots,x_n) = \frac{\delta}{\delta\varphi(x_1)}\cdots\frac{\delta}{\delta\varphi(x_n)}\Gamma[\varphi] \tag{1.78}$$

encode the 1PI n-point correlation functions since

$$\Gamma^{(n)}(x_1,\dots,x_n)\bigg|_{J=0} = \langle \phi(x_1)\dots\phi(x_n)\rangle_{1\mathrm{PI}}. \tag{1.79}$$

Exercise 1.3.6 By varying (1.76) with respect to $\varphi(x)$ and using (1.77) show that the quantum equation of motion is given by

$$\frac{\delta\Gamma[\varphi]}{\delta\varphi(x)} = J(x). \tag{1.80}$$

Exercise 1.3.7 Using the chain rule show that

$$\frac{\delta}{\delta J(x)} = \int d^d y \frac{\delta^2 W[J]}{\delta J(x)\delta J(y)}\frac{\delta}{\delta\varphi(y)} \tag{1.81}$$

as well as

$$\Gamma^{(2)}(x,y) = \left(\frac{\delta^2 W[J]}{\delta J(x)\delta J(y)} \right)^{-1}. \tag{1.82}$$

Therefore $\Gamma^{(2)}|_{J=0}$ is the inverse of the exact propagator.

For practical purposes it is convenient to perform a Fourier transformation. Let us define the *vertex functions* $\Gamma^{(n)}(p_1,\ldots,p_n)$ by

$$(2\pi)^d \delta^d \left(\sum_{i=1}^{n} p_i \right) \Gamma^{(n)}(p_1,\ldots,p_n) = \prod_{k=1}^{n} \int d^d x_k \, e^{-ix_k p_k} \Gamma^{(n)}(x_1,\ldots,x_n) \tag{1.83}$$

with all momenta p_k chosen to be ingoing. We may rewrite the effective action as

$$\Gamma[\varphi] = \frac{1}{2} \int \frac{d^d p}{(2\pi)^d} \, \varphi(-p) \left(p^2 + m^2 - \Pi(p^2) \right) \varphi(p)$$

$$+ \sum_{n=3}^{\infty} \frac{1}{n!} \int \frac{d^d p_1}{(2\pi)^d} \cdots \int \frac{d^d p_n}{(2\pi)^d} (2\pi)^d \delta^d (p_1 + \cdots + p_n)$$

$$\cdot \, \Gamma^{(n)}(p_1,\ldots,p_n) \, \varphi(p_1)\ldots\varphi(p_n), \tag{1.84}$$

where $(p^2 + m^2 - \Pi(p^2))^{-1} = \Gamma^{(2)}$ is the exact propagator and $\Pi(p^2)$ is the *self-energy*.

Exercise 1.3.8   Consider ϕ^4 theory with the Feynman rules given in exercise 1.3.4. In order to determine $\Gamma^{(n)}$ we strip off the external legs, i.e. we do not take into account the factors $(k^2 + m^2)^{-1}$ for the external lines. Show that the one-loop contribution to the 1PI expression $\Gamma^{(2)}_{\text{1-loop}}$ in momentum space, as represented by the Feynman diagram

is given by

$$\Gamma^{(2)}_{\text{1-loop}}(p,-p) = \frac{ig}{2} \int \frac{d^d k}{(2\pi)^d} \frac{1}{k^2 + m^2 - i\delta}. \tag{1.85}$$

After Wick rotation $k_0 \to ik_0$ to Euclidean space, we may set $\delta \to 0$. Performing the integral (1.85) yields

$$\Gamma^{(2)}_{\text{1-loop}}(p,-p) = -\frac{g}{2} \frac{\Gamma\left(1 - \frac{d}{2}\right)}{(4\pi)^{d/2}} m^{(d-2)/2}. \tag{1.86}$$

Show that (1.86) is divergent for $d = 4$. Regularise the expression by *dimensional regularisation*, i.e. by setting $d = 4 - \epsilon$, and show that the singular and finite contributions for $\epsilon \mapsto 0$ are given by

$$\Gamma^{(2)}_{\text{1-loop}}(p,-p) \sim \frac{1}{2} \frac{g}{16\pi^2} m^2 \left(\frac{2}{\epsilon} + 1 - \ln m^2 \right) (e^{-\gamma} 4\pi)^{\epsilon/2}. \tag{1.87}$$

1.4 Wick rotation and statistical mechanics

The path integral as formulated in the last section is very difficult to evaluate numerically. This is due to the fact that the weight factor e^{iS} for the field configurations is not positive definite and oscillates rapidly. By performing a *Wick* rotation we analytically continue the path integral from real times t to imaginary *Euclidean times* $\tau = it$. The generating functional of the Euclidean theory then reads

$$Z[J] = \int \mathcal{D}\phi \, e^{-S_{\mathrm{E}} + \int \mathrm{d}^d x J(x)\phi(x)} \tag{1.88}$$

with the *Euclidean action* S_{E} given by $S_{\mathrm{E}} = \int \mathrm{d}^{d-1}\vec{x}\,\mathrm{d}\tau\, \mathcal{L}_E$,

$$\mathcal{L}_{\mathrm{E}} = \frac{1}{2}\partial^\mu \phi \partial_\mu \phi + \frac{1}{2}m^2 \phi^2 - \mathcal{L}_{\mathrm{int}}[\phi]. \tag{1.89}$$

Note that in the Euclidean action the indices μ are raised and lowered by the Kronecker δ instead of η. Moreover the sum over μ in the Lagrangian (1.89) is implicitly assumed and the kinetic term reads

$$\partial^\mu \phi \partial_\mu \phi = \left(\frac{\partial \phi}{\partial \tau}\right)^2 + (\nabla \phi)^2. \tag{1.90}$$

In the Euclidean version of the path integral, the weight factor $e^{-S_{\mathrm{E}}}$ is strongly damped and positive definite. Therefore the convergence properties of the path integral, and in particular the generating functional (1.88), are more obvious.

Since the weight factor $e^{-S_{\mathrm{E}}}$ is positive definite and therefore can be normalized to one, we may think of it as a density matrix. This connection can be made more precise. The thermodynamical partition function $Z(\beta)$ of a quantum mechanical system at temperature $T = \beta^{-1}$ with Hamiltonian $\hat{H}$ reads

$$Z(\beta) \equiv \mathrm{tr}\, e^{-\beta \hat{H}} = \sum_n \langle n | e^{-\beta \hat{H}} | n \rangle, \tag{1.91}$$

where the set of states $|n\rangle$ forms a basis of the Hilbert space. Note that in quantum field theory, a transition amplitude from an inital state $\langle \phi_{\mathrm{i}}, t = 0 |$ at time $t = 0$ to a final state $|\phi_{\mathrm{f}}, t = t_{\mathrm{f}}\rangle$ is given by the path integral (1.45). For the partition function we thus have to sum over all transition amplitudes with $\phi_{\mathrm{i}} = \phi_{\mathrm{f}}$. Therefore using path integrals, the partition function of a thermal field theory with temperature $T = 1/\beta$, is given by

$$Z(\beta) = \int_{\mathrm{periodic}} \mathcal{D}\phi \, e^{-S_{\mathrm{E}}}, \tag{1.92}$$

where we have restricted the path integral to those configurations which satisfy the periodicity condition $\phi(\tau, \vec{x}) = \phi(\tau + \beta, \vec{x})$. In other words, the path integral of a Euclidean quantum field theory living on a d-dimensional spacetime with topology of a cylinder of circumference β can be viewed as a thermal average of a quantum statistical system in $d - 1$ spatial dimensions. In particular, the generating functional $Z[J = 0]$ of the

d-dimensional Euclidean quantum field theory corresponds to the partition function of the quantum statistical system. The other generating functionals, $W[J]$ and $\Gamma[\varphi]$, given by

$$e^{-W[J]} = Z[J] \tag{1.93}$$

together with (1.76) in the Euclidean theory, also have nice interpretations in terms of statistical physics: $W[J]$ is just the free energy of the statistical system. $\Gamma[\varphi]$, being a Legendre transformation of the free energy, is then identified as the Gibbs free energy in statistical physics.

1.5 Regularisation and renormalisation

In general, local quantum field theories are inherently afflicted by divergences which seem to affect the perturbative expansion. These divergences such as the zero-point energy may be traced back to the infinite number of degrees of freedom present per finite volume. The most prominent of these infinities occur at short distances and high momenta, i.e. in the ultraviolet regime (UV).[2]

These apparent UV divergences may be dealt with consistently using the procedures of *regularisation* and *renormalisation*. *Regularisation* means that the divergences are removed by a suitable procedure, the simplest being to introduce a cut-off Λ in momentum space. Alternatively, *dimensional regularisation* may be used. In this case, the divergences appear as poles in ϵ, where in four dimensions, for instance, ϵ is a regulator defined by modifying the spacetime dimension to $d = 4 - \epsilon$. In the regularised theory obtained by a suitable regularisation procedure, *renormalisation* may be performed. While the *bare parameters* are infinite, the *renormalised parameters* depend on a mass scale. As a consequence, the renormalised (or physical) parameters run if we change this mass scale. The running of the mass and the coupling constants is determined by the *renormalisation group equation*.

1.5.1 Regularisation and renormalisation

Let us now discuss regularisation and renormalisation for the simple example of ϕ^4 theory to one loop. Again we will be brief, referring to standard textbooks on quantum field theory for further details. Nevertheless, it is essential to recall the basic structure here. Later on in section 5.5, we will consider *holographic renormalisation* in the context of the AdS/CFT correspondence, which is an essential feature in the context of gauge/gravity duality. It will be instructive to compare this to the standard procedure in quantum field theory, which we discuss here.

As an illustrative example we consider ϕ^4 theory in four dimensions, whose action is given by (1.62). The starting point is the classical action $\mathcal{S}[\phi]$. Let us consider perturbation theory to first order. Using Feynman diagrams, we calculate the vertex functions $\Gamma^{(n)}$ to

---

[2] In massless theories, infrared (IR) divergences may also appear.

one-loop order in perturbation theory, for which we have to adopt a regularisation procedure. We choose dimensional regularisation. We already determined $\Gamma^{(2)}$ in exercise 1.3.4. Neglecting the finite parts, we find the following divergent contributions to $\Gamma^{(2)}$ and $\Gamma^{(4)}$,

$$\Gamma^{(2)}_{\text{1-loop,div}}(p,-p) = \frac{gm^2}{16\pi^2\epsilon}, \tag{1.94}$$

$$\Gamma^{(4)}_{\text{1-loop,div}}(p_1,p_2,p_3,p_4) = \frac{3g^2}{16\pi^2\epsilon}. \tag{1.95}$$

When using a cut-off Λ instead of dimensional regularisation, $1/\epsilon$ is replaced by $\ln\Lambda$ in the above expressions.

It is now crucial to note that ϕ^4 theory in four dimensions is a *renormalisable* quantum field theory. The essential property of such theories is that the divergences of the type encountered above may be removed consistently by adding new terms to the Lagrangian, which are of the same form as those present in the original Lagrangian, just with divergent coefficients. The new terms are called *counterterms*. In our example of ϕ^4 theory in four dimensions, they take the form

$$\mathcal{L}_{\text{ct}} = -\frac{A}{2}\partial^\mu\phi\partial_\mu\phi - \frac{B}{2}\phi^2 - \frac{C}{4!}\phi^4, \tag{1.96}$$

with coefficients A, B, C which are fixed below. Each of these terms creates additional vertices which contribute to the vertex functions. At tree level, we have

$$\Gamma^{(2)}_{\text{ct;tree}}(p,-p) = -Ap^2 - B,$$
$$\Gamma^{(4)}_{\text{ct;tree}} = -C. \tag{1.97}$$

To first order in perturbation theory, the divergences in $\Gamma^{(n)}_{\text{1-loop}}$ are cancelled by

$$A = 0, \quad B = \frac{gm^2}{16\pi^2\epsilon}, \quad C = \frac{3g^2}{16\pi^2\epsilon}. \tag{1.98}$$

With this choice, $\Gamma^{(n)}_{\text{1-loop}} + \Gamma^{(n)}_{\text{ct;tree}}$ has no pole in ϵ as $\epsilon \to 0$; setting $\epsilon = 0$ yields a finite result. To higher loop order, A must also have non-trivial contributions to cancel the divergences. Note also that the coefficients in (1.96) may be modified by additional finite terms, which leads to a certain amount of arbitrariness, to be fixed below.

To all orders in perturbation theory, the approach outlined above is performed systematically by using *multiplicative renormalisation*. For *renormalisable* theories, all infinities can be reabsorbed into a finite number of coupling constants and masses. Finite results for the renormalised vertex functions are obtained in the limit of $\epsilon \to 0$ for dimensional regularisation, or equivalently for infinite cut-off, $\Lambda \to \infty$, in cut-off regularisation. The key feature of *renormalisability* is that a finite number of counterterms is needed at each order in perturbation theory.

For our example of ϕ^4 theory in four dimensions, we now introduce *renormalised perturbation theory*. Renormalisability implies that all counterterms are of the same form as the contributions to the original Lagrangian. This means that $\mathcal{L}_{\text{ct}}$ is of the form (1.96) with A, B, C now all-order coefficients given by power series in g and in $1/\epsilon$

for dimensional regularisation, or in Λ for cut-off regularisation. We can define a *bare* Lagrangian $\mathcal{L}_{\text{bare}}$ to all orders in perturbation theory by

$$\mathcal{L}_{\text{bare}} = \mathcal{L} + \mathcal{L}_{\text{ct}}, \tag{1.99}$$

which we write as

$$\mathcal{L}_{\text{bare}} = -\frac{1}{2}\partial^\mu\phi_0\partial_\mu\phi_0 - \frac{1}{2}m_0^2\phi_0^2 - \frac{1}{4!}g_0\phi_0^4, \tag{1.100}$$

with

$$\phi_0 = Z_\phi^{1/2}\phi, \qquad Z_\phi = 1 + A. \tag{1.101}$$

Here ϕ_0 is the *bare field* and ϕ is the *renormalised* field. ϕ_0 and ϕ are related by the multiplicative *field renormalisation factor* Z_ϕ. Moreover, we have

$$m_0^2 = \frac{m^2 + B}{Z_\phi} \equiv \frac{m^2 + \delta m^2}{Z_\phi}, \qquad g_0 = \frac{g + C}{Z_\phi^2} \equiv \frac{gZ_g}{Z_\phi^2}. \tag{1.102}$$

Z_g/Z^2 is the coupling renormalisation and δm^2 determines the mass renormalisation. The *bare* parameters ϕ_0, m_0, g_0 may be expressed in a power series in g with higher and higher poles in ϵ. We include only the divergent terms as given to one-loop order by (1.98) to the counterterm Lagrangian (1.96), and do not consider any further finite contributions. This procedure is referred to as the *minimal subtraction (MS) scheme*. In the MS scheme we obtain

$$m_0^2 = m^2\left(1 + \frac{g}{16\pi^2\epsilon}\right), \qquad g_0 = g\left(1 + \frac{3g}{16\pi^2\epsilon}\right), \qquad Z_\phi = 1 + \mathcal{O}(g^2). \tag{1.103}$$

In addition, since the mass dimension of the coupling constant also has to be zero in $d = 4 - \epsilon$ dimensions, we have to introduce an arbitrary mass scale μ and replace $g \mapsto g\mu^\epsilon$ in the Lagrangian. This also affects C in (1.98) which has to be replaced by $C \mapsto C\mu^\epsilon$ and therefore (1.102) reads

$$g_0 = \mu^\epsilon\frac{gZ_g}{Z_\phi^2}. \tag{1.104}$$

Note that μ is not a parameter of the original theory since we have introduced it while performing dimensional analysis for renormalisation. This implies that the bare fields and bare parameters do not depend on μ.

To summarise, the bare fields and bare parameters do not depend on μ, but are infinite as we saw in (1.103). In contrast, the renormalised field ϕ and the renormalised parameters m and g are finite by construction, but depend on μ. The dependence of the parameters on μ is determined by the renormalisation group which we discuss in section 1.5.2.

Using (1.99) we may determine $\Gamma^{(n)}(p_1,\ldots,p_n;m,g)$, the renormalised 1PI vertex functions in momentum space, order by order in perturbation theory, in terms of the unrenormalised (or bare) 1PI vertex function $\Gamma_0^{(n)}(p_1,\ldots,p_n;m_0,g_0)$,

$$\Gamma_0^{(n)}(p_1,\ldots,p_n) = \Gamma^{(n)}(p_1,\ldots,p_n) + \Gamma_{\text{ct}}^{(n)}(p_1,\ldots,p_n) \tag{1.105}$$

and obtain a finite result for $\Gamma^{(n)}(p_1, \ldots, p_n; m, g)$ which is independent of ϵ. For a fixed mass scale μ, the finite renormalised parameters m and g may be read off from $\Gamma^{(2)}$ and $\Gamma^{(4)}$. For example, for m we have to impose

$$\Gamma^{(2)}(p, -p)|_{p^2 = -m^2} = 0, \tag{1.106}$$

$$\frac{\partial}{\partial p^2} \Gamma^{(2)}(p, -p)|_{p^2 = -m^2} = 1. \tag{1.107}$$

1.5.2 The renormalisation group

In the last section we saw that the renormalised parameters, for example the mass m and the coupling constant g for a ϕ^4 theory, will change if we change the mass scale μ. If we change μ from μ_1 to μ_2, the parameters change according to a finite renormalisation transformation denoted by $\mathcal{R}_{\mu_1 \mu_2}$. Suppose we have a third mass scale μ_3 then these transformations satisfy

$$\mathcal{R}_{\mu_1 \mu_2} \mathcal{R}_{\mu_2 \mu_3} = \mathcal{R}_{\mu_1 \mu_3}. \tag{1.108}$$

In other words, the finite transformations $\mathcal{R}$ form a group.[3] In the following we derive a relation, the renormalisation group equation, which allows us to determine these finite transformations.

Let us consider the bare, connected n-point functions $G_{0;c}^{(n)}$, given in momentum space by

$$G_{0;c}^{(n)}(p_1, \ldots, p_n) = \langle \phi_0(p_1) \ldots \phi_0(p_n) \rangle_c = Z_\phi^{n/2} \langle \phi(p_1) \ldots \phi(p_n) \rangle_c$$
$$= Z_\phi^{n/2} G_c^{(n)}(p_1, \ldots, p_n), \tag{1.109}$$

where we have used (1.101). $G_c^{(n)}$ is the renormalised connected n-point function which is given in terms of the renormalised field ϕ and the renormalised parameters m and g while $G_{0;c}^{(n)}$ is given in terms of ϕ_0 and the bare parameters. For the vertex functions $\Gamma^{(n)}$ we stripped off the external legs and therefore

$$\Gamma_0^{(n)}(p_1, \ldots, p_n) = Z_\phi^{-n/2} \Gamma^{(n)}(p_1, \ldots, p_n). \tag{1.110}$$

Note that the left-hand side of (1.110) is independent of the mass scale μ and consequently

$$0 = \mu \frac{\mathrm{d}}{\mathrm{d}\mu} \Gamma_0^{(n)}(p_1, \ldots, p_n) = \mu \frac{\mathrm{d}}{\mathrm{d}\mu} \left(Z_\phi^{-n/2} \Gamma^{(n)}(p_1, \ldots, p_n) \right). \tag{1.111}$$

Using the chain rule we obtain

$$\left(\mu \frac{\partial}{\partial \mu} + \beta \frac{\partial}{\partial g} + m\gamma_m \frac{\partial}{\partial m} - n\gamma \right) \Gamma^{(n)}(p_1, \ldots, p_n) = 0, \tag{1.112}$$

where β, γ_m and γ are given by

$$\beta \equiv \mu \frac{\partial g}{\partial \mu}, \qquad \gamma_m \equiv \frac{\mu}{m} \frac{\partial m}{\partial \mu}, \qquad \gamma \equiv \frac{\mu}{2 Z_\phi} \frac{\partial Z_\phi}{\partial \mu}. \tag{1.113}$$

---

[3] Technically speaking, the renormalisation group is just a semigroup due to the lack of an inverse element: information gets lost when integrating out degrees of freedom in moving to lower energies. This cannot be recovered in a unique fashion, i.e. a quantum field theory may not necessarily have a unique UV completion.

The derivatives are to be taken for fixed bare parameters. Equation (1.112) is the *renormalisation group equation* or *RG equation*. γ is the *anomalous dimension* of the field ϕ, while γ_m is the anomalous dimension of the mass parameter m. Note the similarity between the definitions of γ_m and β. From now on, we view the mass as a generalised coupling constant. In order to obtain an equation analogous to (1.112) for the connected Green function $G_c^{(n)}$, we have to switch the sign in front of the $n\gamma$ term in equation (1.112), as is seen by comparing (1.109) and (1.110).

Exercise 1.5.1   Show that the effective action as given by (1.84) satisfies

$$\left(\mu\frac{\partial}{\partial\mu} + \beta\frac{\partial}{\partial g} + m\gamma_m\frac{\partial}{\partial m} - \gamma\int d^dx\,\varphi(x)\frac{\delta}{\delta\varphi(x)}\right)\Gamma[\varphi,m,g,\mu] = 0. \quad (1.114)$$

Exercise 1.5.2   Using (1.113), determine β, γ_m and γ to first non-trivial order in g for ϕ^4 theory. For $\beta(g)$ you will find

$$\beta(g) = \frac{3g^2}{16\pi^2} + \mathcal{O}(g^3). \quad (1.115)$$

1.5.3   Composite operators and sources

Composite operators consisting of products of elementary quantum fields require special attention as far as renormalisation is concerned since the multiplication of quantum fields with the same spacetime argument leads to additional divergences. In general, composite operators need *additive renormalisation*. Let us explain this using the operator $\phi(x)^2$ as an example. In order to obtain the bare operator $(\phi(x)^2)_0$ we have to apply multiplicative renormalisation generalising (1.101) as well as adding a constant D, i.e.

$$\left(\phi(x)^2\right)_0 = Z_{\phi^2}\phi^2 + D. \quad (1.116)$$

We can unify additive renormalisation with multiplicative renormalisation if we allow for matrices of the form

$$\begin{pmatrix} \phi^2(x) \\ 1 \end{pmatrix}_0 = \begin{pmatrix} Z_{\phi^2} & D \\ 0 & 1 \end{pmatrix}\begin{pmatrix} \phi^2(x) \\ 1 \end{pmatrix}. \quad (1.117)$$

This can be generalised immediately to any other composite operators $\mathcal{O}$ built out of local polynomials of the elementary fields. If $\mathcal{O}^J$ is a basis of such operators and $\mathcal{O}_0^I$ are the corresponding bare operators, the transformation analogous to (1.117) reads

$$\mathcal{O}_0^I = Z^I{}_J\mathcal{O}^J, \quad (1.118)$$

where $Z^I{}_J$ can be arranged such that it is a block upper triangular matrix. Even though I and J numerate infinitely many fields, for a fixed I only a finite number of J exist with $Z^I{}_J \neq 0$. For the renormalised correlation functions of such composite operators we have

$$G^{J_1\ldots J_n}(p_1,\ldots,p_n;g,\mu) = \langle\mathcal{O}^{J_1}(p_1)\ldots\mathcal{O}^{J_n}(p_n)\rangle_c, \quad (1.119)$$

which depend on all coupling constants g as well as on the mass scale μ. Generalising (1.109), the relation between renormalised and bare correlation functions reads

$$G_0^{I_1\ldots I_n}(p_1,\ldots,p_n;g_0) = Z^{I_1}{}_{J_1}\ldots Z^{I_n}{}_{J_n}G^{J_1\ldots J_n}(p_1,\ldots,p_n;g,\mu) \quad (1.120)$$

and therefore the RG equation is given by

$$\left(\mu\frac{\partial}{\partial\mu} + \beta\frac{\partial}{\partial g}\right)G^{I_1\ldots I_n} + \sum_{k=1}^{n}\sum_{J}\gamma^I_J G^{I_1\ldots I_{k-1}J I_{k+1}\ldots I_n} = 0, \tag{1.121}$$

where the coefficients $\beta(g)$ and γ^I_J are

$$\beta(g) = \mu\frac{\partial g}{\partial\mu}, \qquad \gamma^I_J = \mu\sum_{K}(Z^{-1})^I{}_K\frac{\partial Z^K_J}{\partial\mu}, \tag{1.122}$$

evaluated again for fixed bare parameters. While the diagonal entries of γ^I_J are the anomalous dimension, the off-diagonal entries of γ^I_J give rise to operator mixing, as can be seen in (1.121).

In order to calculate correlation functions of the form (1.119) in a convenient way using the path integral, we add sources for the composite operators to the action, just as we did for the elementary fields before. For instance, for a composite scalar operator $\mathcal{O}$, a conserved current J_μ and for the energy-momentum tensor, we may define

$$S' \equiv S + \int d^d x \left(\mathcal{J}(x)\mathcal{O}(x) + A^\mu(x)J_\mu(x) - \frac{1}{2}g^{\mu\nu}(x)T_{\mu\nu}(x)\right), \tag{1.123}$$

such that the generating functional $W[\mathcal{J}, A^\mu, g^{\mu\nu}]$ associated to S' gives rise to

$$\langle\mathcal{O}(x)\rangle = \frac{\delta W}{\delta\mathcal{J}(x)}, \quad \langle J_\mu(x)\rangle = \frac{\delta W}{\delta A^\mu(x)}, \quad \langle T_{\mu\nu}(x)\rangle = -2\frac{\delta W}{\delta g^{\mu\nu}(x)}, \tag{1.124}$$

where the factor of (-2) is a matter of convention.

The sources for the composite operators may be viewed as generalised couplings, which we denote by $\mathcal{J}^I$. From (1.121), we obtain for the generating functional

$$\left(\mu\frac{\partial}{\partial\mu} + \beta\frac{\partial}{\partial g} + \int d^d x\, \gamma^I_J\mathcal{J}^J(x)\frac{\delta}{\delta\mathcal{J}^I}\right)W[\mathcal{J}^I, g, \mu] = 0, \tag{1.125}$$

with the anomalous dimension matrix γ^I_J as in (1.122). Note that the generalised couplings $\mathcal{J}^I$ may be dimensionful.

1.5.4 The Wilsonian renormalisation group

An alternative approach to renormalisation and the renormalisation group was taken by K. Wilson. Whereas in the canonical approach to renormalisation presented in the previous section, the cut-off Λ appeared as a tool for regularisation, in the Wilsonian approach, Λ is a scale of physical significance. This is motivated by statistical mechanics and by condensed matter physics, where such a scale is provided by the lattice spacing.

The starting point for the Wilsonian renormalisation group is again the generating functional,

$$Z[J] = \int\mathcal{D}\phi\, e^{-S_{\mathrm{E}} + \int d^d x J(x)\phi(x)} \tag{1.126}$$

where we have performed a Wick rotation to Euclidean signature. However, in the Wilsonian approach we impose an ultraviolet cut-off Λ by restricting the number of integration variables in the path integral, i.e. we perform the integration in the path integral only over $\phi(k)$ with $|k| < \Lambda$,

$$Z[J] = \int \mathcal{D}\phi_{|k|<\Lambda} e^{-\mathcal{S}_E^{\text{eff}}[\phi;\Lambda]+\int d^d x J(x)\phi(x)}, \tag{1.127}$$

where the measure of the path integral is given by

$$\mathcal{D}\phi_{|k|<\Lambda} = \prod_{|k|<\Lambda} d\phi(k) \tag{1.128}$$

and the *Wilsonian effective action* $\mathcal{S}_E^{\text{eff}}[\phi;\Lambda]$ is determined by

$$e^{-\mathcal{S}_E^{\text{eff}}[\phi;\Lambda]} = \int \mathcal{D}\phi_{|k|>\Lambda} e^{-\mathcal{S}_E[\phi]}. \tag{1.129}$$

For ϕ^4 theory, the Wilsonian effective action $\mathcal{S}_E^{\text{eff}}[\phi;\Lambda]$ is given by an effective Lagrangian in position space expressed as an expansion in local operators,

$$\mathcal{L}_E^{\text{eff}} = \frac{Z(\Lambda)}{2} \partial_\mu \phi \partial^\mu \phi + \frac{m^2(\Lambda)}{2} \phi^2 + \frac{g(\Lambda)}{4!} \phi^4 + \mathcal{O}\left(\frac{1}{\Lambda^2}\right), \tag{1.130}$$

where $Z(\lambda), m^2(\Lambda)$ and $g(\Lambda)$ are all finite functions of Λ. Moreover, the term $\mathcal{O}(1/\Lambda^2)$ in (1.130) represents higher order terms such as ϕ^{2n} with $n \geq 3$ or terms including derivatives. These terms arise from one-loop quantum corrections and compensate for the removal of the large k Fourier components in (1.127) by inducing interactions among the remaining Fourier modes $\phi(k)$, which previously were mediated by fluctuations of the large-k modes.

The Wilsonian approach now consists of studying what happens if we lower the cut-off from Λ to $b\Lambda$ with $b < 1$. The degrees of freedom with high momenta between $b\Lambda$ and Λ are *integrated out*, i.e.

$$Z[J] = \int \mathcal{D}\phi_{|k|<b\Lambda} e^{-\mathcal{S}_E^{\text{eff}}[\phi;b\Lambda]+\int d^d x J(x)\phi(x)}, \tag{1.131}$$

where $\mathcal{S}_E^{\text{eff}}[\phi,b\Lambda]$ involves only the Fourier components $\phi(k)$ with $|k| < b\Lambda$ and we set $J(k) = 0$ for $k > b\Lambda$.

We see that modes of the form $\phi(k)$ with $b\Lambda < |k| < \Lambda$ are no longer present explicitly in the partition function (1.131) since they do not appear in the new effective Lagrangian describing the physics at the new lower scale $b\Lambda$, but their physics is encoded in modifications of the physical parameters such as Z, m and g (as well as the higher order terms suppressed in (1.130)) at the new cut-off scale. The procedure of integrating out high momenta degrees of freedom leads to a *coarse-graining* and a reduction of the number of degrees of freedom.

The procedure of integrating out high momenta degrees of freedom is associative, i.e. we may first integrate out degrees of freedom with momenta between $b_1\Lambda$ and Λ and then with momenta between $b_2\Lambda$ and $b_1\Lambda$ where $b_2 < b_1$. Alternatively, we may integrate out directly all degrees of freedom with momenta between $b_2\Lambda$ and Λ. Note that integrating out high momenta degrees of freedom is irreversible, so the renormalisation group is really only a half group.

Using this procedure we obtain for the coupling constant $g(\Lambda)$

$$\frac{dg(\Lambda)}{d\ln\Lambda} = \beta(g(\Lambda)), \tag{1.132}$$

where for ϕ^4 theory

$$\beta(g(\Lambda)) = \frac{3}{16\pi^2}g(\Lambda)^2 + \cdots. \tag{1.133}$$

In particular, the Wilsonian approach gives us a new interpretation of the β function: the β function measures how the coupling constant $g(\Lambda)$ varies if we integrate out high-energy modes. For ϕ^4 theory, the β function as given by (1.133) is positive and thus g increases with Λ. Solving the differential equation (1.132), we obtain

$$\ln\Lambda = \int_{g_0}^{g(\Lambda)} \frac{dg'}{\beta(g')}, \tag{1.134}$$

Where the integration constant is reabsorbed into g_0. For $\beta(g)$ as given by (1.133), the coupling constant $g(\Lambda)$ becomes arbitrarily large at a finite energy scale $\Lambda = \Lambda_{max}$. This is known as a *Landau pole*.[4] For other field theories in which the β function is positive and scales as $\beta(g) \sim g^{\alpha}$ with $\alpha \leq 1$ for large g, then $\Lambda_{max} \to \infty$.

The points in the space spanned by the coupling constants at which the β function vanishes are referred to as *fixed points*. For field theories in which β has a zero $\beta(g_*) = 0$ and is positive for $g < g_*$, then $g \to g_*$ for $\Lambda \to \infty$. In this case g_* corresponds to a *UV fixed point*. On the other hand, if $\beta(g)$ is negative for small g, then g decreases with Λ and $g \to 0$ for $\Lambda \to \infty$. In this case $g = 0$ is a UV fixed point and the theory is said to be *asymptotically free*. In four dimensions, non-Abelian gauge theories which we will discuss in section 1.7 are candidates for asymptotically free theories. On the other hand, ϕ^4 theory in four dimensions has an *IR fixed point* at $g = 0$ for $\Lambda \to 0$.

Consider a field theory at a UV fixed point. Since b as defined above satisfies $b < 1$, those coupling parameters that are multiplied by negative powers of b grow, while those that are multiplied by positive powers of b decay during the Wilson renormalisation procedure. If the Lagrangian contains growing coefficients, the associated operators will eventually drive the Lagrangian away from the fixed point. Such operators are referred to as *relevant*. Examples of relevant terms in the Lagrangian are mass terms. On the other hand, interaction terms which have a dimensionless coupling correspond to *marginal* operators: in order to determine whether its coefficient grows or decays under the RG procedure, higher order corrections have to be considered. Finally, operators whose coefficients die away under the RG procedure are *irrelevant*.

1.5.5   Relevant, marginal and irrelevant operators

More information about RG fixed points may be obtained by linearising the RG equation around the fixed point and studying its eigenvalues. Generally, the RG equation as given

---

[4] Of course, it is questionable whether we can still use the one-loop result for the β function since the coupling constant is large.

by (1.111)–(1.113) introduces a flow in the space of couplings, parametrised by the renormalisation scale μ. $\beta^I = 0$ for all I corresponds again to a fixed point g_*^I of the renormalisation group, where the couplings are independent of μ. Near the fixed point $g^I = g_*^I$ for all I, we may consider the linearised equation for the β functions, which reads

$$\mu \frac{\partial}{\partial \mu} \left(g^I(\mu) - g_*^I \right) = \sum_J M^I{}_J \left(g^J(\mu) - g_*^J \right), \qquad (1.135)$$

where M is the matrix

$$M^I{}_J = \left(\frac{\partial \beta^I}{\partial g^J} \right)_{g^J = g_*^J}. \qquad (1.136)$$

The solution for $g^I(\mu)$ can be expanded in eigenvectors of M,

$$g^I(\mu) = g_*^I + \sum_J c_J V_J{}^I \mu^{\lambda_J}, \qquad (1.137)$$

where V_J is an eigenvector of M with eigenvalue λ_J. This equation shows that the couplings approach an ultraviolet (UV) fixed point at $\mu \to \infty$ if the $c_J = 0$ for all eigenvectors with $\lambda_J > 0$. The corresponding operators drive the RG flow away from the fixed point and are thus *relevant* operators. This is the case, for instance, for mass operators. On the other hand, the operators with negative eigenvalue are *irrelevant*. As long as perturbation theory is valid, perturbations around the fixed point involving irrelevant operators will die away as $\mu \to \infty$. For *marginal* couplings, i.e. those for which the eigenvalue vanishes, further analysis of the quantum corrections is necessary.

This argument may be generalised to the generalised couplings of (1.125). Consider the case of the RG fixed point $\beta(g) = 0$ in (1.125). Then we may define a generalised β function for the couplings $\mathcal{J}^I$ and linearise around the fixed point as in (1.135). With the values of the generalised couplings given by $\mathcal{J}_*^I$ at the fixed point, the generalised β function reads, linearising around the fixed point,

$$\beta^I(\mathcal{J}) \equiv \mu \frac{\partial}{\partial \mu} \left(\mathcal{J}^I - \mathcal{J}_*^I \right) = \Delta_{(I)} \left(\mathcal{J}^I - \mathcal{J}_*^I \right) + \gamma^I{}_J \left(\mathcal{J}^J - \mathcal{J}_*^J \right), \qquad (1.138)$$

with $\Delta_{(I)}$ the canonical dimension of the coupling $\mathcal{J}^I$. This is the analogue to (1.135) and a similar analysis concerning the eigenvectors and eigenvalues may be applied. For scale invariant theories, the anomalous dimension matrix $\gamma^I{}_J$ may generically be diagonalised. Then, relevant, irrelevant and marginal directions in eigenvector space are determined by the dimensions $\Delta_{(I)} + \gamma_{(I)}$. Note that since $\mathcal{J}^I$ couples to the operator $\mathcal{O}_I$, the canonical dimensions of operator and source are related by $\Delta_{\mathcal{O}^I} = d - \Delta_{(I)}$ for a field theory in d dimensions. An example of a relevant operator is a mass term for a scalar $m^2 \phi^2$ for which at the UV fixed point $\Delta_{(I)} = 2$.

1.6 Dirac fermions

In addition to scalar fields, we may also consider other types of fields such as Dirac fermions or gauge fields. In the following two sections we discuss fermions and gauge fields in more detail. A systematic approach to determining all possible types of fields is postponed to chapter 3, where we investigate the representations of the Lorentz algebra.

1.6.1 Classical theory of fermions

A *Dirac field* $\Psi(x)$ transforms under Lorentz transformations $x \mapsto x' = \Lambda x$ as

$$\Psi(x) \mapsto \Psi'(x) = \exp\left(-\frac{1}{8}\omega_{\mu\nu}[\gamma^\mu, \gamma^\nu]\right) \Psi(\Lambda^{-1}x), \qquad (1.139)$$

where $\omega_{\mu\nu}$ are antisymmetric and γ^μ are *Dirac matrices* satisfying the *Clifford algebra*

$$\{\gamma^\mu, \gamma^\nu\} = -2\eta^{\mu\nu}\mathbb{1}. \qquad (1.140)$$

In appendix B.2.2, we show that such Dirac matrices γ^μ exist for any spacetime dimension d and we give an explicit construction. The Lagrangian of a free Dirac field $\Psi(x)$ with mass m is given by

$$\mathcal{L}_{\text{free}} = i\bar{\Psi}\slashed{\partial}\Psi - m\bar{\Psi}\Psi, \qquad (1.141)$$

where $\slashed{\partial} = \gamma^\mu \partial_\mu$ and $\bar{\Psi} = \Psi^\dagger B$ with $B = \gamma^0$. We consider $\Psi(x)$ and $\bar{\Psi}(x)$ as independent fields. Since $\mathcal{L}_{\text{free}}$ does not depend on $\partial_\mu \bar{\Psi}$, the equation of motion for $\bar{\Psi}$ derived from (1.141) reads

$$(-i\slashed{\partial} + m)\Psi(x) = 0. \qquad (1.142)$$

Exercise 1.6.1   Derive the equation of motion for $\Psi(x)$. What is the relation to (1.142)?

Exercise 1.6.2   Show by acting with $i\slashed{\partial} + m$ on the left of equation (1.142) and using (1.140), that each component of $\Psi(x)$ satisfies the Klein–Gordon equation (1.13).

Although most of the results presented here can be generalised to arbitrary dimensions, we restrict ourselves to $d = 4$ in the following. Then, a convenient basis for the Dirac matrices can be expressed in terms of $\sigma^\mu = (-\mathbb{1}, \vec{\sigma})$ and $\bar{\sigma}^\mu = (-\mathbb{1}, -\vec{\sigma})$ where $\vec{\sigma}$ is a vector built out of the usual Pauli matrices σ^i. The Dirac matrices γ^μ read

$$\gamma^\mu = \begin{pmatrix} 0 & \sigma^\mu \\ \bar{\sigma}^\mu & 0 \end{pmatrix}. \qquad (1.143)$$

Note that γ^0 is Hermitian while γ^k with $k \in \{1, 2, 3\}$ are anti-Hermitian.

Exercise 1.6.3   Verify that the explicit representation of the Dirac matrices, equation (1.143), satisfies the Clifford algebra (1.140).

Exercise 1.6.4   Show that γ_5, defined by $\gamma_5 = i\gamma^0\gamma^1\gamma^2\gamma^3$, takes the form

$$\gamma_5 = \begin{pmatrix} \mathbb{1} & 0 \\ 0 & -\mathbb{1} \end{pmatrix} \qquad (1.144)$$

in this basis. Using γ_5 we may introduce left- and right-handed Weyl fermions as linear combinations involving the projection operators $\frac{1}{2}(\mathbb{1} \pm \gamma_5)$,

$$\Psi_L = \left(\frac{\mathbb{1} + \gamma_5}{2}\right)\Psi, \quad \Psi_R = \left(\frac{\mathbb{1} - \gamma_5}{2}\right)\Psi. \tag{1.145}$$

Exercise 1.6.5 Rewrite the Lagrangian (1.141) in terms of Ψ_L and Ψ_R. For a left-handed Weyl fermion $\Psi = \begin{pmatrix} \psi \\ 0 \end{pmatrix}$ show that the Lagrangian is given by

$$\mathcal{L}_{\text{free}} = -i\psi^\dagger \bar{\sigma}^\mu \partial_\mu \psi. \tag{1.146}$$

The classical Dirac Lagrangian (1.141) has a global $U(1)$ symmetry corresponding to the transformation

$$\Psi \mapsto \Psi' = e^{i\alpha}\Psi, \qquad \bar{\Psi} \mapsto \Psi' = e^{-i\alpha}\bar{\Psi}. \tag{1.147}$$

The corresponding conserved current reads $J^\mu = \bar{\Psi}\gamma^\mu\Psi$. For the case of a massless Dirac fermion, the Lagrangian (1.141) with $m = 0$ is also invariant under

$$\Psi \mapsto \Psi' = e^{i\alpha\gamma_5}\Psi, \quad \bar{\Psi} \mapsto \Psi' = \bar{\Psi}e^{i\alpha\gamma_5}, \qquad J_5^\mu = \bar{\Psi}\gamma^\mu\gamma_5\Psi, \tag{1.148}$$

where J_5^μ is the corresponding conserved current.

Exercise 1.6.6 Using the (massless) Dirac equation, show that both the vector and the axial current, J^μ and J_5^μ, are conserved at the classical level.

In four dimensions we may also introduce a *Majorana field*. In terms of a charge conjugation matrix $\mathcal{C}$ satisfying $\mathcal{C}^{-1}\gamma^\mu\mathcal{C} = -(\gamma^\mu)^T$, the charge conjugate Ψ^c of a Dirac field Ψ is defined by

$$\Psi^c = (\mathcal{B}\mathcal{C})^*\Psi^*. \tag{1.149}$$

By definition, a *Majorana field* is its own charge conjugate, i.e. it satisfies $\Psi = \Psi^c$. Starting from the Lagrangian

$$\mathcal{L}_{\text{free}} = \frac{i}{2}\bar{\Psi}\gamma^\mu\partial_\mu\Psi - \frac{m}{2}\bar{\Psi}\Psi, \tag{1.150}$$

subject to the condition $\Psi = \Psi^c$ for a Majorana field, we can get rid of $\bar{\Psi}$ in (1.150) and write

$$\mathcal{L}_{\text{free}} = \frac{i}{2}\Psi^T\mathcal{C}^T\gamma^\mu\partial_\mu\Psi - \frac{m}{2}\Psi^T\mathcal{C}^T\Psi. \tag{1.151}$$

In the following we will quantise fermionic fields within the path integral approach.

1.6.2 Path integral formulation for fermions

As in the case of a scalar field, we may define a path integral for fermions by integrating over all possible field configurations $\Psi(x)$ and $\bar{\Psi}(x)$. These field configurations, however, will anticommute with themselves, i.e. we have to treat $\Psi(x)$ as a *Grassmann number* valued field.[5] Allowing for Grassmann valued sources η and $\bar{\eta}$ we may define the generating functional

$$Z_0[\eta, \bar{\eta}] = N \int \mathcal{D}\Psi \mathcal{D}\bar{\Psi} e^{i \int \mathrm{d}^d x \left(\mathcal{L}_{\text{free}} + \bar{\eta}\Psi + \bar{\Psi}\eta \right)}, \tag{1.152}$$

where N is chosen such that $Z_0[\eta, \bar{\eta}]\big|_{\eta=\bar{\eta}=0} = 1$. From this we can deduce correlation functions

$$\langle \Psi_{\alpha_1}(x_1) \ldots \bar{\Psi}_{\beta_1}(y_1) \ldots \rangle = \frac{1}{i} \frac{\delta}{\delta \bar{\eta}_{\alpha_1}(x_1)} \ldots i \frac{\delta}{\delta \eta_{\beta_1}(y_1)} \ldots Z_0[\eta, \bar{\eta}]\big|_{\eta=\bar{\eta}=0}. \tag{1.153}$$

Since Ψ and $\bar{\Psi}$ are anticommutating, we pick up a minus sign every time we interchange two fields.

Since the Lagrangian (1.141) is quadratic in the field, we may perform the integration over Ψ and $\bar{\Psi}$ in equation (1.152) using the integration rules reviewed in appendix A. We obtain

$$Z_0[\eta, \bar{\eta}] = e^{i \int \mathrm{d}^d x \int \mathrm{d}^d y \, \bar{\eta}(x) \Delta_{\text{F}}(x-y) \eta(y)} \tag{1.154}$$

where Δ_{F} is the Feynman propagator for a Dirac spinor,

$$\Delta_{\text{F}}(x - y) = \int \frac{\mathrm{d}^d p}{(2\pi)^d} \frac{-\not{p} + m}{p^2 + m^2 - i\epsilon} e^{ip(x-y)}. \tag{1.155}$$

1.7 Gauge theory

1.7.1 Classical gauge theory

In physics it is important to understand symmetries and how to make such symmetries local. The procedure – known as *gauging* – introduces a connection, the gauge field. In this section we briefly review this concept and study gauge theories in detail. It is crucial to understand the dynamics of these gauge theories since all four fundamental forces in the standard model of elementary particle physics, i.e. the strong and weak interactions as well as electromagnetism and gravity, can be formulated in terms of gauge theories.

Abelian gauge theory

We begin by considering a gauge theory based on an Abelian $U(1)$ symmetry group. Consider a free complex scalar field $\phi(x)$ with mass m. As discussed in section 1.2, the action is invariant under the global $U(1)$ transformation $\phi(x) \mapsto \phi'(x) = e^{i\alpha}\phi(x)$, where α is a real parameter. An interesting question is whether we can promote α to be spacetime dependent, i.e. whether we can consider *local* $U(1)$ transformations of the form

$$\phi(x) \mapsto \phi'(x) = e^{i\alpha(x)}\phi(x). \tag{1.156}$$

---

[5] Here, we assume that the reader is familiar with Grassmann numbers. Properties of the Grassmann numbers are reviewed in appendix A.

First we realise that the derivative of $\phi(x)$ does not transform according to (1.156) since

$$\partial_\mu \phi'(x) = \partial_\mu\left(e^{i\alpha(x)}\phi(x)\right) \neq e^{i\alpha(x)}\partial_\mu\phi(x). \tag{1.157}$$

However, by introducing a *connection* or *gauge field* $A = A_\mu dx^\mu$ we may define a covariant derivative D_μ,

$$D_\mu\phi(x) \equiv \left(\partial_\mu + iA_\mu\right)\phi(x) \tag{1.158}$$

such that

$$D_\mu\phi'(x) = D_\mu\left(e^{i\alpha(x)}\phi(x)\right) = e^{i\alpha(x)}D_\mu\phi(x). \tag{1.159}$$

Equation (1.159) only holds provided that the connection $A = A_\mu dx^\mu$ transforms as

$$A_\mu(x) \mapsto A'_\mu(x) = A_\mu(x) - \partial_\mu\alpha(x) \tag{1.160}$$

under the local $U(1)$ transformation (1.156).

Let us now construct an action which is invariant under (1.156) and (1.160). Starting from the Lagrangian given by (1.16), we replace the partial derivative ∂_μ by the covariant derivative D_μ as given by (1.158). This procedure, known as *minimal coupling*, leads to

$$S = -\int d^dx \left((D_\mu\phi)^*D^\mu\phi + m^2\phi^*\phi\right). \tag{1.161}$$

By construction this action is invariant under (1.156). So far the gauge field A_μ is non-dynamical. However, using the covariant derivative D_μ and commutators thereof, such as $[D_\mu, D_\nu] = F_{\mu\nu}$, we may construct the field strength tensor

$$F_{\mu\nu} \equiv \partial_\mu A_\nu - \partial_\nu A_\mu, \tag{1.162}$$

which is unaffected by gauge transformations of A_μ since $\partial_{[\mu}\partial_{\nu]}\alpha(x) = 0$. Since $F_{\mu\nu}$ is a first derivative of the gauge field, the kinetic term of the gauge field A_μ is given by

$$S = -\frac{1}{4g^2}\int d^dx\, F_{\mu\nu}(x)F^{\mu\nu}(x). \tag{1.163}$$

Equation (1.163) is the action of a pure Abelian gauge theory. Its equation of motion reads

$$\partial_\mu F^{\mu\nu} = 0. \tag{1.164}$$

In addition to (1.164), we have a Bianchi identity which is given by

$$\partial_\mu F_{\nu\rho} + \partial_\rho F_{\mu\nu} + \partial_\nu F_{\rho\mu} = 0. \tag{1.165}$$

Note that (1.165) is satisfied using the definition of the field strength tensor (1.162). The equation of motion (1.164) and the Bianchi identity (1.165) give rise to the source-free Maxwell equations.

Exercise 1.7.1 Under a Lorentz transformation $x \mapsto x' = \Lambda x$, the gauge field A transforms as

$$A^\mu(x) \mapsto A'^\mu(x) = \Lambda^\mu_\nu A^\nu(\Lambda^{-1}x). \tag{1.166}$$

Canonically normalised kinetic term for gauge fields

Due to the factor $1/g^2$, the kinetic term in (1.163) is not canonically normalised. Redefining the gauge field by $A_\mu = g\mathcal{A}_\mu$ we obtain the canonically normalised action

$$S = -\frac{1}{4} \int d^d x \, \mathcal{F}_{\mu\nu}(x) \mathcal{F}^{\mu\nu}(x) \qquad (1.173)$$

with $\mathcal{F}_{\mu\nu} = \partial_\mu \mathcal{A}_\nu - \partial_\nu \mathcal{A}_\mu$. The corresponding covariant derivative is given by

$$D_\mu = \partial_\mu + ig\mathcal{A}_\mu. \qquad (1.174)$$

Show that the action (1.163) is invariant under (1.166) and derive the following infinitesimal transformation law using $\Lambda^\mu{}_\nu = \delta^\mu{}_\nu + \omega^\mu{}_\nu$ with $\omega_{\mu\nu} = -\omega_{\nu\mu}$,

$$\delta A_\mu = \omega_{\nu\lambda} x^\nu \partial^\lambda A_\mu - \omega_\mu{}^\nu A_\nu. \qquad (1.167)$$

Using (1.166) for translations, the energy-momentum tensor for an Abelian gauge theory (1.163) can be constructed by generalising (1.26) to gauge fields. This gives

$$\Theta_{\mu\nu} = \frac{1}{g^2} \left(\partial_\nu A_\lambda F_\mu{}^\lambda - \frac{1}{4} \eta_{\mu\nu} F_{\alpha\beta} F^{\alpha\beta} \right). \qquad (1.168)$$

Clearly, the energy-momentum tensor $\Theta_{\mu\nu}$ is not symmetric. However, we can obtain a symmetric energy-momentum tensor using the procedure given in section 1.2, by defining

$$T_{\mu\nu} = \Theta_{\mu\nu} + \partial^\lambda f_{\lambda\mu\nu}, \quad f_{\lambda\mu\nu} = -\frac{1}{g^2} A_\nu F_{\mu\lambda}. \qquad (1.169)$$

We obtain for the canonical energy-momentum tensor $T_{\mu\nu}$

$$T_{\mu\nu} = \frac{1}{g^2} \left(F_{\mu\lambda} F_\nu{}^\lambda - \frac{1}{4} \eta_{\mu\nu} F_{\alpha\beta} F^{\alpha\beta} \right) \qquad (1.170)$$

which is symmetric in μ and ν and in four dimensions is also traceless.

Exercise 1.7.2 By considering the infinitesimal Lorentz transformation (1.167), derive the conservation of the current

$$N^{\nu\kappa\lambda} = x^\kappa T^{\lambda\nu} - x^\lambda T^{\kappa\nu}, \qquad \partial_\nu N^{\nu\kappa\lambda} = 0. \qquad (1.171)$$

Use the result of exercise 1.2.3 to conclude that $T_{\mu\nu}$ is symmetric in μ and ν.

Pure Abelian gauge theory is a free theory. In order to consider interactions, we combine the actions (1.161) and (1.163),

$$S = \int d^d x \left(-\frac{1}{4g^2} F_{\mu\nu} F^{\mu\nu} - (D_\mu \phi)^* D^\mu \phi - m^2 \phi^* \phi \right), \qquad (1.172)$$

where D_μ is given by (1.158). The action (1.172) describes interactions between a complex scalar field ϕ with the gauge field A and therefore is known as *scalar electrodynamics*.

Exercise 1.7.3 We may also minimally couple the gauge field to a Dirac spinor with action
(1.141). Show that the Lagrangian is given by

$$\mathcal{L} = -\frac{1}{4g^2} F^{\mu\nu} F_{\mu\nu} + i\bar{\Psi} \slashed{D} \Psi - m\bar{\Psi}\Psi , \qquad (1.175)$$

where $\slashed{D} = \gamma^\mu D_\mu$ and D_μ is defined in (1.158). This well-studied example describes
(quantum) electrodynamics (QED), i.e. the interactions of electrons, given by the
Dirac spinor Ψ, with light as described by the $U(1)$ gauge field A_μ.

Non-Abelian gauge theory

So far we have discussed how to promote an Abelian global symmetry transformation
such as (1.35) to a local symmetry of the action. We may also apply this procedure to
a non-Abelian global symmetry transformation. In exercise 1.2.6 we studied an example
involving $U(N)$ or $O(2N)$ non-Abelian symmetries. Let us generalise this example and
consider a field transforming in the *fundamental representation* of a compact Lie group [6]
such as $SU(N), SO(N)$ or $USp(N)$. In particular, we may view the field ϕ^i as an element of
an N-dimensional vector space on which the Lie group acts as $\phi(x) \mapsto \phi'(x) = U(x)\phi(x)$.
In terms of components,

$$\phi^j(x) \ \mapsto \ \phi^{j\prime}(x) = \ \left(e^{i\alpha^a(x)T_a}\right)^j{}_k \phi^k(x) \equiv U^j{}_k(x)\phi^k(x), \qquad j,k = 1,2,...,N. \quad (1.176)$$

Here $U(x) = \exp(i\alpha^a(x)T_a)$ is an element of the Lie group $\mathcal{G}$ and T_a are the generators of
the corresponding Lie algebra with commutation relations

$$[T_a, T_b] = if_{ab}{}^c T_c. \qquad (1.177)$$

For example, for $SU(N)$ the corresponding Lie algebra is $\mathfrak{su}(N)$ whose generators T_a in
the fundamental representation are traceless Hermitian $N \times N$ matrices. Taking $\alpha^a(x)$ to
be infinitesimal in (1.176), the field ϕ transforms as

$$\phi^j(x) \mapsto \phi^{j\prime}(x) = \phi^j(x) + i\alpha^a(x) (T_a)^j{}_k \phi^k(x). \qquad (1.178)$$

As in the Abelian case, the derivative of ϕ does not transform as ϕ itself under the non-
Abelian symmetry (1.176). Defining a gauge covariant derivative D_μ by

$$(D_\mu)^i{}_j \equiv \delta^i{}_j \partial_\mu + i A^a_\mu (T_a)^i{}_j \qquad (1.179)$$

and imposing

$$A_\mu(x) \mapsto A'_\mu(x) = A^U_\mu(x) \equiv U(x)A_\mu U(x)^\dagger - iU(x)\partial_\mu U(x)^\dagger \qquad (1.180)$$

as transformation law for A_μ under (1.176), we can check explicitly that $D_\mu\phi$ transforms in
the same way as ϕ under a non-Abelian gauge transformation (1.176). Therefore the action

---

[6] For more details on Lie groups and Lie algebras please consult appendix B, where the classical Lie groups
$SU(N), SO(N)$ and $USp(N)$ as well as the representations of the corresponding Lie algebras are discussed.
Note that for $USp(N)$, N has to be even.

for ϕ^i is gauge invariant provided we replace ∂_μ by D_μ. Defining the field strength tensor by $F_{\mu\nu} = -i[D_\mu, D_\nu]$, or equivalently in terms of the gauge field by

$$
\begin{aligned}
F_{\mu\nu} \equiv F^a_{\mu\nu} T_a & \equiv \partial_\mu A_\nu - \partial_\nu A_\mu + i[A_\mu, A_\nu] \\
& = \left(\partial_\mu A^a_\nu - \partial_\nu A^a_\mu - f_{bc}{}^a A^b_\mu A^c_\nu\right) T_a,
\end{aligned}
\tag{1.181}
$$

we may also form a gauge invariant action with at most two derivatives,

$$
S[A] = -\frac{1}{4g^2} \int d^d x \, F^{a\mu\nu} F^a_{\mu\nu}.
\tag{1.182}
$$

In contrast to Abelian gauge theories, non-Abelian gauge theories are interacting theories since the action also gives rise to cubic and quartic interaction terms, as may be seen by inserting (1.181) into (1.182).

Using a redefinition of the gauge field as in box 1.2, we can normalise the action (1.182) canonically. Note that in this case $D_\mu = \partial_\mu - igA_\mu$ and therefore it is convenient to define the field strength tensor by $F_{\mu\nu} = -(i/g)[D_\mu, D_\nu]$. As a consequence of the redefinition of the gauge field, the coupling constant g also appears in (1.181). In $F^a_{\mu\nu} F^{a\mu\nu}$, the cubic term is proportional to g, while the quartic term is proportional to g^2.

Exercise 1.7.4   Show that the gauge transformation (1.180) for the field strength tensor reads

$$
F_{\mu\nu}(x) \mapsto F'_{\mu\nu}(x) = U(x)F_{\mu\nu}(x)U^\dagger(x)
\tag{1.183}
$$

and that the action (1.182) is indeed gauge invariant.

We may rewrite the Lagrangian (1.182) in terms of a trace using

$$
\mathrm{Tr}(T_a T_b) = C(\mathbf{R})\delta_{ab}.
\tag{1.184}
$$

Here, $C(\mathbf{R})$ is a real number which depends on the representation $\mathbf{R}$ of the gauge group and also on the normalisation of the generators T_a. We may also define the index $T(\mathbf{R})$ of a particular representation $\mathbf{R}$ of the Lie algebra by $T(\mathbf{R}) = C(\mathbf{R})/C(\mathbf{fund})$ where **fund** represents the fundamental representation. A discussion of Lie algebras and their representations is found in appendix B.

For the fundamental representation of $\mathfrak{su}(N)$, usually denoted by its dimension, $\mathbf{N}$, $C(\mathbf{N}) = 1/2$ and therefore the action (1.182) may be written as

$$
S[A] = -\frac{1}{2g^2} \int d^d x \, \mathrm{Tr}(F^{\mu\nu} F_{\mu\nu}).
\tag{1.185}
$$

So far we have considered the matter field ϕ as an N-dimensional vector with components ϕ^k. In addition to such matter fields transforming in the fundamental representation of the gauge group, we may also consider fields in other higher-dimensional representations of the gauge group. This may be done by generalising the transformation laws (1.176) and (1.178) as well as the covariant derivative (1.179) to generators T_a belonging to higher dimensional representations of the Lie algebra.

For example, in the *adjoint representation* of the gauge group the fields Φ can be written as $\Phi = \Phi^a T_a$. This means that Φ are matrices with components $\Phi^i{}_j$ which transform as

$$\Phi^i{}_j \;\equiv\; \Phi^a \, (T_a)^i{}_j \;\mapsto\; \Phi'^i{}_j = \left(e^{i\alpha^b T^b}\right)^i{}_k \Phi^a \, (T_a)^k{}_l \left(e^{-i\alpha^c T_c}\right)^l{}_j, \tag{1.186}$$

or $\Phi(x) \mapsto U(x)\Phi(x)U(x)^\dagger$ for short. Note that this is precisely the transformation law of $F_{\mu\nu}$ derived in exercise 1.7.4. In other words, the field strength tensor transforms in the adjoint representation of the gauge group.[7] For a field Φ in the adjoint representation, the covariant derivative is

$$D_\mu \Phi = \partial_\mu \Phi + i[A_\mu, \Phi]. \tag{1.187}$$

Exercise 1.7.5 Show that the equation of motion for the action (1.182) may be written as

$$D_\mu F^{\mu\nu} = 0, \tag{1.188}$$

where D_μ acts on $F^{\mu\nu}$ as given by (1.187). Moreover, show that the Bianchi identity

$$D_\mu F_{\nu\rho} + D_\rho F_{\mu\nu} + D_\nu F_{\rho\mu} = 0 \tag{1.189}$$

is automatically satisfied.

Exercise 1.7.6 In four dimensions, a term of the form

$$S^{\text{top}}[A] = \int \mathrm{d}^4 x \, \frac{\vartheta}{32\pi^2} F^a_{\mu\nu} \tilde{F}^{a\mu\nu} \tag{1.190}$$

may also be added to the action, with $\tilde{F}^{a\mu\nu} = \frac{1}{2}\epsilon^{\mu\nu\rho\sigma} F^a_{\rho\sigma}$ the dual field strength tensor. $\epsilon^{\mu\nu\rho\sigma}$ is the totally antisymmetric tensor in four dimensions normalised such that $\epsilon^{0123} = -1$. Show that (1.190) is also gauge invariant but does not contribute to the equations of motion. Therefore, in most cases we will set $\vartheta = 0$.

Exercise 1.7.7 Show that we may minimally couple the non-Abelian gauge field to a Dirac spinor Ψ (transforming under the representation **R** of the gauge group) with Lagrangian

$$\mathcal{L} = -\frac{1}{4g^2} F^{a\mu\nu} F^a_{\mu\nu} + i\bar{\Psi}\slashed{D}\Psi - m\bar{\Psi}\Psi, \tag{1.191}$$

where $\slashed{D} = \gamma^\mu D_\mu$. D_μ is defined in (1.179) where the generators T_a have to be in the representation **R**.

1.7.2 Quantisation of gauge theories

There are some additional features to be taken into account when quantising a gauge theory. This is due to the fact that gauge fields related by a gauge transformation are physically equivalent. In fact, the equation of motion (1.188) does not have a unique solution for specified boundary and initial conditions. This leads to difficulties for instance when defining the propagator, for which the equation of motion has to be inverted. The crucial

---

[7] A colloquial expression is to state that the *gauge field* itself transforms in the adjoint representation, although strictly speaking this is not correct since the transformation law (1.180) also involves an inhomogeneous term of the form $U(x)\partial_\mu U(x)^\dagger$.

point is that the physical dynamical variables of a gauge theory really belong to the equivalence class of gauge fields modulo gauge transformations,

$$\mathcal{A}/\mathcal{G} = \left\{ A_\mu \sim A'_\mu : A_\mu, A'_\mu \in \mathcal{A}, \exists\, U \in \mathcal{G} \text{ with } A'_\mu = A^U_\mu \right\}, \qquad (1.192)$$

where A^U_μ is the gauge transformation of A_μ for the group element U of (1.176). The quantisation procedure has to be defined on $\mathcal{A}/\mathcal{G}$, thus eliminating all redundant unphysical degrees of freedom. In the path integral approach, this involves finding the correct integration measure on $\mathcal{A}/\mathcal{G}$. In general, in the path integral we split the measure on $\mathcal{A}$ into

$$Z[J] = \int_{\mathcal{A}} \mathcal{D}A\, e^{i\mathcal{S}[A]+i\int d^dx J^\mu A_\mu} = \left(\int_{\mathcal{G}} \mathcal{D}U \right) \left(\int_{\mathcal{A}/\mathcal{G}} \widetilde{\mathcal{D}A}\, e^{i\mathcal{S}[A]+i\int d^dx J^\mu A_\mu} \right), \quad (1.193)$$

where $\mathcal{D}U = \prod_x dU(x)$ is the measure on the group space $\mathcal{G}$. The integral over $\mathcal{D}U$ will give a number which we may ignore by redefining the generating functional $Z[J]$. In the following the task is to construct the measure $\widetilde{\mathcal{D}A}$.

Ideally, in the path integral on $\mathcal{A}/\mathcal{G}$, each physical gauge field configuration should be taken into account only once, or in other words, each gauge orbit should be intersected only once. This requires a gauge fixing condition $f(A_\mu)$, such that the equation $f(A^U_\mu) = 0$ has a unique solution U_0 for a given A_μ. To compensate for the choice of f, we define $\Delta_f[A_\mu]$ such that

$$\Delta_f[A_\mu] \int_{\mathcal{G}} \mathcal{D}U\, \delta\left(f(A^U_\mu) \right) = 1, \qquad (1.194)$$

where $\delta(f(A^U_\mu))$ is the delta function. Assuming that f has a single zero for $U = U_0$ we find

$$\Delta_f[A_\mu] = \det M_f, \qquad \text{where} \quad M_f(x,y) = \left. \frac{\delta f(x)}{\delta U(y)} \right|_{U=U_0}. \qquad (1.195)$$

Exercise 1.7.8   Using the invariance of $\mathcal{D}U$, i.e. $\mathcal{D}(U\tilde{U}) = \mathcal{D}U$ for a given $\tilde{U}$, show that Δ_f is gauge invariant.

Inserting (1.194) into the path integral (1.193) we get

$$Z[J] = \int_{\mathcal{A}} \mathcal{D}A\, \Delta_f[A_\mu] \int_{\mathcal{G}} \mathcal{D}U\delta\left(f(A^U_\mu) \right)\, e^{i\mathcal{S}[A]+i\int d^dx\, J^\mu A_\mu}$$

$$= \left(\int_{\mathcal{G}} \mathcal{D}U \right) \left(\int_{\mathcal{A}} \mathcal{D}A\, \Delta_f[A_\mu]\delta\left(f(A_\mu) \right)\, e^{i\mathcal{S}[A]+i\int d^dx\, J^\mu A_\mu} \right). \qquad (1.196)$$

In order to derive the second line we used that $\mathcal{D}A$, $\Delta_f[A_\mu]$ and $\mathcal{S}[A]$ are invariant under $A \to A^U$. Note also that (1.196) has the desired product structure (1.193), and we can therefore identify $\int \widetilde{\mathcal{D}A}$ by

$$\int_{\mathcal{A}/\mathcal{G}} \widetilde{\mathcal{D}A} \equiv \int_{\mathcal{A}} \mathcal{D}A\, \Delta_f[A_\mu]\delta\left(f(A_\mu) \right). \qquad (1.197)$$

The contribution $\Delta_f[A_\mu]$ may be written in terms of a path integral over two Grassmann fields[8] $c(x)$ and $\bar{c}(x)$,

$$\Delta_f[A_\mu] = \det M_f = \int \mathcal{D}c\mathcal{D}\bar{c} \exp\left(i\int d^dx\, d^dy\, \bar{c}(x)M_f(x,y)c(y)\right) \equiv \int \mathcal{D}c\mathcal{D}\bar{c}\, e^{iS_{\text{gh}}}. \tag{1.198}$$

The fields $c(x)$, $\bar{c}(x)$ are referred to as *ghost* fields which may be viewed as spurious fields contributing to the Feynman graphs.

Moreover, we may convert $\delta(f(A_\mu))$ into an exponential factor. To see this, we slightly generalise the gauge fixing condition to

$$f(A_\mu(x)) = B(x), \tag{1.199}$$

where $B(x)$ is a function of spacetime which does not depend on $A_\mu(x)$ explicitly. Δ_f is still defined by (1.194). Inserting a constant of the form

$$\text{Constant} = \int \mathcal{D}B \exp\left(-\frac{i}{2\xi}\int d^dx\, B^2\right), \tag{1.200}$$

where ξ is the *gauge parameter*, into the generating functional (1.196) and ignoring overall constants we obtain

$$Z[J] = \int \mathcal{D}A_\mu \mathcal{D}c\mathcal{D}\bar{c}\mathcal{D}B\, e^{iS[A]+iS_{\text{gh}}+i\int d^dx\left(J^\mu A_\mu - \frac{1}{2\xi}B^2(x)\right)} \delta(f(A_\mu) - B). \tag{1.201}$$

We may then perform the integral over $\mathcal{D}B$ and obtain

$$Z[J] = \int \mathcal{D}A_\mu \mathcal{D}c\mathcal{D}\bar{c} \exp\left(iS_{\text{eff}}[A,c,\bar{c}] + i\int d^dx\, J^\mu A_\mu\right), \tag{1.202}$$

with

$$S_{\text{eff}}[A,c,\bar{c}] = S[A] + S_{\text{gf}} + S_{\text{gh}}, \quad S_{\text{gf}} = -\frac{1}{2\xi}\int d^dx\, \left(f_a(A_\mu)\right)^2, \tag{1.203}$$

and S_{gh} as given by (1.198).

In many cases, it is sufficient to consider a very simple choice for the gauge, which is the linear covariant gauge given by $f(A_\mu) = \partial_\mu A^\mu$, and thus $M_f^{ab}(x,y) = -\partial^\mu D_\mu^{ab}\delta^d(x-y)$. With this gauge choice, the Feynman rules for the canonically normalised field introduced in box 1.2 are derived straightforwardly. In the case of gauge group $U(N)$, we have for the propagators of the gauge field and of the ghosts

$$\Delta_{A,\mu\nu,ab}(p) = \left(\frac{\eta_{\mu\nu}}{p^2 - i\epsilon} + (\xi - 1)\frac{p_\mu p_\nu}{(p^2 - i\epsilon)^2}\right)\delta_{ab},$$
$$\Delta_{\text{gh},ab}(p) = \frac{1}{p^2 - i\epsilon}\delta_{ab}. \tag{1.204}$$

Exercise 1.7.9 Derive the remaining Feynman rules for a non-Abelian gauge theory including the rules involving ghost fields. In the non-Abelian case, the non-linear contribution to $F_{\mu\nu}$ gives rise to interactions with the vertices shown in figure 1.1 for the canonically normalised field of box 1.2.

---

[8] For more details on Grassmann fields see appendix A.

Figure 1.1   Feynman rules for the three- and four-point vertex for canonically normalised gauge field.

Exercise 1.7.10   In the case of $U(1)$ Abelian gauge theory, the formalism simplifies considerably. In particular show that for any gauge fixing condition (1.199), the linear response matrix M_f as given by (1.195) will be independent of A_μ and thus it does not contribute to the path integral over A_μ. In other words, the Faddeev–Popov ghosts decouple from the gauge field and may be absorbed into the overall normalisation.

Based on this formalism, it is now possible to evaluate the renormalisation behaviour of gauge theories. We summarise its main features in section 1.7.3 below.

1.7.3   Renormalisation of gauge theories

In this section we are mainly interested in non-Abelian gauge theories in four spacetime dimensions and we discuss their renormalisation. The gauge coupling is dimensionless and thus satisfies a necessary condition for renormalisability.

To calculate the β function, we have to consider graphs with two, three and four external legs, involving both gauge field and ghost propagators. If we additionally couple N_f Dirac spinors transforming in the representation **R** of the gauge group minimally to the non-Abelian gauge field as discussed in exercise 1.7.7, the one-loop β function reads

$$\beta(g) = -\frac{g^3}{16\pi^2}\left(\frac{11}{3}C(\mathbf{adj}) - \frac{4}{3}N_f\,C(\mathbf{R})\right), \qquad (1.205)$$

where $C(\mathbf{R})$ is given by (1.184) and **adj** denotes the adjoint representation of the gauge group. For the Lie algebra $\mathfrak{su}(N)$, $C(\mathbf{adj}) = N$ while $C(\mathbf{fund}) = 1/2$ and we obtain the well-known result

$$\beta(g) = -\frac{g^3}{48\pi^2}\left(11N - 2N_f\right) \qquad (1.206)$$

for the β function of $SU(N)$ Yang–Mills theory coupled to N_f Dirac spinors in the fundamental representation. For $N = 3$, the one-loop β function is negative unless $N_f > 16$ implying that $g = 0$ is a stable UV fixed point for $\mu \to \infty$. This means that the theory becomes free at very high energies, i.e. *asymptotically free* as defined on page 26.

Exercise 1.7.11   Consider an Abelian $U(1)$ gauge theory coupled to a Dirac fermion Ψ as discussed in exercise 1.7.3. Show that the one-loop β function is given by

$$\beta(g) = \frac{g^3}{12\pi^2}. \qquad (1.207)$$

Note that the β function is positive, implying that $g = 0$ is an IR stable fixed point as $\mu \to 0$; thus the theory known as QED is free at low energies, while at high energies the coupling constant diverges.

1.7.4 Wilson loops in gauge theories

In addition to local operators constructed from elementary fields such as $F_{\mu\nu}$, there is also an important class of non-local operators in a gauge theory, the *Wilson loops*. These are non-local gauge invariant operators which describe the parallel transport for a quark along a closed path $\mathcal{C}$. Under a gauge transformation the quark field Ψ transforms in the fundamental representation of the gauge group. In what follows, we assume that its motion does not source the gauge field. In other words, we view the quark as an infinitely heavy test particle. The quark field Ψ picks up a phase factor $W(\mathcal{C})$ around the closed path $\mathcal{C}$,

$$\Psi(x + \mathcal{C}) = W(\mathcal{C})\Psi(x). \tag{1.208}$$

For pure Yang–Mills theory with gauge group $SU(N)$, the Wilson loop operator $\mathcal{W}(\mathcal{C})$ in the fundamental representation is defined by

$$\mathcal{W}(\mathcal{C}) = \frac{1}{N}\mathrm{Tr}W(\mathcal{C}) = \frac{1}{N}\mathrm{Tr}\left(\mathcal{P}\exp\left[i\oint_{\mathcal{C}}\mathrm{d}x^{\mu}A_{\mu}\right]\right), \tag{1.209}$$

where the trace is over the fundamental representation of the gauge group. Using a particular parametrisation $x^{\mu}(s)$ for the closed path $\mathcal{C}$ with $s \in [0, 1]$, the exponent reads

$$i\oint_{\mathcal{C}}\mathrm{d}x^{\mu}A_{\mu} = i\int_{0}^{1}\mathrm{d}s\frac{\mathrm{d}x^{\mu}}{\mathrm{d}s}A_{\mu}\left(x(s)\right). \tag{1.210}$$

Note that $A_{\mu} = A_{\mu}^{a}T_{a}$ are Lie algebra valued fields and therefore do not commute in general. This leads to an ambiguity in the definition of the exponential function in terms of its Taylor series. This ambiguity is cured by imposing the path ordering prescription denoted by $\mathcal{P}$: gauge fields $A(x(s))$ are ordered such that higher values of the parameter s along the path appear on the left, i.e. for example

$$\mathcal{P}\left(A(x(s_1))A(x(s_2))\right) = \begin{cases} A(x(s_1))A(x(s_2)) & \text{for } s_1 > s_2, \\ A(x(s_2))A(x(s_1)) & \text{for } s_1 < s_2. \end{cases} \tag{1.211}$$

Exercise 1.7.12 We generalise the definition of $W(\mathcal{C})$ to a path with starting point x and end point y and denote it by $W(x; y)$. $W(x; y)$ is a *Wilson line*. Show that $W(x; y)$ transforms as

$$W(x; y) \mapsto U(y)W(x; y)U^{\dagger}(x) \tag{1.212}$$

under a gauge transformation of the form (1.180). Use this result to show that $\mathcal{W}(\mathcal{C})$ is gauge invariant for a closed path $\mathcal{C}$.

Exercise 1.7.13 Show that $\mathcal{W}(\mathcal{C})^{\dagger} = \mathcal{W}(-\mathcal{C})$, where $-\mathcal{C}$ denotes the opposite orientation of the path $\mathcal{C}$.

The vacuum expectation value of the Wilson loop operator $\langle\mathcal{W}(\mathcal{C})\rangle$ for a certain contour $\mathcal{C}$, such as the rectangular path in figure 1.2 encodes interesting information about the gauge theory. For example, it provides an order parameter for *confinement* and *deconfinement*, an important property of non-Abelian gauge theories which we will discuss extensively

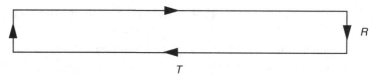

Figure 1.2 Rectangular Wilson loop which extends a time T in the t direction and a distance R in the x^1 direction. We consider the limit $T \gg R$ and R fixed.

in chapter 13. As an example of the Wilson loop, we choose the contour $\mathcal{C}$ to be a $T \times R$ rectangular path in the (t, x^1)-plane as shown in figure 1.2. In the limit $R \ll T$, this configuration corresponds to a static *quark–antiquark pair* at a fixed distance R: the temporal sides of the rectangle are the worldlines of the two particles. Since the rectangle is oriented, one of the temporal sides corresponds to a particle propagating forward in time, i.e. a quark Ψ, while the other temporal side corresponds to a particle propagating backwards in time, or an antiparticle, i.e. the antiquark $\bar{\Psi}$. Thus, only the potential term in the Yang–Mills action contributes and we have

$$\langle \mathcal{W}[\mathcal{C}(R, T)] \rangle = A(R) \exp[-TV(R)], \qquad (1.213)$$

with $V(R)$ the quark–antiquark potential and R-dependent normalisation $A(R)$. Consequently, from the Wilson loop we obtain the quark–antiquark potential by virtue of

$$V(R) \equiv - \lim_{T \to \infty} \frac{\ln \langle \mathcal{W}[\mathcal{C}(R, T)] \rangle}{T}. \qquad (1.214)$$

A more formal argument for the validity of (1.213) is that when a complete basis of eigenstates is inserted into a quark correlator then, in the limit of very large T, the lowest energy eigenvalue dominates in the exponential $\exp(-\mathcal{S})$, which is precisely the potential.

Exercise 1.7.14 Calculate $\langle \mathcal{W}(\mathcal{C}) \rangle$ for the rectangular path discussed above within $U(1)$ gauge theory. Identify the static potential from your result.

1.7.5 Large N expansion in gauge theory

As first pointed out by Gerard 't Hooft in 1974, non-Abelian gauge theories simplify considerably in the limit $N \to \infty$. This limit also enters gauge/gravity duality in a crucial way. The large N limit is motivated by an expansion used in statistical mechanics, where the number of field components is taken to be large and an expansion in the inverse of this number is performed. The expansion of non-Abelian gauge theory in $1/N$ rearranges the Feynman diagrams in such a way that they correspond to a string theory expansion with string coupling $1/N$. This suggests that non-Abelian gauge theories are equivalent to string theories, at least at large N. A particular virtue of the AdS/CFT correspondence is to make this mapping between field theory and string theory precise for a well-defined class of examples.

Consider $SU(N)$ Yang–Mills theory. Its β function is given by (1.206) with $N_{\mathrm{f}} = 0$. Naively, in the $N \to \infty$ limit, the β function diverges. However, if $\lambda \equiv g^2 N$ is kept fixed while taking the large N limit, then the renormalisation group equation for λ has finite coefficients, and from (1.206) we obtain

$$\mu \frac{\mathrm{d}\lambda}{\mathrm{d}\mu} = -\frac{11}{24\pi^2}\lambda^2 + \mathcal{O}(\lambda^3). \tag{1.215}$$

λ is referred to as the *'t Hooft coupling*. Thus the limit $N \to \infty$ with $\lambda = g^2 N$ kept fixed exists and is non-trivial since the corresponding field theory is not free as we can see from (1.215). In particular, the effective coupling constant in the large N limit is not g which goes to zero but rather λ.

To illustrate the relation between a field theory expansion in the large N limit and a string theory expansion, let us consider a toy model involving just a scalar field Φ in the adjoint representation of the gauge group, with $\Phi = \Phi^a T_a$. More explicitly, writing out the indices of the matrices T_a, we have

$$\Phi^i{}_j \equiv \Phi^a (T_a)^i{}_j. \tag{1.216}$$

For the generators T_a themselves, we may in principle take any representation. Here, we take T_a to be in the fundamental representation.[9] Moreover we assume that the interaction vertices of this matrix model mimic Yang–Mills theory: in a canonical normalisation, the cubic vertex is proportional to g and the quartic vertex is proportional to g^2, as in figure 1.1. The Lagrangian of this toy matrix model then reads

$$\mathcal{L} = -\frac{1}{2}\mathrm{Tr}\left(\partial_\mu \Phi \, \partial^\mu \Phi\right) + g\,\mathrm{Tr}\left(\Phi^3\right) + g^2\,\mathrm{Tr}\left(\Phi^4\right). \tag{1.217}$$

A rescaling $\tilde{\Phi} = g\Phi$ turns this Lagrangian into

$$\mathcal{L} = \frac{1}{g^2}\left[-\frac{1}{2}\mathrm{Tr}\left(\partial_\mu \tilde{\Phi}\,\partial^\mu \tilde{\Phi}\right) + \mathrm{Tr}\left(\tilde{\Phi}^3\right) + \mathrm{Tr}\left(\tilde{\Phi}^4\right)\right]. \tag{1.218}$$

To have a well-defined $N \to \infty$ limit, it is convenient to introduce the 't Hooft coupling

$$\lambda \equiv g^2 N \tag{1.219}$$

as above. If we send $N \to \infty$ at constant λ, the coefficient of (1.218) diverges, but the number of components in the fields diverges as well. In fact, a subtle cancellation mechanism between the two infinities will take place. To see this at work, we analyse particular Feynman graphs in the 't Hooft limit. Let us first concentrate on theories with gauge group $U(N)$. In the notation (1.216), the free propagator of the field $\tilde{\Phi}$ has the structure [10]

$$\left\langle \tilde{\Phi}^i{}_j(x)\,\tilde{\Phi}^k{}_l(y)\right\rangle = \delta^i{}_l \delta^k{}_j \frac{g^2}{4\pi^2(x-y)^2}, \tag{1.221}$$

---

[9] We choose the generators in such a way that $\mathrm{Tr}\,(T_a T_b) = \delta_{ab}$, which corresponds to $C(N) = 1$ in (1.184) for the group $U(N)$. This is a different convention than used in our discussion of non-Abelian gauge theory. The new convention is generally used both for large N theory and for string theory, as we will see below.

[10] Note that for the field Φ of (1.217), the free propagator reads

$$\left\langle \Phi^i{}_j(x)\,\Phi^k{}_l(y)\right\rangle = \delta^i{}_l \delta^k{}_j \frac{1}{4\pi^2(x-y)^2}, \tag{1.220}$$

without factors of the coupling constant g in the numerator.

The T_a in (1.222) form a complete set of matrices, implying that

$$\sum_{a=1}^{N^2} (T_a)^i_{\ j}(T_a)^k_{\ l} \propto \delta^i_{\ l}\,\delta^k_{\ j}.$$ (1.223)

The proportionality constant is fixed by considering $j = k$ and $i = l$ and summing over i and j. With

$$\sum_{a=1}^{N^2} \mathrm{Tr}(T_a T_a) = C(\mathbf{N}) \cdot \delta_{aa} = 1 \cdot N^2$$ (1.224)

we obtain (1.222).

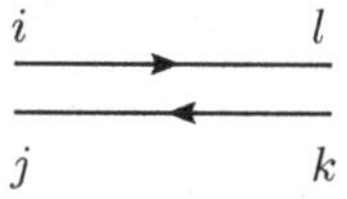

Figure 1.3   Double line notation of a field in the adjoint representation of the gauge group.

which is found using the $\mathfrak{u}(N)$ completeness relation

$$\sum_{a=1}^{N^2} (T_a)^i_{\ j}(T_a)^k_{\ l} = \delta^i_{\ l}\,\delta^k_{\ j},$$ (1.222)

as described in box 1.3. The spacetime dependence of the propagator in (1.221) is the appropriate one for a scalar field in four dimensions. Then the expression (1.221) for the propagator suggests a *double line notation* as shown in figure 1.3. The arrow on each line points from an upper to a lower index. Feynman diagrams then become networks of double lines. As can be read off from the Lagrangian (1.218), vertices scale as $g^{-2} = \frac{N}{\lambda}$, while propagators being the inverse of the kinetic term scale as $g^2 = \frac{\lambda}{N}$. Moreover, the sum over indices in a trace contributes a factor of N for each closed loop. If we introduce the shorthand notation (V, E, F) for the numbers of vertices, propagators (edges) and loops (faces) respectively, a Feynman diagram with V vertices, E propagators and F loops is proportional to

$$\mathrm{diagram}(V, E, F) \ \sim \ N^{V-E+F}\lambda^{E-V} \ = \ N^\chi\,\lambda^{E-V}.$$ (1.225)

The power of the expansion parameter N is precisely the *Euler characteristic*

$$\chi \ \equiv \ V - E + F \ = \ 2 - 2g,$$ (1.226)

related to the number of handles of the surface (the *genus*) g.

Any physical quantity in this theory may be expressed in an expansion of N and g. For example, the partition function Z and the generating functional W for connected Green's functions read

Figure 1.4 Two different vacuum amplitudes. The left diagram has genus zero, i.e. is a planar diagram and scales as N^2, the right diagram has genus one and scales as N^0. For the left diagram, $E = 3, F = 3, V = 2$, while for the right diagram, $E = 6, F = 2, V = 4$.

Figure 1.5 Propagator and interaction vertex for closed strings.

$$iW = \ln Z = \sum_{g=0}^{\infty} N^{2-2g} \sum_{i=0}^{\infty} c_{g,i}\, \lambda^i = \sum_{g=0}^{\infty} N^{2-2g} f_g(\lambda) \qquad (1.227)$$

with $f_g(\lambda)$ a polynomial in the 't Hooft coupling. For large N, the series is clearly dominated by surfaces of minimal genus $g = 0$, the so-called *planar diagrams*. As an example let us compare the vacuum amplitudes shown in figure 1.4. The graph shown on the left is planar and of order N^2, while the graph on the right is non-planar of genus one and scales as N^0, as may been seen from the counting of edges, faces and vertices as given in the caption of figure 1.4.

A crucial point in view of correspondences between quantum field theory and string theory is that the large N expansion is formally the same as a perturbation expansion of closed oriented strings with string coupling $1/N$ as we will see in chapter 4. The basic building blocks of this string expansion, the propagator and the cubic interaction vertex of a closed string, are shown in figure 1.5. The equivalence with the double line large N field theory expansion can be demonstrated by considering again the graphs of figure 1.4. By a smooth, topologically trivial transformation, the field theory graphs are mapped to corresponding string diagrams which are of the same genus. This is shown in figure 1.6. On the left we have the single line Feynman diagrams, in the middle we have the corresponding double line Feynman diagrams, and on the right we have the mapping to a sphere for genus zero and to a torus for genus one of the corresponding Feynman diagrams, respectively.

In the simple toy model considered here, it is not known which string theory fits the field theory perturbative series. For $\mathcal{N} = 4$ Super Yang–Mills theory, however, which is introduced in Chapter 3, the AdS/CFT correspondence tells us which string theory leads to the correct expansion: ten-dimensional type IIB superstring theory on $AdS_5 \times S^5$.

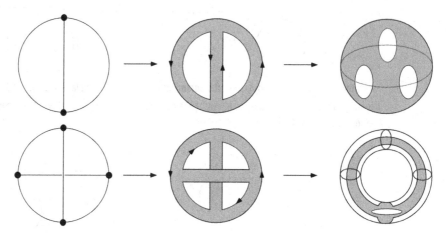

The double line field theory graphs are mapped to a string theory graph of the same genus.

In addition to the vacuum diagrams discussed above, we may also consider correlation functions involving *single-trace* gauge invariant operators of the form

$$\mathcal{O}_j(x) \equiv \frac{1}{N} \text{Tr} \left(\prod_{i=1}^{j} \Phi_i(x) \right). \tag{1.228}$$

These correlation functions are obtained by adding source terms for these operators to the original action,

$$\mathcal{S} \to \mathcal{S}' = \mathcal{S} + N \sum_j g_j \mathcal{O}_j. \tag{1.229}$$

Then, with W the generating functional for connected Green's functions computed from $\mathcal{S}'$, we have

$$\langle \mathcal{O}_1(x_1) \mathcal{O}_2(x_2) \cdots \mathcal{O}_n(x_n) \rangle_\text{c} = \frac{1}{i^{n-1} N^n} \frac{\delta^n W}{\delta g_1 \cdots \delta g_n} \bigg|_{g_j=0}. \tag{1.230}$$

The leading contribution to this correlator comes from planar diagrams with n operator insertions. Since W scales as N^2 in the planar limit, the correlator in (1.230) scales as N^{2-n}, such that the two-point function ($n = 2$) is canonically normalised and the three-point function ($n = 3$) scales as $1/N$. Therefore, $1/N$ may be viewed as a coupling constant which plays the same role as g_s in the string theory perturbative expansion.

Note that for large N, the three-point functions are suppressed compared to the two-point functions. Thus it seems that the theory becomes effectively semi-classical but still non-trivial. The semi-classical nature of the large-N limit can also be observed if we take into account the disconnected Feynman diagrams. For two-point functions of single trace operators (1.228), the disconnected diagrams dominate and we have

$$\langle \mathcal{O}_1(x_1) \mathcal{O}_2(x_2) \rangle = \langle \mathcal{O}_1(x_1) \rangle \langle \mathcal{O}_2(x_2) \rangle + \mathcal{O}(N^{-2}). \tag{1.231}$$

This is referred to as the *factorisation* property and holds for a general set of gauge invariant operators $\mathcal{O}_1, \ldots, \mathcal{O}_n$, and not only for single-trace operators.

So far we have considered the gauge group $U(N)$. However, the arguments presented above can be generalised to other gauge groups, for example $SU(N)$. For the gauge group $SU(N)$, the completeness relation (1.222) reads

$$\sum_{a=1}^{N^2-1} (T_a)^i{}_j (T_a)^k{}_l = \delta^i{}_l \delta^k{}_j - \frac{1}{N} \delta^i{}_j \delta^k{}_l. \tag{1.232}$$

Thus in the $N \to \infty$ limit, the term removing the trace in (1.232) vanishes, and the propagator is again of the form (1.221). Consequently, the structure of the planar diagrams is analogous to the $U(N)$ case. For $SO(N)$ or $USp(N)$ theories, the adjoint representation may be written as a product of two fundamental representations rather than a product of a fundamental and an anti-fundamental representation. Since the fundamental representation is real, there are no arrows on the propagators and we expect the planar diagrams obtained to be associated to a non-orientable string theory.

1.8 Symmetries, Ward identities and anomalies

1.8.1 Ward identities

The presence of a symmetry in a quantum field theory leads to relations between the correlation functions. These are known as the *Ward identities*. The Ward identities are the implications of Noether's theorem in quantum theory.

Let us consider the generating functional $Z[J]$ and change the variables $\phi(x) \mapsto \tilde{\phi}(x) = \phi(x) + \delta\phi(x)$, where $\delta\phi(x)$ is an arbitrary infinitesimal shift. The generating functional $Z[J]$ is invariant under this shift. Assuming that the measure is invariant, i.e. $\mathcal{D}\phi = \mathcal{D}\tilde{\phi}$, we obtain

$$0 = \delta Z[J] = i \int \mathcal{D}\phi \, e^{i(S + \int d^d x J(x)\phi(x))} \int d^d x \left(\frac{\delta S}{\delta \phi(x)} + J(x) \right) \delta\phi(x). \tag{1.233}$$

Taking functional derivatives with respect to $J(x_i)$ and subsequently setting J to zero, we obtain the Schwinger–Dyson equations

$$0 = i \left\langle \frac{\delta S}{\delta \phi(x)} \phi(x_1) \ldots \phi(x_n) \right\rangle + \sum_{j=1}^{n} \langle \phi(x_1) \ldots \phi(x_{j-1}) \delta(x - x_j) \phi(x_{j+1}) \ldots \phi(x_n) \rangle. \tag{1.234}$$

In particular we see that the classical equations of motion $\delta S / \delta \phi(x) = 0$ are satisfied up to contact terms when inserted into correlation functions.

Exercise 1.8.1 Solve the Schwinger–Dyson equations for $n = 1$ for a free real scalar field with mass m.

Let us apply the Schwinger–Dyson equations to continuous symmetry transformations $\phi(x) \mapsto \phi(x) + \delta\phi(x)$. The variation of the Lagrangian reads, without using the equations of motion,

$$
\begin{aligned}
\delta\mathcal{L} &= \frac{\partial\mathcal{L}}{\partial\phi(x)}\delta\phi(x) + \frac{\partial\mathcal{L}}{\partial(\partial_\mu\phi(x))}\partial_\mu\delta\phi(x) \\
&= \partial_\mu\left(\frac{\partial\mathcal{L}}{\partial(\partial_\mu\phi(x))}\delta\phi(x)\right) + \frac{\delta S}{\delta\phi(x)}\delta\phi(x),
\end{aligned}
\tag{1.235}
$$

where we have used equation (1.10). Suppose $\delta\phi$ corresponds to a symmetry which leaves the Lagrangian invariant, $\delta\mathcal{L} = 0$. Then the Noether current given by (1.22) with $X^\mu = 0$ satisfies

$$
\partial_\mu \mathcal{J}^\mu(x) = \frac{\delta S}{\delta\phi(x)}\delta\phi(x).
\tag{1.236}
$$

Inserting this equation into the Schwinger–Dyson equations, we obtain Ward identities of the form

$$
\partial_\mu\langle\mathcal{J}^\mu(x)(x_1)\dots\phi(x_n)\rangle - i\sum_{j=1}^{n}\langle\phi(x_1)\dots\delta\phi(x_j)\delta(x-x_j)\dots\phi(x_n)\rangle = 0.
\tag{1.237}
$$

Therefore the classical statement $\partial_\mu\mathcal{J}^\mu(x) = 0$ is true up to contact terms if we insert the statement into correlation functions. The contact terms depend on the symmetry transformation through the presence of $\delta\phi(x)$. If $\delta\phi$ does not involve time derivatives and if we know the Noether charge $\mathcal{Q} = \int \mathrm{d}^{d-1}\vec{x}\,\mathcal{J}^t$ we can reconstruct $\delta\phi$ using

$$
[\hat{\mathcal{Q}}, \hat{\phi}(x)] = i\delta\hat{\phi}(x).
\tag{1.238}
$$

Exercise 1.8.2   Verify (1.238) using the commutation relations (1.39).

1.8.2 Anomalies

In deriving the Schwinger–Dyson equation (1.234), we have assumed that the path integral measure is invariant under the symmetry transformation. However, this is not necessarily the case. Symmetries present at the classical level may be broken by *anomalies* when the theory is quantised.

An example of a global symmetry which is conserved in the classical theory but broken in the quantised theory is the axial symmetry introduced in (1.148). It turns out that the path integral measure is not invariant under this symmetry. Consider the Lagrangian of four-dimensional Abelian gauge theory coupled to a massless Dirac field given by (1.175) with $m = 0$. In the classical theory, this Lagrangian is invariant under the global $U(1)_A$ symmetry with transformations given by (1.148). In the classical theory, the corresponding axial current $J_5^\mu = \bar{\Psi}\gamma^\mu\gamma_5\Psi$ is divergence free, $\partial_\mu J_5^\mu = 0$. In the quantised theory, however, the path integral measure is not invariant under this symmetry. Equivalently, in the perturbative expansion the calculation of the contributions corresponding to the two triangle graphs shown in figure 1.7 gives rise to an anomalous contribution to the

Figure 1.7   Triangle Feynman graphs giving rise to the axial anomaly.

divergence of the axial current of the form

$$\partial_\mu \langle J_5^\mu \rangle = -\frac{g^2}{16\pi^2}\epsilon^{\mu\nu\rho\sigma}F_{\mu\nu}F_{\rho\sigma}, \tag{1.239}$$

which makes use of $4i\epsilon_{\mu\nu\sigma\rho} = \text{Tr}(\gamma_\mu\gamma_\nu\gamma_\sigma\gamma_\rho\gamma_5)$. This is known as the Adler–Bell–Jackiw or ABJ anomaly. Its origin may also be traced back to the fact that the measure of the path integral is no longer invariant under the axial symmetry. It can be shown that its coefficient is one-loop exact, i.e. there are no further contributions at higher orders in perturbation theory.

It is important to note that the axial symmetry which becomes anomalous according to (1.239) is a global symmetry. A gauge symmetry, on the other hand, may not be anomalous since this would correspond to an inconsistent theory. Using the fact that the conserved vector current operator is obtained by functionally varying the generating functional with respect to the gauge field, the anomaly (1.239) may also be written as an anomalous Ward identity for the three-point function obtained by varying (1.239) twice with respect to the gauge field,

$$\partial_\mu^x \langle J_5^\mu(x)J^\nu(y)J^\omega(z)\rangle = -\frac{g^2}{4\pi^2}\epsilon^{\alpha\nu\gamma\omega}\partial_\alpha\delta^{(4)}(x-y)\partial_\gamma\delta^{(4)}(x-z), \tag{1.240}$$

where J_5^μ, J^ν are defined in (1.147), (1.148).

For non-Abelian symmetries, we have to include a factor of

$$d_{abc}^{\mathbf{R}} = \text{Tr}\left(T_a^{\mathbf{R}}\left\{T_b^{\mathbf{R}}, T_c^{\mathbf{R}}\right\}\right) \tag{1.241}$$

in (1.240), where $\mathbf{R}$ denotes the representation of the massless fermions. It is convenient to define an anomaly coefficient $r(\mathbf{R})$ relative to the fundamental representation by

$$d_{abc}^{\mathbf{R}} \equiv r(\mathbf{R})d_{abc}^{\text{fund}} = r(\mathbf{R})\text{Tr}\left(T_a^{\text{fund}}\left\{T_b^{\text{fund}}, T_c^{\text{fund}}\right\}\right). \tag{1.242}$$

By definition, $r(\text{fund}) = 1$. It turns out that for the axial anomaly – and more generally for anomalous global symmetries – the anomaly coefficient in the IR is the same as in the UV, although the degrees of freedom in the IR are different from those in the UV. This property is known as *'t Hooft anomaly matching* which is summarised in box 1.4. 't Hooft anomaly matching states that the anomalies of chiral and axial currents must be the same in both the UV and IR limits of the renormalised theory. This provides an important constraint on the anomaly coefficients in the IR.

Box 1.4	't Hooft anomaly matching

't Hooft anomaly matching states that the anomalies of chiral and axial currents must be the same in both the UV and IR limits of the renormalised theory. The argument is as follows. Consider a global symmetry with group G. The associated current has an anomalous divergence involving background fields with coefficient r_{UV}. Now promote G to a local gauge symmetry. For consistency of the theory, the anomaly is now required to vanish. This is achieved by adding spectator fermions to the original theory for which the anomaly coefficient is r_S such that $r_{UV} + r_S = 0$. These additional fermions are chosen to be singlets of G. The gauge symmetry G must persist in the IR, such that also $r_{IR} + r_S = 0$. This implies $r_{UV} = r_{IR}$ which is the statement of anomaly matching. In the limit when the gauge coupling is set to zero and G becomes global again, the spectator fermions and the background fields decouple while leaving the current three-point functions unchanged, such that $r_{UV} = r_{IR}$ remains valid for the global symmetry.

1.9 Further reading

Weinberg has written a beautiful exhaustive series of books on quantum field theory [1, 2, 3]. A compact introduction is the book by Ryder [4]. A comprehensive book geared towards elementary particle physics is that by Peskin and Schroeder [5], while Zinn-Justin takes a view on applications in statistical mechanics [6]. An introduction to formal concepts of quantum field theory may be found in [7] and in [8].

In our introduction to renormalisation, we followed [9] which also provides a short overview of many field theory concepts. Quantisation and renormalisation of gauge theories is discussed for instance in [10]. The original paper of Faddeev and Popov on non-Abelian gauge theories is [11]. A seminal paper which introduces the Wilson loop among other important concepts is [12]. The triangle anomaly was found in [13] and [14]. 't Hooft proposed anomaly matching in [15]. An alternative argument is given in [16]. Anomaly matching is reviewed in [17].

A useful set of lecture notes for many aspects of quantum field theory and renormalisation is [18].

References

[1] Weinberg, Steven. 1995. *The Quantum Theory of Fields*. Vol. 1: *Foundations*. Cambridge University Press.

[2] Weinberg, Steven. 1996. *The Quantum Theory of Fields*. Vol. 2: *Modern Applications*. Cambridge University Press.

[3] Weinberg, Steven. 2000. *The Quantum Theory of Fields*. Vol. 3: *Supersymmetry*. Cambridge University Press.

[4] Ryder, L. H. 1985. *Quantum Field Theory*. Cambridge University Press.

[5] Peskin, Michael E., and Schroeder, Daniel V. 1995. *An Introduction to Quantum Field Theory*. Addison-Wesley.

[6] Zinn-Justin, Jean. 1989. *Quantum Field Theory and Critical Phenomena*, 4th edition 2002. Clarendon Press, Oxford.

[7] Kugo, T. 1997. *Gauge Field Theory* (in German). Springer, Berlin.

[8] Flory, Mario, Helling, Robert C., and Sluka, Constantin. 2012. How I learned to stop worrying and love QFT. ArXiv:1201.2714.

[9] Srednicki, M. 2007. *Quantum Field Theory*. Cambridge University Press.

[10] Cheng, T. P., and Li, L. F. 1985. *Gauge Theory of Elementary Particle Physics*. Clarendon, Oxford Science Publications.

[11] Faddeev, L. D., and Popov, V. N. 1967. Feynman diagrams for the Yang-Mills field. *Phys. Lett.*, **B25**, 29–30.

[12] Wilson, Kenneth G. 1974. Confinement of quarks. *Phys. Rev.*, **D10**, 2445–2459.

[13] Adler, Stephen L. 1969. Axial vector vertex in spinor electrodynamics. *Phys. Rev.*, **177**, 2426–2438.

[14] Bell, J. S., and Jackiw, R. 1969. A PCAC puzzle: $\pi^0 \rightarrow \gamma\gamma$ in the sigma model. *Nuovo Cimento.*, **A60**, 47–61.

[15] 't Hooft, Gerard. 1980. Naturalness, chiral symmetry, and spontaneous chiral symmetry breaking. *NATO Adv. Study Inst. Ser. B Phys.*, **59**, 135.

[16] Frishman, Y., Schwimmer, A., Banks, Tom, and Yankielowicz, S. 1981. The axial anomaly and the bound state spectrum in confining theories. *Nucl. Phys.*, **B177**, 157.

[17] Harvey, Jeffrey A. 2005. TASI 2003 Lectures on Anomalies. ArXiv:hep-th/0509097.

[18] Osborn, Hugh. 2013. Advanced Quantum Field Theory. Available at www.damtp.cam.ac.uk/user/ho/.

2

Elements of gravity

Einstein's theory of gravity is based on two physical assumptions: *gravity is geometry* and *matter curves spacetime*. In particular this means that on the one hand, particles follow *geodesics* in curved spacetime; the resulting motion appears to an observer as the effect of gravity. On the other hand, matter is also a source of spacetime curvature and hence of gravity.

Spacetime is modelled in terms of differentiable manifolds. This is reasonable since physics does not depend on the coordinate system chosen and therefore has to be invariant under general coordinate transformations. Hence in section 2.1 below, we review basic concepts of manifolds and introduce tensors. Moreover, we construct covariant derivatives. Einstein's equations of gravity are formulated in section 2.2, while their solutions are discussed in the subsequent sections: maximally symmetric spacetimes in 2.3 and black hole solutions in 2.4. Finally, the energy-momentum tensor in curved space has to satisfy important conditions, the *energy conditions*, which we summarise in section 2.5.

2.1 Differential geometry

2.1.1 Manifolds, tangent and cotangent spaces

Let us consider a d-dimensional differentiable real manifold $\mathcal{M}$. A coordinate system x is a map from a subset of $\mathcal{M}$ to $\mathbb{R}^d$. Of course, for a manifold $\mathcal{M}$, a unique or a preferred coordinate system does not exist. However, the change from one coordinate system x to another coordinate system x', which is defined for the same subset of $\mathcal{M}$ for simplicity, has to be smooth. In other words, the transition functions $x \circ x'^{-1}$ and $x' \circ x^{-1}$ have to be differentiable maps from $\mathbb{R}^d$ to $\mathbb{R}^d$. At each point $p \in \mathcal{M}$, we may define $T_p(\mathcal{M})$ as the space of *tangent vectors*. This vector space $T_p(\mathcal{M})$ is d-dimensional and a particular set of basis vectors is given by $\partial_\mu = \partial/\partial x^\mu$, where x^μ are the coordinates. Therefore any vector $V \in T_p(\mathcal{M})$ may be written as

$$V = V^\mu \partial_\mu. \tag{2.1}$$

Besides the tangent space $T_p(M)$ for every point $p \in \mathcal{M}$, we may also define the corresponding *cotangent space* $T_p^*(\mathcal{M})$ consisting of all linear maps from $T_p(\mathcal{M})$ into $\mathbb{R}$. The basis ∂_μ of the tangent space $T_p(\mathcal{M})$ induces a dual basis $\mathrm{d}x^\nu$ of the cotangent

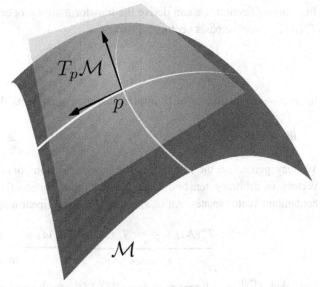

A manifold $\mathcal{M}$ (dark grey). The grey plane $T_p(\mathcal{M})$ is the tangent space at the point $p \in \mathcal{M}$.

space $T_p^*(\mathcal{M})$ by virtue of

$$\mathrm{d}x^\nu(\partial_\mu) = \delta_\mu^\nu. \tag{2.2}$$

Therefore, as in the case of the tangent space, every vector of the cotangent space $W \in T_p^*(\mathcal{M})$ may be written as

$$W = W_\nu \mathrm{d}x^\nu. \tag{2.3}$$

2.1.2 Covariance and tensors

Since physics does not depend on the coordinates which we have chosen, it is essential that all physical statements are independent of the choice of coordinates. A change of coordinates $x \mapsto x'$ induces a change of basis in the tangent space. In particular, basis vectors $\partial_\nu = \partial/\partial x^\nu$ transform into $\partial_\nu' = \partial/\partial x'^\nu$ according to

$$\partial_\mu' = \frac{\partial x^\nu}{\partial x'^\mu}\,\partial_\nu \tag{2.4}$$

where we have used the chain rule. Under a coordinate transformation $x \mapsto x'$, any vector $V \in T_p(\mathcal{M})$ is invariant,

$$V = V^\nu\,\partial_\nu = V'^\mu\,\partial_\mu'. \tag{2.5}$$

However, since the basis vectors transform as (2.4), we have to impose the following transformation properties on the components V^μ,

$$V'^\mu = V^\nu\,\frac{\partial x'^\mu}{\partial x^\nu}. \tag{2.6}$$

In a similar fashion, we can derive the transformation property of cotangent vectors $W \in T_p^*(\mathcal{M})$. Since the basis vectors transform as

$$\mathrm{d}x'^{\mu} = \frac{\partial x'^{\mu}}{\partial x^{\nu}} \, \mathrm{d}x^{\nu} \tag{2.7}$$

under $x \mapsto x'$, invariance of W implies that the components W_{μ} have to transform as

$$W = W_{\nu} \, \mathrm{d}x^{\nu} = W'_{\mu} \, \mathrm{d}x'^{\mu} = W'_{\mu} \frac{\partial x'^{\mu}}{\partial x^{\nu}} \, \mathrm{d}x^{\nu} \quad \Rightarrow \quad W'_{\mu} = W_{\nu} \frac{\partial x^{\nu}}{\partial x'^{\mu}}. \tag{2.8}$$

We may generalise the transformation law of the components of vectors and cotangent vectors to arbitrary tensors. Consider the tensor product of tangent vector spaces or contangent vector spaces. An (r,s) tensor $T^{(r,s)}$ is a linear map of the form

$$T^{(r,s)} : \underbrace{T_p^*(\mathcal{M}) \times \cdots \times T_p^*(\mathcal{M})}_{r \text{ times}} \times \underbrace{T_p(\mathcal{M}) \times \cdots \times T_p(\mathcal{M})}_{s \text{ times}} \to \mathbb{R}. \tag{2.9}$$

Note that $T^{(1,0)}$ is a linear map from $T_p^*(\mathcal{M})$ into the real numbers and therefore is an element of the tangent space $T_p(\mathcal{M})$. Similarly, $T^{(0,1)}$ is an element of the cotangent space. For generic r and s, $T^{(r,s)}$ may be expressed in components as

$$T^{(r,s)} = T^{\mu_1 \dots \mu_r}{}_{\nu_1 \dots \nu_s} \, \partial_{\mu_1} \otimes \cdots \otimes \partial_{\mu_r} \otimes \mathrm{d}x^{\nu_1} \otimes \cdots \otimes \mathrm{d}x^{\nu_s}, \tag{2.10}$$

where the components $T^{\mu_1 \dots \mu_r}{}_{\nu_1 \dots \nu_s}$ are functions of $p \in \mathcal{M}$.

Exercise 2.1.1   Show that under a transformation $x \mapsto x'$ the components of the (r,s) tensor transform as

$$T'^{\mu_1 \dots \mu_r}{}_{\nu_1 \dots \nu_s} = \frac{\partial x'^{\mu_1}}{\partial x^{\rho_1}} \cdots \frac{\partial x'^{\mu_r}}{\partial x^{\rho_r}} \frac{\partial x^{\sigma_1}}{\partial x'^{\nu_1}} \cdots \frac{\partial x^{\sigma_s}}{\partial x'^{\nu_s}} T^{\rho_1 \dots \rho_r}{}_{\sigma_1 \dots \sigma_s}. \tag{2.11}$$

Exercise 2.1.2   Show that if we multiply an (r_1, s_1) tensor S and an (r_2, s_2) tensor T, the resulting tensor $S \cdot T$ is of type $(r_1 + r_2, s_1 + s_2)$.

Given an (r,s) tensor T of the form (2.10) with components $T^{\mu_1 \dots \mu_r}{}_{\nu_1 \dots \nu_s}$, we may symmetrise or antisymmetrise separately in the lower (or upper) indices. For a tensor with only lower indices, symmetrisation over the indices $\nu_1, \dots, \nu_n$ (with $n \leq s$) is denoted by

$$T_{(\nu_1 \dots \nu_n) \nu_{n+1} \dots \nu_s} = \frac{1}{n!} \left(T_{\nu_1 \dots \nu_n \nu_{n+1} \dots \nu_s} + \text{permutations of } \nu_1 \dots \nu_n \right). \tag{2.12}$$

For antisymmetrising in the lower indices $\nu_1, \dots, \nu_n$ (with $n \leq s$), we write

$$T_{[\nu_1 \dots \nu_n] \nu_{n+1} \dots \nu_s} = \frac{1}{n!} \left(T_{\nu_1 \dots \nu_n \nu_{n+1} \dots \nu_s} \pm \text{alternating permutations of } \nu_1 \dots \nu_n \right). \tag{2.13}$$

Moreover, for (anti-)symmetrising in non-adjacent indices, we place vertical bars $|\dots|$ around the indices which are not (anti-)symmetrised. For example, $(\mu_1 \mu_2 | \mu_3 | \mu_4)$ means that we symmetrise in μ_1, μ_2 and μ_4, but not in μ_3.

2.1.3 Metric and vielbeins

An important role is played by the *metric*, a particular $(0, 2)$ tensor g which at each point $p \in \mathcal{M}$ is a non-degenerate symmetric bilinear form g, i.e.

$$g : \begin{cases} T_p(\mathcal{M}) \times T_p(\mathcal{M}) & \to & \mathbb{R} \\ (u, v) & \mapsto & g(u, v), \end{cases} \tag{2.14}$$

with $g(u, v) = g(v, u)$. Non-degeneracy means that for every $u \neq 0$, there exists a vector v such that $g(u, v) \neq 0$. In other words, there is no vector (besides the zero vector) which is orthogonal to every other vector in $T_p(\mathcal{M})$. Note that as explained at the beginning of section 1.1, we do not impose positivity of the metric. This is somewhat contrary to the usual definition in mathematics. It allows us to define spacelike, timelike and null vectors.

The metric may be expressed in terms of the basis vectors $dx^\mu \otimes dx^\nu$ of $T_p(\mathcal{M}) \times T_p(\mathcal{M})$ using the components $g_{\mu\nu}(x)$ as

$$ds^2 \equiv g_{\mu\nu}(x) \, dx^\mu \otimes dx^\nu. \tag{2.15}$$

We will suppress $\otimes$ and write $g_{\mu\nu} dx^\mu dx^\nu$ as a shorthand notation. Equation (2.15) introduces the notion of an *(infinitesimal) line element*. If $g_{\mu\nu}$ has only positive eigenvalues the manifold is *Riemannian* while if it has one negative eigenvalue the manifold is *Lorentzian*. For Lorentzian manifolds, the infinitesimal line element ds^2 determines whether a vector dx^μ, viewed as an infinitesimal distance between points on a manifold, is spacelike, timelike or lightlike depending on the sign of ds^2.

In the case of flat d-dimensional Minkowski space, the metric components $g_{\mu\nu}$ are given by $\eta_{\mu\nu}$. Let us give other examples of Lorentzian manifolds, restricted to four spacetime dimensions for simplicity.

- One example is the *Schwarzschild* metric for a black hole of mass M,

$$ds^2 = -\left(1 - \frac{2GM}{r}\right) dt^2 + \frac{dr^2}{1 - \frac{2GM}{r}} + r^2\left(d\theta^2 + \sin^2\theta \, d\phi^2\right), \tag{2.16}$$

  where G is the Newton constant.
- Another example is the *de Sitter* metric describing an accelerated expansion of the universe in the inflationary phase,

$$ds^2 = -dt^2 + e^{2Ht}\left((dx_1)^2 + (dx_2)^2 + (dx_3)^2\right). \tag{2.17}$$

As in the case of flat spacetime, we can use the metric with components $g_{\mu\nu}$ and its inverse $g^{\mu\nu}$ to lower and raise indices. In particular, a vector $V \in T_p(\mathcal{M})$ with components $V = V^\mu \partial_\mu$ may be mapped to $W = W_\mu dx^\mu \in T_p^*(\mathcal{M})$ using $W_\mu = g_{\mu\nu} V^\nu$. In other words, the metric induces a natural isomorphism between the tangent space $T_p(\mathcal{M})$ and the cotangent space $T_p^*(\mathcal{M})$.

Instead of ∂_μ and dx^μ (with $\mu \in \{0, \ldots, d-1\}$) as basis vectors for $T_p(\mathcal{M})$ and $T_p^*(\mathcal{M})$, we can use another set of basis vectors, for example e_a where $a \in \{0, \ldots, d-1\}$ for $T_p(\mathcal{M})$. Here we use Latin indices such as a instead of Greek indices to remind us that in general it is a non-coordinate basis. A convenient choice for a non-coordinate basis e_a of $T_p(\mathcal{M})$ is such that $g(e_a, e_b) = \eta_{ab}$, or, in terms of components,

$$g_{\mu\nu}\, e_a^\mu\, e_b^\nu = \eta_{ab}. \tag{2.18}$$

Here, η_{ab} represents the components of the Minkowski metric. e_a^μ are the components of e_a with respect to the basis ∂_μ, i.e. $e_a = e_a^\mu \partial_\mu$. The set of basis vectors e_a with $a \in \{0, \ldots, d-1\}$ is referred to as an *vielbein*. Since the components e_a^μ describe a change of basis, they may be inverted,

$$e_a^\mu\, e_\nu^a = \delta^\mu{}_\nu, \qquad e_a^\mu\, e_\mu^b = \delta^b{}_a. \tag{2.19}$$

Therefore (2.18) may be rephrased as

$$g_{\mu\nu} = \eta_{ab}\, e_\mu^a\, e_\nu^b, \tag{2.20}$$

and we may consider the vielbein as the 'square-root' of the metric. Using e_μ^a and e_a^μ any (r, s) tensor can be converted into a tensor with Latin indices. For example, $(1, 0)$ and $(0, 1)$ tensors transform as

$$V^a = e_\mu^a\, V^\mu, \qquad V_a = e_a^\mu\, V_\mu \tag{2.21}$$

and similarly for any other (r, s) tensor. For tensors with more than one index, we do not have to transform both indices. For example, we may consider objects such as $V^a{}_\nu$ which are given in terms of $V^a{}_\nu$ by

$$V^a{}_\nu = e_\mu^a\, V^\mu{}_\nu. \tag{2.22}$$

Note also that due to (2.18) the metric $g_{\mu\nu}$ expressed in the frame basis is flat. Therefore the Greek indices are referred to as *curved* space indices and the Latin indices are referred to as *flat* indices.

The choice of the orthonormal basis e_a is not unique. We may choose another basis e_a' satisfying (2.18). However, $e_a(x)$ may be related to $e_b'(x)$ by

$$e_a'(x) = \Lambda_a{}^b(x)\, e_b(x), \qquad \text{or equivalently} \qquad e'^a(x) = \Lambda^a{}_b(x)\, e^b(x), \tag{2.23}$$

where the matrices $\Lambda_a{}^b(x)$ and $\Lambda^a{}_b(x)$ have to satisfy

$$\Lambda_a{}^c(x)\, \Lambda_b{}^d(x)\, \eta^{ab} = \eta^{cd}, \qquad \text{or equivalently} \qquad \Lambda^a{}_c(x)\, \Lambda^b{}_d(x)\, \eta_{ab} = \eta_{cd}. \tag{2.24}$$

Note the similarity of (2.24) with (1.4) in chapter 1. Here, the matrices $\Lambda^a{}_c(x)$ are spacetime dependent and therefore we refer to (2.23) as a *local Lorentz transformation*. Under this transformation, the objects V^a and V_a defined by (2.21) transform as

$$V'^a = \Lambda^a{}_b(x)\, V^b \qquad \text{and} \qquad V_a' = \Lambda_a{}^b(x)\, V_b, \tag{2.25}$$

respectively. Besides the local Lorentz transformations we still have the freedom to perform coordinate transformations $x \mapsto x'$. Performing both transformations at the same time, tensors with mixed flat and curved indices, such as $V^a{}_\nu$ defined in (2.22), transform according to

$$V'^a{}_\nu = \Lambda^a{}_b(x)\, \frac{\partial x^\rho}{\partial x'^\nu}\, V^b{}_\rho. \tag{2.26}$$

It is straightforward to generalise this rule for any (r, s) tensor with mixed indices.

2.1.4 Covariant derivative

With the exception of scalar fields, the partial derivative of any (r, s) tensor is no longer a tensor. For example, let us consider a one-form $W = W_\nu dx^\nu$, i.e. a $(0, 1)$ tensor, and take the derivative with respect to ∂_μ. From the index structure, we expect that $\partial_\mu W_\nu$ should transform as a $(0, 2)$ tensor. However, under a change of coordinates $\partial_\mu W_\nu$ transforms as

$$\partial'_\mu W'_\nu = \frac{\partial}{\partial x'^\mu} W'_\nu = \frac{\partial x^\rho}{\partial x'^\mu} \frac{\partial}{\partial x^\rho} \left(\frac{\partial x^\sigma}{\partial x'^\nu} W_\sigma \right)$$

$$= \frac{\partial x^\rho}{\partial x'^\mu} \frac{\partial x^\sigma}{\partial x'^\nu} \left(\frac{\partial}{\partial x^\rho} W_\sigma \right) + W_\sigma \frac{\partial x^\rho}{\partial x'^\mu} \frac{\partial^2 x^\sigma}{\partial x^\rho \partial x'^\nu}. \tag{2.27}$$

While the first term in the second line of (2.27) is the expected transformation law of a $(0, 2)$ tensor, the second, inhomogeneous term spoils it. We encountered a similar situation in section 1.7 when promoting a global symmetry to a local symmetry in the context of gauge theory. There, the derivative of the field does not transform in the same way as the field itself. To achieve this, the introduction of a covariant derivative was necessary. Here we proceed in the same way. The covariant derivatives, denoted by ∇_μ, should satisfy the following properties:

- ∇_μ maps (r, s) tensors to $(r, s + 1)$ tensors,
- ∇_μ is linear, $\nabla_\mu(T + S) = \nabla_\mu T + \nabla_\mu S$,
- ∇_μ satisfies the Leibniz rule $\nabla_\mu(S \cdot T) = (\nabla_\mu S) \cdot T + S \cdot (\nabla_\mu T)$.

These properties imply that the covariant derivative will be the standard derivative plus a correction term, the *connection*. The connection in general relativity consists of a $d \times d$ matrix $(\Gamma_\mu)^\nu{}_\lambda \equiv \Gamma^\nu_{\mu\lambda}$ for each coordinate labelled by μ. Acting on vectors V^μ and one-forms W_μ, the covariant derivative is given by

$$\nabla_\mu V^\nu = \partial_\mu V^\nu + \Gamma^\nu_{\mu\lambda} V^\lambda, \tag{2.28a}$$

$$\nabla_\mu W_\nu = \partial_\mu W_\nu - \Gamma^\lambda_{\mu\nu} W_\lambda. \tag{2.28b}$$

The action of ∇_μ on a generic (r, s) tensor is obtained in generalisation of (2.28a) and (2.28b). For each upper index we obtain a connection term as in (2.28a), while for each lower index we obtain a connection term as in (2.28b). The opposite signs in front of the correction terms in (2.28a) and (2.28b) make sure that the contraction $V^\mu W_\mu$ transforms as a scalar, $\nabla_\nu(V^\mu W_\mu) = \partial_\nu(V^\mu W_\mu)$. Here we made the additional reasonable assumptions that for scalars ϕ, the covariant derivative acts as the partial derivative, i.e. $\nabla_\mu \phi = \partial_\mu \phi$, and that the covariant derivative commutes with contractions.

The *Christoffel* or *Levi-Civita connection* is the unique connection which is symmetric in the lower indices, $\Gamma^\rho_{\mu\nu} = \Gamma^\rho_{\nu\mu}$, and which is *metric compatible*, $\nabla_\mu g_{\nu\rho} = 0$. In terms of the metric the Christoffel connection is given by

$$\Gamma^\lambda_{\mu\nu} = \frac{1}{2} g^{\lambda\rho} \left(\partial_\mu g_{\nu\rho} + \partial_\nu g_{\rho\mu} - \partial_\rho g_{\mu\nu} \right). \tag{2.29}$$

A connection symmetric in its lower indices is also referred to as *torsion free*. We restrict our discussion to such connections from now on. So far, the covariant derivative is defined only for tensors with curved indices. On tensors with Latin indices, such as those defined by (2.21), the covariant derivative ∇_μ acts as

$$\nabla_\mu V^a = \partial_\mu V^a + \omega_\mu{}^a{}_b V^b \tag{2.30a}$$

$$\nabla_\mu V_a = \partial_\mu V_a - \omega_\mu{}^b{}_a V_b. \tag{2.30b}$$

Here, $\omega_\mu{}^a{}_b$ is the *spin connection*. We may raise and lower the flat indices a and b in the spin connection by η, for example $\omega_{\mu ab} = \eta_{ac}\, \omega_\mu{}^c{}_b$. For a metric compatible connection, the spin connection $\omega_{\mu ab}$ is antisymmetric in a and b.

The generalisation to tensors with more than one index and with mixed indices (such as (2.22)) is straightforward. For each upper Latin index we add a spin connection term of the form (2.30a), while for each lower Latin index we add a spin connection term of the form (2.30b). In the case of Greek indices the connection terms are still given by (2.28a) and (2.28b).

The spin connection is determined by $\nabla_\mu e_\nu^a = 0$, which in terms of the connection reads

$$\nabla_\mu e_\nu^a = \partial_\mu e_\nu^a + \omega_\mu{}^a{}_b e_\nu^b - \Gamma_{\mu\nu}^\lambda e_\lambda^a = 0. \tag{2.31}$$

For a given connection $\Gamma_{\mu\nu}^\lambda$, solving (2.31) for the spin connection $\omega_\mu{}^a{}_b$ we obtain

$$\omega_\mu{}^a{}_b = e_\lambda^a e_b^\nu \Gamma_{\mu\nu}^\lambda - e_b^\nu \partial_\mu e_\nu^a. \tag{2.32}$$

Sometimes it is tedious first to calculate the connection and then to determine the spin connection using (2.32). For the torsion-free case, we will derive an equation in exercise 2.1.7 which allows us to determine the spin connection without computing the connection $\Gamma_{\mu\nu}^\lambda$ beforehand.

2.1.5 Lie derivative

The covariant derivative ∇_μ is not the only possible way to define a derivative. Another possibility is given by the *Lie derivative* $\mathscr{L}_V$ along a vector field $V = V^\mu(x)\,\partial_\mu$. For example, the Lie derivative acts on the scalar $\phi(x)$ by

$$\mathscr{L}_V \phi(x) = V^\rho(x)\partial_\rho \phi(x). \tag{2.33}$$

Note that $\mathscr{L}_V \phi$ is again a scalar field. We may extend the definiton of a Lie derivative to arbitrary tensor fields. For example, applied to vector fields and one-forms the Lie derivative reads

$$\mathscr{L}_V U^\mu = V^\rho\,\partial_\rho U^\mu - (\partial_\rho V^\mu)\,U^\rho, \tag{2.34}$$

$$\mathscr{L}_V W_\mu = V^\rho\,\partial_\rho W_\mu + (\partial_\mu V^\rho)\,W_\rho. \tag{2.35}$$

This can be generalised in a straightforward way to any (r, s) tensor field. For example, we can apply the Lie derivative to generic tensors of rank two such as $T^\mu{}_\nu$ or $T_{\mu\nu}$

$$\mathscr{L}_V T^\mu{}_\nu = V^\rho\,\partial_\rho T^\mu{}_\nu - (\partial_\rho V^\mu)\,T^\rho{}_\nu + (\partial_\nu V^\rho)\,T^\mu{}_\rho, \tag{2.36}$$

$$\mathcal{L}_V T_{\mu\nu} = V^\rho \, \partial_\rho T_{\mu\nu} + (\partial_\mu V^\rho) \, T_{\rho\nu} + (\partial_\nu V^\rho) \, T_{\mu\rho}. \tag{2.37}$$

Note that in order to define the Lie derivative we did not need to specify a connection $\Gamma^\rho_{\mu\nu}$.

Exercise 2.1.3 Show that the Lie derivative of a vector field U along another vector field V as given by (2.34) may be rewritten as the commutator of the two vector fields, also known as the *Lie bracket*,

$$\mathcal{L}_V U^\mu \equiv \left[V, U \right]^\mu = V^\nu \partial_\nu U^\mu - U^\nu \partial_\nu V^\mu, \tag{2.38}$$

and satisfies $\mathcal{L}_V U = -\mathcal{L}_U V$.

Exercise 2.1.4 Show that in the case of a symmetric connection, $\Gamma^\lambda_{\mu\nu} = \Gamma^\lambda_{\nu\mu}$, we may replace the partial derivative ∂ by a covariant derivative ∇ in the definition of the Lie derivative, i.e. in (2.34), (2.35), (2.36) and (2.37).

Exercise 2.1.5 Show that the metric $g_{\mu\nu}$ satisfies

$$\mathcal{L}_V g_{\mu\nu} = \nabla_\mu V_\nu + \nabla_\nu V_\mu, \tag{2.39}$$

where ∇_μ is the covariant derivative with Christoffel connection (2.29). This is precisely the transformation law of the metric under an infinitesimal coordinate transformation. Hint: Apply exercise 2.1.4 to (2.37).

The Lie derivative becomes important if we consider infinitesimal coordinate transformations. Under a transformation of the form $x^\mu \mapsto x'^\mu = x^\mu - \xi^\mu(x)$, an (r,s) tensor T transforms as (2.11). Keeping only terms which are first order in ξ, the transformation law (2.11) may be expressed as

$$\delta T^{\mu_1 \dots \mu_r}{}_{\nu_1 \dots \nu_s}(x) \equiv T'^{\mu_1 \dots \mu_r}{}_{\nu_1 \dots \nu_s}(x) - T^{\mu_1 \dots \mu_r}{}_{\nu_1 \dots \nu_s}(x) = \mathcal{L}_\xi T^{\mu_1 \dots \mu_r}{}_{\nu_1 \dots \nu_s}(x). \tag{2.40}$$

In particular, we are interested in those vector fields V which leave the metric invariant, i.e.

$$\mathcal{L}_V g_{\mu\nu} = 0. \tag{2.41}$$

Vector fields satisfying (2.41) are *Killing vector fields*. Using exercise 2.1.5, we may characterise them by the property $\nabla_\mu V_\nu + \nabla_\nu V_\mu = 0$. Killing vector fields also allow a definition of stationary spacetimes. A spacetime is *stationary* if it has a timelike Killing vector field, at least asymptotically at large distances.

2.1.6 Differential forms, volume form and Hodge dual

A particularly important subset of tensors is given by the $(0, p)$ tensors [1]

$$\omega^{(p)} : \left\{ \begin{array}{ccc} T_p(\mathcal{M}) \times \dots \times T_p(\mathcal{M}) & \to & \mathbb{R} \\ (v^{(1)}, \dots, v^{(p)}) & \mapsto & \omega^{(p)}\left(v^{(1)}, \dots, v^{(p)}\right) \end{array} \right\}, \tag{2.42}$$

which are antisymmetric if we pairwise interchange $v^{(i)}$ and $v^{(j)}$,

$$\omega^{(p)}\left(v^{(1)}, \dots, v^{(i)}, \dots, v^{(j)}, \dots, v^{(p)}\right) = -\omega^{(p)}\left(v^{(1)}, \dots, v^{(j)}, \dots, v^{(i)}, \dots, v^{(p)}\right). \tag{2.43}$$

---

[1] Note the confusing notation in the following equation. p denotes an integer number, with $0 \le p \le d$, while p in $T_p(\mathcal{M})$ specifies a point on the manifold $\mathcal{M}$.

These antisymmetric $(0,p)$ tensors $\omega^{(p)}$ are referred to as *differential forms*. If we want to indicate the rank of the form, we write *p-form* instead. Note that due to the antisymmetric property, $p \leq d$ where d is the dimension of the manifold $\mathcal{M}$. The vector space of all p-forms of a manifold $\mathcal{M}$ is denoted by $\Lambda^{(p)}(\mathcal{M})$. Note that covectors are one-forms and, by definition, scalar fields are zero-forms.

Exercise 2.1.6   Show that $\Lambda^{(p)}(\mathcal{M})$ is a vector space of dimension $\frac{d!}{p!(d-p)!}$.

In analogy to (2.10), any p-form $\omega^{(p)}$ may be expressed in terms of the coordinate basis $dx^{\mu_1} \wedge \cdots \wedge dx^{\mu_p}$ by

$$\omega^{(p)} = \frac{1}{p!}\omega_{\mu_1...\mu_p}dx^{\mu_1} \wedge \cdots \wedge dx^{\mu_p}. \tag{2.44}$$

Here, the basis elements $dx^{\mu_1} \wedge \cdots \wedge dx^{\mu_p}$ are defined by

$$dx^{\mu_1} \wedge \cdots \wedge dx^{\mu_p} = \sum_{\sigma \in S_p} \text{sgn}(\sigma)\, dx^{\mu_{\sigma(1)}} \otimes \cdots \otimes dx^{\mu_{\sigma(p)}}. \tag{2.45}$$

The basis elements $dx^{\mu_1} \wedge \cdots \wedge dx^{\mu_p}$ are therefore totally antisymmetrised. In (2.45), S_p is the set of permutations of $\{1,\ldots,p\}$. Any permutation σ can be decomposed into a product of transpositions, i.e. pairwise interchanges of entries. $\text{sgn}(\sigma)$ is the sign of a permutation σ, which is $+1$ if the number of transpositions is even and -1 otherwise.

We further define the *wedge product* of a p-form $\omega^{(p)}$ and a q-forms $\omega^{(q)}$ by constructing the antisymmetrised product of the p- and q-form, $\omega^{(p)} \wedge \omega^{(q)}$. The resulting object is a $(p+q)$-form. Note that the wedge product defined in this way is not commutative since in general

$$\omega^{(p)} \wedge \omega^{(q)} = (-1)^{pq}\omega^{(q)} \wedge \omega^{(p)}. \tag{2.46}$$

Instead of using the coordinate basis, we may also use the frame basis $\{e^a\}$. In this basis, any p-form $\omega^{(p)}$ of the form (2.44) reads

$$\omega^{(p)} = \frac{1}{p!}\omega_{a_1...a_p}e^{a_1} \wedge \cdots \wedge e^{a_p}, \tag{2.47}$$

where, as usual, the coefficients $\omega_{a_1...a_p}$ are related to $\omega_{\mu_1...\mu_p}$ by

$$\omega_{a_1...a_p} = \omega_{\mu_1...\mu_p}e_{a_1}^{\mu_1} \cdots e_{a_p}^{\mu_p}. \tag{2.48}$$

The *exterior derivative* d acts on p-forms $\omega^{(p)}$ by

$$d\omega^{(p)} = \frac{1}{p!}\left(\partial_\mu\omega_{\mu_1...\mu_p}\right)dx^\mu \wedge dx^{\mu_1} \wedge \cdots \wedge dx^{\mu_p}. \tag{2.49}$$

The resulting object $d\omega^{(p)}$ is a $(p+1)$-form. The exterior derivative is a linear operation and in addition satisfies the following conditions

$$d\left(\omega^{(p)} \wedge \omega^{(q)}\right) = d\omega^{(p)} \wedge \omega^{(q)} + (-1)^p\omega^{(p)} \wedge d\omega^{(q)}, \tag{2.50}$$

$$d^2\omega^{(p)} = 0, \tag{2.51}$$

for any p-form $\omega^{(p)}$ and any q-form $\omega^{(q)}$. Note also that in order to define the exterior derivative by (2.49), we did not have to specify a connection.

Exercise 2.1.7   Show that for a given symmetric connection, (2.31) may be written as

$$\mathrm{d}e^a + \omega^a{}_b \wedge e^b = 0, \qquad (2.52)$$

where $\omega^a{}_b = \omega_\mu{}^a{}_b \mathrm{d}x^\mu$. Note that in (2.52), the connection $\Gamma^\lambda_{\mu\nu}$ drops out and thus this equation may be used to determine the spin connection in the torsion-free case. In particular, there is no need to calculate $\Gamma^\lambda_{\mu\nu}$ explicitly.

Canonical volume form

The differential forms play a crucial role in defining a canonical volume for curved space-times. To see this let us consider a d-dimensional manifold with vielbeins $e^0, \ldots, e^{d-1}$. Note that the space of differential forms of rank d is one dimensional. For example, in the local frame we may consider the d-form $e^0 \wedge \cdots \wedge e^{d-1}$ which is unique up to multiplication by arbitrary functions. Let us refer to this d-form as the canonical volume form, given by

$$\mathrm{dVol} = e^0 \wedge \cdots \wedge e^{d-1}. \qquad (2.53)$$

To manipulate dVol further, it is convenient to define the totally antisymmetric Levi-Civita symbol in the local frame basis,

$$\epsilon_{a_1 \ldots a_d} = \begin{cases} +1 & : \ (a_1, \ldots, a_d) \text{ is even permutation of } (0, 1, \ldots, d-1), \\ -1 & : \ (a_1, \ldots, a_d) \text{ is odd permutation of } (0, 1, \ldots, d-1), \\ 0 & : \ \text{otherwise}. \end{cases} \qquad (2.54)$$

As usual, the Latin indices a are raised by η^{ab} and therefore $\epsilon^{01\ldots(d-1)} = -1$. For consistency, to define the totally antisymmetric tensor in coordinate basis, we have to insert appropriate factors of e where $e \equiv \det(e^a_\mu) = \sqrt{-\det g}$,

$$\epsilon_{\mu_1 \ldots \mu_d} \equiv \frac{1}{e} \epsilon_{a_1 \ldots a_d} e^{a_1}_{\mu_1} \ldots e^{a_d}_{\mu_d}, \qquad (2.55)$$

$$\epsilon^{\mu_1 \ldots \mu_d} \equiv e \, \epsilon^{a_1 \ldots a_d} e^{\mu_1}_{a_1} \ldots e^{\mu_d}_{a_d}. \qquad (2.56)$$

Both antisymmetric tensors, $\epsilon_{\mu_1 \ldots \mu_d}$ and $\epsilon^{\mu_1 \ldots \mu_d}$ take values in $\{0, \pm 1\}$ and therefore they are usually referred to as tensor densities. Note also that we cannot obtain $\epsilon^{\mu_1 \ldots \mu_d}$ by raising the indices of $\epsilon_{\mu_1 \ldots \mu_d}$ with the metric g. We should think of (2.55) and (2.56) as two separate definitions. This also implies that it makes no sense to consider the antisymmetric tensors with raised and lowered indices, such as $\epsilon_{\mu_1 \ldots \mu_p}{}^{\nu_{p+1} \ldots \nu_d}$.

A further way to write the canonical volume dVol is obtained by using (2.55),

$$\begin{aligned} \mathrm{dVol} &= \frac{1}{d!} \epsilon_{a_1 \ldots a_d} e^{a_1} \wedge \cdots \wedge e^{a_d} \\ &= \frac{1}{d!} e \, \epsilon_{\mu_1 \ldots \mu_d} \mathrm{d}x^{\mu_1} \wedge \cdots \wedge \mathrm{d}x^{\mu_d} \\ &= e \, \mathrm{d}x^0 \wedge \cdots \wedge \mathrm{d}x^{d-1} \equiv \mathrm{d}^d x \sqrt{-g}. \end{aligned} \qquad (2.57)$$

Therefore the canonical volume in coordinate basis is given by $\mathrm{d}^d x \sqrt{-g}$. It is straightforward to check that this form is indeed invariant under coordinate transformations $x \mapsto x'$.

Therefore if we want to integrate a scalar field $\phi(x)$ over the d-dimensional manifold $\mathcal{M}$ the appropriate measure is given by $d^d x \sqrt{-g}$ and hence the integral reads

$$\int_{\mathcal{M}} d\mathrm{Vol}(x)\phi(x) = \int_{\mathcal{M}} d^d x \sqrt{-g(x)}\, \phi(x). \tag{2.58}$$

Exercise 2.1.8   Verify that the Christoffel connection (2.29) satisfies

$$\Gamma^\mu_{\mu\nu} = \frac{1}{\sqrt{-g}}\, \partial_\nu \sqrt{-g}, \tag{2.59}$$

and that a covariant divergence of a vector may be written as

$$\nabla_\mu V^\mu = \frac{1}{\sqrt{-g}}\, \partial_\mu \left(\sqrt{-g}\, V^\mu\right). \tag{2.60}$$

Exercise 2.1.9   Using (2.60) show that Stokes' theorem also holds in curved spacetime,

$$\int_{\mathcal{N}} d^d x \sqrt{-g}\nabla_\mu V^\mu = \int_{\partial\mathcal{N}} d^{d-1}x \sqrt{-\gamma}\, n_\mu V^\mu, \tag{2.61}$$

for a region $\mathcal{N} \subseteq \mathcal{M}$ of the manifold with boundary $\partial\mathcal{N}$. n^μ is the vector normal to $\partial\mathcal{N}$ pointing outwards and γ_{ij} is the induced metric on $\partial\mathcal{N}$.

Hodge dual

The result of exercise 2.1.6 implies that the vector spaces of p-forms and of $(d-p)$-forms have the same dimension. We may therefore establish a linear one-to-one map between $\Lambda^p(\mathcal{M})$ and $\Lambda^{d-p}(\mathcal{M})$, the *Hodge dual*, by

$$* : \left\{ \begin{array}{ccc} \Lambda^p(\mathcal{M}) & \to & \Lambda^{d-p}(\mathcal{M}) \\ \omega^{(p)} & \mapsto & {}^*\omega^{(p)} \end{array} \right\}. \tag{2.62}$$

We may thus define a map using the local frame basis,

$$* \left(e^{a_1} \wedge \cdots \wedge e^{a_p}\right) = \frac{1}{(d-p)!}\epsilon_{b_1 \ldots b_{d-p}}{}^{a_1 \ldots a_p} e^{b_1} \wedge \cdots \wedge e^{b_{d-p}}. \tag{2.63}$$

Any p-form $\omega^{(p)}$ can be expressed in the frame basis and therefore we may generalise the Hodge dual to any p-form by

$$* \omega^{(p)} = \frac{1}{p!}\omega_{a_1 \ldots a_p}{}^*(e^{a_1} \wedge \cdots \wedge e^{a_p}). \tag{2.64}$$

The components of the Hodge dual ${}^*\omega^{(p)}$ are then given by

$$\left({}^*\omega^{(p)}\right)_{b_1 \ldots b_{d-p}} = \frac{1}{p!}\epsilon_{b_1 \ldots b_{d-p}}{}^{a_1 \ldots a_p}\, \omega_{a_1 \ldots a_p}. \tag{2.65}$$

Instead of working in the frame basis we can also use the coordinate basis. Since $dx^\mu = e^\mu_a e^a$ and using (2.55), the duality map (2.63) for the coordinate basis reads

$$*(dx^{\mu_1} \wedge \cdots \wedge dx^{\mu_p}) = \frac{1}{(d-p)!}e\, \epsilon_{\nu_1 \ldots \nu_{d-p}\sigma_1 \ldots \sigma_p}g^{\mu_1\sigma_1} \cdots g^{\mu_p\sigma_p}\, dx^{\nu_1} \wedge \cdots \wedge dx^{\nu_{d-p}}$$
$$\tag{2.66}$$

and hence the components of the Hodge dual $*\omega^{(p)}$ are

$$\left(*\omega^{(p)}\right)_{\nu_1\ldots\nu_{d-p}} = \frac{1}{p!}\, e\, \epsilon_{\nu_1\ldots\nu_{d-p}\sigma_1\ldots\sigma_p} g^{\mu_1\sigma_1} \ldots g^{\mu_p\sigma_p} \omega_{\mu_1\ldots\mu_p}. \tag{2.67}$$

Exercise 2.1.10   Show that the Hodge dual of the function $f(x) = 1$ is given by

$$*f = \mathrm{d}^d x \sqrt{-g}. \tag{2.68}$$

Exercise 2.1.11   Verify that acting twice with the Hodge dual on a p-form $\omega^{(p)}$ returns the same p-form $\omega^{(p)}$ up to a sign, i.e.

$$**\omega^{(p)} = (-1)^{p(d-p)}\omega^{(p)} \qquad \text{or} \qquad **\omega^{(p)} = (-1)^{p(d-p)+1}\omega^{(p)}, \tag{2.69}$$

in the case of Riemannian or Lorentzian manifolds, respectively. In the case of a Lorentzian manifold the term -1 may be traced back to $\det\eta = -1$.

For any p-form $\omega^{(p)}$, the Hodge dual $*\omega^{(p)}$ is a $(d-p)$-form and hence $*\omega^{(p)} \wedge \omega^{(p)}$ may be integrated over the manifold $\mathcal{M}$. Indeed it is straightforward to show that

$$\int_{\mathcal{M}} *\omega^{(p)} \wedge \omega^{(p)} = \frac{1}{p!} \int \mathrm{d}^d x \sqrt{-g}\, \omega^{\mu_1\ldots\mu_p} \omega_{\mu_1\ldots\mu_p}. \tag{2.70}$$

2.1.7 Curvature and parallel transport

In order to introduce the notion of *curvature*, we define the *parallel transport* of a vector V along a path $x^\mu(\lambda)$ by a vanishing covariant derivative

$$\frac{\mathrm{d}x^\rho}{\mathrm{d}\lambda}\nabla_\rho V^\mu \equiv \frac{\mathrm{d}V^\mu}{\mathrm{d}\lambda} + \Gamma^\mu_{\rho\sigma}\frac{\mathrm{d}x^\rho}{\mathrm{d}\lambda}V^\sigma = 0. \tag{2.71}$$

A *geodesic* is a curve $x^\mu(\lambda)$ along which the tangent vector $V^\mu = \mathrm{d}x^\mu/\mathrm{d}\lambda$ is parallel transported. It therefore satisfies the *geodesic equation*

$$\frac{\mathrm{d}^2 x^\mu}{\mathrm{d}\lambda} + \Gamma^\mu_{\rho\sigma}\frac{\mathrm{d}x^\rho}{\mathrm{d}\lambda}\frac{\mathrm{d}x^\sigma}{\mathrm{d}\lambda} = 0. \tag{2.72}$$

In general, parallel transport of a vector along a closed loop in a curved spacetime will lead to a different vector than before. For example, in the configuration of figure 2.2, consider

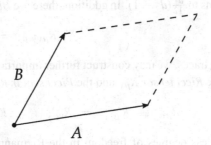

Figure 2.2   Parallel transport in curved spacetime.

the parallel transport of a vector V first along A and then along B and compare it to the vector first parallely transported along B and then along A. The difference δV between both parallely transported vectors has to be proportional to V, A and B, i.e.

$$\delta V^\rho = R^\rho{}_{\mu\alpha\beta} V^\mu A^\alpha B^\beta. \tag{2.73}$$

The parallel transport around such a closed loop may be viewed as taking the commutator of covariant derivatives. This commutator then measures the difference between first parallely transporting a vector the way $A \to B$ and then the other way $B \to A$. In this way, we can express the *Riemann curvature tensor $R^\rho{}_{\mu\alpha\beta}$* defined by (2.73) in terms of the connection

$$\left[\nabla_\alpha, \nabla_\beta\right] V^\rho = \partial_\alpha(\nabla_\beta V^\rho) + \Gamma^\rho_{\alpha\mu} \nabla_\beta V^\mu - \Gamma^\sigma_{\alpha\beta} \nabla_\sigma V^\rho - (\beta \leftrightarrow \alpha)$$

$$\equiv R^\rho{}_{\mu\alpha\beta} V^\mu - T^\sigma_{\alpha\beta} \nabla_\sigma V^\rho. \tag{2.74}$$

In particular, the Riemann tensor $R^\rho{}_{\mu\alpha\beta}$ and the torsion tensor $T^\sigma_{\mu\nu}$ for a given connection Γ are

$$R^\rho{}_{\mu\alpha\beta} = \partial_\alpha \Gamma^\rho_{\beta\mu} - \partial_\beta \Gamma^\rho_{\alpha\mu} + \Gamma^\rho_{\alpha\sigma} \Gamma^\sigma_{\beta\mu} - \Gamma^\rho_{\beta\sigma} \Gamma^\sigma_{\alpha\mu} \tag{2.75}$$

and

$$T^\sigma_{\mu\nu} = -2\Gamma^\sigma_{[\mu\nu]}. \tag{2.76}$$

Note that for connections $\Gamma^\sigma_{\mu\nu}$ which are symmetric in μ and ν, the torsion tensor $T^\sigma{}_{\mu\nu}$ has to vanish. Such connections are therefore *torsion free*. For torsion-free connections, the Riemann tensor measures the deviation of the commutator of covariant derivatives from its flat space value, and therefore the curvature of spacetime. Its purely lower case version

$$R_{\mu\nu\alpha\beta} = g_{\mu\sigma} R^\sigma{}_{\nu\alpha\beta} \tag{2.77}$$

has various symmetries in its indices, namely

$$R_{\mu\nu\alpha\beta} = -R_{\mu\nu\beta\alpha} = -R_{\nu\mu\alpha\beta} = R_{\alpha\beta\mu\nu}, \tag{2.78a}$$

$$R_{\mu\alpha\beta\gamma} + R_{\mu\beta\gamma\alpha} + R_{\mu\gamma\alpha\beta} = 0. \tag{2.78b}$$

This reduces the number of algebraically independent components in d spacetime dimensions to $\frac{d^2}{12}(d^2 - 1)$. In addition, there are *Bianchi identities*

$$\nabla_{[\lambda} R_{\mu\nu]\alpha\beta} = 0. \tag{2.79}$$

By taking traces, we may construct further important tensorial quantities from the Riemann tensor: the *Ricci tensor $R_{\mu\nu}$* and the *Ricci scalar R*,

$$R_{\mu\nu} \equiv R^\lambda{}_{\mu\lambda\nu} = R_{\nu\mu}, \qquad R \equiv R^\mu{}_\mu = g^{\mu\nu} R_{\mu\nu}. \tag{2.80}$$

The traceless degrees of freedom in the Riemann tensor which are absent in (2.80) are collected in the *Weyl tensor*

A *complex manifold* of complex dimension n is defined in analogy to a real manifold using complex local coordinate systems in $\mathbb{C}^n$. Its transition maps are required to be biholomorphic, i.e. the map and its inverse are both holomorphic functions. The complex local coordinates are denoted by z^k, $k = 1, \ldots, n$, while their complex conjugates are $\bar{z}^{\bar{k}}$, $\bar{k} = 1, \ldots, n$. A complex manifold admits a complex structure J which is a $(1, 1)$ tensor satisfying

$$J_k^{\ l} = i\delta_k^{\ l}, \quad J_{\bar{k}}^{\ \bar{l}} = -i\delta_{\bar{k}}^{\ \bar{l}}, \quad J_k^{\ \bar{l}} = J_{\bar{l}}^{\ k} = 0 \tag{2.83}$$

in complex coordinates. For a complex Riemannian manifold, the metric in complex local coordinates is given by

$$ds^2 = g_{kl}dz^k dz^l + g_{\bar{k}\bar{l}}d\bar{z}^{\bar{k}}d\bar{z}^{\bar{l}} + g_{\bar{k}l}d\bar{z}^{\bar{k}}dz^l + g_{k\bar{l}}dz^k d\bar{z}^{\bar{l}}. \tag{2.84}$$

To ensure that the metric is real, we have to impose the conditions

$$g_{\bar{k}\bar{l}} = (g_{kl})^* \quad \text{and} \quad g_{\bar{k}l} = (g_{k\bar{l}})^*. \tag{2.85}$$

A special case of a complex Riemannian manifold is a Hermitian manifold, for which

$$g_{kl} = g_{\bar{k}\bar{l}} = 0. \tag{2.86}$$

$$C_{\mu\nu\alpha\beta} \equiv R_{\mu\nu\alpha\beta} - \frac{2}{d-2}\left(g_{\mu[\alpha}R_{\beta]\nu} - g_{\nu[\alpha}R_{\beta]\mu}\right) + \frac{2}{(d-1)(d-2)}g_{\mu[\alpha}g_{\beta]\nu}R. \tag{2.81}$$

This has the same symmetries (2.78a) and (2.78b) as the original Riemann tensor, with in addition $C^\mu_{\ \nu\alpha\mu} = 0$. The Bianchi identities (2.79) of the full Riemann tensor imply

$$\nabla^\mu R_{\mu\nu} = \frac{1}{2}\nabla_\nu R \tag{2.82}$$

at the level of Ricci tensor and scalar.

Parallel transport of a vector around a closed loop leads to a rotation of the vector as compared to its original direction. Therefore a group structure may be associated with closed loop parallel transport, which in the case of an n-dimensional Riemannian manifold is $O(n)$, or $SO(n)$ if the manifold is orientable. Note that in addition to real manifolds, there are also complex manifolds as introduced in box 2.1.

2.1.8 Covariantising field theories

In chapter 1 we considered field theories in flat spacetime. Here we generalise field theories to curved spacetime backgrounds. Let us describe a recipe of how to generalise actions formulated in flat Minkowski spacetime to general spacetimes. This recipe corresponds to the minimal way to couple fields to curved spacetime, and is therefore referred to as *minimal coupling*. The recipe involves the following steps.

- Replace the derivative ∂_μ by the covariant derivative ∇_μ which is given in terms of the Christoffel connection (2.29).
- Replace the flat metric $\eta_{\mu\nu}$ by the curved spacetime metric $g_{\mu\nu}$.
- Replace the integration measure $\int d^d x$ by $\int d^d x \sqrt{-g}$.

Let us illustrate this recipe using two examples. First consider a scalar field ϕ. The action in flat spacetime was constructed in section 1.1. Using the recipe of minimal coupling mentioned above, the action for a scalar field in curved spacetime reads

$$S = \int d^d x \sqrt{-g} \left(-\frac{1}{2} g^{\mu\nu} \partial_\mu \phi \partial_\nu \phi - \frac{1}{2} m^2 \phi^2 + \mathcal{L}_{\text{int}}[\phi] \right). \tag{2.87}$$

Note that in the action (2.87) we have used partial derivatives instead of covariant derivatives since both act on scalars. Second, let us translate the energy-momentum conservation, $\partial_\mu T^{\mu\nu} = 0$, to curved spacetime. Using the covariant derivative ∇_μ given by the Christoffel connection (2.29), the equation for energy conservation in curved spacetime reads

$$\nabla_\mu T^{\mu\nu} = 0. \tag{2.88}$$

In addition to the recipe for minimally coupling fields to gravity discussed above, we may also write down additional terms coupling the fields ϕ to the curvature tensor $R_{\mu\nu\rho\sigma}$. These terms are examples of non-minimal couplings of the scalar field to the metric. Moreover, these terms violate the strong equivalence principle, since the curvature tensor does not vanish in local inertial frames. For example, for a scalar field ϕ, we may add the term

$$S_{R\phi^2} = -\xi \int d^d x \sqrt{-g} R \phi^2. \tag{2.89}$$

Since the term is quadratic in ϕ, it may be viewed as an additional mass term which need not be constant.

Exercise 2.1.12 By calculating the equations of motion of a scalar field ϕ with the additional interaction term $S_{R\phi^2}$, show that this term induces a mass correction for ϕ proportional to the Ricci scalar R.

Exercise 2.1.13 Consider the action of a free massless scalar field theory and add the term (2.89) with $\xi = (d-2)/((8(d-1))$. Show that the resulting action of the scalar field is invariant under *conformal transformations*,

$$g_{\mu\nu} \mapsto \tilde{g}_{\mu\nu} = \Omega^{-2}(x) g_{\mu\nu}. \tag{2.90}$$

It is also straightforward to couple minimally (non-)Abelian gauge theories to curved spacetime. Using the recipe presented above, the action for a (non-)Abelian gauge field reads

$$S[A] = -\frac{1}{2g_{\text{YM}}^2} \int d^d x \sqrt{-g} g^{\mu\rho} g^{\nu\sigma} \text{Tr}(F_{\rho\sigma} F_{\mu\nu}). \tag{2.91}$$

Although we have to replace the partial derivatives by covariant derivatives, the field strength tensor $F_{\mu\nu}$ is still given by the flat space expression (1.181) since, for torsion-free theories, the connection term drops out when antisymmetrising.

More work is needed in order to couple spinors to curved spacetime. The gamma matrices in curved spacetime γ^μ have to satisfy

$$\{\gamma^\mu, \gamma^\nu\} = -2g^{\mu\nu}\mathbb{1}. \tag{2.92}$$

Using the gamma matrices Γ^a of flat spacetime, i.e. satisfying

$$\{\Gamma^a, \Gamma^b\} = -2\eta^{ab}\mathbb{1}, \tag{2.93}$$

it is straightforward to construct γ^μ using vielbeins,

$$\gamma^\mu = e_a^\mu \Gamma^a. \tag{2.94}$$

We further define $\Gamma^{ab} \equiv \frac{1}{2}[\Gamma^a, \Gamma^b]$. In order to generalise the Lagrangian (1.141) to curved spacetime, we also have to replace the partial derivative ∂_μ by the covariant derivative ∇_μ given by

$$\nabla_\mu = \partial_\mu + \frac{1}{4}\omega_{\mu ab}\Gamma^{ab}, \tag{2.95}$$

where $\omega_{\mu ab}$ are the components of the spin connection. With these ingredients, the action for a Dirac spinor minimally coupled to curved spacetime reads

$$S = \int d^d x \sqrt{-g}\left(i\bar{\Psi}\slashed{\nabla}\Psi - m\bar{\Psi}\Psi\right), \tag{2.96}$$

where $\slashed{\nabla} = \gamma^\mu\nabla_\mu$ and $\bar{\Psi} = \Psi^\dagger\gamma^0$.

2.2 Einstein's field equations

In the previous section we have seen how to describe curved spacetimes and how to generalise field theories to those spacetimes. In this section, we introduce Einstein's field equations for gravity. A first guess for the field equations of gravity, relating geometry and matter, is '$R_{\mu\nu} \propto T_{\mu\nu}$'. However, this turns out to be incorrect, since energy conservation $\nabla^\mu T_{\mu\nu} = 0$ would then imply $\nabla^\mu R_{\mu\nu} = 0$, which is not true in general, as is seen from (2.82). However, the *Einstein tensor* defined by $G_{\mu\nu} = R_{\mu\nu} - \frac{1}{2}Rg_{\mu\nu}$ satisfies $\nabla^\mu G_{\mu\nu} = 0$, and therefore Einstein introduced the field equations

$$R_{\mu\nu} - \frac{1}{2}Rg_{\mu\nu} + \Lambda g_{\mu\nu} = \kappa^2 T_{\mu\nu}, \tag{2.97}$$

which give rise to the correct gravitational law in the Newtonian approximation. Note that the *Einstein equations* (2.97) relate the matter part, given by $T_{\mu\nu}$ on the right-hand side, to the spacetime geometry data on the left-hand side. The parameter Λ is the *cosmological constant* and κ is related to Newton's gravitational constant G by $\kappa^2 = 8\pi G$.

The Einstein equations may also be derived from an action principle. The appropriate action of the gravitational system is given by

$$S[g_{\mu\nu}, \phi] = S_{\text{EH}}[g_{\mu\nu}] + S_{\text{matter}}[g_{\mu\nu}, \phi], \tag{2.98}$$

where $\mathcal{S}_{\text{matter}}$ is the matter part as constructed in section 2.1.8. ϕ denotes any possible field, which is not necessarily a scalar field. We consider only covariant derivatives and curvature tensors that are determined by the Christoffel connection with respect to the metric $g_{\mu\nu}$. $\mathcal{S}_{\text{EH}}[g_{\mu\nu}]$ is the *Einstein–Hilbert action* with cosmological constant Λ,

$$\mathcal{S}_{\text{EH}}[g_{\mu\nu}] = \frac{1}{2\kappa^2} \int d^d x \sqrt{-g} \left(R - 2\Lambda \right). \tag{2.99}$$

By varying the action S with respect to the metric $g_{\mu\nu}$ we obtain

$$\frac{\delta \mathcal{S}_{\text{EH}}}{\delta g^{\mu\nu}} = \frac{\sqrt{-g}}{2\kappa^2} \left(R_{\mu\nu} - \frac{1}{2} g_{\mu\nu} R + \Lambda g_{\mu\nu} \right). \tag{2.100}$$

Defining the energy-momentum tensor $T_{\mu\nu}$ in curved spacetime by

$$T_{\mu\nu} \equiv -\frac{2}{\sqrt{-g}} \frac{\delta \mathcal{S}_{\text{matter}}}{\delta g^{\mu\nu}}, \tag{2.101}$$

we can rederive Einstein's field equations (2.97). In section 1.2 we derived the canonical energy-momentum tensor using Noether's theorem. Here, we have used an alternative definition of the energy-momentum tensor in equation (2.101). This new definition leads to the same results for scalar fields and has many advantages. First of all, the definition (2.101) is by construction symmetric in the indices μ and ν. Therefore we do not have to add additional terms, as was the case in the canonical approach. Second, the canonical energy-momentum tensor cannot be generalised to arbitrary curved spacetimes. Finally, for gauge fields the definition (2.101) is always gauge invariant (since the action and metric are), whereas in the canonical approach this does not have to be the case, as we have seen in section 1.7.

Exercise 2.2.1   Eliminate R of Einstein's field equations (2.97) by computing the trace of (2.97).

Exercise 2.2.2   Use the alternative definition (2.101) for the energy-momentum tensor and determine $T_{\mu\nu}$ for a real scalar field with mass m and interacting Lagrangian $\mathcal{L}_{\text{int}}$. What happens to the trace of the energy-momentum tensor if we consider a massless free theory and in addition if we include the term (2.89) with $\xi = (d-2)/(8(d-1))$?

2.3 Maximally symmetric spacetimes

Let us discuss solutions of Einstein's equations with $T_{\mu\nu} = 0$, i.e. vacuum solutions without any matter. In particular, we are interested in those spacetimes with maximal symmetry. The symmetries of spacetime are given by Killing vector fields X which by definition satisfy $\mathcal{L}_X g_{\mu\nu} = 0$. A Killing vector field is linear dependent if it can be written as a linear combination of other Killing vector fields with constant coefficients.

The question arises how many linear independent Killing fields, i.e. isometries, may a manifold have. For example, the isometries of d-dimensional Minkowski space are given by Lorentz transformations and translations in space and time. Therefore we have d

translational isometries and $d(d-1)/2$ rotational isometries including boosts, i.e. in total $d(d+1)/2$ isometries. It can be shown that a manifold of dimension d can only have at most $d(d+1)/2$ linearly independent Killing vector fields. The spacetimes which satisfy this bound are referred to as *maximally symmetric spacetimes*. Minkowski spacetime is therefore an example of a maximally symmetric spacetime.

For a maximally symmetric spacetime, the curvature has to be the same everywhere. If we think in terms of translational and rotational isometries, it is obvious that the curvature has to be the same at each point of the manifold and in each direction. Therefore we should be able to express the Riemann tensor in terms of the Ricci scalar. Due to the symmetries of the Riemann tensor given by (2.78a) and (2.78b), we have

$$R_{\mu\nu\rho\sigma} = \frac{R}{d(d-1)} \left(g_{\nu\sigma} g_{\mu\rho} - g_{\nu\rho} g_{\mu\sigma} \right). \tag{2.102}$$

Therefore we see that we may classify maximally symmetric spacetimes according to their dimension, the value of the Ricci scalar as well as whether the spacetime manifold is Riemannian or Lorentzian.

Let us first consider the case of Riemannian manifolds, in which the maximally symmetric spacetimes are locally Euclidean, spherical or hyperbolic. The line element of these spaces may be written in a compact way,

$$ds^2 = \frac{d\chi^2}{1 - k\chi^2} + \chi^2 d\Omega_{d-1}^2 \equiv dK_d^2, \tag{2.103}$$

where $k \in \{0, \pm 1\}$ and $d\Omega_{d-1}^2$ is the line element of a unit sphere S^{d-1}. $d\Omega_{d-1}^2$ may be defined iteratively by

$$d\Omega_1 = d\theta_1, \qquad d\Omega_j^2 = d\theta_j^2 + \sin^2 \theta_j \, d\Omega_{j-1}^2, \tag{2.104}$$

where $\theta_1 \in [0, 2\pi[$ and $\theta_j \in [0, \pi[$ for $j \in \{2, \ldots, d-1\}$.

For $k = 0$, we obtain Euclidean space in spherical coordinates where χ is the radial coordinate. For $k = 1$ after the coordinate transformation $\chi = \sin\phi$, where $\phi \in [0, \pi[$, the line element (2.103) reads

$$ds^2 = d\phi^2 + \sin^2 \phi \, d\Omega_{d-1}^2, \tag{2.105}$$

which corresponds to a unit sphere. In the case of $k = -1$, we can use $\chi = \sinh\psi$ with $\psi \in [0, \infty[$, to get the line element of a hyperboloid,

$$ds^2 = d\psi^2 + \sinh^2 \phi \, d\Omega_{d-1}^2. \tag{2.106}$$

Also in the case of a Lorentzian manifold we find three maximally symmetric spacetimes depending on the sign of the Ricci scalar R. For $R = 0$, the maximally symmetric spacetime is Minkowski spacetime, which we discuss first in section 2.3.1. For $R < 0$ the maximally symmetric spacetime is *Anti-de Sitter (AdS) space* whose properties we review in section 2.3.2. For $R > 0$, the maximally symmetric spacetime is de Sitter space which we do not discuss in detail here.

Such maximally symmetric spacetimes may occur as solutions of Einstein's equation (2.97) in the vacuum, i.e. without any matter content $T_{\mu\nu}$. Multiplying (2.97) with $g^{\mu\nu}$ and setting $T_{\mu\nu} = 0$, we obtain $R = 2d\Lambda/(d - 2)$. Therefore the cosmological constant has to be positive or negative in the case of de Sitter or Anti-de Sitter space, respectively.

2.3.1 Minkowski spacetime

d-dimensional Minkowski spacetime is a solution to the vacuum Einstein equations with $\Lambda = 0$. We use coordinates such that the line element ds is given by

$$ds^2 = \eta_{\mu\nu} dx^\mu dx^\nu. \tag{2.107}$$

Let us study the causal structure of this spacetime. In general, the causal structure may be visualised by a *conformal diagram*, which is also referred to as a *Penrose diagram*. A *conformal diagram* is defined by the following two properties. To study the causal structure of spacetime, we have to use coordinates that vary in a finite range only. Furthermore, null geodesics should always remain straight lines at angles of $\pm 45°$.

Consider first two-dimensional Minkowski spacetime with metric $ds^2 = -dt^2 + dx^2$ where $-\infty < t, x < \infty$. Introducing light-cone coordinates of the form $u = t - x$ and $v = t + x$ and mapping these to a finite interval through $\tilde{u} = \arctan(u)$, $\tilde{v} = \arctan(v)$, the metric reads

$$ds^2 = -\frac{1}{\cos^2 \tilde{u} \cos^2 \tilde{v}} d\tilde{u} \, d\tilde{v} = \frac{1}{4 \cos^2 \tilde{u} \cos^2 \tilde{v}} \left(-d\tilde{t}^2 + d\tilde{x}^2 \right), \tag{2.108}$$

where we have introduced $\tilde{t}$ and $\tilde{x}$ as $\tilde{t} = \frac{1}{2}(\tilde{u} + \tilde{v})$ and $\tilde{x} = \frac{1}{2}(\tilde{v} - \tilde{u})$. In this way we have mapped two-dimensional Minkowski spacetime into a finite region given by $-\pi < \tilde{t} + \tilde{x} < \pi$ and $-\pi < \tilde{t} - \tilde{x} < \pi$. Note that the resulting metric (2.108) is conformal to Minkowski spacetime. Since null geodesics are invariant under conformal transformations, the null geodesics are still straight lines at $\pm 45°$. Therefore we may use the coordinates $\tilde{t}$ and $\tilde{x}$ to draw the conformal diagram of Minkowski space. In the conformal diagram 2.3 there are various infinities present. First of all, there are three special points which are referred to as

$$i^+ \equiv \text{future timelike infinities,}$$
$$i^- \equiv \text{past timelike infinities,}$$
$$i^0 \equiv \text{spacelike infinities.}$$

Timelike geodesics begin at i^- and end at i^+, whereas spacelike geodesics begin and end at i^0. Furthermore, we may define

$$\mathscr{I}^+ \equiv \text{future null infinities connecting } i^0, i^+,$$
$$\mathscr{I}^- \equiv \text{past null infinities connecting } i^0, i^-.$$

All null geodesics originate from $\mathscr{I}^-$ and reach $\mathscr{I}^+$.

For higher dimensional Minkowski spacetime, the conformal diagram looks slightly different. We start from the metric

$$ds^2 = -dt^2 + d\vec{x}^2 = -dt^2 + dr^2 + r^2 d\Omega_{d-2}^2, \tag{2.111}$$

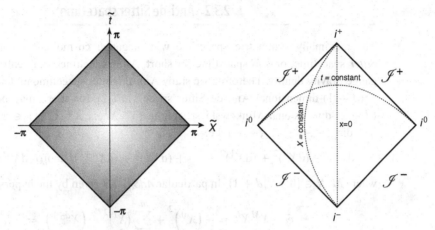

Figure 2.3   Conformal diagram of two-dimensional Minkowski spacetime.

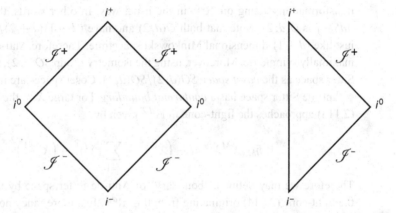

Figure 2.4   Conformal diagram of two-dimensional Minkowski spacetime (left) and of d-dimensional Minkowski spacetime for $d > 2$ (right). For the d-dimensional case, a sphere S^{d-2} has to be added at each point.

where r is the radial direction, $r^2 = \vec{x}^2$. Note that now r is restricted to values $r \geq 0$. Repeating the same analysis as in the two-dimensional case, the conformal diagram of d-dimensional Minkowski spacetime is given by figure 2.4. To draw this figure we have suppressed a sphere S^{d-2} which we have to attach at each point of the conformal diagram.

Moreover, we are interested in spacetimes which are 'deformed' in the interior but asymptote to Minkowski spacetime. One example is the Schwarzschild black hole which we discuss in section 2.4. To make more precise what we mean by 'asymptote to Minkowski spacetime' we introduce the notion of *asymptotically flat spacetimes*. By definition, an asymptotically flat spacetime shares future null infinity $\mathscr{I}^+$, spacelike infinity i^0 and past null infinity $\mathscr{I}^-$ with the causal diagram of Minkowski spacetime, but not necessarily timelike infinities $i^\pm$.

2.3.2 Anti-de Sitter spacetime

The maximally symmetric spacetime with negative cosmological constant is *Anti-de Sitter* spacetime, or *AdS* spacetime for short. AdS spacetime is of central importance for gauge/gravity duality. Therefore we study Anti-de Sitter spacetime in detail.[2]

$(d + 1)$-dimensional Anti-de Sitter space, AdS_{d+1} for short, may be embedded into $(d + 2)$-dimensional Minkowski spacetime $(X^0, X^1, ..., X^d, X^{d+1}) \in \mathbb{R}^{d,2}$, with metric $\bar{\eta} = \text{diag}(-, +, +, \ldots, +, -)$, i.e.

$$ds^2 = -(dX^0)^2 + (dX^1)^2 + \cdots + (dX^d)^2 - (dX^{d+1})^2 \equiv \bar{\eta}_{MN} dX^M dX^N, \quad (2.112)$$

where $M, N \in \{0, \ldots, d + 1\}$. In particular AdS_{d+1} is given by the hypersurface

$$\bar{\eta}_{MN} X^M X^N = -\left(X^0\right)^2 + \sum_{i=1}^{d} \left(X^i\right)^2 - \left(X^{d+1}\right)^2 = -L^2 \quad (2.113)$$

inside $\mathbb{R}^{d,2}$. In (2.113) L is the radius of curvature of the Anti-de Sitter space, as we will see later. Note that the hypersurface given by (2.113) is invariant under $O(d, 2)$ transformations acting on $\mathbb{R}^{d,2}$ in the usual way. In other words, the isometry group of AdS_{d+1} is $O(d, 2)$. Note that both $O(d, 2)$ and have $(d + 1)(d + 2)/2$ Killing generators, just like $(d + 1)$-dimensional Minkowski spacetime. Therefore Anti-de Sitter space is also maximally symmetric. Moreover, using the isometry group $SO(d, 2)$, we may write Anti-de Sitter space as the *coset space* $SO(d, 2)/SO(d, 1)$. Coset spaces are introduced in box 2.2.

Anti-de Sitter space has a *conformal boundary*. For large X^M, the hyperboloid given by (2.113) approaches the light-cone in $\mathbb{R}^{d,2}$ given by

$$\bar{\eta}_{MN} X^M X^N = -\left(X^0\right)^2 + \sum_{i=1}^{d} \left(X^i\right)^2 - \left(X^{d+1}\right)^2 = 0. \quad (2.114)$$

Therefore we may define a 'boundary' of Anti-de Sitter space by the set of all lines on the light-cone (2.114) originating from $0 \in \mathbb{R}^{d,2}$. In a more fancy notation, the conformal boundary of AdS_{d+1}, denoted by ∂AdS_{d+1}, is given by the set of points

$$\partial AdS_{d+1} = \left\{ [X] | X \in \mathbb{R}^{d,2}, X \neq 0, \bar{\eta}_{MN} X^M X^N = 0 \right\}, \quad (2.115)$$

Box 2.2   **Coset spaces**

A maximally symmetric spacetime may be represented as a *coset space*. The coset is obtained by modding out the isometry group of the spacetime $\mathcal{M}$ by the *stabiliser group* for each point $p \in \mathcal{M}$. The stabiliser group contains those isometries which leave p invariant. For example, for S^2, the isometry group is $SO(3)$. Each point p on S^2 is invariant under the rotations around the axis connecting the centre of the sphere to p. Thus S^2 is given by the coset space $SO(3)/SO(2)$. Similarly, in d dimensions, S^d corresponds to $SO(d + 1)/SO(d)$. A further example is Minkowski space which is given by the Poincaré group modded out by the Lorentz group, $ISO(d, 1)/SO(d, 1)$. Finally, AdS_{d+1} corresponds to $SO(d, 2)/SO(d, 1)$.

---

[2] For notational consistency with later chapters, we define $(d + 1)$-dimensional Anti-de Sitter space instead of d-dimensional Anti-de Sitter space.

where we identify $[X]$ with $[\tilde{X}]$ if $(X^0, X^1, \ldots, X^{d+1}) = \lambda(\tilde{X}^0, \tilde{X}^1, \ldots, \tilde{X}^{d+1})$ for a real number λ. To see the topology of the conformal boundary ∂AdS_{d+1}, we can represent any element $[X]$ of ∂AdS_{d+1} by the point X satisfying

$$\sum_{i=1}^{d} \left(X^i\right)^2 = 1. \tag{2.116}$$

Since X also has to satisfy (2.114) we further obtain

$$\left(X^0\right)^2 + \left(X^{d+1}\right)^2 = 1. \tag{2.117}$$

We therefore conclude that the conformal boundary of AdS_{d+1} is $(S^1 \times S^{d-1})/\mathbb{Z}_2$. Note that in this expression, we have to divide by $\mathbb{Z}_2$: $X \in \mathbb{R}^{d,2}$ and $-X$ satisfying (2.116) and (2.117) are different points in $S^1 \times S^{d-1}$, while according to the identification involved in (2.115), they are the same point in ∂AdS_{d+1}.

How should we think about the space ∂AdS_{d+1}? It turns out that ∂AdS_{d+1} is a compactification of d-dimensional Minkowski spacetime. To verify this, consider a point $X \neq 0$ satisfying (2.114). Introducing coordinates (u, v) by

$$u = X^{d+1} + X^d, \qquad v = X^{d+1} - X^d, \tag{2.118}$$

we may rewrite (2.114) as

$$uv = \eta_{\mu\nu} X^\mu X^\nu, \tag{2.119}$$

where μ and ν take values in $\{0, \ldots, d-1\}$ and $\eta_{\mu\nu}$ is the diagonal matrix with entries $\mathrm{diag}(-1, 1, \ldots, 1)$. Whenever $v \neq 0$ we can rescale the X so that $v = 1$. For given X^μ with $\mu \in \{0, \ldots, d-1\}$ we can solve (2.119) for u. Therefore for $v \neq 0$ we have d-dimensional Minkowski spacetime. The points with $v = 0$ correspond to infinities which we added to d-dimensional Minkowski spacetime. Analysing (2.119) we see that we added a light-cone to our Minkowski spacetime. We see in chapter 3 that this is necessary to define conformal transformations. This also explains why ∂AdS_{d+1} is a conformal compactification of d-dimensional Minkowski spacetime.

Let us study different coordinate systems for AdS_{d+1}. For example we may use the parametrisation

$$
\begin{aligned}
X^0 &= L \cosh \rho \cos \tau, \\
X^{d+1} &= L \cosh \rho \sin \tau, \\
X^i &= L \, \Omega_i \sinh \rho, \qquad \text{for } i = 1, \ldots, d,
\end{aligned}
\tag{2.120}
$$

where Ω_i with $i = 1, \ldots, d$ are angular coordinates satisfying $\sum_i \Omega_i^2 = 1$. In other words Ω_i parametrise a $(d-1)$-dimensional sphere S^{d-1}. The remaining coordinates take the ranges $\rho \in \mathbb{R}_+$ and $\tau \in [0, 2\pi[$. The coordinates (ρ, τ, Ω_i) are referred to as *global coordinates* of AdS_{d+1} since all points of the hypersurface (2.113) are taken into account exactly once.

Figure 2.5 displays AdS_2 spacetime embedded into $\mathbb{R}^{1,2}$. The coordinates of $\mathbb{R}^{1,2}$ are X^0, X^1 and X^2. Using the coordinates (2.120), we may extend ρ to $\rho \in \mathbb{R}$ capturing the effect of $\Omega^1 = \pm 1$. In particular, for AdS_2 the spatial section of the conformal boundary consists of two points, since $S^0 = \{\pm 1\}$.

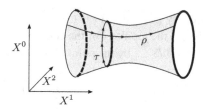

Figure 2.5 Schematic picture of AdS_2 embedded in $\mathbb{R}^{1,2}$.

Inserting (2.120) into (2.112) yields the metric

$$\mathrm{d}s^2 = L^2 \left(-\cosh^2 \rho \, \mathrm{d}\tau^2 + \mathrm{d}\rho^2 + \sinh^2 \rho \, \mathrm{d}\Omega_{d-1}^2 \right). \tag{2.121}$$

It features a timelike Killing vector ∂_τ on the whole manifold, so τ may be called the global time coordinate. In this parametrisation, only the maximal compact subgroup $SO(2) \times SO(d)$ of the isometry group $SO(2,d)$ of AdS_{d+1} is manifest. While $SO(2)$ generates translations in τ, $SO(d)$ acts by rotating the X^i coordinates (with $i = 1, \ldots, d$).

To investigate the conformal boundary of AdS space in global coordinates, it is convenient to introduce a new coordinate θ by $\tan \theta = \sinh \rho$. Then the metric (2.121) becomes that of the *Einstein static universe* $\mathbb{R} \times S^d$,

$$\mathrm{d}s^2 = \frac{L^2}{\cos^2 \theta} \left(-\mathrm{d}\tau^2 + \mathrm{d}\theta^2 + \sin^2 \theta \, \mathrm{d}\Omega_{d-1}^2 \right). \tag{2.122}$$

However, since $0 \leq \theta < \frac{\pi}{2}$, this metric covers only half of $\mathbb{R} \times S^d$. The causal structure remains unchanged when scaling this metric to get rid of the overall factor. Further, we may add the point $\theta = \frac{\pi}{2}$ corresponding to spatial infinity. Then the compactified spacetime is given by

$$\mathrm{d}s^2 = -\mathrm{d}\tau^2 + \mathrm{d}\theta^2 + \sin^2 \theta \, \mathrm{d}\Omega_{d-1}^2, \qquad 0 \leq \theta \leq \frac{\pi}{2}, \qquad 0 \leq \tau < 2\pi. \tag{2.123}$$

If we specify boundary conditions on $\mathbb{R} \times S^{d-1}$ at $\theta = \frac{\pi}{2}$, then the Cauchy problem is well posed. As one can easily read off from (2.123), the $\theta = \frac{\pi}{2}$ boundary of conformally compactified AdS_{d+1} is identical to the conformal compactification of d-dimensional Minkowski spacetime.

Note that the timelike coordinate τ is periodic in 2π and hence Anti-de Sitter spacetime has closed timelike curves. To avoid inconsistencies, we should consider the universal covering of Anti-de Sitter space by unwrapping the timelike circle. This is done by taking $\tau \in \mathbb{R}$ without identifying points. The universal covering is denoted by $\widetilde{AdS}_{d+1}$.

Exercise 2.3.1 In global coordinates (2.121), consider a radially directed light ray starting from $\rho = \rho_0$ with proper time $\tau(\rho_0) = 0$. Find the trajectory $\tau(\rho)$ for such a light ray. What is the coordinate time for a geodesic to go from ρ_0 to the boundary and come back? What is the proper time measured by a stationary observer's clock at ρ_0 for this trajectory?

Exercise 2.3.2 Determine the behaviour of a massive geodesic in the radial direction of AdS space in global coordinates (2.121). Show that a massive geodesic never reaches the conformal boundary of AdS space.

Let us introduce another useful parametrisation of the hyperboloid (2.113) using the coordinates $t \in \mathbb{R}, \vec{x} = (x^1, \ldots, x^{d-1}) \in \mathbb{R}^{d-1}$ as well as $r \in \mathbb{R}_+$. The parametrisation in these coordinates is given by

$$X^0 = \frac{L^2}{2r}\left(1 + \frac{r^2}{L^4}\left(\vec{x}^2 - t^2 + L^2\right)\right), \tag{2.124}$$

$$X^i = \frac{rx^i}{L} \quad \text{for } i \in \{1, \ldots, d-1\}, \tag{2.125}$$

$$X^d = \frac{L^2}{2r}\left(1 + \frac{r^2}{L^4}\left(\vec{x}^2 - t^2 - L^2\right)\right), \tag{2.126}$$

$$X^{d+1} = \frac{rt}{L}. \tag{2.127}$$

Due to the restriction $r > 0$, we cover only one-half of the AdS_{d+1} spacetime. These *local* coordinates are referred to as *Poincaré patch* coordinates. In the Poincaré patch, the metric of AdS_{d+1} space reads

$$ds^2 = \frac{L^2}{r^2}dr^2 + \frac{r^2}{L^2}\left(-dt^2 + d\vec{x}^2\right) \equiv \frac{L^2}{r^2}dr^2 + \frac{r^2}{L^2}\left(\eta_{\mu\nu}dx^\mu dx^\nu\right), \tag{2.128}$$

where $\mu = 0, \ldots, d$, $x^0 = t$ and $\eta_{\mu\nu} = \text{diag}(-1, +1, \ldots, +1)$. An explicit calculation of the Ricci scalar for AdS_{d+1} gives $R = -\frac{d(d+1)}{L^2}$, i.e. the curvature is indeed negative and constant. This also confirms that L is the radius of curvature.

Exercise 2.3.3 Calculate the Christoffel symbols $\Gamma^\mu_{\rho\sigma}$ of the metric (2.128) as well as the Riemann tensor $R^\mu_{\nu\rho\sigma}$ in the Poincaré patch of AdS_{d+1}. Check that the Ricci scalar R is given by $R = -\frac{d(d+1)}{L^2}$ and that Anti-de Sitter space is maximally symmetric since (2.102) is satisfied. Moreover, confirm that Anti-de Sitter space satisfies Einstein's field equations (2.97) with $T_{\mu\nu} = 0$. The cosmological constant Λ is given by

$$\Lambda = -\frac{d(d-1)}{2L^2}. \tag{2.129}$$

We may view $(d + 1)$-dimensional Anti-de Sitter spacetime in the Poincaré patch as flat spacetime, parametrised by the coordinates $t, \vec{x}$, plus an extra warped direction, which is denoted by r. For a fixed value of r, the d-dimensional transverse spacetime is flat Minkowski spacetime, i.e. $\mathbb{R}^{d-1,1}$. A cartoon of Anti-de Sitter space in these coordinates is shown in figure 2.6. The horizontal direction in figure 2.6 displays the radial direction r of Anti-de Sitter spacetime. To the right of figure 2.6, r is zero, whereas to the left, r asymptotes to infinity. As we will see, these are two special values of the radial direction.

For $r \to 0$, i.e. to the right of figure 2.6, we have a degenerate Killing horizon, also known as a Poincaré horizon. A *Killing horizon* is a null hypersurface uniquely defined by $k_\mu k^\mu = 0$, where k_μ is a Killing vector. Note that the Poincaré horizon is only a coordinate singularity, not a curvature singularity: on the other side of the horizon, i.e. for $r < 0$, there is another Poincaré patch, which is needed to cover the whole of AdS spacetime.

Note that the metric (2.128) has a second order pole for $r \to \infty$, i.e. g_{ii} diverges quadratically for $r \to \infty$. Indeed, it is possible to show that any metric of asymptotically Anti-de Sitter spaces always has such a quadratic divergence for a particular value r_* of the radial direction. The slice of spacetime for fixed $r = r_*$ is the *conformal boundary* of

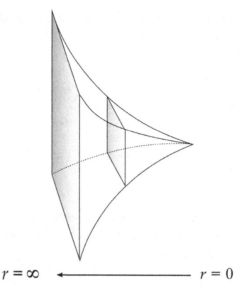

$r = \infty$ $\longleftarrow$ $r = 0$

Figure 2.6 A cartoon of Anti-de Sitter spacetime.

AdS space. In the coordinates used above in (2.128), the conformal boundary is located at $r \to \infty$.

In order to continue the metric to the boundary of AdS space, we have to ensure finiteness by multiplying the metric by a *defining function* $g(r, t, \vec{x})$, which is constructed as follows. $g(r, t, \vec{x})$ has to be a positive smooth function of the coordinates r, t and $\vec{x}$. Moreover, $g(r, t, \vec{x})$ must have a second order zero at $r = \infty$. An example of a defining function $g(r, t, \vec{x})$ is given by $g(r, t, \vec{x}) = (L^2/r^2)\omega(t, \vec{x})$, where ω is a smooth and positive function of $\vec{x}$ and t. Multiplying the metric (2.128) with g and taking the limit $r \to \infty$, we may define a finite boundary metric given by $ds^2_{\partial AdS} = \omega(t, \vec{x})(-dt^2 + d\vec{x}^2)$. Different choices of $\omega(t, \vec{x})$, or more generally different choices of $g(r, t, \vec{x})$, give rise to different boundary metrics. Therefore the bulk metric determines a class of boundary metrics which are related by conformal transformations. This class is referred to as *conformal structure*, i.e. an equivalence class of metrics related to each other by conformal transformations. Hence the boundary of Anti-de Sitter spacetimes may be referred to as conformal.

Whereas in the defining equation (2.113), the isometry group $SO(d, 2)$ of AdS_{d+1} is obvious, only the following subgroups of $SO(d, 2)$ are manifest for the metric in Poincaré coordinates,

- $ISO(d - 1, 1)$, i.e. all Poincaré transformations acting on the coordinates $(t, \vec{x})$,
- $SO(1, 1)$ acting on coordinates $t, \vec{x}$ and r as

$$(t, \vec{x}, r) \mapsto (\lambda t, \lambda \vec{x}, r/\lambda). \tag{2.130}$$

We see that we can identify the elements of $ISO(d - 1, 1)$ with the Poincaré transformations on the conformal boundary of AdS space. How do the other generators of the isometry group of AdS_{d+1} act on the conformal boundary of AdS space? It can be shown that the isometry group $SO(d, 2)$ acts on the boundary as the conformal group of Minkowski space.

In particular, the subgroup $SO(1, 1)$ is identified with the dilatation D of the conformal symmetry group of $\mathbb{R}^{d-1,1}$.

Sometimes it is more convenient to invert the r-coordinate by defining $z = L^2/r$. As opposed to the *r-coordinates* given by $(r, t, \vec{x})$, the conformal boundary in the *z-coordinates* $(z, t, \vec{x})$ is located at $z = 0$, whereas the Poincaré horizon is at $z \to \infty$. It is easy to verify that the metric in z-coordinates reads

$$\mathrm{d}s^2 = \frac{L^2}{z^2}\left(\mathrm{d}z^2 - \mathrm{d}t^2 + \mathrm{d}\vec{x}^2\right) = \frac{L^2}{z^2}\left(\mathrm{d}z^2 + \eta_{\mu\nu}\mathrm{d}x^\mu\mathrm{d}x^\nu\right). \tag{2.131}$$

Exercise 2.3.4 Use the coordinate transformation $z = \exp(-r/L)$ to rewrite the metric (2.131) as

$$\mathrm{d}s^2 = \mathrm{d}r^2 + L^2\, e^{2r/L}\, \eta_{\mu\nu}\mathrm{d}x^\mu\mathrm{d}x^\nu. \tag{2.132}$$

Note that the conformal boundary is at $r \to \infty$ while the horizon is located at $r \to -\infty$.

Exercise 2.3.5 Show that defining $\rho = z^2$, we obtain for the metric (2.131)

$$\mathrm{d}s^2 = L^2\left(\frac{\mathrm{d}\rho^2}{4\rho^2} + \frac{1}{\rho}\,\eta_{\mu\nu}\mathrm{d}x^\mu\mathrm{d}x^\nu\right), \tag{2.133}$$

which is referred to as the *Fefferman–Graham metric*.

The conformal diagram or Penrose diagram of AdS space may be obtained from a conformal map just as discussed for Minkowski space in section 2.3.1. Consider the global coordinates (2.121). A conformal transformation maps AdS_2 to the space $[0, 2\pi[\times[-\pi/2, \pi/2]$. The conformal diagram of AdS_2 is shown in figure 2.7, from which the conformal diagram of AdS_{d+1} may be obtained by adding a sphere S^{d-1} to each

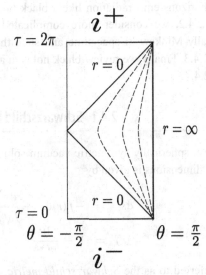

Figure 2.7 Conformal diagram of AdS_2. The Poincaré coordinates cover the triangular region shown. The dashed lines correspond to constant finite values of the Poincaré coordinate r. $r = 0$ and $r = \infty$ correspond to the two legs of the right-angled triangle and to its hypotenuse, respectively.

point. For the universal covering $\widetilde{AdS}_{d+1}$, τ is decompactified, such that the coordinate range becomes $\mathbb{R} \times [-\pi/2, \pi/2]$. The Poincaré coordinates cover the triangular region shown, while the global coordinates cover the entire conformal diagram. To formulate the AdS/CFT correspondence, we also need to define a Euclidean signature version of $(d+1)$-dimensional Anti-de Sitter space. To define Euclidean AdS_{d+1}, we simply Wick rotate the component X^0 in the defining equation (2.113). Therefore the isometry group of Euclidean AdS_{d+1} is given by $SO(d+1, 1)$ instead of $SO(d, 2)$. For the metric in global coordinates given by (2.121) we have to use $\tau_E = i\tau$, where τ_E is the Euclidean time and thus we have

$$ds^2 = L^2 \left(\cosh^2 \rho \, d\tau_E^2 + d\rho^2 + \sinh^2 \rho \, d\Omega_{d-1}^2 \right). \tag{2.134}$$

For the metric in Poincaré coordinates we just have to replace $\eta_{\mu\nu}$ by $\delta_{\mu\nu}$ in equations (2.128) and (2.131), with δ the standard Kronecker symbol.

2.4 Black holes

An interesting class of solutions of Einstein's equations of motion are black holes which by definition have at least one event horizon. An *event horizon* is a boundary in spacetime beyond which events cannot influence an outside observer. For example, Minkowski spacetime has no event horizon since all inextendible null curves start at $\mathscr{I}^-$ and terminate at $\mathscr{I}^+$.

We first discuss asymptotically flat Schwarzschild black holes in section 2.4.1. We realise that, as a result of the event horizon, the black hole may have a non-zero temperature, the Hawking temperature T_H. Indeed, quantum field theory in curved spacetime predicts that event horizons emit radiation like a black body with a finite temperature T_H. Then, in section 2.4.2, we consider more complicated charged and rotating black holes in asymptotically Minkowski spacetime and state the laws of black hole thermodynamics in section 2.4.3. Finally we review black holes in asymptotically Anti-de Sitter spacetime in section 2.4.4.

2.4.1 Schwarzschild black hole

The simplest, spherically symmetric vacuum solution to Einstein's equations (2.97) with $\Lambda = 0$ in d dimensions is given by

$$ds^2 = -f(r)dt^2 + \frac{dr^2}{f(r)} + r^2 \, d\Omega_{d-2}^2, \tag{2.135}$$

$$f(r) = 1 - \frac{2\mu}{r^{d-3}} \tag{2.136}$$

which is referred to as the *Schwarzschild metric*. Historically, for $d = 4$ where $f(r) = 1 - 2\mu/r$, this was the first non-trivial solution to Einstein's equations, found in 1916. $d\Omega_{d-2}$ is the infinitesimal angular element in $d - 2$ dimensions. μ in (2.136) is related to the mass of a black hole,

$$\mu = \frac{8\pi GM}{(d-2)\text{Vol}(S^{d-2})}, \qquad \text{Vol}(S^{d-2}) = \frac{2\pi^{\frac{d-1}{2}}}{\Gamma(\frac{d-1}{2})}, \qquad (2.137)$$

with G the Newton constant and $\text{Vol}(S^{d-2})$ the volume of the sphere S^{d-2}. The parameter M represents the mass of a *black hole* centred at the (spatial) origin. According to *Birkhoff's theorem*, this is the unique time independent spherically symmetric solution to Einstein's equations in the vacuum. Obviously, there are two special values $r = 0$ and $r = r_h$ for the radial coordinate. The origin $r = 0$ is a singularity since the curvature becomes infinite there. This is a curvature singularity and not just a coordinate singularity since this divergence occurs in any coordinate system.

The second special value $r = r_h$ is given by $f(r_h) = 0$. This condition gives $r_h = (2\mu)^{1/(d-3)}$ which is referred to as the *Schwarzschild radius*. In the special case of $d = 4$ dimensions, we have $r_h = 2GM$. For any number of dimensions, the curvature is finite at r_h. We will see that at this radius, there is an event horizon of a *black hole*.

In order to obtain the causal structure of the black hole spacetime, we have to extend the range of the coordinates. For simplicity, let us consider the case of a four-dimensional Schwarzschild black hole with coordinates (t, r, Ω_2) where Ω_2 are the coordinates of S^2 which we suppress from now on. Therefore the metric in the (t, r) subspace reads

$$ds^2 = -\left(1 - \frac{2GM}{r}\right) dt^2 + \left(1 - \frac{2GM}{r}\right)^{-1} dr^2. \qquad (2.138)$$

Considering radial null curves for which $ds^2 = 0$, we see that the light-cones close up if we approach the horizon $r \to 2GM$.

To study the causal structure, it is convenient to introduce the *tortoise coordinate* $r^*(r)$ satisfying

$$dr^* = \frac{dr}{1 - \frac{2GM}{r}}. \qquad (2.139)$$

Integrating both sides we obtain

$$r^* = r - 2GM + 2GM \ln\left(\frac{r}{2GM} - 1\right). \qquad (2.140)$$

Note that the tortoise coordinate r^* is only defined for $r \geq 2GM$. Moreover, the horizon located at $r = 2GM$ is pushed to minus infinity in tortoise coordinates, $r^* \to -\infty$. In the *light-cone tortoise coordinates* u and v,

$$v \equiv t + r^*, \qquad u \equiv t - r^*, \qquad (2.141)$$

the line element (2.138) reads

$$ds^2 = \left(1 - \frac{2GM}{r(u,v)}\right)(-dt^2 + dr^{*2}) = -\left(1 - \frac{2GM}{r(u,v)}\right) du \, dv. \qquad (2.142)$$

Note that none of the metric components is infinite for $r \to 2GM$. In fact, in this limit g_{tt} and $g_{r^*r^*}$ both vanish.

Curves with $v = $ constant correspond to infalling radial null geodesics while curves with $u = $ constant correspond to outgoing null geodesics. Instead of using the light-cone tortoise

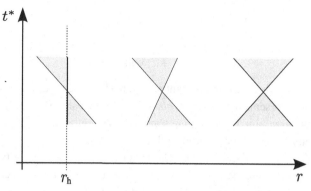

Figure 2.8 Light-cones tilt and narrow when approaching the horizon $r \rightarrow r_{\mathrm{h}} = 2GM$. Here (r, v) are the ingoing Eddington–Finkelstein coordinates, with $t^* = v - r$ on the vertical axis, such that the diagonals with $v = $ constant lines at $-45°$.

coordinates (u, v) we may use the original radial coordinate r as well as one of the two light-cone coordinates, say v. In the coordinates (r, v) known as infalling Eddington–Finkelstein coordinates the line element (2.138) reads

$$\mathrm{d}s^2 = -\left(1 - \frac{2GM}{r}\right)\mathrm{d}v^2 + 2\mathrm{d}v\,\mathrm{d}r. \tag{2.143}$$

Note that the horizon is still located at the finite value $r = 2GM$ and none of the metric components diverges. Therefore we explicitly see that the horizon $r_{\mathrm{h}} = 2GM$ is just a coordinate singularity since we can extend our coordinate system by using Eddington–Finkelstein coordinates. Studying radial null curves in these coordinates we see that the light-cones do not close up but become tilted as we see in figure 2.8.

We can further extend the coordinates by using $(\tilde{u}, \tilde{v})$

$$\tilde{u} \equiv -4GM \exp\left(-\frac{u}{4GM}\right), \qquad \tilde{v} \equiv 4GM \exp\left(\frac{v}{4GM}\right), \tag{2.144}$$

which are referred to as *Kruskal–Szekeres light-cone coordinates*. In these coordinates the metric (2.142) reads

$$\mathrm{d}s^2 = \frac{2GM}{r(\tilde{u}, \tilde{v})} \exp\left(1 - \frac{r(\tilde{u}, \tilde{v})}{2GM}\right) \mathrm{d}\tilde{u}\,\mathrm{d}\tilde{v}, \tag{2.145}$$

where the radius r is now implicitly defined by

$$\tilde{u}\tilde{v} = -(4GM)^2 \left(\frac{r(\tilde{u}, \tilde{v})}{2GM} - 1\right) \exp\left(\frac{r(\tilde{u}, \tilde{v})}{2GM} - 1\right). \tag{2.146}$$

The Kruskal–Szekeres coordinates as defined in (2.144) range from $-\infty < \tilde{u} < 0$ and $0 < \tilde{v} < +\infty$ covering the exterior of the black hole, i.e. $r > 2GM$ in the original coordinates (t, r). The metric (2.146) is also defined for $\tilde{u} > 0$ or $\tilde{v} < 0$ and therefore we may extend the Kruskal–Szekeres coordinates to $\tilde{u}, \tilde{v} \in (-\infty, \infty)$. In this context the question arises whether we may further extend the spacetime. It turns out that this cannot be done and therefore the Kruskal–Szekeres coordinates are the maximal analytic extension of the Schwarzschild solution.

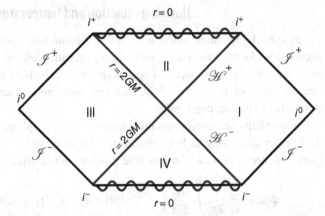

Figure 2.9 Conformal diagram of a maximally extended asymptotically flat Schwarzschild black hole. The Kruskal–Szekeres light-cone coordinates $(\tilde{u}, \tilde{v})$ are mapped to a finite interval using $\bar{u} = \arctan \tilde{u}$ and $\bar{v} = \arctan \tilde{v}$.

The conformal diagram for the maximally extended asymptotically flat Schwarzschild black hole is given in figure 2.9. If $\tilde{u}\tilde{v} < 0$, we have $r > 2GM$ which describes the exterior of the black hole. The coordinate patch (labelled I) has $-\infty < \tilde{u} < 0$ and $0 < \tilde{v} < +\infty$ and corresponds to the exterior of the Schwarzschild spacetime. Another coordinate patch, region III, is given by $0 < \tilde{u} < +\infty$ and $-\infty < \tilde{v} < 0$. Moreover, we have two regions, namely II and IV, which correspond to $r < 2GM$. Region II contains the future singularity at $r = 0$ while region IV contains the past singularity at $r = 0$.

The Kruskal–Szekeres coordinates $(\tilde{t}, \tilde{r})$ given by $\tilde{u} = \tilde{t} - \tilde{r}$ and $\tilde{v} = \tilde{t} + \tilde{r}$ have a number of useful properties. First, in the $(\tilde{t}, \tilde{r})$ subspace, radial null curves are given by straight lines at $\pm 45°$ angles, $\tilde{t} = \pm \tilde{r} + \text{constant}$. In particular, the null curves $\tilde{t} = \pm \tilde{r}$ correspond to event horizons which are located at $r = 2GM$ in the original coordinates. Moreover, surfaces of constant r are given by hyperbolae $\tilde{t}^2 - \tilde{r}^2 = \text{constant}$ in the $(\tilde{t}, \tilde{r})$ plane, while surfaces of constant t become straight lines through the origin.

We have two event horizons given by the lines through the origin at an angle of $\pm 45°$ relative to the $\tilde{t}, \tilde{r}$ axes: the past horizon $\mathcal{H}^-$ at $\tilde{v} = 0$ and the future horizon $\mathcal{H}^+$ at $\tilde{u} = 0$. Note that the conformal diagram shares the asymptotic structure of flat Minkowski spacetime, consisting of $\mathcal{J}^+$, i^0 and $\mathcal{J}^-$. Therefore we refer to the Schwarzschild black hole as asymptotically flat. Within the conformal diagram of the maximally extended Schwarzschild spacetime as shown in figure 2.9, there are four different physical objects encoded. The conformal diagram has two singularities located at $r = 0$, one in the future, one in the past, as well as two horizons and two asymptotically flat regimes. The fully extended Schwarzschild solution contains (a) two *white holes* that share the past singularity IV, but have different asymptotically flat spacetime regions I or III; and (b) two *black holes* with common future singularity II, but different spacetime regions I and III. The tortoise coordinates (t, r^*) parametrise only part of the region I of the Kruskal diagram.

However, when we describe black holes which result from the collapse of a matter distribution, the relevant diagram contains only part of the regions I and II, and we also have to omit the past horizon $\mathcal{H}^-$.

Hawking radiation and temperature

Let us consider quantum fields in curved spacetime with special emphasis on black hole geometries. As we saw above, there is no globally defined timelike Killing vector. In particular, the vector ∂_t becomes spacelike for $r < 2GM$. Thus there is no unique mode decomposition of the quantum field. Different observers will decompose the quantum fields $\hat{\phi}$ in different momentum modes.

For simplicity let us suppress the coordinates of the sphere S^2 and reduce the quantum field theory effectively to 1+1 dimensions. An observer located at rest far away from the black hole decomposes a massless field $\hat{\phi}$ using the coordinates (t, r), in the following way,

$$\hat{\phi}(u, v) = \frac{1}{2\pi} \int_0^\infty \frac{d\Omega}{2\Omega} \left(e^{-i\Omega u} \hat{b}(\Omega) + e^{i\Omega u} \hat{b}^\dagger(\Omega) + \text{left-moving} \right), \qquad (2.147)$$

while an observer freely falling through the horizon will use Kruskal–Szekeres coordinates and therefore finds

$$\hat{\phi}(\tilde{u}, \tilde{v}) = \frac{1}{2\pi} \int_0^\infty \frac{d\omega}{2\omega} \left(e^{-i\omega \tilde{u}} \hat{a}(\omega) + e^{i\omega \tilde{u}} \hat{a}^\dagger(\omega) + \text{left-moving} \right). \qquad (2.148)$$

The corresponding vacua of the two observers are given by

$$\hat{b}(\Omega)|0_B\rangle = 0 \quad \text{and} \quad \hat{a}(\omega)|0_K\rangle = 0, \qquad (2.149)$$

where $|0_B\rangle$ is the *Boulware vacuum* while $|0_K\rangle$ is the *Kruskal vacuum* state. From the point of view of the observer at rest far away from the black hole, the Kruskal vacuum state $|0_K\rangle$ contains particles. This may be seen from the fact that the modes $\hat{a}(\omega)$ and $\hat{b}(\Omega)$ and their Hermitian conjugates are related by a *Bogolyubov transformation*. In particular,

$$\hat{b}(\Omega) = \int_0^\infty d\omega \left[\alpha_{\Omega\omega} \hat{a}(\omega) - \beta_{\Omega\omega} \hat{a}^\dagger(\omega) \right]. \qquad (2.150)$$

Determining the coefficients $\alpha_{\Omega\omega}$, $\beta_{\Omega\omega}$ we find that the observer at rest at infinity sees particles with a thermal spectrum

$$n(\Omega) = \left(\exp\left(\frac{\Omega}{T_H} \right) - 1 \right)^{-1}, \qquad (2.151)$$

where T_H is the Hawking temperature.

Reintroducing the sphere S^2 by generalising the arguments above to 3+1 dimensions using spherical harmonics, we have to correct the thermal spectrum (2.151) by a greybody factor $\Gamma_l(\Omega) < 1$. Moreover, instead of bosonic fields we may consider fermionic fields. Also in this case, the Hawking effect occurs. However, for half-integer spin fields the Hawking radiation is a Fermi–Dirac distribution rather than the Bose–Einstein distribution. For the spectrum of the Hawking radiation, we thus have

$$n_\Omega = \Gamma_l(\Omega) \left(\exp\left(\frac{\Omega}{T_H} \right) \pm 1 \right)^{-1}, \qquad (2.152)$$

where the plus/minus sign corresponds to fermionic/bosonic fields, respectively.

It remains to determine the Hawking temperature T_H, which can be achieved by applying the methods outlined above. However, we may also determine T_H purely from the metric using Euclidean quantum gravity, in which the canonical partition function for gravity at temperature $T = 1/\beta$ is given by

$$Z(\beta) = \int \mathcal{D}g \, e^{-\mathcal{S}[g]}, \tag{2.153}$$

where the integral is performed over all Riemannian metrics satisfying certain asymptotic fall-off conditions. The right-hand side also depends on β via the asymptotic periodicity which has to be satisfied by all geometries (cf. chapter 1). We may approximate the partition function using the saddle point approximation as

$$Z(\beta) \simeq e^{-\mathcal{S}^*}. \tag{2.154}$$

The saddle point geometries are solutions of the classical equations of motion of the action and thus have to be regular. $\mathcal{S}^*$ denotes the value of the classical action for these saddle point geometries.

The action $\mathcal{S}$ to be used in (2.154) is the Einstein–Hilbert action (2.99) (with $\Lambda = 0$ and in Euclidean signature) supplemented by a boundary term $\mathcal{S}_{\text{bdy}}$, the *Gibbons–Hawking* term,

$$\mathcal{S}_{\text{bdy}} = -\frac{1}{8\pi G} \int_{\partial \mathcal{M}} \mathrm{d}^{d-1}x \sqrt{\gamma} K. \tag{2.155}$$

Here, K is the trace of the *extrinsic curvature* and γ is the induced metric on the boundary $\partial \mathcal{M}$. In terms of the induced metric and an outward-pointing unit vector normal to the boundary, we have

$$K = \gamma^{\mu\nu} \nabla_\mu n_\nu. \tag{2.156}$$

The boundary term (2.155) may be motivated as follows. The Ricci tensor involves two derivatives acting on the metric component $g_{\mu\nu}$. However, it is assumed in general that the Lagrangian contains terms at most of first order in the derivatives, see the discussion below (1.7) in chapter 1. Of course, we can always integrate the Einstein–Hilbert Lagrangian by parts to remove the second derivatives. If the manifold $\mathcal{M}$ is not compact and has a boundary $\partial \mathcal{M}$, we are left with a boundary term of the form (2.155).

In other words, only if we supplement the Einstein–Hilbert action with the Gibbons–Hawking term, i.e. if we consider

$$\mathcal{S} = -\frac{1}{16\pi G} \int_{\mathcal{M}} \mathrm{d}^d x \sqrt{g} R - \frac{1}{8\pi G} \int_{\partial \mathcal{M}} \mathrm{d}^{d-1}x \sqrt{\gamma} K, \tag{2.157}$$

is the resulting action an extremum under variations of the metric provided that the variations vanish at the boundary. However, the variation of normal derivatives of the metric does not have to vanish at the boundary $\partial \mathcal{M}$. Note the different signs in (2.157) and (2.99), which are due to changing to Euclidean signature.

Using (2.154), we may calculate the entropy S via

$$S = \ln Z - \beta \frac{\partial \ln Z}{\partial \beta} = \beta \frac{\partial \mathcal{S}^*}{\partial \beta} - \mathcal{S}^*. \tag{2.158}$$

Let us apply this procedure. First we have to identify those black hole geometries which are regular after Wick rotation $\tau = it$ to Euclidean space. We compactify the Euclidean timelike direction on a circle. Instead of (2.135), we allow for a more general metric of the form

$$ds^2 = f(r)d\tau^2 + g^{-1}(r)dr^2 + r^2 d\Omega_{d-2}^2, \tag{2.159}$$

where we assume that both $f(r)$ and $g(r)$ have a first order zero at $r = r_h$, i.e. $f(r_h) = 0$ but $f'(r_h) \neq 0$ and similarly for $g(r)$. Then close to $r = r_h$ we may expand $f(r)$ and $g(r)$ as

$$f(r) \simeq f'(r_h)(r - r_h) + \mathcal{O}((r - r_h)^2), \qquad g(r) \simeq g'(r_h)(r - r_h) + \mathcal{O}((r - r_h)^2). \tag{2.160}$$

The fact that f and g vanish at $r = r_h$ imposes a constraint on the periodicity in the τ direction, which ensures regularity of the Euclidean space: This space should be a genuine stationary point of the Einstein–Hilbert action and hence has to be regular. Inserting the expansions (2.160) into (2.159) we obtain for $r \simeq r_h$

$$ds^2 = f'(r_h)(r - r_h)d\tau^2 + \frac{1}{g'(r_h)(r - r_h)}dr^2 + \ldots \tag{2.161}$$

$$\equiv \rho^2 d\phi^2 + d\rho^2 + \ldots, \tag{2.162}$$

where the dots stand for the regular angular part and

$$\rho^2 = \frac{4(r - r_h)}{g'(r_h)}, \qquad \phi = \frac{1}{2}\sqrt{g'(r_h)f'(r_h)}\,\tau. \tag{2.163}$$

From (2.162) we may interpret (ρ, ϕ) as polar coordinates. To avoid a conical singularity, ϕ must be of period 2π, i.e. $\phi \sim \phi + 2\pi$. This implies

$$\tau \sim \tau + \frac{4\pi}{\sqrt{f'(r_h)g'(r_h)}}. \tag{2.164}$$

As explained in section 1.4, τ has to be periodic with periodicity $1/T$ where T is the corresponding temperature. Therefore we may read off the Hawking temperature as

$$T_H = \frac{\sqrt{f'(r_h)g'(r_h)}}{4\pi}. \tag{2.165}$$

For the special case $f'(r_h) = g'(r_h)$ we obtain

$$T_H = \frac{|f'(r_h)|}{4\pi}. \tag{2.166}$$

Exercise 2.4.1   Suppose $f(r)$ has a double zero at $r = r_h$, i.e. $f(r_h) = f'(r_h) = 0$ but $f''(r_h) \neq 0$, while $g(r_h) \neq 0$ in the metric (2.159). Show that the spacetime is regular provided that τ is perodic with periodicity $1/T_H$, where

$$T_H = \frac{\sqrt{2f''(r_h)g(r_h)}}{4\pi}. \tag{2.167}$$

Exercise 2.4.2   Using L'Hospital's rule, show that (2.165) and (2.167) may be derived from

$$T_H = \frac{1}{4\pi}\sqrt{\frac{g(r_h)}{f(r_h)}}f'(r_h). \tag{2.168}$$

For the Schwarzschild black hole considered above, we have

$$T_H = \frac{2\mu(d-3)r_h^{2-d}}{4\pi} = \frac{d-3}{4\pi}(2\mu)^{-\frac{1}{d-3}}. \tag{2.169}$$

Using this expression, we may write the Hawking temperature in terms of the mass M of the black hole. We see that the Hawking temperature increases when μ and thus the mass decreases. Therefore, the heat capacity $\partial M/\partial T$ is negative. In particular, for $d = 4$ and $\mu = GM$ we have

$$T_H = \frac{1}{8\pi GM}, \text{ or equivalently } M = \frac{1}{8\pi GT_H} = \frac{\beta}{8\pi G} \tag{2.170}$$

Alternatively, we may determine the Hawking temperature T_H of any generic Killing horizon by

$$T_H = \frac{\kappa}{2\pi} \tag{2.171}$$

where κ is the surface gravity defined by

$$\kappa k^\mu = k^\nu \nabla_\nu k^\mu. \tag{2.172}$$

Here k^μ is the Killing vector associated with the Killing horizon. Equation (2.172) should be evaluated at the horizon. For asymptotically flat spacetimes we have to choose the Killing vector such that it satisfies

$$k^\mu k_\mu \to -1 \quad \text{as} \quad r \to \infty. \tag{2.173}$$

To determine the entropy of the black hole solution, we may use the saddle point method of Euclidean quantum gravity as outlined above. In particular, the partition function is given by (2.154). In this equation $\mathcal{S}^*$ is the on-shell value of (2.157) for the regular geometries constructed above. However, note that $\mathcal{S}^*$ is infinite for these cases. We may regularise the expression by subtracting the action $\mathcal{S}^*[g^0]$ of a reference background with metric g^0. A natural choice for this reference background is the Minkowski vacuum, which may be regarded as the ground state for asymptotically flat boundary conditions. For vacuum solutions for which $R_{\mu\nu} = 0$, the bulk term in (2.157) vanishes and we are left with the boundary term

$$\mathcal{S}^*[g] - \mathcal{S}^*[g^0] = -\frac{1}{8\pi G}\int_{\partial \mathcal{M}} d^{d-1}x\,(\sqrt{\gamma}K - \sqrt{\gamma^0}K^0). \tag{2.174}$$

The integral is taken at the asymptotic boundary of spacetime, where both metrics have to be taken to coincide asymptotically. To evaluate (2.174) explicitly, in the case of a four-dimensional black hole solution given by (2.135) with $d\Omega_2^2 = d\theta^2 + \sin^2\theta\,d\phi^2$, we take the boundary to be a spherical shell at large radius $r = R$. With

$$\sqrt{\gamma} = r^2\left(1 - \frac{2GM}{r}\right)^{1/2}\sin\theta, \tag{2.175}$$

and recalling that Euclidean time is periodic with period $1/T$, in 3+1 dimensions we have

$$\int_{\partial \mathcal{M}} d^3x\,\sqrt{\gamma}K = \frac{4\pi}{T}(2R - 3GM) \tag{2.176}$$

for the contribution from the Schwarzschild metric. To evaluate K, we have used that $\sqrt{\gamma}K = n^\mu \partial_\mu \sqrt{\gamma}$, where n^μ is the outward radial unit normal vector at the boundary. For the Minkowski contribution, the temperature is arbitrary. We fix it such that asymptotically, the metric matches the Schwarzschild solution. This implies that we have to match the length of the circles of Euclidean time,

$$\int_0^{1/T} d\tau \sqrt{\gamma_{\tau\tau}} = \int_0^{1/T_0} d\tau \sqrt{\gamma_{\tau\tau}^0}. \tag{2.177}$$

With $K^0 = 2/r$, $\sqrt{\gamma^0} = r^2\sin\theta$ for Minkowski space, the result for the second contribution to (2.174) is then

$$\int_{\partial\mathcal{M}} d^3x \sqrt{\gamma^0}K^0 = \frac{8\pi}{T}R\left(1 - \frac{2GM}{R}\right)^{1/2}, \tag{2.178}$$

and in the limit $R \to \infty$ we have

$$\mathcal{S}^*[g] - \mathcal{S}^*[g^0] = \frac{M}{2T}. \tag{2.179}$$

With the free energy given by $F = -T\ln Z = T(\mathcal{S}^*[g] - \mathcal{S}^*[g^0])$, and the inverse temperature $\beta \equiv 1/T$ with $k_B = 1$, we find using conventional thermodynamical relations and (2.170) that the energy is given by

$$E = \frac{\partial(\beta F)}{\partial\beta} = \frac{\beta}{8\pi G} = M, \tag{2.180}$$

and the entropy by

$$S = \left(\beta\frac{\partial}{\partial\beta} - 1\right)(\beta F) = \frac{\beta^2}{16\pi G}. \tag{2.181}$$

Equation (2.181) may be rewritten in terms of the area $A = 4\tilde{\pi}r_h^2$ of the black hole horizon,

$$S = \frac{A}{4G}. \tag{2.182}$$

This is the important result of Bekenstein and Hawking according to which the entropy of a black hole is given by the area of its horizon.

2.4.2 Charged and rotating black holes

So far we have considered solutions for the vacuum Einstein equations as obtained from the Einstein–Hilbert action. Now we consider the case where the action also contains an Abelian gauge field, which provides a matter contribution,

$$\mathcal{S} = \int d^d x \sqrt{-g}\left(\frac{1}{2\kappa^2}R - \frac{1}{4}F_{\mu\nu}F^{\mu\nu}\right). \tag{2.183}$$

The solution for the metric following from the associated equations of motion is known as the *Reissner–Nordström solution*. In d dimensions, it takes the form

$$ds^2 = -f(r)dt^2 + f^{-1}(r)dr^2 + r^2 d\Omega_{d-2}^2, \tag{2.184}$$

where the factor $f(r)$ in the black hole metric now takes the form

$$f(r) = 1 - \frac{2\mu}{r^{d-3}} + \frac{\theta^2}{r^{2(d-3)}}. \tag{2.185}$$

The parameters μ and θ are related to the mass M and charge Q of the black hole, respectively, by virtue of

$$M = \frac{(d-2)\text{Vol}(S^{d-2})}{8\pi G}\mu, \qquad \text{Vol}(S^{d-2}) = \frac{2\pi^{(d-1)/2}}{\Gamma\left(\frac{d-1}{2}\right)}, \tag{2.186}$$

$$Q^2 = \frac{(d-2)(d-3)\text{Vol}(S^{d-2})}{8\pi G}\theta^2. \tag{2.187}$$

For $\mu^2 > \theta^2$, there are two different zeros for f, the *outer horizon* $r_{h,+}$ and the *inner horizon* $r_{h,-}$, which are given by

$$r_{h,\pm} = \left(\mu \pm \sqrt{\mu^2 - \theta^2}\right)^{\frac{1}{d-3}}. \tag{2.188}$$

In this case, the black hole horizon is given by the outer horizon. The Hawking temperature and entropy are given by

$$T = \frac{|f'(r_{h,+})|}{4\pi} = \frac{d-3}{4\pi r_{h,+}}\left(1 - \left(\frac{r_{h,-}}{r_{h,+}}\right)^{d-3}\right), \tag{2.189}$$

$$S = \frac{A_+}{4G} = \frac{\text{Vol}(S^{d-2})r_{h,+}^{d-2}}{4G}. \tag{2.190}$$

For $\mu^2 = \theta^2$ we have $r_{h,+} = r_{h,-}$, which implies $T = 0$ and

$$f(r) = \left(1 - \left(\frac{r_{h,+}}{r}\right)^{d-3}\right)^2. \tag{2.191}$$

This case is the *extremal* Reissner–Nordström black hole. Defining

$$\tilde{r}^{d-3} = r^{d-3} - r_{h,+}^{d-3} \tag{2.192}$$

such that the horizon is at $\tilde{r} = 0$, we may rewrite the metric of the extremal Reissner–Nordström black hole as

$$ds^2 = -H^{-2}dt^2 + H^{2/(d-3)}d\vec{x}^2, \qquad \vec{x} \in \mathbb{R}^{d-1}, \tag{2.193}$$

with $d\vec{x}^2 = d\tilde{r}^2 + \tilde{r}^2 d\Omega_{d-2}^2$, $|\vec{x}| \equiv \tilde{r}$ and

$$H = 1 + \left(\frac{r_{h,+}}{|\vec{x}|}\right)^{d-3}. \tag{2.194}$$

H is a harmonic function. Consequently, we may also consider a *multicentre solution* of the form

$$H = 1 + \sum_i \frac{q_i}{|\vec{x} - \vec{x}_i|^{d-3}}. \tag{2.195}$$

These solutions are stable since in the extremal case $\mu^2 = \theta^2$, the electric and gravitational forces cancel each other.

For $\mu^2 < \theta^2$, there is no horizon present and the geometry has a naked singularity. In supergravity theories, we always have $M \geq |Q|$ due to the BPS bound $M = |Q|$. This implies that naked singularities are absent in these theories.

In addition to the charged black holes, there are also rotating black holes known as *Kerr black holes*. Black holes which in addition to charge and mass have an angular momentum J are referred to as *Kerr–Newman* black holes. While rotating black holes have also been studied in the context of gauge/gravity duality, we do not discuss them further in this book.

2.4.3 Black hole thermodynamics

The thermal properties of black holes may be summarised in four laws which have analogues in the corresponding four laws of standard thermodynamics. As we shall see later, many of the features of black hole thermodynamics have a very natural reinterpretation in the context of the AdS/CFT correspondence.

The four laws of black hole thermodynamics read as follows. The *zeroth law* of black hole thermodynamics states that the surface gravity κ is constant over the horizon. This implies thermal equilibrium. We have checked this explicitly for Schwarzschild black holes, but, as is less trivial, it is indeed constant for charged black holes in any dimension. Due to the zeroth law, the surface gravity corresponds to temperature. The same applies to the electrostatic potential Φ and the angular velocity Ω of a charged or rotating black hole. The *first law* states energy conservation: the change in the mass M of the black hole is related to the change in its area A, angular momentum J and charge Q by

$$dM = \frac{\kappa}{8\pi G} \delta A + \Omega \delta J + \Phi \delta Q. \qquad (2.196)$$

For $J = Q = 0$, and relating κ to the Hawking temperature of the Schwarzschild black hole, $T_{\mathrm{H}} = \kappa/(2\pi)$, we obtain

$$dM = \frac{1}{4G} T_{\mathrm{H}} \delta A \equiv T_{\mathrm{H}} \delta S_{\mathrm{BH}} \quad \Rightarrow \quad S_{\mathrm{BH}} = \frac{A}{4G} \qquad (2.197)$$

with *Bekenstein–Hawking entropy* S_{BH}. The *second law* states that the total entropy of a system consisting of a black hole and matter contributions never decreases, i.e.

$$dS_{\mathrm{tot}} = dS_{\mathrm{matter}} + dS_{\mathrm{BH}} \geq 0. \qquad (2.198)$$

The *third law* corresponds to Nernst's law: it is impossible to reduce the surface gravity κ to zero by a finite sequence of operations (e.g. by absorbing matter). These four laws are analogous to the four laws of standard thermodynamics.

2.4.4 Asymptotically AdS black holes

In order to obtain the black hole metric in asymptotically global Anti-de Sitter space, we simply have to add r^2/L^2 to the function $f(r)$ defined in (2.136). For $r \to \infty$, $f(r) \to r^2/L^2$ and hence the metric is asymptotically AdS. For later convenience, we consider $(d + 1)$-dimensional Anti-de Sitter space, and therefore the line element now reads

$$ds^2 = -f(r)dt^2 + \frac{dr^2}{f(r)} + r^2 d\Omega_{d-1}^2, \tag{2.199}$$

$$f(r) = 1 - \frac{2\mu}{r^{d-2}} + \frac{r^2}{L^2}, \tag{2.200}$$

with L the AdS curvature radius. This is referred to as the AdS–Schwarzschild black hole. In this case,

$$\mu = \frac{8\pi GM}{(d-1)\text{Vol}(S^{d-1})}, \tag{2.201}$$

where M is the mass of the AdS–Schwarzschild black hole, measured relative to the AdS ground state. Note that (2.201) coincides with (2.137), except that we now consider the case of $d + 1$ dimensions. There is again an event horizon at $r = r_h$, where r_h is the larger root of $f(r_h) = 0$. As for the black hole in asymptotically flat space, this event horizon is associated with a finite temperature.

Exercise 2.4.3   Calculate the Hawking temperature of the AdS–Schwarzschild black hole and show that it is given by

$$T = \frac{dr_h^2 + (d-2)L^2}{4\pi L^2 r_h}. \tag{2.202}$$

Show that T has a minimum as function of r_h. Determine the minimal temperature $T_{\min}$. This implies that black holes exist only for temperatures larger than $T_{\min}$.

Black holes with $r_h \ll L$ are referred to as *small black holes*. These have thermodynamic properties which are similar to those of the Schwarzschild black hole in asymptotically flat space, since, roughly speaking, we may neglect the r^2/L^2 term in $f(r)$ given by (2.200). In this limit, the Hawking temperature (2.202) is given by

$$T = \frac{d-2}{4\pi} \frac{1}{r_h}. \tag{2.203}$$

Hence, the temperature of the small black holes decreases with increasing r_h, while the mass M increases. Thus the heat capacity $C \sim \partial M/\partial T$ of the small black hole is negative. On the other hand, there are also *large black holes* for which $r_h \gg L$. Consequently, we may drop the constant 1 in (2.200) in this case. For large black holes, the Hawking temperature is given by

$$T = \frac{dr_h}{4\pi L^2} \tag{2.204}$$

and thus increases with increasing r_h. Since the mass M also increases with r_h, the heat capacity of large black holes is positive. This implies that the black hole can be in equilibrium with its own Hawking radiation. The idea is that since AdS space may be viewed as a box, the radiation emitted by the black hole is reflected and will be reabsorbed by the black hole. This is possible due to the behaviour of null geodesics in AdS space, as explored in exercise 2.3.1. If emission and absorption rates coincide, thermal equilibrium is reached.

Using (2.103), we may generalise the AdS–Schwarzschild metric to non-spherical event horizons,

$$ds^2 = -f(r)dt^2 + \frac{dr^2}{f(r)} + r^2 dK_{d-1}^2,$$ (2.205)

$$f(r) = k - \frac{2\mu}{r^{d-2}} + \frac{r^2}{L^2}.$$ (2.206)

k may take the values $1, 0, -1$. For $k = 1$, we recover the AdS–Schwarzschild metric as given by (2.199). For $k = 0$, dK_{d-1}^2 reduces to the metric of flat Euclidean space and is no longer compact. The corresponding spacetime is referred to as a *black brane*. Its thermodynamics is similar to that of the large black holes. For the case $k = -1$, dK_{d-1}^2 corresponds to a hyperbolic space and the black holes are referred to as topological. Note that the generalisation (2.205) is possible only in asymptotically Anti-de Sitter space, not in asymptotically flat space.

2.5 Energy conditions

In addition to the vacuum solutions to Einstein's equations considered so far, there are also solutions corresponding to a specific matter distribution, which enters in the Einstein equations by virtue of the energy-momentum tensor $T_{\mu\nu}$. From the matter field action, we may determine $T_{\mu\nu}$ from (2.101). However, sometimes it is not desirable to specify a particular matter system in the form of a Lagrangian and an associated energy-momentum tensor, since a general theory of gravity and its phenomena should be maximally independent of any assumptions concerning non-gravitational physics. For instance, this applies to the proof of important theorems for black holes, such as no-hair theorems and black hole thermodynamics.

However, in order to obtain sensible results we have to impose certain criteria on the form of the energy-momentum tensor which are met by relevant matter theories realised in nature. Such criteria are given by *energy conditions*. Let us list these conditions for a d-dimensional gravitational system.

- *Null energy condition*: the null energy condition holds if, for any arbitrary null vector ζ^μ,

$$T_{\mu\nu}\zeta^\mu\zeta^\nu \geq 0.$$ (2.207)

- *Weak energy condition*: the weak energy condition holds if, for any arbitrary time-like vector ξ^μ,

$$T_{\mu\nu}\xi^\mu\xi^\nu \geq 0.$$ (2.208)

Note that in the case of a future-directed timelike vector ξ^μ, $T_{\mu\nu}\xi^\mu\xi^\nu$ is the energy density of matter as measured by an observer whose relativistic velocity is given by ξ. According to the weak energy condition, this energy density should be non-negative.

- *Strong energy condition*: for $d > 2$, the strong energy condition holds if, for any timelike vector ξ^μ,

$$\left(T_{\mu\nu} - \frac{1}{d-2} g_{\mu\nu} T\right) \xi^\mu \xi^\nu \geq 0. \tag{2.209}$$

- *Dominant energy condition*: the dominant energy condition is satisfied if, for any null vector ζ^μ,

$$T_{\mu\nu} \zeta^\mu \zeta^\nu \geq 0 \quad \text{and} \quad T^{\mu\nu} \zeta_\mu \text{ is a non-spacelike vector.} \tag{2.210}$$

Exercise 2.5.1 Assuming that the Einstein equations are given by

$$R_{\mu\nu} - \frac{1}{2} R \, g_{\mu\nu} = \kappa^2 T_{\mu\nu}, \tag{2.211}$$

where we include a possible cosmological constant as a term in $T_{\mu\nu}$ which is proportional to $g_{\mu\nu}$, show that we may rewrite the strong energy condition as

$$R_{\mu\nu} \xi^\mu \xi^\nu \geq 0. \tag{2.212}$$

2.6 Further reading

Below in [1, 2, 3, 4, 5] we give a list of very useful introductions to general relativity, and in particular to black holes embedded in flat space. Quantum field theory in curved spacetime and its application to black holes is treated in the books [6, 7, 8, 9]. A recent study of stationary black holes is [10]. Higher dimensional black holes are reviewed in [11]. The Gibbons–Hawking term was introduced in [12]. In addition, the thermodynamics of the AdS black hole is discussed in [13].

References

[1] Carroll, Sean M. 2004. *Spacetime and Geometry: An Introduction to General Relativity*. Addison-Wesley.

[2] Wald, R. M. 1984. *General Relativity. Physics, Astrophysics*. University of Chicago Press.

[3] Schutz, Bernard F. 1985. *A First Course in General Relativity*. Cambridge University Press.

[4] Stephani, Hans. 2004. *Relativity: An Introduction to Special and General Relativity*. Cambridge University Press.

[5] Mukhanov, Viatcheslav, and Winitzki, Sergei. 2007. *Introduction to Quantum Effects in Gravity*. Cambridge University Press.

[6] Hawking, S. W., and Ellis, G. F. R. 1973. *The Large Scale Structure of Space-Time*. Cambridge University Press.

[7] Birrell, N. D., and Davies, P. C. W. 1982. *Quantum Fields in Curved Space*. Cambridge Monographs on Mathematical Physics. Cambridge University Press.

[8] Fulling, S. A. 1989. *Aspects of Quantum Field Theory in Curved Space-Time*. Cambridge University Press.

[9] Wald, Robert M. 1995. *Quantum Field Theory in Curved Space-Time and Black Hole Thermodynamics*. University of Chicago Press.

[10] Chrusciel, Piotr T., Costa, Joao Lopes, and Heusler, Markus. 2012. Stationary black holes: uniqueness and beyond. *Living Rev. Relativity*, **15**, 7.

[11] Emparan, Roberto, and Reall, Harvey S. 2008. Black holes in higher dimensions. *Living Rev. Relativity*, **11**, 6.

[12] Gibbons, G. W., and Hawking, S. W. 1977. Action integrals and partition functions in quantum gravity. *Phys. Rev.*, **D15**, 2752–2756.

[13] Witten, Edward. 1998. Anti-de Sitter space, thermal phase transition, and confinement in gauge theories. *Adv. Theor. Math. Phys.*, **2**, 505–532.

3 Symmetries in quantum field theory

In this chapter we introduce symmetries of quantum field theories which will be important later on: conformal symmetry and supersymmetry. First, as a guiding example, we discuss the Lorentz and Poincaré symmetries. We work out the corresponding algebras and show how fields transform under their representations. In particular we discuss the tensor and spinor representations of the Lorentz algebra, building on the concepts introduced in chapter 1. Moreover, we consider massless and massive states within the Poincaré algebra and discuss their consequences. An important property of the symmetries is to constrain the correlation functions.

We then discuss the *Coleman–Mandula theorem* which states that under certain reasonable circumstances, the largest possible bosonic symmetry algebra of a quantum field theory is the Poincaré algebra plus some internal symmetries. The Coleman–Mandula theorem may be bypassed by extending the Poincaré algebra. First, instead of just the Lorentz transformations and translations, we consider theories which are invariant under conformal transformations. Second, we add spinorial charges to the Poincaré algebra which satisfy anticommutation relations. This is the basic idea behind supersymmetry. Finally, we combine both extensions of the Poincaré algebra and study superconformal theories. For each extended symmetry, we use the experience gained from the Lorentz and Poincaré algebras to discuss the representations of the corresponding extended algebra and the transformation laws of fields. Moreover, we look at how these constrain the correlation functions.

3.1 Lorentz and Poincaré symmetry

In chapter 1 we saw that Lorentz transformations are of the form $x^\mu \mapsto x^{\mu\prime} = \Lambda(\omega)^\mu{}_\nu x^\nu$ and leave the length element (1.1) invariant provided that (1.4) is satisfied. Infinitesimally we may expand $\Lambda(\omega)$ as

$$\Lambda(\omega)^\mu{}_\nu = \delta^\mu{}_\nu + \eta^{\mu\rho}\omega_{\rho\nu}, \tag{3.1}$$

where $\omega_{\rho\nu}$ is antisymmetric under the exchange of the two indices ρ and ν. The finite transformations are easily reconstructed by exponentiating the infinitesimal form. To do so, it is convenient to introduce the generators $J_{\mu\nu}$, which are $d \times d$ matrices such that

$$\Lambda(\omega)^\mu{}_\nu = \delta^\mu{}_\nu + \frac{i}{2}\omega^{\rho\sigma}\left(J_{\rho\sigma}\right)^\mu{}_\nu. \tag{3.2}$$

The components of $J_{\rho\sigma}$ are specified by

$$\left(J_{\rho\sigma}\right)^{\mu}{}_{\nu} = i\left(\eta_{\rho\nu}\delta^{\mu}_{\sigma} - \eta_{\sigma\nu}\delta^{\mu}_{\rho}\right). \tag{3.3}$$

In particular, $J_{\rho\sigma}$ satisfy the commutation relations of the Lie algebra $\mathfrak{so}(d-1,1)$

$$[J_{\mu\nu}, J_{\rho\sigma}] = i\left(\eta_{\mu\rho}J_{\nu\sigma} + \eta_{\nu\sigma}J_{\mu\rho} - \eta_{\nu\rho}J_{\mu\sigma} - \eta_{\mu\sigma}J_{\nu\rho}\right). \tag{3.4}$$

The finite form of the Lorentz transformations (3.2) is given by

$$\Lambda(\omega) = \exp\left(\frac{i}{2}\omega_{\mu\nu}J^{\mu\nu}\right). \tag{3.5}$$

The generators J_{kl} with $k,l = 1,\ldots,d-1$ correspond to rotations, whereas J_{0k} are generators of boosts. Note that not all the generators $J_{\mu\nu}$ can be Hermitian due to the non-compactness of the Lorentz group $SO(d-1,1)$. Indeed, the rotation generators may be chosen to be Hermitian, whereas the boost generators are anti-Hermitian,

$$\left(J_{kl}\right)^{\dagger} = J_{kl}, \qquad \left(J_{0k}\right)^{\dagger} = -J_{0k}. \tag{3.6}$$

A question which remains to be addressed is how the different Lorentz covariant fields, local symmetry currents as well as conserved charges of a field theory transform under finite-dimensional representations of the Lorentz algebra. Let ϕ be a field with n components, i.e. we can think of ϕ as a column vector with components $\phi^a, a = 1,\ldots,n$. Under an infinitesimal Lorentz transformation (3.1) the field ϕ transforms as

$$\delta\phi^a = \frac{i}{2}\omega_{\mu\nu}\left(\mathcal{J}^{\mu\nu}\right)^a{}_b \phi^b. \tag{3.7}$$

Here, the $\mathcal{J}^{\mu\nu}$ have to satisfy the Lorentz algebra (3.4). We can think of $\mathcal{J}^{\mu\nu}$ for fixed μ and ν as an $n \times n$ matrix (note that n does not have to be identical to d) and ϕ as a column vector with n entries. For non-infinitesimal Lorentz transformations, the transformation rule (3.7) has to be exponentiated and hence reads

$$\phi'^a(x) = D(\Lambda(\omega))^a{}_b\, \phi^b(\Lambda^{-1}x) \qquad \text{with} \quad D(\Lambda(\omega)) = \exp\left(\frac{i}{2}\omega_{\mu\nu}\mathcal{J}^{\mu\nu}\right). \tag{3.8}$$

The only difference between $D(\Lambda(\omega))$ and $\Lambda(\omega)$ is the replacement of $J^{\mu\nu}$ with $\mathcal{J}^{\mu\nu}$. Therefore in order to classify all possible transformation laws of the form (3.8) we have to study all possible choices of $\mathcal{J}^{\mu\nu}$ in more detail. To do this we employ the language of representation theory which is reviewed in appendix B.

Since the $\mathcal{J}^{\mu\nu}$ satisfy the commutation relations (3.4) the matrices $\mathcal{J}^{\mu\nu}$ form a *representation* of the Lorentz algebra. In particular, we are only interested in irreducible representations since these correspond to elementary fields. In the next section we discuss important finite-dimensional irreducible representations of the Lorentz algebra $\mathfrak{so}(d-1,1)$.

3.1.1 Tensor representations

The simplest representation is the *trivial*, *singlet* or *scalar* representation. The associated vector space is one dimensional and its elements are denoted by ϕ. Finally, $\mathcal{J}$ is given by $\mathcal{J}_1^{\rho\sigma} = 0$. With this assignment, the Lorentz algebra (3.4) is trivially satisfied. This is the representation corresponding to the scalar field considered in section 1.1.

Next we consider the *vector* representation, which has dimension d and therefore is usually denoted by $\mathbf{d}$. The field ϕ has d components, ϕ^ρ, $\rho = 0, \ldots, d-1$ and the components of the $d \times d$ matrices $\mathcal{J}_{\mathbf{d}}^{\rho\sigma}$ are given by (3.3), i.e.

$$\left(\mathcal{J}_{\mathbf{d}}^{\rho\sigma}\right)^\mu_{\ \nu} = i\left(\delta^\rho_\nu \eta^{\mu\sigma} - \delta^\sigma_\nu \eta^{\mu\rho}\right). \tag{3.9}$$

We already know an example of a field which transforms under this vector representation, namely the vector fields A_μ introduced in section 1.7.

So far we have found the irreducible representations of the Lorentz algebra corresponding to scalar fields and vector fields. What does the representation of a field $\phi_{\mu_1 \ldots \mu_n}$ with n indices look like? In order to construct such a representation we have to consider a rank n tensor product of the vector representation. The resulting representations are in general reducible since they can be decomposed into partially symmetrised or antisymmetrised tensors. For simplicity let us consider the case $n = 2$, i.e. a field with two indices, $\phi_{\mu\nu}$. We know that we can decompose the field into a symmetrised part, $\phi_{(\mu\nu)}$, and an antisymmetrised part, $\phi_{[\mu\nu]}$, as defined in equations (2.12) and (2.13). In the language of representations, we can decompose the rank two tensor product representation $\mathbf{d} \otimes \mathbf{d}$ into a direct sum of a symmetric rank two representation, $\mathbf{d} \otimes_S \mathbf{d}$ of dimension $\frac{1}{2}d(d+1)$ and an antisymmetric rank two representation $\mathbf{d} \otimes_A \mathbf{d}$ with dimension $\frac{1}{2}d(d-1)$,

$$\mathbf{d} \otimes \mathbf{d} = (\mathbf{d} \otimes_S \mathbf{d}) \oplus (\mathbf{d} \otimes_A \mathbf{d}). \tag{3.10}$$

Note that in general neither $\mathbf{d} \otimes_S \mathbf{d}$ nor $\mathbf{d} \otimes_A \mathbf{d}$ is irreducible. Let us first consider the symmetric rank two representation $\mathbf{d} \otimes_S \mathbf{d}$. To decompose the representation further into a sum of irreducible representations, we make use of the invariant tensors. Given any reducible representation, a smaller one can be obtained by contracting the tensors of the representation with the invariant tensors. Let us demonstrate this for the Lorentz algebra.

For any orthogonal group $SO(p,q)$, and therefore in particular for the Lorentz group $SO(d-1,1)$, the metric $\eta_{\mu\nu}$ (or its inverse $\eta^{\mu\nu}$) and the totally antisymmetric tensor are the only two invariant tensors. Contracting the symmetric tensor $\phi_{(\mu\nu)}$ with the totally antisymmetric tensor gives zero, such that we only have to consider $\eta^{\mu\nu}\phi_{(\mu\nu)}$, which is the trace of the rank two tensor. This implies that the symmetric rank two tensor can be decomposed into a traceless symmetric rank two tensor, denoted by $\mathbf{S}$ and its trace part,

$$\mathbf{d} \otimes_S \mathbf{d} = \mathbf{1} \oplus \mathbf{S}. \tag{3.11}$$

Starting from a generic rank two tensor $\phi_{\rho\sigma}$, which does not necessarily have to be symmetric in ρ and σ, we may use the projection operator

$$P^{\rho\sigma}_{\mu\nu} = \frac{1}{2}\left(\delta^\rho_\mu \delta^\sigma_\nu + \delta^\sigma_\mu \delta^\rho_\nu\right) - \frac{1}{d}\eta_{\mu\nu}\eta^{\rho\sigma} \tag{3.12}$$

to find the components of $\phi_{\rho\sigma}$ within the symmetric traceless part $\mathbf{S}$.

Exercise 3.1.1 Show that for a general rank two tensor ϕ with components $\phi_{\mu\nu}$, the projected tensor $P^{\rho\sigma}_{\mu\nu}\phi_{\rho\sigma}$ is symmetric and traceless. Moreover prove that $P^{\rho\sigma}_{\mu\nu}$ is a projection operator by checking $P^{\rho\sigma}_{\alpha\beta}P^{\alpha\beta}_{\mu\nu} = P^{\rho\sigma}_{\mu\nu}$.

Box 3.1	(Anti-)self-dual tensors in Euclidean spacetime

In the case of Euclidean spacetime the Hodge dual satisfies $^*(^*\phi) = +\phi$ and therefore we may impose the conditions

$$^*\phi = \phi \qquad \text{or} \qquad ^*\phi = -\phi, \tag{3.16}$$

defining self-dual and anti-self-dual tensors in the representation $\mathbf{3}^\pm$, respectively. In this case, the representations $\mathbf{3}^\pm$ are no longer mapped to each other under complex conjugation.

Let us now consider the rank two antisymmetric tensors, $\phi_{[\mu\nu]}$, which are antisymmetric in the indices μ and ν and thus transform in the representation $\mathbf{d} \otimes_A \mathbf{d}$. It is convenient to think about the rank two antisymmetric tensor as a two-form of the form $\phi = \phi_{[\mu\nu]} \mathrm{d}x^\mu \wedge \mathrm{d}x^\nu$.

If we use the invariant tensor $\eta^{\mu\nu}$ to contract the indices μ and ν, the result vanishes. Hence you might conclude that the representation $\mathbf{d} \otimes_A \mathbf{d}$ is already irreducible. However, in four-dimensional Minkowski spacetime, we may use the totally antisymmetric tensor $\epsilon_{\mu\nu\rho\sigma}$, given by equation (2.54) and normalised such that $\epsilon^{0123} = -1$, to relate the two-form ϕ to another rank two antisymmetric tensor $^*\phi$, the *Hodge dual* of ϕ, by

$$^*\phi^{[\mu\nu]} = \frac{1}{2}\epsilon^{\mu\nu\rho\sigma}\phi_{[\rho\sigma]}. \tag{3.13}$$

In particular notice that $^*(^*\phi) = -\phi$ in agreement with (2.69). Using the Hodge dual we may impose two different projection conditions

$$^*\phi = i\phi \qquad \text{or} \qquad ^*\phi = -i\phi. \tag{3.14}$$

An antisymmetric tensor satisfying (3.14) with the plus (or minus) sign is a *self-dual* (or *anti-self-dual*) tensor, respectively. The (anti-)self-dual tensors give rise to three-dimensional irreducible representations $\mathbf{3}^+$ and $\mathbf{3}^-$. Note that both representations are complex and that under complex conjugation both representations map into each other, i.e. $\left(\mathbf{3}^\pm\right)^* = \mathbf{3}^\mp$. This is no longer true if we use four-dimensional Euclidean spacetime, as pointed out in box 3.1.

To summarise, we may decompose a rank two antisymmetric tensor into its self-dual and anti-self-dual parts in four spacetime dimensions,

$$\mathbf{4} \otimes_A \mathbf{4} = \mathbf{3}^+ \oplus \mathbf{3}^-. \tag{3.15}$$

3.1.2 Spinor representations

In addition to tensor representations, the Lorentz group admits a further class of irreducible representations, the spinor representations.[1] These may be constructed by using the *Clifford algebra*,

$$\gamma_\mu\gamma_\nu + \gamma_\nu\gamma_\mu \equiv \{\gamma_\mu, \gamma_\nu\} = -2\eta_{\mu\nu}\mathbb{1}. \tag{3.17}$$

---

[1] Mathematically speaking, spinors are representations of the *spin group* which is the double cover of the Lorentz group. This implies that spinors are *projective representations* of the Lorentz group.

The matrices γ_μ are the *Dirac gamma matrices*. Using the anti-commutation relations (3.17) we conclude that $(\gamma_0)^2 = \mathbb{1}$ and $(\gamma_k)^2 = -\mathbb{1}$ for $k \in \{1, 2, \ldots, d-1\}$. Therefore γ_0 has eigenvalues ± 1, while γ_k has eigenvalues $\pm i$ and thus γ_0 may be chosen to be Hermitian while γ_k is anti-Hermitian,

$$(\gamma_0)^\dagger = \gamma_0, \qquad (\gamma_k)^\dagger = -\gamma_k. \tag{3.18}$$

Using the Dirac gamma matrices it is possible to construct a representation of the Lorentz algebra

$$\mathcal{J}^{\mu\nu} = \frac{i}{4}\left[\gamma^\mu, \gamma^\nu\right], \tag{3.19}$$

the *Dirac spinor representation* of the Lorentz algebra. Here, we have raised the indices of γ_μ with $\eta^{\mu\nu}$, i.e. $\gamma^\mu = \eta^{\mu\nu}\gamma_\nu$.

Exercise 3.1.2 Show explicitly that (3.19) is a representation of the Lorentz algebra, i.e. that the commutation relations (3.4) are satisfied.

For general spacetime dimension d it is possible to construct the Dirac matrices γ^μ, see appendix B.2.2. In fact for even d, we can find up to similarity transformations one complex irreducible representation of the Clifford algebra. In contrast, for odd d we find two inequivalent complex irreducible representations of the Clifford algebra. For both even and odd dimensions d, the irreducible representations are of complex dimension $2^{\lfloor d/2 \rfloor}$.

Let us first restrict our discussion to $d = 4$ and consider the irreducible representation of complex dimension four satisfying the Clifford algebra (3.17). The question arises whether the generators (3.19) form a reducible or irreducible representation of the Lorentz group. In order to answer that question we have to collect some basic facts about the Dirac gamma matrices.

Up to similarity transformations, the Dirac gamma matrices γ_μ form a unique irreducible representation of the Clifford algebra (3.17). Hence other sets of possible gamma matrices such as $\{-\gamma_\mu\}$, or $\{\pm\gamma_\mu^T\}, \{\pm\gamma_\mu^*\}$ and $\{\pm\gamma_\mu^\dagger\}$ have to be related to $\{\gamma_\mu\}$ by a similarity transformation. For example, γ_μ is related to $\gamma_\mu^\dagger$ by $\mathcal{B}$,

$$\mathcal{B}\gamma_\mu\mathcal{B}^{-1} = (\gamma_\mu)^\dagger, \tag{3.20}$$

where $\mathcal{B}$ is given by

$$\mathcal{B} = \gamma^0. \tag{3.21}$$

Here we have chosen the phase of $\mathcal{B}$ such that $\mathcal{B}^2 = \mathbb{1}$ and $\mathcal{B} = \mathcal{B}^* = \mathcal{B}^\dagger$. Moreover, the similarity transformation which takes $-\gamma_\mu$ into γ_μ is given by γ_5

$$\gamma_5\gamma_\mu\gamma_5^{-1} = -\gamma_\mu, \tag{3.22}$$

where

$$\gamma_5 = i\gamma^0\gamma^1\gamma^2\gamma^3. \tag{3.23}$$

Exercise 3.1.3 Prove the similarity transformations (3.20) and (3.22) provided that $\mathcal{B}$ and γ_5 are given by (3.21) and (3.23).

Exercise 3.1.4   Show that γ_5 given by (3.23) has the following properties

$$\{\gamma_5, \gamma_\mu\} = 0, \qquad \gamma_5^2 = \mathbb{1}, \qquad \gamma_5 = \gamma_5^\dagger. \tag{3.24}$$

Exercise 3.1.5   Further show that γ_5 is traceless and satisfies $[\gamma_5, \mathcal{J}^{\mu\nu}] = 0$ where $\mathcal{J}^{\mu\nu}$ is given by (3.19).

Exercise 3.1.6   Using the gamma matrices (1.143) show that $\mathcal{J}^{\mu\nu}$ is given by

$$\mathcal{J}^{\mu\nu} = \begin{pmatrix} \sigma^{\mu\nu} & 0 \\ 0 & \bar{\sigma}^{\mu\nu} \end{pmatrix}, \tag{3.25}$$

where

$$\sigma^{\mu\nu} = \frac{i}{4}\left(\sigma^\mu\bar{\sigma}^\nu - \sigma^\nu\bar{\sigma}^\mu\right), \qquad \bar{\sigma}^{\mu\nu} = \frac{i}{4}\left(\bar{\sigma}^\mu\sigma^\nu - \bar{\sigma}^\nu\sigma^\mu\right). \tag{3.26}$$

In order to relate $-\gamma_\mu^{\mathrm{T}}$ to γ_μ we introduce the matrix $\mathcal{C}$,

$$\mathcal{C}\gamma_\mu\mathcal{C}^{-1} = -\gamma_\mu^{\mathrm{T}}, \tag{3.27}$$

which can be chosen such that

$$\mathcal{C}\mathcal{C}^\dagger = \mathbb{1}, \qquad \mathcal{C} = -\mathcal{C}^{\mathrm{T}}. \tag{3.28}$$

$\mathcal{C}$ is known as the *charge conjugation matrix*. Using (3.20) and (3.28) we can also relate γ_μ^* to γ_μ,

$$\gamma_\mu^* = -\mathcal{B}\mathcal{C}\gamma_\mu(\mathcal{B}\mathcal{C})^{-1}. \tag{3.29}$$

Using these similarity transformations it is possible to define projection conditions on the spinors. In particular, we see that the Dirac spinor is reducible under the Lorentz algebra. Let us define two different projection conditions.

- **Weyl spinors** Because $\gamma_5^2 = \mathbb{1}$ the eigenvalues of γ_5 are ± 1. Moreover, γ_5 is traceless and therefore has two eigenvalues $+1$ and two eigenvalues -1. Let us choose a basis in which γ_5 is diagonal. We know that $\mathcal{J}^{\mu\nu}$ commutes with γ_5. Therefore in the eigenbasis of γ_5, the Dirac representation $\mathcal{J}^{\mu\nu}$ is block diagonal with two 2×2 blocks and is therefore reducible. Indeed we can project the Dirac spinor Ψ onto complex two component left- and right-handed Weyl spinors, ψ_L and ψ_R, defined by

$$\Psi_{\mathrm{L}} = \begin{pmatrix} \psi_{\mathrm{L}} \\ 0 \end{pmatrix} = \mathcal{P}_+\Psi, \qquad \Psi_{\mathrm{R}} = \begin{pmatrix} 0 \\ \psi_{\mathrm{R}} \end{pmatrix} = \mathcal{P}_-\Psi, \qquad \mathcal{P}_\pm = \frac{1}{2}\left(\mathbb{1} \pm \gamma_5\right). \tag{3.30}$$

Note that the two Weyl representations, denoted by $\mathbf{2_L}$ and $\mathbf{2_R}$ are inequivalent. Under complex conjugation, $\mathbf{2_L}$ and $\mathbf{2_R}$ transform into each other.
- **Majorana spinors** Since γ_μ^* and γ_μ are related by $\mathcal{B}\mathcal{C}$, we can also derive a relation between $\mathcal{J}^{\mu\nu}$ and $(\mathcal{J}^{\mu\nu})^*$ of the form

$$\mathcal{B}\mathcal{C}\,\mathcal{J}^{\mu\nu}\,(\mathcal{B}\mathcal{C})^{-1} = -\left(\mathcal{J}^{\mu\nu}\right)^*. \tag{3.31}$$

Therefore the complex conjugated Dirac spinor Ψ^* transforms in the same way as $\mathcal{BC}\Psi$ under Lorentz transformations and we can impose the reality condition

$$\Psi^* = \mathcal{BC}\Psi. \tag{3.32}$$

This projection condition also defines a two-dimensional representation, the *Majorana spinor*. Since $\Psi^{**} = \Psi$, the projector $\mathcal{BC}$ has to satisfy $(\mathcal{BC})^*\mathcal{BC} = \mathbb{1}$.

Exercise 3.1.7 Show that in four-dimensional Minkowski spacetime $(\mathcal{BC})^*\mathcal{BC} = \mathbb{1}$ is indeed satisfied and thus we can define Majorana spinors. Hint: Take $\mathcal{BC} = i\gamma^2$ for the gamma matrices (1.143).

Note that Majorana spinors and Weyl spinors correspond to two different representations of the Lorentz algebra. In particular, these two representations are not equivalent. However, since they have the same dimension – both have four real components – we can find a one-to-one map between the components of Weyl and Majorana spinors.

Let us generalise these results to d-dimensional Minkowski spacetime. In even spacetime dimensions we can always define the projection operators $\mathcal{P}_\pm$ leading to left- and right-handed Weyl spinors of complex dimension $2^{d/2-1}$, respectively. The Majorana condition may be modified to

$$\gamma_\mu^* = \mp\mathcal{BC}\gamma_\mu(\mathcal{BC})^{-1}, \qquad \Psi^* = \mathcal{BC}\Psi, \tag{3.33}$$

where we still have to satisy $(\mathcal{BC})^*\mathcal{BC} = \mathbb{1}$. In the case of the upper sign in (3.33), the spinor is called a Majorana spinor, while with the lower sign it is a pseudo-Majorana spinor. Such (pseudo-)Majorana spinors exist in dimensions $d = 0, 1, 2, 3, 4 \pmod 8$ and they possess half of the degrees of freedom of the Dirac spinor Ψ.

In even dimensions we may ask further whether the Weyl and the Majorana conditions are compatible, i.e. whether the corresponding projection operators commute. This is only possible in $d \equiv 2 \pmod 8$ dimensions where we can define such *Majorana–Weyl* spinors. In odd spacetime dimensions we can only impose Majorana or pseudo-Majorana conditions. In table 3.1 all possible types of spinors in $d \leq 11$ dimensional Minkowski spacetime are listed as well as the real dimension of the smallest irreducible representation.

3.1.3 An alternative way in four dimensions

The discussion of representations of the Lorentz group performed in the last two subsections can be applied to any dimensions. If we want to classify the representations of the Lorentz algebra in four dimensions there is a more direct way which will be discussed in this section. The generators $J^{\mu\nu}$ of the Lorentz algebra can be grouped into the boosts K_i and the rotations J_i given by

$$K_i = J^{0i}, \qquad J_i = \frac{1}{2}\epsilon_{ijk}J^{jk} \qquad \text{with} \quad i,j,k \in \{1,2,3\}, \tag{3.34}$$

where index summation is understood and $\epsilon_{ijk} = \epsilon_{0ijk}$. Introducing the generators L_i and R_i by

$$L_k = \frac{1}{2}(J_k + iK_k), \qquad R_k = \frac{1}{2}(J_k - iK_k), \tag{3.35}$$

d	Dimension (real)	Weyl	Majorana	Pseudo-Majorana	Majorana–Weyl
2	1	•	•	•	•
3	2		•		
4	4	•	•		
5	8				
6	8	•			
7	16				
8	16	•		•	
9	16			•	
10	16	•	•	•	•
11	32		•		

Table 3.1 Types of spinors in d-dimensional Minkowski spacetime

Table 3.2 Irreducible Lorentz group representations in $d = 4$

(j_L, j_R)	Representation	Field	Name
$(0,0)$	$\mathbf{1}$	ϕ	scalar
$(\frac{1}{2},0)$	$\mathbf{2}_L$	ψ_L	left-handed Weyl spinor
$(0,\frac{1}{2})$	$\mathbf{2}_R$	ψ_R	right-handed Weyl spinor
$(\frac{1}{2},\frac{1}{2})$	$\mathbf{4}$	ϕ_μ	vector
$(1,0)$	$\mathbf{3}^+$	$\phi^+{}_{[\mu\nu]}$	antisymmetric self-dual tensor
$(0,1)$	$\mathbf{3}^-$	$\phi^-{}_{[\mu\nu]}$	antisymmetric anti-self-dual tensor
$(1,1)$	$\mathbf{9}$	$P^{\rho\sigma}_{\mu\nu}\phi_{\rho\sigma}$	symmetric traceless tensor

the Lorentz algebra (3.4) can be rewritten in the form

$$\left[L_i, L_j\right] = i\epsilon_{ijk}L_k, \qquad \left[R_i, R_j\right] = i\epsilon_{ijk}R_k, \qquad \left[L_i, R_j\right] = 0. \qquad (3.36)$$

We have rewritten the commutation relations in terms of two commuting $\mathfrak{su}(2)$ algebras which we denote by $\mathfrak{su}(2)_L$ and $\mathfrak{su}(2)_R$, respectively. In order to study the representations of the Lorentz algebra, we have only to study representations of $\mathfrak{su}(2)$. The representations of $\mathfrak{su}(2)$, denoted by $\mathbf{j}$, are labelled by a half-integer j and are $(2j + 1)$-dimensional.

Therefore we may classify the representations of $\mathfrak{so}(3, 1)$ by two half-integers, j_L and j_R. Among other representations we obtain the list of irreducible representations given in table 3.2.

3.1.4 Poincaré algebra and particle states

Let us now extend the Lorentz algebra to the Poincaré algebra. In addition to the generators $J^{\mu\nu}$ of Lorentz transformations, we also have to consider P_μ generating infinitesimal translations. The generators P^μ and $J_{\rho\sigma}$ have to satisfy the commutation relations (3.4) as well as

$$[J_{\mu\nu}, P_\rho] = i\left(\eta_{\mu\rho}P_\nu - \eta_{\nu\rho}P_\mu\right), \qquad [P_\mu, P_\nu] = 0. \tag{3.37}$$

In other words, P_ρ transforms as a vector under Lorentz transformations and the momenta commute. The corresponding Poincaré group is a semi-direct product of translations and Lorentz transformations. Note that the Poincaré group is not compact. In particular the boosts and the translations are non-compact transformations.

In quantum field theory, we use unitary representations of the symmetry groups. However, besides the trivial representation, a non-compact group does not have unitary finite-dimensional representations. Therefore the representations have to be labelled by continuous parameters. In this case we can label the representations of the Poincaré algebra by the momentum p^μ.

A strategy for classifying all unitary representations is to choose a frame in which the non-compact transformations are fixed, such that we only have to deal with the compact transformations. In this frame we classify all possible representations of the compact generators. For simplicity, we consider the Poincaré algebra in four spacetime dimensions. The different infinite-dimensional unitary representations correspond to massive and massless particle states.

For massive particles, we can always boost to a frame such that $p^\mu = (m, 0, 0, 0)$. The *little group* is given by those transformations which leave this momentum vector p^μ invariant, i.e. in this case $SO(3)$. The representations for $SO(3)$ are labelled by a half-integer s, the spin of the field. The representation has dimension $2s + 1$. In order to determine the spin of the massive field, it is convenient to introduce the *Pauli–Lubanski* vector W with components W_σ given by

$$W_\sigma = \frac{1}{2}\epsilon_{\mu\nu\rho\sigma}J^{\mu\nu}P^\rho, \tag{3.38}$$

and its square $W^2 = W_\sigma W^\sigma$. It can be shown that P^μ, W^2 and one component of the Pauli–Lubanski vector W_σ are commuting. Instead of the four operators P^μ, $\mu \in \{0, \ldots, 3\}$, we can also use $P^2 = P_\mu P^\mu$ and the three spatial components P^i, $i \in \{1, 2, 3\}$. In summary, massive particle states can be classified according to their mass $m^2 = -P_\mu P^\mu$, their spatial momentum P^i, their spin $W^2 = m^2 s(s + 1)$, and one of the spin components $W_3 = ms_3$ where s_3 can take values in $\{-s, -s+1, \ldots, s-1, s\}$. The corresponding eigenstates of the massive particle are thus determined by $|p^\mu, s, s_3\rangle$.

Let us now consider massless particles. In this case it is not possible to boost into a frame where all spatial components of p^μ are zero. Instead it is possible to boost to a frame with $p^\mu = (E, 0, 0, E)$. The little group of p^μ is generated by $N_1 = J_{10} + J_{13}$, $N_2 = J_{20} + J_{23}$ and by J_{12}. Since N_1 and N_2 are non-compact generators, they must be trivially realised in any unitary finite-dimensional representation. Thus a unitary finite-dimensional representation is labelled by just one number, the so-called helicity λ, which is the eigenvalue of the generator J_{12} corresponding to rotations around the x^3-axis. It turns out that the helicity λ has to be (half-)integer and thus the states of the massless particles are denoted by $|p^\mu, \lambda\rangle$.

It is straightforward to generalise the argument to $d \neq 4$ spacetime dimensions. For a massive particle we may boost to the rest-frame and hence the little group is $SO(d-1)$, while for massless particles the little group contains $SO(d-2)$ but not $SO(d-1)$.

3.1.5 Ward identities

As discussed in section 1.8, continuous symmetries will impose restrictions on the n-point correlation functions. Here we investigate how they are constrained by Lorentz transformations and translations.

Let us first investigate how symmetries act on the quantum fields $\phi(x)$ which do not necessarily have to be scalar fields. For Lorentz transformations, the change of the field $\phi(x)$ is given by

$$\delta\phi(x) = \tilde{\phi}(x) - \phi(x) = e^{i/2\,\omega_{\mu\nu}\mathcal{J}^{\mu\nu}}\phi(\Lambda^{-1}x) - \phi(x), \tag{3.39}$$

and therefore the corresponding infinitesimal transformation at $x = 0$ reads

$$\delta\phi(0) = \frac{i}{2}\omega_{\mu\nu}\mathcal{J}^{\mu\nu}\phi(0). \tag{3.40}$$

In quantum field theory we have to promote the generator $\mathcal{J}^{\mu\nu}$ of Lorentz transformations to an operator $\hat{\mathcal{J}}^{\mu\nu}$ as explained in section 1.8. In particular, using (1.238), the infinitesimal change $\delta\phi(x)$ may be written in terms of a commutator involving $\hat{\mathcal{J}}^{\mu\nu}$,

$$\delta\phi(0) = -\frac{i}{2}\omega_{\mu\nu}[\hat{\mathcal{J}}^{\mu\nu}, \phi(0)]. \tag{3.41}$$

Thus we conclude

$$[\hat{\mathcal{J}}^{\mu\nu}, \phi(0)] = -\mathcal{J}^{\mu\nu}\phi(0). \tag{3.42}$$

Here, for pedagogical reasons we have introduced the hat on $\mathcal{J}^{\mu\nu}$ to emphasise that we do not mean the generator $\mathcal{J}^{\mu\nu}$ acting on spacetime but the corresponding operator acting on the Hilbert space of quantum fields.

Exercise 3.1.8   Generalise the commutator (3.42) to fields evaluated at an arbitrary spacetime point x by showing that

$$[\hat{\mathcal{J}}^{\mu\nu}, \phi(x)] = -\mathcal{J}^{\mu\nu}\phi(x) + i\left(x^\mu\partial^\nu - x^\nu\partial^\mu\right)\phi(x). \tag{3.43}$$

For translations the same procedure may be applied. The infinitesimal transformation of $\phi(x)$ reads

$$\delta\phi(x) = -a^\mu\partial_\mu\phi(x) \tag{3.44}$$

and thus we obtain for the commutator

$$[\hat{P}_\mu, \phi(x)] = -i\partial_\mu\phi(x). \tag{3.45}$$

For now on, we will omit the hat on the operators. Moreover, in analogy to $\mathcal{J}^{\mu\nu}$ acting on the fields we may define $\mathcal{P}_\mu = -i\partial_\mu$ acting on the fields $\phi(x)$ of the Hilbert space.

| Box 3.2 | Coleman–Mandula theorem |

The Coleman–Mandula theorem states that for quantum field theories, Poincaré (i.e. spacetime) and internal symmetries can only be combined in a direct product symmetry group. This implies that in theories with a mass gap, all conserved quantities are Lorentz scalars. The proof of this theorem is obtained by considering the S-matrix of the theory and its transformation properties under Lie algebras.

One way to bypass this theorem is to consider conformal field theories, in which there is neither a mass gap nor an S-matrix. Conformal symmetry is a non-trivial extension of Poincaré symmetry. A further way around this theorem is to consider supersymmetry, which involves a Lie superalgebra instead of a Lie algebra.

Exercise 3.1.9 Show that the operators $\mathcal{P}^\rho$ and $\mathcal{J}^{\mu\nu} - i\,(x^\mu \partial^\nu - x^\nu \partial^\mu)$ indeed satisfy the commutation relations (3.37).

Exercise 3.1.10 Show that the operator $\mathcal{T}(a) = \exp(-i\hat{P}_\mu a^\mu)$ acts on $\phi(x)$ as

$$\mathcal{T}(a)^{-1}\phi(x)\mathcal{T}(a) = \phi(x-a). \tag{3.46}$$

The Ward identities corresponding to Lorentz transformations and translations are given by the general expression (1.237), where $\delta\phi$ is given by (3.41) and (3.44). We see that the n-point correlation functions $\langle \phi(x_1) \dots \phi(x_n)\rangle$ can only depend on the differences $(x_i - x_j)^2$. In particular, the one-point function has to be a constant and the two-point function is of the form

$$\langle \phi(x_1)\phi(x_2)\rangle = f\left((x_1 - x_2)^2\right), \tag{3.47}$$

with f an arbitrary function.

3.1.6 Beyond the Poincaré algebra

We saw that in order to classify the particle states, it is important to know the symmetry algebra. In the last section we considered the Poincaré algebra which we can also extend by an internal symmetry. Then the particle states are classified by their momentum, mass, spin (or helicity in the massless case) and charge from the internal symmetry.

In this context an important question is whether we can further extend the Poincaré algebra. There is a powerful theorem, the *Coleman–Mandula theorem* stated in box 3.2, which states that under certain – important but reasonable – assumptions and in a theory with non-trivial scattering in more than 1+1 dimensions, the only conserved quantities transforming as tensors under the Lorentz group are the energy-momentum vector P^μ and the generator of Lorentz transformations $J^{\rho\sigma}$, as well as possible internal symmetries which commute with P^μ and $J^{\rho\sigma}$.

Can we bypass the Coleman–Mandula theorem? Indeed there are two possibilities. If the theory has only massless particles, the Poincaré algebra is extended to the conformal algebra. Besides Lorentz transformations and translations, the theory is also invariant under angle preserving transformations. This will impose restrictions on the dynamics of the theory. We will study the consequences in more detail in section 3.2.

Moreover, we do not have to restrict ourselves to quantities transforming in a tensor representation of the Lorentz group. We can also consider conserved quantities transforming in the spinorial representation giving rise to a spinorial charge Q_α and impose anticommutation relations for those generators. Then the Poincaré algebra is extended to what is known as supersymmetry algebra. This will be investigated in section 3.3.

3.2 Conformal symmetry

3.2.1 Conformal algebra

In Euclidean spacetime, the Poincaré group can be extended to the conformal group. This group consists of transformations preserving angles. In Minkowski spacetime we can define conformal transformations as the most general locally causality preserving transformations, i.e. spacelike (timelike) separated points are mapped to spacelike (timelike) separated points. In particular, lightlike separated points will remain lightlike separated.

Allowing for a non-trivial line element $ds^2 = g_{\mu\nu}(x)dx^\mu dx^\nu$ with metric components $g_{\mu\nu}$, conformal transformations can be viewed as those transformations which leave the metric $g_{\mu\nu}$ invariant up to an arbitrary (but positive!) spacetime dependent scale factor, i.e. conformal transformations are those transformations $x \mapsto \tilde{x} = f(x)$ for which

$$g_{\mu\nu}(x) \mapsto \Omega(x)^{-2}g_{\mu\nu}(x) \equiv e^{2\sigma(x)}g_{\mu\nu}(x). \tag{3.48}$$

Therefore conformal transformations change the length of an infinitesimal spacetime interval by $ds'^2 = e^{2\sigma(x)}ds^2$ but they leave angles invariant locally and preserve the causal structure.

Let us now determine the conformal transformations in the case of a flat spacetime metric, i.e. for $g_{\mu\nu} = \eta_{\mu\nu}$. For an infinitesimal transformation $x^\mu \mapsto \tilde{x}^\mu = x^\mu + \epsilon^\mu(x)$, this implies that the metric transforms as

$$\eta_{\mu\nu} \mapsto \eta_{\mu\nu} + \partial_\mu\epsilon_\nu + \partial_\nu\epsilon_\mu. \tag{3.49}$$

Using the definition (3.48), an infinitesimal conformal transformation has to satisfy

$$\partial_\mu\epsilon_\nu + \partial_\nu\epsilon_\mu = 2\sigma(x)\eta_{\mu\nu}, \tag{3.50}$$

where we have used $\Omega(x) = 1 - \sigma(x) + \mathcal{O}(\sigma^2)$. Contracting the indices of both sides with $\eta^{\mu\nu}$, we obtain $\partial \cdot \epsilon = \partial_\mu\epsilon^\mu = \sigma(x) \cdot d$ in d dimensions. Therefore the infinitesimal transformation is conformal if $\epsilon(x)$ satisfies

$$\left(\eta_{\mu\nu}\partial_\rho\partial^\rho + (d-2)\partial_\mu\partial_\nu\right)\partial \cdot \epsilon = 0. \tag{3.51}$$

Note that equation (3.51) simplifies if we set $d = 2$. This will have dramatic consequences. Therefore, from now on, we have to distinguish in this section between $d = 2$ and $d > 2$. Let us first consider the case $d > 2$.

Table 3.3 Conformal transformations in $d > 2$ dimensions

Name	$\epsilon^\mu(x)$	$\sigma(x)$	Operator
Translation	a^μ	0	P_μ
Lorentz transformations	$\omega^\mu{}_\nu x^\nu$, $\omega_{\mu\nu} = -\omega_{\nu\mu}$	0	$J_{\mu\nu}$
Dilatation	λx^μ	λ	D
Special conformal transformation	$b^\mu x^2 - 2(b \cdot x)x^\mu$	$-2(b \cdot x)$	K_μ

Conformal algebra in $d > 2$ dimensions

For $d > 2$, the conformal Killing equation (3.51) is solved if $\epsilon(x)$ is at most of second order in x. Therefore we may write

$$\epsilon^\mu(x) = a^\mu + \omega^\mu{}_\nu x^\nu + \lambda x^\mu + b^\mu x^2 - 2(b \cdot x)x^\mu, \tag{3.52}$$

where $(b \cdot x)$ and x^2 are the shorthand notations for $b_\mu x^\mu$ and $x_\mu x^\mu$. For ϵ^μ given by (3.52), we have $\sigma = \lambda - 2b \cdot x$. We note that the parameters $a^\mu, \omega^\mu{}_\nu, \lambda$ and b^μ have a finite number of components. The conformal algebra and the associated symmetry group are thus finite dimensional. The geometric interpretation of the parameters is given in table 3.3.

The generators corresponding to a_μ and $\omega_{\mu\nu}$ are the momentum vector P^μ and $J^{\mu\nu}$. In addition we have the new operators D, corresponding to dilatations parametrised by λ, and special conformal transformations K^μ. The conformal algebra consisting of $J^{\mu\nu}, P^\mu, D$ and K^μ is given by the commutation relations of the Poincaré algebra, (3.4) and (3.37), as well as

$$[J_{\mu\nu}, K_\rho] = i\left(\eta_{\mu\rho}K_\nu - \eta_{\nu\rho}K_\mu\right),$$
$$[D, P_\mu] = iP_\mu, \qquad [D, K_\mu] = -iK_\mu, \qquad [D, J_{\mu\nu}] = 0, \tag{3.53}$$
$$[K_\mu, K_\rho] = 0, \qquad [K_\mu, P_\nu] = -2i\left(\eta_{\mu\nu}D - J_{\mu\nu}\right).$$

Let us analyse the commutation relations of the conformal algebra in more detail. In particular we see that the generators $J^{\mu\nu}$ form a subalgebra, the Lorentz algebra, $\mathfrak{so}(d - 1, 1)$. In fact, the generators of the conformal algebra can be grouped in such a way that the conformal algebra is the algebra $\mathfrak{so}(d, 2)$. The generators of $\mathfrak{so}(d, 2)$ – denoted by $\bar{J}_{AB} = -\bar{J}_{BA}$, where A and B run from 0 to $d + 1$ – have to satisfy an algebra such as (3.4) with $\eta = \text{diag}(-1, 1, \ldots, 1)$ replaced by $\bar{\eta} = \text{diag}(-1, 1, \ldots, 1, 1, -1)$. In particular, the generators $\bar{J}_{\mu\nu} \equiv J_{\mu\nu}$ with $\mu, \nu \in \{0, \ldots, d - 1\}$ are the generators of the Lorentz group satisfying the usual commutation relations (3.4).

In order to map the conformal algebra (3.53) to the algebra $\mathfrak{so}(d, 2)$ we have to construct the map between the remaining $\mathfrak{so}(d, 2)$ generators $\bar{J}_{\mu d}, \bar{J}_{\mu(d+1)}$ (with $\mu \in \{0, \ldots, d - 1\}$) and $\bar{J}_{d(d+1)}$ and the generators D, P_μ, K_μ of the conformal algebra. The generator $\bar{J}_{d(d+1)}$ transforms as a scalar under the Lorentz transformations $\mathfrak{so}(d - 1, 1)$ and hence has to commute with $J_{\mu\nu}$ where $\mu, \nu \in \{0, \ldots, d - 1\}$. Thus we may identify

$$\bar{J}_{d(d+1)} = -D. \tag{3.54}$$

Moreover, we know that $\bar{J}_{\mu d}$ and $\bar{J}_{\mu (d+1)}$ are vectors under the Lorentz transformation $\mathfrak{so}(d-1,1)$ and hence they have to be related to P_μ and K_μ. It is straightforward – see exercise 3.2.1 – to work out the precise identification,

$$\bar{J}_{\mu d} = \frac{1}{2}\left(K_\mu - P_\mu\right), \qquad \bar{J}_{\mu (d+1)} = \frac{1}{2}\left(P_\mu + K_\mu\right), \tag{3.55}$$

where $\mu \in \{0, \ldots, d-1\}$.

Exercise 3.2.1 Show that under the identifications (3.54) and (3.55) the conformal algebra can be written in terms of an $\mathfrak{so}(d,2)$.

Exercise 3.2.2 If we consider conformal transformations of flat spacetime with signature (p, q), i.e. η has p times the eigenvalue $+1$ and q times the eigenvalue -1, the conformal algebra is given by $\mathfrak{so}(p+1, q+1)$. In particular, show that the conformal algebra for d-dimensional Euclidean spacetime is $\mathfrak{so}(d+1, 1)$.

Let us now consider finite transformations in addition to the infinitesimal ones considered above. In particular we are interested in scale transformations (with parameter λ which is now finite rather than infinitesimal) and special conformal transformations (with parameter b^μ)

$$x^\mu \mapsto \lambda x^\mu, \tag{3.56}$$

$$x^\mu \mapsto \frac{x^\mu + b^\mu x^2}{1 + 2b \cdot x + b^2 x^2}. \tag{3.57}$$

For finite conformal transformations, it is useful to introduce the *inversion*

$$x^\mu \to x'^\mu = \frac{x^\mu}{x^2}. \tag{3.58}$$

Note that the special conformal transformations and the inversion are not globally defined. For the inversion, points satisfying $x^2 = 0$ are mapped to infinity which is not part of the flat Euclidean or Minkowski spacetime. The same is true for the special conformal transformation. For a given vector b^μ all points x with $1 + 2b \cdot x + b^2 x^2 = 0$ are mapped to infinity. In order to define conformal transformations globally, we have to add points to our flat spacetime. In technical terms we have to consider the conformal compactifications of $\mathbb{R}^d$ or $\mathbb{R}^{d-1,1}$. In the case of flat Euclidean space $\mathbb{R}^d$ we only have to add one point called 'infinity' since the equation $x^2 \equiv \delta_{\mu\nu} x^\mu x^\nu = 0$ is satisfied only for $x = 0$. However, for flat Minkowski spacetime $\mathbb{R}^{d-1,1}$ we have to add all points satisfying $x^2 \equiv \eta_{\mu\nu} x^\mu x^\nu = 0$ which is the light-cone of the point $x = 0$ to $\mathbb{R}^{d-1,1}$ to obtain the conformal compactification.

The transformation (3.58) is not connected to the identity and thus is an element of $O(d, 2)$ rather than $SO(d, 2)$. However, a combination of transformations involving an even number of inversions again gives rise to conformal transformation connected to the identity, as exemplified in the following exercise.

Exercise 3.2.3 Show that the special conformal transformation can be decomposed into a inversion $x^\mu \mapsto x'^\mu = \frac{x^\mu}{x^2}$, a translation $x'^\mu \mapsto x''^\mu = x'^\mu + b^\mu$ and another inversion $x''^\mu \mapsto x'''^\mu = \frac{x''^\mu}{x''^2}$.

Exercise 3.2.4 Show that the inversion is not connected to the identity, i.e. it cannot be written as $x'^\mu = x^\mu + \epsilon^\mu(x)$ with an infinitesimal ϵ^μ.

In Euclidean signature, the conformal group generated by translations, rotations, dilatations and special conformal transformations is $SO(d + 1, 1)$. Any conformal transformation can be generated by combining inversions with rotations and translations, but only a combination of two, or any even number of, inversions may be an element of the conformal group $SO(d + 1, 1)$. The group generated by rotations, translations and an arbitrary number of inversions is $O(d + 1, 1)$.

For any conformal transformation we may define $\mathcal{R}^\mu{}_\rho(x)$ by

$$\mathcal{R}^\mu{}_\rho(x) = \Omega(x) \frac{\partial x'^\mu}{\partial x^\rho}. \tag{3.59}$$

$\mathcal{R}^\mu{}_\rho(x)$ is a local Lorentz transformation since

$$\mathcal{R}^\mu{}_\rho(x) \mathcal{R}^\nu{}_\sigma(x) \eta_{\mu\nu} = \eta_{\rho\sigma}. \tag{3.60}$$

In the case of Euclidean signature, we have to replace $\eta_{\mu\nu}$ by $\delta_{\mu\nu}$ and thus $\mathcal{R}^\nu{}_\sigma(x)$ is a local orthogonal rotation belonging to $O(d)$. This will prove to be useful for the construction of conformal correlation functions below.

For an inversion $x'^\mu = x^\mu/x^2$, we have

$$\Omega(x) = x^2, \tag{3.61}$$

and the local orthogonal rotation (3.59) is given by $\mathcal{R}^{\mu\nu}(x) = I^{\mu\nu}(x)$ with

$$I^{\mu\nu}(x) = \delta^{\mu\nu} - 2\frac{x^\mu x^\nu}{x^2}. \tag{3.62}$$

The inversion matrix I of (3.62) plays the important role of a parallel transport. For two points x, y we have

$$I^{\mu\nu}(x' - y') = \mathcal{R}^\mu{}_\alpha(x)\mathcal{R}^\nu{}_\beta(y)I^{\alpha\beta}(x - y), \tag{3.63}$$

which implies

$$(x' - y')^2 = \frac{(x - y)^2}{\Omega(x)\Omega(y)}. \tag{3.64}$$

Exercise 3.2.5 Show that $\det I = -1$ for I defined in (3.62).

Also, using the inversion we may define a vector which transforms homogeneously under conformal transformations. For three points x, y, z, this vector is constructed from the difference of the inversions of $(x - z)_\mu$ and $(y - z)_\mu$. This vector is denoted as Z_μ and is defined at the point z by

$$Z^\mu = \frac{(x - z)^\mu}{(x - z)^2} - \frac{(y - z)^\mu}{(y - z)^2}, \tag{3.65}$$

which implies

$$Z^2 = \frac{(x - y)^2}{(x - z)^2(y - z)^2}. \tag{3.66}$$

For any conformal transformation, the vector Z^μ transforms covariantly,

$$Z'^\mu = \Omega(z)\mathcal{R}^\mu{}_\alpha(z)Z^\alpha. \tag{3.67}$$

At the points x, y, similar covariant vectors X^μ, Y^μ may be defined by cyclic permutation.

Conformal algebra in $d = 2$ dimensions

In the special case of two dimensions, the condition (3.51) specifying infinitesimal conformal transformation takes the simple form

$$\partial_0\epsilon_1 = -\partial_1\epsilon_0, \qquad \partial_0\epsilon_0 = \partial_1\epsilon_1. \tag{3.68}$$

These equations are easily identified as the Cauchy–Riemann differential equation of complex analysis in Euclidean spacetime where it is convenient to introduce complex coordinates $z = x^0 + ix^1, \bar{z} = x^0 - ix^1$. Then $\epsilon = \epsilon^0 + i\epsilon^1$ is a function of z, i.e. holomorphic, while $\bar{\epsilon} = \epsilon^0 - i\epsilon^1$ depends only on $\bar{z}$ and thus is anti-holomorphic. We may expand $\epsilon(z)$ and $\bar{\epsilon}(\bar{z})$ as

$$\epsilon(z) = -\sum_{n\in\mathbb{Z}}\epsilon_n z^{n+1}, \quad \bar{\epsilon}(\bar{z}) = -\sum_{n\in\mathbb{Z}}\bar{\epsilon}_n \bar{z}^{n+1}. \tag{3.69}$$

Thus the infinitesimal transformation given by $z \mapsto z' = z + \epsilon(z)$ and $\bar{z} \mapsto \bar{z}' = \bar{z} + \bar{\epsilon}(\bar{z})$ is conformal. The generators of a conformal transformation where only $\epsilon_n, \bar{\epsilon}_n \neq 0$ are

$$l_n = -z^{n+1}\partial_z, \qquad \bar{l}_n = -\bar{z}^{n+1}\partial_{\bar{z}}. \tag{3.70}$$

Exercise 3.2.6   Show that the commutation relations of the generators $l_n, \bar{l}_m$ are given by

$$[l_n, l_m] = (m - n)l_{m+n}, \quad [\bar{l}_n, \bar{l}_m] = (m - n)\bar{l}_{m+n}, \quad [l_n, \bar{l}_m] = 0. \tag{3.71}$$

In particular note that the generator $\{l_{-1}, l_0, l_1\}$ and its complex conjugate generate the finite-dimensional subalgebra $\mathfrak{sl}(2, \mathbb{R}) \oplus \mathfrak{sl}(2, \mathbb{R})$. This subalgebra corresponds to *global* conformal transformation. This subset is equivalent to the conformal transformations also present in more than two dimensions, as given by (3.52). The transformations for all other n are referred to as *local* conformal transformations. These do not have any counterpart in higher dimensions.

Let us only consider the generators l_n and their commutation relations. In the quantum theory, the corresponding operators L_n satisfy slightly different commutation relations,

$$[L_n, L_m] = (m - n)L_{m+n} + \frac{c}{12}(m^3 - m)\delta_{m+n,0}, \tag{3.72}$$

which is known as the *Virasoro algebra*. Here, the coefficient c is referred to as the *central charge*. Note that the additional term in the commutation relations (3.72) is a quantum effect and that it is multiplied by the identity operator which trivially commutes with all other operators L_n. Thus the Virasoro algebra is referred to as the *central extension* of the algebra (3.71).

3.2.2 Field transformations

The fields in a conformal field theory (CFT) transform in irreducible representations of the conformal algebra. In order to construct the transformation representations for general dimensions, we use the method of *induced representations*. First, we analyse the transformation properties of the fields ϕ at $x = 0$. Then, with the help of the momentum vector P^μ, we may shift the argument of the field to an arbitrary point x in order to obtain the general transformation rule. We already used this method above for the Poincaré algebra. For the Lorentz transformations we have postulated that

$$\left[J_{\mu\nu}, \phi(0)\right] = -\mathcal{J}_{\mu\nu}\phi(0), \tag{3.73}$$

where $\mathcal{J}_{\mu\nu}$ is a finite-dimensional representation of the Lorentz group determining the spin for the field $\phi(0)$. For the conformal algebra, in addition we postulate commutation relations with the dilatation operator D,

$$[D, \phi(0)] = -i\Delta\phi(0). \tag{3.74}$$

This relation implies that ϕ has the scaling dimension Δ, i.e. under dilatations $x \mapsto x' = \lambda x$ it transforms as

$$\phi(x) \mapsto \phi'(x') = \lambda^{-\Delta}\phi(x). \tag{3.75}$$

In particular, a field ϕ which transforms covariantly under an irreducible representation of the conformal algebra has a fixed scaling dimension and is therefore an eigenstate of the dilatation operator D. Moreover, in a conformal algebra it is sufficient to consider particular fields, the *conformal primary* fields, which satisfy the commutation relation

$$\left[K_\mu, \phi(0)\right] = 0. \tag{3.76}$$

By applying the commutation relations of D with P_μ and K_μ to the eigenstates of D, we see that P_μ increases the scaling dimension while K_μ decreases it. As discussed in box 3.3, in a unitary CFT, there is a lower bound on the scaling dimension of the fields. This implies that any conformal representation must contain operators of lowest dimension which due to (3.76) are annihilated by K_μ at $x^\nu = 0$. In a given irreducible multiplet of the conformal algebra, the conformal primary fields are thus fields of lowest scaling dimension determined by the relation (3.76). All other fields, the *conformal descendants* of ϕ, are obtained by acting with P_μ on the conformal primary fields.

So far we have considered the transformation properties of ϕ at $x^\mu = 0$. Using the operator $\mathcal{T}(x)$ as introduced in (3.46) we may write $\phi(x) = \mathcal{T}(x)\phi(0)\mathcal{T}^{-1}(x)$ and thus we

---

Box 3.3   **Unitarity bound for conformal field theories**

Consider the subalgebra $\mathfrak{so}(1,1) \oplus \mathfrak{so}(3,1)$ of the four-dimensional conformal algebra, corresponding to dilatations and Lorentz transformations. This allows us to label representations of the conformal algebra by (Δ, j_L, j_R), with Δ the scaling dimension and j_L, j_R the Lorentz quantum numbers given in table 3.2. For any quantum field theory, unitarity implies that all states in a representation have positive norm. This imposes bounds on the unitary representations. Let us consider the compact subalgebra $\mathfrak{so}(2) \oplus \mathfrak{so}(4)$ of $\mathfrak{so}(4,2)$. Unitary representations of this group are labelled by (Δ, j_L, j_R) and have to satisfy the constraints

$$\Delta \geq 1 + j_L \quad \text{for } j_R = 0, \qquad \Delta \geq 1 + j_R \quad \text{for } j_L = 0,$$

$$\Delta \geq 2 + j_L + j_R \quad \text{for both } j_L, j_R \neq 0. \tag{3.77}$$

Examples are as follows: scalars must have $\Delta \geq 1$, vectors must have $\Delta \geq 3$, symmetric traceless tensors must have $\Delta \geq 4$. These bounds are saturated for a free scalar field ϕ satisfying the equation of motion, for a conserved current J_μ and for a conserved symmetric traceless tensor $T_{\mu\nu}$. In d dimensions, the bound for scalars generalises to

$$\Delta \geq \frac{d-2}{2}. \tag{3.78}$$

---

can deduce the commutation relations for a conformal primary field $\phi(x)$,

$$\begin{aligned}
\left[P_\mu, \phi(x)\right] &= -i\partial_\mu\phi(x) \equiv \mathcal{P}_\mu\phi(x), \\
[D, \phi(x)] &= -i\Delta\phi(x) - ix^\mu\partial_\mu\phi(x) \equiv \mathcal{D}\phi(x), \\
\left[J_{\mu\nu}, \phi(x)\right] &= -\mathcal{J}_{\mu\nu}\phi(x) + i(x_\mu\partial_\nu - x_\nu\partial_\mu)\phi(x) \equiv \tilde{\mathcal{J}}_{\mu\nu}\phi(x), \\
\left[K_\mu, \phi(x)\right] &= \left(i(-x^2\partial^\mu + 2x_\mu x^\rho\partial_\rho + 2x_\mu\Delta) - 2x^\nu\mathcal{J}_{\mu\nu}\right)\phi(x) \equiv \mathcal{K}_\mu\phi(x).
\end{aligned} \tag{3.79}$$

Exercise 3.2.7   Use a general infinitesimal conformal transformation specified by $\epsilon(x)$ as in (3.52) to show that the transformations (3.79) may be summarised in the form

$$\delta_\epsilon\phi(x) = -\mathscr{L}_\nu\phi(x), \quad \mathscr{L}_\nu = \epsilon(x)\cdot\partial + \frac{\Delta}{d}\partial\cdot\epsilon(x) - \frac{i}{2}\partial_{[\mu}\epsilon_{\nu]}(x)\mathcal{J}_{\mu\nu}. \tag{3.80}$$

Exercise 3.2.8   Show that the operators $\mathcal{P}_\mu, \mathcal{D}, \mathcal{K}_\mu$ and $\tilde{\mathcal{J}}_{\mu\nu}$ defined by (3.79) form a representation of the conformal algebra.

In two dimensions, there is a distinction between *primary* and *quasi-primary* conformal fields according to their transformation properties. This is discussed in box 3.4.

3.2.3 Energy-momentum tensor and CFT

According to the Noether theorem reviewed in chapter 1, every continuous symmetry is associated with a conserved current. For translations, the conserved current is the energy-momentum tensor $T_{\mu\nu}$, while for Lorentz transformations, the conserved current is given

| Box 3.4 | (Quasi-)primary fields in two-dimensional CFTs |

In two dimensions, we have to distinguish two different kinds of primary fields. Quasi-primary fields satisfy the commutation relations (3.79) while *primary* fields transform under a two-dimensional conformal transformation $z \mapsto f(z), \bar{z} \mapsto \bar{f}(\bar{z})$ as

$$\phi(z, \bar{z}) \mapsto \phi'(z, \bar{z}) = \left(\frac{\partial f}{\partial z}\right)^h \left(\frac{\partial \bar{f}}{\partial \bar{z}}\right)^{\bar{h}} \phi(f(z), \bar{f}(\bar{z})). \tag{3.81}$$

The quasi-primary fields are those for which this transformation rule applies only to global transformations, for which $n = \{-1, 0, 1\}$ in (3.71).

by $N_{\mu\nu\rho} = x_\nu T_{\mu\rho} - x_\rho T_{\mu\nu}$. The associated Noether charges are

$$P_\nu = \int d^{d-1}x \, T^0{}_\nu$$

$$M_{\nu\rho} = \int d^{d-1}x \, (x_\nu T^0{}_\rho - x_\rho T^0{}_\nu). \tag{3.82}$$

The remaining conformal transformations, i.e. the scale transformation and the special conformal transformations, also give rise to conserved currents $J_{(D)\mu}$ and $J_{(K)\mu\nu}$, respectively,

$$J_{(D)\mu} = x^\nu T_{\mu\nu}, \qquad J_{(K)\mu\nu} = x^2 T_{\mu\nu} - 2x_\nu x^\rho T_{\mu\rho}. \tag{3.83}$$

The corresponding generators are

$$D = \int d^{d-1}x \, x^\rho T^0{}_\rho, \tag{3.84}$$

$$K_\nu = \int d^{d-1}x \, \left(x^2 T^0{}_\nu - 2x_\nu x^\rho T^0{}_\rho\right).$$

These symmetries impose restrictions on the energy-momentum tensor. In exercise 1.2.3, we have seen that due to Lorentz and translation invariance, there exists an *improved* energy-momentum tensor which has to be symmetric in its indices, i.e. $T^{\mu\nu} = T^{\nu\mu}$. If the theory considered is also invariant under scale transformations, its energy-momentum tensor has to be traceless, $T^\mu{}_\mu = 0$, since

$$0 = \partial^\nu J_{(D)\nu} = \partial^\nu (x^\rho T_{\nu\rho}) = (\partial^\nu x^\rho) T_{\nu\rho} + x^\rho \partial^\nu T_{\nu\rho} = T^\rho{}_\rho. \tag{3.85}$$

The dilatation charge – or its associated current, respectively – generates scale transformations. Therefore, tracelessness of the energy-momentum tensor guarantees scale invariance of the classical field theory.

Exercise 3.2.9 In (2.101) in chapter 2, we defined the energy-momentum tensor (in Lorentz signature) by

$$T_{\mu\nu} = -\frac{2}{\sqrt{-g}} \frac{\delta S}{\delta g^{\mu\nu}}, \tag{3.86}$$

where $\mathcal{S} = \int d^d x \sqrt{-g}\, \mathcal{L}$ is the classical action. We have seen that $T_{\mu\nu}$ is symmetric and also gauge invariant by construction. Show that $T_{\mu\nu}$ is traceless if the action $\mathcal{S}$ of the field theory is scale invariant.

3.2.4 Correlation functions

Conformal symmetry imposes significant restrictions on the possible form of the quantum field theory correlation functions introduced in (1.47). In particular, it determines the form of the two- and three-point correlation functions up to a manageable number of parameters. This applies both to the case of $d = 2$ and to the case of $d > 2$, though in the latter case there is generally more freedom remaining.

As discussed in section 1.8, the invariance $\delta \mathcal{S} = 0$ of the action under conformal transformations on the classical level leads to a Ward identity for correlation functions of the form

$$\sum_{i=1}^{n} \langle \phi_1(x_1)\phi_2(x_2) \ldots \delta\phi_i(x_i) \ldots \phi_n(x_n) \rangle = 0, \tag{3.87}$$

where $\delta\phi$ is given by (3.80). In particular we obtain the *dilatation Ward identity*

$$\sum_{i=1}^{n} \left(x_i^\mu \frac{\partial}{\partial x_i^\mu} + \Delta_i \right) \langle \phi_1(x_1)\phi_2(x_2) \ldots \phi_i(x_i) \ldots \phi_n(x_n) \rangle = 0, \tag{3.88}$$

where Δ_i is the scaling dimension of the field ϕ_i. The Ward identities associated with dilatations and special conformal transformations constrain the spacetime dependence of the correlation functions.

For example, using the invariance under dilatations, the two-point function of two scalar conformal primary operators ϕ_1 and ϕ_2 with scaling dimensions Δ_1 and Δ_2 transforms as

$$\langle \phi_1(x_1)\phi_2(x_2) \rangle = \lambda^{\Delta_1 + \Delta_2} \langle \phi_1(\lambda x_1)\phi_2(\lambda x_2) \rangle. \tag{3.89}$$

Since the correlation function $\langle \phi_1(x_1)\phi_2(x_2) \rangle$ can only depend on $(x_1 - x_2)^2$ due to Poincaré invariance, we conclude

$$\langle \phi_1(x_1)\phi_2(x_2) \rangle = \frac{C_{\phi_1\phi_2}}{(x_1 - x_2)^{\Delta_1 + \Delta_2}}. \tag{3.90}$$

Here $(x_1 - x_2)^{\Delta_1 + \Delta_2}$ is an abbreviation for $((x_1 - x_2)^2)^{(\Delta_1 + \Delta_2)/2}$ which we use from now on. We may further constrain the correlation function by applying an inversion: the two-point function is zero unless both fields have the same scaling dimension Δ. Moreover, since the constant $C_{\phi_1\phi_2}$ appearing in the two-point function is real and symmetric under the exchange of ϕ_1 and ϕ_2, i.e. $C_{\phi_1\phi_2} = C_{\phi_2\phi_1}$, we can diagonalise the constant C in the space of scalar primary operators $\mathcal{O}$ such that C is only non-zero for conjugated operators $\mathcal{O}$ and $\bar{\mathcal{O}}$. Finally, by redefining the operators $\mathcal{O}$ and $\bar{\mathcal{O}}$, we can set $C = 1$ and thus obtain for a scalar conformal primary operator $\mathcal{O}$ of scaling dimension Δ,

$$\langle \mathcal{O}(x_1)\bar{\mathcal{O}}(x_2) \rangle = \frac{1}{(x_1 - x_2)^{2\Delta}}. \tag{3.91}$$

In the same way, we can show that for three scalar conformal primary operators $\mathcal{O}_i$ ($i = 1, 2, 3$) with scaling dimension Δ_i, the three-point function reads

$$\langle \mathcal{O}_1(x_1) \mathcal{O}_1(x_2) \mathcal{O}_3(x_3) \rangle$$
$$= \frac{C_{\mathcal{O}_1\mathcal{O}_2\mathcal{O}_3}}{(x_1 - x_2)^{\Delta_1+\Delta_2-\Delta_3} (x_2 - x_3)^{-\Delta_1+\Delta_2+\Delta_3} (x_1 - x_3)^{\Delta_1-\Delta_2+\Delta_3}} \quad (3.92)$$

with constants $C_{\mathcal{O}_1\mathcal{O}_2\mathcal{O}_3}$ determined by the field content. Four-point correlators $\langle \mathcal{O}_1(x_1) \mathcal{O}_2(x_2) \mathcal{O}_3(x_3) \mathcal{O}_4(x_4) \rangle$ are less constrained by the symmetry since they involve dimensionless *cross ratios* $\frac{(x_1-x_2)^2}{(x_3-x_4)^2}$ and $\frac{(x_1-x_3)^2}{(x_2-x_4)^2}$.

Exercise 3.2.10 Show that the correlation functions (3.91) and (3.92) satisfy the Ward identities associated with dilatations and special conformal transformations.

Let us now consider general conformal primary operators $\mathcal{O}^i$ in more than two dimensions, using Euclidean signature. We use the index i to denote components in a space on which a representation of $O(d)$ (or $O(d - 1, 1)$ in Minkowski signature) acts. Examples of $\mathcal{O}^i$ are a vector current J^μ or the energy-momentum tensor $T^{\mu\nu}$. Applying the procedure outlined above on a case-by-case basis to all these operators is very tedious. However, we can also use the method of induced representations in the case of conformal transformations. The matrix $\mathcal{R}$ as defined in (3.59) gives rise to a local Lorentz transformation or local rotation in Minkowski or Euclidean signature respectively, whose representations we have already studied in section 3.1. Thus a general conformal primary operator transforms as

$$\mathcal{O}^i(x) \mapsto \mathcal{O}'^i(x') = \Omega(x)^\Delta D(\mathcal{R}(x))^i{}_j \, \mathcal{O}^j(x), \quad (3.93)$$

where $\Omega(x)$ is the scale factor defined in (3.48) and Δ is the conformal dimension. $D(\mathcal{R}(x))$ is the appropriate representation of the local Lorentz transformation.

It is now straightforward to construct conformally covariant expressions for the two-point functions of conformal primary operators. For the field $\mathcal{O}$ transforming as in (3.93) and its associated conjugate field $\overline{\mathcal{O}}$, which transforms in the conjugate representation, i.e. $\overline{\mathcal{O}}_i(x) \mapsto \Omega(x)^\Delta \overline{\mathcal{O}}_j(x) \left(D(\mathcal{R}(x))^{-1} \right)^j{}_i$, then, for $\mathcal{O}, \overline{\mathcal{O}}$ in irreducible representations of $O(d)$, we may write in general

$$\langle \mathcal{O}^i(x) \overline{\mathcal{O}}_j(y) \rangle = \frac{C_{\mathcal{O}}}{(x - y)^{2\Delta}} D(I(x - y))^i{}_j, \quad (3.94)$$

where $C_{\mathcal{O}}$ is an overall constant scale factor which we can set to one by redefining the operators.

Applying this result to the conserved vector current J_μ we have

$$\langle J_\mu(x) J_\nu(y) \rangle = \frac{C_J}{(x - y)^{2(d-1)}} I_{\mu\nu}(x - y). \quad (3.95)$$

Moreover, for the energy-momentum tensor $T_{\mu\nu}$ the general result (3.94) implies

$$\langle T_{\mu\nu}(x) T_{\sigma\rho}(y) \rangle = \frac{C_T}{(x - y)^{2d}} \mathcal{I}^T_{\mu\nu,\sigma\rho}(x - y), \quad (3.96)$$

where

$$\mathcal{I}^T_{\mu\nu,\rho\sigma}(x - y) = I_{\mu\alpha}(x - y) I_{\nu\beta}(x - y) P^{\alpha\beta}_{\rho\sigma}, \quad (3.97)$$

Box 3.5 **Energy-momentum tensor two-point function in $d=2$**

In two dimensions, it is convenient to introduce complex coordinates such that $T = T_{zz}, \bar{T} = \bar{T}_{\bar{z}\bar{z}}$. In this case the general result (3.96) is in agreement with the well-known expression for the two-dimensional case,

$$\langle T(z)T(w) \rangle = \frac{c/2}{(z-w)^4}. \tag{3.98}$$

Here c is the Virasoro central charge of (3.72). Consistency with the Virasoro algebra is obtained by expanding the energy-momentum tensor in a Laurent series in the Virasoro generators,

$$T(z) = \sum_{n \in \mathbb{Z}} z^{-n-2} L_n, \quad L_n = \frac{1}{2\pi i} \oint dz \, z^{n+1} T(z), \tag{3.99}$$

and calculating the commutator $[L_m, L_n]$.

with P the projection operator onto the space of symmetric traceless tensors defined in (3.12). $\mathcal{I}^T$ represents the corresponding inversion tensor. Since $\partial_\mu J^\mu$ is a scalar and $\partial_\mu T^{\mu\nu}$ is a vector, J_μ and $T_{\mu\nu}$ have dimensions $d-1$ and d, respectively. This ensures that (3.95) and (3.96) automatically satisfy the required conservation equations. For the case of two dimensions, the expression (3.96) simplifies if complex coordinates are used. This is discussed in box 3.5.

Three-point function and operator product expansion

The general formula for a conformally covariant three-point function for conformal primary fields in d dimensions is straightforward to construct using the vector Z defined in (3.65) and appropriate representations D of the inversion matrix I given in (3.62), which is a representation of $O(d)$. The most general expression for the three-point function of three arbitrary conformal primary operators reads

$$\langle \mathcal{O}_1^i(x) \, \mathcal{O}_2^j(y) \, \mathcal{O}_3^k(z) \rangle = \frac{D_{1\,i'}^i(I(x-z)) D_{2\,j'}^j(I(y-z)) \, t^{i'j'k}(Z)}{(x-z)^{2\Delta_1} \, (y-z)^{2\Delta_2}}, \tag{3.100}$$

where D_1, D_2 are appropriate $O(d)$ representations acting on the operators $\mathcal{O}_1^i$, $\mathcal{O}_2^j$, respectively. Moreover, t^{ijk} is homogeneous in Z, i.e.

$$t^{ijk}(\lambda Z) = \lambda^{\Delta_3 - \Delta_1 - \Delta_2} \, t^{ijk}(Z), \tag{3.101}$$

and has to satisfy

$$D_{1\,i'}^i(R) \, D_{2\,j'}^j(R) \, D_{3\,k'}^k(R) \, t^{i'j'k'}(Z) = t^{ijk}(RZ) \tag{3.102}$$

for all $R \in O(d)$. These conditions which ensure that t is a homogeneous function covariant under $O(d)$ transformations are sufficient to guarantee that (3.100) satisfies the conformal ward identities (3.87). This is due in particular to the fact that the parallel transport relation (3.63) extends to arbitrary representations D.

Note that (3.100) seems not to be symmetric under the exchange of the three operators. However, this is only apparent as demonstrated in the exercise below.

Exercise 3.2.11 Show that the expression (3.100) is symmetric under the exchange of the three operators, by explicitly performing the following steps.

(i) Check the following relations

$$I_\mu{}^\alpha(x-z)Z_\alpha = -\frac{(x-y)^2}{(z-y)^2}X_\mu, \tag{3.103}$$

$$I_\mu{}^\alpha(x-z)I_{\alpha\nu}(z-y) = I_{\mu\nu}(x-y) + 2(x-y)^2 X_\mu Y_\nu, \tag{3.104}$$

$$I_\sigma{}^\alpha(y-z)I_{\alpha\mu}(z-x) = I_\sigma{}^\alpha(y-x)I_{\alpha\mu}(X). \tag{3.105}$$

(ii) With the help of these relations, we obtain the equivalent expression

$$\langle \mathcal{O}_1^i(x)\,\mathcal{O}_2^j(y)\,\mathcal{O}_3^k(z)\rangle = \frac{D_{2\,j'}^j(I(y-x))D_{3\,k'}^k(I(z-x))\,t^{j'k'i}(X)}{(x-y)^{2\Delta_2}\,(x-z)^{2\Delta_3}}, \tag{3.106}$$

with

$$t^{jki}(X) = (X^2)^{\Delta_1-\Delta_3}D_{2\,j'}^j(I(X))\,t^{ij'k}(-X). \tag{3.107}$$

(iii) Show that for bosonic fields, exchange symmetry requires

$$t^{ijk}(Z) = t^{jik}(-Z) = D^i_{\,i'}(I(Z))\,t^{ki'j}(-Z). \tag{3.108}$$

For particular operators, the explicit form of the three-point function is found by obtaining the most general expression for $t(Z)$ satisfying all the conditions given. For example, for three scalar conformal primary operators $\mathcal{O}_1$, $\mathcal{O}_2$ and $\mathcal{O}_3$, the tensor $t(Z)$ is given by

$$t(Z) = C_{\mathcal{O}_1\mathcal{O}_2\mathcal{O}_3}\left(\frac{(x-z)\,(y-z)}{(x-y)}\right)^{\Delta_1+\Delta_2-\Delta_3} \tag{3.109}$$

and thus the three-point function (3.100) simplifies to (3.92).

For generic conformal primary operators, $t(Z)$ has a direct significance since it represents the leading term in the operator product expansion (OPE). The leading contribution of the operator $\bar{\mathcal{O}}_{3k}$ to the operator product of $\mathcal{O}_1^i(x)\mathcal{O}_2^j(y)$ as $x \to y$ is given by

$$\mathcal{O}_1^i(x)\mathcal{O}_2^j(y) \sim \frac{1}{C_{\mathcal{O}_3}}t^{ijk}(x-y)\,\bar{\mathcal{O}}_{3k}(y), \tag{3.110}$$

where $C_{\mathcal{O}_3}$ is the normalisation constant (3.94) of the two-point function $\langle \mathcal{O}_3\bar{\mathcal{O}}_3\rangle$.

As an explicit example, we obtain the three-point function of three vector currents. The application of the general formalism as detailed above gives in this case

$$\langle J_\mu(x)J_\nu(y)J_\omega(z)\rangle = \frac{1}{(x-z)^{2d-2}(y-z)^{2d-2}}\,I_\mu{}^\alpha(x-z)I_\nu{}^\beta(y-z)\,t_{\alpha\beta\omega}(Z), \tag{3.111}$$

where $t_{\mu\nu\omega}(Z)$ contains two independent forms with parameters a and b, respectively,

$$t_{\mu\nu\omega}(Z) = a\,\frac{Z_\mu Z_\nu Z_\omega}{Z^{d+2}} + b\,\frac{1}{Z^d}(Z_\mu\delta_{\nu\omega} + Z_\nu\delta_{\mu\omega} - Z_\omega\delta_{\mu\nu}). \tag{3.112}$$

The explicit expression for the energy-momentum tensor three-point function $\langle T_{\mu\nu}(x)T_{\sigma\rho}(y)T_{\alpha\beta}(z)\rangle$ is more involved. It has three independent forms in general. In three dimensions, the number of independent forms reduces to two and in two dimensions to one, again the Virasoro central charge.

Four-point functions

Even in a conformal field theory, four-point functions are less constrained than two- and three-point functions. This is due to the fact that with four coordinates, it is possible to construct two dimensionless invariants, the cross ratios. These are given by

$$\eta = \frac{x_{12}^2 x_{34}^2}{x_{13}^2 x_{24}^2}, \qquad \xi = \frac{x_{14}^2 x_{23}^2}{x_{13}^2 x_{24}^2}, \tag{3.113}$$

where $x_{ij}^2 \equiv (x_i - x_j)^2$ with $i,j = 1,\dots,4$. A four-point function of scalar conformal primary operators $\mathcal{O}_i$ of conformal dimension Δ_i takes the general form

$$\langle \mathcal{O}(x_1)\mathcal{O}(x_2)\mathcal{O}(x_3)\mathcal{O}(x_4) \rangle = \frac{1}{x_{12}^{\Delta_1+\Delta_2} x_{34}^{\Delta_3+\Delta_4}} F(\eta,\xi), \tag{3.114}$$

with $F(\eta,\xi)$ a function of the two cross ratios which is unconstrained except for the requirement that the four-point function must be invariant under the exchange of any two of the operators involved.

The function $F(\eta,\xi)$ is related to the OPE coefficients introduced in (3.110) via the *double OPE* obtained by taking the simultaneous short-distance limit on two pairs of operators, for instance $x_1 \to x_2$ and $x_3 \to x_4$,

$$\langle \mathcal{O}_{\Delta_1}(x_1)\mathcal{O}_{\Delta_2}(x_2)\mathcal{O}_{\Delta_3}(x_3)\mathcal{O}_{\Delta_4}(x_4) \rangle = \sum_{\Delta\Delta'} \frac{c_{\Delta_1\Delta_2\Delta}}{x_{12}^{\Delta_1+\Delta_2-\Delta}} \frac{1}{x_{13}^{\Delta+\Delta'}} \frac{c_{\Delta_3\Delta_4\Delta'}}{x_{34}^{\Delta_3+\Delta_4-\Delta'}}, \tag{3.115}$$

where the $c_{\Delta_i\Delta_j\Delta_k}$ are determined by the three-point function (3.92) or the OPE (3.109).

3.2.5 Renormalisation of CFTs

A necessary condition for a field theory to be conformal is its scale invariance. Therefore a necessary condition for conformal symmetry in a renormalised theory is that all β functions vanish. Nevertheless, the anomalous dimensions γ may still be non-zero. Let us consider the example of a two-point function of a primary operator of classical scale dimension Δ. For $\beta \equiv 0$, the RG equation discussed in chapter 1 reduces to

$$\left(\mu \frac{\partial}{\partial \mu} + 2\gamma \right) \langle \mathcal{O}(x)\overline{\mathcal{O}}(y) \rangle_c = 0. \tag{3.116}$$

From dimensional analysis, also in the renormalised theory the two-point function is of engineering dimension 2Δ and satisfies

$$\left(\mu \frac{\partial}{\partial \mu} - (x-y) \cdot \frac{\partial}{\partial(x-y)} \right) \langle \mathcal{O}(x)\overline{\mathcal{O}}(y) \rangle_c = 2\Delta \langle \mathcal{O}(x)\overline{\mathcal{O}}(y) \rangle_c. \tag{3.117}$$

Equation (3.116) then implies

$$-(x-y) \cdot \frac{\partial}{\partial(x-y)} \langle \mathcal{O}(x)\overline{\mathcal{O}}(y) \rangle_c = 2(\Delta+\gamma)\langle \mathcal{O}(x)\overline{\mathcal{O}}(y) \rangle_c. \tag{3.118}$$

Solutions of this equation for the renormalised two-point function are of the general form

$$\langle \mathcal{O}(x)\overline{\mathcal{O}}(y)\rangle_c = \frac{f(g)}{(x-y)^{2\Delta}\left((x-y)^2\mu^2\right)^{\gamma}}, \tag{3.119}$$

where for regularity, the dimensionful scale μ has to be introduced, for example by using an appropriate regularisation scheme. Due to (3.74) and (3.79), γ is the quantum correction to the eigenvalue of the dilatation operator. This may be rewritten as

$$\langle \mathcal{O}(x)\overline{\mathcal{O}}(y)\rangle_c \sim \frac{f(g)}{(x-y)^{2\Delta_0}}\left(1-\gamma\,\ln\mu^2(x-y)^2+\ldots\right). \tag{3.120}$$

3.2.6 Conformal Ward identities and trace anomaly

In section 1.8, we introduced Ward identities which are relations between correlation functions originating from a continuous symmetry. We now consider Ward identities for conformal symmetry. For this purpose it is appropriate to couple the quantum field theory considered to a classical, non-propagating curved space background in which the metric is the source for an operator insertion of the energy-momentum tensor,

$$\langle T_{\mu\nu}(x)\rangle = -\frac{2}{\sqrt{g}}\frac{\delta W}{\delta g^{\mu\nu}(x)}, \tag{3.121}$$

where we consider the Euclidean signature case. Here W is the generating functional for connected Green's functions. From the quantum field theory point of view, (3.121) ensures that $T_{\mu\nu}$ is a well-defined regularised expression for a composite operator. For a conformally invariant theory we expect

$$0 = \delta_\sigma W = \int d^d x \frac{\delta W}{\delta g^{\mu\nu}}\delta_\sigma g^{\mu\nu} = -\int d^d x \sqrt{g}\sigma(x)\langle T^\mu{}_\mu\rangle, \tag{3.122}$$

which implies

$$\langle T^\mu{}_\mu\rangle = 0. \tag{3.123}$$

This equation has to be read with caution though, since it implies that just the vacuum expectation value of $T^\mu{}_\mu$ vanishes, and not the operator $T^\mu{}_\mu$ itself. As we discussed in section 1.8.2, the necessity for regularisation and renormalisation within the quantisation procedure may lead to contributions within the generating functional which break a symmetry which was present at the classical level. These lead to additional contributions to the Ward identities, the *anomalies*. This happens in particular for conformal symmetry, leading to non-trivial contributions to the trace of the energy-momentum tensor involving the curvature. Since the energy-momentum tensor has scaling dimension d, the conformal anomaly in d dimensions is a scalar of dimension d.

Let us begin by considering the two-dimensional case. Here, the conformal anomaly reads

$$\langle T^\mu{}_\mu(x)\rangle = \frac{c}{24\pi}R, \tag{3.124}$$

with R the Ricci scalar, which has dimension 2. c is the central charge also present in the Virasoro algebra (3.72). The conformal anomaly (3.124) is a topological density since

$$\int d^2x \sqrt{g}\, R = 2\pi \chi, \tag{3.125}$$

where χ is the Euler number.

Naively, the anomaly (3.124) vanishes on flat space where the curvature vanishes. However, the anomaly has implications for the correlation functions of a conformal field theory even on flat space. This becomes evident when varying both sides of (3.124) a second time with respect to the metric. Then, (3.124) has important consequences for the energy-momentum tensor two-point function. Varying (3.124) with respect to $g^{\sigma\rho}(y)$ gives rise to

$$\langle T^\mu{}_\mu(x) T_{\sigma\rho}(y)\rangle = -\frac{c}{12\pi}(\partial_\sigma \partial_\rho - \delta_{\sigma\rho}\partial^2)\delta^2(x-y) \tag{3.126}$$

where we have used that in an expansion around flat space,

$$R \sim -(\partial_\sigma \partial_\rho - \delta_{\sigma\rho}\partial^2)\delta g^{\sigma\rho}. \tag{3.127}$$

It is important to note that (3.126) also holds for conformal field theory on flat space.

It is now straightforward to show that the two-dimensional energy-momentum tensor two-point function given by (3.96) or (3.98) in complex coordinates satisfies the Ward identity (3.126). This may be seen as follows. In two dimensions, the general result for the energy-momentum tensor two-point function (3.96) may be rewritten as

$$\langle T_{\mu\nu}(x) T_{\sigma\rho}(y)\rangle = -\frac{c}{48\pi^2} S^x_{\mu\nu} S^y_{\sigma\rho} \ln((x-y)^2\mu^2), \qquad S_{\mu\nu} = \partial_\mu \partial_\nu - \delta_{\mu\nu}\partial^2. \tag{3.128}$$

Using complex coordinates in which $T(z) = -2\pi T_{zz}(x)$ it is straightforward, starting from (3.128), to recover the standard complex coordinate result (3.98). Moreover, (3.128) satisfies the the Ward identity (3.126) as may be seen by carefully taking two of the four derivatives in (3.128), noting that the expression $1/x^2$ is singular as distribution in two dimensions. In fact, for general dimensions we have

$$\frac{1}{x^{2\lambda}} \sim \frac{1}{d+2n-2\lambda} \frac{1}{2^{2n}n!} \frac{2\pi^{d/2}}{\Gamma(d/2+n)}(\partial^2)^n\delta^{(d)}(x). \tag{3.129}$$

In two dimensions this implies

$$\frac{1}{2}S_\mu{}^\mu \ln(x^2\mu^2) = \frac{1}{2}(-\partial^2)\ln(x^2\mu^2)$$
$$= -\frac{1}{2}\partial^\mu \frac{2x_\mu}{x^2} = 2\pi\delta^{(2)}(x). \tag{3.130}$$

When taking the second derivative, the pole in the denominator is cancelled by a factor $d-2$ in the numerator, such that the calculation leaves a finite result. For the two-point function we thus have

$$\langle T^\mu{}_\mu(x) T_{\sigma\rho}(y)\rangle = -\frac{c}{12\pi}S_{\sigma\rho}\delta^{(2)}(x-y), \tag{3.131}$$

in agreement with the Ward identity (3.126).

Exercise 3.2.12   Show that the two-dimensional energy-momentum tensor two-point function given by (3.96) or (3.98) in complex coordinates satisfies the Ward identity (3.126), by explicitly performing the calculation outlined above.

There are no conformal anomalies in odd dimensions since it is impossible to construct scalars of odd dimension using just the curvature. In four dimensions, the conformal anomaly takes the form

$$\langle T^{\mu}{}_{\mu} \rangle = \frac{c}{16\pi^2} C^{\mu\nu\sigma\rho} C_{\mu\nu\sigma\rho} - \frac{a}{16\pi^2} E. \tag{3.132}$$

Here $C_{\mu\nu\sigma\rho}$ is the Weyl tensor, which in general d dimensions is given by (2.81). The Weyl tensor has the index symmetries

$$C_{\mu\nu\sigma\rho} = C_{[\mu\nu][\sigma\rho]}, \quad C_{\mu[\sigma\rho\nu]} = 0, \quad C_{\mu\sigma\rho\mu} = 0. \tag{3.133}$$

Note that the Weyl tensor vanishes for $d \leq 3$ dimensions. The second term in (3.132) involves the Euler topological density E, which in four dimensions by given by

$$E = \frac{1}{4} \varepsilon_{\alpha\beta\gamma\delta} \varepsilon_{\mu\nu\sigma\rho} R^{\alpha\beta\mu\nu} R^{\gamma\delta\sigma\rho} = R^{\mu\nu\sigma\rho} R_{\mu\nu\sigma\rho} - 4 R^{\mu\nu} R_{\mu\nu} + R^2. \tag{3.134}$$

The Euler density satisfies

$$\int \mathrm{d}^4 x \sqrt{g} E = 4\pi \chi, \tag{3.135}$$

where χ is the Euler number and $\varepsilon_{\alpha\beta\gamma\delta}$ is the totally antisymmetric symbol in four dimensions. Since E is a topological density, this anomaly contribution parallels the Ricci scalar in two dimensions which is also a topological density. The term involving the square of the Weyl tensor is absent from the anomaly in $d \leq 3$ where the Weyl tensor vanishes.

In four dimensions, the conformal anomaly gives rise to anomalous terms in the Ward identity relating the energy-momentum tensor three- and two-point functions. Varying (3.132) twice with respect to the metric, we obtain

$$\langle T_{\mu}{}^{\mu}(x) T_{\sigma\rho}(y) T_{\alpha\beta}(z) \rangle = 2 \left(\delta^4(x-y) + \delta^4(x-z) \right) \langle T_{\sigma\rho}(y) T_{\alpha\beta}(z) \rangle$$
$$- 32c\, \mathcal{E}^C_{\sigma\epsilon\eta\rho,\alpha\gamma\delta\beta}\, \partial^{\varepsilon} \partial^{\eta} \delta^4(x-y) \partial^{\gamma} \partial^{\delta} \delta^4(x-z)$$
$$+ 4a \left(\varepsilon_{\sigma\alpha\varepsilon\kappa} \varepsilon_{\rho\beta\eta\lambda} \partial^{\kappa} \partial^{\lambda} \left(\partial^{\varepsilon} \delta^4(x-y) \partial^{\eta} \delta^4(x-z) \right) + \sigma \leftrightarrow \rho \right), \tag{3.136}$$

where $\mathcal{E}^C$ is a projector which projects onto tensors which have the same index symmetry as the Weyl tensor (3.133). This projector may be constructed from the product $\delta_{\alpha\mu} \delta_{\beta\nu} \delta_{\gamma\sigma} \delta_{\varepsilon\rho}$ with suitable permutations and suitable traces subtracted.

In four dimensions, the conformal anomaly leads to additional contributions to the Ward identity involving the three-point function which are present even on flat space. There is a linear relation between the three independent coefficients in the energy-momentum tensor three-point function and the parameters a, c of the conformal anomaly. The third independent form in the energy-momentum tensor three point function is anomaly free. Moreover, the anomaly coefficient c is proportional to the coefficient of the energy-momentum tensor two-point function, while a may not be related to the two-point function.

This feature is very different from the two-dimensional case where the coefficient of the topological density contribution to the anomaly is proportional to the coefficient of the two-point function.

3.3 Supersymmetry

By introducing conformal symmetry in the preceding sections, we have discussed one possibility for extending the Poincaré algebra. Conformal symmetry is realised in massless quantum field theories with dimensionless coupling constants if no scale is generated via the renormalisation process.

A further way to bypass the Coleman–Mandula theorem is to allow for one or more spinor *supercharges* Q^a, where a specifies the number of independent supersymmetries present, i.e. $a = 1, \ldots, \mathcal{N}$. In this way we obtain a new symmetry algebra, the *supersymmetry algebra*, which also involves anticommutation relations.

3.3.1 Supersymmetry algebra

In the case of four dimensions, in which we are mostly interested, it is convenient to use the Weyl notation and have a left-handed spinor Q_α^a and its right-handed counterpart $\overline{Q}_{a\dot\alpha} = (Q_\alpha^a)^*$ where the $SL(2, \mathbb{C})$ indices $\alpha, \dot\alpha$ take values $1, 2$ and $a = 1, \ldots, \mathcal{N}$ counts the number of independent supersymmetries.

The two-component Weyl spinor notation is related to the Dirac four-spinor notation by

$$Q_{\mathrm{D}}^a = \begin{pmatrix} Q_\alpha^a \\ \overline{Q}^{a\dot\alpha} \end{pmatrix}, \qquad \gamma^\mu = \begin{pmatrix} 0 & \sigma^\mu_{\alpha\dot\beta} \\ \bar{\sigma}^{\mu\dot\alpha\beta} & 0 \end{pmatrix}, \qquad (3.137)$$

as in (1.143), where $\sigma^\mu = (-\mathbb{1}, \sigma^i)$ and $\bar{\sigma}^\mu = (-\mathbb{1}, -\sigma^i)$ are four vectors of 2×2 matrices with the standard Pauli matrices σ^i as their spatial entries.

Simple supersymmetry algebra

Let us first restrict ourselves to the case of one supercharge, i.e. $\mathcal{N} = 1$. The supercharges obey commutation relations of a *graded* Lie algebra, sometimes also called a *superalgebra*. Besides the usual generators of a Lie algebra – which we will refer to as bosonic generators from now on – we also have fermionic generators. While bosonic generators have grade 0, the fermionic generators have grade $+1$. To assign the grade to a product of fields we simply add the grades of the individual fields modulo 2. In particular, the product of two fermionic generators is a bosonic generator, since $1 + 1 = 0 \,(\mathrm{mod}\,2)$ while the product of a bosonic and a fermionic generator is always fermionic.

The (anti-)commutation relation of two generators, denoted by $\mathcal{O}_1$ and $\mathcal{O}_2$, with grades g_1 and g_2 is given by

$$[\mathcal{O}_1, \mathcal{O}_2] = \mathcal{O}_1\mathcal{O}_2 - (-1)^{g_1 g_2}\mathcal{O}_2\mathcal{O}_1. \qquad (3.138)$$

In particular, the notation of the bracket $[\cdot, \cdot\}$ suggests that it can be either a commutator or an anticommutator. To be precise, in the case of two fermionic generators the bracket is an anticommutator, while in all other cases it is a commutator.

It turns out that the supercharges are the fermionic generators of such a graded algebra. However, the way in which a graded Lie algebra can be related to the symmetries of quantum field theory is very restricted, since its structure has to be compatible with the Poincaré algebra and other internal symmetries. The most general supersymmetry algebra for one supercharge Q (with components Q_α and $\overline{Q}_{\dot\alpha}$ in Weyl notation) reads

$$
\begin{aligned}
&[Q_\alpha, J^{\mu\nu}] = (\sigma^{\mu\nu})_\alpha{}^\beta Q_\beta, &&[\overline{Q}_{\dot\alpha}, J^{\mu\nu}] = \epsilon_{\dot\alpha\dot\beta}(\bar\sigma^{\mu\nu})^{\dot\beta}{}_{\dot\gamma}\overline{Q}^{\dot\gamma}, \\
&[Q_\alpha, P^\mu] = 0, &&[\overline{Q}_{\dot\alpha}, P^\mu] = 0, &&(3.139)\\
&\{Q_\alpha, \overline{Q}_{\dot\alpha}\} = 2\sigma^\mu_{\alpha\dot\alpha}P_\mu, &&\{Q_\alpha, Q_\beta\} = \{\overline{Q}_{\dot\alpha}, \overline{Q}_{\dot\beta}\} = 0,
\end{aligned}
$$

where we have raised and lowered the spinor indices by $\epsilon_{\dot\alpha\dot\beta}$ and $\epsilon_{\alpha\beta}$ – for more details consult appendix B, in particular section B.2.2. These commutation relations have to be supplemented by the commutation relations of the Poincaré algebra. The first line of (3.139) states that Q_α and $Q_{\dot\alpha}$ are left- and right-handed spinors, transforming in $(1/2, 0)$ and $(0, 1/2)$ of $\mathfrak{su}(2)_L \oplus \mathfrak{su}(2)_R$, respectively. The second line of (3.139) is a consequence of a Jacobi identity. This follows from the fact that the only term consistent with the index structure of $[Q_\alpha, P^\mu]$ is $(\sigma^\mu)_{\alpha\dot\alpha}\overline{Q}^{\dot\alpha}$. Using the Jacobi identity involving P^μ, P^ν and Q_α, we see that this term has to be absent, which implies the second line of (3.139).

For the third line of (3.139) there is a similar argument. The only term consistent with the index structure of the anticommutator $\{Q_\alpha, \overline{Q}_{\dot\alpha}\}$ is $\sigma^\mu_{\alpha\dot\alpha}P_\mu$. The proportionality constant is chosen to be two, which also fixes the normalisation of Q_α and $\overline{Q}_{\dot\alpha}$. For the anticommutator $\{Q_\alpha, Q_\beta\}$, we also write down all terms with the appropriate index structure. Then, by considering the Jacobi identity involving P_μ, Q_α and Q_β, we may show that this term has to be absent and therefore the anticommutator $\{Q_\alpha, Q_\beta\}$ has to vanish.

Moreover, in the case of one supercharge there is also a $U(1)$ automorphism of the supersymmetry algebra known as R-symmetry,

$$
Q_\alpha \mapsto Q'_\alpha = e^{i\alpha}Q_\alpha, \qquad \overline{Q}_{\dot\alpha} \mapsto \overline{Q}'_{\dot\alpha} = e^{-i\alpha}\overline{Q}_{\dot\alpha}. \tag{3.140}
$$

The corresponding generator of this $U(1)$ automorphism is denoted by R and the non-vanishing commutation relations read

$$
[Q_\alpha, R] = Q_\alpha, \qquad [\overline{Q}_{\dot\alpha}, R] = -\overline{Q}_{\dot\alpha}. \tag{3.141}
$$

Extended supersymmetry algebra

Let us now consider more than one Dirac supercharge Q_D^a with $a = 1, \ldots, \mathcal{N}$, or equivalently Q_α^a and $\overline{Q}_{b\dot\beta}$ in Weyl notation. As in the case of simple supersymmetry with $\mathcal{N} = 1$, the supercharges transform as left- and right-handed spinors under Lorentz transformations and commute with the generators of translations. Therefore the first two lines of the algebra (3.139) are valid after attaching indices a and b to appropriate places. The third line of the supersymmetry algebra (3.139) has to be modified and becomes

$$\{Q_\alpha^a, \overline{Q}_{b\dot\beta}\} = 2\sigma_{\alpha\dot\beta}^\mu P_\mu \delta_b^a, \qquad \{Q_\alpha^a, Q_\beta^b\} = \varepsilon_{\alpha\beta} Z^{ab}, \qquad \{\overline{Q}_{a\dot\alpha}, \overline{Q}_{b\dot\beta}\} = \varepsilon_{\dot\alpha\dot\beta} \bar{Z}_{ab}.$$
(3.142)

Here, the new ingredients are the numbers $Z^{ab}, \bar{Z}_{ab}$ referred to as *central charges* since Z^{ab} and $\bar{Z}_{ab}$ commute with all the other generators of the supersymmetry algebra. In other words, Z^{ab} generate the centre of the supersymmetry algebra. In order to respect the anticommutator's symmetry, the central charges Z^{ab} need to be antisymmetric, $Z^{ab} = -Z^{ba}$, and also have to satisfy $Z^{ab} = (\bar{Z}^\dagger)_{ab}$ due to $\overline{Q}_{a\dot\alpha} = (Q_\alpha^a)^*$

As in the case of $\mathcal{N} = 1$ supersymmetry, the supersymmetry algebra (3.142) is invariant under global phase rotations of the supercharges Q_α^a and $\overline{Q}_{a\dot\alpha}$ of the form

$$Q_\alpha^a \mapsto Q_\alpha^{a'} = R^a{}_b Q_\alpha^b, \qquad \overline{Q}_{a\dot\alpha} \mapsto \overline{Q}_{a\dot\alpha}' = \overline{Q}_{b\dot\alpha}(R^\dagger)^b{}_a,$$
(3.143)

where $R^a{}_b$ are components of an $\mathcal{N} \times \mathcal{N}$ matrix R. This symmetry of the supersymmetry algebra is known as R-symmetry. The R-symmetry is a global non-Abelian symmetry, and in in four spacetime dimensions, the symmetry group satisfies $R \in U(\mathcal{N})$. Note that Q_α^a transforms in the fundamental representation $\mathcal{N}$ of $U(\mathcal{N})$ while $\overline{Q}_{a\dot\alpha}$ transforms in the complex conjugate representation $\bar{\mathcal{N}}$. This also explains why we label Q_α^a with an upper index a, while $\overline{Q}_{a\dot\alpha}$ has a lower index a.

Let us denote the generators of the transformation (3.143) by T^j. The commutation relations involving T^j read

$$[Q_\alpha^a, T^j] = B^{ja}{}_b Q_\alpha^b, \qquad [\overline{Q}_{a\dot\alpha}, T^j] = -B^j{}_a{}^b \overline{Q}_{b\dot\alpha}, \qquad [T^j, T^k] = if^{jk}{}_l T^l.$$
(3.144)

The components $B^{ja}{}_b$ and $B^j{}_a{}^b$ satisfy $(B^{j\dagger})^a{}_b = (B^j)_a{}^b$. Since Z^{ab} is also an $\mathcal{N} \times \mathcal{N}$ matrix, we can define coefficients $A^{ab}{}_j$ such that

$$Z^{ab} = A^{ab}{}_j T^j,$$
(3.145)

where we also have to satisfy $B^{ia}{}_b A_j{}^{bc} = -A_j{}^{ab} B^i{}_c{}^b$. This condition puts constraints on the maximal possible R-symmetry group in four dimensions. For example, if $Z_{ab} = 0$, the maximal R-symmetry group is $U(\mathcal{N})$. On the other hand for $Z_{ab} \neq 0$, the maximal R-symmetry is a subgroup of $U(\mathcal{N})$, the *compact symplectic group* $USp(\mathcal{N})$, which is discussed in detail in appendix B. So far we have considered the classical level not taking into account quantum effects. On the quantum level, part or all of these R-symmetries may be broken by anomalies.

Representations of the supersymmetry algebra

In this section we determine the irreducible representations of the supersymmetry algebra (3.139). Since the supersymmetry algebra is an extension of the Poincaré algebra, we can label the different states of an irreducible representation in terms of Lorentz quantum numbers.

As in the case of the Poincaré algebra, it is useful to determine the Casimir operators of the theory. The Poincaré algebra had two Casimir operators, $P^2 = P_\mu P^\mu$ and $W^2 = W_\mu W^\mu$ where W_μ is the Pauli–Lubanski vector defined in (3.38).

Exercise 3.3.1 Show that $W^2 = W_\mu W^\mu$ can be written in the form $W^2 = C_{\mu\nu}C^{\mu\nu}$ with $C_{\mu\nu} = W_\mu P_\nu - W_\nu P_\mu$.

For the supersymmetry algebra, P^2 is still a Casimir operator but W^2 is not. However, it is possible to modify the definition of $C_{\mu\nu}$,

$$\tilde{C}_{\mu\nu} = \tilde{W}_\mu P_\nu - \tilde{W}_\nu P_\mu, \qquad \text{with} \quad \tilde{W}_\mu = W_\mu - \frac{1}{4}\overline{Q}_{a\dot\alpha}\bar\sigma_\mu^{\dot\alpha\alpha}Q_\alpha^a, \tag{3.146}$$

where a sum over a is implicitly assumed. With this modified definition of $\tilde{C}_{\mu\nu}$ it turns out that $W^2 = \tilde{C}_{\mu\nu}\tilde{C}^{\mu\nu}$ is again a Casimir operator.

Before we start constructing massless and massive representations of the supersymmetry algebra let us collect some basic facts valid for any representation.

- The mass of all fields in a supersymmetry multiplet is the same since $P^\mu P_\mu$ is a Casimir operator of the supersymmetry algebra.
- In the case of a gauge theory, the generators of the gauge group commute with the supercharges and therefore all fields in an irreducible supersymmetry multiplet are in the same representation of the gauge group.
- In any supersymmetry multiplet, the number of bosonic degrees of freedom, n_B, equals the number of fermionic degrees of freedom, n_F, i.e. $n_B = n_F$.

In order to prove the last statement we introduce the fermion number operator $(-)^F$, defined by

$$(-)^F |B\rangle = |B\rangle, \qquad (-)^F |F\rangle = -|F\rangle, \tag{3.147}$$

where $|B\rangle$ represents a bosonic state while $|F\rangle$ is fermionic. Since the supersymmetry charges Q_α and $\overline{Q}_{\dot\alpha}$ turn a bosonic state into a fermion state (and vice versa), Q_α and $\overline{Q}_{\dot\alpha}$ have to anticommute with $(-)^F$. Since $n_B - n_F = \text{Tr}\left((-)^F\right)$ where the trace is taken over the states of an irreducible supermultiplet, we have to show $\text{Tr}\left((-)^F\right) = 0$. We guide the reader through the proof in the following exercise.

Exercise 3.3.2 Show that $\text{Tr}\left((-)^F\{Q_\alpha, \overline{Q}_{\dot\alpha}\}\right) = 0$ by using the cyclicity property of the trace. Calculate $\text{Tr}\left((-)^F\{Q_\alpha, \overline{Q}_{\dot\alpha}\}\right)$ directly using (3.139) and conclude that $\text{Tr}\left((-)^F\right) = 0$.

We now turn to representations of the supersymmetry algebra. In analogy to the representation of the Poincaré algebra, these representations give rise to supersymmetry multiplets of fields.

Massless representation

Since the Poincaré algebra is part of the supersymmetry algebra, massless states $|p^\mu, \lambda\rangle$ are labelled by the momentum p^μ, which are the eigenvalues p^μ of the momentum operator P^μ and the helicity λ. Since the states are massless, we may choose a lightlike frame such as $p^\mu = (E, 0, 0, E)$. The helicity λ is the eigenvalue of the generator J_{12} of the little group of p^μ.

For the choice of p^μ, the Casimir operators $P^\mu P_\mu$ and $\tilde{C}_{\mu\nu}\tilde{C}^{\mu\nu}$ are zero and the anticommutation relation evaluated for p^μ reads

$$\{Q^a_\alpha, \overline{Q}_{b\dot\beta}\} = 2\,\delta^a_b\,\sigma^\mu_{\alpha\dot\beta}\,P_\mu = 2\,\delta^a_b\,E\,(-\sigma^0 + \sigma^3)_{\alpha\dot\beta} = 4\,\delta^a_b\,E\begin{pmatrix} 1 & 0 \\ 0 & 0 \end{pmatrix}_{\alpha\dot\beta}. \qquad (3.148)$$

Let us now act with Q^a_2 on a state of a massless particle $|p^\mu, \lambda\rangle$. Since $\{Q^a_2, \overline{Q}^b_2\} = 0$ we have

$$\langle p^\mu, \lambda| \overline{Q}^a_2 Q^b_2 |\tilde{p}^\mu, \tilde{\lambda}\rangle = 0, \qquad (3.149)$$

which implies that Q^a_2 has to be realised trivially, i.e. $Q^a_2|p^\mu, \lambda\rangle$ is zero for any $a = 1, \ldots, \mathcal{N}$. The components Q_1 of the supercharge satisfy $\{Q_1, \overline{Q}_{\dot 1}\} = 4E$, so defining creation and annihilation operators a and $a^\dagger$ via

$$a^b = \frac{Q^b_1}{2\sqrt{E}} \qquad a^\dagger_b = \frac{\overline{Q}_{b\dot 1}}{2\sqrt{E}}, \qquad (3.150)$$

we obtain anticommutation relations

$$\{a^b, a^\dagger_c\} = \delta^b_c, \qquad \{a^b, a^c\} = \{a^\dagger_b, a^\dagger_c\} = 0. \qquad (3.151)$$

Using the commutation relation,

$$[Q^a_\alpha, J_{12}] = (\sigma_{12})_\alpha{}^\beta Q^a_\beta = \frac{1}{2}(\sigma_3)_\alpha{}^\beta Q^a_\beta, \qquad (3.152)$$

where we have used (3.26). Specialising (3.152) to $\alpha = 1$ and acting with states $|p^\mu, \lambda\rangle$ on it, we conclude that $Q^a_1|p^\mu, \lambda\rangle$ has helicity $\lambda - \frac{1}{2}$. Hence, Q^b_1 and thus a^b lower the helicity by $\frac{1}{2}$. By similar reasoning, we find that the helicity of $\overline{Q}_{b\dot 1}|p^\mu, \lambda\rangle$ is $\lambda + \frac{1}{2}$ and thus $\overline{Q}_{b\dot 1}$, or equivalently $a^\dagger_b$, raise the helicity by $\frac{1}{2}$.

To construct supersymmetry multiplets, we start with a vacuum state of minimum helicity λ which is referred to as $|\Omega\rangle$. Let us first focus on simple supersymmetry, i.e. $\mathcal{N} = 1$. By definition, Q_1 annihilates $|\Omega\rangle$, i.e. $Q_1|\Omega\rangle = 0$ since otherwise $|\Omega\rangle$ would not have lowest helicity. By acting with $Q_{\dot 1}$ we can raise the helicity by $\frac{1}{2}$. However, note that $\overline{Q}^2_{\dot 1} = 0$ and hence the complete multiplet consists of the two particle states

$$|\Omega\rangle = |p^\mu, \lambda\rangle, \qquad a^\dagger |\Omega\rangle = |p^\mu, \lambda + \tfrac{1}{2}\rangle. \qquad (3.153)$$

If we want to realise such a multiplet in a relativistic quantum field theory, we have to add the CPT conjugated states which have opposite chirality to ensure CPT invariance,

$$|p^\mu, \pm\lambda\rangle, \qquad |p^\mu, \pm(\lambda + \tfrac{1}{2})\rangle. \qquad (3.154)$$

Examples multiplets obtained in this way are the $\mathcal{N} = 1$ *chiral* multiplet with $\lambda = 0$, and the $\mathcal{N} = 1$ *vector* multiplet with $\lambda = \frac{1}{2}$. The vector multiplets are referred to as *gauge* multiplets if they take values in a gauge algebra.

Let us generalise these results to massless representations of extended supersymmetry algebras. The only difference is that in this case, up to $\mathcal{N}$ different creation operators $\overline{Q}_{a\dot 1}$ with $a = 1, \ldots, \mathcal{N}$ may be applied to the vacuum state $|\Omega\rangle$. In this way we can construct $2^\mathcal{N}$ different states. The massless supersymmetry multiplets with $|\lambda| \leq 1$ are listed in table 3.4.

Table 3.4	Massless supersymmetry multiplets with $|\lambda| \leq 1$					
Helicity ≤ 1	$\mathcal{N} = 1$ gauge $+$ CPT	$\mathcal{N} = 1$ chiral $+$ CPT	$\mathcal{N} = 2$ gauge $+$ CPT	$\mathcal{N} = 2$ hyper	$\mathcal{N} = 3$ gauge $+$ CPT	$\mathcal{N} = 4$ gauge
1	1	0	1	0	1	1
1/2	1	1	2	2	3 + 1	4
0	0	1 + 1	1 + 1	4	3 + 3	6
−1/2	1	1	2	2	1 + 3	4
−1	1	0	1	0	1	1
Bosonic + fermionic degrees of freedom	2 + 2	2 + 2	4 + 4	4 + 4	8 + 8	8 + 8

Massive representation with vanishing central charge

In the case of massive representations, we may consider the rest-frame of the particle in which the momentum p^μ reads $p^\mu = (m, 0, 0, 0)$. Here m is the mass of the particle. Therefore acting with particle states $|p^\mu, s, s_3\rangle$ (as considered in section 3.1.4) on the supersymmetry algebra (3.142), we obtain

$$\{Q_\alpha^a, \overline{Q}_{b\dot\beta}\} = 2m \, \delta_b^a \, (\sigma_0)_{\alpha\dot\beta} = 2m \, \delta_b^a \begin{pmatrix} 1 & 0 \\ 0 & 1 \end{pmatrix}_{\alpha\dot\beta}, \tag{3.155}$$

where we have set $Z = 0$ since we first restrict our attention to representations with vanishing central charge. In contrast to massless representations we cannot conclude that $Q_2^a |p^\mu, s, s_3\rangle = 0$ for all $a = 1, \ldots, \mathcal{N}$. Therefore we expect to have more states in massive representations than in the massless representations. In particular, we may define the creation and annihilation operators

$$a_\alpha^b = \frac{Q_\alpha^b}{\sqrt{2m}}, \qquad \left(a^\dagger\right)_{\dot\alpha}^a = \frac{\overline{Q}_{\dot\alpha}^a}{\sqrt{2m}}. \tag{3.156}$$

Again, a lowers the spin, while $a^\dagger$ raises the spin. Since the vacuum by definition has lowest spin, it is annihilated by a_α^b, i.e. $a_\alpha^a |\Omega\rangle = 0$. Acting with the creation operators $\left(a^\dagger\right)_{\dot\alpha}^a$ and products thereof, we can construct states with higher spin. Since $b = 1, \ldots, \mathcal{N}$ and $\dot\alpha \in \{\dot1, \dot2\}$, we have in total $2\mathcal{N}$ creation operators giving rise to $2^{2\mathcal{N}}$ states for a generic massive representation, as opposed to $2^{\mathcal{N}}$ states for a massless representation.

Massive representation with non-vanishing central charge

In the case of a non-vanishing central charge Z, it turns out that the massive representation is shortened, i.e. some entries vanish. Since, by definition, the central charges commute with all generators, we may choose a basis in which the central charges are diagonal with eigenvalues q^i. These eigenvalues may be arranged in an antisymmetric $\mathcal{N} \times \mathcal{N}$ matrix. For $\mathcal{N} = 2$, we define the components of the antisymmetric Z^{ab} to be

$$Z^{ab} = \begin{pmatrix} 0 & q_1 \\ -q_1 & 0 \end{pmatrix}. \tag{3.157}$$

More generally, if $\mathcal{N} > 2$ with $\mathcal{N}$ even, we have

$$Z^{ab} = \begin{pmatrix} 0 & q_1 & 0 & 0 & 0 & \cdots \\ -q_1 & 0 & 0 & 0 & 0 & \cdots \\ 0 & 0 & 0 & q_2 & 0 & \cdots \\ 0 & 0 & -q_2 & 0 & 0 & \cdots \\ 0 & 0 & 0 & 0 & \ddots \\ \vdots & \vdots & \vdots & \vdots & & \ddots \\ & & & & & 0 & q_{\frac{\mathcal{N}}{2}} \\ & & & & & -q_{\frac{\mathcal{N}}{2}} & 0 \end{pmatrix}, \tag{3.158}$$

with *central charges* q_a. Note that for $\mathcal{N}$ odd, the last row and the last column of this matrix consist of the entry 0. For now let us restrict ourselves to $\mathcal{N}$ even. Using the linear combinations $\tilde{Q}^j_{\alpha\pm} \equiv (Q^{2j-1}_\alpha \pm (Q^{2j}_\alpha)^\dagger)$ for $j = 1, \ldots, \mathcal{N}/2$ the only non-zero anticommutators in the supersymmetry algebra read

$$\begin{aligned} \{\tilde{Q}^i_{\alpha+}, (\tilde{Q}^j_{\beta+})^\dagger\} &= \delta^i_j \delta^\beta_\alpha (2m + q_j) \\ \{\tilde{Q}^i_{\alpha-}, (\tilde{Q}^j_{\beta-})^\dagger\} &= \delta^i_j \delta^\beta_\alpha (2m - q_j). \end{aligned} \tag{3.159}$$

For unitary particle representations, both sides of this relation must be positive and hence $|q_j| \leq 2m$ for $j = 1, \ldots, \mathcal{N}/2$. A special case arises when the right-hand side vanishes, i.e. for $|q_j| = 2m$. This is the *Bogomolnyi–Prasad–Sommerfield bound* or *BPS bound*. If k of the q_i are equal to $\pm 2m$, there are $2\mathcal{N} - 2k$ creation operators and $2^{2(\mathcal{N}-k)}$ states. These states are referred to as $1/2^k$ *BPS multiplets*, i.e. there are *one-half BPS multiplets*, *one-quarter BPS multiplets*, and so on. The possible BPS multiplets are summarised in table 3.5. BPS states play an important role in physics. These states and the corresponding bounds were first found in *soliton* (monopole) solutions of Yang–Mills systems, which are localised finite energy solutions of the classical equations of motion. In this case the bound is an inequality between the energy of the monopole solution and the corresponding charge. Since the equality is satisfied for BPS states, these states corrrespond to the lightest charged particles, and these particles are stable. In chapter 2, we saw that BPS states also appear when discussing charged black holes. The BPS states correspond to extremal

Table 3.5 BPS multiplets

$k = 0$	$2^{2\mathcal{N}}$ states	long multiplet
$0 < k < \frac{\mathcal{N}}{2}$	$2^{2(\mathcal{N}-k)}$ states	short multiplet
$k = \frac{\mathcal{N}}{2}$	$2^{\mathcal{N}}$ states	ultrashort multiplet

black holes. These black hole solutions are BPS states for extended supergravity theories. Moreover, BPS states are important in understanding strong–weak coupling dualities in supersymmetric field theory. Note that by definition, BPS states are short multiplets. The size of the multiplet is not expected to change if we dial the coupling constant from weak to strong coupling, assuming there is no phase transition at a finite coupling constant. Finally, BPS states will also play a crucial role in string theory. Some of the extended objects known as *D-branes* are BPS states.

3.3.2 A first supersymmetric field theory: toy model

So far we have extended the Poincaré algebra to the supersymmetric algebra. Of course, then the question arises whether we can find a field theory which is invariant under the supersymmetry transformations and whose associated Noether charges give rise to the supersymmetry algebra. In this section we discuss the simplest possible supersymmetric field theory in four dimensions based on the $\mathcal{N} = 1$ supersymmetric chiral multiplet which has a left-handed Weyl fermion with components ψ_α and complex scalar fields ϕ.

For simplicity let us first consider only massless, non-interacting fields. Then the Lagrangian reads [2]

$$\mathcal{L} = -\partial_\mu \phi^* \partial^\mu \phi - i\bar{\psi}\bar{\sigma}^\mu \partial_\mu \psi. \tag{3.160}$$

The Lagrangian (3.160) is invariant under the supersymmetry transformations

$$\delta_\epsilon \phi = \sqrt{2}\epsilon\psi, \qquad \delta_\epsilon \psi_\alpha = \sqrt{2}i \left(\sigma^\mu \bar{\epsilon}\right)_\alpha \partial_\mu \phi, \tag{3.161}$$

where ϵ (with components ϵ_α) is an infinitesimal, anticommuting, two-component Weyl fermion parameterising the supersymmetry transformation. In particular ϵ does not depend on the spacetime variables x^μ and thus (3.161) is a global symmetry of the Lagrangian (3.160).

Exercise 3.3.3 Show that the Lagrangian (3.160) is invariant under the supersymmetry transformation (3.161). By allowing for a spacetime dependent $\epsilon(x)$ in (3.161) construct the associated supercurrent.

Exercise 3.3.4 Construct the Weyl Spinor Noether charges $\mathcal{Q}$ associated to the supersymmetry and show by using the equal time (anti-)commutation relations (1.39) that

$$i\delta_\epsilon \phi = \left[\epsilon\mathcal{Q} + \bar{\epsilon}\overline{\mathcal{Q}}, \phi\right], \qquad i\delta_\epsilon \psi = \left[\epsilon\mathcal{Q} + \bar{\epsilon}\overline{\mathcal{Q}}, \psi\right]. \tag{3.162}$$

Let us now derive the supersymmetry algebra. For this we consider two infinitesimal supersymmetry transformations ϵ and η of the form (3.161). It may be shown that for the scalar field ϕ we have

$$\left[\delta_\epsilon, \delta_\eta\right]\phi = 2i(\eta\sigma^\mu\bar{\epsilon} - \epsilon\sigma^\mu\bar{\eta})\partial_\mu\phi, \tag{3.163}$$

while for the Weyl fermion ψ we obtain

$$\left[\delta_\epsilon, \delta_\eta\right]\psi = 2i(\eta\sigma^\mu\bar{\epsilon} - \epsilon\sigma^\mu\bar{\eta})\partial_\mu\psi - 2i(\bar{\epsilon}\bar{\sigma}^\mu\partial_\mu\psi)\eta + 2i(\bar{\eta}\bar{\sigma}^\mu\partial_\mu\psi)\epsilon. \tag{3.164}$$

---

[2] Here, ψ is a Weyl fermion and we use the shorthand notation $\bar{\psi} = \psi^\dagger$.

Exercise 3.3.5 Derive the relations (3.163) and (3.164).

Since the equations of motion imply $\bar{\sigma}^\mu \partial_\mu \psi = 0$, we conclude that

$$\left[\delta_\epsilon, \delta_\eta\right] = 2i(\eta \sigma^\mu \bar{\epsilon} - \epsilon \sigma^\mu \bar{\eta})\partial_\mu, \tag{3.165}$$

i.e. the commutator applied to any field is just a translation by the vector $\eta \sigma^\mu \bar{\epsilon}$. Note that in order to derive that result we had to use the equations of motion. Therefore the algebra closes only on-shell.

So far we have considered a non-interacting massless theory of fermions and bosons. The next step is to include interactions. In the presence of interactions, the equations of motion will be non-linear. Since, as we saw, part of the equations of motion arises from the commutator of two supersymmetry transformations, the supersymmetry transformations will generically be non-linear for an interacting theory. However, they may be kept linear in the fields by introducing an additional non-dynamical complex scalar field F. This field may be viewed as an auxiliary Lagrange multiplier which can be *integrated out*, i.e. it may be eliminated using its non-dynamical equation of motion. Therefore, even when introducing the auxiliary fields, the theory still has two bosonic and two fermionic degrees of freedom on-shell.

The free part of the Lagrangian,

$$\mathcal{L}_{\text{kin}} = -\partial_\mu \phi^* \partial^\mu \phi - i\bar{\psi}\bar{\sigma}^\mu \partial_\mu \psi + F^* F, \tag{3.166}$$

may be supplemented by a term

$$\mathcal{L}_{\text{mass}} = m\left(-\frac{1}{2}\psi\psi + \bar{\psi}\bar{\psi} + F\phi + F^*\phi^*\right) \tag{3.167}$$

generating a mass m for the fermion as well as for the scalar field ϕ if we integrate out F. Moreover, we may also allow for interactions by considering the term

$$\mathcal{L}_{\text{int}} = g\left(\phi^2 F + \phi^{*2} F^* - \psi\psi\phi - \bar{\psi}\bar{\psi}\phi\right). \tag{3.168}$$

The total Lagrangian $\mathcal{L} = \mathcal{L}_{\text{kin}} + \mathcal{L}_{\text{mass}} + \mathcal{L}_{\text{int}}$ is invariant under the supersymmetry transformations

$$\delta_\epsilon \phi = \sqrt{2}\epsilon\psi, \qquad \delta_\epsilon \psi_\alpha = +\sqrt{2}\epsilon_\alpha F + \sqrt{2}i(\sigma^\mu \bar{\epsilon}_\alpha)\partial_\mu \phi, \qquad \delta_\epsilon F = \sqrt{2}i\bar{\epsilon}\bar{\sigma}^\mu \partial_\mu \psi. \tag{3.169}$$

In the Lagrangians given by equations (3.166), (3.167) and (3.168), the auxiliary field F appears without its derivatives. F is thus non-dynamical and may be eliminated using its equation of motion, which is referred to as integrating out F. Using F as given above, the equations of motion are

$$F^* = -m\phi - g\phi^2 \tag{3.170}$$

and therefore the *on-shell* version of the Lagrangian reads

$$\mathcal{L}_{\text{on-shell}} = -\partial^\mu \phi^* \partial_\mu \phi - i\bar{\psi}\bar{\sigma}^\mu \partial_\mu \psi - \frac{1}{2}m\psi\psi - \frac{1}{2}m\bar{\psi}\bar{\psi} - g\phi\psi\psi - g^*\phi^*\bar{\psi}\bar{\psi}. \tag{3.171}$$

Note that in particular the supersymmetry transformations of the on-shell theory are non-linear in the fields,

$$\delta_\epsilon \phi = \sqrt{2}\epsilon\psi, \qquad \delta_\epsilon\psi = \sqrt{2}i(\sigma^\mu\bar\epsilon)\partial_\mu\phi - \sqrt{2}(m\phi^* + g\phi^{*2})\epsilon. \qquad (3.172)$$

The results presented in this section demonstrate that there is an interacting field theory involving the fields of the $\mathcal{N} = 1$ supersymmetric chiral multiplet. This is referred to as the *Wess–Zumino model*. Since we constructed a supersymmetric theory by hand, the question arises whether there is a more elegant formulation of supersymmetric theories for which it is guaranteed that the action is invariant under the supersymmetry transformations. This is achieved by the $\mathcal{N} = 1$ superspace formalism which we discuss in the next section.

3.3.3 $\mathcal{N} = 1$ superspace formalism

A convenient way to write supersymmetry multiplets and Lagrangians of supersymmetric theories is obtained by introducing fermionic coordinates in addition to the well-known bosonic ones. In particular, for $\mathcal{N} = 1$ supersymmetry in (3+1)-dimensional flat space, we add left- and right-handed Weyl spinor coordinates θ_α and $\bar\theta_{\dot\alpha}$ to our ordinary Minkowski space $\mathbb{R}^4$. In this way we obtain the *superspace* $\mathbb{R}^{4|4}$. The coordinates of this space are denoted by z^A, with

$$z^A = \left(x^\mu, \theta_\alpha, \bar\theta_{\dot\alpha}\right). \qquad (3.173)$$

In order to obtain the group corresponding to the supersymmetry algebra, we have to exponentiate the generators of the algebra. In particular, we may define the operator

$$G(x, \theta, \bar\theta) = e^{-ix_\mu P^\mu + i\theta Q + i\bar\theta\bar Q}, \qquad (3.174)$$

with the scalar products given by $\theta^\alpha Q_\alpha = \theta^\alpha \epsilon_{\alpha\beta} Q^\beta$, $\bar\theta_{\dot\alpha}\bar Q^{\dot\alpha} = \bar\theta_{\dot\alpha}\epsilon^{\dot\alpha\dot\beta}\bar Q_{\dot\beta}$. Using the Baker–Campbell–Hausdorff formula as well as the (anti-)commutation relations between P_μ, Q and $\bar Q$, we obtain the product of two such operators,

$$G(0, \xi, \bar\xi)G(x, \theta, \bar\theta) = G(x^\mu + i\theta\sigma^\mu\bar\xi - i\xi\sigma^\mu\bar\theta, \theta + \xi, \bar\theta + \bar\xi). \qquad (3.175)$$

Writing the indices explicitly, we have, for instance, $\theta\sigma^\mu\bar\xi = \theta^\alpha\sigma^\mu_{\alpha\dot\alpha}\bar\xi^{\dot\alpha}$.

Using equation (3.175) we find the action of $g(\xi, \bar\xi) = G(0, \xi, \bar\xi)$ acting on superspace coordinates $(x^\mu, \theta, \bar\theta)$,

$$g(\xi, \bar\xi): \quad (x^\mu, \theta, \bar\theta) \mapsto (x^\mu + i\theta\sigma^\mu\bar\xi - i\xi\sigma^\mu\bar\theta, \theta + \xi, \bar\theta + \bar\xi), \qquad (3.176)$$

which may be represented by the differential operator $\xi Q + \bar\xi\bar Q = \xi^\alpha Q_\alpha + \bar\xi_{\dot\alpha}\bar Q^{\dot\alpha}$ with

$$Q_\alpha = \frac{\partial}{\partial\theta^\alpha} - i\sigma^\mu_{\alpha\dot\alpha}\bar\theta^{\dot\alpha}\partial_\mu, \qquad (3.177)$$

$$\bar Q^{\dot\alpha} = \frac{\partial}{\partial\bar\theta_{\dot\alpha}} - i\theta^\alpha\sigma^\mu_{\alpha\dot\beta}\epsilon^{\dot\beta\dot\alpha}\partial_\mu. \qquad (3.178)$$

If we consider the multiplication of $G(x, \theta, \bar\theta)$ by $G(0, \xi, \bar\xi)$ from the right instead of left multiplication as in (3.175), then the differential operator $\xi D + \bar\xi\bar D = \xi^\alpha D_\alpha + \bar\xi_{\dot\alpha}\bar D^{\dot\alpha}$ has to be used, with

$$\mathcal{D}_\alpha = \frac{\partial}{\partial \theta^\alpha} + i\sigma^\mu_{\alpha\dot\alpha}\bar\theta^{\dot\alpha}\partial_\mu, \tag{3.179}$$

$$\bar{\mathcal{D}}_{\dot\alpha} = -\frac{\partial}{\partial \bar\theta^{\dot\alpha}} - i\theta^\alpha\sigma^\mu_{\alpha\dot\alpha}\partial_\mu. \tag{3.180}$$

Exercise 3.3.6   Show that (3.177) acting on superspace coordinates $z = (x, \theta, \bar\theta)$ gives rise to the superspace transformation (3.176).

Exercise 3.3.7   Show that

$$\begin{array}{ll}
\{\mathcal{Q}_\alpha, \bar{\mathcal{Q}}_{\dot\alpha}\} = 2i\sigma^\mu_{\alpha\dot\alpha}\partial_\mu, & \{\mathcal{Q}_\alpha, \mathcal{Q}_\beta\} = \{\bar{\mathcal{Q}}_{\dot\alpha}, \bar{\mathcal{Q}}_{\dot\beta}\} = 0, \\
\{\mathcal{D}_\alpha, \bar{\mathcal{D}}_{\dot\alpha}\} = -2i\sigma^\mu_{\alpha\dot\alpha}\partial_\mu, & \{\mathcal{D}_\alpha, \mathcal{D}_\beta\} = \{\bar{\mathcal{D}}_{\dot\alpha}, \bar{\mathcal{D}}_{\dot\beta}\} = 0, \\
\{\mathcal{D}_\alpha, \mathcal{Q}_\beta\} = \{\mathcal{D}_\alpha, \bar{\mathcal{Q}}_{\dot\beta}\} = 0, & \{\bar{\mathcal{D}}_{\dot\alpha}, \mathcal{Q}_\beta\} = \{\bar{\mathcal{D}}_{\dot\alpha}, \bar{\mathcal{Q}}_{\dot\beta}\} = 0.
\end{array} \tag{3.181}$$

General superfields

Let us consider a general superfield $\mathbf{F}(x, \theta, \bar\theta)$ which maps a point $(x, \theta, \bar\theta)$ of the superspace to $\mathbf{F}(x, \theta, \bar\theta)$. Note that $\mathbf{F}$ does not have to be a scalar in superspace but can also carry vector or spinor indices. The superfield $\mathbf{F}(x, \theta, \bar\theta)$ can be expanded in powers of θ and $\bar\theta$. The coefficients of that expansion are the fields of the corresponding supersymmetry multiplet. Due to the anticommutativity of θ and $\bar\theta$, the expansion of $F(x, \theta, \bar\theta)$ truncates at order $\theta^2\bar\theta^2$,

$$\begin{aligned}
\mathbf{F}(x, \theta, \bar\theta) = {}& f^{(1)}(x) + \theta f^{(2)}(x) + \bar\theta\bar f^{(3)}(x) + \theta^2 f^{(4)}(x) + \bar\theta^2 f^{(5)}(x) \\
& + \theta\sigma^\mu\bar\theta f^{(6)}_\mu + \theta^2\bar\theta\bar f^{(7)} + \bar\theta^2\theta f^{(8)} + \theta^2\bar\theta^2 f^{(9)}(x),
\end{aligned} \tag{3.182}$$

where $f^{(1)}(x)$, $f^{(4)}(x)$, $f^{(5)}(x)$, $f^{(9)}(x)$ are scalars, $f^{(2)}(x)$, $f^{(8)}(x)$ and $f^{(3)}(x)$, $f^{(7)}(x)$ are left- and right-handed Weyl spinors and $f^{(6)}(x)$ is a vector field. To pick a certain component field of $\mathbf{F}$, we write $\mathbf{F}_{|\dots}$, i.e. in order to pick $\bar f^{(3)}$ we write $\mathbf{F}_{|\bar\theta} = f^{(3)}$ and for $f^{(9)}$ we write $\mathbf{F}_{|\theta^2\bar\theta^2} = f^{(9)}$. The component field $f^{(9)}(x)$ multiplied by $\theta^2\bar\theta^2$ is usually referred to as a *D-term*, while the component fields $f^{(4)}(x)$ and $f^{(5)}(x)$ in front of θ^2 and $\bar\theta^2$ are *F-terms*.

The supersymmetry transformation δ_ϵ of the superfield $\mathbf{F}$ is defined by acting with supersymmetry transformations on the individual component fields

$$\begin{aligned}
\delta_\epsilon\mathbf{F}(x, \theta, \bar\theta) = {}& \delta_\epsilon f^{(1)}(x) + \theta\delta_\epsilon f^{(2)}(x) + \bar\theta\delta_\epsilon\bar f^{(3)}(x) + \theta^2\delta_\epsilon f^{(4)}(x) + \bar\theta^2\delta_\epsilon f^{(5)}(x) \\
& + \theta\sigma^\mu\bar\theta\delta_\epsilon f^{(6)}_\mu + \theta^2\bar\theta\delta_\epsilon\bar f^{(7)} + \bar\theta^2\theta\delta_\epsilon f^{(8)} + \theta^2\bar\theta^2\delta_\epsilon f^{(9)}(x),
\end{aligned} \tag{3.183}$$

and may be realised using the operators $\mathcal{Q}$ and $\bar{\mathcal{Q}}$ as defined in equations (3.177) and (3.177)

$$\delta_\epsilon\mathbf{F}(x, \theta, \bar\theta) = \left(\epsilon\mathcal{Q} + \bar\epsilon\bar{\mathcal{Q}}\right)\mathbf{F}(x, \theta, \bar\theta). \tag{3.184}$$

By expanding the superfield $\mathbf{F}$ we find the supersymmetry transformation law for the component fields. Note that the component fields of the general superfield $\mathbf{F}(x, \theta, \bar\theta)$ do not fit into an $\mathcal{N} = 1$ supersymmetric multiplet, since there are too many degrees of freedom. In other words, the component fields of the superfield $\mathbf{F}$ do not transform under an irreducible representation of the supersymmetry algebra. By imposing conditions of the general superfield $\mathbf{F}$ we can find the superfield analogue of the $\mathcal{N} = 1$ chiral multiplet and of the $\mathcal{N} = 1$ vector multiplet. Let us first study the chiral multiplet.

Chiral superfield

The chiral superfield denoted by $\Phi(x, \theta, \bar{\theta})$ is determined by the constraint

$$\bar{\mathcal{D}}_{\dot{\alpha}} \Phi(x, \theta, \bar{\theta}) = 0. \tag{3.185}$$

In order to find the component fields, it is convenient to introduce new superspace coordinates y_-^μ and y_+^μ, which are related to x^μ by

$$y_\pm^\mu = x^\mu \pm i\theta\sigma^\mu\bar{\theta}. \tag{3.186}$$

The coordinate y_+ satisfies

$$\bar{\mathcal{D}}_{\dot{\alpha}} y_+^\mu = 0. \tag{3.187}$$

Moreover, since in addition $\bar{\mathcal{D}}_{\dot{\alpha}}\theta = 0$, the superfield $\Phi(x, \theta, \bar{\theta})$ satisfying (3.185) may be written as an arbitrary function of y_+ and θ,

$$
\begin{aligned}
\Phi(x, \theta, \bar{\theta}) &= \phi(y_+) + \sqrt{2}\theta\psi(y_+) + \theta^2 F(y_+) \\
&= \phi(x) + i\theta\sigma^\mu\bar{\theta}\partial_\mu\phi(x) + \frac{1}{4}\theta^2\bar{\theta}^2\partial_\rho\partial^\rho\phi(x) \\
&\quad + \sqrt{2}\theta\psi(x) - \frac{i}{\sqrt{2}}\theta^2\partial_\mu\psi(x)\sigma^\mu\bar{\theta} + \theta^2 F(x).
\end{aligned}
\tag{3.188}
$$

Here, $\phi(x)$ is a complex scalar field, while ψ is a left-handed Weyl spinor. Moreover, F is an auxiliary complex scalar. In the last two lines of (3.188) we have rewritten the fields ϕ, ψ and F in x-coordinates by using a Taylor expansion. The use of the coordinate y_- corresponds to the *chiral representation* in which the supersymmetry derivatives take the asymmetric representation

$$\mathcal{D}_\alpha = \frac{\partial}{\partial\theta^\alpha} + 2i\sigma^\mu_{\alpha\dot{\alpha}}\bar{\theta}^{\dot{\alpha}}\frac{\partial}{\partial y_+^\mu}, \qquad \bar{\mathcal{D}}_{\dot{\alpha}} = -\frac{\partial}{\partial\bar{\theta}^{\dot{\alpha}}}. \tag{3.189}$$

Exercise 3.3.8 Show that if Φ_1 and Φ_2 are chiral superfields, then $\Phi_1 + \Phi_2$ and $\Phi_1\Phi_2$ are also chiral superfields.

Exercise 3.3.9 Use the supersymmetry transformation (3.184) of the superfield to determine the supersymmetry transformations of the component fields,

$$
\begin{aligned}
\delta_\epsilon \phi(x) &= \sqrt{2}\epsilon\psi, \\
\delta_\epsilon \psi(x) &= \sqrt{2}i(\sigma^\mu\bar{\epsilon})\partial_\mu\phi(x) + \sqrt{2}\epsilon_\alpha F(x), \\
\delta_\epsilon F(x) &= \sqrt{2}i\bar{\epsilon}\bar{\sigma}^\mu\partial_\mu\psi(x).
\end{aligned}
\tag{3.190}
$$

Do they agree with (3.169)?

We may also introduce an anti-chiral multiplet $\Phi^\dagger$ satisfying

$$\mathcal{D}_\alpha \Phi^\dagger = 0. \tag{3.191}$$

$\Phi^\dagger$ has a similar expansion in terms of y_+ and $\bar\theta$ as given in equation (3.188) for the chiral case. While this expansion is involved in the chiral representation used above, it becomes simple again using the *anti-chiral* representation obtained by conjugation.

Exercise 3.3.10   Show that the complex conjugate of $\mathcal{D}$ is $\bar{\mathcal{D}}$. Moreover, argue that if Φ is a chiral superfield, then its complex conjugate is an anti-chiral superfield. Determine the expansion of $\Phi^\dagger$.

Exercise 3.3.11   Check that the $\theta^2 \bar\theta^2$ component of the product superfield $\Phi^\dagger \Phi$ is given by

$$\left(\Phi^\dagger \Phi \right)_{\theta^2 \bar\theta^2} = \frac{1}{4} \left(-2\partial_\mu \phi^* \partial^\mu \phi + \phi^* \partial_\rho \partial^\rho \phi + \phi \partial_\rho \partial^\rho \phi^* - 2i\bar\psi \sigma^\mu \overleftrightarrow{\partial}_\mu \psi + 4F^* F \right). \tag{3.192}$$

Note that $\Phi^\dagger \Phi$ is a real superfield.

Vector superfield

Let us consider a second type of superfield, the *vector superfield*. The vector field $V(x, \theta, \bar\theta)$ is obtained from the general superfield (3.182) by imposing the covariant reality condition

$$V(x, \theta, \bar\theta) = V^\dagger(x, \theta, \bar\theta). \tag{3.193}$$

The most general superfield satisfying this reality condition is given by

$$\begin{aligned}
V(x, \theta, \bar\theta) = {}& C(x) + i\theta \chi(x) - i\bar\theta \bar\chi(x) \\
& + \frac{i}{2}\theta^2 (M(x) + iN(x)) - \frac{i}{2}\bar\theta^2 (M(x) - iN(x)) - \theta \sigma^\mu \bar\theta A_\mu(x) \\
& + i\theta \bar\theta^2 \left(\bar\lambda(x) + \frac{i}{2}\bar\sigma^\mu \partial_\mu \chi(x) \right) - i\bar\theta^2 \theta \left(\lambda(x) + \frac{i}{2}\sigma^\mu \partial_\mu \chi(x) \right) \\
& + \frac{1}{2}\theta^2 \bar\theta^2 \left(D(x) + \frac{1}{2}\partial_\rho \partial^\rho C(x) \right).
\end{aligned} \tag{3.194}$$

Here we have eight bosonic degrees of freedom (complex scalar fields $C(x)$, $N(x)$, $M(x)$ and vector field $A_\mu(x)$) as well as eight fermionic degrees of freedom ($\chi(x)$, $\bar\chi(x)$, $\lambda(x)$, $\bar\lambda(x)$). In the following we will see that we can define a gauge transformation such that a few fields can be set to zero. Let us consider a general chiral field $\Phi(x, \theta, \bar\theta)$ and its anti-chiral field $\Phi^\dagger(x, \theta, \bar\theta)$. From (3.188), their sum is given by

$$\begin{aligned}
\Phi + \Phi^\dagger = {}& \phi(x) + \phi^*(x) + \sqrt{2}\theta \psi(x) + \sqrt{2}\bar\theta \bar\psi(x) + \theta^2 F(x) + \bar\theta^2 F^*(x) \\
& + i\theta \sigma^\mu \bar\theta \partial_\mu(\phi(x) - \phi^*(x)) + \frac{i}{\sqrt{2}}\theta^2 \bar\theta \bar\sigma^\mu \partial_\mu \psi(x) + \frac{i}{\sqrt{2}}\bar\theta^2 \theta \sigma^\mu \partial_\mu \bar\psi(x) \\
& + \frac{1}{4}\theta^2 \bar\theta^2 \partial_\rho \partial^\rho(\phi(x) + \phi^*(x)).
\end{aligned} \tag{3.195}$$

There is a gauge transformation

$$V \mapsto V + \Phi + \Phi^\dagger, \tag{3.196}$$

such that $C(x) = N(x) = M(x) = \chi(x) = \bar{\chi}(x) = 0$. In this gauge, the *Wess–Zumino gauge*, or WZ gauge for short, the expansion of $V_{WZ}(x, \theta, \bar{\theta})$ reads

$$V_{WZ}(x, \theta, \bar{\theta}) = -\theta \sigma^\mu \bar{\theta} A_\mu(x) + i\theta^2 \bar{\theta} \bar{\lambda}(x) - i\bar{\theta}^2 \theta \lambda(x) + \frac{1}{2}\theta^2 \bar{\theta}^2 D(x). \tag{3.197}$$

In the Wess–Zumino gauge, only the gauge field A_μ, the gaugino λ as well as the auxiliary field D appears.

Exercise 3.3.12   Show that

$$V_{WZ}^2 = -\frac{1}{2}\theta^2 \bar{\theta}^2 A_\mu(x) A^\mu(x), \qquad V_{WZ}^n = 0 \quad \text{for } n \geq 3. \tag{3.198}$$

Exercise 3.3.13   By decomposing the gauge transformation (3.196) into components, show that it corresponds to a canonical gauge transformation.

The superfield V may be viewed as the supersymmetric generalisation of the Yang–Mills potential. The generalisation of the Yang–Mills field strength is encoded in the gauge invariant chiral (or anti-chiral, respectively) superfields W_α and $\bar{W}_{\dot{\alpha}}$,

$$W_\alpha = -\frac{1}{4}\bar{\mathcal{D}}^2 \mathcal{D}_\alpha V, \qquad \bar{W}_{\dot{\alpha}} = -\frac{1}{4}\mathcal{D}^2 \bar{\mathcal{D}}_{\dot{\alpha}} V. \tag{3.199}$$

The component expansions for the field strengths W_α and $\bar{W}_{\dot{\alpha}}$ are given using $y_+ \equiv x + i\theta\sigma\bar{\theta}$ and $y_- \equiv x - i\theta\sigma\bar{\theta}$,

$$W_\alpha = -i\lambda_\alpha(y_-) + \left[\delta_\alpha{}^\beta D(y_-) - \frac{i}{2}(\sigma^\mu \bar{\sigma}^\nu)_\alpha{}^\beta F_{\mu\nu}(y_-)\right]\theta_\beta + \theta\theta\sigma^\mu_{\alpha\dot{\alpha}}\partial_\mu \bar{\lambda}^{\dot{\alpha}}(y_-),$$

$$\bar{W}_{\dot{\alpha}} = i\bar{\lambda}_{\dot{\alpha}}(y_+) + \left[\epsilon_{\dot{\alpha}\dot{\beta}}D(y_+) + \frac{i}{2}\epsilon_{\dot{\alpha}\dot{\gamma}}(\bar{\sigma}^\mu \sigma^\nu)_{\dot{\beta}}{}^{\dot{\gamma}}F_{\mu\nu}(y_+)\right]\bar{\theta}^{\dot{\beta}} - \epsilon_{\dot{\alpha}\dot{\beta}}\bar{\theta}\bar{\theta}\bar{\sigma}^{\mu\dot{\beta}\alpha}\partial_\mu \lambda_\alpha(y_+),$$

$$\tag{3.200}$$

where we have introduced the field strength tensor $F_{\mu\nu}$ associated with the gauge field A_μ,

$$F_{\mu\nu} \equiv \partial_\mu A_\nu - \partial_\nu A_\mu. \tag{3.201}$$

Exercise 3.3.14   Show that $\bar{\mathcal{D}}_{\dot{\beta}}W_\alpha = \mathcal{D}_\beta \bar{W}_{\dot{\alpha}} = 0$ and that W_α is invariant under the gauge transformation (3.196).

Exercise 3.3.15   Prove that W_α and $\bar{W}_{\dot{\alpha}}$ satisfy the constraint $\mathcal{D}^\alpha W_\alpha = \bar{\mathcal{D}}_{\dot{\alpha}}\bar{W}^{\dot{\alpha}}$.

Exercise 3.3.16   Confirm that

$$W^\alpha W_\alpha|_{\theta^2} = -2i\lambda(x)\sigma^\mu \partial_\mu \bar{\lambda}(x) - \frac{1}{2}F^{\mu\nu}F_{\mu\nu} + D^2 + \frac{i}{4}\epsilon_{\mu\nu\rho\sigma}F^{\mu\nu}F^{\rho\sigma}. \tag{3.202}$$

For non-Abelian theories, the supersymmetric field strengths take the form

$$W_\alpha = -\frac{1}{4}\bar{\mathcal{D}}\bar{\mathcal{D}}(e^{-V}\mathcal{D}_\alpha e^V), \qquad \bar{W}_{\dot{\alpha}} = \frac{1}{4}\mathcal{D}\mathcal{D}(e^V \bar{\mathcal{D}}_{\dot{\alpha}}e^{-V}). \tag{3.203}$$

These transform as

$$W_\alpha \mapsto e^{-i\Lambda}W_\alpha e^{i\Lambda}, \qquad \bar{W}_{\dot{\alpha}} \mapsto e^{-i\bar{\Lambda}}\bar{W}_{\dot{\alpha}}e^{i\bar{\Lambda}} \tag{3.204}$$

under gauge transformations, with Λ chiral and $\bar{\Lambda}$ anti-chiral.

3.3.4 Action in $\mathcal{N} = 1$ superspace formalism

In this section we find a prescription to construct a Lagrangian $\mathcal{L}$ which is invariant under supersymmetry up to a total derivative. We will see that the $\mathcal{N} = 1$ superspace formalism introduced above is very convenient. Let us start with the simplest case of only chiral superfields.

An action for $\mathcal{N} = 1$ chiral superfields

We aim to find a Lagrangian $\mathcal{L}(\Phi^k)$ depending on the $\mathcal{N} = 1$ chiral superfields Φ^k as well as first derivatives thereof, such that under an $\mathcal{N} = 1$ supersymmetry transformation δ_ϵ the Lagrangian is a total derivative.

The most general Lagrangian for chiral superfields Φ^k with these properties may be written as

$$\mathcal{L} = K(\Phi^k, \Phi^{k\dagger})_{|\theta^2\bar{\theta}^2} + \left(W(\Phi^k)_{|\theta^2} + W^\dagger(\Phi^{k\dagger})_{|\bar{\theta}^2} \right), \tag{3.205}$$

where $K(\Phi^k, \Phi^{k\dagger})$ is a real function of Φ^k and $\Phi^{k\dagger}$, known as the *Kähler potential*. W is the *superpotential* which only depends on Φ^k, not on $\Phi^{k\dagger}$, and thus may be viewed as a holomorphic function. Note that $W^\dagger$ depends only on $\Phi^{k\dagger}$ and therefore is anti-holomorphic. In the Lagrangian (3.205), only the D-term of the Kähler potential, i.e. the $\theta^2\bar{\theta}^2$ component of $K(\Phi^k, \Phi^{k\dagger})$ and the F-terms, i.e. the θ^2 component of $W(\Phi^k)$ and the $\bar{\theta}^2$ component of $W^\dagger(\Phi^{k\dagger})$ enter.

Instead of restricting the Kähler potential to the term $\theta^2\bar{\theta}^2$ and the superpotential W to the term θ^2, we may introduce integrals over the Grassmann variable and make use of the identities

$$\int \mathrm{d}^2\theta \, \theta^2 = 1, \qquad \int \mathrm{d}^4\theta \, \theta^2 \, \bar{\theta}^2 \equiv \int \mathrm{d}^2\theta \mathrm{d}^2\bar{\theta} \, \theta^2 \, \bar{\theta}^2 = 1, \tag{3.206}$$

in order to write

$$K(\Phi^k, \Phi^{k\dagger})_{|\theta^2\bar{\theta}^2} = \int \mathrm{d}^4\theta \, K(\Phi^k, \Phi^{k\dagger}), \qquad W(\Phi^k)_{|\theta^2} = \int \mathrm{d}^2\theta \, W(\Phi^k). \tag{3.207}$$

Then the action $\mathcal{S} = \int \mathrm{d}^4x \, \mathcal{L}$ reads

$$\mathcal{S} = \int \mathrm{d}^4x \, \mathrm{d}^4\theta \, K(\Phi^k, \Phi^{k\dagger}) + \int \mathrm{d}^4x \left(\int \mathrm{d}^2\theta \, W + \text{h.c.} \right). \tag{3.208}$$

Exercise 3.3.17 Show by explicit calculation that for the vector superfield V and the chiral superfield Φ, we have

$$\int \mathrm{d}^2\theta \, \Phi(x, \theta, \bar{\theta}) = F(x), \qquad \int \mathrm{d}^2\theta \, \mathrm{d}^2\bar{\theta} \, V(x, \theta, \bar{\theta}) = D(x), \tag{3.209}$$

with F, D given by the component expansions (3.188) and (3.197). This confirms that the superspace integration projects out the F- and D-terms.

For just one chiral field Φ, a renormalisable theory is obtained in the following way. The most general renormalisable Kähler potential K and the most general renormalisable superpotential W for one chiral superfield Φ are given by

$$K = \Phi^\dagger \Phi, \qquad W = \frac{m}{2}\Phi^2 + \frac{g}{3}\Phi^3. \tag{3.210}$$

This choice corresponds to the *Wess–Zumino model*. In this renormalisable model, the superpotential W may in principle also contain a term linear in Φ. However, this term may be set to zero by an appropriate field redefinition.

Let us further investigate the supersymmetric action

$$S = \int \mathrm{d}^4x\, \mathrm{d}^4\theta\, \Phi^\dagger \Phi. \tag{3.211}$$

Note that this action is invariant under global $U(1)$ transformations of the form $\Phi \mapsto e^{i\alpha}\Phi$ and $\Phi^\dagger \mapsto e^{-i\alpha}\Phi^\dagger$. In the case of more than one chiral field, this global transformation might be extended to a non-Abelian transformation.

It is also possible to promote this global symmetry to a local one. For example we may consider a non-Abelian gauge transformation for a chiral superfield Φ,

$$\Phi^j \mapsto \left(e^{i\alpha^a(x)T_a}\right)^j{}_k \Phi^k. \tag{3.212}$$

Defining $\Omega(x) = \alpha^a(x)T_a$ and promoting it to a chiral superfield, we generalise (3.212) to

$$\Phi \mapsto e^{i\Omega(x)}\Phi. \tag{3.213}$$

As in the non-supersymmetric case, the kinetic term (3.211) for Φ is not invariant under (3.213) unless we introduce a vector superfield V which transforms as

$$e^V \mapsto e^{i\Omega^\dagger}e^V e^{-i\Omega} \tag{3.214}$$

under the non-Abelian gauge transformation (3.213). If we modify the kinetic term (3.211) for the chiral superfield Φ to

$$S_{\text{non-Abelian}} = \int \mathrm{d}^4x \int \mathrm{d}^4\theta\, \mathrm{Tr}\left(\Phi^\dagger e^V \Phi e^{-V}\right) \tag{3.215}$$

then indeed the action is invariant under the local non-Abelian gauge symmetry. V may also be made dynamical, as we now discuss.

Action for the $\mathcal{N} = 1$ gauge vector superfield

The supersymmetric version of the Yang–Mills action is obtained using the supersymmetric field strengths as given by (3.200) and (3.203) for the Abelian and non-Abelian cases, respectively. It reads

$$S = \frac{1}{4g_{\text{YM}}^2}\int \mathrm{d}^4x\left(\int \mathrm{d}^2\theta\, \mathrm{Tr}\left(W^\alpha W_\alpha\right) + \int \mathrm{d}^2\bar\theta\, \mathrm{Tr}\left(\bar{W}_{\dot\alpha}\bar{W}^{\dot\alpha}\right)\right). \tag{3.216}$$

Using (3.202) we can indeed verify that $\mathcal{S}$ is a superymmetric generalisation of (1.185). Introducing the complex coupling constant τ,

$$\tau = \frac{\vartheta}{2\pi} + i\frac{4\pi}{g_{\text{YM}}^2},$$

(3.217)

we may further generalise (3.216) by allowing a non-vanishing ϑ term, such that we have

$$\mathcal{S} = \frac{1}{8\pi^2} \int \mathrm{d}^4x \operatorname{Im} \operatorname{Tr} \left(\tau \int \mathrm{d}^2\theta \operatorname{Tr}(W^\alpha W_\alpha) \right).$$

(3.218)

3.3.5 Renormalisation of supersymmetric theories

The renormalisation of supersymmetric theories follows the renormalisation principles and methods presented in chapter 1. Nevertheless, supersymmetric theories have special properties under renormalisation. In particular cases, cancellations occur between bosonic and fermionic propagating degrees of freedom. This leads to *non-renormalisation theorems*. In particular, within perturbation theory the superpotential contributions to the superspace action, which are either chiral or anti-chiral, are unaffected by the renormalisation process. On the other hand, non-perturbative corrections to the superpotential are possible.

Renormalisation of the Wess–Zumino model

Let us demonstrate the implications of the non-renormalisation theorems by considering the quantisation of the Wess–Zumino model in superspace, as given by (3.210). We have similar renormalisation relations between bare and renormalised couplings and fields as those introduced in chapter 1 for ϕ^4 theory in four dimensions,

$$g_0 = gZ_g, \quad \Phi_0 = \Phi Z.$$

(3.219)

It can be shown order by order in perturbation theory that for the Wess–Zumino model,

$$Z_g Z^3 = 1.$$

(3.220)

This non-renormalisation theorem implies that

$$g_0 {\Phi_0}^3 = g\Phi^3.$$

(3.221)

Note that only the product (3.220) is not renormalised, while the coupling and the field are when taken separately. This also implies that the kinetic term in the action which involves an integral over the entire superspace is renormalised. As explained in chapter 1, the β and γ functions are obtained from the coupling and field renomalisations, respectively. It follows from (3.220) that

$$\beta(g) - 3\gamma(g)g = 0.$$

(3.222)

This relation between β and γ functions is special to supersymmetric theories and a direct consequence of the non-renormalisation theorem.

Renormalisation of supersymmetric gauge theories

Similarly, supersymmetry also imposes constraints on the gauge theory β function. An expression for the gauge β function $\beta(g)$ is well known for $\mathcal{N} = 1$ theories to all orders in perturbation theory. It is given by the *NSVZ β function*, named after Novikov, Shifman, Vainshtein and Zakharov,

$$\beta(g) = -\frac{g^3}{8\pi^2} \frac{3\,C(\mathbf{adj}(G)) - \sum_A C(\mathbf{R_A})\,(1 - 2\,\gamma_A)}{1 - g^2\,C(\mathbf{adj}(G))/(8\pi^2)}. \tag{3.223}$$

Here, γ_A denotes the anomalous dimension of the superfield Φ^A, which is in the representation $\mathbf{R_A}$. The group theory factor C is defined in (1.184).

The NSVZ beta function may be derived by using a non-perturbative instanton argument. It is a renormalisation scheme dependent result which has been verified to fourth order by explicit perturbative calculations after a suitable redefinition of the couplings.

3.3.6 Maximally supersymmetric Yang–Mills theory in $d = 4$

In four spacetime dimensions, the largest amout of supersymmetry with a particle multiplet representation of spin ≤ 1 is $\mathcal{N} = 4$, corresponding to sixteen preserved Poincaré supercharges. Theories with more supersymmetry generators will involve a spin two field and thus gravity. This may be seen as follows. Each supercharge $Q_\alpha^a, \overline{Q}_{a\dot{\alpha}}$ changes the spin of the state it acts on by $1/2$. All massless states with helicities between -1 and 1 are generated by acting with no more than $\mathcal{N}_{\max} = 4$ different supercharges. Therefore $\mathcal{N} = 4$ supersymmetric field theories in four spacetime dimensions are *maximally supersymmetric* or *maximally extended*. Since any multiplet has to include particles of spin one, all the $\mathcal{N} = 4$ supersymmetric field theories must be constructed only from the gauge multiplet discussed in section (3.3.3). Therefore all particles are massless and there will be no central charges.

The massless $\mathcal{N} = 4$ supersymmetric gauge multiplet contains a gauge field $A_\mu(x)$, four Weyl fermions $\lambda_\alpha^a(x)$ ($a \in \{1, 2, 3, 4\}$) as well as six real scalars $\phi^i(x)$ ($i \in \{1, \ldots, 6\}$). The field content of the $\mathcal{N} = 4$ supersymmetry multiplet as well as the quantum numbers of the elementary fields under the R-symmetry group $SU(4)_R$ are summarised in table 3.6.

There are two ways to obtain the Lagrangian of $\mathcal{N} = 4$ Super Yang–Mills theory and the corresponding supersymmetry transformations. On the one hand, $\mathcal{N} = 4$ supersymmetric field theory is also $\mathcal{N} = 1$ supersymmetric by default, and we may use $\mathcal{N} = 1$ superspace formalism to write down the theory. In order to obtain the full $\mathcal{N} = 4$ supersymmetry with

Table 3.6 Field content of the $\mathcal{N} = 4$ supersymmetry multiplet

	Field	Range	Representation of $SU(4)_R$
Vector	A_μ		**1** singlet
Weyl fermions	λ_α^a	$a = 1, 2, 3, 4$	**4** fundamental
Real scalars	ϕ^i	$i = 1, 2, \ldots, 6$	**6** antisymmetric

R-symmetry group $SU(4)_R$, the coupling constants and the superpotential of the $\mathcal{N} = 1$ formulation have to preserve certain constraints. Below, we explain how to obtain the Lagrangian and the supersymmetry transformations using this approach. On the other hand, $\mathcal{N} = 4$ Super Yang–Mills theory may also be obtained from the dimensional reduction of $\mathcal{N} = 1$ Super Yang–Mills theory in ten dimensions. This parent theory is the unique supersymmetric theory in ten dimensions which has spin one as its highest spin state. We also demonstrate how this reduction works in detail. This derivation will be helpful later on in understanding how the AdS/CFT correspondence arises geometrically.

$\mathcal{N} = 4$ Super Yang–Mills theory in $\mathcal{N} = 1$ superspace

The $\mathcal{N} = 1$ superspace formulation of $\mathcal{N} = 4$ Super Yang–Mills theory requires three chiral superfields Φ_i, $i = 1, 2, 3$, as well as the gauge superfield V with field strength W_α. $\mathcal{D}_\alpha, \bar{\mathcal{D}}^{\dot{\alpha}}$ are superderivatives acting on these superfields. The unique field theory action with $\mathcal{N} = 4$ supersymmetry is given by, with $Tr(T_a T_b) = \delta_{ab}$,

$$
\mathcal{S}_{\mathcal{N}=4} = \int \mathrm{d}^4x \, \mathrm{Tr} \left[\int \mathrm{d}^4\theta \, \Phi^{i\dagger} e^V \Phi^i e^{-V} + \frac{1}{8\pi} \mathrm{Im}\left(\tau \int \mathrm{d}^2\theta \, W_\alpha W^\alpha \right) \right.
$$
$$
\left. + \left(ig_{\mathrm{YM}} \frac{\sqrt{2}}{3!} \int \mathrm{d}^2\theta \, \epsilon_{ijk} \Phi^i [\Phi^j, \Phi^k] + \mathrm{h.c.} \right) \right], \tag{3.224}
$$

where τ is complex gauge coupling (3.217) and W_α is the chiral spinor field constructed from the vector field V, given by (3.204).

The precise dynamics of $\mathcal{N} = 4$ supersymmetric Yang–Mills theory is almost entirely dictated by supersymmetry and the large R-symmetry group at the level of a renormalisable Lagrangian. Besides choosing the gauge group, we have the freedom to adjust the Yang–Mills gauge coupling g_{YM} and the ϑ parameter. Note that the ϑ parameter breaks CP invariance and may be set to zero.

Writing out the superfields in component fields using the notation of table 3.6, the action (3.225) gives

$$
\mathcal{L} = \mathrm{Tr}\left(-\frac{1}{2 g_{\mathrm{YM}}^2} F_{\mu\nu} F^{\mu\nu} + \frac{\vartheta}{16\pi^2} F_{\mu\nu} \tilde{F}^{\mu\nu} - i\bar{\lambda}^a \bar{\sigma}^\mu D_\mu \lambda_a \right.
$$
$$
- \sum_i D_\mu \phi^i \, D^\mu \phi^i + g_{\mathrm{YM}} \sum_{a,b,i} C^{ab}{}_i \, \lambda_a [\phi^i, \lambda_b]
$$
$$
\left. + g_{\mathrm{YM}} \sum_{a,b,i} \bar{C}_{iab} \bar{\lambda}^a [\phi^i, \bar{\lambda}^b] + \frac{g_{\mathrm{YM}}^2}{2} \sum_{i,j} [\phi^i, \phi^j]^2 \right), \tag{3.225}
$$

where $F_{\mu\nu} = \partial_\mu A_\nu - \partial_\nu A_\mu + i[A_\mu, A_\nu]$ is the field strength tensor and D_μ is the covariant derivative acting on the adjoint fields by $D_\mu \cdot = \partial_\mu \cdot + i[A_\mu, \cdot]$. Moreover, $\tilde{F}_{\mu\nu} = \frac{1}{2}\varepsilon_{\mu\nu\lambda\rho} F^{\lambda\rho}$. The $C^{ab}{}_i$ are Clebsch–Gordan coefficients that couple two **4** representations to a **6** representation of $\mathfrak{su}(4)_R$. We may view these coefficients as six-dimensional generalisation of the four-dimentinal matrices $\sigma^\mu_{\dot{\alpha}\alpha}$. The Lagrangian (3.225) is invariant under supersymmetry transformations given by

$$\delta_\epsilon \phi^i = \left[\epsilon_a^\alpha Q_\alpha^a , \phi^i \right] = \epsilon_a^\alpha C^{iab} \lambda_{\alpha b},$$

$$\delta_\epsilon \lambda_{\beta b} = \left[\epsilon_a^\alpha Q_\alpha^a , \lambda_{\beta b} \right] = F_{\mu\nu}^+ \epsilon_{\alpha b} (\sigma^{\mu\nu})^\alpha_{\ \beta} + \left[\phi^i , \phi^j \right] \epsilon_{\beta a} (C_{ij})^a_{\ b},$$

$$\delta_\epsilon \bar{\lambda}^b_{\ \dot\beta} = \left[\epsilon_a^\alpha Q_\alpha^a , \bar{\lambda}^b_{\ \dot\beta} \right] = C_i^{ab} \epsilon_a^\alpha \bar\sigma^\mu_{\alpha\dot\beta} D_\mu \phi^i, \qquad (3.226)$$

$$\delta_\epsilon A_\mu = \left[\epsilon_a^\alpha Q_\alpha^a , A^\mu \right] = \epsilon_a^\alpha (\sigma_\mu)_\alpha^{\ \dot\beta} \bar{\lambda}^a_{\dot\beta}.$$

Note that $F_{\mu\nu}^+$ is the self-dual part $\frac{1}{2}(F_{\mu\nu} + \tilde{F}_{\mu\nu})$ of the field strength, and the constants $(C_{ij})^a_{\ b}$ are related to bilinears in Clifford Dirac matrices of $SO(6)_R$.

$\mathcal{N} = 4$ Super Yang–Mills theory from dimensional reduction

The Lagrangian and the supersymmetry transformations of $\mathcal{N} = 4$ Super Yang–Mills theory may be obtained from a dimensional reduction of $\mathcal{N} = 1$ Super Yang–Mills theory in ten dimensions. The action of $\mathcal{N} = 1$ Super Yang–Mills theory in ten dimensions is

$$S_{10D} = \int \mathrm{d}^{10}x \, \mathrm{Tr} \left(-\frac{1}{2} F_{mn} F^{mn} + \frac{i}{2} \bar\Psi \Gamma^m D_m \Psi \right) \qquad (3.227)$$

where Γ^m are Dirac matrices in ten dimensions. F_{mn} is the field strength tensor as defined in (1.181). As outlined in box 1.2, we rescale the gauge fields by a factor of the coupling constant g such that the field strength tensor is given by

$$F_{mn} = \partial_m A_n - \partial_n A_m + ig[A_m, A_n]. \qquad (3.228)$$

In (3.227), Ψ represents a Majorana–Weyl fermion which has sixteen real independent components according to table 3.1. Both F_{mn} and Ψ transform in the adjoint representation of the gauge group and thus the covariant derivative D_m of Ψ reads

$$D_m \Psi = \partial_m \Psi + ig[A_m, \Psi]. \qquad (3.229)$$

The action (3.227) is invariant under the supersymmetry transformations

$$\delta_\epsilon A_m = i\bar\epsilon \Gamma_m \Psi, \qquad (3.230)$$

$$\delta_\epsilon \Psi = \Gamma_{mn} F^{mn} \Psi, \qquad (3.231)$$

with ϵ the anticommuting parameter of the transformation.

For the dimensional reduction of this theory to four dimensions à la Kaluza–Klein on a six-dimensional torus T^6, which we now perform explicitly for this theory, split the index m into two ranges μ and i, which run from 0 to 3 and from 1 to 6, respectively. Moreover, the ten-dimensional gauge field $A = A_m dx^m$ decomposes into a four-dimensional gauge field with components A_μ and into ϕ^i which are the last six components of the gauge field:

$$A_m = \left(A_\mu(x^\nu), \phi_i(x^\nu) \right). \qquad (3.232)$$

We also assume that the fields A_μ and ϕ_i depend only on the first four coordinates x^μ with $\mu = 0, \ldots, 3$, and are independent of the remaining coordinates.

Note that ϕ^i transforms trivially under a Lorentz transformation of the four-dimensional coordinates x^μ and thus is a real scalar field in four spacetime dimensions while A_μ

transforms as a vector. Using the decomposition of A_m, it is straightforward to work out the components $F_{\mu i}$ and F_{ij},

$$F_{\mu i} = \partial_\mu \phi_i + ig[A_\mu, \phi_i] = D_\mu \phi_i, \qquad F_{ij} = ig[\phi_i, \phi_j], \tag{3.233}$$

where D_μ is the covariant derivative in four dimensions. Thus the F^2 contribution to the Lagrangian (3.227) reads

$$-\frac{1}{2} \mathrm{Tr}\left(F_{mn} F^{mn}\right) = \mathrm{Tr}\left(-\frac{1}{2} F_{\mu\nu} F^{\mu\nu} - D_\mu \phi^i D^\mu \phi^i + \frac{1}{2} g^2 [\phi_i, \phi_j][\phi^i, \phi^j]\right). \tag{3.234}$$

These terms are also present in the Lagrangian of $\mathcal{N} = 4$ super Yang–Mills theory. To derive the other contributions to (3.225), we have to reduce the kinetic term $i\bar{\Psi}\Gamma^m D_m \Psi$ for the Majorana–Weyl fermion Ψ on $\mathbb{R}^{3,1} \times T^6$. Here, Γ^m are the ten-dimensional Dirac matrices, which we decompose into Dirac matrices γ^μ of the four-dimensional spacetime as well as $\hat{\gamma}^i$ which are the gamma matrices of T^6. Choosing a convenient basis for these gamma matrices, for example in terms of the 't Hooft symbols, we indeed can dimensionally reduce the term to (3.225).

Exercise 3.3.18   Perform the dimensional reduction of the kinetic term of the fermions, $i\bar{\Psi}\Gamma^m D_m \Psi$, explicitly.

Properties of $\mathcal{N} = 4$ supersymmetric Yang–Mills theory

Let us list here several important facts about $\mathcal{N} = 4$ supersymmetric Yang–Mills theory in four spacetime dimensions.

- Since the coupling constant is dimensionless and all fields are massless, the action of $\mathcal{N} = 4$ Super Yang–Mills theory is scale invariant on the classical level. The engineering mass dimensions of the fields are $[A_\mu] = 1$, $[\lambda] = 3/2$, $[\phi^i] = 1$.
- It is quite remarkable that the theory is also scale invariant after quantisation. This is connected to the fact that $\mathcal{N} = 4$ Super Yang–Mills theory is believed to be a UV finite theory and that the β function vanishes exactly to all orders in perturbation theory. In fact, scale invariance is part of a larger symmetry, the conformal symmetry group $SO(4, 2)$. Moreover, the Lagrangian is also invariant under $\mathcal{N} = 4$ supersymmetry with R-symmetry group $SU(4)_R$. In section 3.4 we will study the consequences of combining supersymmetry and conformal symmetry into a larger symmetry, *superconformal* symmetry. Using this superconformal symmetry, we can classify the spectrum of all states of $\mathcal{N} = 4$ Super Yang–Mills theory.
- Using perturbative quantisation, it can be shown that $\mathcal{N} = 4$ Super Yang–Mills theory does not have UV divergences in the correlation functions of elementary fields. Since also the corrections of instantons are finite, the theory is believed to be UV finite.
- Furthermore, $\mathcal{N} = 4$ Super Yang–Mills theory is invariant under the *S-duality* group $SL(2, \mathbb{Z})$ acting on the complex coupling constant τ as

$$\tau \to \frac{a\tau + b}{c\tau + d}, \qquad ad - bc = 1, \qquad a, b, c, d \in \mathbb{Z}. \tag{3.235}$$

This S-duality is remarkable since it implies a strong–weak duality: the coupling constant is given by $\tau = 4\pi i/g_{YM}^2$. Let us now apply the S-duality transformation with $b = -c = 1, a = d = 0$. This transformation changes the coupling constant g_{YM} to $4\pi/g_{YM}$.

- The $\mathcal{N} = 4$ Super Yang–Mills theory has two different classes of vacua. Since the scalar potential must vanish in the supersymmetric ground state and each interaction term $[\phi^i, \phi^j]^2$ is non-negative, the scalar fields have to be constant and have to satisfy $[\phi^i, \phi^j] = 0$ for any pair of indices $i, j \in \{1, \ldots, 6\}$. This condition can be satisfied in two different ways. Either the vacuum expectation values of ϕ^i vanish, which corresponds to the *superconformal phase*, or there exists at least one scalar ϕ^i for which the vacuum expectation value is non-zero. The latter case is referred to as the *Coulomb phase*. In this phase, conformal invariance is broken since a length scale $\langle \phi^i \rangle$ is introduced. Moreover, the gauge symmetry is also broken down to a subgroup. For example, for gauge group $SU(N)$ the gauge symmetry may generically be broken to $U(1)^{N-1}$.

3.4 Superconformal symmetry

In this section we study the consequences if a supersymmetric theory is also conformal. Then the symmetry algebra is extended to the superconformal algebra. This enlarged symmetry puts stringent conditions on the spectrum of the theory. In particular, we can reveal properties of the spectrum by studying representations of the superconformal algebra.

First we will discuss $\mathfrak{su}(2, 2|\mathcal{N})$ which is the superconformal algebra for an $\mathcal{N}$ extended supersymmetric conformal theory in four spacetime dimensions. We will work out part of the representations of this algebra. Finally, we comment on the special case $\mathcal{N} = 4$ and study in detail $\mathfrak{su}(2, 2|N)$ and $\mathfrak{psu}(2, 2|N)$, which is the symmetry algebra of $\mathcal{N} = 4$ Super Yang–Mills theory.

3.4.1 Superconformal algebra

The generators of the superconformal algebra can be grouped into the generators of the conformal group, i.e. $J_{\mu\nu}, P_\mu, D$ and K_μ as well as the (Poincaré) supercharges Q_α^a and $\bar{Q}_{\dot{\alpha}}^a$. It turns out that these are not all the generators of the superconformal group. In fact, to ensure closure of the superconformal algebra, we have to introduce further fermionic supercharges which we denote by S_α^a and $\bar{S}_{\dot{\alpha}}^a$. While the Poincaré supercharges Q_α^a and $\bar{Q}_{a\dot{\alpha}}$ correspond to the fermionic superpartners of P_μ, the special conformal supercharges S_α^a and $S_{a\dot{\alpha}}$ are the fermionic superpartners of K_μ.

The (anti-)commutation relations of the extended superconformal algebra $\mathfrak{su}(2, 2|\mathcal{N})$ in four spacetime dimensions are explicitly given in appendix B.3.2. In particular, the (anti-)commutation relations involving S_α^a and $S_{a\dot{\alpha}}$ read

$$\begin{aligned}
\{Q_\alpha^a, Q_\beta^b\} &= \{S_{\alpha a}, S_{\beta b}\} = \{Q_\alpha^a, \bar{S}_{\dot\beta}^b\} = 0, \\
\{Q_\alpha^a, \overline{Q}_{\dot\beta b}\} &= 2\,(\sigma^\mu)_{\alpha\dot\beta}\, P_\mu\, \delta_b^a, \\
\{S_\alpha^a, \bar{S}_{\dot\beta b}\} &= 2\,(\sigma^\mu)_{\alpha\dot\beta}\, K_\mu\, \delta_b^a, \\
\{Q_\alpha^a, S_{\beta b}\} &= \varepsilon_{\alpha\beta}\,(\delta_b^a D + R_b^a) + \frac{1}{2}\,\delta_b^a\, J_{\mu\nu}\,(\sigma^{\mu\nu})_{\alpha\beta}.
\end{aligned} \tag{3.236}$$

3.4.2 Representations of the superconformal algebra

Here we consider non-vanishing local operators $\mathcal{O}(x)$ constructed from the elementary fields of the conformal theory. In the case of a gauge theory we consider only gauge invariant operators. These operators $\mathcal{O}$ are characterised by the conformal dimension Δ and the spin $\mathcal{J}_{\mu\nu}$,

$$[D, \mathcal{O}(0)] = -i\Delta\mathcal{O}(0), \qquad [J_{\mu\nu}, \mathcal{O}(0)] = -\mathcal{J}_{\mu\nu}\mathcal{O}(0). \tag{3.237}$$

A particularly important subset of operators are the *superconformal primary operators* $\mathcal{O}$. In a given superconformal multiplet of $\mathfrak{su}(2,2|\mathcal{N})$, these have the lowest dimension, denoted by Δ. According to (3.236), the special conformal supersymmetry generators S_α^a and $S_{a\dot\alpha}$ lower the dimension. In addition, unitarity imposes a lower bound on the dimension of the operators. This implies that the superconformal primary operators $\mathcal{O}$ have to satisfy

$$[S_\alpha^a, \mathcal{O}\} = 0, \qquad [\bar{S}_{a\dot\alpha}, \mathcal{O}\} = 0 \tag{3.238}$$

for all $a = 1, \ldots, \mathcal{N}$ and $\alpha \in \{1, 2\}$. Commutation and anticommutation brackets in (3.238) have to be chosen depending on the bosonic or fermionic nature of the operator $\mathcal{O}$. In section 3.2.2, we discussed a similar set of fields, the conformal primary operators which have the lowest dimension in a given representation of the conformal group. Since $\{S, \bar{S}\} \sim K_\mu$, superconformal primaries are conformal primaries, but not vice versa.

Starting from the superconformal primary operator, we may construct descendants of the superconformal primary operator by applying any generator of the superconformal algebra. For example, by applying P_μ to $\mathcal{O}$ we obtain a descendant $[P_\mu, \mathcal{O}(x)] = -i\partial_\mu\mathcal{O}(x)$ whose dimension Δ is increased by 1. We may also apply P_μ as well as other generators of the superconformal algebra more than once in order to obtain new descendants. The superconformal primary operator and its descendants correspond to an irreducible representation of $\mathfrak{su}(2,2|\mathcal{N})$ for $\mathcal{N} < 4$, and of $\mathfrak{psu}(2,2|4)$ for $\mathcal{N} = 4$.

A special kind of descendants of the superconformal primary operator $\mathcal{O}$ are *superdescendants* $\mathcal{O}'$ defined by

$$\mathcal{O}' = [Q, \mathcal{O}\}. \tag{3.239}$$

The dimension of the superdescendant operator is increased by $1/2$, i.e. $\Delta_{\mathcal{O}'} = \Delta_{\mathcal{O}} + \frac{1}{2}$. These superdescendant operators are important since they are conformal primary operators,

$$[K_\mu, \mathcal{O}'] = [K_\mu, [Q, \mathcal{O}\}] = 0 \tag{3.240}$$

where we have used the Jacobi identity as well as the commutator $[K_\mu, Q] \sim S$. Each of these superdescendants gives rise to a *Verma module*, i.e. essentially a conformal multiplet, and all Verma modules are linked to each other by supersymmetry transformations Q.

The superconformal primary operators are in one-to-one correspondence with irreducible representations of $\mathfrak{su}(2, 2|\mathcal{N})$. Of particular interest is a subset of superconformal primary operators, the *chiral primary operators*. In addition to (3.238) the chiral primary operators are also annihilated by at least one of the Q_α^a,

$$\{Q_\alpha^a, \mathcal{O}\} = 0 \tag{3.241}$$

for at least one $a \in \{1, \ldots, \mathcal{N}\}$ and one $\alpha \in \{1, 2\}$. According to the definition of section 3.3.1, they are thus BPS operators. The multiplet formed by chiral primary operators is smaller – though still infinite – than the multiplets formed by superconformal primary operators which are not chiral primary operators. The chiral primary operators $\mathcal{O}$ are important since their conformal dimension Δ does not receive any quantum corrections since the conformal dimension Δ is related to the spin and to the quantum numbers of the R-symmetry group. Schematically, this may be seen as follows,

$$0 = [\{S, Q\}, \mathcal{O}(0)] = [L + D + R, \mathcal{O}(0)] \sim (\Delta + R + \mathcal{J})\mathcal{O}(l) \tag{3.242}$$

where we have not specified the indices on S and Q, for simplicity. Note that the argument only works for those Q_α^a which annihilate $\mathcal{O}$. The relation (3.242) implies that Δ has to be a function of the spin, which is encoded in $\mathcal{J}$, and of the R-symmetry quantum numbers.

3.4.3 Superconformal operators in $\mathcal{N} = 4$ theory

In the case of $\mathcal{N} = 4$ Super Yang–Mills theory in four spacetime dimensions, the superconformal representations discussed above are realised in terms of gauge invariant composite operators involving the fields of the $\mathcal{N} = 4$ Super Yang–Mills theory Lagrangian. The elementary fields of this theory are the scalars ϕ^i, the fermions $\psi_\alpha^a, \bar{\psi}_{\dot{\alpha}a}$ and the gauge field A_μ. Under a gauge transformation, the scalars, fermions and the field strength tensor $F_{\mu\nu}$ as well as covariant derivatives of these fields transform covariantly. Gauge–invariant operators are obtained by taking the trace of a product of such covariant fields evaluated at the same space–time point, i.e. for instance

$$\mathcal{O}(x) = \mathrm{Tr}(\phi^i \ldots \phi^j)(x). \tag{3.243}$$

These local operators are *single-trace operators*. Let us give some examples of these operators.

Of central importance are those single-trace operators involving only scalars ϕ^i which are of the form

$$\mathcal{O}(x) = \mathrm{Str}\big(\phi^{\{i_1} \phi^{i_2} \ldots \phi^{i_k\}}\big)(x). \tag{3.244}$$

Here, Str stands for the *symmetrised trace* of the gauge algebra, which for the scalars $\phi^i = \phi^{ia} T_a$ in the adjoint representation is given by the sum over all permutations,

$$\mathrm{Str}\,(T_{a_1} \cdots T_{a_n}) = \sum_{\text{all permutations}\,\sigma} \mathrm{Tr}(T_{\sigma(a_1)} \cdots T_{\sigma(a_n)}). \tag{3.245}$$

This symmetrisation ensures that (3.244) is totally symmetric. Moreover, the curly brackets for the field indices in (3.244) denote that all traces are removed. This ensures that the resulting operators correspond to an irreducible representation of the superconformal algebra. An example is the single-trace operator of dimension $\Delta = 2$,

$$\text{Str}(\phi^{\{i}\phi^{j\}}) = \text{Tr}(\phi^i\phi^j) - \frac{1}{6}\delta^{ij}\text{Tr}(\phi^k\phi^k). \tag{3.246}$$

The operators (3.244) are chiral primary operators and one-half BPS states of the superconformal algebra. Since the dimension of the scalars Φ^i is one in four dimensions, their dimension is $\Delta = k$ with k the number of scalar fields present. Since these operators correspond to entries in a unitary superconformal multiplet, they are protected when renormalised, such that in agreement with (3.242) they do not acquire an anomalous dimension.

On the other hand, there are also unprotected single-trace operators such as the *Konishi operator*, which for $\mathcal{N} = 4$ Super Yang–Mills theory reads

$$K = \text{Tr}(\phi^i\phi^i). \tag{3.247}$$

Let us look at the $\mathfrak{su}(4)$ representations these operators belong to. Unitary representations of the superconformal algebra $\mathfrak{su}(2,2|4)$ are labelled by the quantum numbers of the maximal bosonic subalgebra of $\mathfrak{su}(2,2|4)$. This subalgebra is the direct product of the Lorentz algebra $\mathfrak{so}(1,3)$, the dilations $\mathfrak{so}(1,1)$ and the $\mathfrak{su}(4)_R$ R-symmetry algebra. The corresponding quantum numbers are spin labels $s_\pm$ for the Lorentz algebra, the scale dimension Δ for dilatations and the three Dynkin labels $[r_1, r_2, r_3]$ for $\mathfrak{su}(4)_R$. The Dynkin labels are defined in appendix B.1.1. They determine the dimension of the $\mathfrak{su}(4)$ representation, as explained in appendix B.2.1.

The chiral primary or one-half BPS operators given by (3.244), are built from a k-fold symmetric product of Φ^i. Since the scalars Φ^i transform in the $[0,1,0]$ representation of $\mathfrak{su}(4)_R$, the chiral primary operator of the from (3.244) has to be in $[0,k,0]$. According to the general result (B.34) in appendix B.2.1, the dimension of the associated representation is

$$\dim[0,k,0] = \frac{1}{12}(k+1)(k+2)^2(k+3). \tag{3.248}$$

The simplest example of a chiral primary operator is the case $k = 2$, with dimension $\dim[0,2,0] = 20$. Historically, this is referred to as the representation **20'** of $\mathfrak{su}(4)_R$. The associated operator in $\mathcal{N} = 4$ theory is the one given in (3.246).

Descendants are obtained by applying the $\mathcal{N} = 4$ supersymmetry transformations (3.236) to the fields in the chiral primary operators. To obtain the corresponding $\mathfrak{su}(4)_R$ representations, we recall that the $Q's$ transform in the representation **4** of $\mathfrak{su}(4)_R$. For instance, when applying a Q generator to the scalar field ϕ^i in the **6** representation, we obtain $\mathbf{4} \otimes \mathbf{6} \rightarrow \bar{\mathbf{4}} \oplus \mathbf{20}$. However, in agreement with the supersymmetry algebra representation (3.226), i.e. $[Q,\phi] \sim \lambda$, the coefficient of the **20** is absent due to multiplet shortening.

In addition to the 1/2 BPS operators, there are also 1/4 and 1/8 BPS operators. A summary of representations corresponding to BPS operators of $\mathfrak{su}(4)_R$ is given in table 3.7. The Konishi operator (3.247) is an example of a non-BPS operator.

Table 3.7	Superconformal BPS operators for $\mathfrak{su}(4)_R$	
Operator	$\mathfrak{su}(4)_R$	Dimension
1/2 BPS	$[0, k, 0]$, $\quad k \geq 2$	$\Delta = k$
1/4 BPS	$[l, k, l]$, $\quad l \geq 1$	$\Delta = k + 2l$
1/8 BPS	$[l, k, l + 2m]$, $\quad k \geq 2$	$\Delta = k + 2l + 3m$

The single-trace operators are the leading ones in the large N limit. In addition, there are also subleading *multi-trace operators*, which are given by products of single-trace operators. For instance, the 1/4 and 1/8 BPS operators introduced above are realised by multi-trace operators.

3.5 Further reading

For Lorentz and Poincaré symmetry in field theory, any book on quantum field theory is recommended. A particularly elegant approach may be found in [1].

Early results for conformal correlation functions in general dimensions include [2, 3]. A complete discussion of conformal two- and three-point functions, their Ward identities and their relations to anomalies in more than two dimensions may be found in [4, 5]. The seminal paper on conformal field theory in two dimensions is [6]. A review book on two-dimensional conformal field theory is [7]. Lecture notes on this subject include [8, 9].

A standard reference on supersymmetry is [10]. The Coleman–Mandula theorem was stated in [11]. The property of S-duality in Yang–Mills theory was conjectured by Montonen and Olive [12] and also in [13, 14]. This conjecture was further supported for $\mathcal{N} = 4$ Super Yang–Mills theory by Osborn [15].

The NSVZ β function was proposed in [16]. It was shown to be equivalent to a four-loop computation in the DRED renormalisation scheme in [17].

For a detailed discussion of superconformal symmetry and correlation functions see [18]. BPS operators in $\mathcal{N} = 4$ theory are reviewed in [19]. In [20], how to obtain the action of $\mathcal{N} = 4$ from dimensional reduction is reviewed.

References

[1] Weinberg, Steven. 1995. *The Quantum Theory of Fields*. Vol. 1: *Foundations*. Cambridge University Press.

[2] Mack, G., and Salam, Abdus. 1969. Finite component field representations of the conformal group. *Ann. Phys.*, **53**, 174–202.

[3] Schreier, E. J. 1971. Conformal symmetry and three-point functions. *Phys. Rev.*, **D3**, 980–988.

[4] Osborn, H., and Petkou, A. C. 1994. Implications of conformal invariance in field theories for general dimensions. *Ann. Phys.*, **231**, 311–362.

[5] Erdmenger, J., and Osborn, H. 1997. Conserved currents and the energy momentum tensor in conformally invariant theories for general dimensions. *Nucl. Phys.*, **B483**, 431–474.

[6] Belavin, A. A., Polyakov, A. M., and Zamolodchikov, A. B. 1984. Infinite conformal symmetry in two-dimensional quantum field theory. *Nucl. Phys.*, **B241**, 333–380.

[7] Di Francesco, P., Mathieu, P., and Senechal, D. 1997. *Conformal Field Theory*. Springer, New York.

[8] Ginsparg, Paul H. 1988. Applied conformal field theory. ArXiv:hep-th/9108028.

[9] Blumenhagen, Ralph, and Plauschinn, Erik. 2009. *Introduction to Conformal Field Theory*. Lecture Notes in Physics, Vol. 779, Springer.

[10] Wess, J., and Bagger, J. 1992. *Supersymmetry and Supergravity*. Princeton University Press.

[11] Coleman, Sidney R., and Mandula, J. 1967. All possible symmetries of the S-matrix. *Phys. Rev.*, **159**, 1251–1256.

[12] Montonen, C., and Olive, David I. 1977. Magnetic monopoles as gauge particles? *Phys. Lett.*, **B72**, 117.

[13] Goddard, P., Nuyts, J., and Olive, David I. 1977. Gauge theories and magnetic charge. *Nucl. Phys.*, **B125**, 1.

[14] Witten, Edward, and Olive, David I. 1978. Supersymmetry algebras that include topological charges. *Phys. Lett.*, **B78**, 97.

[15] Osborn, Hugh. 1979. Topological charges for $\mathcal{N} = 4$ supersymmetric gauge theories and monopoles of spin 1. *Phys. Lett.*, **B83**, 321.

[16] Novikov, V. A., Shifman, Mikhail A., Vainshtein, A. I., and Zakharov, Valentin I. 1983. Exact Gell-Mann-Low function of supersymmetric Yang-Mills theories from instanton calculus. *Nucl. Phys.*, **B229**, 381.

[17] Jack, I., Jones, D. R. T., and North, C. G. 1997. Scheme dependence and the NSVZ Beta function. *Nucl. Phys.*, **B486**, 479–499.

[18] Park, Jeong-Hyuck. 1999. Superconformal symmetry and correlation functions. *Nucl. Phys.*, **B559**, 455–501.

[19] D'Hoker, Eric, and Freedman, Daniel Z. 2002. Supersymmetric gauge theories and the AdS/CFT correspondence. TASI 2001 School Proceedings. ArXiv:hep-th/0201253.

[20] Park, Jeong-Hyuck, and Tsimpis, Dimitrios. 2007. Topological twisting of conformal supercharges. *Nucl. Phys.*, **B776**, 405–430.

4   Introduction to superstring theory

In this chapter we introduce and review those important developments in string theory which led to the discovery of the AdS/CFT correspondence and which are a necessary prerequisite for understanding the chapters which follow, in particular the motivation of the AdS/CFT conjecture itself. To illustrate the ideas and to keep the arguments as simple as possible, we first consider bosonic string theory and quantise open and closed strings in Minkowski spacetime. Later we generalise the ideas to superstring theory. Moreover, inherently non-perturbative objects, the *branes*, are introduced which play an important role in the AdS/CFT duality.

4.1 Bosonic string theory

4.1.1 From point particles to strings

The basic idea behind string theory is to consider one-dimensional extended strings as the fundamental objects rather than point particles. Such a string sweeps out a (1+1)-dimensional *worldsheet* in spacetime and not just a worldline as is the case for pointlike particles.

The worldsheet Σ is parametrised by two coordinates, the proper time τ and the spatial extent σ of the string. The coordinate σ takes values in the interval $[0, \sigma_0]$, where σ_0 will be chosen later in a convenient way. The embedding of the worldsheet of the fundamental string into the *target spacetime* is given by functions $X^M(\tau, \sigma)$ as shown in figure 4.1. Here we assume the target spacetime is D-dimensional Minkowski spacetime with metric η_{MN} and generalise it to curved spacetimes later on.

The physics depends only on the embedding into target spacetime and not on the parametrisation of the worldsheet. The simplest parametrisation invariant action for strings is the Nambu–Goto action,

$$\mathcal{S}_{\mathrm{NG}} = -\frac{1}{2\pi\alpha'} \int_\Sigma \mathrm{d}^2\sigma \sqrt{-\det\left(\partial_\alpha X^M \, \partial_\beta X^N \eta_{MN}\right)}, \tag{4.1}$$

where we define $\mathrm{d}^2\sigma = \mathrm{d}\sigma^0 \mathrm{d}\sigma^1$ with $(\sigma^0, \sigma^1) \equiv (\tau, \sigma)$. α' is related to the string length l_{s} by $\alpha' = l_{\mathrm{s}}^2$ and we refer to $\tau_{F1} = \frac{1}{2\pi\alpha'}$ as the tension of the fundamental string.

Due to the square-root in (4.1), it is difficult to deal with the Nambu–Goto action. For example, it is very complicated to quantise the theory specified by the action (4.1). We may

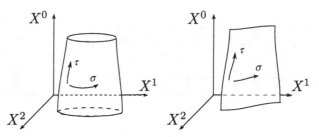

Figure 4.1 Embedding of strings into D-dimensional target spacetime (here $D = 3$). There are two types of strings with different worldsheet topologies as discussed later on: closed and open strings. For closed strings the worldsheet is a cylinder, while for open strings it is a strip.

get rid of the square root in (4.1) by introducing a worldsheet metric $h_{\alpha\beta}(\sigma)$ as an auxiliary field. The dynamics of the string is then given by the *Polyakov action*

$$S_P = -\frac{1}{4\pi\alpha'} \int_\Sigma d^2\sigma \, \sqrt{-h} \, h^{\alpha\beta} \, \partial_\alpha X^M \, \partial_\beta X^N \eta_{MN}, \tag{4.2}$$

where $h = \det(h_{\alpha\beta})$ and $h^{\alpha\beta}$ is the inverse matrix of $h_{\alpha\beta}$, i.e. $h^{\alpha\beta} h_{\beta\gamma} = \delta^\alpha_\beta$. Using the equations of motion for $h_{\alpha\beta}$, $\delta S_P/\delta h^{\alpha\beta} = 0$, we conclude that the worldsheet energy-momentum tensor $T_{\alpha\beta}$ has to vanish,

$$T_{\alpha\beta} \equiv -\frac{4\pi\alpha'}{\sqrt{-h}} \frac{\delta S_P}{\delta h^{\alpha\beta}} = \partial_\alpha X^M \partial_\beta X^N \eta_{MN} - \frac{1}{2} h_{\alpha\beta} h^{\rho\sigma} \partial_\rho X^M \partial_\sigma X^N \eta_{MN} = 0. \tag{4.3}$$

We thus may eliminate the worldsheet metric from the Polyakov action (4.2) and obtain the Nambu–Goto action (4.1). The equation $T_{\alpha\beta} = 0$ puts constraints on the dynamical fields X^M of the Polyakov action which are known as *Virasoro constraints*. Therefore both actions are equivalent at the classical level. From now on we will work with the Polyakov action (4.2) in view of quantising the theory. First let us analyse the symmetries which are preserved by the Polyakov action.

- *D-dimensional Poincaré transformations of target spacetime.* The action (4.2) is invariant under

$$X^M \mapsto X'^M = \Lambda^M_N X^N + a^M \qquad \text{and} \qquad \delta h^{\alpha\beta} = 0, \tag{4.4}$$

  where Λ^M_N and a^M are Lorentz transformations and spacetime translations of the D-dimensional target spacetime, respectively.

- *Reparametrisations of the worldsheet.* The action (4.2) is invariant under reparametrisations of the worldsheet given by $\sigma^\alpha \mapsto \sigma'^\alpha = f^\alpha(\sigma)$. In particular $X^M(\tau,\sigma)$ and $h_{\alpha\beta}(\tau,\sigma)$ transform according to

$$h_{\alpha\beta}(\tau,\sigma) = \frac{\partial f^\gamma}{\partial\sigma^\alpha} \frac{\partial f^\delta}{\partial\sigma^\beta} h_{\gamma\delta}(\tau',\sigma') \qquad \text{and} \qquad X'^M(\tau',\sigma') = X^M(\tau,\sigma). \tag{4.5}$$

- *Weyl transformations.* The action (4.2) is invariant under

$$h_{\alpha\beta}(\tau,\sigma) \mapsto e^{2\omega(\tau,\sigma)} h_{\alpha\beta}(\tau,\sigma) \qquad \text{and} \qquad X'^M(\tau,\sigma) = X^M(\tau,\sigma). \tag{4.6}$$

The local symmetries may be used to choose a convenient gauge in which the worldsheet metric is diagonal. From now on we choose the *conformal gauge*

$$h_{\alpha\beta} = e^{2\omega(\tau,\sigma)}\eta_{\alpha\beta}, \qquad \text{with} \qquad \eta = \begin{pmatrix} -1 & 0 \\ 0 & 1 \end{pmatrix}. \qquad (4.7)$$

In this gauge, the Polyakov action $\mathcal{S}_P$ reads

$$\mathcal{S}_P = \frac{1}{4\pi\alpha'}\int d^2\sigma \left(\partial_\tau X^M \partial_\tau X^N - \partial_\sigma X^M \partial_\sigma X^N\right)\eta_{MN} \qquad (4.8)$$

and the equations of motion for $X^M(\tau,\sigma)$ are given by a relativistic wave equation,

$$(\partial_\tau^2 - \partial_\sigma^2)X^M = \partial_+\partial_- X^M = 0, \qquad (4.9)$$

where we have introduced light-cone coordinates $\sigma^\pm = \tau \pm \sigma$ as well as the corresponding derivatives $\partial_\pm = \partial/\partial\sigma^\pm$. The equations of motion have to be supplemented by the the boundary condition

$$\partial_\sigma X^M \delta X_M|_0^{\sigma_0} = 0. \qquad (4.10)$$

Moreover we have to impose the Virasoro constraints (4.3) in addition, which in the conformal gauge (4.7) read

$$T_{++} = \partial_+ X^M \partial_+ X_M = 0, \quad T_{--} = \partial_- X^M \partial_- X_M = 0, \quad T_{+-} = T_{-+} = 0. \qquad (4.11)$$

4.1.2 String spectrum in Minkowski spacetime

We now study classical string solutions. It is straightforward to solve (4.9) by decomposing $X^M(\tau,\sigma)$ into *left- and right-moving modes*, $X_{(L)}^M$ and $X_{(R)}^M$ which depend only on σ^+ and σ^-, respectively, i.e.

$$X^M(\tau,\sigma) = X_{(L)}^M(\sigma^+) + X_{(R)}^M(\sigma^-). \qquad (4.12)$$

Both, the left- and right-moving modes, $X_{(L)}^M$ and $X_{(R)}^M$ can be decomposed into a Fourier series of the form

$$X_{(L)}^M(\sigma^+) = \frac{\tilde{x}_0^M}{2} + \frac{\alpha'}{2}\tilde{p}^M \sigma^+ + i\sqrt{\frac{\alpha'}{2}}\sum_{n\neq 0}\frac{\tilde{\alpha}_n^M}{n}e^{-in\sigma^+},$$

$$\qquad\qquad\qquad\qquad\qquad\qquad\qquad\qquad\qquad\qquad\qquad\qquad (4.13)$$

$$X_{(R)}^M(\sigma^-) = \frac{x_0^M}{2} + \frac{\alpha'}{2}p^M \sigma^- + i\sqrt{\frac{\alpha'}{2}}\sum_{n\neq 0}\frac{\alpha_n^M}{n}e^{-in\sigma^-}.$$

The constants x_0^M and $\tilde{x}_0^M$ can be related to the centre of mass position of the string, $(x_0^M + \tilde{x}_0^M)/2$. Moreover, p^M and $\tilde{p}^M$ are the momenta of the modes and thus the centre of mass momentum is given by $(p^M + \tilde{p}^M)/2$. Later on it is convenient to introduce α_0^M and $\tilde{\alpha}_0^M$ via $\alpha_0^M = \sqrt{\frac{\alpha'}{2}}p^M$ and similarly for $\tilde{\alpha}_0^M$. Note also that X^M has to be real and therefore $\alpha_{-n}^M = (\alpha_n^M)^*$ and $\tilde{\alpha}_{-n}^M = (\tilde{\alpha}_n^M)^*$.

There are different possibilities for satisfying the boundary conditions (4.10) which may be rephrased in terms of the topology of the worldsheet Σ. For free strings, the worldsheet has the topology either of a cylinder in the case of closed strings, or of a strip in the case of open strings.

Let us first discuss closed strings, for which we set $\sigma_0 = 2\pi$, i.e. the coordinate σ describing the spatial extension is in the interval $\sigma \in [0, 2\pi[$. The embedding functions $X^M(\tau, \sigma)$ satisfy the periodic boundary conditions

$$X^M(\tau, 0) = X^M(\tau, 2\pi), \qquad \partial_\sigma X^M(\tau, 0) = \partial_\sigma X^M(\tau, 2\pi) \tag{4.14}$$

and also $h_{\alpha\beta}(\tau, 0) = h_{\alpha\beta}(\tau, 2\pi)$. Due to the periodic identification, the boundary condition (4.10) is automatically satisfied. The left- and right-moving modes satisfy the boundary conditions if we set $p^M = \tilde{p}^M$. Moreover, it is convenient to set $x_0^M = \tilde{x}_0^M$.

Let us now consider open strings, for which it is convenient to set $\sigma_0 = \pi$, i.e. the spatial extension of the string is parametrised by $\sigma \in [0, \pi]$, where the two endpoints are given by $\sigma = 0$ and $\sigma = \pi$. Let $\bar{\sigma}$ be either $\bar{\sigma} = 0$ or $\bar{\sigma} = \pi$, i.e. one of the two endpoints. The boundary term (4.10) which has to vanish allows for two different boundary conditions:

- Neumann boundary conditions

$$\partial_\sigma X^M(\tau, \bar{\sigma}) = 0; \tag{4.15}$$

- Dirichlet boundary conditions

$$\delta X^M(\tau, \bar{\sigma}) = 0, \tag{4.16}$$

which means that the endpoint given by $\sigma = \bar{\sigma}$ of the string is fixed to $\bar{x}_0^M$,

$$X^M(\tau, \bar{\sigma}) = \bar{x}_0^M. \tag{4.17}$$

Both boundary conditions may be implemented for each string endpoint independently and for each target spacetime dimension. The only exception is the time direction for which we have to impose Neumann boundary conditions. For the space directions, we may impose either Dirichlet or Neumann boundary conditions for the string endpoint given by $\bar{\sigma} = 0$, and also either Dirichlet or Neumann boundary conditions for the other string endpoint given by $\bar{\sigma} = \pi$. In the case of Neumann boundary conditions for both string endpoints, the boundary condition is referred to as NN, while if we impose Dirichlet boundary conditions for both string endpoints, the boundary condition is referred to as DD. In the case of mixed Neumann and Dirichlet boundary conditions for the two endpoints, we speak of DN or ND boundary conditions for the two endpoints, depending on whether the endpoint given by $\bar{\sigma} = 0$ satisfies Dirichlet or Neumann boundary conditions, respectively.

In case of NN boundary conditions for the target spacetime coordinate X^M, we may decompose $X^M(\tau, \sigma)$ satisfying (4.9) into modes of the form

$$X^M(\tau, \sigma) = x_0^M + 2\alpha' p^M \tau + i\sqrt{2\alpha'} \sum_{n \neq 0} \frac{\alpha_n^M}{n} e^{-in\tau} \cos(n\sigma). \tag{4.18}$$

Here, we may also define $\alpha_0^M = \sqrt{\frac{\alpha'}{2}} p^M$. Note that the mode expansion for NN boundary conditions allows for a centre of mass motion given by x_0^M and p^M. In addition, p^M is the total momentum in the closed string since

$$p^M = \int_0^\pi d\sigma \, \Pi^M(\tau, \sigma), \quad \text{with canonical momentum} \quad \Pi^M(\tau, \sigma) = \frac{\partial_\tau X^M(\tau, \sigma)}{2\pi \alpha'}. \quad (4.19)$$

In particular, for NN boundary conditions the total momentum p^M is conserved. In contrast to the closed string we only have one set of oscillators α_n^M, which satisfy the reality condition $\alpha_{-n}^M = (\alpha_n^M)^*$.

Next, let us consider DD boundary conditions for X^M. Due to the boundary conditions, X^M has to satisfy $X^M(\tau, 0) = x_i^M$ and $X^M(\tau, \pi) = x_f^M$ where x_i^M and x_f^M are the coordinates of the string endpoints. The mode expansion is given by

$$X^M(\tau, \sigma) = x_i^M + \frac{1}{\pi}(x_f^M - x_i^M)\sigma + \sqrt{2\alpha'} \sum_{n \neq 0} \frac{\alpha_n^M}{n} e^{-in\tau} \sin(n\sigma). \quad (4.20)$$

Obviously, the momentum of the open string p^M given by (4.19) is not conserved for DD boundary conditions. This is not a surprise, however, since by imposing Dirichlet boundary conditions we have broken translational invariance in this direction, and momentum is no longer conserved. Where does the momentum flow to? Since the open string endpoints end on two hypersurfaces parametrised by $x^M = x_i^M$ and by $x^M = x_f^M$, these hypersurfaces have to absorb the momentum of the open string and therefore have to be dynamical. Note that in the case $x_i^M = x_f^M$, we only have one hypersurface on which both string endpoints end. These dynamical objects on which the open string endpoints end are referred to as *Dirichlet branes*, or *D-branes* for short. The D-branes are extended in those directions in which we impose Neumann boundary conditions, and are transverse to those directions in which we impose Dirichlet boundary conditions.

Exercise 4.1.1 Work out the mode expansion for X^M for ND boundary conditions, i.e. for Neumann boundary conditions for $\bar\sigma = 0$ and Dirichlet boundary conditions for $\bar\sigma = \pi$ with $X^M(\tau, \pi) = x_f^M$.

We have solved the classical equations of motion and the boundary conditions. But we are not yet done. In addition, the Virasoro constraints (4.11) have to be satisfied. For the closed string, introducing

$$T_{++} = \alpha' \sum_m \tilde{L}_m e^{-im\sigma^+}, \qquad T_{--} = \alpha' \sum_m L_m e^{-im\sigma^-}, \quad (4.21)$$

the constraints (4.11) amount to

$$\tilde{L}_m = L_m = 0 \quad \text{for all} \quad m, \quad (4.22)$$

where

$$\tilde{L}_m = \frac{1}{2} \sum_n \tilde\alpha_n^M \tilde\alpha_{m-n,M}, \qquad L_m = \frac{1}{2} \sum_n \alpha_n^M \alpha_{m-n,M}. \quad (4.23)$$

In the case of an open string, we only have one type of oscillating mode, say α_n^M, and we have to implement $L_m = 0$.

So far our discussion has been entirely classical. Let us now quantise the open and closed strings by imposing canonical equal-time time commutation relations

$$[X^M(\tau,\sigma), \Pi^N(\tau,\sigma')] = i\eta^{MN}\delta(\sigma - \sigma'). \tag{4.24}$$

For example, for the open string satisfying NN boundary conditions in all D target spacetime dimensions, we then conclude using the mode expansion (4.18) that the only non-vanishing commutation relations are given by

$$[x_0^M, p^N] = i\eta^{MN} \qquad \text{and} \qquad [\alpha_m^M, \alpha_n^N] = m\eta^{MN}\delta_{m,-n}. \tag{4.25}$$

Defining creation operators a_m^M and annihilation operators $a_m^{M\dagger}$ by

$$a_m^M = \frac{1}{\sqrt{m}}\alpha_m^M \qquad \text{and} \qquad a_m^{M\dagger} = \frac{1}{\sqrt{m}}\alpha_{-m}^M \tag{4.26}$$

for all $m > 0$, we then may express the commutation relations (4.25) in the form

$$[a_m^M, a_n^{N\dagger}] = \eta^{MN}\delta_{mn}, \qquad [a_m^M, a_n^N] = [a_m^{M\dagger}, a_n^{N\dagger}] = 0. \tag{4.27}$$

Each string mode characterised by m and M gives rise to a Hilbert space of the harmonic oscillator with the exception of a_m^0 and $a_m^{0\dagger}$. In that case the commutation relations satisfy $[a_m^0, a_m^{0\dagger}] = -1$ and therefore the associated Hilbert space contains negative norm states. In order to obtain a sensible quantum theory we have to show that these negative norm states decouple from the theory. Indeed this happens and is a consequence of the Virasoro constraints, as we now show.

Negative norm states already arise in gauge theories and may be cured by fixing the gauge symmetries. In string theory the relevant gauge symmetries are diffeomorphisms and Weyl symmetries which we have fixed by using the conformal gauge (4.7) giving rise to Virasoro constraints (4.11). These Virasoro constraints may be solved straightforwardly in light-cone coordinates in which we will work from now on. Defining $X^\pm = X^0 \pm X^{D-1}$ and setting

$$X^+ = x_0^+ + 2\alpha'p^+\tau, \tag{4.28}$$

we can solve the Virasoro constraints by expressing X^- as a function of X^i ($i = 1, \ldots, D-2$) and X^+ up to a constant x_0^-. The dynamical degrees of freedom are therefore p^+, x_0^- as well as p^i, x_0^i and $a_m^i, a_m^{i\dagger}$ for $i = 1, \ldots, D-2$, satisfying the commutation relations (4.27) with $M \to i, N \to j$ and $\eta^{MN} \to \delta^{ij}$. Defining the vacuum $|0, k\rangle$ by

$$p^M|0, k\rangle = k^M|0, k\rangle, \qquad a_m^i|0, k\rangle = 0 \tag{4.29}$$

we may form a general state $|N, k\rangle$ by acting with $a_m^{i\dagger}$ on $|0, k\rangle$, i.e.

$$|N, k\rangle = \left[\prod_{i=1}^{D-2}\prod_{n=1}^{\infty}\frac{(a_n^{i\dagger})^{N_{in}}}{\sqrt{N_{in}!}}\right]|0, k\rangle. \tag{4.30}$$

Here, N_{in} is the occupation number for each mode, i.e. $|N, k\rangle$ satisfies

$$a_n^{i\dagger}a_n^i|N, k\rangle = N_{in}|N, k\rangle \tag{4.31}$$

without summing over i and n on the left-hand side of equation (4.31). Let us now implement the Virasoro constraints (4.11). For this purpose, we rewrite L_n for $n \neq 0$ in terms of the creation and annihilation operators $a_m^{i\dagger}$ and a_m^i. For L_0 we have to take care of normal ordering ambiguities. Using (4.26) we may write L_0 as

$$L_0 = \alpha' p^M p_M + \sum_{i=1}^{D-2} \sum_{n=1}^{\infty} n a_n^{i\dagger} a_n^i \equiv \alpha' p^M p_M + N. \tag{4.32}$$

Since a physical state $|\psi\rangle$ has to solve the Virasoro constraint, we require

$$(L_0 - a)|\psi\rangle = 0, \qquad L_n|\psi\rangle = 0 \quad \text{for all } n \in \mathbb{N}, \tag{4.33}$$

where a arises from ordering ambiguities and is given by

$$a = \frac{D-2}{2} \sum_{n \geq 0} n. \tag{4.34}$$

Naively this sum is infinite. However, using ζ-function regularisation, a is given by

$$a = -\frac{D-2}{24}. \tag{4.35}$$

Exercise 4.1.2 Regularise the infinite sum $\sum_{n=1}^{\infty} n$ by $\sum_{n=1}^{\infty} n e^{-\epsilon n}$ and show that

$$\sum_{n=1}^{\infty} n e^{-\epsilon n} \sim \frac{1}{\epsilon^2} - \frac{1}{12} + \mathcal{O}(\epsilon^2)$$

for small ϵ. How do we obtain (4.35)?

Using (4.33) we may determine the mass $M^2 = -k^M k_M$ of the string state (4.31) to be

$$M^2 = \frac{1}{\alpha'} \left(N + \frac{2-D}{24} \right), \tag{4.36}$$

where N is defined implicitly in (4.32). Let us now determine the open string spectrum of the lowest mass states. The lightest state is the vacuum $|k, 0\rangle$ with mass $M^2 = (2-D)/(24\alpha')$. For $D > 2$ the mass squared is negative, i.e. $|k, 0\rangle$ is tachyonic and therefore the vacuum is unstable. It is not known whether bosonic string theory has a stable vacuum. However, the vacuum in superstring theory is stable. Let us ignore this problem for now and consider the first excited state,

$$a_1^{i\dagger}|k, 0\rangle, \qquad \text{with} \quad M^2 = \frac{D-26}{24}. \tag{4.37}$$

All the states of the form (4.37) transform as a vector under the rotation group $SO(D-2)$ of the transverse space. This implies that the string excitation has to be massless, and therefore bosonic string theory in Minkowski spacetime is consistent only in $D = 26$ dimensions. The argument goes as follows. As argued in section 3.1.4, for the case of $D = 3$, the little group for massless representations in $\mathbb{R}^{3,1}$ is $ISO(2)$, the group of motions in two-dimensional Euclidean space. A subgroup of $ISO(2)$ is $SO(2)$ which corresponds to the rotations in the two-dimensional Euclidean space. In contrast, massive representations have $SO(3)$ as little group. This result can be generalised to arbitrary D using the same arguments as presented in section 3.1.4. If the little group contains $SO(D-2)$ and not

$SO(D-1)$, the states forming the representation have to be massless. Since the states $a_1^{i\dagger}|k,0\rangle$ transform as a vector under $SO(D-2)$, they have to be massless and therefore the dimension has to be $D=26$.

Although the classical theory is invariant under Lorentz transformations for any D as argued in (4.4), this is not the case for the quantised theory of bosonic strings unless $D=26$. In other words, only for $D=26$ are the anomalies absent.

It is tempting to identify the modes $a_1^{i\dagger}|k,0\rangle$ with gauge bosons. In section 4.4.1 we will see that this is indeed the case.

Exercise 4.1.3   In light-cone gauge, quantise the open string which has d ND directions. It may be assumed that there are at least two NN directions. Construct the string states and show that their mass is given by

$$M^2 = \frac{1}{\alpha'}\left(\sum_{i=1}^{D-2}\sum_{n=1}^{\infty} nN_{in} + a\right) + \left(\frac{\Delta x}{2\pi\alpha'}\right)^2,\tag{4.38}$$

where a is

$$a = -\frac{D-2}{24} + d.\tag{4.39}$$

The last term in (4.38) represents the contribution to the mass of a stretched open string between two D-branes separated by a distance Δx.

Let us now consider closed strings. The quantisation may be performed analogously to the open string case using light-cone quantisation. The only difference is that the mode expansion of the functions X^M, as given for instance by (4.18), now involves *two* modes α_n^i and $\tilde{\alpha}_n^i$ with $i = 1,\ldots,D-2$. L_n and $\tilde{L}_n$ as defined in (4.23) give rise to two copies of Virasoro algebras. As in the case of the open string, see (4.32), L_0 and $\tilde{L}_0$ suffer from an operator ordering ambiguity. Therefore the Virasoro constraints for a physical state $|\psi\rangle$ read

$$L_n|\psi\rangle = \tilde{L}_n|\psi\rangle = 0 \quad \text{for all } n \in \mathbb{N},\tag{4.40}$$

$$(L_0 - a)|\psi\rangle = (\tilde{L}_0 - \tilde{a})|\psi\rangle = 0,\tag{4.41}$$

where a and $\tilde{a}$ are given by

$$a = \tilde{a} = -\frac{D-2}{12}.\tag{4.42}$$

Consider a physical state $|\psi\rangle$ of the form

$$|N,\tilde{N},k\rangle = \left[\prod_{i=1}^{D-2}\prod_{n=1}^{\infty}\frac{\left(a_n^{i\dagger}\right)^{N_{in}}\left(\tilde{a}_n^{i\dagger}\right)^{\tilde{N}_{in}}}{\sqrt{N_{in}!\,\tilde{N}_{in}!}}\right]|0,0,k\rangle,\tag{4.43}$$

where $|0,0,k\rangle$ is an eigenstate of p^M with eigenvalue k^M and is annihilated by all operators a_n^i and $\tilde{a}_n^i$. The mass of the string state (4.43) is given by

$$M^2 = \frac{2}{\alpha'}\left(\sum_{i=1}^{D-2}\sum_{n=1}^{\infty}\left(nN_{in} + n\tilde{N}_{in}\right) + \frac{2-D}{12}\right).\tag{4.44}$$

The N_{in}, $\tilde{N}_{in}$ are the eigenvalues of $a_{in}^\dagger a_{in}$, $\tilde{a}_{in}^\dagger \tilde{a}_{in}$, respectively. In order to be physical, the string state $|\psi\rangle$ given by (4.43) has to satisfy the level matching condition

$$\left(L_0 - \tilde{L}_0 - a + \tilde{a}\right)|\psi\rangle = 0 \tag{4.45}$$

for the Virasoro generators L_0, $\tilde{L}_0$. Using (4.42) and defining

$$N \equiv \sum_{i=1}^{D-2}\sum_{n=1}^{\infty} n N_{in}, \qquad \tilde{N} \equiv \sum_{i=1}^{D-2}\sum_{n=1}^{\infty} n \tilde{N}_{in}, \tag{4.46}$$

the level matching condition is equivalent to

$$N - \tilde{N} = 0. \tag{4.47}$$

Note that the level matching condition couples the oscillators a_{in} and $\tilde{a}_{in}$ in a subtle way.

Let us now consider the spectrum of the lowest mass closed string states. The vacuum $|0, 0, k\rangle$ is the lightest state with $M^2 = \frac{2-D}{6\alpha'}$ which is again tachyonic. The first excited states are of the form

$$a_1^i \tilde{a}_1^j |0, 0, k\rangle, \qquad \text{with mass} \quad M^2 = \frac{26 - D}{6\alpha'}. \tag{4.48}$$

Note that for $D = 26$, these states are a massless rank two tensor which may be decomposed into a symmetric traceless tensor, an antisymmetric tensor and a scalar. In section 4.1.4, the symmetric traceless tensor is identified with the graviton, while the scalar is the *dilaton*. The antisymmetric tensor gives rise to the *Kalb–Ramond field*.

4.1.3 String perturbation theory

So far we have considered only non-interacting strings and we have discussed their different mass states in detail. Let us now allow for interactions among the strings. Assuming a small coupling constant between strings, the idea of the perturbative expansion in terms of Feynman diagrams, similar to those introduced within quantum field theory, may be carried over to string theory in a natural way.

So far, we have considered worldsheets with the topology of a cylinder for closed strings and world sheets with the topology of a strip for open strings. To describe interactions of strings, we have to use more involved worldsheet topologies. For example, in order to allow for splitting and joining of closed strings, additional handles have to be added to the worldsheet. An example of a closed string decaying into two closed strings is shown on the left-hand side of figure 4.2.

In the path integral approach, which is again very useful here, we have to sum over different worldsheet topologies which connect initial and final string configurations, weighted by a factor involving the exponential of the action. For closed strings, the sum contains all two-dimensional oriented surfaces without boundaries, while for open strings, boundaries have to be included. In both cases, the worldsheets are characterised by their topology, in particular by their genus g which corresponds to the number of handles of a surface. The path integral over all worldsheets Σ decomposes into a sum over genera g and

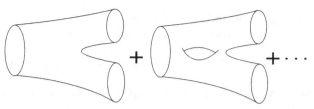

Figure 4.2   The lowest and the first-order contribution to the interaction vertex in the string perturbation expansion. The diagrams may be viewed as string splitting when moving from left to right, and to joining of strings when moving from right to left.

an integral over the worldsheets Σ_g with genus g, giving rise to a partition function of the form

$$Z = \int_\Sigma \mathcal{D}X^M \mathcal{D}h_{\alpha\beta}\, e^{-\mathcal{S}'_P} = \sum_{g=0}^\infty \int_{\Sigma_g} \mathcal{D}X^M \mathcal{D}h_{\alpha\beta}\, e^{-\mathcal{S}'_P} \tag{4.49}$$

after Wick rotation to Euclidean space. The action in (4.49) is given by

$$\mathcal{S}'_P = \mathcal{S}_P - \lambda\chi, \qquad \chi = \frac{1}{4\pi}\int_\Sigma \mathrm{d}^2\sigma\, \sqrt{h}R_{(h)}. \tag{4.50}$$

Here $\mathcal{S}_P$ is the Polyakov action (4.2) and $R_{(h)}$ is the Ricci scalar of the worldsheet metric $h_{\alpha\beta}$. It is important to note that χ is the Euler number and therefore χ is a topological term not contributing to the equations of motion. The Euler number is related to the genus of the worldsheet Σ_g by $\chi = 2 - 2g$. Consequently, the partition function (4.49) may be rewritten as

$$Z = \sum_{g=0}^\infty e^{-\lambda(2-2g)}\int_{\Sigma_g} \mathcal{D}X^M \mathcal{D}h_{\alpha\beta}\, e^{-\mathcal{S}_P}. \tag{4.51}$$

In this expression, the factor $e^{-\lambda(2-2g)}$ gives a different weight to different topologies. Adding a handle to any worldsheet reduces the Euler number by two. As shown in figure 4.2, the process which is described by adding a handle corresponds to emitting and reabsorbing a closed string. The closed string coupling may thus be identified with $g_{\text{closed}} = e^\lambda$. An analogous argument for open strings shows that $g^2_{\text{open}} = e^\lambda$, such that

$$g_s \equiv g_{\text{closed}} = g^2_{\text{open}} = e^\lambda. \tag{4.52}$$

Later, we will see that the string coupling is fixed by the vacuum expectation value of the dilaton field.

Although the idea of summing over all worldsheets is simple, it is difficult to define and perform this expansion in practice. A simplification is obtained by taking the limit where the string sources are located at infinity. This corresponds to an S-matrix element with specified incoming and outgoing strings which are on-shell. Using conformal transformations, the corresponding worldsheets may be transformed to compact surfaces with n points removed. These points or *punctures* correspond to the external string states. Within the path integral approach, scattering amplitudes are obtained by summing over all surfaces with n punctures and by integrating against the external string state wave functions at these punctures. On the worldsheet, these external string states are represented by *vertex operators*.

4.1.4 Bosonic string theory in background fields: emergence of gravity

Up to now, we have considered the propagation of open and closed strings in Minkowski spacetime. By coupling the fundamental string to the massless closed string excitations, which involve the graviton, strings propagating through curved background spacetime can be described. In particular, the symmetric traceless part of the state $\alpha_{-1}^{(M} \tilde{\alpha}_{-1}^{N)} |0, 0, k\rangle$ may be identified with the metric of the target spacetime g_{MN}.

The Polyakov action in a curved target spacetime becomes

$$S = -\frac{1}{4\pi\alpha'} \int d^2\sigma \sqrt{-h}\, h^{\alpha\beta}\, \partial_\alpha X^M\, \partial_\beta X^N\, g_{MN}(X) . \tag{4.53}$$

In addition, we have a *Kalb–Ramond field* B_{MN} antisymmetric in the indices M and N, and a dilaton ϕ associated with the remaining massless closed string states $\alpha_{-1}^{[M} \tilde{\alpha}_{-1}^{N]} |0, 0, k\rangle$ and $\alpha_{-1}^M \tilde{\alpha}_{-1M} |0, 0, k\rangle$. Their action reads

$$S_{B,\phi} = -\frac{1}{4\pi\alpha'} \int d^2\sigma \sqrt{-h}\, \left(\epsilon^{\alpha\beta}\, \partial_\alpha X^M\, \partial_\beta X^N\, B_{MN}(X) + \alpha'\, R_h\, \phi(X) \right), \tag{4.54}$$

where $R_{(h)}$ denotes the Ricci scalar on the worldsheet. By comparison with the string theory perturbative expansion, we identify the string coupling as $g_s = e^\phi$. To ensure Weyl invariance of the quantised theory, we have to impose tracelessness of the worldsheet energy-momentum tensor. The trace of the worldsheet energy-momentum tensor reads

$$T^\alpha{}_\alpha = -\frac{1}{2\alpha'}\, \beta_{MN}^g\, h^{\alpha\beta}\, \partial_\alpha X^M\, \partial_\beta X^N - \frac{1}{2\alpha'}\, \beta_{MN}^B\, \epsilon^{\alpha\beta}\, \partial_\alpha X^M\, \partial_\beta X^N - \frac{1}{2}\, \beta^\phi\, R_{(h)} \tag{4.55}$$

where the β functions are given by

$$\begin{aligned}
\beta_{MN}^g &= -\alpha' \left(R_{MN} + 2\nabla_M \nabla_N \phi - \frac{1}{4}\, H_{MLR}\, H_N{}^{LR} \right) \\
\beta_{MN}^B &= \alpha' \left(-\frac{1}{2}\, \nabla^L H_{LMN} + \nabla^\lambda \phi\, H_{LMN} \right) \\
\beta^\phi &= \alpha' \left(\frac{D-26}{6\alpha'} - \frac{1}{2}\, \nabla^2 \phi + \nabla_M \phi\, \nabla^M \phi - \frac{1}{24}\, H_{MNL}\, H^{MNL} \right)
\end{aligned} \tag{4.56}$$

to order α'. Using differential forms, we may define a field strength $H = dB$ for the Kalb–Ramond field, with

$$H_{MNL} \equiv \partial_M B_{NL} + \partial_N B_{LM} + \partial_L B_{MN} \tag{4.57}$$

in component notation.

The string theory in this background is Weyl invariant only if the energy-momentum tensor is traceless, $T^\alpha{}_\alpha = 0$ and thus $\beta_{MN}^g = \beta_{MN}^B = \beta^\phi = 0$. Note that the vanishing of the β functions imposes restrictions on the target spacetime. For example, for a constant dilaton ϕ and for a vanishing Kalb–Ramond field, we deduce that the dimension of the target spacetime has to be $D = 26$ and the target spacetime metric $g_{MN}(x)$ has to satisfy the vacuum Einstein equation.

Remarkably, the vanishing of the β functions (4.56) may be derived as equations of motion from the target spacetime action

$$S = \frac{1}{2\tilde{\kappa}^2} \int \mathrm{d}^D X \sqrt{-g}\, e^{-2\phi} \left(R + 4\nabla_M \phi\, \nabla^M \phi \right.$$
$$\left. - \frac{1}{12}\, H_{MNR}\, H^{MNR} - \frac{2(D-26)}{3\alpha'} + \mathcal{O}(\alpha') \right), \quad (4.58)$$

where R and ∇_M are the Ricci scalar and the covariant derivatives associated with the target spacetime metric g_{MN}. Therefore we may view (4.58) as an effective action for the massless closed string states ϕ, g_{MN}, B_{MN}.

As already mentioned above, the string coupling constant is given by the expectation value of the dilaton $g_s = e^\phi$. Moreover, the massless rank two symmetric tensor field g_{MN} may be identified with the graviton, since g_{MN} has to satisfy the equations of motion $\beta_{MN}^g = 0$, which also follow immediately from the effective action. The first term in (4.58) is an Einstein–Hilbert term coupled to a dilaton. Therefore g_{MN} is identified with the target spacetime metric.

Note that the sign of the kinetic term of the dilaton field ϕ in (4.58) is opposite to the usual convention. By rescaling the metric for $D > 2$,

$$\tilde{g}_{MN} = e^{\frac{4}{D-2}(\phi_0 - \phi)} g_{MN}, \quad (4.59)$$

the Einstein–Hilbert term of the action (4.58) may be normalised canonically and the prefactor of the kinetic term turns negative,

$$S = \frac{1}{2\kappa^2} \int \mathrm{d}^D X \sqrt{-\tilde{g}} \left(\tilde{R} - \frac{4}{D-2}\nabla_M \tilde{\phi} \nabla^M \tilde{\phi} - \frac{1}{12} e^{-\frac{8}{D-2}\tilde{\phi}} H_{MNR} H^{MNR} \right.$$
$$\left. - \frac{2(D-26)}{3\alpha'} e^{\frac{4}{D-2}\tilde{\phi}} + \mathcal{O}(\alpha') \right), \quad (4.60)$$

with $\tilde{\phi} = \phi - \phi_0$ where ϕ_0 is the asymptotic value of the dilaton field. Note that $\kappa = \tilde{\kappa} e^{\phi_0} = \tilde{\kappa} g_s$ and κ may be identified with $\sqrt{8\pi G}$ where G is the Newton constant. Looking at the part involving the Ricci scalar $\tilde{R}$, which is determined by the rescaled metric $\tilde{g}_{MN}$, we see that we have removed the factor involving the dilaton ϕ in the Einstein–Hilbert part of the action (4.60). Whereas the action written in terms of the original fields is the *string-frame action*, the canonically normalised action (4.60) is referred to as the *Einstein-frame action*.

4.2 Superstring theory

Bosonic string theory which we have described so far has two major shortcomings. First, it contains tachyons in both the open string and the closed string sectors, i.e. states of negative mass squared. Second, the bosonic string lacks fermionic degrees of freedom necessary to model the particles observed in nature.

Fermionic degrees of freedom are obtained naturally by introducing supersymmetry. The supersymmetrised Polyakov action for the string position X^M and its worldsheet superpartner Ψ^M in conformal gauge $h_{\alpha\beta}(\tau,\sigma) = e^{2\omega(\tau,\sigma)}\eta_{\alpha\beta}$ is given by

$$S = -\frac{1}{4\pi\alpha'}\int d^2\sigma\, \eta^{\alpha\beta}\big(\partial_\alpha X^M\, \partial_\beta X^N + i\bar\Psi^M\,\gamma_\alpha\,\partial_\beta\Psi^N\big)g_{MN}(X). \tag{4.61}$$

Here, we consider again a flat D-dimensional target spacetime, i.e. $g_{MN} = \eta_{MN}$. The Ψ^M are spinors on the two-dimensional worldsheet. According to table 3.1, these may be chosen to be Majorana spinors, i.e. $\Psi^M = \big(\psi_-^M, \psi_+^M\big)^{\mathrm{T}}$ with two real components $\psi_\pm^M$. At the same time, these spinors Ψ^M are vectors in the target spacetime, as indicated by the index M. The γ^α denote worldsheet Dirac matrices for which one possible representation is

$$\gamma^0 = \begin{pmatrix} 0 & -i \\ i & 0 \end{pmatrix}, \quad \gamma^1 = \begin{pmatrix} 0 & i \\ i & 0 \end{pmatrix}. \tag{4.62}$$

Thus the fermionic part of the action (4.61) may be rewritten as

$$S_{\mathrm{f}} = \frac{i}{2\pi\alpha'}\int d^2\sigma\, \big(\psi_-^M\,\partial_+\psi_{-M} + \psi_+^M\,\partial_-\psi_{+M}\big). \tag{4.63}$$

The equations of motion describe left- and right-moving waves just like in the bosonic sector,

$$\partial_+\psi_-^M = \partial_-\psi_+^M = 0. \tag{4.64}$$

The total action is invariant under the worldsheet supersymmetry transformations $\delta_\epsilon X^M = \bar\epsilon\Psi^M$ and $\delta_\epsilon\Psi^M = \gamma^\alpha\partial_\alpha X^M\epsilon$, where the parameter ϵ is an infinitesimal constant Majorana spinor.

Integrating the action (4.63) by parts, we obtain the boundary term

$$\delta S_{\mathrm{f}} = \frac{i}{4\pi\alpha'}\int d\tau\, \big(\psi_-^M\,\delta\psi_{-M} - \psi_+^M\,\delta\psi_{+M}\big)\Big|_{\sigma=0}^{\sigma=\pi}, \tag{4.65}$$

which imposes boundary conditions. As in the bosonic case we will discuss two now different types of strings satisfying the boundary conditions, i.e. open and closed strings. The analysis for X^M is the same as beforehand. Therefore we only discuss the fermionic fields $\psi_\pm^M$.

4.2.1 Open superstrings

In the open string sector, the contributions in (4.65) arising from $\sigma = 0$ and $\sigma = \pi$ have to vanish separately. This is equivalent to

$$\psi_-^M\,\delta\psi_{-M} - \psi_+^M\,\delta\psi_{+M}\Big|_{\sigma=0,\pi} = 0 \Leftrightarrow \delta\big(\psi_{+M}\big)^2\Big|_{\sigma=0,\pi} = \delta\big(\psi_{-M}\big)^2\Big|_{\sigma=0,\pi} = 0. \tag{4.66}$$

Since the overall sign of the spinor components may be chosen arbitrarily, we impose $\psi_+^M(\tau,0) = \psi_-^M(\tau,0)$, then the boundary condition at $\sigma = \pi$ leaves two options corresponding to the *Neveu–Schwarz* (or NS) and the *Ramond* (or R) *sectors* of the theory,

$$
\begin{aligned}
\text{R}: \quad & \psi_+^M(\tau, \pi) = +\psi_-^M(\tau, \pi), \\
\text{NS}: \quad & \psi_+^M(\tau, \pi) = -\psi_-^M(\tau, \pi).
\end{aligned}
\tag{4.67}
$$

These boundary conditions give rise to the Fourier expansions

$$
\begin{aligned}
\text{R}: \quad & \psi_\mp^M(\tau, \pi) = \frac{1}{\sqrt{2}} \sum_{n \in \mathbb{Z}} d_n^M \, e^{-in\sigma_\mp}, \\
\text{NS}: \quad & \psi_\mp^M(\tau, \pi) = \frac{1}{\sqrt{2}} \sum_{r \in \mathbb{Z}-\frac{1}{2}} b_r^M \, e^{-ir\sigma_\mp},
\end{aligned}
\tag{4.68}
$$

with Grassmann-valued Fourier modes d_n, b_r. We now proceed as in the case of the bosonic string by promoting the Fourier modes to operators and imposing (anti-)commutation relations. For d_n and b_r the anticommutation relations read

$$
\{d_m^M, d_n^N\} = \eta^{MN} \delta_{m,-n}, \qquad \{b_r^M, b_s^N\} = \eta^{MN} \delta_{r,-s}.
\tag{4.69}
$$

These modes may be used to construct the states of the theory. For the NS sector, the ground state is defined by $b_r^M |0\rangle_{\text{NS}} = 0$ for $r > 0$, and the modes b_r^M with $r < 0$ are creation operators. In the R sector, we also define the ground state such that it is annihilated by the d_m^M with $m > 0$. However, in the R sector this ground state is degenerate since $\{d_m^M, d_0^N\} = 0$ for $m > 0$, i.e. the d_0^M take a ground state into another ground state. Note that d_0^M satisfy the algebra (4.69) and thus we may represent d_0^M by the gamma matrices of the target spacetime, Γ^M. Hence the ground state in the R sector is a spacetime spinor with spacetime spin one-half.

The higher string states are created by acting with creation operators b_r and d_m with $r, m < 0$ on the ground state of the NS and R sectors, respectively. The modes b_r and d_m anti-commute and hence we can apply each creation operator only once.

As in the bosonic case we can now work out the masses of these states. For example, the excited state[1] $b_{-1/2}^i |0\rangle_{\text{NS}} = 0$ in the NS sector has a mass

$$
M^2 = \frac{1}{\alpha'} \left(\frac{1}{2} - \frac{D-2}{16} \right)
\tag{4.70}
$$

and transforms as a vector in $SO(D-2)$. However, we would have expected $SO(D-1)$ unless the state is massless. Hence we conclude $D = 10$ for superstring theories. Note that this immediately implies that the NS ground state $|0\rangle_{\text{NS}}$ is tachyonic with $M^2 = -1/2\alpha'$. In the R sector, the vacuum is massless but still contains both chiralities. In table 4.1 the massless open string states are classified in ten dimensions according to their $SO(8)$ representation. $\mathbf{8_v}$ denotes the fundamental representation of $SO(8)$ while $\mathbf{8}$ and $\mathbf{8'}$ are the irreducible spinorial representations of opposite chirality.

In order to get rid of the tachyon in the NS sector as well as one of the chiralities in the R sector, we introduce the fermion number $\exp(i\pi F)$ which counts how often a fermionic creation operator is applied to the vacuum. We keep only those states with an odd number

---

[1] Note that here we replaced the index M of $b_{-1/2}^M$ by $i = 1, \ldots D - 2$. Technically speaking, we have again introduced light-cone coordinates in target spacetime to solve the Virasoro constraints.

Table 4.1 Lowest supersymmetric open string states			
Sector	$\exp(i\pi F)$	$SO(8)$ rep	m^2
NS	+	$\mathbf{8_v}$	0
NS	−	$\mathbf{1}$	$-1/2\alpha'$
R	+	$\mathbf{8}$	0
R	−	$\mathbf{8'}$	0

of creation operators applied to $|0\rangle_{NS}$. In the R sector we have a choice whether we keep only those states with an even or with an odd number of creation operators acting on the vacuum. Depending on the choice, the physical massless string state in the R sector has a definite chirality and is real, i.e. it is a Majorana–Weyl spinor.

This truncation prescription, known as *GSO projection* due to Gliozzi, Scherk and Olive, projects out all tachyonic states and furthermore leaves an equal number of fermions and bosons at each mass level. Thus it paves the way for target spacetime supersymmetry. For example, at the massless level we are left with $\{b^i_{-1/2}|0\rangle_{NS}, |0\rangle_R\}$ which we may identify as an $\mathcal{N} = 1$ supersymmetric gauge multiplet, where $b^i_{-1/2}|0\rangle_{NS}$ are the gauge bosons, while the spacetime spinor $|0\rangle_R$ is the gaugino.

4.2.2 Closed superstrings

The closed sector of superstring theory may be constructed in four different ways. Each of left and right movers may be taken from open string NS and R sectors. From a spacetime point of view, we have the following statistics for the states: the NS-NS and R-R sectors are spacetime bosons, while the NS-R and R-NS sectors are spacetime fermions.

For the closed string, we thus obtain the lowest states from two copies of the open string states as given by table 4.1. The lowest closed string states are given in table 4.2, where + and − again correspond to the fermion number $\exp(i\pi F)$. The representations of table 4.2 have the following properties. The $\mathbf{28}$, $\mathbf{56_t}$ and $\mathbf{35_\pm}$ represent two-, three- and four-forms, where the four-form satisfies a self-duality condition. The $\mathbf{35}$ is a symmetric traceless tensor of rank two. The $\mathbf{56}$ and $\mathbf{56'}$ are vector spinors and will correspond to gravitinos.

The GSO projection gives rise to four consistent closed superstring theories in ten dimensions. For gauge/gravity duality, two of them, referred to as *type IIA* and *type IIB* superstring theory, are of particular interest, as we will see below. These contain the following sectors,

$$\text{Type IIA}: \quad (NS+, NS+), \quad (R+, NS+), \quad (NS+, R-), \quad (R+, R-),$$
$$\text{Type IIB}: \quad (NS+, NS+), \quad (R+, NS+), \quad (NS+, R+), \quad (R+, R+).$$

Table 4.2 Lowest supersymmetric closed string states

Sector	$SO(8)$ representation	m^2
$(\text{NS}\oplus, \text{NS}\oplus)$	$\mathbf{8_v} \otimes \mathbf{8_v} = \mathbf{1} \oplus \mathbf{28} \oplus \mathbf{35}$	0
$(\text{NS}-, \text{NS}-)$	$\mathbf{1}$	$-1/2\alpha'$
$(\text{R}\oplus, \text{R}\oplus)$	$\mathbf{8} \otimes \mathbf{8} = \mathbf{1} \oplus \mathbf{28} \oplus \mathbf{35_+}$	0
$(\text{R}-, \text{R}-)$	$\mathbf{8'} \otimes \mathbf{8'} = \mathbf{1} \oplus \mathbf{28} \oplus \mathbf{35_-}$	0
$(\text{R}\oplus, \text{R}-)$	$\mathbf{8} \otimes \mathbf{8'} = \mathbf{8_v} \oplus \mathbf{56_t}$	0
$(\text{NS}\oplus, \text{R}\oplus)$	$\mathbf{8_v} \otimes \mathbf{8} = \mathbf{8'} \oplus \mathbf{56}$	0
$(\text{NS}\oplus, \text{R}-)$	$\mathbf{8_v} \otimes \mathbf{8'} = \mathbf{8} \oplus \mathbf{56'}$	0

Since type IIB has fermion number $+1$ in every entry, it has a chiral structure. Using the results of table 4.2, this corresponds to the representations

$$\text{Type IIA}: \quad \mathbf{1} \oplus \mathbf{8_v} \oplus \mathbf{28} \oplus \mathbf{56_t} \oplus \mathbf{35} \oplus \mathbf{8} \oplus \mathbf{8'} \oplus \mathbf{56} \oplus \mathbf{56'}, \tag{4.71}$$

$$\text{Type IIB}: \quad \mathbf{1}^2 \oplus \mathbf{28}^2 \oplus \mathbf{35} \oplus \mathbf{35_+} \oplus \mathbf{8'}^2 \oplus \mathbf{56}^2. \tag{4.72}$$

Type IIA theory has spinors $\mathbf{8}$, $\mathbf{8'}$ as well as $\mathbf{56}$, $\mathbf{56'}$ which have different chirality. On the other hand, type IIB theory is chiral as noted above. The $\mathbf{56}$, $\mathbf{56'}$ representations correspond to gravitinos, which are the superpartners of the graviton which corresponds to the $\mathbf{35}$ representation.

The NS-NS sector contains the fields ϕ, B_{MN}, g_{MN}. These correspond to the $\mathbf{1}$, $\mathbf{28}$ and $\mathbf{35}$ representations of $SO(8)$. We encountered these fields previously in the bosonic string theory. On the other hand, the 'mixed' NS-R, R-NS sectors contain SUSY superpartners such as the gravitino and dilatino. The R-R sector is more complicated due to the degenerate ground state. There are two possible inequivalent R-R ground states which differ by chirality, corresponding to type IIA and type theory IIB superstring theory. In type IIB, left- and right-moving sectors have the same chirality, which leads to a scalar $C_{(0)}$ and antisymmetric tensor fields $C_{(2)}$ and $C_{(4)}$ of rank two and four at the massless level. Type IIA theory with R-R ground states of opposite chiralities gives rise to $C_{(1)}$ and $C_{(3)}$ antisymmetric tensor fields.

As emerged in the 1980s and 1990s there are in fact three further consistent superstring theories known as type I and heterotic string theories, with gauge groups $SO(32)$ and $E_8 \times E_8$. They are connected with each other and with the type II models by a web of dualities. For the purpose of this book, however, it is sufficient to focus on type II theories.

4.2.3 The low-energy effective action: supergravity

The low-energy effective action for the type II superstring theories is obtained from the massless closed superstring states listed in table 4.2. This action corresponds to the action of *supergravity*. We obtain actions for type II supergravity in the string frame, writing out only the bosonic part of the supergravity actions. Supergravity may also be constructed independently of string theory by requiring local supersymmetry. This naturally encompasses general relativity. Supergravity theories may be formulated in any dimension less than or equal to $D = 11$. Many supergravity theories in less than $D = 11$ may be obtained by compactifying the unique eleven-dimensional theory whose UV completion is referred to as *M-theory*.

Type IIB supergravity

The fields of type IIB supergravity are obtained by identifying the fields corresponding to the representations in (4.72). These are listed in table 4.3. Due to its field content, the theory is chiral and violates parity. The bosonic part of the type IIB supergravity action reads, in string frame,

$$S_{\text{IIB}} = \frac{1}{2\tilde{\kappa}_{10}^2} \left[\int d^{10}X \sqrt{-g} \left(e^{-2\phi} \left(R + 4\partial_M \phi \partial^M \phi - \frac{1}{2}|H_{(3)}|^2 \right) \right. \right.$$
$$\left. - \frac{1}{2}|F_{(1)}|^2 - \frac{1}{2}|\tilde{F}_{(3)}|^2 - \frac{1}{4}|\tilde{F}_{(5)}|^2 \right)$$
$$\left. - \frac{1}{2} \int C_{(4)} \wedge H_{(3)} \wedge F_{(3)} \right], \tag{4.73}$$

where we use the notation

$$\int d^{10}X \sqrt{-g} \, |F_{(p)}|^2 = \frac{1}{p!} \int d^{10}X \sqrt{-g} \, g_{M_1 N_1} \cdots g_{M_p N_p} \bar{F}^{M_1 \cdots M_p} F^{N_1 \cdots N_p} \tag{4.74}$$

and $\bar{F}_{(p)}$ denotes the complex conjugate [2] of $F_{(p)}$. Moreover, $\tilde{\kappa}_{10}$ is the ten-dimensional gravitational constant,

$$2\tilde{\kappa}_{10}^2 = (2\pi)^7 \alpha'^4. \tag{4.75}$$

As in the bosonic case, in order to identify the ten-dimensional Newton constant, we have to take the asymptotic value ϕ_0 of the dilaton into account and consider instead κ_{10} given by

$$2\kappa_{10}^2 = 2\tilde{\kappa}_{10}^2 g_s^2 = (2\pi)^7 \alpha'^4 g_s^2, \tag{4.76}$$

Table 4.3 Field content of type IIB supergravity		
Field	$SO(8)$ representation	Physical property
g_{MN}	$\mathbf{35}$	metric (graviton)
$C_{(0)} + i\exp(-\phi)$	$\mathbf{1}^2$	axion-dilaton
$B_{(2)}, C_{(2)}$	$\mathbf{28}^2$	two-form
$C_{(4)}$	$\mathbf{35}_+$	self-dual four-form
$\Psi_{I\alpha}{}^M, \quad I = 1, 2$	$\mathbf{56}'^2$	Majorana–Weyl gravitinos
$\lambda_{I\alpha}, \quad I = 1, 2$	$\mathbf{8}'^2$	Majorana–Weyl dilatinos

---

[2] The RR form fields discussed here are assumed to be real and thus the complex conjugation is trivial.

which is related to the ten-dimensional Newton constant G_{10} by $\kappa_{10}^2 = 8\pi G_{10}$. The field strength tensors in (4.73) are given by

$$F_{(p)} = \mathrm{d}C_{(p-1)}, \quad H_{(3)} = \mathrm{d}B_{(2)}, \quad \tilde{F}_{(3)} = F_{(3)} - C_{(0)}H_{(3)}, \tag{4.77}$$

$$\tilde{F}_{(5)} = F_{(5)} - \frac{1}{2}C_{(2)} \wedge H_{(3)} + \frac{1}{2}B_{(2)} \wedge F_{(3)},$$

with d the exterior derivative. In addition, we have to impose the self-duality constraint

$$^*\tilde{F}_{(5)} = \tilde{F}_{(5)}. \tag{4.78}$$

Type IIA supergravity

For type IIA supergravity, the field content is again given by the representations of (4.71). The bosonic part of the corresponding action reads

$$S_{\mathrm{IIA}} = \frac{1}{2\kappa_{10}^2}\Big[\int \mathrm{d}^{10}x\sqrt{-g}\Big(e^{-2\phi}\Big(R + 4\partial_M\phi\partial^M\phi - \frac{1}{2}|H_{(3)}|^2\Big) - \frac{1}{2}|F_{(2)}|^2 - \frac{1}{2}|\tilde{F}_{(4)}|^2\Big)$$

$$- \frac{1}{2}\int B \wedge F_{(4)} \wedge F_{(4)}\Big], \tag{4.79}$$

where

$$\tilde{F}_{(4)} = \mathrm{d}A_{(3)} - A_{(1)} \wedge F_{(3)}. \tag{4.80}$$

An important fact about this action is that it may be derived from eleven-dimensional supergravity by dimensional reduction. Eleven-dimensional supergravity is unique in the sense that it is the only (local) supersymmetric theory in eleven dimensions containing only massless particles of spin ≤ 2. In particular, it contains two bosonic fields, the metric G_{MN} and a three-form potential $A_{(3)} = A_{MNR}\mathrm{d}x^M \wedge \mathrm{d}x^N \wedge \mathrm{d}x^R$. While the metric has 44 physical states, the three-form potential has 84, thus adding up to 128 states. The bosonic part of the action of eleven-dimensional supergravity is given by

$$S_{11} = \frac{1}{2\kappa_{11}^2}\left[\int \mathrm{d}^{11}x\sqrt{-g}\left(R - \frac{1}{2}|F_{(4)}|^2\right) - \frac{1}{6}\int A_{(3)} \wedge F_{(4)} \wedge F_{(4)}\right], \tag{4.81}$$

where $F_{(4)} = \mathrm{d}A_{(3)}$ and $2\kappa_{11}^2 = (2\pi)^8 l_{\mathrm{p}}{}^9$, where l_{p} is the Planck length in the eleven-dimensional theory.

Exercise 4.2.1 Reduce eleven-dimensional supergravity as given by (4.81) to ten dimensions by a Kaluza–Klein reduction on a circle with radius $R_{11} = g_{\mathrm{s}}^{2/3}l_{\mathrm{p}} = g_{\mathrm{s}}l_{\mathrm{s}}$ and show that you obtain type IIA supergravity. In particular you may decompose the eleven-dimensional metric with components $g_{\overline{M}\overline{N}}$ into

$$\mathrm{d}s^2 = g_{\overline{MN}}\mathrm{d}x^{\overline{M}}\mathrm{d}x^{\overline{N}}$$

$$= \exp\left(-\frac{2}{3}\phi\right)g_{MN}^{(10)}(x)\mathrm{d}x^M\mathrm{d}x^N + \exp\left(\frac{4}{3}\phi\right)\left(\mathrm{d}x^{10} + C_M\mathrm{d}x^M\right)^2, \tag{4.82}$$

where $\overline{M}, \overline{N} = 0, \ldots, 10$ and $M, N = 0, \ldots 9$. $g^{(10)}$ is the metric of the ten-dimensional theory, ϕ is the dilaton and $C_{(1)} = C_M dx^M$ is the R-R one-form. Moreover, we may decompose the three-form $A_{(3)}$ of eleven-dimensional supergravity into the three-form R-R form field $C_{(3)}$ and the Kalb–Ramond field of type IIA superstring theory by $A_{MNP} = C_{MNP}$ and $A_{MN10} = B_{MN}$.

4.2.4 String theory on Calabi–Yau manifolds

An approach to constructing string theories with low-energy behaviour similar to the field theories of the standard model of elementary particles is to compactify six of the ten dimensions on a *Calabi–Yau manifold*. These are particular compact *Kähler manifolds* as defined in box 4.1, with a trivial canonical bundle.

The Calabi–Yau theorem, proved by Yau following an earlier conjecture by Calabi, states that for a given Calabi-Yau manifold with trivial canonical bundle, there exists a unique metric for which the Ricci scalar vanishes, i.e. which is Ricci flat. Moreover, this is equivalent to stating that in n complex dimensions, such a Kähler manifold admits a non-vanishing holomorphic n-form Ω, or a global holonomy contained in $SU(n)$.

In string theory, the required trivial canonical bundle is obtained from imposing supersymmetry. In fact, the condition of vanishing infinitesimal supersymmetry transformations requires the existence of a covariantly constant spinor,

$$\nabla_m \epsilon = 0. \tag{4.85}$$

The spinor ϵ provides the required bundle. In six real (or three complex) dimensions, the most relevant case in string theory when compactifying ten-dimensional space down to four dimensions, the Kähler form J and the holomorphic three-form Ω may be constructed explicitly from the spinor ϵ by virtue of

$$J_{k\bar{l}} = -i\epsilon^\dagger \Gamma_k \Gamma_{\bar{l}} \epsilon, \tag{4.86}$$

$$\Omega = \Omega_{jkl} dz^j \wedge dz^k \wedge dz^l, \qquad \Omega_{jkl} = \epsilon^T \Gamma_j \Gamma_k \Gamma_l \epsilon, \tag{4.87}$$

Box 4.1	Kähler manifolds

A *Kähler manifold* of complex dimension n is a Hermitian complex manifold as defined in box 2.1 for which the Kähler form

$$J = g_{k\bar{l}} dx^k \wedge d\bar{z}^l \tag{4.83}$$

is closed, $dJ = 0$. This implies that locally, we may write

$$g_{k\bar{l}} = \frac{\partial}{\partial z^k} \frac{\partial}{\partial \bar{z}^l} K(z, \bar{z}), \tag{4.84}$$

i.e. $J = \partial \bar{\partial} K$. K is referred to as the Kähler potential. For a closed Kähler form, parallel transport does not mix holomorphic and anti-holomorphic indices. The holonomy of an n-dimensional Kähler manifold is $U(n)$.

where the Γ_i are six-dimensional Dirac gamma matrices.

Examples of Calabi–Yau manifolds include the following.

- The complex plane $\mathbb{C}$ and the two-torus T^2 in two (real) dimensions. There are no further examples in two dimensions since any compact Riemann surface, except the torus T^2, cannot have a vanishing Ricci scalar and thus is not a Calabi–Yau manifold.
- In four (real) dimensions there are two compact Calabi–Yau manifolds: the four-torus T^4 and $K3$. Examples of non-compact Calabi-Yau manifolds are $\mathbb{C}^2 \times T^2$ and $\mathbb{C}^4$.
- In six (real) dimensions there are many Calabi–Yau three-folds known. In fact, the number of Calabi–Yau three-folds may even be infinite since at present there is no mathematical proof for the finiteness of the number. Note that a non-trivial solution to (4.85) requires a geometry with non-differentiable points which have non-trivial (i.e. non-contractible) cycles.

4.3 Web of dualities

The different types of string theory, such as type IIA and type IIB string theories, as well as type I and heterotic string theories, are related to each other by different kinds of *dualities*. Prominent examples of dualities are S-duality and T-duality, which we introduce here.

4.3.1 T-Duality

T-Duality (or target space duality) denotes the equivalence between two superstring theories compactified on different background spacetimes. Let us consider type II superstring theory compactified on a circle, i.e. the coordinate X^9 is periodically identified in the following way,

$$X^9 \sim X^9 + 2\pi R. \tag{4.88}$$

T-Duality of closed strings

First let us consider closed strings. The embedding function $X^9(\tau, \sigma)$ has to satisfy the periodicity condition

$$X^9(\tau, \sigma + 2\pi) = X^9(\tau, \sigma) + 2m\pi R, \tag{4.89}$$

where R is the radius of the circle and m is an arbitrary integer. The number m counts how often the closed string winds around the compactified direction X^9 and is therefore called the winding number.

In the non-compactified directions, the mode decomposition (4.13) for the right- and left-moving modes can be used subject to $p_{(R)}^M = p_{(L)}^M$. In the compactified direction the same mode decomposition may be applied, however now with $p_{(R)}^9 \neq p_{(L)}^9$. Omitting the oscillatory terms, we have the decomposition

$$X_{(R)}^9(\tau - \sigma) = \frac{1}{2}x_0^M + \alpha' p_{(R)}^9(\tau - \sigma) + \cdots,$$

$$X_{(L)}^9(\tau + \sigma) = \frac{1}{2}x_0^M + \alpha' p_{(L)}^9(\tau + \sigma) + \cdots. \qquad (4.90)$$

Since $X^9 = X_{(L)}^9 + X_{(R)}^9$, the periodicity condition reads

$$\alpha'(p_{(L)}^9 - p_{(R)}^9) = mR. \qquad (4.91)$$

Since the X^9 direction is compactified, the centre of mass momentum $p_{(R)}^9 + p_{(L)}^9$ is quantised in units of $1/R$, i.e.

$$p_{(L)}^9 + p_{(R)}^9 = \frac{n}{R}. \qquad (4.92)$$

Thus $p_{(R)}^9$ and $p_{(L)}^9$ are given by

$$p_{(L)}^9 = \frac{1}{2}\left(\frac{n}{R} + \frac{mR}{\alpha'}\right), \qquad (4.93)$$

$$p_{(R)}^9 = \frac{1}{2}\left(\frac{n}{R} - \frac{mR}{\alpha'}\right). \qquad (4.94)$$

We are now interested in the spectrum of the closed string states. First of all, the level matching condition (4.45) for the closed string is modified,

$$\bar{N} - N = nm, \qquad (4.95)$$

and the mass formula for string states reads

$$M^2 = \left(\frac{mR}{\alpha'}\right)^2 + \left(\frac{n}{R}\right)^2 + \frac{2}{\alpha'}\left(N + \bar{N} - 2\right). \qquad (4.96)$$

However, this is not the whole story. The closed string sector has a remarkable symmetry. Considering the mass formula, it turns out that the closed string spectrum for a compactification with radius R is identical to the closed string spectrum for a compactification with radius $\tilde{R} = \alpha'/R$ if we interchange the winding number m and momentum number n,

$$R \leftrightarrow \tilde{R} = \frac{\alpha'}{R}, \qquad (4.97)$$

$$(n, m) \leftrightarrow (m, n). \qquad (4.98)$$

Although here we have described the proof of T-duality only for free strings, it can be shown that T-duality of closed strings is an exact symmetry at the quantum level also if interactions are included.

In fact it is not possible to distinguish between both compactifications. Note that if R is large, then the dual radius $\tilde{R}$ is small. This is a remarkable feature, which is not present in usual field theories of pointlike particles. Since T-duality exchanges the winding number on the circle with the quantum number of the corresponding (discrete) momentum, it is clear that this symmetry has no counterpart in ordinary point-particle field theory, as the ability of closed strings to wind around the compact dimension is essential.

T-duality of open strings

At first sight, it seems that T-duality does not apply to theories with open strings, since open strings do not have a winding sector. However, T-duality may be restored in the open string sector with the help of *D-branes* which are hyperplanes where open strings end. By applying T-duality, not only the radius of the compactified dimension changes, but also the dimension of the D-brane.

To see this, let us consider the propagation of open bosonic strings in a spacetime which is compactified in the X^9 direction. Furthermore we assume for simplicity that we have a space-filling D9-brane, i.e. the endpoints of the string can move freely. As it was in the case of closed strings, the centre of mass momentum in the compactified direction is quantised, i.e. $p^9 = n/R$, and contributes terms of the form n^2/R^2 to the mass formula of string states. However, this contribution changes if we apply the T-duality rules of closed strings only. Since the dual radius is $\tilde{R} = \alpha'/R$, the contribution to the mass formula changes to $n^2\tilde{R}^2/\alpha'^2$.

T-duality may be restored in the open string sector by considering D-branes. Instead of the D9-brane described above, consider now a D8-brane in the dual theory, which does not wrap the X^9-direction. Due to the Dirichlet boundary conditions, we have no momentum states in the compact direction. In addition, the endpoints of the open string must remain attached to points with $x^9 = x_0^9 + 2\pi n\tilde{R}$, where x_0^9 is the position of the D8-brane in the compactified direction. Therefore we get winding states in the dual theory which contribute to the mass formula by

$$\left(\frac{n\tilde{R}}{\alpha'}\right)^2 = \left(\frac{n}{R}\right)^2. \tag{4.99}$$

This is precisely the contribution of the momentum states in the original theory with a space-filling D9-brane.

Therefore T-duality is an exact symmetry of the open string sector, if the dimension of the D-brane is also changed. This means that the type of boundary conditions of open strings (Neumann or Dirichlet) has to be exchanged in the direction in which T-duality is performed.

As an example consider a D8-brane stretched along the coordinates $X^0, X^1, \ldots, X^8$. In these directions Neumann boundary conditions for open strings are imposed. Moreover, in the X^9-direction, open strings will satisfy Dirichlet boundary conditions. Assuming that the X^8 and X^9 directions are compactified on circles with radii R_8 and R_9 respectively, we can apply T-duality to both compact directions. Performing a T-duality transformation along X^9, the open strings in the dual theory in the X^9-direction also satisfy Neumann boundary conditions. Therefore in the dual theory, a D9-brane exists and the radii of the two compactified directions are given by R_8 and α'/R_9, respectively. Applying a T-duality along X^8 instead, the open strings no longer satisfy Neumann boundary conditions in the X^8-direction. Therefore we are left with a D7-brane in the dual theory, which is compactified on circles with radii α'/R_8 and R_9.

It turns out that a given theory and its T-dual have different chirality in the right-moving Ramond sector. This means that T-duality reverses the relative chiralities of the right- and left-moving ground states. Thus, if we start with type IIA theory which is non-chiral and T-dualise, we obtain type IIB theory which is chiral, and vice versa. Moreover, T-duality relates the different R-R forms $C_{(p)}$ of type IIA and IIB theories to each other.

Applying T-duality in curved spacetime leads to a change in the background fields, i.e. in the metric, Kalb–Ramond field, dilaton and R-R form fields. Performing a T-duality transformation along the X^9-direction, the new background fields, which are denoted by a tilde, are given in terms of the original fields by

$$\tilde{g}_{99} = \frac{1}{g_{99}}, \quad \tilde{g}_{9M} = \frac{B_{9M}}{g_{99}}, \qquad \tilde{g}_{MN} = g_{MN} + \frac{B_{9M}B_{9N} - g_{9M}g_{9N}}{g_{99}},$$
$$\tilde{B}_{9M} = -\tilde{B}_{M9} = \frac{g_{9M}}{g_{99}}, \qquad \tilde{B}_{MN} = B_{MN} + \frac{g_{9M}B_{9N} - B_{9M}B_{9N}}{g_{99}}, \qquad (4.100)$$

for $M, N \neq 9$ and similar rules for the R-R form fields $C_{(p)}$. These relations are referred to as *Buscher rules*.

4.3.2 S-duality

S-Duality is a *strong–weak coupling duality* in the sense that a superstring theory in the weak coupling regime is mapped to another strongly coupled superstring theory. S-Duality relates the string coupling constant g_s to $1/g_s$ in the same way that T-duality maps the radius of the compactified dimension R to α'/R.

The most prominent example where S-duality is present is type IIB superstring theory. This theory is mapped to itself under S-duality. This is due to the fact that S-duality is a special case of the $SL(2, \mathbb{Z})$ symmetry of type IIB superstring theory. In the massless spectrum of type IIB superstring theory, the scalars ϕ and $C_{(0)}$ and the two-form potentials $B_{(2)}$ and $C_{(2)}$ are present in pairs. Arranging the R-R scalar $C_{(0)}$ and the dilaton ϕ in a complex scalar $\tau = C_{(0)} + i \exp(-\phi)$, the $SL(2, \mathbb{R})$ symmetry of the equations of motion of type IIB supergravity (see (4.73)) acts as

$$\tau \mapsto \frac{a\tau + b}{c\tau + d}, \qquad (4.101)$$

with the real parameters a, b, c, d satisfying $ad - bc = 1$. Moreover, the R-R two-form potential $C_{(2)}$ and the NS-NS $B_{(2)}$ transform according to

$$\begin{pmatrix} B_{(2)} \\ C_{(2)} \end{pmatrix} \mapsto \begin{pmatrix} d & -c \\ -b & a \end{pmatrix} \begin{pmatrix} B_{(2)} \\ C_{(2)} \end{pmatrix}. \tag{4.102}$$

Due to charge quantisation, this symmetry group breaks down to $SL(2, \mathbb{Z})$ of the full superstring theory. A particular case of the above symmetry is S-duality. If the R-R scalar $C_{(0)}$ vanishes, the coupling constant $g_s = \exp(\phi)$ of type IIB superstring theory may be mapped to $1/g_s$ by the $SL(2, \mathbb{Z})$ transformation with $a = d = 0$, and $b = -c = 1$, i.e.

$$\phi \mapsto -\phi, \qquad B_{(2)} \mapsto C_{(2)}, \quad C_{(2)} \mapsto -B_{(2)}. \tag{4.103}$$

Note that the $SL(2, \mathbb{Z})$ duality of type IIB superstring theory is a strong–weak coupling duality relating different regimes of the *same* theory.

Since the NS-NS field $B_{(2)}$ couples to the fundamental string, the fundamental string carries one unit of $B_{(2)}$ charge, but is not charged under the R-R two-form field $C_{(2)}$. However, there are also solitonic strings which are charged under the R-R two-form field $C_{(2)}$, but not under the Kalb–Ramond field $B_{(2)}$. These objects are D1-branes as we will see in section 4.4.2. Under S-duality, a fundamental string is transformed into a D1-brane and vice versa. Moreover, a general $SL(2, \mathbb{Z})$ transformation maps the fundamental string into a bound state called a (p, q) string, carrying p units of NS-NS charge and q units of R-R charge. The same is true for the magnetic dual of the fundamental string, the NS5-brane, which will be discussed in section 4.4.4 in more detail. There exist bound states, with p units of NS-NS charge and q units of R-R charge, which are denoted by (p, q) NS5-brane.

4.4 D-branes and other non-perturbative objects

So far we have quantised open and closed (super-)strings and derived the low-energy effective action of closed string theory. In this section we study non-perturbative objects, such as D-branes in string theory. These objects may be viewed from two different point of view. We can view D-branes as hyperplanes where open strings can end. The open strings may deform the D-brane and may lead to non-trivial gauge fields on it. Thus the D-brane not only encodes the boundary conditions in a geometric way but is rather a dynamical object which we will study in section 4.4.1.

However, we can also view D-branes as very massive objects curving the surrounding spacetime. In this picture, D-branes correspond to non-trivial soliton-like solutions in string theory or its low-energy limit supergravity. This will be discussed in section 4.4.2

4.4.1 Low-energy effective action of D-branes

The analysis of the low-energy effective action of closed strings of section 4.2.3 can be repeated in the open string sector, in which the open string boundary conditions are specified by hyperplanes, the D-branes. The endpoint of a string is charged and thus couples to a gauge field A with field strength tensor F living on the D-brane. Imposing that the worldsheet energy-momentum tensor vanishes, we obtain constraints on the form and the dynamics of the gauge field. As in the case of closed strings, these constraints can be rewritten in terms of equations of motion of a Dp-brane action. In the following, we will discuss this action.

Let ξ^a denote the coordinates for the worldvolume of a Dp-brane. For the case of the fundamental string, this reduces to $\xi^0 = \tau$ and $\xi^1 = \sigma$. In direct analogy to the string worldsheet area action, the bosonic part of the Dp-brane action is given by

$$S_{\text{DBI}} = -\tau_p \int d^{p+1}\xi \, e^{-\phi} \sqrt{-\det(P[g]_{ab} + P[B]_{ab} + 2\pi\alpha' F_{ab})}, \qquad (4.104)$$

where $P[g]$ and $P[B]$ denote the pullback of the NS-NS sector bulk fields g_{MN} and B_{MN},

$$P[g]_{ab} = \frac{\partial X^M}{\partial \xi^a} \frac{\partial X^N}{\partial \xi^b} g_{MN}. \qquad (4.105)$$

Moreover, F_{ab} are the components of a U(1) gauge field A living on the brane. The action (4.104) is known as the *Dirac–Born–Infeld action* or, *DBI action*. Its prefactor τ_p reads

$$\tau_p = (2\pi)^{-p}\alpha'^{-(p+1)/2}. \qquad (4.106)$$

Let us consider a few simple examples. If we consider a constant dilaton ϕ with $e^\phi = g_s$ as well as a vanishing Kalb–Ramond field B and a gauge field F vanishing to zero, we see that the Dp-brane action (4.104) reduces to

$$S_{\text{DBI}} = -\frac{\tau_p}{g_s} \int d^{p+1}\xi \sqrt{-\det(P[g]_{ab})}, \qquad (4.107)$$

and thus the D-brane tends to minimise its volume. Hence the prefactor τ_p/g_s is viewed as a tension and we can think of the DBI action as a generalisation of the worldsheet action of strings to higher dimensions. However, unlike fundamental strings, D-branes are non-perturbative objects since the tension and therefore the energy scales as $1/g_s$.

The Dp-brane also has a gauge field F living on it. To investigate its dynamics, consider the embedding of a Dp-brane into flat space with $B = 0$ and with a constant dilaton $e^\phi = g_s$. Expanding the DBI action (4.104) using

$$\det(1 + M) = 1 - \frac{1}{2}\text{Tr}(M^2) + \cdots \qquad (4.108)$$

for antisymmetric matrices M, we obtain to lowest non-trivial order in α',

$$S_{\text{DBI}} = -(2\pi\alpha')^2 \frac{\tau_p}{4g_s} \int d^{p+1}\xi F_{ab} F^{ab}. \qquad (4.109)$$

This implies that the DBI action for a single Dp-brane is a generalisation of Yang–Mills theory with gauge group $U(1)$. From (4.109) we can read off the Yang–Mills coupling constant g_{YM} as

$$g_{\text{YM}}^2 = \frac{g_s}{\tau_p (2\pi\alpha')^2} = (2\pi)^{p-2} g_s \alpha'^{\frac{p-3}{2}}. \qquad (4.110)$$

There are also non-trivial couplings to the R-R forms. The R-R forms $C_{(p)}$ define charges for D-branes in a natural way. In analogy to an electrically charged point particle whose worldline Σ_1 couples to the pullback of the gauge field one-form A by virtue of

$$S_0 = \mu_0 \int_{\Sigma_1} P[A], \tag{4.111}$$

a general $(p+1)$-form $C_{(p+1)}$ couples to surfaces Σ_{p+1} of dimension $(p+1)$ by virtue of the diffeomorphism invariant action

$$S_p = \mu_p \int_{\Sigma_{p+1}} P[C_{(p+1)}], \quad \text{with} \quad \mu_p = \frac{\tau_p}{g_s}. \tag{4.112}$$

This action is invariant under Abelian gauge transformations of rank p, $\delta C_{(p+1)} = d\lambda_p$, where λ_p is a p-form. The full action corresponding to a Dp-brane involves a *Chern–Simons term*, $S = S_{\text{DBI}} \pm S_{\text{CS}}$,

$$S_{\text{CS}} = \mu_p \int \sum_q P[C_{(q+1)}] \wedge e^{P[B]+2\pi\alpha'F}, \tag{4.113}$$

describes the interaction of the R-R fields $C_{(q+1)}$ with the NS-NS field B. The exponential of the two-form $\mathcal{F} \equiv P[B] + 2\pi\alpha'F$ has to be understood in terms of the wedge product. The integral in (4.113) picks the appropriate $(p+1)$-form.

Coincident D-branes: Chan–Paton factors

So far we have seen that open strings on one Dp-brane are described by a $U(1)$ gauge theory. In order to generalise this to non-Abelian gauge theories, *Chan–Paton factors* are introduced on a stack of N coincident Dp-branes. Chan–Paton factors are non-dynamical degrees of freedom from the worldsheet point of view, which are assigned to the endpoints of the string. These factors label the open strings that connect the various coincident D-branes. For example, the Chan–Paton factor λ_{ij} labels strings stretching from brane i to brane j, with $i,j \in \{1,\ldots,N\}$. The resulting matrix λ is an element of a Lie algebra. It turns out that the only Lie algebra consistent with open string scattering amplitudes is $U(N)$ in the case of oriented strings, where N is the number of coincident D-branes. Therefore λ can be chosen as a Hermitian matrix and λ_{ij} are the corresponding entries of the matrix.

Although the Chan–Paton factors are global symmetries of the worldsheet action, the symmetry turns out to be local in the target spacetime. The theory of open strings ending on coincident D-branes can effectively be described by a non-Abelian gauge theory.

So far we have considered only *oriented* strings. This means that a left to right direction on the string may be defined unambiguously. This is obvious since we parametrise the spatial extent by σ. *Unoriented* strings are constructed by imposing the worldsheet parity transformation Ω,

$$\Omega: \sigma \to \sigma_0 - \sigma, \tag{4.114}$$

where $\sigma_0 = 2\pi$ for closed strings and $\sigma_0 = \pi$ for open strings. This transformation, which changes the orientation of the worldsheet, is a global symmetry of string theory. We may truncate the theory consistently by using only Ω-invariant string states. We can view Ω as being an *O-plane*. If we consider both O-planes and D-branes, the low-energy gauge theory may have gauge group $SO(N)$ or $USp(N)$.

4.4.2 D-branes in supergravity

We now look for solution - like solutions of the supergravity equations of motion. A Dp-brane is a BPS solution of ten-dimensional supergravity, i.e. it perserves half of the Poincaré supercharges Q_α of the background. It has a $(p+1)$-dimensional flat hypersurface with Poincaré invariance group $\mathbb{R}^{p+1} \times SO(p, 1)$. The transverse space is then of dimension $D - p - 1$.

A Dp-brane in ten dimensions has symmetries $\mathbb{R}^{p,1} \times SO(p, 1) \times SO(9 - p)$. An ansatz which solves the equations of motion of type IIB supergravity is

$$\mathrm{d}s^2 = H_p(r)^{-1/2} \eta_{\mu\nu}\, \mathrm{d}x^\mu \mathrm{d}x^\nu + H_p(r)^{1/2}\, \mathrm{d}y^i \mathrm{d}y^i, \tag{4.115}$$

$$e^\phi = g_{\mathrm{s}} H_p(r)^{(3-p)/4}, \tag{4.116}$$

$$C_{(p+1)} = \left(H_p(r)^{-1} - 1 \right) \mathrm{d}x^0 \wedge \mathrm{d}x^1 \wedge \cdots \wedge \mathrm{d}x^p, \tag{4.117}$$

$$B_{MN} = 0, \tag{4.118}$$

where x^μ with $\mu = 0, \ldots, p$ are the coordinates on the brane worldvolume and y^i with $i = p+1, \ldots, 9$ denote the coordinates perpendicular to the brane. Moreover, r is defined by $r^2 = \sum_{i=p+1}^{9} y_i^2$. Plugging this ansatz into the equations of motion of type IIB supergravity, the equations of motion imply in particular that

$$\Box H_p(r) = 0 \tag{4.119}$$

for $r \neq 0$. In other words, $H_p(r)$ has to be a harmonic function and therefore can be written as

$$H_p(r) = 1 + \left(\frac{L_p}{r} \right)^{7-p}. \tag{4.120}$$

The constant 1 is chosen such that far away from the brane, i.e. for $r \to \infty$, ten-dimensional Minkowski spacetime is recovered.

In order to determine L_p in the ansatz (4.120), we have to determine the charge of the Dp-brane solution. This charge may be calculated by integrating the R-R flux through the $(8 - p)$-dimensional sphere at infinity, which surrounds the pointlike charge in the $(9 - p)$-dimensional transversal space,

$$Q = \frac{1}{2\kappa_{10}^2} \int_{S^{8-p}} {}^*F_{(p+2)}, \tag{4.121}$$

where $*$ is the ten-dimensional Hodge operator. For the solution (4.115) the charge Q is given by $Q = N \cdot \mu_p$, i.e. the charge Q encodes the total number of Dp-branes.

Calculating the right-hand side of (4.121) and setting $Q = N$, the characteristic length L_p is found to be

$$L_p^{7-p} = (4\pi)^{(5-p)/2} \Gamma\left(\frac{7-p}{2}\right) g_s N \alpha'^{(7-p)/2}.\tag{4.122}$$

In particular, for N D3-branes which play a major role in the AdS/CFT correspondence we obtain the relation

$$L_3^4 = 4\pi g_s N \alpha'^2.\tag{4.123}$$

In type IIA/B superstring theory, Dp-branes with p even or odd, respectively, are stable since R-R gauge potentials $C_{(p+1)}$ are present to which Dp-branes couple. These branes are referred to as BPS since their mass (energy) is proportional to their charge Q,

$$M = \mathrm{Vol}(\mathbb{R}^{p,1}) \cdot N \cdot \mu_p.\tag{4.124}$$

Since type IIA and type IIB supergravity have different $C_{(p)}$ forms, the dimensionality of possible Dp-branes in the two theories also differs. The possible D-branes are listed in table 4.4. The forms $C_{(6)}$ and $C_{(8)}$ are the Hodge dual of the forms $C_{(4)}$, $C_{(2)}$, respectively. D5- and D7-branes couple magnetically as prescribed by (4.112) to the Hodge duals of $C_{(4)}$, $C_{(2)}$, respectively.

In addition, there are also *near-extremal* non-BPS p-branes with solution

$$ds^2 = H_p(r)^{-1/2}\left(-f(r)dt^2 + dx^i dx^i\right) + H_p(r)^{1/2}\left(\frac{dr^2}{f(r)} + r^2 d\Omega_{8-p}^2\right),\tag{4.125}$$

$$f(r) = 1 - \frac{r_h^{7-p}}{r^{7-p}},\tag{4.126}$$

where we have used the polar coordinates $dy^j dy^j \equiv dr^2 + r^2 d\Omega_{8-p}^2$. The dilaton and R-R forms are of the same form as in the D-brane case. These near-extremal branes also have $Q = N \cdot \mu_p$, but their mass is no longer proportional to Q. The parameter r_h plays the role of a horizon since $f(r_h) = 0$.

Table 4.4 Branes in type IIA and type IIB string theory		
type IIB	$\leftrightarrow$	D(-1), D1, D3, D5, D7 branes
type IIA	$\leftrightarrow$	D0, D2, D4, D6, D8 branes

4.4.3 M-branes in M-theory

Since eleven-dimensional supergravity has only an antisymmetric tensor field $A_{(3)}$ of rank three, the possibility for realising branes is very restricted. The consistent supergravity solutions are a 2-brane, referred to as an M2-brane, and its magnetic dual, an M5-brane.

A stack of N coincident M2-branes in flat spacetime sources the fields

$$
\begin{aligned}
\mathrm{d}s^2 &= H(r)^{-2/3}\eta_{\mu\nu}\mathrm{d}x^\mu\mathrm{d}x^\nu + H(r)^{1/3}\left(\mathrm{d}r^2 + \mathrm{d}r^2\mathrm{d}\Omega_7^2\right), \\
A_{(3)} &= H(r)^{-1}\mathrm{d}x^0 \wedge \mathrm{d}x^1 \wedge \mathrm{d}x^2,
\end{aligned}
\tag{4.127}
$$

where $\mu, \nu = 0, 1, 2$ label the worldvolume coordinates of the M2-branes. The function $H(r)$ and the characteristic length scale L are given by

$$
H(r) = 1 + \frac{L^6}{r^6}, \qquad L^6 = 32\pi^2 N l_p^6. \tag{4.128}
$$

Moreover, there are M5-branes which are the magnetic dual of the M2-branes. The corresponding supergravity solution is given by

$$
\mathrm{d}s^2 = H(r)^{-1/3}\eta_{\mu\nu}\mathrm{d}x^\mu\mathrm{d}x^\nu + H(r)^{2/3}(\mathrm{d}r^2 + r^2\mathrm{d}\Omega_4^2), \tag{4.129}
$$

$$
A_{(6)} = H(r)^{-1}\mathrm{d}x^0 \wedge \mathrm{d}x^1 \wedge \cdots \wedge \mathrm{d}x^5, \tag{4.130}
$$

where again $H(r)$ is harmonic and reads

$$
H(r) = 1 + \frac{L^3}{r^3}, \tag{4.131}
$$

where $L^3 = \pi N l_p^3$ and $A_{(6)}$ is the magnetic dual of $A_{(3)}$.

4.4.4 Further supergravity solutions

In addition to the classical supergravity solutions given by D-branes and M-branes, there are many more solutions of classical supergravity. Here we restrict our discussion to those which are charged under the Kalb–Ramond field $B_{(2)}$, the fundamental string, also denoted by F1, as well as its magnetic dual, the NS5-brane. Both objects are solutions of both type IIA and type IIB supergravity.

In Einstein frame, the fundamental string solution or *F1-string* is given by

$$
\mathrm{d}s^2 = H_1(r)^{-3/4}\eta_{\mu\nu}\mathrm{d}x^\mu\mathrm{d}x^\nu + H_1(r)^{1/4}(\mathrm{d}r^2 + r^2\mathrm{d}\Omega_3^2), \tag{4.132}
$$

$$
e^\phi = H_1(r)^{-1/2}g_s, \tag{4.133}
$$

$$
B_{(2)} = \left(H_1(r)^{-1} - 1\right)\mathrm{d}x^0 \wedge \mathrm{d}x^1, \tag{4.134}
$$

where the function $H_1(r)$ reads

$$
H_1(r) = 1 + \frac{L^6}{r^6}, \qquad L^6 = 32\pi^2 g_s^2 \alpha'^3. \tag{4.135}
$$

The magnetic dual of the F-string is the *NS5-brane*, whose supergravity solution in Einstein frame reads

$$ds^2 = H_5(r)^{-1/4}\eta_{\mu\nu}dx^\mu dx^\nu + H_5(r)^{3/4}(dr^2 + r^2 d^2\Omega_3^2), \tag{4.136}$$

$$e^\phi = H_5(r)^{1/2}g_s, \tag{4.137}$$

$$B_{(6)} = \left(H_5(r)^{-1} - 1\right)dx^0 \wedge \cdots \wedge dx^5, \tag{4.138}$$

where the function $H_5(r)$ is given by

$$H_5(r) = 1 + \frac{L^2}{r^2}, \qquad L^2 = N\alpha' \tag{4.139}$$

and N is the number of NS5-branes.

4.5 Further reading

There are a number of standard textbooks and reviews on string theory, which include [1, 2, 3, 4, 5]. A very pedagogical introduction to string theory is [6]. Moreover, [7] is a textbook on supergravity and [8] is a textbook on D-branes. A discussion of Calabi–Yau manifolds and their use in string theory is given in [9]. We refer to these books for references to the original literature. Here we just note the following original references. The GSO projection was developed in [10]. T-duality for open strings is discussed in [11]. D-branes were proposed in [12].

References

[1] Polchinski, J. 1998. *String Theory*. Vol. 1: *An Introduction to the Bosonic String*. Cambridge University Press.

[2] Polchinski, J. 1998. *String Theory*. Vol. 2: *Superstring Theory and Beyond*. Cambridge University Press.

[3] Becker, K., Becker, M., and Schwarz, J. H. 2007. *String Theory and M-Theory: A Modern Introduction*. Cambridge University Press.

[4] Kiritsis, Elias. 2007. *String Theory in a Nutshell*. Princeton University Press.

[5] Blumenhagen, Ralph, Lüst, Dieter, and Theisen, Stefan. 2013. *Basic Concepts of String Theory*. Springer.

[6] Zwiebach, B. 2004. *A First Course in String Theory*. Cambridge University Press.

[7] Freedman, Daniel Z., and Van Proeyen, Antoine. 2012. *Supergravity* Cambridge University Press.

[8] Johnson, C. V. 2003. *D-Branes*. Cambridge University Press.

[9] Ibanez, Luis E., and Uranga, Angel M. 2012. *String Theory and Particle Physics: An Introduction to String Phenomenology*. Cambridge University Press.

[10] Gliozzi, F., Scherk, Joel, and Olive, David I. 1977. Supersymmetry, supergravity theories and the dual spinor model. *Nucl. Phys.*, **B122**, 253–290.

[11] Dai, Jin, Leigh, R. G., and Polchinski, Joseph. 1989. New connections between string theories. *Mod. Phys. Lett.*, **A4**, 2073–2083.

[12] Polchinski, Joseph. 1995. Dirichlet branes and Ramond-Ramond charges. *Phys. Rev. Lett.*, **75**, 4724–4727.

PART II

GAUGE/GRAVITY DUALITY

5 The AdS/CFT correspondence

In theoretical physics, important new results have often been found by realising that two different concepts are related to each other at a deep and fundamental level. Examples of such relations are *dualities* which relate two seemingly different quantum theories to each other by stating that the theories are in fact equivalent. In particular, the Hilbert spaces and the dynamics of the two theories agree. From a mathematical point of view, this means that the theories are identical. However, from a physical point of view, their descriptions may differ, for instance there may be different Lagrangians for the two theories. The duality examples mentioned in box 5.1 either relate quantum field theories together, or they relate string theories together. The Anti-de Sitter/Conformal Field Theory correspondence (AdS/CFT), however, is a new type of duality which relates a quantum field theory on flat spacetime to a string theory. This is particularly remarkable since string theory is a very promising candidate for a consistent theory of quantum gravity. Naively, quantum field theory on flat spacetime does not appear to be a theory of quantum gravity. However, the AdS/CFT correspondence, being a duality, implies that the two theories are equivalent. This explains why many scientists think that the AdS/CFT correspondence, discovered by Maldacena in 1997, is one of the most exciting discoveries in modern theoretical physics in the last two decades.

Moreover, the AdS/CFT correspondence is an important realisation of the *holographic principle*. This principle states that in a gravitational theory, the number of degrees of freedom in a given volume V scales as the surface area ∂V of that volume, as described in box 5.2. The theory of quantum gravity involved in the AdS/CFT correspondence is defined on a manifold of the form $AdS \times X$, where AdS is the Anti-de Sitter space and X is a compact space. The quantum field theory may be thought of as being defined on the conformal boundary of this Anti-de Sitter space.

In a particular limit, the AdS/CFT correspondence is an example of a strong–weak coupling duality. If the field theory is strongly coupled, the dual gravity theory is classical

---

Box 5.1   **Examples of dualities**

Many remarkable examples of dualities have been discussed in the previous chapters, for instance the Montonen–Olive duality of $\mathcal{N} = 4$ Super Yang–Mills theory and string dualities such as T- and S-duality, as well as their generalisations. A further well-studied example in two-dimensional quantum field theory relates the massive Thirring model and the sine–Gordon model. In this case, fermionic degrees of freedom are mapped to bosonic degrees of freedom by a procedure referred to as *bosonisation*.

Box 5.2 | **Holographic principle**

In the context of semi-classical considerations for quantum gravity, the holographic principle asserts that the information stored in a volume V_{d+1} is encoded in its boundary area A_d measured in units of the Planck area l_p^d. This is motivated by the *Bekenstein bound*. The Bekenstein bound states that the maximum amount of entropy stored in a volume is given by $S = A_d/(4G)$, with A_d measured in Planck units and G the Newton constant. The name 'holographic principle' alludes to the fact that this principle is similar to a hologram as known from optics, where the information contained in a volume is stored on a surface.

---

and weakly curved. For that reason the AdS/CFT correspondence is a very promising approach to the study of strongly coupled field theories. Certain questions within strongly coupled quantum field theories become computationally tractable on the gravity side and also conceptionally clearer.

5.1 The AdS/CFT correspondence: a first glance

The AdS/CFT correspondence [1] relates gravity theories on asymptotically Anti-de Sitter spacetimes to conformal field theories. There are many specific examples. For simplicity, here we restrict our discussion to the most prominent example which relates $\mathcal{N} = 4$ Super Yang–Mills theory in 3+1 dimensions and IIB superstring theory on $AdS_5 \times S^5$.

The strongest form of the AdS_5/CFT_4 correspondence states that

---

$\mathcal{N} = 4$ Super Yang–Mills (SYM) theory
with gauge group $SU(N)$ and Yang–Mills coupling constant g_{YM}

is *dynamically equivalent* to

type IIB superstring theory
with string length $l_s = \sqrt{\alpha'}$ and coupling constant g_s
on $AdS_5 \times S^5$ with radius of curvature L and N units of $F_{(5)}$ flux on S^5.

The two free parameters on the field theory side, i.e. g_{YM} and N, are mapped to the free parameters g_s and $L/\sqrt{\alpha'}$ on the string theory side by

$$g_{YM}^2 = 2\pi g_s \qquad \text{and} \qquad 2g_{YM}^2 N = L^4/\alpha'^2.$$

Within this duality, the string theory is defined on the product spacetime $AdS_5 \times S^5$ involving five-dimensional Anti-de Sitter space and a five-dimensional sphere. These both have the same radius L. Type IIB string theory on $AdS_5 \times S^5$ is referred to as the 'AdS side' of the AdS/CFT correspondence. The two free parameters on the AdS side are the string coupling g_s and the dimensionless ratio L^2/α', where $\alpha' = l_s^2$ with l_s the string length. Note that only the ratio $L/\sqrt{\alpha'}$ is important rather than the characteristic length scale of AdS space, i.e. the radius of curvature L, and the string length l_s separately.

The field theory under consideration is the $\mathcal{N} = 4$ $SU(N)$ Super Yang–Mills theory introduced in chapter 3. This field theory is conformally invariant and thus is denoted as the 'CFT side' of the correspondence. Its parameters are the rank of the gauge group N and the coupling constant g_{YM}^2. To establish the correspondence, the parameters of the two sides are identified as follows,

$$g_{\text{YM}}^2 = 2\pi g_s \quad \text{and} \quad 2g_{\text{YM}}^2 N = L^4/\alpha'^2. \tag{5.1}$$

Note that while the first of these equations involves g_{YM}, the second involves the 't Hooft coupling $\lambda = g_{\text{YM}}^2 N$.

What is the meaning of the statement that the two theories are *dynamically equivalent*? The correspondence states that the two theories, i.e. $\mathcal{N} = 4$ Super Yang–Mills theory and type IIB string theory on $AdS_5 \times S^5$, are identical and therefore describe the same physics from two very different perspectives. In particular, if the AdS/CFT conjecture holds, all the physics of one description is mapped onto all the physics of the other. This is very peculiar since in this way, we can map a possible candidate for a theory of quantum gravity, i.e. type IIB string theory, to a field theory without any gravitational degrees of freedom. Moreover, the AdS/CFT correspondence is a realisation of the *holographic principle* as described in box 5.2: the information of the five-dimensional theory obtained from Kaluza–Klein reduction of type IIB string theory on S^5 is mapped to a four-dimensional theory which lives on the conformal boundary of the five-dimensional spacetime.

Although the strongest form of the AdS_5/CFT_4 correspondence as stated above is very interesting and stimulates new ideas, it is very difficult to perform explicit calculations for generic values of the parameters. Therefore it is necessary to lessen the strength, but not the importance, of the proposed AdS/CFT correspondence by taking certain limits on both sides. We will see that in this way, we obtain more tractable forms of the AdS_5/CFT_4 correspondence. A duality between two theories as proposed above is most useful if we obtain new insights into the non-perturbative behaviour, i.e. into the strong coupling dynamics of one theory from the computable weak coupling perturbative behaviour of the other. This will be our guiding principle for restricting the correspondence to particular parameter regimes.

Since currently string theory is best understood in the perturbative regime, it is useful to specialise the string theory side of the correspondence to weak coupling, i.e. to $g_s \ll 1$, while keeping $L/\sqrt{\alpha'}$ constant. At leading order in g_s, the AdS side then reduces to *classical* string theory, in the sense that we take only tree level diagrams into account within string perturbation theory, not the entire string genus expansion. The string length l_s as measured in units of L is kept constant. This is referred to as the *strong form* of the AdS/CFT correspondence. What is the corresponding limit on the CFT side? Using the map between parameters as stated in (5.1), we see that $g_{\text{YM}} \ll 1$ while $g_{\text{YM}}^2 N$ stays finite. In other words, we have to take the large N limit $N \to \infty$ for fixed λ, which is known as the *'t Hooft limit*. As pointed out in chapter 1, this corresponds to the planar limit of the gauge theory. In this respect, AdS/CFT is a concrete realisation of the idea of 't Hooft

Table 5.1 Different forms of the AdS/CFT correspondence		
	$\mathcal{N} = 4$ Super Yang–Mills theory	IIB theory on $AdS_5 \times S^5$
Strongest form	any N and λ	Quantum string theory, $g_s \neq 0$, $\alpha'/L^2 \neq 0$
Strong form	$N \to \infty$, λ fixed but arbitrary	Classical string theory, $g_s \to 0$, $\alpha'/L^2 \neq 0$
Weak form	$N \to \infty$, λ large	Classical supergravity, $g_s \to 0$, $\alpha'/L^2 \to 0$

that the planar limit of a quantum field theory is a string theory. We conclude that a $1/N$ expansion on the field theory side can be mapped to an expansion in the genus of the string worldsheet on the string theory side since $1/N \propto g_s$ for fixed λ.

In the 't Hooft limit, there is only one free parameter on both sides: on the field theory side we can tune the 't Hooft coupling λ, whereas on the string theory side the radius of curvature $L/\sqrt{\alpha'}$ is a free parameter. The two parameters are related by $L^4/\alpha'^2 = 2\lambda$. Since we are interested in strongly coupled field theories, we take the limit $\lambda \to \infty$ on the field theory side, which corresponds to $\sqrt{\alpha'}/L \to 0$. The string length is then very small compared to the radius of curvature. Therefore, for $\sqrt{\alpha'}/L \to 0$ we obtain the point-particle limit of type IIB string theory, which is given by type IIB supergravity on $AdS_5 \times S^5$. This leads to a strong/weak duality in the sense that *strongly* coupled $\mathcal{N} = 4$ Super Yang–Mills is mapped to type IIB supergravity on *weakly* curved $AdS_5 \times S^5$ space. Due to the special limit taken, this is referred to as the *weak form* of the AdS/CFT conjecture. The three different forms of the AdS/CFT correspondence are summarised in table 5.1.

5.2   D3-branes and their two faces

In this section we motivate how the particular example of the AdS/CFT correspondence stated in the preceding section arises within the framework of superstring theory. In particular, we restrict our arguments to the weak form of the correspondence and describe the decoupling limit that is essential for the AdS/CFT correspondence. As explained in chapter 4, superstring theory is much more than just a theory of closed strings. Besides fundamental strings, superstring theory also contains various non-perturbative solitonic higher dimensional objects known as Dirichlet branes, or D-branes for short. D-branes may be viewed from two different perspectives: the *open string* and the *closed string* perspectives. Which perspective is the right one depends on the value of the string coupling constant g_s, which controls the interaction strength between open and closed strings (see Figure 5.1).

- **Open string perspective.** D-branes may be visualised as higher dimensional objects where open strings can end. Since we have to treat strings as small perturbations, this perspective is only reliable if the coupling constant for open and closed strings is small,

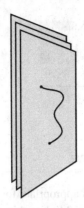

Figure 5.1 D-branes: open string perspective (left) versus closed string perspective (right).

i.e. for $g_s \ll 1$. Moreover, if we neglect massive string excitations, i.e. for low energies $E \ll \alpha'^{-1/2}$, the dynamics of the open strings is described by a supersymmetric gauge theory living on the worldvolume of the D-branes. The gauge field A_μ corresponds to open string excitations parallel to the D-brane while open string excitations transversal to the D_p-brane are scalar fields from the worldvolume point of view. In the case of N coincident D-branes, as explained in section 4.4.1, the gauge group is $U(N)$. Then the effective coupling constant is given by $g_s N$ and the open string perspective is reliable for $g_s N \ll 1$.

- **Closed string perspective.** D-branes may be viewed as solitonic solutions of the low-energy limit of superstring theory, i.e. of supergravity. We may consider D-branes as sources of the gravitational field which curves the surrounding spacetime. The characteristic length scale L should be large in order to ensure weak curvature and the validity of the supergravity approximation. In the case of N coincident D-branes, L^4/α'^2 is proportional to $g_s N$. Therefore the closed string perspective is reliable only for $g_s N \gg 1$.

When applied to a stack of N D3-branes in flat spacetime, these two perspectives on D-branes allow us to motivate the AdS_5/CFT_4 correspondence, which relates four-dimensional $\mathcal{N} = 4$ Super Yang–Mills theory to type IIB superstring theory in $AdS_5 \times S^5$. In the following discussion we concentrate on the weak form of the conjecture as stated in table 5.1.

The stack of N coincident D3-branes extends along the spacetime directions x^0, x^1, x^2 and x^3, and is transversal to the other six spatial directions $x^4, \ldots, x^9$. Without loss of generality, we may describe the embedding of the stack of D3-branes into ten-dimensional flat spacetime by $x^4 = \cdots = x^9 = 0$. The embedding of the D3-branes is visualised in table 5.2, where the directions with Neumann boundary conditions are represented by $\bullet$, and directions with Dirichlet boundary conditions are denoted by $-$.

	0	1	2	3	4	5	6	7	8	9
Table 5.2 Embedding of N coincident D3-branes in flat ten-dimensional spacetime.										
N D3	•	•	•	•	–	–	–	–	–	–

5.2.1 Open string perspective

First, let us consider the open string perspective which is appropriate for $g_s N \ll 1$. We study type IIB superstring theory in flat (9+1)-dimensional Minkowski spacetime where we also embed N coincident D3-branes as given by table 5.2. This configuration breaks half of the thirty-two supercharges of type IIB superstring theory in flat spacetime.

Perturbative string theory in this background consists of two kinds of strings: open strings beginning and ending on the D3-branes and closed strings. Open strings may be viewed as excitations of the (3+1)-dimensional hyperplane, while closed strings are excitations of the (9+1)-dimensional flat spacetime.

Now consider the N D3-branes in flat spacetime at energies $E \ll \alpha'^{-1/2}$. In other words, we take only massless excitations into account and ignore all other stringy excitations since they have energies of order $\alpha'^{-1/2}$. Since the setup preserves supersymmetry, the string modes may be arranged in supermultiplets. To be precise, sixteen of the thirty-two supercharges are preserved. The massless closed string states give rise to a ten-dimensional $\mathcal{N} = 1$ supergravity multiplet. The massless open string excitations may be grouped into a four-dimensional $\mathcal{N} = 4$ supermultiplet which consists of a gauge field A_μ and six real scalar fields ϕ^i as well as fermionic superpartners. According to their transformation properties under worldvolume rotations of the D3-branes, the bosonic massless open string excitations longitudinal to the D3-branes give rise to a gauge field A_μ and the bosonic massless open string excitations transversal to the D3-branes are described by six real scalar fields ϕ^i. Since we consider N coincident D3-branes, the open strings between different branes are massless. As discussed in chapter 4, all the open string modes are therefore valued in the adjoint representation of the gauge group $U(N)$.

The complete effective action for all massless string modes may be written as

$$S = S_{\text{closed}} + S_{\text{open}} + S_{\text{int}}, \tag{5.2}$$

where S_{closed} contains the closed string modes, S_{open} the open string modes and S_{int} the interactions between open and closed string modes. S_{closed} is the action of ten-dimensional supergravity plus some higher derivative terms. As explained in section 4.1.4, schematically S_{closed} reads

$$S_{\text{closed}} = \frac{1}{2\kappa^2} \int d^{10}x \sqrt{-g} e^{-2\phi} (R + 4\partial_M \phi \partial^M \phi)) + \cdots$$
$$\sim -\frac{1}{2} \int d^{10}x \partial_M h \partial^M h + \mathcal{O}(\kappa), \tag{5.3}$$

where κ is given by $2\kappa^2 = (2\pi)^7 \alpha'^4 g_s^2$. g_{MN} and ϕ are the metric and the dilaton, respectively. The second line in (5.3) shows the schematic form of the lowest-order contribution in metric fluctuations h, which is obtained by expanding $g = \eta + \kappa h$. Note that h is multiplied by a factor κ in the expansion to ensure canonical normalisation of the kinetic term for h in the action. In (5.3), the terms involving field strength tensors of the R-R form fields as well as fermionic fields such as the gravitino are not explicitly shown.

The actions $\mathcal{S}_{\text{open}}$ and $\mathcal{S}_{\text{int}}$ can be derived from the Dirac–Born–Infeld action and the Wess–Zumino term. The Dirac–Born–Infeld action for a single D3-brane reads

$$\mathcal{S}_{\text{DBI}} = -\frac{1}{(2\pi)^3 \alpha'^2 g_s} \int d^4x \, e^{-\phi} \sqrt{-\det(\mathcal{P}[g] + 2\pi\alpha' F)}, \tag{5.4}$$

where we have also set the Kalb–Ramond field to zero for simplicity. The world-volume fields are the coordinates x^μ, where $\mu \in \{0,1,2,3\}$. The six coordinates transverse to the D3-brane are labelled by x^i. Moreover, we introduce six real scalar fields ϕ^i which may be identified with the transverse coordinates x^{i+3} by $x^{i+3} = 2\pi\alpha' \phi^i$. Thus the pullback $\mathcal{P}$ of the metric to the worldvolume is given by

$$\mathcal{P}[g]_{\mu\nu} = g_{\mu\nu} + (2\pi\alpha') \left(g_{i+3\,\nu} \partial_\mu \phi^i + g_{\mu j+3} \, \partial_\nu \phi^j \right) + (2\pi\alpha')^2 g_{i+3\,j+3} \partial_\mu \phi^i \partial_\nu \phi^j. \tag{5.5}$$

Expanding $e^{-\phi}$ and $g = \eta + \kappa h$, we find to leading order in α',

$$\mathcal{S}_{\text{open}} = -\frac{1}{2\pi g_s} \int d^4x \left(\frac{1}{4} F_{\mu\nu} F^{\mu\nu} + \frac{1}{2} \eta^{\mu\nu} \partial_\mu \phi^i \partial_\nu \phi^i + \mathcal{O}(\alpha') \right) \tag{5.6}$$

$$\mathcal{S}_{\text{int}} = -\frac{1}{8\pi g_s} \int d^4x \, \phi F_{\mu\nu} F^{\mu\nu} + \cdots. \tag{5.7}$$

An example of a term which is present in $\mathcal{S}_{\text{open}}$ at higher order in α' is given by $\alpha'^2 F^4$. For $\mathcal{S}_{\text{int}}$, we have shown only one term explicitly which is of the form ϕF^2. This term suggests that a dilaton can decay into two gauge bosons on the D3-branes.

So far, we have discussed the low-energy effective actions $\mathcal{S}_{\text{open}}$ and $\mathcal{S}_{\text{int}}$ for the case of a single D3-brane. Generalising the action to the case of N coincident D3-branes, the scalars and gauge fields are $U(N)$ valued, $\phi^i = \phi^{ia} T_a$, $A_\mu = A_\mu^a T_a$, and we have to trace over the gauge group to ensure gauge invariance. This implies for instance that the gauge kinetic term becomes $F_{\mu\nu}^a F^{a\mu\nu}$. Moreover, we have to replace the partial derivatives by the covariant derivatives and we have to add a scalar potential V of the form

$$V = \frac{1}{2\pi g_s} \sum_{i,j} \text{Tr} \left[\phi^i, \phi^j \right]^2 \tag{5.8}$$

to the action $\mathcal{S}_{\text{open}}$ to lowest order in α'.

Let us now naively take the limit $\alpha' \to 0$. Then we find that $\mathcal{S}_{\text{open}}$ is just the bosonic part of the action of $\mathcal{N} = 4$ Super Yang–Mills theory provided that we identify

$$2\pi g_s = g_{\text{YM}}^2. \tag{5.9}$$

All other terms in $\mathcal{S}_{\text{open}}$ are of order α' or higher, and therefore vanish for $\alpha' \to 0$. Moreover, since in this limit also $\kappa \propto \alpha'^2 \to 0$, we observe that $\mathcal{S}_{\text{closed}}$ is just the action of free supergravity in (9+1)-dimensional Minkowski spacetime. Finally, $\mathcal{S}_{\text{int}}$ also vanishes in the limit $\alpha' \to 0$, i.e. open and closed strings decouple. Note that this is not obvious for $\mathcal{S}_{\text{int}}$ as given in (5.7). However, in the same way as for h in (5.3), we also have to rescale the dilaton ϕ by κ in (5.7) for canonical normalisation. Therefore S_{int} is of order κ and vanishes for $\alpha' \to 0$.

To summarise, we have seen that in the naive limit $\alpha' \to 0$, the open and closed strings decouple. While the dynamics of open strings give rise to $\mathcal{N} = 4$ Super Yang–Mills theory, the closed strings are effectively described by supergravity in flat (9+1)-dimensional spacetime.

Suppose we start with $N+1$ D3-branes in flat (9+1)-dimensional spacetime and separate one of the branes from the other N coincident branes, say in the x^9 direction. While the N coincident D3-branes are located at $x^9 = 0$, the other brane is at $x^9 = r$. Considering only massless modes, this system is no longer described by a $U(N + 1)$ gauge theory, but by a $U(N) \times U(1)$ theory as discussed in section 4.4.1. Indeed, this system may be viewed as being in a Higgs phase characterised by $\langle \phi^9 \rangle = r/(2\pi\alpha')$. In the decoupling limit $\alpha' \to 0$ we have to keep all field theory quantities fixed, i.e. in particular $\langle \phi^9 \rangle$. Therefore the correct decoupling limit, the so-called *Maldacena limit*, is given by

$$\alpha' \to 0 \qquad \text{with} \qquad u = \frac{r}{\alpha'} \quad \text{kept fixed,} \tag{5.10}$$

where r is any distance.

5.2.2 Closed string perspective

In order to motivate the AdS$_5$/CFT$_4$ correspondence, let us now interchange the two limits, i.e. the strong coupling and the low-energy limit. Consider N D3-branes in the strongly coupled limit $g_s N \to \infty$. In this limit, we have to take the closed string perspective. The N D3-branes may be viewed as massive charged objects sourcing various fields of type IIB supergravity, and therefore also of type IIB string theory. In this background, closed strings of type IIB superstring theory will propagate.

The supergravity solution of N D3-branes preserving $SO(3, 1) \times SO(6)$ isometries of $\mathbb{R}^{9,1}$ and half of the supercharges of type IIB supergravity, i.e. sixteen out of the thirty-two supercharges, is given by

$$\begin{aligned} ds^2 &= H(r)^{-1/2}\eta_{\mu\nu}\,\mathrm{d}x^\mu \mathrm{d}x^\nu + H(r)^{1/2}\delta_{ij}\,\mathrm{d}x^i \mathrm{d}x^j, \\ e^{2\phi(r)} &= g_s^2, \\ C_{(4)} &= \left(1 - H(r)^{-1}\right)\mathrm{d}x^0 \wedge \mathrm{d}x^1 \wedge \mathrm{d}x^2 \wedge \mathrm{d}x^3 + \cdots, \end{aligned} \tag{5.11}$$

where $\mu, \nu = 0, \ldots, 3$ and $i, j = 4, \ldots, 9$. The radial coordinate r is defined by $r^2 = \sum_{i=4}^{9} x_i^2$, The $\cdots$ in the expression for $C_{(4)}$ stand for terms which ensure self-duality of the associated field strength tensor $F_{(5)} = \mathrm{d}C_{(4)}$. In the present discussion, these terms do not play an important role and we omit them.

Inserting the ansatz (5.11) into the equations of motion of type IIB supergravity, we find that $H(r)$ has to be harmonic, $\Box_g H(r) = 0$ for $r \neq 0$. This equation is solved by

$$H(r) = 1 + \left(\frac{L}{r}\right)^4. \tag{5.12}$$

Here, L is just a constant and we cannot determine L using supergravity. However, from string theory we know that the flux of the $F_{(5)}$ through the sphere S^5 has to be quantised, since this flux counts the number of coincident D3-branes. Using this argument, we deduced in (4.122) that

$$L^4 = 4\pi g_s N \alpha'^2. \tag{5.13}$$

The background consists of two different regions for small r and large r, respectively. If $r \gg L$, then $H(r)$ can be approximated by $H(r) \sim 1$ and the metric (5.11) reduces to ten-dimensional flat spacetime. On the other hand, $r \ll L$ corresponds to the *near-horizon region* or *throat*, in which $H(r)$ is given by $H(r) \sim L^4/r^4$ and the metric reads

$$
\begin{aligned}
ds^2 &= \frac{r^2}{L^2}\eta_{\mu\nu}\,dx^\mu dx^\nu + \frac{L^2}{r^2}\delta_{ij}\,dx^i dx^j \\
&= \frac{L^2}{z^2}\left(\eta_{\mu\nu}\,dx^\mu dx^\nu + dz^2\right) + L^2 ds_{S^5}^2.
\end{aligned}
\tag{5.14}
$$

In the second line we have introduced a new coordinate $z = L^2/r$ as well as spherical coordinates $(r, \Omega_5) \in \mathbb{R}_+ \times S^5$ instead of $(x^4, \ldots, x^9) \in \mathbb{R}^6$ by

$$\delta_{ij}\,dx^i dx^j = dr^2 + r^2 ds_{S^5}^2, \tag{5.15}$$

where $ds_{S^5}^2$ is the metric on S^5 with unit radius. The first terms in the second line of (5.14) correspond to AdS$_5$.

We thus have two different types of closed strings: closed strings propagating in flat ten-dimensional spacetime and closed strings propagating in the near-horizon region. When taking the low-energy limit (5.10), both types of closed strings decouple from each other. This may be seen as follows. Consider a string excitation with energy $\sqrt{\alpha'}E_r$ measured in string units at a fixed radial position r. Though string states in the throat may have large energies $\sqrt{\alpha'}E_r \gg 1$, they should not be integrated out at low energies since the energy E_∞ as measured by an observer at infinity is given by

$$E_\infty = \sqrt{-g_{00}}E_r = H(r)^{-1/4}E_r. \tag{5.16}$$

Since in the low-energy limit, we consider string states in the throat where $r \ll L$, $H(r)$ takes the form $H(r) \sim L^4/r^4$. We find that although the energy of a string excitation E_r close to the throat $r = 0$ might be large, E_∞ is very small, since

$$\sqrt{\alpha'}E_\infty \sim \frac{r}{L}\sqrt{\alpha'}E_r \to 0 \tag{5.17}$$

for $\sqrt{\alpha'}E_r$ fixed, but $r \ll L$. The observer at infinity therefore sees two different low-energy modes: the supergravity modes propagating in flat ten-dimensional spacetime and string excitations in the throat region, which corresponds to an $AdS_5 \times S^5$ spacetime.

To summarise, the background consists of two different regions: a near-horizon region and an asymptotically flat region. The dynamics of the closed strings in asymptotically flat spacetime are described by type IIB supergravity in ten-dimensional flat spacetime, while the strings in the throat region are described by fluctuations about the $AdS_5 \times S^5$ solution of IIB supergravity. When taking the low-energy limit (5.10), both types of closed strings decouple from each other.

In this limit we have

$$\frac{L^4}{r^4} = 4\pi g_s N \frac{\alpha'^2}{r^4} = 4\pi g_s N \underbrace{\frac{\alpha'^4}{r^4}}_{\text{constant}} \cdot \underbrace{\alpha'^{-2}}_{\to \infty} \to \infty, \tag{5.18}$$

i.e. we effectively zoom into the near-horizon region. Therefore the Maldacena limit (5.10) is also referred to as the *near-horizon limit*.

Due to (5.18), we can approximate $H(r)$ by $H(r) \simeq L^4/r^4$ in the near-horizon limit. Thus we obtain for the metric and the four-form potential $C_{(4)}$

$$ds^2 = \frac{r^2}{L^2} \eta_{\mu\nu} dx^\mu dx^\nu + \frac{L^2}{r^2} dr^2 + L^2 d\Omega_5^2, \tag{5.19}$$

$$C_{(4)} = \frac{r^4}{L^4} dx^0 \wedge dx^1 \wedge dx^2 \wedge dx^3 + \cdots. \tag{5.20}$$

As may be read off from (5.19), the D3-brane metric (5.11) reduces to $AdS_5 \times S^5$ in the near-horizon limit. The radius of the sphere S^5 and of AdS_5 are equal and are given by $L^4 = 4\pi g_s N \alpha'^2$. Since AdS_5 and S^5 are both maximally symmetric spacetimes, the curvature of AdS_5 and S^5 factors are given by

$$\begin{aligned} AdS_5 : \quad R_{mlns} &= -\frac{1}{L^2} \left(g_{mn} g_{ls} - g_{ms} g_{ln} \right), & R_{mn} &= -\frac{4}{L^2} g_{mn}, \\ S^5 : \quad R_{\alpha\gamma\beta\delta} &= +\frac{1}{L^2} \left(g_{\alpha\beta} g_{\gamma\delta} - g_{\alpha\delta} g_{\gamma\beta} \right), & R_{\alpha\beta} &= +\frac{4}{L^2} g_{\alpha\beta}, \end{aligned} \tag{5.21}$$

where Latin indices denote AdS_5 coordinates and Greek indices denote S^5 coordinates. The Ricci scalars of AdS space and of the sphere are given by $R_{AdS_5} = -20/L^2$ and by $R_{S^5} = 20/L^2$, respectively. Thus the Ricci scalar R for $AdS_5 \times S^5$ reads $R = R_{AdS_5} + R_{S^5} = 0$.

Strictly speaking, the Maldacena limit (5.10) requires the use of coordinates which are kept fixed. A way to achieve this is to introduce the coordinate $u = \frac{r}{\alpha'}$ and to rewrite the metric and the four-form potential in terms of u. This gives

$$\frac{1}{\alpha'} ds^2 = \frac{u^2}{\tilde{L}^2} \eta_{\mu\nu} dx^\mu dx^\nu + \frac{\tilde{L}^2}{u^2} du^2 + \tilde{L}^2 d\Omega_5^2, \tag{5.22}$$

$$\frac{1}{\alpha'^2} C_{(4)} = \frac{u^4}{\tilde{L}^4} dx^0 \wedge dx^1 \wedge dx^2 \wedge dx^3 + \cdots, \tag{5.23}$$

with $\tilde{L}^2 = L^2/\alpha'$.

5.2.3 Combining both perspectives

In both pictures, the open and the closed string perspectives, we found two decoupled effective theories in the low-energy limits.

- Closed string perspective: type IIB supergravity on $AdS_5 \times S^5$ and type IIB supergravity on $\mathbb{R}^{9,1}$.
- Open string perspective: $\mathcal{N} = 4$ Super Yang–Mills theory on flat four-dimensional spacetime and type IIB supergravity on $\mathbb{R}^{9,1}$.

The two perspectives should be equivalent descriptions of the same physics, and type IIB supergravity on $\mathbb{R}^{9,1}$ is present in both perspectives. This suggests that the other two theories should also be identified. Therefore Maldacena conjectured that $\mathcal{N} = 4$ Super Yang–Mills theory in four dimensions is equivalent to type IIB supergravity on $AdS_5 \times S^5$, although the fundamental degrees of freedom on the two sides are very different. Relaxing the low-energy limit leads to the conjecture that $\mathcal{N} = 4$ Super Yang–Mills theory in four dimensions is equivalent to type IIB string theory on $AdS_5 \times S^5$.

One obvious puzzle remains. We argued that N coincident D3-branes give rise to an $\mathcal{N} = 4$ gauge multiplet in the adjoint representation of $U(N)$. However, in box 5.52 on page 180 we stated that $\mathcal{N} = 4$ Super Yang–Mills theory with gauge group $SU(N)$ (and not $U(N)$!) is equivalent to type IIB string theory on $AdS_5 \times S^5$. It turns out that the overall $U(1) \subset U(N)$ degrees of freedom decouple from the $SU(N)$ degrees of freedom. The $U(1)$ degrees of freedom correspond to *singleton* fields in the gravity theory which are only located at the boundary and cannot propagate into the bulk of AdS_5.

5.2.4 Comparison of symmetries

A first obvious check of the conjecture is to see whether the symmetries agree on both sides. In chapter 3 we discussed the symmetries of $\mathcal{N} = 4$ Super Yang–Mills theory. For example, $\mathcal{N} = 4$ Super Yang–Mills theory is conformal with a vanishing β function. The conformal group in four dimensions is $SO(4, 2)$. Moreover, the theory preserves $\mathcal{N} = 4$ supersymmetry, i.e. there are sixteen Poincaré supercharges which may be grouped into four spinors Q_α^a, where $a = 1, \ldots, 4$ and $\alpha = 1, \ldots, 4$. Since the theory is conformal, in addition to these sixteen Poincaré supercharges there are also sixteen superconformal supercharges, denoted by S_α^a. All of these symmetries form the the supergroup $PSU(2, 2|4)$, under which $\mathcal{N} = 4$ Super Yang–Mills theory is invariant. Details of this supergroup will be discussed in chapter 7. Its superalgebra is given in appendix B. The bosonic subgroup of $PSU(2, 2|4)$ is given by $SU(2, 2) \sim SO(4, 2)$ and $SU(4) \sim SO(6)$. The fermionic part of the supergroup $PSU(2, 2|4)$ is generated by the Poincaré supercharges Q_α^a and the superconformal supercharges S_α^a.

Let us consider the symmetries of string theory on $AdS_5 \times S^5$. First, at the level of geometry, the theory is invariant under the isometry groups of AdS_5 and S^5, which are given by $SO(4, 2)$ and $SO(6)$, respectively. These coincide with the bosonic subgroups of $PSU(2, 2|4)$. Moreover, as we will see in chapter 7, string theory on $AdS_5 \times S^5$ preserves this $PSU(2, 2|4)$ symmetry. Consequently, the symmetries of $\mathcal{N} = 4$ Super Yang–Mills theory in flat spacetime and of type IIB string theory on $AdS_5 \times S^5$ coincide.

5.3 Field-operator map

The AdS/CFT correspondence proposes a map between two different theories. As we will see in this section, this map provides a precise one-to-one relation between operators in $\mathcal{N} = 4$ Super Yang–Mills theory and the spectrum of type IIB string theory on

$AdS_5 \times S^5$. This one-to-one map or *dictionary* arises from the fact that the symmetries on the two sides of the correspondence coincide, which allows field theory operators in certain representations of the $PSU(2, 2|4)$ symmetry to be mapped to string states on $AdS_5 \times S^5$ in the same representation. We begin by establishing this map for the weak form of the AdS/CFT correspondence, where $\mathcal{N} = 4$ Super Yang–Mills operators are mapped to supergravity fields. In particular, this field-operator map allows the AdS/CFT correspondence to be formulated as a relation between generating functionals in field theory and supergravity.

5.3.1 Representations for field theory operators

Our goal is to work out the precise dictionary between objects of the two equivalent theories, in particular between representations of the common symmetry groups. We will relate field theory operators to supergravity fields which transform in the same representation of the superconformal algebra $\mathfrak{su}(2, 2|4)$ or its bosonic subalgebra $\mathfrak{so}(6) \oplus \mathfrak{so}(4, 2)$. This provides a one-to-one map between gauge invariant operators in $\mathcal{N} = 4$ Super Yang–Mills theory and classical fields in IIB supergravity on $AdS_5 \times S^5$.

The field theory operators for which the map is established have to be gauge invariant, which implies that they have to be composite operators. An important class of such operators are the 1/2 BPS or chiral primary operators introduced in chapter 3. We recall from section 3.4.3 that in $SU(N)$ $\mathcal{N} = 4$ Super Yang–Mills theory in four dimensions, the scalar 1/2 BPS operator $\mathcal{O}_\Delta$ of conformal dimension Δ belongs to a representation of $\mathfrak{su}(4)$ with Dynkin labels $[0, \Delta, 0]$, or equivalently to an $\mathfrak{so}(6)$ representation with Dynkin labels $[\Delta, 0, 0]$. In terms of the elementary fields, it takes the explicit form

$$\mathcal{O}_\Delta(x) \equiv \text{Str}\big(\phi^{i_1}(x)\,\phi^{i_2}(x)\ldots\phi^{i_\Delta}(x)\big) \;=\; C^\Delta_{i_1\ldots i_\Delta}\,\text{Tr}\big(\phi^{i_1}(x)\,\phi^{i_2}(x)\ldots\phi^{i_\Delta}(x)\big). \quad (5.24)$$

Here, ϕ^i are the elementary scalar fields of $\mathcal{N} = 4$ Super Yang–Mills theory transforming in the representation **6** of $\mathfrak{so}(6) \cong \mathfrak{su}(4)$ and $C^\Delta_{i_1\ldots i_\Delta}$ belongs to the totally symmetric rank Δ tensor representation of $\mathfrak{so}(6)$. We take the $C^\Delta_{i_1\ldots i_\Delta}$ to be orthonormal. The trace in (5.24) is taken over colour indices. Recall that all the fields transform in the adjoint representation of $SU(N)$). The normalisation is chosen such that all planar graphs scale with N^2. The operators of (5.24) are single-trace operators.

5.3.2 The dual fields of supergravity

On the supergravity side, there are fields in the same representations of $PSU(2, 2|4)$ as given for the operators on the field theory side in the preceding section. These supergravity fields are obtained by decomposing all fields of IIB supergravity into Kaluza–Klein towers on S^5, i.e. by expanding the fields in a complete set of *spherical harmonics* $Y^I(\Omega_5)$ of S^5.

The spherical harmonics form a basis of the space of all functions on S^5. They correspond to irreducible representations of $\mathfrak{so}(6)$, or equivalently of $\mathfrak{su}(4)$. A function on S^5 can be regarded as a restriction on the coordinates x^i of the space $\mathbb{R}^6$ into which

S^5 is embedded. Then, each totally symmetric traceless tensor $C^I_{i_1\ldots i_l}$ of rank l defines a spherical harmonic by

$$Y^I = C^I_{i_1\ldots i_l} x^{i_1}\ldots x^{i_l} \tag{5.25}$$

where we take the tensors C to be orthonormal, i.e.

$$C^I_{i_1\ldots i_l} C^{J\,i_1\ldots i_l} = \delta^{IJ}. \tag{5.26}$$

The spherical harmonics Y^I transform in the representation $[0, l, 0]$ of $SU(4)$ or equivalently in the representation $[l, 0, 0]$ of $SO(6)$. Restricting $\mathbb{R}^6$ to a sphere S^5 with radius L, it can be shown that

$$\Box_{S^5} Y^I = -\frac{1}{L^2} l(l+4) Y^I. \tag{5.27}$$

For a field φ, suppressing any Lorentz indices, we have the Kaluza–Klein expansion

$$\varphi(x, z, \Omega_5) = \sum_{I=0}^{\infty} \varphi^I(x, z)\, Y^I(\Omega_5), \tag{5.28}$$

where (x^μ, z) with $\mu = 0, 1, 2, 3$ denotes the coordinates on AdS_5 and Ω_5 denotes the coordinates on S^5. Inserting the ansatz (5.28) into the ten-dimensional supergravity equations of motion determines the masses and couplings of the AdS_5 fields $\varphi^I(x)$. The decomposition of the IIB supergravity fields into spherical harmonics is an involved calculation [2], therefore we restrict our discussion to some examples here. Moreover, in this section, we consider only the linearised case which provides the masses of the $\varphi^I(x, z)$. The next order which provides cubic couplings will be considered in chapter 6.

As an example we consider those fluctuations which are dual to 1/2 BPS operators. For this purpose, we recall the relevant supergravity background from section 4.2.3. In particular, type IIB supergravity contains a self-dual five-form field F. It enters the ten-dimensional equations of motion for the graviton via

$$R_{MN} = \frac{1}{3!} F_{MABCD} F_N{}^{ABCD}, \tag{5.29}$$

where the capital letters denote ten-dimensional indices. In the $AdS_5 \times S^5$ background solution, the five-form takes particularly simple values: along the legs of AdS space, the five-form $F_{(5)}$ is proportional to the volume form of AdS_5 while along the legs of the sphere, $F_{(5)}$ is proportional to the volume form of S^5. To be precise, we have

$$\bar{F}_{m_1\ldots m_5} = \frac{4}{L}\sqrt{-g_{AdS_5}}\,\varepsilon_{m_1\ldots m_5}\,, \qquad \bar{F}_{\alpha_1\ldots \alpha_5} = \frac{4}{L}\sqrt{g_{S^5}}\,\varepsilon_{\alpha_1\ldots \alpha_5}, \tag{5.30}$$

where the AdS_5 indices are denoted by m_i, $i = 1, 2, \ldots, 5$ and the S^5 indices by $\alpha_i, 1 = 1, 2, \ldots, 5$. Consider now fluctuations of the metric and the R-R five-form around that background, i.e.

$$g_{MN} = \bar{g}_{MN} + h_{MN}\,, \qquad F = \bar{F} + \delta F, \tag{5.31}$$

where $\bar{g}_{MN}$ is the metric of $AdS_5 \times S^5$ and $\bar{F}$ is given in (5.30). In the following, we are restricted to those modes dual to scalar 1/2 BPS operators and thus we consider fluctuations of the S^5 parts of the trace of the metric h_2 and of the four-form $a_{\alpha\beta\gamma\delta}$, defined by

$$h_{\alpha\beta} = h_{(\alpha\beta)} + \frac{h_2}{5}\bar{g}_{\alpha\beta}, \qquad \delta F_{\alpha\beta\gamma\delta\epsilon} = \nabla_{[\alpha}a_{\beta\gamma\delta\epsilon]}. \tag{5.32}$$

For the fields of (5.32), the Kaluza–Klein expansion introduced above reads

$$h_2(z, x, \Omega_5) = \sum h_2^I(z, x)Y^I(\Omega_5), \qquad a_{\alpha\beta\gamma\delta}(z, x, \Omega_5) = \sum b^I(z, x)\epsilon_{\alpha\beta\gamma\delta\epsilon}\nabla^\epsilon Y^I(\Omega_5), \tag{5.33}$$

where (z, x) denote the coordinates of AdS_5 and Ω_5 the coordinates of S^5. To linear order in the fluctuations, inserting this ansatz into the IIB supergravity equations of motion (5.31) gives two coupled equations for the coefficients b^I and h_2^I. These may easily be diagonalised and thus decouple. In particular, this involved procedure leads to the equation

$$\Box_{AdS_5}s^I(z, x) = \frac{1}{L^2}l(l-4)s^I(z, x) \tag{5.34}$$

for the new variable

$$s^I = \frac{1}{20(l+2)}\left(h_2^I - 10(l+4)b^I\right). \tag{5.35}$$

Equation (5.34) is a Klein–Gordon equation for a scalar field of mass $m^2L^2 = l(l-4)$.

Correspondingly, to quadratic order the Kaluza–Klein decomposition of the supergravity fields gives the following dimensionally reduced supergravity action for the s^I modes,

$$S = -\frac{4N^2}{(2\pi)^5L^8}\int d^4x\, dz\, \sqrt{-\bar{g}}\,\frac{A_I}{2}\left(\bar{g}^{mn}\partial_m s^I \partial_n s^I + l(l-4)(s^I)^2\right). \tag{5.36}$$

Here, the prefactor is obtained from writing the ten-dimensional gravitational coupling (4.76) in field theory quantities using the AdS/CFT identification (5.1),

$$\frac{1}{16\pi G_{10}} \equiv \frac{1}{2\kappa_{10}^2} = \frac{4N^2}{(2\pi)^5L^8}. \tag{5.37}$$

The constant A_I is determined from the ten-dimensional IIB supergravity action to be

$$A_I = 32\,\frac{l(l-1)(l+2)}{l+1}\,Z(l)\,, \qquad Z(l)\delta^{IJ} \equiv \int_{S^5} d\Omega\, Y^I(\Omega)Y^J(\Omega), \tag{5.38}$$

where $Z(\ell)$ evaluates to $Z(\ell) = \pi^3/(2^{k-1}(k+1)(k+2))$.

5.3.3 Field-operator map: representations

By comparing the field theory results of section 5.3.1 with the supergravity results of section 5.3.2, we see that when identifying $l = \Delta$, the s^I scalar fields of supergravity as defined in (5.35) are in the same representation $[0, \Delta, 0]$ of $SU(4)$ as the one-half BPS field theory operators $\mathcal{O}^\Delta$ of (5.24). It is thus natural to expect that s^Δ and $\mathcal{O}^\Delta$ are mapped into each other by the proposed holographic dictionary. This map may be extended to superconformal descendants of $\mathcal{O}^\Delta$. These are mapped to appropriate descendants of the s^Δ. Similarly, the holographic dictionary relates the energy-momentum tensor to AdS metric fluctuations, as well as the R-symmetry current to a gauge field fluctuation in the five-dimensional AdS supergravity obtained from the Kaluza–Klein reduction,

Table 5.3 Mass dimension relations	
Type of field	**Relation between m and Δ**
scalars, massive spin two fields	$m^2 L^2 = \Delta(\Delta - d)$
massless spin two fields	$m^2 L^2 = 0, \Delta = d$
p-form fields	$m^2 L^2 = (\Delta - p)(\Delta + p - d)$
spin 1/2, spin 3/2	$|m|L = \Delta - d/2$
rank s symmetric traceless tensor	$m^2 L^2 = (\Delta + s - 2)(\Delta - s + 2 - d)$

$$h^{\mu\nu} \longleftrightarrow T_{\mu\nu}, \qquad A^\mu \longleftrightarrow J_\mu. \tag{5.39}$$

A further example is the map between the dilaton ϕ and $T_{\mathrm{r}}(F^2)$ plus its supersymmetric completion to $\mathscr{L}_{\mathcal{N}=4}$. Note that non-BPS operators, such as the Konishi operator introduced in section 3.4.3, are not dual to supergravity modes present in the supergravity limit. They are expected to be dual to genuine string modes.

General d dimensions

So far we have considered the explicit example of the duality between $\mathcal{N} = 4$ theory in four dimensions and supergravity on $AdS_5 \times S^5$. However, it is also possible to motivate the AdS$_{d+1}$/CFT$_d$ correspondence in dimensions different from $d = 4$, for instance for M2- or M5-branes or for the D1/D5-brane system. Let us consider the case of general d. The Klein–Gordon equation (5.34) is then replaced by

$$\Box_{\mathrm{AdS}}\phi = \frac{1}{L^2}\Delta(\Delta - d)\phi \tag{5.40}$$

for a generic scalar ϕ of mass

$$m^2 L^2 = \Delta(\Delta - d). \tag{5.41}$$

A similar analysis involving the Kaluza–Klein reduction may also be performed for fields of different spin. This gives different relations between Δ and m as listed in table 5.3.

5.3.4 Field-operator map: boundary asymptotics

The proposed field-operator map, which is based on symmetry arguments above, may be made more explicit by considering the boundary behaviour of the supergravity fields. Let us consider a scalar ϕ dual to a primary operator. Its action is given by (5.36). For simplicity, in the present context it is sufficient to consider a toy model action of the form

$$S = -\frac{C}{2}\int dz\, d^d x \sqrt{-g}\,(g^{mn}\partial_m\phi\,\partial_n\phi + m^2\phi^2), \tag{5.42}$$

with $C \propto N^2$ and the mass m given by (5.41). It is convenient to work in AdS Poincaré coordinates, for which the metric is given by

$$ds^2 = g_{mn}dx^m dx^n = \frac{L^2}{z^2} \left(dz^2 + \eta_{\mu\nu} dx^\mu dx^\nu \right). \tag{5.43}$$

The Klein–Gordon equation for the scalar ϕ reads

$$(\Box_g - m^2)\phi = 0, \qquad \Box_g \phi = \frac{1}{\sqrt{-g}} \, \partial_m \left(\sqrt{-g} \, g^{mn} \, \partial_n \phi \right), \tag{5.44}$$

with

$$\Box_g \big|_{\text{AdS}} = \frac{1}{L^2} \left(z^2 \, \partial_z^2 - (d-1) z \, \partial_z + z^2 \eta_{\mu\nu} \partial^\mu \partial^\nu \right). \tag{5.45}$$

For AdS_{d+1} space with coordinates (5.43), it is convenient to perform a Fourier decomposition in the x^μ directions and to consider a plane wave ansatz of the form $\phi(z,x) = e^{ip^\mu x_\mu} \phi_p(z)$. Then, the Klein–Gordon equation for the modes $\phi_p(z)$ reads

$$z^2 \partial_z^2 \phi_p(z) - (d-1)z \partial_z \phi_p(z) - (m^2 L^2 + p^2 z^2)\phi_p(z) = 0, \tag{5.46}$$

where $p^2 \equiv \eta_{\mu\nu} p^\mu p^\nu$. This equation has two independent solutions which are characterised by their asymptotics as $z \to 0$,

$$\phi_p(z) \sim \begin{cases} z^{\Delta_+} & \text{normalisable,} \\ z^{\Delta_-} & \text{non-normalisable,} \end{cases} \tag{5.47}$$

where $\Delta_\pm$ are the roots of $m^2 L^2 = \Delta(\Delta - d)$ given by

$$\Delta_\pm = \frac{d}{2} \pm \sqrt{\frac{d^2}{4} + m^2 L^2}. \tag{5.48}$$

By definition, $\Delta_+ \geq \Delta_-$ and $\Delta_- = d - \Delta_+$. Near the boundary, i.e. for $z \to 0$, we can expand $\phi(z,x)$ as

$$\phi(z,x) \sim \phi_{(0)}(x) z^{\Delta_-} + \phi_{(+)}(x) z^{\Delta_+} + \cdots, \tag{5.49}$$

where $\cdots$ stands for subleading terms in z. The non-normalisable fields define associated boundary fields by virtue of

$$\phi_{(0)}(x) \equiv \lim_{z \to 0} \phi(z,x) \, z^{-\Delta_-} = \lim_{z \to 0} \phi(z,x) \, z^{\Delta_+ - d}. \tag{5.50}$$

By dimensional analysis, we may identify the normalisable AdS mode $\phi_{(+)}$ as vacuum expectation value for a dual scalar field theory operator $\mathcal{O}$ of dimension $\Delta \equiv \Delta_+$, and the non-normalisable modes $\phi_{(0)}$ as source for this operator. Equation (5.41) then provides a relation between the conformal dimension of the field theory operator and the mass of the dual supergravity field.

Let us explain the nomenclature *normalisable* versus *non-normalisable* in (5.47) in more detail. A solution is normalisable if the action evaluated on this solution is finite. Let us check this for the action (5.42) for a field $\phi = \phi(z)$, for which the action becomes

$$S = -\frac{C L^{d-1}}{2} \int dz \, d^d x \, \frac{1}{z^{d+1}} \left(z^2 \, \partial_z \phi \partial_z \phi + m^2 L^2 \phi^2 \right). \tag{5.51}$$

In flat space, fields with negative m^2 have an upside-down potential which leads to an instability. In $(d+1)$-dimensional Anti-de Sitter space, however, scalar fields are still stable even for negative m^2 if their mass satisfies

$$m^2 L^2 \geq -d^2/4, \qquad (5.52)$$

provided the fluctuations have the asymptotic behaviour $\phi \sim z^{\Delta_\pm}$ as discussed in the main text. This is the *Breitenlohner–Freedman bound*.

A straightforward way to obtain the bound is to consider the action (5.51), introducing a new coordinate $y = \ln z$ and rescaling the scalar, $\phi = z^{d/2}\varphi$. Up to a boundary term, the action (5.51) then becomes

$$S = -\frac{CL^{d-1}}{2} \int dy\, d^d x \left(\partial_y \varphi \partial_y \varphi + \left(m^2 L^2 + \frac{1}{4} d^2 \right) \varphi^2 \right). \qquad (5.53)$$

This may be interpreted as the action of a scalar field φ with effective mass squared $m_{\text{eff}}^2 L^2 = m^2 L^2 + \frac{1}{4} d^2$ in flat spacetime. However, in flat spacetime the scalar field theory is only consistent for $m_{\text{eff}}^2 \geq 0$. Thus we obtain the Breitenlohner–Freedman bound (5.52).

Exercise 5.3.1 Near the boundary where $\phi \sim z^\Delta$, show that the action (5.51) is finite when integrating from $z = 0$ to $z = \varepsilon$ provided that $\Delta \geq d/2$.

The result of this exercise implies, together with (5.48), that Δ_+ leads to the normalisable mode. Note that when $\Delta = \Delta_+ > d$, the dual operator on the field theory side is irrelevant and the mass of the supergravity field is positive, $m > 0$. For $\Delta_+ = d$, the dual operator is marginal and the mass m vanishes. Naively, we expect that m^2 has to be positive or zero to give rise to a consistent theory. However, this is not true in Anti-de Sitter spacetimes: we note that m^2 may be negative in AdS space, while satisfying $m^2 L^2 \geq -d^2/4$. This lower bound is referred to as the *Breitenlohner–Freedman bound* [3, 4] and is described in box 5.3. Thus supergravity fields with mass $0 > m^2 L^2 \geq -d^2/4$ are dual to field theory operators of conformal dimension $\Delta \equiv \Delta_+ \geq d/2$ but $\Delta < d$.

The situation is more intricate, however. By an integration by parts, as commonly performed below in section 5.4 for the calculation of correlation functions, and by removing the finite boundary term which arises, the action (5.51) becomes

$$S' = -\frac{CL^{d-1}}{2} \int dz\, d^d x \frac{1}{z^{d+1}} \left(-z^2 \phi\, \partial_z^2 \phi + (d-1)z\, \phi \partial_z \phi + m^2 L^2 \phi^2 \right). \qquad (5.54)$$

The boundary term discarded is non-zero if $\Delta \leq d/2$. This implies that in this case, the action has changed compared with the original definition (5.51).

Exercise 5.3.2 Repeat exercise 5.3.1 for the action (5.54) and show that it is finite when integrating from $z = 0$ to $z = \epsilon$ provided that $\Delta \geq (d-2)/2$.

This new bound $\Delta \geq (d-2)/2$ corresponds to the *unitarity bound* of a scalar field in quantum field theory as discussed in box 3.3. Thus in order to describe dual field theory operators with conformal dimension Δ satisfying $(d - 2)/2 \leq \Delta < d/2$, we have to identify Δ_- as the conformal dimension $\Delta \equiv \Delta_-$. Thus the identification of the source and the vacuum expectation value of the field theory operator can be interchanged for supergravity fields in the mass range

$$-\frac{d^2}{4} < m^2 L^2 \leq -\frac{d^2}{4} + 1. \tag{5.55}$$

Either the source is identified with the leading contribution $\phi_{(0)}$ of the near-boundary expansion of the corresponding supergravity field and the vacuum expectation value with the subleading contribution $\phi_{(+)}$, or vice versa.

This is consistent with the fact that for $m^2 L^2$ satisfying (5.55), there are two consistent ways of imposing boundary conditions on the supergravity fields, while for larger m^2 there is only one.

5.4 Correlation functions

In sections 5.3.3 and 5.3.4, we have shown that there is a one-to-one correspondence or *dictionary* between field theory operators $\mathcal{O}$ and gravity fields ϕ in the same representation of the symmetry group or isometry, respectively. Within this correspondence the boundary value $\phi_{(0)}$ may be interpreted as a source for the field operator $\mathcal{O}$, as discussed within field theory in chapter 1. This suggests a duality between generating functionals on both sides of the correspondence.

5.4.1 Field-operator map for generating functionals

The field-operator map leads to a map between the generating functionals in the following way. The quantum field theory is defined on the d-dimensional conformal boundary of the $(d + 1)$-dimensional AdS space. The generating functional $W[\phi_{(0)}]$ for connected Green's functions of composite field theory operators $\mathcal{O}$ is given in terms of the source fields $\phi_{(0)}$, which we introduce by adding source terms to the action $\mathcal{S}$

$$\mathcal{S}' = \mathcal{S} - \int d^d x\, \phi_{(0)}(x)\, \mathcal{O}(x) \tag{5.56}$$

and compute the partition function $Z[\phi_{(0)}]$ for the action $\mathcal{S}'$. In Euclidean signature we have

$$Z[\phi_{(0)}] = e^{-W[\phi_{(0)}]} = \left\langle \exp\left(\int d^d x\, \phi_{(0)}(x)\, \mathcal{O}(x) \right) \right\rangle_{\text{CFT}}, \tag{5.57}$$

where the fields $\phi_{(0)}$ play the role of the sources.

The AdS side of the duality is governed by an action $\mathcal{S}_{\text{sugra}}[\phi]$, where ϕ are fields in five-dimensional Anti-de Sitter spacetime. The action $\mathcal{S}_{\text{sugra}}[\phi]$ can be derived from a Kaluza–Klein reduction of ten-dimensional type IIB supergravity on $AdS_5 \times S^5$. The AdS/CFT conjecture states that precisely this classical supergravity action is the generating functional for a connected Green's function of composite gauge invariant operators $\mathcal{O}$. To be precise, $S_{\text{sugra}}[\phi]$ is related to the generating functional $W[\phi_{(0)}]$ by

$$W[\phi_{(0)}] = S_{\text{sugra}}[\phi]\Big|_{\lim_{z\to 0}(\phi(z,x)\,z^{\Delta-d})=\phi_{(0)}(x)}, \tag{5.58}$$

where we have assumed that the composite operator $\mathcal{O}$ on the field theory side has dimension Δ and its source is given by $\phi_{(0)}$. In other words, the field theory generating functional as given by (5.57) is identified with a classical action on $(d+1)$-dimensional Anti-de Sitter space, subject to the boundary condition that the $(d+1)$-dimensional fields ϕ assume the boundary values $\phi_{(0)}$ in agreement with (5.50). Thus (5.58) is a central result which formulates the AdS/CFT conjecture in a precise way for local gauge invariant operators and their dual sources. Note that (5.58) is a very non-trivial statement since it equates a generating functional of a field theory in the large N limit with a generating functional for a gravity theory.

This identification of generating functionals may also be formulated for the strongest form of the AdS/CFT correspondence, as summarised in box 5.4. Moreover, the AdS/CFT correspondence for generating functionals may also be given in Lorentzian signature, which is required in particular for implementing causality. This will be discussed in detail in chapters 11 and 12.

The map between generating functionals as given in (5.58) or equivalently in (5.59) is the starting point for the holographic calculation of correlation functions of composite gauge invariant operators. Introducing for all composite operators $\mathcal{O}_i$ on the field theory side the corresponding sources $\phi_{(0)}^i$, we obtain connected correlation functions from the generating functional $W[\phi_{(0)}^i]$ by taking derivatives with respect to the sources $\phi_{(0)}^i$,

$$\langle \mathcal{O}_1(x_1)\,\mathcal{O}_2(x_2)\ldots\mathcal{O}_n(x_n)\rangle_{\text{CFT,c}} = -\frac{\delta^n W}{\delta\phi_{(0)}^1(x_1)\,\delta\phi_{(0)}^2(x_2)\ldots\delta\phi_{(0)}^n(x_n)}\Big|_{\phi_{(0)}^i=0}. \tag{5.62}$$

Each composite operator $\mathcal{O}_i$ corresponds to a gravity field $\phi^i(z,x)$. Thus the AdS/CFT correspondence states that correlation functions for local gauge invariant operator $\mathcal{O}$ on the gravity side are obtained as follows.

- Determine the bulk field ϕ which is dual to the operator $\mathcal{O}$ of dimension Δ and compute S_{sugra} by reducing type IIB supergravity on the sphere S^5.
- Solve the supergravity equations of motion for ϕ, subject to the boundary condition $\phi(z,x) \sim z^{d-\Delta}\phi_{(0)}(x)$ for $z \to 0$.
- Insert the solution $\tilde{\phi}$ into the supergravity action, subject to the appropriate boundary behaviour.
- Take variational derivatives with respect to the source $\phi_{(0)}$ to obtain connected correlation functions.

5.4.2 Witten diagrams and AdS propagators

To make explicit use of the recipe given above, we note that the AdS/CFT correspondence for generating functionals as given by (5.58) implies that the calculation of correlation

Box 5.4	AdS/CFT for generating functionals

For the strongest form of the AdS/CFT correspondence, the precise statement of the correspondence as discussed in this section equates the partition function of type IIB string theory with the generating functional of CFT correlation functions. Let ϕ be a general field propagating in the bulk, not necessarily a scalar, dual to an operator of dimension Δ, with possible indices suppressed. Near the boundary at $z \to 0$, let ϕ take the asymptotic leading behaviour $\phi(z,x) \sim z^{d-\Delta}\phi_{(0)}(x)$. In the strong form, the AdS/CFT correspondence is the statement that

$$\left\langle \exp\left(\int d^d x\, \mathcal{O}\phi_{(0)} \right) \right\rangle_{CFT} = Z_{string}\Big|_{\lim_{z \to 0}(\phi(z,x)\, z^{\Delta-d})=\phi_{(0)}(x)}, \tag{5.59}$$

in Euclidean signature. In the partition function Z_{string}, we integrate over all possible field configurations for ϕ. Note that Z_{string} is not known explicitly.

For the weak form of the correspondence, a *saddle point* to the superstring partition function Z_{string} is given by type IIB supergravity. Thus we may approximate the string partition function by

$$Z_{string}\Big|_{\lim_{z \to 0}(\phi(z,x)\, z^{\Delta-d})=\phi_{(0)}(x)} \approx e^{-S_{sugra}}\Big|_{\lim_{z \to 0}(\tilde{\phi}(z,x)\, z^{\Delta-d})=\phi_{(0)}(x)}, \tag{5.60}$$

where $\tilde{\phi}$ denotes the solution of type IIB supergravity with leading asymptotic behaviour $z^{d-\Delta}\phi_{(0)}$ near the conformal boundary at $z=0$. This amounts to a *saddle point approximation*. In the weak form, the AdS/CFT correspondence therefore equates

$$\left\langle \exp\left(\int d^d x\, \mathcal{O}\phi_{(0)} \right) \right\rangle_{CFT} = e^{-S_{sugra}}\Big|_{\lim_{z \to 0}(\tilde{\phi}(z,x)\, z^{\Delta-d})=\phi_{(0)}(x)}. \tag{5.61}$$

The on-shell bulk action S_{sugra} acts as the generating functional for connected correlation functions involving the operator $\mathcal{O}$.

functions amounts to computing tree level diagrams on the gravity side. These tree level diagrams in AdS space are referred to as *Witten diagrams* for which we can develop a pictorial language which we refer to as Feynman rules. Examples of Witten diagrams are shown in figure 5.2. Let us give the corresponding Feynman rules.

- The external sources $\phi_{(0)}(x)$ of composite gauge invariant operators $\mathcal{O}$ on the field theory side are located at the conformal boundary of AdS space, which is represented by the circle in figure 5.2. The bulk of AdS spacetime is given by the interior of the circle.
- Propagators depart from the external sources either to another boundary point or to an interior interaction point (in which case they are called *bulk-to-boundary propagators*).
- The structure of the interior interaction points is governed by the interaction terms in the supergravity action. So far we have neglected these terms. In chapter 6 we derive these interaction terms by explicitly performing the Kaluza–Klein reduction of type IIB supergravity on S^5.
- Two interior interaction points may be connected by a *bulk-to-bulk propagator*.

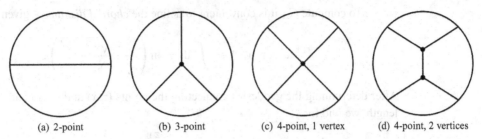

(a) 2-point   (b) 3-point   (c) 4-point, 1 vertex   (d) 4-point, 2 vertices

Figure 5.2 Examples of tree level *Witten diagrams* in AdS spacetime. The circle denotes the boundary of AdS space and its interior corresponds to the bulk of AdS space. The vertices given by the dots are thus in the bulk of AdS space. Diagrams (a)–(c) involve bulk-to-boundary propagators only, while diagram (d) also contains a bulk-to-bulk propagator.

The explicit form of both the bulk-to-bulk and the bulk-to-boundary propagators is obtained from Green's functions of operator $\Box_g - m^2$ in AdS space, subject to the appropriate boundary conditions. We now derive these propagators for scalar supergravity fields. To ensure regularity of the generating functional as in (5.58), we consider the Euclidean case and use the Euclidean AdS metric in $d+1$ dimensions,

$$\mathrm{d}s^2 = \frac{L^2}{z^2}\left(\mathrm{d}z^2 + \delta_{\mu\nu}\mathrm{d}x^\mu \mathrm{d}x^\nu\right), \tag{5.63}$$

where $x \in \mathbb{R}^d$. In the following, we consider a scalar field ϕ_Δ of mass $m^2 L^2 = \Delta(\Delta - d)$ which according to the AdS/CFT dictionary is dual to a scalar operator $\mathcal{O}$ of dimension Δ.

Suppose we would like to solve the source-free equation of motion for ϕ_Δ, $(\Box_g - m^2)\phi_\Delta(z,x) = 0$ subject to the boundary condition $\phi(z,x) = \phi_{(0),\Delta}(x)z^{d-\Delta}$ for $z \to 0$. As in classical electrodynamics, we can reformulate this problem by using the integral kernel K_Δ, which we refer to as the bulk-to-boundary propagator,

$$\phi_\Delta(z,x) = \int_{\partial\mathrm{AdS}} \mathrm{d}^d y\, K_\Delta(z,x;y)\, \phi_{(0),\Delta}(y), \tag{5.64}$$

where $\phi_{(0),\Delta}(y)$ depends only on the boundary variable y^μ. In the same spirit, we can write the solution of the Klein–Gordon equation with source $J(z,x)$, i.e. $(\Box_g - m^2)\phi_\Delta(z,x) = J(z,x)$ by the bulk-to-bulk propagator G_Δ

$$\phi_\Delta(z,x) = \int_{\mathrm{AdS}} \mathrm{d}w\, \mathrm{d}^d y\, \sqrt{g}\, G_\Delta(z,x;w,y)\, J(w,y), \tag{5.65}$$

where the coordinates (z,x) denote a point with bulk coordinate z and boundary coordinates x^μ, while (w,y) denotes a point with bulk coordinate w and boundary coordinates y^μ. Thus the bulk-to-bulk propagator G_Δ has to satisfy

$$\left(\Box_g - m^2\right) G_\Delta(z,x;w,y) = \frac{\delta(z-w)\delta^d(x-y)}{\sqrt{g}}, \tag{5.66}$$

where the action of the Laplacian $\Box_g$ on scalar fields is in general given by (5.45). It turns out that (5.66) is a hypergeometric equation.

Note that the Green function $G_\Delta(z,x;w,y)$ should respect the isometries of AdS space, and thus G_Δ can only depend on the distance $d(z,x;w,y)$ between the points (z,x) and

(w, y). To compute G_Δ it is convenient to define the *chordal distance* ξ given by

$$d(z, x; w, y) \equiv \int\limits_{(z,x)}^{(w,y)} ds = \ln\left(\frac{1 + \sqrt{1 - \xi^2}}{\xi}\right). \tag{5.67}$$

After determining the geodesics connecting the points (z, x) and (w, y) and computing its length, we find for ξ,

$$\xi = \frac{2zw}{z^2 + w^2 + (x - y)^2}. \tag{5.68}$$

The Green function solving (5.66) is then given by a hypergeometric function in the argument ξ, namely

$$G_\Delta(z, x; w, y) = G_\Delta(\xi) = \frac{C_\Delta}{2^\Delta (2\Delta - d)} \, \xi^\Delta \cdot {}_2F_1\left(\frac{\Delta}{2}, \frac{\Delta + 1}{2}; \Delta - \frac{d}{2} + 1; \xi^2\right)$$
$$\text{with} \quad C_\Delta = \frac{\Gamma(\Delta)}{\pi^{d/2}\Gamma(\Delta - \frac{d}{2})}. \tag{5.69}$$

In order to obtain the bulk-to-boundary propagator K_Δ, we use the explicit expression for G_Δ and we put one of its points to the boundary, i.e. we take the limit $w \to 0$. To be precise, the bulk-to-boundary propagator K_Δ is given by

$$K_\Delta(z, x; y) = \lim_{w \to 0} \frac{2\Delta - d}{w^\Delta} G_\Delta(z, x; w, y) \tag{5.70}$$

in terms of the bulk-to-bulk propagator G_Δ. Performing the limit (5.70) explicitly, we obtain

$$K_\Delta(z, x; y) = C_\Delta \left(\frac{z}{z^2 + (x - y)^2}\right)^\Delta. \tag{5.71}$$

Note that K_Δ is regular in the interior, i.e. for $z \to \infty$. Moreover, it diverges as

$$\lim_{z \to 0} \left(z^{\Delta - d} K_\Delta(z, x; y)\right) = \delta^d(x - y) \tag{5.72}$$

near the boundary $z \to 0$ and thus corresponds to a delta-distribution like source. Note that (5.72) justifies the choice of the factor $2\Delta - d$ in (5.70) a posteriori. Thus for a given source $\phi_{(0),\Delta}$, the solution to the source-free equations of motion is given by

$$\phi_\Delta(z, x) = \frac{\Gamma(\Delta)}{\pi^{d/2}\Gamma(\Delta - \frac{d}{2})} \int_{\partial\text{AdS}} d^d y \left(\frac{z}{z^2 + (x - y)^2}\right)^\Delta \phi_{(0),\Delta}(y) \tag{5.73}$$

and thus $\phi_{(+)}(x)$ as defined by the expansion (5.49) is given by

$$\phi_{(+)}(x) = \lim_{z \to 0} \left(z^{-\Delta}\phi_\Delta(z, x)\right) = \frac{\Gamma(\Delta)}{\pi^{d/2}\Gamma(\Delta - \frac{d}{2})} \int_{\partial\text{AdS}} d^d y \, (x - y)^{-2\Delta} \, \phi_{(0),\Delta}(y). \tag{5.74}$$

Exercise 5.4.1 Show that you can verify (5.70) also by using Green's second identity,

$$\int_M dz \, d^d x \sqrt{g} \left(\phi(\Box_g - m^2)\psi - \psi(\Box_g - m^2)\phi\right) = \int_{\partial M} d^d x \sqrt{\gamma}(\phi\partial_n\psi - \psi\partial_n\phi), \tag{5.75}$$

where γ is the determinant of the induced metric on ∂M and ∂_n is the derivative normal to the boundary, i.e. in our case $\partial_n \sim \partial_z$.

In chapter 6, we will use the propagators introduced here to perform checks on the AdS/CFT correspondence as stated quantitatively in (5.58). Generically, the on-shell supergravity action diverges. These divergences arise from the infinite volume of AdS space, i.e. they are long-distance or infrared (IR) divergences. In the field theory, we have a similar phenomenon, namely pure contact terms in correlation functions. These terms, which are unphysical since they are scheme dependent, arise due to short-distance ultraviolet (UV) divergences.

Thus we see that in the AdS/CFT correspondence, IR divergences on the gravity side are connected to UV divergences on the field theory side. To make the AdS/CFT correspondence meaningful, we have to regulate and renormalise these divergences. As an example of a consistent procedure to achieve this, we consider the calculation of a scalar two-point function in the next section.

5.4.3 Two-point function

Let us calculate the two-point function $\langle \mathcal{O}(x)\mathcal{O}(y) \rangle$ of composite gauge invariant operators $\mathcal{O}$ of the d-dimensional field theory. The corresponding contribution is given by the Witten diagram in figure 5.2(a). Although this seems to be straightforward, we will see that the calculation of the two-point function requires careful treatment of potential divergences at the boundary [5].

For simplicity, consider again a scalar operator $\mathcal{O}$ with conformal dimension Δ on the field theory side which is dual to a scalar field ϕ in the $(d + 1)$-dimensional gravity theory. Since the Witten diagram of interest does not contain any interaction vertices in the bulk, only quadratic terms in the scalar field ϕ in the bulk gravity action are important and we can neglect the self-interactions. Moreover, note that the scalar field couples to the metric since it contributes to the energy-momentum tensor in the Einstein equation. However, here we consider the *probe limit* in which the contribution of the scalar to the energy-momentum tensor is neglected – since such contributions are precisely the interaction vertices which are not of importance in this analysis.

Thus in Euclidean signature, the relevant part of the action $\mathcal{S}[\phi]$ reads

$$\mathcal{S}[\phi] = \frac{C}{2} \int dz\, d^d x\, \sqrt{g}\, \left(g^{mn}\, \partial_m \phi\, \partial_n \phi\, +\, m^2\, \phi^2 \right), \qquad (5.76)$$

where we have to adjust the mass of the scalar such that it satisfies $m^2 L^2 = \Delta(\Delta - d)$. Note that in the probe limit, Anti-de Sitter spacetime is still a solution. In the following, we will use the explicit metric (5.63). The equation of motion of the action (5.76) reads

$$(\Box_g - m^2)\phi = 0, \qquad \Box_g \phi = \frac{1}{\sqrt{g}} \partial_m(\sqrt{g} g^{mn} \partial_n \phi). \qquad (5.77)$$

In order to proceed we have to find the solution to this equation of motion subject to the boundary condition (5.49) for any source $\phi_{(0)}(x)$. We obtain this solution for example by integrating (5.64) with (5.71). Then, we insert this solution into the action (5.76) and determine the *on-shell action* $\mathcal{S}[\phi]$. Since by construction the solution ϕ satisfies the equations of motion, $\mathcal{S}[\phi]$ is just a boundary term of the form

$$\mathcal{S}[\phi] = -\frac{C}{2} \int d^d x \sqrt{g} g^{zz} \, \phi(z,x) \, \partial_z \phi(z,x) \Big|_{z=\epsilon}. \tag{5.78}$$

The integrand of (5.78) has to be evaluated at both limits of integration, i.e. for $z \to \infty$ and $z = 0$. However, imposing regularity in the interior ensures that the integrand vanishes for $z \to \infty$, as we will see in detail below when considering the explicit solutions. At the lower limit, $z = 0$, the expression $\sqrt{g} g^{zz} = (L/z)^{d-1}$ is divergent and thus we have to regularise $\mathcal{S}[\phi]$, for example by omitting the region $0 < z < \epsilon$ and imposing all boundary conditions at $z = \epsilon$ where ϵ is small.

Since we restrict z to $z \geq \epsilon$, it is no longer possible to use the isometries of AdS spacetime in order to find the solution ϕ, as we did in the preceding section. Instead, we perform a Fourier transform along the boundary coordinates x while keeping the radial direction z in configuration space. The Fourier transformation reads

$$\phi(z,x) = \int \frac{d^d p}{(2\pi)^d} \, e^{ip \cdot x} \, \phi(z,p), \tag{5.79}$$

where p is the momentum along the field theory directions and $p \cdot x = \delta_{\mu\nu} p^\mu x^\nu$. Let us assume that the functions $\phi(z,p)$ satisfy (5.46) with $p^2 = \delta_{\mu\nu} p^\mu p^\nu$, where $|p| = \sqrt{p^2}$ is real due to the Euclidean signature. Then $\phi(z,x)$ as given by (5.79) also satisfies the equations of motion derived from (5.76).

While in section 5.3.4, we only studied the asymptotic boundary behaviour of equation (5.46), here the full solution is required. We note that (5.46) is a *Bessel equation* which has two independent solutions in terms of modified Bessel functions,

$$\phi(z,p) = A_p z^{d/2} K_\nu(z|p|) + B_p z^{d/2} I_\nu(z|p|), \tag{5.80}$$

where $\nu = \Delta - d/2 = \sqrt{d^2/4 + m^2 L^2}$. Since $I_\nu(z)$ diverges exponentially for $z \to \infty$, imposing regularity in the interior implies that we have to omit this solution by setting $B_p = 0$. In contrast, $K_\nu(z)$ decays exponentially for $z \to \infty$ and thus is regular in the interior of AdS space. In particular, for this solution the integrand of (5.78) vanishes for $z \to \infty$. Note in addition that for $\nu \neq 0$ we have $K_\nu(z) \sim z^{-\nu}$ for $z \to 0$, and thus $\phi(z,p)$ has the correct boundary behaviour for $z \to 0$,

$$\phi(z,p) \sim z^{d/2-\nu} A_p = z^{d-\Delta} A_p, \tag{5.81}$$

if A_p are related to the Fourier modes of the source $\phi_{(0)}(x)$, which we will denote by $\phi_{(0)}(p)$. Since we have to match $\phi(z,p)$ to $\phi_{(0)}(p)$ at $z = \epsilon$, the correct normalised solution for $\phi(z,p)$ reads

$$\phi(z,p) = \frac{z^{d/2} K_\nu(z|p|)}{\epsilon^{d/2} K_\nu(\epsilon|p|)} \, \phi_{(0)}(p) \, \epsilon^{d-\Delta}. \tag{5.82}$$

We determine the on-shell action by inserting the solution into (5.78) and we obtain

$$\mathcal{S}[\phi] = -\frac{C L^{d-1}}{2\epsilon^{d-1}} \int \frac{d^d p}{(2\pi)^d} \frac{d^d q}{(2\pi)^d} \, (2\pi)^d \delta^d(p+q) \, \phi(z,p) \, \partial_z \phi(z,q) \Big|_{z=\epsilon}. \tag{5.83}$$

Using (5.82) we can express $\phi(z, p)$ in terms of $\phi_{(0)}(p)$ and thus the action depends only on $\phi_{(0)}$, i.e. $\mathcal{S}[\phi_{(0)}]$. Moreover, using (5.58) and (5.62), we obtain for the two-point functions for the dual CFT operators,[1]

$$
\begin{aligned}
\langle \mathcal{O}(p)\, \mathcal{O}(q) \rangle_\epsilon &= -(2\pi)^{2d} \frac{\delta^2 \mathcal{S}[\phi_{(0)}]}{\delta\phi_{(0)}(-p)\, \delta\phi_{(0)}(-q)} \\
&= -\frac{(2\pi)^d \delta^d(p+q)\, CL^{d-1}}{\epsilon^{2\Delta-d-1}} \left. \frac{\mathrm{d}}{\mathrm{d}z} \ln\!\left(z^{d/2}\, K_\nu(z|p|) \right) \right|_{z=\epsilon} \\
&= -\frac{(2\pi)^d \delta^d(p+q)\, CL^{d-1}}{\epsilon^{2\Delta-d}} \left(\frac{d}{2} + \frac{\epsilon|p|\, K_\nu'(\epsilon|p|)}{K_\nu(\epsilon|p|)} \right), \qquad (5.84)
\end{aligned}
$$

where we still have to take the limit $\epsilon \to 0$ to obtain the two-point function. Thus we have to expand the Bessel function $K_\nu(u)$ for small arguments u. The form of the expansion depends on whether ν is a positive integer or not. Here, we discuss the case where ν is a positive integer. This includes the cases where the associated CFT operator $\mathcal{O}$ has conformal dimension $\Delta = \nu + d/2$. Thus the conformal dimension is an integer in even dimension, which for example is true for the chiral primary operators $\mathcal{O}_k$ given by (5.24).

The expansion of the Bessel function $K_\nu(u)$ for $u \to 0$ and $\nu \in \mathbb{N}$ has the schematic form [2]

$$
K_\nu(u) \sim u^{-\nu}\left(a_0 + a_1\, u^2 + \mathcal{O}(u^4)\right) + u^\nu \ln u\, \left(b_0 + b_1\, u^2 + \mathcal{O}(u^4)\right), \qquad (5.85)
$$

where the coefficients a_i and b_i are functions of ν. In the following we do not need the exact values of a_i and b_i but rather only the quotient

$$
\frac{2\nu b_0}{a_0} = \frac{(-1)^{\nu+1}}{2^{2(\nu-1)}\Gamma(\nu)^2}. \qquad (5.86)
$$

Using the expansion (5.86) in (5.84), we obtain

$$
\langle \mathcal{O}(p)\, \mathcal{O}(q) \rangle_\epsilon = (2\pi)^d\, \delta^d(p+q)\, CL^{d-1}\Bigg(\frac{\beta_0 + \beta_1\, \epsilon^2\, |p|^2 + \cdots + \beta_\nu\, (\epsilon|p|)^{2(\nu-1)}}{\epsilon^{2\Delta-d}} \\
- \frac{2\,\nu\, b_0}{a_0}\, |p|^{2\nu} \ln(\epsilon|p|)(1+\mathcal{O}(\epsilon^2)) \Bigg),
$$
$$
\qquad (5.87)
$$

where the coefficients β_i are ratios of a_k and b_k and thus functions of ν. For example, $\beta_0 = \nu - d/2$. The terms on the first line of (5.87), involving powers of the momentum, correspond to scheme dependent contact terms: their Fourier transform gives terms of the form $\sim \Box^m \delta^d(x-y)$.

Thus we concentrate on the second line of (5.87). In the limit $\epsilon \to 0$, only the first term in the second line, involving the logarithm of the momentum, contributes. Thus using

---

[1] Note that

$$
\int \mathrm{d}^d x\, \mathcal{O}(x)\phi_{(0)}(x) = \int \frac{\mathrm{d}^d p}{(2\pi)^d} \phi_{(0)}(-p)\mathcal{O}(p)
$$

with our convention and thus we have to take derivatives with respect to $(2\pi)^{-d}\phi_{(0)}(-p)$ to insert operators $\mathcal{O}(p)$ into correlation functions.

[2] In the case of non-integer ν, we just have to replace $u^\nu \ln u$ by u^ν in the expansion (5.85).

(5.86), we obtain the correct non-local result for the correlator

$$\langle \mathcal{O}(p)\,\mathcal{O}(q) \rangle = -(2\pi)^d \delta^d(p+q)\, CL^{d-1} \frac{(-1)^{\nu+1}}{2^{2(\nu-1)}\Gamma(\nu)^2}\, |p|^{2\nu}\, \ln(\epsilon|p|). \qquad (5.88)$$

Transforming the non-local contribution $\propto |p|^{2\nu} \ln|p|$ back to position space yields the ϵ-independent result [6, 7]

$$\langle \mathcal{O}(x)\,\mathcal{O}(y) \rangle = CL^{d-1} \frac{\Gamma(\Delta)}{\Gamma(\Delta-d/2)}\, \frac{2\Delta-d}{\pi^{d/2}\,|x-y|^{2\Delta}}, \qquad (5.89)$$

which agrees with the spatial dependence expected from conformal field theory, see (3.94). For a given theory, we only have to determine C. For example, for the chiral primary operators $\mathcal{O}(x)$ of $\mathcal{N}=4$ Super Yang–Mills theory as given by (5.24), the coefficient C may be read off from (5.36).

5.5 Holographic renormalisation

As we have seen in the preceding section, calculation of two-point functions from the propagation of the dual supergravity fields through AdS space is straightforward in principle. However, there are divergences present which are associated with the boundary behaviour of the supergravity fields. These require regularisation in general. In the calculation of the two-point function in section 5.4.3, we used a cut-off regularisation.

There is a general consistent method available for dealing with near-boundary divergences, which is known as *holographic renormalisation* [8, 9]. This method is of great importance for performing explicit tests of the AdS/CFT correspondence. This includes the calculation of correlation functions and of anomalies. We therefore describe holographic renormalisation in detail. As a simple example we begin by considering holographic renormalisation of a scalar field.

5.5.1 Scalar

Let us revist the calculation of the two-point function $\langle \mathcal{O}(x)\mathcal{O}(y) \rangle$ as given by the Witten diagram figure 5.2(a). Thus we only have to consider the action as given by (5.76) and the corresponding equation of motion (5.77). For holographic renormalisation, we use a slightly different metric of AdS space. A very convenient choice is the *Fefferman–Graham* coordinates as obtained in (2.133),

$$ds^2 = g_{mn}dx^m dx^n = L^2 \left(\frac{d\rho^2}{4\rho^2} + \frac{1}{\rho}\delta_{\mu\nu}dx^\mu dx^\nu \right). \qquad (5.90)$$

In these coordinates, the AdS boundary is located at $\rho = 0$. In order to solve (5.77), we consider an ansatz of the form

$$\phi(\rho, x) = \rho^{(d-\Delta)/2} \bar{\phi}(\rho, x), \tag{5.91}$$

$$\bar{\phi}(\rho, x) = \phi_{(0)}(x) + \rho \phi_{(2)}(x) + \rho^2 \phi_{(4)}(x) + \cdots, \tag{5.92}$$

i.e. we perform an expansion of ϕ about the AdS boundary at $\rho = 0$. According to the results of section 5.3.4, the boundary term $\phi_{(0)}$ corresponds to the source for the dual scalar operator in the field theory. Inserting (5.92) into (5.77) gives

$$0 = [(m^2 L^2 - \Delta(\Delta - d))\bar{\phi}(\rho, x) \tag{5.93}$$
$$- \rho(\Box_0 \bar{\phi}(\rho, x) + 2(d - 2\Delta + 2)\partial_\rho \bar{\phi}(\rho, x) + 4\rho \partial_\rho^2 \bar{\phi}(\rho, x))],$$

with $\Box_0 \equiv \delta^{\mu\nu} \partial_\mu \partial_\nu$ the Laplace operator at the d-dimensional boundary. This equation is now solved order by order in ρ. To lowest order, by just setting $\rho = 0$, we obtain the well-known relation for the scalar mass, $m^2 L^2 = \Delta(\Delta - d)$. Inserting this into (5.93), we obtain

$$\Box_0 \bar{\phi}(\rho, x) + 2(d - 2\Delta + 2)\partial_\rho \bar{\phi}(\rho, x) + 4\rho \partial_\rho^2 \bar{\phi}(\rho, x) = 0. \tag{5.94}$$

Setting $\rho = 0$, this implies

$$\phi_{(2)}(x) = \frac{1}{2(2\Delta - d - 2)} \Box_0 \phi_{(0)}(x) \tag{5.95}$$

for the second coefficient in the expansion (5.92).

Exercise 5.5.1 Perform the calculation of $\phi_{(2)}(x)$ explicitly, following the steps given above.

Similarly, we may obtain the higher order coefficients by differentiating (5.94) with respect to ρ an appropriate number of times and subsequently setting $\rho = 0$. In this way we obtain

$$\phi_{(2n)} = \frac{1}{2n(2\Delta - d - 2n)} \Box_0 \phi_{(2n-2)}. \tag{5.96}$$

Thus we can solve recursively for $\phi_{(2n)}$ with $n \in \mathbb{N}$. Note, however, that this procedure stops if the denominator in (5.96) vanishes, i.e. for $2\Delta - d - 2k = 0$ with $k \in \mathbb{N}$. This can only happen for integer conformal dimensions Δ in even dimensions or for half integer conformal dimensions Δ in odd dimensions.[3] In both cases, a logarithmic term has to be introduced at order ρ^k in the expansion (5.92) to obtain a solution. As an example, let us consider the case $2\Delta - d - 2 = 0$, i.e. for $k = 1$, in which the expansion about the boundary is given by

$$\bar{\phi}(\rho, x) = \phi_{(0)} + \rho(\phi_{(2)} + \ln \rho \, \chi_{(2)}) + \cdots. \tag{5.97}$$

Inserting this into (5.93) again we now obtain

$$\chi_{(2)} = -\frac{1}{4} \Box_0 \phi_{(0)}, \tag{5.98}$$

---

[3] In even dimensions, we have to be careful for instance of chiral primary operators since their dimension is integer.

while $\phi_{(2)}$ is no longer determined by $\phi_{(0)}$ in a simple manner by expanding the equation of motion near the conformal boundary. However, $\phi_{(2)}$ is still determined if we know the full solution with the appropriate boundary conditions. In particular, $\phi_{(2)}(x)$ can be related to $\phi_{(0)}(x)$ in a non-local way using infinitely many derivatives.

Let us discuss the more general case $2\Delta - d - 2k = 0$ with k an integer. In this case, we find

$$\chi_{(2k)} = -\frac{1}{2^{2k}\Gamma(k)\Gamma(k+1)}(\Box_0)^k\phi_{(0)}. \tag{5.99}$$

In this case, it is $\phi_{(2k)}$ which is no longer determined by the equation of motion.

Next we aim to evaluate the action (5.76) on the asymptotic boundary expansion solution we just constructed. This requires regularisation for which we introduce a cut-off at $\rho = \epsilon$. Since the equation of motion is satisfied, the bulk contribution to the action vanishes. However, as in section 5.4.3 – see (5.78) – a non-zero boundary contribution remains, obtained by integration by parts,

$$S_{\text{reg}} = -\frac{C}{2}\int d^dx\sqrt{g}g^{\rho\rho}\phi\partial_\rho\phi\Big|_{\rho=\epsilon}, \tag{5.100}$$

where ϕ is the solution to the equation of motion with appropriate boundary conditions. In particular, ϕ can be expanded near the boundary as in (5.91) and thus S_{reg} reads

$$S_{\text{reg}} = -CL^{d-1}\int d^dx\, \rho^{-\Delta+\frac{d}{2}}\left(\frac{1}{2}(d-\Delta)\bar{\phi}(\rho,x)^2 + \rho\bar{\phi}(\rho,x)\partial_\rho\bar{\phi}(\rho,x)\right)\Big|_{\rho=\epsilon}$$
$$= CL^{d-1}\int d^dx\left(\epsilon^{-\Delta+\frac{d}{2}}a_{(0)} + \epsilon^{-\Delta+\frac{d}{2}+1}a_{(2)} + \cdots - \ln\epsilon\, a_{(2\Delta-d)}\right), \tag{5.101}$$

with coefficients $a_{(2n)}$, which are local functions, depending only on the boundary source $\phi_{(0)}$. These coefficients are given by

$$a_{(0)} = -\frac{1}{2}(d-\Delta)\phi_{(0)}^2, \tag{5.102}$$

$$a_{(2)} = -(d-\Delta+1)\phi_{(0)}\phi_{(2)} = -\frac{d-\Delta+1}{2(2\Delta-d-2)}\phi_{(0)}\Box_0\phi_{(0)}, \tag{5.103}$$

$$a_{(2\Delta-d)} = -\frac{d}{2^{2k+1}\Gamma(k)\Gamma(k+1)}\phi_{(0)}(\Box_0)^k\phi_{(0)}. \tag{5.104}$$

These coefficients have important applications. For instance, they play a crucial role in the computation of the conformal anomaly.

Note that since $\Delta > d/2$, the on-shell value of the action (5.101) diverges if we take $\epsilon \to 0$. We will subtract these divergences by introducing a counterterm. In order to obtain diffeomorphism invariant counterterms in the bulk at $\rho = \epsilon$ we have to invert the expansion (5.92). Instead of expanding $\phi(\epsilon, x)$ in a power series of $\phi_{(0)}, \phi_{(2)}, \ldots$, we solve for $\phi_{(0)}$ and $\phi_{(2)}, \ldots$ in terms of $\phi(\epsilon, x)$ and derivatives $\Box_\gamma\phi(\epsilon, x)$. Here, γ is the induced metric $\gamma_{\mu\nu} = L^2\delta_{\mu\nu}/\epsilon$ on the hyperplane given by $\rho = \epsilon$. The Laplace operator $\Box_\gamma$ is given by

$\Box_\gamma = \gamma^{\mu\nu} \partial_\mu \partial_\nu$. Thus for $\Delta \neq 1 + d/2$ we have, to second order in ϵ,

$$\phi_{(0)}(x) = \epsilon^{-(d-\Delta)/2} \left(\phi(\epsilon, x) - \frac{1}{2(2\Delta - d - 2)} \Box_\gamma \phi(\epsilon, x) \right),$$

$$\phi_{(2)}(x) = \epsilon^{-(d-\Delta)/2-1} \frac{1}{2(2\Delta - d - 2)} \Box_\gamma \phi(\epsilon, x). \tag{5.105}$$

Let us now construct the action for the counterterms. The counterterms should cancel the divergences of the on-shell action (5.101). Assuming that the conformal dimension satisfies $d/2 < \Delta < d/2 + 1$, only the terms $a_{(0)}$ and $a_{(2)}$ give rise to divergent terms which we subtract. Using these expressions, the counterterm action (5.101) may be written in the form

$$S_{\text{ct}} = \frac{C}{L} \int d^d x \sqrt{\gamma} \left(\frac{d - \Delta}{2} \phi^2(\epsilon, x) + \frac{1}{2(2\Delta - d - 2)} \phi(\epsilon, x) \Box_\gamma \phi(\epsilon, x) \right) + \cdots. \tag{5.106}$$

If $\Delta > d/2 + 1$ we have to add higher derivative terms which are summarised in the dots. For particular values of the conformal dimension, i.e. for $\Delta = d/2 + k$ for some $k \in \mathbb{N}$, there may also be contributions involving logarithms $\ln(\epsilon)$ in S_{ct}. In particular, for $k = 1$, we have to modify (5.106) by replacing the prefactor of $\phi \Box_\gamma \phi$ by $-(1/4) \ln \epsilon$.

Defining the renormalised action S_{ren} by

$$S_{\text{ren}} = \lim_{\epsilon \to 0} S_{\text{sub}}, \qquad \text{where} \qquad S_{\text{sub}} = S_{\text{reg}} + S_{\text{ct}}, \tag{5.107}$$

holographic renormalisation provides a systematic approach to treat divergences of the on-shell action. In particular, using the renormalised action S_{ren}, the one-point function may be computed by

$$\langle \mathcal{O}(x) \rangle_s = -\frac{\delta S_{\text{ren}}}{\delta \phi_{(0)}(x)}. \tag{5.108}$$

The subscript s indicates that we keep the dependence on the sources and do not set them to zero after taking the derivative. Using the definition of S_{sub}, see (5.107), as well as (5.91), it is sometimes convenient to write (5.108) in a covariant way,

$$\langle \mathcal{O}(x) \rangle_s = - \lim_{\epsilon \to 0} \left(\frac{L^d}{\epsilon^{\Delta/2}} \frac{1}{\sqrt{\gamma}} \frac{\delta S_{\text{sub}}}{\delta \phi(\epsilon, x)} \right), \tag{5.109}$$

where we have used the metric γ_{ij} on the hyperplane $\rho = \epsilon$.

After constructing the counterterms, let us evaluate the action S_{ren}. It is convenient to use the metric (5.63) of the Poincaré patch of AdS space instead of Fefferman–Graham coordinates[4] since we already know the solution (5.73) in terms of the source which we denote here by $\phi_{(0)}(x)$. This solution satisfies the boundary expansion (5.49) (with $\Delta_+ \equiv \Delta$ and thus $\Delta_- = d - \Delta$) which we explicitly insert into the regularised action S_{reg}. The

---

[4] To construct the counterterms, it was convenient to use Fefferman–Graham coordinates. However, the final result S_{ct} is written in a manifest covariant way and thus we can use any coordinate system.

divergent terms in $\mathcal{S}_{\text{reg}}$ are cancelled by the counterterm action. The relevant terms which are finite for $\epsilon \to 0$ read

$$\mathcal{S}_{\text{reg}} = -\frac{dC\,L^{d-1}}{2} \int d^dx\, \phi_{(0)}(x)\phi_{(+)}(x). \tag{5.110}$$

Inserting the boundary expansion (5.49) into the counterterm action, we obtain for the terms which are finite for $\epsilon \to 0$,

$$\mathcal{S}_{\text{ct}} = \frac{C}{L} \int d^dx\, \sqrt{\gamma}\, \frac{d - \Delta}{2}\phi^2(\epsilon, x) = CL^{d-1}(d - \Delta) \int d^dx\, \phi_{(0)}(x)\phi_{(+)}(x) \tag{5.111}$$

and thus we have

$$\mathcal{S}_{\text{sub}} = \frac{CL^{d-1}}{2}(d - 2\Delta)\frac{\Gamma(\Delta)}{\pi^{d/2}\Gamma(\Delta - \frac{d}{2})} \int d^dx \int d^dy\, \frac{\phi_{(0)}(x)\phi_{(0)}(y)}{(x - y)^{2\Delta}} + \cdots , \tag{5.112}$$

where we used (5.74) for $\phi_{(+)}(x)$. Moreover, in (5.112) the dots denote terms which are also finite when taking $\epsilon \to 0$, and which may be removed by adding finite counterterms. Using (5.108), we obtain

$$\langle \mathcal{O}(x)\mathcal{O}(y) \rangle = CL^{d-1}(2\Delta - d)\frac{\Gamma(\Delta)}{\pi^{d/2}\Gamma(\Delta - \frac{d}{2})}\frac{1}{(x - y)^{2\Delta}}, \tag{5.113}$$

which is precisely (5.89).

Let us go one step back. Using (5.108) or equivalently (5.109) we obtain for the one-point function

$$\langle \mathcal{O}(x) \rangle_s = CL^{d-1}(2\Delta - d)\phi_{(+)}(x) + \mathcal{C}(\phi_{(0)}(x)) \tag{5.114}$$

which may be checked by taking another functional derivative with respect to $\phi_{(0)}(y)$ and using the explicit solution (5.74) for $\phi_{(+)}$.

In (5.114), $\mathcal{C}$ denotes a local function of the source $\phi_{(0)}$ as well as derivatives thereof [5] which will give rise to contact terms in the correlation functions. It is now straightforward to calculate higher order correlation functions by virtue of

$$\langle \mathcal{O}(x_1)\mathcal{O}(x_2) \ldots \mathcal{O}(x_n) \rangle = CL^{d-1}(2\Delta - d)\frac{\delta\phi_{(+)}(x_1)}{\delta\phi_{(0)}(x_2) \cdots \delta\phi_{(0)}(x_n)}\bigg|_{\phi_{(0)}=0}, \tag{5.115}$$

omitting the contact terms involving δ functions.

5.5.2 Metric

A similar boundary expansion as presented above for a scalar field may also be performed for the metric due to the celebrated *Fefferman–Graham theorem* of differential geometry [10, 11].

The starting point is to consider the gravity action in Euclidean signature,

$$S = -\frac{1}{16\pi G} \int d^{d+1}x\, \sqrt{g}\left(R + \frac{d(d - 1)}{L^2}\right) - \frac{1}{8\pi G} \int d^dx\, \sqrt{\gamma}K, \tag{5.116}$$

---

[5] The function $\mathcal{C}$ may also depend on $\chi_{(2k)}$ in the case of operators with conformal dimension $\Delta = d/2 + k$.

with a Gibbons–Hawking boundary term as introduced in (2.153).

In the following we allow for deviations of the metric from Anti-de Sitter spacetime and consider *asymptotically AdS manifolds*.[6] The metric of such a manifold reads

$$ds^2 = \tilde{g}_{mn}dx^m dx^n = L^2 \left(\frac{d\rho^2}{4\rho^2} + \frac{1}{\rho} g_{\mu\nu}(\rho, x)dx^\mu dx^\nu \right). \tag{5.117}$$

Note that the metric (5.117) generalises the metric (2.133) considered in section 2.3.2, where $g_{\mu\nu}(\rho, x) = \eta_{\mu\nu}$. Here, $g_{\mu\nu}(\rho, x)$ depends on ρ and may generate a boundary curvature. In the following we pick a boundary metric $g_{(0)\,\mu\nu}$ to be a representative of the conformal equivalence class or conformal structure. In most cases, $g_{(0)\,\mu\nu}$ will be the flat metric.

The Fefferman–Graham theorem states that when the Einstein equations of motion are satisfied, then $g_{\mu\nu}(\rho, x)$ has a boundary expansion of the form

$$g_{\mu\nu}(\rho, x) = g_{(0)\mu\nu}(x) + \rho\, g_{(2)\mu\nu}(x) + \rho^2\, g_{(4)\mu\nu}(x) + \cdots \tag{5.118}$$

in odd boundary dimensions. If the boundary is of even dimension, additional logarithmic terms appear:

$$g_{\mu\nu}(\rho, x) = g_{(0)\mu\nu}(x) + \rho\, g_{(2)\mu\nu}(x) + \cdots + \rho^{d/2} \ln \rho\, h_{(d)\mu\nu}(x) + \mathcal{O}(\rho^{\frac{d}{2}+1}). \tag{5.119}$$

In both cases, the tensors $g_{(k)\mu\nu}(x)$ are constructed from the boundary metric $g_{(0)\mu\nu}(x)$, its curvature and its covariant derivative. They depend on x, but not on ρ, and may be calculated explicitly by inserting the metric (5.117) into the $(d+1)$-dimensional equations of motion.

Exercise 5.5.2 Calculate $g_{(2)}$ explicitly by inserting the metric (5.117) into the Einstein equation. The result is

$$g_{(2)\mu\nu}(x) = \frac{L^2}{d-2} \left(R_{\mu\nu} - \frac{1}{2(d-1)} R g_{(0)\mu\nu} \right), \tag{5.120}$$

which is a conformally covariant tensor.

Using this boundary expansion, holographic renormalisation of the $(d+1)$-dimensional Einstein–Hilbert action (5.116) may be performed. Inserting the metric (5.117) with the expansion (5.119) into the action (5.116) gives rise to a boundary expansion of the gravity action, which takes the form

$$S = -\frac{1}{16\pi G} \int d^{d+1}x \sqrt{\det g^{(0)}} \left(\epsilon^{-d/2} a_{(0)} + \epsilon^{-d/2+1} a_{(2)} + \cdots - \ln \epsilon\, a_{(d)} \right) + S_{\text{finite}}, \tag{5.121}$$

where we have introduced a cut-off at $\rho = \epsilon$. The action contribution S_{finite} summarises the finite contributions. The remaining terms characterise the divergences for $\epsilon \to 0$. The coefficients $a_{(n)}$ are determined in terms of $g_{\mu\nu}^{(0)}$ or its Riemann tensor or contractions thereof. An explicit calculation as outlined above gives

---

[6] Sometimes such manifolds are also referred to as a *conformally compact Einstein manifolds*.

$$a_{(0)} = \frac{2(d-1)}{L},$$

$$a_{(2)} = \frac{L}{2(d-1)}R, \tag{5.122}$$

$$a_{(4)} = \frac{L^3}{2(d-2)^2}\left(R^{\mu\nu}R_{\mu\nu} - \frac{1}{(d-1)}R^2\right).$$

The divergence of $\mathcal{S}$ at the boundary is regularised by adding suitable counterterms $\mathcal{S}_{ct}$ which make $\mathcal{S}_{ren} = \mathcal{S} + \mathcal{S}_{ct}$ finite in the limit $\epsilon \to 0$. The simplest form of regularisation is the minimal subtraction scheme, which amounts simply to subtracting the divergent terms from (5.121). For example, the divergence coming from $a_{(0)}$ can be cured by adding the counterterm

$$\mathcal{S}_{ct} = \frac{1}{8\pi G}\frac{d-1}{L}\int d^d x \sqrt{\gamma}, \tag{5.123}$$

which cancels the infinite volume of AdS space. For $d \geq 4$ we have to add more counterterms which are proportional to $\sqrt{\gamma}R$ or even higher order in R. Holographic renormalisation is essential for the holographic calculation of the boundary conformal anomaly, an important test of the AdS/CFT correspondence which we discuss in the next chapter. Here we proceed by briefly discussing the analogue of the scalar result (5.109) for gravity, which is obtained by inverting the boundary expansion, as done for the scalar field in section 5.5.1. The expectation value of the energy-momentum tensor $\langle T_{\mu\nu}(x)\rangle$ in the presence of a non-trivial metric $g_{\mu\nu}^{(0)}$ as a source reads

$$\langle T_{\mu\nu}(x)\rangle_s = -\frac{2}{\sqrt{\det g_{(0)}}}\frac{\delta\mathcal{S}_{ren}}{\delta g_{(0)}^{\mu\nu}(x)}$$

$$= \lim_{\epsilon\to 0} -\frac{2}{\sqrt{\det g(\epsilon, x)}}\frac{\delta\mathcal{S}_{sub}}{\delta g^{\mu\nu}(x)} = \lim_{\epsilon\to 0}\left(\frac{L^{d-2}}{\epsilon^{d/2-1}}T_{\mu\nu}^{(\gamma)}\right), \tag{5.124}$$

with $\mathcal{S}_{ren}$ and $\mathcal{S}_{sub}$ the action contributions obtained from (5.116) in analogy to the scalar case (5.107). $T_{\mu\nu}^{(\gamma)}$ is the stress tensor obtained at $\rho = \epsilon$ and $\gamma_{\mu\nu}$ is the induced metric at the cut-off $\rho = \epsilon$,

$$\gamma_{\mu\nu}(x) = \frac{L^2}{\epsilon}g_{\mu\nu}(\rho = \epsilon, x). \tag{5.125}$$

For $d = 4$, the explicit result for $T_{\mu\nu}^{(\gamma)}$ of (5.124) is given by [12, 8]

$$T_{\mu\nu}^{(\gamma)} = \frac{1}{8\pi G}\left(K_{\mu\nu} - K\gamma_{\mu\nu} - \frac{3}{L}\gamma_{\mu\nu} - \frac{L}{2}G_{\mu\nu}\right), \tag{5.126}$$

with $G_{\mu\nu} = R_{\mu\nu} - 1/2\gamma_{\mu\nu}R$ the Einstein tensor for the induced metric. $K_{\mu\nu}$ is the extrinsic curvature as defined in (2.156) on the hyperplane at $\rho = \epsilon$.

Holographic renormalisation will be used in chapter 6 to calculate the conformal anomaly, which provides a powerful test of the AdS/CFT correspondence. Here we proceed by discussing Wilson loops, which provide an example for realising the AdS/CFT duality for non-local operators.

5.6 Wilson loops in $\mathcal{N} = 4$ Super Yang–Mills theory

The Wilson loop as introduced in chapter 1 has a very intuitive gravity dual. To study this, let us first construct the Wilson loop operator in $\mathcal{N} = 4$ Super Yang–Mills theory. Since all degrees of freedom in this theory are massless and transform in the adjoint representation of the gauge group, we must find a way naturally to introduce very massive particles transforming in the fundamental representation. This is necessary to consider a quark–antiquark pair as in section 1.7.4. For this purpose it is very useful to think of $\mathcal{N} = 4$ Super Yang–Mills theory as the low-energy worldvolume theory of open strings ending on a stack of N D3-branes in flat space. To introduce massive fundamental particles, we consider $N+1$ D3-branes instead, where the extra D3-brane is separated from the remaining N D3-branes in at least one of the six transverse directions x^{i+3} with $i = 1, \ldots, 6$. The separation of the D3-branes corresponds to giving a vacuum expectation value to the scalars ϕ^i. Let us assume that this separation is given by a vector n with components n^i such that $x^{i+3} = Mn^i$ in the six-dimensional space. M is chosen such that $\delta_{ij}n^i n^j = 1$ and corresponds to the separation distance in the flat six-dimensional spacetime, which we take to be large in the following, i.e. $M \gg 1$.

On the gauge theory side, we can view the system as an $SU(N + 1)$ $\mathcal{N} = 4$ Super Yang–Mills theory where the six scalars $\hat{\phi}^i$ which are valued in $\mathfrak{su}(N + 1)$ may be expressed as

$$\hat{\phi}^i = \begin{pmatrix} \phi^i & \omega^i \\ \omega^{i\dagger} & Mn^i \end{pmatrix}. \tag{5.127}$$

While ϕ^i are the remaining massless scalars of the $SU(N)$ theory, the fields ω^i and $\omega^{i\dagger}$, which are column (row) vectors of length N, respectively, are in the fundamental or anti-fundamental representation of $\mathfrak{su}(N)$. Thus ω^i and $\omega^{i\dagger}$ describe modes of strings stretched between the stack of N D3-branes and the separated $D3$-brane. Moreover, the component Mn^i in (5.127) corresponds to the vacuum expectation value which gives rise to a mass of order M for ω^i and $\omega^{i\dagger}$ by the usual Higgs mechanism (see Figure 5.3). In addition, for large N we may ignore the fields on the single brane.

The procedure described introduces massive fundamental particles to $\mathcal{N} = 4$ Super Yang–Mills theory. These particles correspond to the ground states of the open strings stretching between the stack of N D3-branes and the single separated brane and are referred to as *W-bosons* of the broken gauge group $SU(N + 1) \rightarrow SU(N) \times U(1)$ and their superpartners. The trajectories of these W-bosons around a closed path $\mathcal{C}$ give rise to a phase factor, which is given by the vacuum expectation value of

$$\mathcal{W}(\mathcal{C}) = \frac{1}{N} \text{Tr} \, \mathcal{P} \exp \left(i \oint_{\mathcal{C}} ds \, \left(A_\mu \dot{x}^\mu + |\dot{x}| \Phi_i n^i \right) \right) \tag{5.128}$$

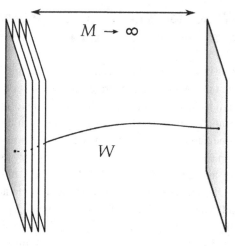

N D3-branes   Single D3-brane

Figure 5.3 Heavy particles ω may be introduced into $\mathcal{N} = 4$ Super Yang–Mills theory by Higgsing the $SU(N + 1)$ gauge group. This corresponds geometrically to separating one D3-brane from the stack of N D3-branes in at least one of the transverse directions. The separation distance is taken to be large, $M \to \infty$.

in Minkowski signature. In Euclidean signature which we use in this section, the Wilson loop operator reads

$$W(\mathcal{C}) = \frac{1}{N} \mathrm{Tr}\, \mathcal{P} \exp \left(\oint_{\mathcal{C}} \mathrm{d}s \, \left(iA_\mu \dot{x}^\mu + |\dot{x}| \Phi_i n^i \right) \right). \qquad (5.129)$$

Note that in Euclidean space, the phase factor also has a real part. In equations (5.128) and (5.129), the n^i, which satisfy $\delta_{ij} n^i n^j = 1$, may be considered as coordinates on S^5, while the x^μ are coordinates on $\mathbb{R}^{3,1}$ or $\mathbb{R}^4$, respectively. Note that the Wilson loop operator given in (5.129) is not the most general one for $\mathcal{N} = 4$ Super Yang–Mills theory. We can consider a generalisation of the form

$$W(\mathcal{C}) = \frac{1}{N} \mathrm{Tr}\, \mathcal{P} \exp \left(\oint_{\mathcal{C}} \mathrm{d}s \, \left(iA_\mu \dot{x}^\mu + \Phi_i \dot{y}^i \right) \right). \qquad (5.130)$$

Setting $\dot{y}^i = |\dot{x}| n^i$, we recover (5.129). This is equivalent to $\dot{x}^2 = \dot{y}^2$, which is also related to supersymmetry. The Wilson loop operator may be viewed as the phase factor of an excited state of the massive open string stretching between the single D3-brane and the stack of N D3-branes.

Exercise 5.6.1   Show that subject to $\dot{x}^2 = \dot{y}^2$, (5.130) is invariant under local supersymmetry transformations, i.e. infinitesimal supersymmetry transformations with parameter $\epsilon = \epsilon(x)$. This implies that for every x^μ, the integrand of the Wilson loop is 1/2 BPS, however the exact form of the conserved supercharges is spacetime dependent.

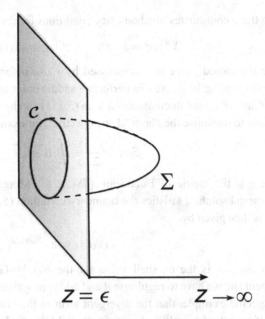

Figure 5.4   Gravity dual of a Wilson loop. The gravity dual of a Wilson loop along $\mathcal{C}$ is given by the minimal surface Σ which shares the closed curve $\mathcal{C}$ at the boundary. This surface corresponds to the worldsheet of a string.

5.6.1 Gravity dual of the $\mathcal{N} = 4$ Super Yang–Mills Wilson loop

The dual gravitational description of the Wilson loop in $\mathcal{N} = 4$ Super Yang–Mills theory is very intuitive. To make a proposal for this gravity dual, let us go back to the derivation of the Wilson loop operator from the system of $N + 1$ D3-branes where one of them is separated from the remaining N. The massive fundamental particles correspond to open strings stretching between the N D3-branes and the separated brane. These fundamental particles are now considered to move on a closed loop on the single D3-brane.

To obtain the gravity dual description, the N D3-branes are replaced by the space $AdS_5 \times S^5$. The single D3-brane is taken to be located at $z = \epsilon, \epsilon \to 0$, corresponding to the conformal boundary of AdS_5. Therefore the expectation value of the Wilson loop operator is dual to the semi-classical partition function of a macroscopic string in $AdS_5 \times S^5$ whose worldsheet Σ ends on the path of the Wilson loop at the boundary. This is shown in figure 5.4. Thus the Wilson loop expectation value is given by the path integral

$$\langle \mathcal{W}(\mathcal{C}) \rangle = \int \mathcal{D}X^\mu \mathcal{D}\theta^i \mathcal{D}h_{\alpha\beta} \exp\left(-\mathcal{S}_{\text{Polyakov}}\right), \tag{5.131}$$

where $\mathcal{S}_{\text{Polyakov}}$ is the Polyakov action of the fundamental string in $AdS_5 \times S^5$. We have to perform the integral over all string embeddings which end on the curve $\mathcal{C}$. To be precise, let us denote the embedding functions by X^μ and X^i along the field theory directions and the six perpendicular directions, respectively. With the radial direction r given by $r^2 = \sum_{i=1}^{g}(X^i)^2$, we may introduce $\theta^i (i = 1, \cdots, 6)$ such that $\sum_{i=1}^{6}(\theta^i)^2 = 1$ and $X^{i+3} = r \cdot \theta^i$.

Using these coordinates, the boundary conditions for the fundamental string read.

$$X^\mu|_{\partial\Sigma} = x^\mu(s), \qquad \theta^i|_{\partial\Sigma} = n^i(s), \; r|_{\partial\Sigma} = 0, \tag{5.132}$$

where the closed curve is parametrised by s. As before, at the large N limit and large 't Hooft coupling limit, we can perform a saddle point approximation to (5.131), in which the X^μ and θ^i do not fluctuate and $h_{\alpha\beta}$ in (5.131) can be integrated out. In other words, we just have to minimise the classical string action, for example the Nambu–Goto action

$$\mathcal{S}_{NG} = \frac{1}{2\pi\alpha'} \int_\Sigma d^2\sigma \sqrt{P[g]_{ab}}, \tag{5.133}$$

where g is the metric of Euclidean $AdS_5 \times S^5$. Moreover, we have to make sure that the classical solution satisfies the boundary condition (5.132). The Wilson loop expection value is then given by

$$\langle \mathcal{W}(\mathcal{C}) \rangle = e^{-\mathcal{S}_{NG,min}}, \tag{5.134}$$

where $\mathcal{S}_{NG,min}$ is the on-shell value for the Nambu-Goto action. As usual $\mathcal{S}_{NG,min}$ is divergent and we have to regularise it and add appropriate counterterms. We will see below in a specific example, that the divergent term in the Nambu–Goto action corresponds to the self-energy of pointlike charges in the field theory. From now on, we will subtract this contribution and the final result for the Wilson loop is

$$\langle \mathcal{W}(\mathcal{C}) \rangle = e^{-\mathcal{S}_{NG,min} - \mathcal{S}_{ct}}. \tag{5.135}$$

Let us see how this works for a specific example and calculate the quark–antiquark potential, which is related to the Wilson loop by (1.213), for $\mathcal{N} = 4$ Super Yang–Mills theory using the AdS/CFT correspondence.

To compute the potential for a pair of static quarks and antiquarks in $\mathcal{N} = 4$ Super Yang–Mills theory, we consider the rectangular Wilson loop with $R \ll T$ as shown in figure 1.2. In this case, the string in the gravity description is just hanging down in the radial direction of AdS space, with its endpoints on the AdS boundary. We take the separation R of the Wilson loop to be extended in the x^1 direction, which in the subsequent we denote by x. The embedding of the string in AdS_5 space is parametrised by a function $z(x)$, while the string does not move on S^5. We write its derivative with respect to x as z'. In Euclidean signature, the induced metric on the string worldsheet is

$$ds^2 = \frac{L^2}{z^2} \left(d\tau^2 + (1 + z'^2) \, dx^2 \right), \tag{5.136}$$

for which the Nambu–Goto action reads

$$S = \frac{1}{2\pi\alpha'} \int d\tau \, dx \sqrt{P[g]_{ab}} = \frac{TL^2}{2\pi\alpha'} \int dx \frac{\sqrt{1 + z'^2}}{z^2}. \tag{5.137}$$

This action reminds us of problems encountered in classical mechanics, in which x plays the role of time and the x-integrand of the Nambu–Goto action defines a Lagrangian $\mathcal{L}(z, z')$. Since this Lagrangian is independent of x, there is a constant of motion, referred to as z_*^2, which can be determined by calculating the first integral. This gives

$$z^2\sqrt{1 + z'^2} = \text{constant} \equiv z_*^2. \tag{5.138}$$

Solving (5.138) for z' and integrating after using separation of variables, we obtain x as a function of z,

$$x = \pm \int_{z_*}^{z} \frac{\zeta^2}{\sqrt{z_*^4 - \zeta^4}} d\zeta. \tag{5.139}$$

The two signs in this equation correspond to the two sides of the hanging string. The boundary conditions for the two endpoints of the string are given by

$$z(x = -R/2) = z(x = R/2) = 0. \tag{5.140}$$

Equations (5.139) and (5.140) relate the constant of motion z_* and the separation R. Performing the integral in (5.139) subject to the boundary conditions (5.140), we obtain

$$z_* = \frac{R}{2\sqrt{2}\pi^{3/2}} \left(\Gamma(\tfrac{1}{4})\right)^2. \tag{5.141}$$

The action is now obtained by inserting the constant of motion (5.138) into the Nambu–Goto action (5.137). Noting that $z(x)$ is double valued, corresponding to the two sides of the string symmetric to each other, we have

$$S_{\text{on-shell}} = 2 \frac{TL^2 z_*^2}{2\pi\alpha'} \int_{\epsilon}^{z_*} \frac{dz}{z^2\sqrt{z_*^4 - z^4}}, \tag{5.142}$$

where we have to introduce a cut-off ϵ to perform the regularisation. The integral yields for small ϵ

$$S_{\text{on-shell}} = \frac{TL^2}{\pi\alpha' z_*} \left(-\frac{\pi^{3/2}\sqrt{2}}{\left(\Gamma(\tfrac{1}{4})\right)^2} + \frac{z_*}{\epsilon} \right) + \mathcal{O}(\epsilon). \tag{5.143}$$

Note that according to (1.213), the on-shell action corresponds to $TV(R)$, with $V(R)$ the quark–antiquark potential. Equation (5.143) can be regularised by subtracting the quark and antiquark masses, which are obtained from the Nambu–Goto action for two parallel strings stretching from $z = \epsilon$ to $z = \infty$. This action yields TV_0, with a potential V_0 due to the quark masses which coincides precisely with the divergent term in (5.143). Finally, we obtain for the quark–antiquark potential

$$\begin{aligned} V_{\bar{q}q} \equiv V - V_0 &= -\frac{L^2}{\alpha'} \frac{4\pi^2}{\left(\Gamma(\tfrac{1}{4})\right)^4} \frac{1}{R} \\ &= -\sqrt{2\lambda} \frac{4\pi^2}{\left(\Gamma(\tfrac{1}{4})\right)^4} \frac{1}{R}, \end{aligned} \tag{5.144}$$

where we have used $L^2 = \sqrt{2\lambda}\alpha'$.

We see that we obtain a potential of Coulomb form, $V(R) \sim 1/R$, as expected by dimensional analysis for a conformal field theory. The appearance of the square-root of the 't Hooft coupling in the exponential function is a non-perturbative effect due to the strong coupling nature of the AdS/CFT correspondence. $\sqrt{2\lambda}$ may be traced back to the appearance of L^2/α' in the string action.

Exercise 5.6.2   Show that the expectation value for one straight Wilson line (closed to a loop at infinity) satisfies $\langle \mathcal{W} \rangle = 1$ with the counterterm advertised above. This is an alternative possibility for fixing the counterterm since the straight Wilson line is a one-half BPS operator and its expectation value has to be 1.

Exercise 5.6.3   Calculate holographically the expectation value of circular Wilson loops with radius R.

5.7 Further reading

The original paper proposing the AdS/CFT duality conjecture is [1]. The field-operator map and the calculation of correlation functions within AdS/CFT were established in [13, 14]. Reviews of the AdS/CFT correspondence are [15, 16, 17].

The duality between sine–Gordon and Thirring models is discussed in [18, 19]. The holographic principle was proposed in [20, 21] and is discussed in the AdS/CFT context in [22].

The mass spectrum of IIB supergravity fields on $AdS_5 \times S^5$ was worked out in [2].

The Breitenlohner–Freedman bound was found in [3, 4], where the consistent AdS boundary conditions are also discussed. The possibility of interchanging the dimensions Δ_+ and Δ_- in a certain parameter range, leading to an interchange of source and vacuum expectation value, was discussed in [23].

In [5], the scalar two-point function was obtained using the AdS/CFT correspondence. The Fourier transform of (5.88) leading to the position-space result (5.89) for the two-point function is reviewed in [7], and was performed in [5] using earlier results given in the appendix of [6].

Holographic renormalisation was established in [8] and is reviewed in [9]. The Fefferman–Graham theorem is given in [10], see also [11].

The holographic dual of the Wilson loop was given in [24, 25]. Further information about holographic Wilson loops may be found for instance in [26] and also in the AdS/CFT review [17]. Wilson loops in $\mathcal{N} = 4$ Super Yang–Mills theory were calculated and shown to match with AdS/CFT expectations in [28]. In exercise 5.6.3 we considered the circular Wilson loop, its vacuum expectation value can be calculated using a matrix model [29], as shown explicitly in [30] using localisation techniques.

References

[1] Maldacena, Juan Martin. 1998. The large N limit of superconformal field theories and supergravity. *Adv. Theor. Math. Phys.*, **2**, 231–252.

[2] Kim, H. J., Romans, L. J., and van Nieuwenhuizen, P. 1985. The mass spectrum of chiral $N = 2\, D = 10$ supergravity on S^5. *Phys. Rev.*, **D32**, 389.

[3] Breitenlohner, Peter, and Freedman, Daniel Z. 1982. Positive energy in Anti-de Sitter backgrounds and gauged extended supergravity. *Phys. Lett.*, **B115**, 197.

[4] Breitenlohner, Peter, and Freedman, Daniel Z. 1982. Stability in gauged extended supergravity. *Ann. Phys.*, **144**, 249.

[5] Freedman, Daniel Z., Mathur, Samir D., Matusis, Alec, and Rastelli, Leonardo. 1999. Correlation functions in the CFT$_d$/AdS$_{(d+1)}$ correspondence. *Nucl. Phys.*, **B546**, 96–118.

[6] Freedman, Daniel Z., Johnson, Kenneth, and Latorre, Jose I. 1992. Differential regularization and renormalization: a new method of calculation in quantum field theory. *Nucl. Phys.*, **B371**, 353–414.

[7] Freedman, Daniel Z., and Van Proeyen, Antoine. 2012. *Supergravity*. Cambridge University Press.

[8] de Haro, Sebastian, Solodukhin, Sergey N., and Skenderis, Kostas. 2001. Holographic reconstruction of space-time and renormalization in the AdS/CFT correspondence. *Commun. Math. Phys.*, **217**, 595–622.

[9] Skenderis, Kostas. 2002. Lecture notes on holographic renormalization. *Class. Quantum Grav.*, **19**, 5849–5876.

[10] Fefferman, C., and Graham, C. R. 1985. Conformal invariants, in 'The Mathematical Heritage of Elie Cartan' (Lyon, 1984). *Asterisque*, 95–116.

[11] Fefferman, C., and Graham, C. R. 2007. The ambient metric. ArXiv:0710.0919.

[12] Balasubramanian, Vijay, and Kraus, Per. 1999. A Stress tensor for Anti-de Sitter gravity. *Commun. Math. Phys.*, **208**, 413–428.

[13] Witten, Edward. 1998. Anti-de Sitter space and holography. *Adv. Theor. Math. Phys.*, **2**, 253–291.

[14] Gubser, S. S., Klebanov, Igor R., and Polyakov, Alexander M. 1998. Gauge theory correlators from noncritical string theory. *Phys. Lett.*, **B428**, 105–114.

[15] D'Hoker, Eric, and Freedman, Daniel Z. 2002. Supersymmetric gauge theories and the AdS/CFT correspondence. TASI 2001 School Proceedings. ArXiv:hep-th/0201253. 3–158.

[16] Aharony, Ofer, Gubser, Steven S., Maldacena, Juan Martin, Ooguri, Hirosi, and Oz, Yaron. 2000. Large N field theories, string theory and gravity. *Phys. Rep.*, **323**, 183–386.

[17] Ramallo, Alfonso V. 2013. Introduction to the AdS/CFT correspondence. ArXiv: 1310.4319.

[18] Coleman, Sidney R. 1975. The quantum Sine-Gordon equation as the massive Thirring model. *Phys. Rev.*, **D11**, 2088.

[19] Mandelstam, S. 1975. Soliton operators for the quantised Sine-Gordon equation. *Phys. Rev.*, **D11**, 3026.

[20] 't Hooft, Gerard. 1993. Dimensional reduction in quantum gravity. ArXiv:gr-qc/9310026.

[21] Susskind, Leonard. 1995. The world as a hologram. *J. Math. Phys.*, **36**, 6377–6396.

[22] Susskind, Leonard, and Witten, Edward. 1998. The holographic bound in anti-de Sitter space. ArXiv:hep-th/9805114.

[23] Klebanov, Igor R., and Witten, Edward. 1999. AdS/CFT correspondence and symmetry breaking. *Nucl. Phys.*, **B556**, 89–114.

[24] Maldacena, Juan Martin. 1998. Wilson loops in large N field theories. *Phys. Rev. Lett.*, **80**, 4859–4862.

[25] Rey, Soo-Jong, Theisen, Stefan, and Yee, Jung-Tay. 1998. Wilson–Polyakov loop at finite temperature in large N gange theory and Anti-de Sitter supergravity. *Nucl. Phys.*, **B527**, 171–186.

[26] Drukker, Nadav, Gross, David J., and Ooguri, Hirosi. 1999. Wilson loops and minimal surfaces. *Phys. Rev.*, **D60**, 125006.

[27] Erickson, J. K., Semenoff, G. W., and Zarembo, K. 2000. Wilson loops in $N = 4$ supersymmetric Yong–Mills theory. *Nucl. Phys.*, **B582**, 155–175.

[28] Drukker, N., and Gross, D. J. 2002. An exact prediction of $N = 4$ SUSYM theory for string theory. *J. Math. Phys.*, **42**, 2896–2914.

[29] Pestun, V. 2012. Localisation of gange theory on a four-sphere and super symmetric Wilson loops. *Commun. Math. Phys.*, **313**, 71–129.

6 Tests of the AdS/CFT correspondence

In the preceding chapter we introduced the conjectured duality of $\mathcal{N} = 4$ Super Yang–Mills theory and IIB string theory on $AdS_5 \times S^5$ as a map between the open and closed string pictures of D3-branes. As we explained, a proof of this duality would require a full non-perturbative understanding of quantised string theory in a curved space background. This is absent at present. This means that, to date, it is not possible to give a proof of the AdS/CFT correspondence.

Nevertheless, some very non-trivial tests of the conjectured duality are possible, for which observables are calculated on both sides and perfect agreement is found. Some of these tests are presented in this chapter. Generally, tests of the correspondence are possible for both the strong form and the weak form as defined in table 5.1, both of which require the $N \to \infty$ limit on the field theory side to ensure a classical calculation on the gravity side. While tests for the strong form with the 't Hooft coupling λ fixed but arbitrary will be discussed in chapter 7, in this chapter we focus on tests for the weak form for which λ is taken to be large. This amounts to calculations in classical supergravity on the gravity side.

An important issue is that in the weak form, the AdS/CFT correspondence maps a quantum field theory in the strongly coupled regime to a gravity theory in the weakly coupled regime. In part III of this book, we will use this property to make non-trivial predictions for strongly coupled field theories which cannot be obtained by standard quantum field theory methods. However, for the tests considered here, our aim is to compare observables calculated perturbatively at weak coupling within quantum field theory to the same QFT observables calculated at strong coupling using the AdS/CFT correspondence. This implies that only calculations for observables *independent of the coupling* may be compared directly to each other. Coupling-dependent field theory observables, on the other hand, may take very different values at weak coupling, as accessible by a perturbative Feynman diagram expansion, and at strong coupling, as accessible by an AdS/CFT computation involving classical supergravity.

Fortunately, supersymmetry and in particular the very special renormalisation properties of $\mathcal{N} = 4$ Super Yang–Mills theory come to our help, since they give rise to a multitude of *non-renormalisation theorems* in generalisation of those discussed in section 3.3.5. In particular, in subsection 3.4.3 we discussed 1/2 BPS operators which are protected with respect to quantum corrections, such that their dimension does not receive quantum corrections. As we will see in the present chapter, in addition there are many non-trivial examples of *correlation functions* of 1/2 BPS operators which do not receive any quantum corrections involving λ.

Non-renormalisation theorems also protect the coefficient of the conformal anomaly in $\mathcal{N} = 4$ Super Yang–Mills theory, which is one-loop exact and independent of λ. The conformal anomaly was introduced in section 3.2.6. The comparison of the value of the anomaly coefficients obtained perturbatively and using the AdS/CFT correspondence constitutes a second very non-trivial check of the correspondence which we also discuss below.

6.1 Correlation function of 1/2 BPS operators

6.1.1 Three-point function of 1/2 BPS operators

An impressive test of the AdS/CFT correspondence is provided by the three-point functions of 1/2 BPS operators in $\mathcal{N} = 4$ Super Yang–Mills theory at large N. The result for this three-point function calculated within perturbative quantum field theory agrees with the same three-point function calculated using the AdS/CFT correspondence. For the latter calculation, Witten diagrams as introduced in section 5.4.2 are used.

This example is a very non-trivial case of the AdS/CFT correspondence at work. Because of its importance, we present the calculations involved in full generality. The key idea is to calculate three-point functions for 1/2 BPS operators in $\mathcal{N} = 4$ Super Yang–Mills theory first perturbatively at weak coupling, and then at strong coupling using the AdS/CFT correspondence. Due to the non-renormalisation theorems, we expect the two calculations to yield identical results, and we will confirm through the explicit calculations that this is indeed the case. The necessary calculations involve several steps. The first step is to ensure that the 1/2 BPS operators are normalised in the same way both in the field theory and in the gravity calculation. This is achieved by normalising the operators with the help of the two-point function. Then the three-point function is calculated perturbatively to lowest order and it is shown that higher order corrections involving λ are absent. Finally we calculate the three-point function on the gravity side and find exact agreement.

Let us summarise how we will proceed to perform this non-trivial test.

- We calculate the two-point function in both approaches to fix the normalisation.
- We calculate the three-point function in Super Yang–Mills theory to lowest order in the coupling.
- We check that this result for the three-point function is not renormalised at higher orders, i.e. we prove a non-renormalisation theorem to show independence of the correlator on the coupling.
- We calculate the three-point function on the gravity side. Its spacetime dependence is obtained from the propagators in the Witten diagrams and the coupling is obtained from the Kaluza–Klein reduction.

6.1.2 Field theory correlation function of one-half BPS operators

We begin by considering the one-half BPS two- and three-point functions to lowest order in perturbation theory on the field theory side of the correspondence. We write the one-half BPS operators of $\mathcal{N} = 4$ Super Yang–Mills operators defined in 3.4.3 in the form [1]

$$\mathcal{O}_k^I = C_{i_1 \ldots i_k}^I \, \mathrm{Tr}(\phi^{i_1} \ldots \phi^{i_k}), \tag{6.1}$$

where the dimension of the operator is given by $\Delta = k$ and the C^I are totally symmetric traceless rank k tensors of $SO(6)$. The $\mathcal{N} = 4$ Super Yang–Mills action (3.225) gives rise to the scalar propagators

$$\langle \phi^{ia}(x) \, \phi^{jb}(y) \rangle = \frac{\delta^{ij} \, \delta^{ab}}{(2\pi)^2 \, (x - y)^2}. \tag{6.2}$$

To lowest order in perturbation theory in the large N limit, the two-point function on the field theory side is given by the Feynman diagram shown in figure 6.1. The composite operators correspond to k legs each. The two operators are thus linked by k scalar propagators. The large N limit implies that only the planar diagram has to be considered. The corresponding two-point function containing a product of k scalar propagators as well the appropriate symmetry factors reads

$$
\begin{aligned}
\langle \mathcal{O}_k^I(x) \, \mathcal{O}_k^J(y) \rangle &= C_{i_1 \ldots i_k}^I \, C_{j_1 \ldots j_k}^J \, \langle \mathrm{Tr}(\phi^{i_1}(x) \ldots \phi^{i_k}(x)) \, \mathrm{Tr}(\phi^{j_1}(y) \ldots \phi^{j_k}(y)) \rangle \\
&= C_{i_1 \ldots i_k}^I \, C_{j_1 \ldots j_k}^J \, \frac{N^k \left(\delta^{i_1 j_1} \delta^{i_2 j_2} \ldots \delta^{i_k j_k} + \text{permutations} \right)}{(2\pi)^{2k} \, (x - y)^{2k}} \\
&= \frac{k \, N^k \, \delta^{IJ}}{(2\pi)^{2k} \, (x - y)^{2k}}.
\end{aligned}
\tag{6.3}
$$

where $(x - y)^{2k}$ is an abbreviation for $((x - y)^2)^k$. The last equality in (6.3) only holds at leading order in N, where only cyclic permutations are taken into account. Moreover, we have used the orthonormality of the tensors $C_{i_1 \ldots i_k}^I$.

Similarly, for the three-point function to lowest order in perturbation theory and in the limit of large N we have [1]

$$\langle \mathcal{O}_{k_1}^I(x) \, \mathcal{O}_{k_2}^J(y) \, \mathcal{O}_{k_3}^K(z) \rangle = \frac{N^{\Sigma/2} \, k_1 \, k_2 \, k_3 \, \langle C^I \, C^J \, C^K \rangle}{N \, (2\pi)^\Sigma \, |x - y|^{2\alpha_3} \, |y - z|^{2\alpha_1} \, |x - z|^{2\alpha_2}}. \tag{6.4}$$

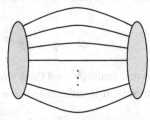

Figure 6.1 Feynman diagram for the two-point correlation function $\langle \mathcal{O}_k^I(x) \, \mathcal{O}_k^J(y) \rangle$ to lowest order in the coupling, $\mathcal{O}(g_{\mathrm{YM}}^0)$, in the planar limit. This is referred to as the *rainbow diagram*.

Note that the spacetime dependence is completely determined by conformal invariance. We use the shorthand notation

$$\Sigma = k_1 + k_2 + k_3, \qquad \alpha_i = \frac{\Sigma}{2} - k_i, \tag{6.5}$$

such that for example $\alpha_1 = (k_2 + k_3 - k_1)/2$ and $\langle C^I C^J C^K \rangle$ denotes a uniquely defined $SO(6)$ tensor contraction of indices determined by the Feynman graph.

In view of the comparison with the gravity result to be obtained below, we define normalised operators

$$\tilde{\mathcal{O}}_k^I \equiv \frac{(2\pi)^k}{N^{k/2}\sqrt{k}} \mathcal{O}_k^I \tag{6.6}$$

for which the two-point function is normalised to one,

$$\langle \tilde{\mathcal{O}}_k^I(x)\, \tilde{\mathcal{O}}_k^J(y) \rangle = \frac{\delta^{IJ}}{(x-y)^{2k}}, \tag{6.7}$$

and the three-point function reads

$$\langle \tilde{\mathcal{O}}_{k_1}^I(x)\, \tilde{\mathcal{O}}_{k_2}^J(y)\, \tilde{\mathcal{O}}_{k_3}^K(z) \rangle = \frac{\sqrt{k_1 k_2 k_3}\, \langle C^I C^J C^K \rangle}{N\, (x-y)^{2\alpha_3}\, (y-z)^{2\alpha_1}\, (x-z)^{2\alpha_2}}. \tag{6.8}$$

This holds in the large N limit. For finite N, non-planar corrections of order $\frac{1}{N^2}$ arise. The result (6.8) will be compared to the gravity calculation of the same correlator below. From now on, we do not write the dimension k characterising the one-half BPS operator $\mathcal{O}_k^I$ explicitly, since it is determined by the $SO(6)$ tensor $C_{i_1 \ldots i_k}^I$, and we use the shorthand notation $\mathcal{O}^I$.

6.1.3 Non-renormalisation theorem

Before we can compare (6.8) to the gravity calculation, we have to demonstrate the absence of higher order quantum corrections of order $\mathcal{O}(\lambda)$ both in $\langle \mathcal{O}\mathcal{O} \rangle$ and in $\langle \mathcal{O}\mathcal{O}\mathcal{O} \rangle$. The argument for this non-renormalisation property [2] which we give in this section holds for any N.

The starting point is the Lagrangian of $\mathcal{N} = 4$ Super Yang–Mills theory in components as given by (3.225). To streamline the argument, it is convenient to introduce complex scalar fields Φ^j

$$\Phi^j \equiv \phi^j + i\phi^{j+3}, \qquad j = 1, 2, 3, \tag{6.9}$$

combining the six real scalar $\mathcal{N} = 4$ theory fields ϕ^j into three complex fields. Due to the supersymmetry present, it is sufficient to consider the two-point function

$$\langle \mathrm{Tr}\left((\Phi^1)^k(x)\right) \mathrm{Tr}\left((\bar{\Phi}^1)^k(y)\right) \rangle = \frac{P_{k,k,0}(N)}{\left(2\pi\,(x-y)\right)^{2k}} \tag{6.10}$$

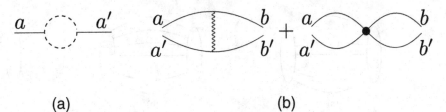

(a)   (b)

Figure 6.2 Possible quantum corrections to scalar propagators in $\mathcal{N} = 4$ Super Yang–Mills theory at order g_{YM}^2. (*a*) Self-energy correction from a fermion loop. (*b*) Exchange contributions: gauge boson exchange and quartic scalar interaction.

with the polynomial $P_{k,k,0}$ in N given by

$$P_{k,k,0}(N) = \sum_{\sigma \in S_k} \text{Tr}\left(T^{a_1} T^{a_2} ... T^{a_k}\right) \text{Tr}\left(T^{a_{\sigma(1)}} T^{a_{\sigma(2)}} ... T^{a_{\sigma(k)}}\right)$$

$$= k \left(\frac{N}{2}\right)^k + \text{lower order in } N. \tag{6.11}$$

We consider possible $\mathcal{O}(g_{YM}^2)$ quantum corrections to the scalar propagators in the rainbow graph figure 6.1. The possible interaction contributions are displayed in figure 6.2. First, there is a self-energy contribution from a fermion loop as shown on the left-hand side of figure 6.2. This corresponds to a term of the form

$$\mathcal{A}^{aa'}(x,y) = \delta^{aa'} N A(x,y) G(x,y), \tag{6.12}$$

where

$$A(x,y) = a_0 + a_1 \ln\left(\mu^2(x-y)^2\right), \tag{6.13}$$

$$G(x,y) = \frac{1}{4\pi^2(x-y)^2}. \tag{6.14}$$

Here, $G(x,y)$ is the scalar propagator and $A(x,y)$ arises from integrating over the vertices according to position space Feynman rules. The precise value of the coefficients a_1, a_2 is not essential for the argument which follows.

Moreover, there are exchange contributions as shown on the left-hand side of figure 6.2. These involve a gauge boson exchange and a quartic scalar interaction. Here, application of position space Feynman rules gives

$$\mathcal{B}^{aa'bb'}(x,y) = \left(f^{pab} f^{pa'b'} + f^{pab'} f^{pa'b}\right) B(x,y) G(x,y)^2, \tag{6.15}$$

where

$$B(x,y) = b_0 + b_1 \ln\left(\mu^2(x-y)^2\right). \tag{6.16}$$

Again, the exact value of the coefficients b_0 and b_1 is not relevant for the argument which follows. The interactions shown in figure 6.2 lead to quantum corrections to the rainbow graph of figure 6.1. These $\mathcal{O}(g_{YM}^2)$ corrections are shown in figure 6.3. It turns out that the three graphs shown in figure 6.3 cancel each other for all N and for all k. This is proved using combinatorics for the matrices T^a, as well as their commutation relation, as we now

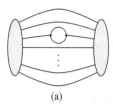   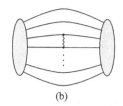   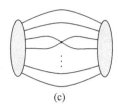

(a)   (b)   (c)

Figure 6.3   Possible corrections to the rainbow graph at order g_{YM}^2 obtained from the interactions shown in figure 6.2.

explain. To begin with, we note a useful trace identity valid for any square matrices K and M_i, which reads

$$\sum_{i=1}^{n} \mathrm{Tr}\left(M_1 \dots M_{i-1}\left[M_i,\, K\right]M_{i+1}\dots M_n\right) = 0. \tag{6.17}$$

The first step is to insert the exchange contributions (6.15) between all pairs of adjacent lines in the rainbow graph, using $[T^a, T^b] = if^{abc}T^c$. The result is, for a fixed $\sigma \in S_k$,

$$\frac{1}{4}\left(-2B(x,y)\right)\mathrm{Tr}\left(T^{a_1}\dots T^{a_k}\right)\sum_{i\neq j=1}^{k}\mathrm{Tr}\left(T^{a_{\sigma(1)}}\dots\left[T^{a_{\sigma(i)}},\, T^p\right]\dots\left[T^{a_{\sigma(j)}},\, T^p\right]\dots T^{a_{\sigma(k)}}\right). \tag{6.18}$$

Next we apply (6.17) to one of the two commutators in (6.18) to find

$$\frac{B(x,y)}{2}\mathrm{Tr}\left(T^{a_1}\dots T^{a_k}\right)\sum_{i=1}^{k}\mathrm{Tr}\left(T^{a_{\sigma(1)}}\dots\left[\left[T^{a_{\sigma(i)}},\, T^p\right],\, T^p\right]\dots T^{a_{\sigma(k)}}\right)$$

$$= \frac{N\,B(x,y)}{2}\mathrm{Tr}\left(T^{a_1}\dots T^{a_k}\right)\sum_{i=1}^{k}\mathrm{Tr}\left(T^{a_{\sigma(1)}}\dots T^{a_{\sigma(i)}}\dots T^{a_{\sigma(k)}}\right). \tag{6.19}$$

The last step follows from the fact that $\left[[\cdot,\, T^p],\, T^p\right]$ is the Casimir operator of the adjoint representation of $SU(N)$, such that $\left[[T^a, T^p], T^p\right] = NT^a$. The final sum over i is then just a sum over k identical terms.

Next we evaluate the self-energy corrections obtained from inserting (6.12) into one of the lines in the rainbow graph. Since there are k such lines, there is an overall factor of k. The sum over all contributions shown in figure 6.3 then gives

$$\frac{kN\left(B(x,y) + 2A(x,y)\right)}{2}\sum_{\sigma\in S_k}\mathrm{Tr}\left(T^{a_1}\dots T^{a_k}\right)\mathrm{Tr}\left(T^{a_{\sigma(1)}}\dots T^{a_{\sigma(k)}}\right)$$

$$= \frac{kN\left(B + 2A\right)P_{k,k,0}(N)}{2}. \tag{6.20}$$

Finally, we note that (6.20) vanishes since $B(x,y) + 2A(x,y) = 0$ due to a non-renormalisation theorem for $\mathcal{N} = 4$ Super Yang–Mills theory. This non-renormalisation theorem follows from the fact that the operator $\mathrm{Tr}\left(\phi^2\right)$ is in the same supersymmetry multiplet as the energy-momentum tensor $T_{\mu\nu}$. It can be shown that the latter is not renormalised, since the energy-momentum tensor is conserved. Therefore by supersymmetry, $\mathrm{Tr}\left(\phi^2\right)$ is protected as well. This implies that $\langle\mathcal{O}_2(x)\mathcal{O}_2(y)\rangle$ does not have any quantum corrections of order $\mathcal{O}(g_{\mathrm{YM}}^2)$, and thus $B(x,y) + 2A(x,y) = 0$. Equation (6.20) then implies that $\langle \mathcal{O}_k(x)\mathcal{O}_k(y)\rangle$ is non-renormalised and independent of g_{YM} (and thus of λ) for all k. Note that this non-renormalisation theorem for the two-point function of $\frac{1}{2}$ BPS operators holds for all values of N.

A similar analysis applies to the three-point functions of $\frac{1}{2}$ BPS operators as well. Four-point functions, however, are renormalised in general, though there are special exceptional cases where they are not, as will be discussed below in section 6.2.

6.1.4 Three-point function on the gravity side

Since we have obtained an exact result for the three-point function of one-half BPS operators on the field theory side, we are now ready to compare this result with its gravity counterpart.

Let us consider three-point functions of scalar fields in AdS spacetimes. The associated Witten tree diagram is displayed in figure 5.2(b). It is specified by three boundary points x, y, z, by three bulk-to-boundary propagators and by a bulk coupling associated with the cubic vertex. This coupling is determined by the Kaluza–Klein reduction on S^5.

Recall from section 5.4.2 that the bulk-to-boundary Green's function in AdS_{d+1} for the scalar field dual to an operator of dimension $\Delta = k$ is given by

$$K_k(w, x; y) = \frac{\Gamma(k)}{\pi^{d/2}\,\Gamma(k-2)} \left(\frac{w}{w^2 + (x-y)^2}\right)^k. \tag{6.21}$$

Due to the defining property $\lim_{w\to 0}\left[w^{k-d} K_k(w,x;y)\right] = \delta^d(x-y)$, we may express the bulk field ϕ in terms of its values at the boundary

$$\phi(w, x) = \frac{\Gamma(k)}{\pi^{d/2}\,\Gamma(k-2)} \int d^d y \left(\frac{w}{w^2 + (x-y)^2}\right)^k \phi_0(y). \tag{6.22}$$

Spatial dependence

To calculate the spatial dependence of the three-point function associated with the Witten diagram in figure 5.2(b), it is useful to use a compact notation for the coordinates which is widely used in the literature. This is introduced as follows. Bulk points are denoted by $(d+1)$-dimensional variables w, which are composed of $(w_0, \vec{w})$ with w_0 the radial coordinate and $\vec{w}$ the d-dimensional coordinate parallel to the boundary. By analogy, the coordinates of the boundary points are denoted by $\vec{x}, \vec{y}, \vec{z}$. Moreover, for the denominator in the bulk-to-boundary propagator we use the notation $(w - \vec{x})^2 \equiv w_0^2 + (\vec{w} - \vec{x})^2$. Using the new notation, (6.21) reads

$$K_k(w, \vec{x}) = \frac{\Gamma(k)}{\pi^{d/2}\,\Gamma(k-2)} \left(\frac{w_0}{w_0^2 + (\vec{w} - \vec{x})^2}\right)^k = \frac{\Gamma(k)}{\pi^{d/2}\,\Gamma(k-2)} \left(\frac{w_0}{(w - \vec{x})^2}\right)^k. \tag{6.23}$$

Using this notation, let us now evaluate [2] the spatial dependence of the three-point function associated with the Witten diagram figure 5.2(b), which is given by

$$A(\vec{x},\vec{y},\vec{z}) \equiv \int dw_0 \, d^d\vec{w} \, \frac{1}{w_0^{d+1}} \left(\frac{w_0}{(w-\vec{x})^2}\right)^{k_1} \left(\frac{w_0}{(w-\vec{y})^2}\right)^{k_2} \left(\frac{w_0}{(w-\vec{z})^2}\right)^{k_3}. \quad (6.24)$$

The prefactor, which is given by the factors of the propagators in (6.21) in combination with the cubic coupling λ_{IJK} arising from the Kaluza–Klein reduction, will be taken care of below.

We proceed by evaluating the integral in (6.24). The integrand involves three factors. To evaluate the integral in closed form, it is necessary to reduce this to two factors. This may be achieved by using an *inversion* as introduced in section 3.2.1. We reexpress the $(d+1)$-dimensional integration variable as

$$w_m = \frac{w'_m}{(w')^2}. \quad (6.25)$$

Similarly, we set

$$\vec{x} = \frac{\vec{x}'}{(\vec{x}')^2}, \qquad \vec{y} = \frac{\vec{y}'}{(\vec{y}')^2}, \qquad \vec{z} = \frac{\vec{z}'}{(\vec{z}')^2} \quad (6.26)$$

for the boundary coordinates. For the bulk-to-boundary propagator (6.23) this leads to

$$K_k(w,\vec{x}) = (\vec{x}')^{2k} \, K_k(w',\vec{x}'). \quad (6.27)$$

The factor $|\vec{x}'|^{2k}$ is very similar to the expressions used for d-dimensional conformal field theory in section 3.2.1, since $(\vec{x}')^{2k} = 1/(\vec{x})^{2k}$. Moreover, the inversion is an isometry of AdS space, such that the volume element is invariant,

$$\frac{d^{d+1}w}{w_0^{d+1}} = \frac{d^{d+1}w'}{(w'_0)^{d+1}}. \quad (6.28)$$

Under the inversions, the three-point function (6.24) transforms as

$$A(\vec{x},\vec{y},\vec{z}) = (\vec{x}')^{2k_1} \, (\vec{y}')^{2k_2} \, (\vec{z}')^{2k_3} \, A(\vec{x}',\vec{y}',\vec{z}'). \quad (6.29)$$

Using the inversion, we can reduce the number of factors in the denominator of (6.24) from three to two, as follows. First, translation invariance allows us to set one argument to zero, i.e. $\vec{z} = 0$. This allows us to write

$$A(\vec{x},\vec{y},\vec{z}) = A(\vec{x}-\vec{z},\vec{y}-\vec{z},0) \equiv A(\vec{u},\vec{v},0). \quad (6.30)$$

Here, the third factor of (6.24) reduces to the simple form

$$\left(\frac{w_0}{(w-\vec{z})^2}\right)^{k_3} = \left(\frac{w_0}{w^2}\right)^{k_3} = (w'_0)^{k_3} \quad (6.31)$$

without denominator. Using the inversion we then obtain

$$A(\vec{u},\vec{v},0) = \frac{1}{|\vec{u}|^{2k_1}\,|\vec{v}|^{2k_2}} \int \frac{d^{d+1}w'}{(w'_0)^{d+1}} \left(\frac{w'_0}{(w'-\vec{u}')^2}\right)^{k_1} \left(\frac{w'_0}{(w'-\vec{v}')^2}\right)^{k_2} (w'_0)^{k_3}. \quad (6.32)$$

By translation invariance of the $\vec{w}$ integration variable, the integral can only depend on the difference $\vec{u}' - \vec{v}'$, and dimensional analysis fixes the power to be $|\vec{u}' - \vec{v}'|^{k_3-k_1-k_2}$. Hence, we have already found the spacetime dependence to be

$$A(\vec{u}, \vec{v}, 0) \sim \frac{(\vec{u}' - \vec{v}')^{k_3-k_1-k_2}}{(\vec{u})^{2k_1}(\vec{v})^{2k_2}}$$

$$= \frac{1}{(\vec{x} - \vec{y})^{k_1+k_2-k_3}(\vec{y} - \vec{z})^{k_2+k_3-k_1}(\vec{z} - \vec{x})^{k_3+k_1-k_2}}$$

$$\equiv f(\vec{x}, \vec{y}, \vec{z}). \tag{6.33}$$

Note that good care has to be taken to restore the w, $\vec{x}$, $\vec{y}$, $\vec{z}$ variables when using the inversion. A useful formula is

$$(\vec{u}' - \vec{v}')^2 = \frac{(\vec{x} - \vec{y})^2}{(\vec{x} - \vec{z})^2(\vec{y} - \vec{z})^2}, \tag{6.34}$$

which is equivalent to (3.65) in section 3.2.1.

With only two factors in the denominator, we can evaluate $A(\vec{u}, \vec{v}, 0)$ in closed form using Feynman parameter methods. The prefactor in $A(\vec{x}, \vec{y}, \vec{z}) = a \cdot f(\vec{x}, \vec{y}, \vec{z})$ is found to be

$$a = -\frac{\Gamma[\frac{1}{2}(k_1 + k_2 - k_3)]\,\Gamma[\frac{1}{2}(k_2 + k_3 - k_1)]\,\Gamma[\frac{1}{2}(k_3 + k_1 - k_2)]\,\Gamma[\frac{1}{2}(\sum_i k_i - d)]}{2\pi^d\,\Gamma[k_1 - \frac{d}{2}]\,\Gamma[k_2 - \frac{d}{2}]\,\Gamma[k_3 - \frac{d}{2}]}. \tag{6.35}$$

We note that Gamma functions in this expression may lead to poles for particular values of the operator scaling dimensions.

Cubic coupling

Next we need to calculate the cubic coupling with which (6.24) enters the three-point function

$$\langle \mathcal{O}^I(\vec{x})\,\mathcal{O}^J(\vec{y})\,\mathcal{O}^K(\vec{z})\rangle = -\frac{4N^2}{(2\pi)^5}\lambda^{IJK} A(\vec{x}, \vec{y}, \vec{z}). \tag{6.36}$$

Here, the indices I, J, K specify representations of $SO(6)$ of dimension k_1, k_2, k_3, respectively. The coupling λ^{IJK} arises from Kaluza–Klein reduction of the type IIB supergravity action on S^5. We outline how to perform this reduction, extending the results presented in section 5.3.2. There we considered the supergravity action only to second order in the fluctuations. Now, to calculate three-point functions, we also need to include the cubic terms to obtain the correct cubic coupling necessary for evaluating the three-point Witten diagram [1].

To obtain the cubic couplings, we have to look at fluctuations about the IIB supergravity background to third order. We decompose the fluctuations in spherical harmonics as described in section 5.3.2. For the S^5, is convenient to work in de-Donder gauge for which

$$\nabla^\alpha h_{\alpha\beta} = \nabla^\alpha a_{\alpha\mu_1\mu_2\mu_3} = 0. \tag{6.37}$$

Inserting the ansatz (5.33) into the ten-dimensional equations of motion leads to diagonalisation and decoupling. As in section 5.3.2, the modes s^I which couple to the field theory

1/2 BPS operators $\mathcal{O}_k^I$ are given by (5.35). To cubic order in the action, these S^5 modes satisfy a five-dimensional equation of motion in AdS space, which in generalisation of (5.34) is of the form

$$\left(\nabla_\mu \nabla^\mu - k(k-4)\right) s^I = \lambda^{IJK} s^J s^K. \tag{6.38}$$

For consistency of the notation in the present chapter, we use k in (6.38), which coincides with l in (5.34), and set $L = 1$. In (6.38), the cubic coupling λ^{IJK} with $SO(6)$ indices I, J, K is given by

$$\lambda^{IJK} = a(k_1, k_2, k_3) \frac{128 \, \Sigma \left((\Sigma/2)^2 - 1\right) \left((\Sigma/2)^2 - 4\right) \alpha_1 \alpha_2 \alpha_3 \left\langle C^I C^J C^K \right\rangle}{(k_1 + 1)(k_2 + 1)(k_3 + 1)}. \tag{6.39}$$

We use the usual shorthand notation $\Sigma = k_1 + k_2 + k_3$ and $\alpha_1 = \frac{k_2 + k_3 - k_1}{2}$ and their cyclic permutations. The numbers $a(k_1, k_2, k_3)$ relate S^5 integrals of spherical harmonics with the unique $SO(6)$ tensors $\left\langle C^I C^J C^K \right\rangle$,

$$\int_{S^5} d\Omega \, Y^I(\Omega) \, Y^J(\Omega) \, Y^K(\Omega) = a(k_1, k_2, k_3) \left\langle C^I \, C^J \, C^K \right\rangle,$$

$$a(k_1, k_2, k_3) = \frac{\pi^3}{(\Sigma/2 + 2)! \, 2^{\Sigma/2 - 1}} \frac{k_1! \, k_2! \, k_3!}{\alpha_1! \, \alpha_2! \, \alpha_3!}. \tag{6.40}$$

Note that for orthonormal C^I, the tensors $\left\langle C^I C^J C^K \right\rangle$ are unique and therefore coincide with the field theory tensors of (6.4). $\left\langle C^I C^J C^K \right\rangle$ is the unique $SO(6)$ invariant which can be formed from $C^I_{i_1 \ldots i_{k_1}}$, $C^J_{i_1 \ldots i_{k_2}}$ and $C^K_{i_1 \ldots i_{k_3}}$. It is obtained by contracting α_1 indices between C^J and C^K, α_2 indices between C^I and C^K and α_3 indices between C^I and C^J.

In extension of (5.36) to cubic order, the Kaluza–Klein decomposition of the supergravity fields gives the following dimensionally reduced supergravity action for the s^I modes,

$$S = \frac{4 N^2}{(2\pi)^5} \int d^5x \, \sqrt{-g} \left[\frac{A_I}{2} \left(-\partial^m s^I \partial_m s^I - k(k-4) (s^I)^2 \right) + \frac{1}{3} \lambda_{IJK} s^I s^J s^K \right]. \tag{6.41}$$

Calculating the two-point function according to the prescription of section 5.4.3 from the action (6.41), we obtain

$$\left\langle \mathcal{O}^I(x) \, \mathcal{O}^J(y) \right\rangle = \frac{4 N^2}{(2\pi)^5} \frac{\pi}{2^{k-7}} \frac{k(k-1)^2(k-2)^2}{(k+1)^2} \frac{\delta^{IJ}}{(x-y)^{2k}}, \tag{6.42}$$

using A_I and $Z(k)$ given by (5.38). We then define normalised operators $\tilde{\mathcal{O}}^I(x)$ dual to $s^I(x)$ such that

$$\left\langle \tilde{\mathcal{O}}^I(x) \tilde{\mathcal{O}}^J(y) \right\rangle = \frac{\delta^{IJ}}{(x-y)^{2k}}. \tag{6.43}$$

The three-point function is computed using λ^{IJK}, as well as the operator normalisation as given by (6.43) and the result (6.33), (6.35) for $A(x, y, z)$. We find

$$\left\langle \tilde{\mathcal{O}}^I(x) \tilde{\mathcal{O}}^J(y) \tilde{\mathcal{O}}^K(z) \right\rangle = \frac{1}{N} \frac{\sqrt{k_1 \, k_2 \, k_3} \left\langle C^I \, C^J \, C^K \right\rangle}{(x-y)^{2\alpha_3} \, (y-z)^{2\alpha_1} \, (z-x)^{2\alpha_2}}. \tag{6.44}$$

Remarkably, this gravitational correlator coincides with the field theory result (6.8)!

Note again that for comparing quantum field theory and supergravity, it is essential to use the two-point function to normalise the operators in the same way on both sides of the correspondence. Also, it is essential to consider observables which are independent of the coupling, since the field theory calculation is performed at weak coupling while the supergravity calculation is dual to a strong coupling result in field theory. Further impressive and very non-trivial tests of the correspondence beyond non-renormalised operators, where the results *do* depend on the coupling, have been obtained in the *integrability* approach. This requires considering the strong form of the AdS/CFT correspondence in the nomenclature of table 5.1. This will be discussed in chapter 7.

6.2 Four-point functions

6.2.1 General case

It is also possible to calculate four-point functions using the AdS/CFT correspondence. Since four-point functions are generically renormalised even for 1/2 BPS operators, a direct comparison between weak coupling field theory results and AdS/CFT is not possible in general, except for some very special cases, some of which we present in section 6.2.2 below.

The Witten diagrams associated to AdS/CFT four-point functions for 1/2 BPS operators involve *contact graphs* with a quartic coupling and four bulk-to-boundary propagators, as well as *exchange graphs* with two cubic couplings and a bulk-to-bulk propagator in addition to the bulk-to-boundary propagators. These Witten diagrams are shown in figure 6.4. The contact graph corresponds to the amplitude

$$D_{\Delta_1\Delta_2\Delta_3\Delta_4}(\vec{x}_1,\vec{x}_2,\vec{x}_3,\vec{x}_4) = \mathcal{G}_{\Delta_1\Delta_2\Delta_3\Delta_4} \int \frac{d^{d+1}z}{z_0^{d+1}} \prod_{i=1}^{4} K_{\Delta_i}(z,\vec{x}_i) \tag{6.45}$$

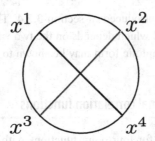 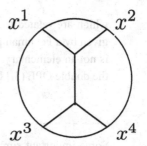

Figure 6.4   Four-point contact graph (left) and exchange graph (right).

and the exchange graph corresponds to

$$S(\vec{x}_1, \vec{x}_2, \vec{x}_3, \vec{x}_4) = \mathcal{G}_{\Delta_3 \Delta_4 \Delta} \int d^{d+1}w \sqrt{g} K_{\Delta_3}(w, \vec{x}_3) K_{\Delta_4}(w, \vec{x}_4) A(w, \vec{x}_1, \vec{x}_2), \quad (6.46)$$

$$A(w, \vec{x}_1, \vec{x}_2) = \mathcal{G}_{\Delta_1 \Delta_2 \Delta} \int d^{d+1}z \, G_\Delta(w, z) K_{\Delta_1}(z, \vec{x}_1) K_{\Delta_2}(z, \vec{x}_2), \quad (6.47)$$

where we have used the same notation for the coordinates as introduced for the calculation of the three-point function in section 6.1.4. The K_{Δ_i} are the bulk-to-boundary propagators as in (5.64) and G_{Δ_i} are the bulk-to-bulk propagators (5.65). The $\mathcal{G}$ are the cubic and quartic couplings obtained from the Kaluza–Klein reduction. For the full four-point function, a sum over all intermediate states in the exchange diagram is necessary. These states and the associated couplings may be obtained from the Clebsch–Gordan expansion of the two $\mathfrak{su}(4)_R$ representations associated with the two bulk-to-boundary propagators meeting at the cubic vertex,

$$[0, \Delta_1, 0] \otimes [0, \Delta_2, 0] = \oplus_{m=1}^{\Delta_2} \oplus_{n=0}^{\Delta_2 - m} [n, \Delta_1 + \Delta_2 - 2m - 2n, n]. \quad (6.48)$$

This formula generalises the examples for $\mathfrak{su}(4)_R$ tensor products discussed in appendix B.2.1. In addition to scalars, this may also involve intermediate vector and tensor states.

The integrals in (6.45), (6.46) and (6.47) are difficult to evaluate in general. However, simplifications occur if the conformal dimensions of the operators involved coincide in pairs, i.e. $\Delta_1 = \Delta_2 = \Delta$, $\Delta_3 = \Delta_4 = \Delta'$. As an example, let us consider the contact graph amplitude (6.45). Applying twice both the techniques introduced in section 6.1.4, i.e. the inversion and Feynman parameter methods, allows us to write the contact graph amplitude in compact form,

$$D_{\Delta\Delta\Delta'\Delta'}(\vec{x}_1, \ldots, \vec{x}_4) = (-1)^{\Delta + \Delta'} \frac{\vec{x}_{12}^{2\Delta'} \vec{x}_{13}^{2\Delta} \vec{x}_{14}^{2\Delta'}}{(\vec{x}^2 + \vec{y}^2)^{\Delta'}}$$
$$\times \left(\frac{\partial}{\partial s}\right)^{\Delta' - 1} \left[s^{\Delta - 1} \left(\frac{\partial}{\partial s}\right)^{\Delta - 1} H(s, t) \right], \quad (6.49)$$

where we have defined the inversion differences

$$\vec{x} \equiv \vec{x}'_{13} - \vec{x}'_{14}, \qquad \vec{y} \equiv \vec{x}'_{13} - \vec{x}'_{12} \quad (6.50)$$

as well as the variables

$$s = \frac{1}{2} \frac{x_{13}^2 x_{24}^2}{x_{12}^2 x_{34}^2 + x_{14}^2 x_{23}^2}, \qquad t = \frac{x_{12}^2 x_{34}^2 - x_{14}^2 x_{23}^2}{x_{12}^2 x_{34}^2 + x_{14}^2 x_{23}^2} \quad (6.51)$$

which are related to the cross ratios introduced in section 3.2.4. The function $H(s, t)$ is given by a Feynman parameter integral which depends on the two variables (6.51). $H(s, t)$ is not an elementary function; its asymptotic form may be shown to be in agreement with the double OPE (3.115).

6.2.2 Extremal correlation functions

Some important simplifications occur for four-point functions with the property that the conformal dimension of one of the 1/2 BPS operators involved is equal to the sum of all

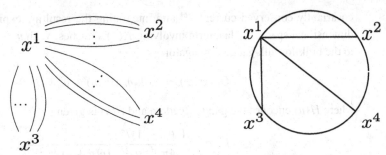

Figure 6.5 Extremal four-point function: field theory Feynman graph at large N (left); AdS/CFT Witten diagram (right).

of the other dimensions, i.e. $\Delta_1 = \Delta_2 + \Delta_3 + \Delta_4$. In this case, the four-point function can be shown to factorise into a product of two-point functions both on the field theory side and on the gravity side [3]. In this special case, the four-point function satisfies a non-renormalisation theorem. The field theory Feynman diagram for this four-point function is shown in figure 6.5 on the left. It corresponds to a four-point function of the factorised form

$$\langle \mathcal{O}_{\Delta_1}(\vec{x}_1)\mathcal{O}_{\Delta_2}(\vec{x}_2)\mathcal{O}_{\Delta_3}(\vec{x}_3)\mathcal{O}_{\Delta_4}(\vec{x}_4)\rangle = A(\Delta_i; N)\prod_{i=2}^{4}\frac{1}{(\vec{x}_1 - \vec{x}_i)^{\Delta_i}}. \qquad (6.52)$$

A similar factorisation arises on the gravity side. In this case, the coupling obtained from the Kaluza–Klein reduction vanishes. On the other hand, there is a singularity when the interaction vertex coincides with the boundary point $\vec{x}_1$ where the operator with conformal dimension Δ_1 is located. A careful analytic continuation shows that the product of coupling and amplitude gives rise to a finite result which again corresponds to a product of three two-point functions. The associated Witten diagram is shown in figure 6.5 on the right.

6.2.3 Vector and tensor propagators

In addition to the scalar propagators introduced above, propagators for vector and tensor fields may also appear as both bulk-to-bulk and bulk-to-boundary propagators.

For the gauge propagator, the relevant part of the action is

$$\mathcal{S}_A = \frac{1}{2\kappa_{d+1}^2}\int d^{d+1}z\sqrt{g}\left(\frac{1}{4}F_{mn}F^{mn} - A_m J^m\right), \qquad (6.53)$$

where we set $L = 1$ and the gauge field A_m sources a conserved bulk current J^m. This leads to

$$-\frac{1}{\sqrt{g}}\partial_p\left(\sqrt{g}g^{pq}\partial_{[q}G_{m]n}(z,z')\right) = g_{mn}\delta(z,z') + \partial_m\partial_n\Lambda(\xi) \qquad (6.54)$$

for the bulk-to-bulk gauge propagator, where the notation $z_m = (z_0,\vec{z})$ is used for the $(d+1)$-dimensional coordinates, with z_0 the radial variable. ξ is the chordal distance as in (5.67). The function Λ reflects the gauge freedom. When multiplying (6.54) with the covariantly conserved current J^m and integrating the resulting expression over $(d+1)$-dimensional space, the last term involving $\Lambda(\xi)$ vanishes. With $u \equiv 1/\xi$, (6.54) gives rise to the bulk-to-bulk gauge propagator

$$G_{mn}(z,z') = -(\partial_m \partial_n u)F(u) + \partial_m \partial_n H(u), \tag{6.55}$$

where $H(u)$ encodes the gauge freedom and $F(u)$ is given by

$$F(u) = \frac{\Gamma((d-1)/2)}{4\pi^{(d+1)/2}} \frac{1}{(u(u+2))^{(d-1)/2}}. \tag{6.56}$$

Similarly, the bulk-to-boundary gauge propagator satisfying

$$A_m(z) = \int d^{d+1}w \sqrt{g}\, G_{m\mu}(w,\vec{x})A_\mu(\vec{x}) \tag{6.57}$$

is given by

$$G_{m\mu}(z,\vec{x}) = C_d \frac{z_0^{d-2}}{(z-\vec{x})^{2(d-1)}} I_{m\mu}(z-\vec{x}) + S_{m\mu}(z,\vec{x}), \tag{6.58}$$

$$C_d = \frac{\Gamma(d)}{2\pi^{d/2}\Gamma(d/2)}, \tag{6.59}$$

where $S_{m\mu}$ expresses the gauge freedom and can be fixed in such a way that $\partial_{\vec{x}_\mu} G_{m\mu}(z,\vec{x}) = 0$, consistent with a conserved boundary current. $I_{m\mu}$ is an inversion tensor as defined in (3.62). The field strength $F_{mn} = \partial_m A_n - \partial_n A_m$ for A_n of (6.57) gives rise to

$$\partial_{[m} G_{n]\mu}(z,\vec{x}) = (d-2)C_d \frac{z_0^{d-3}}{(z-\vec{x})^{2(d-1)}} I_{0[m}(z-\vec{x})I_{n]\mu}(z-\vec{x}), \tag{6.60}$$

which is gauge invariant as expected. The gauge bulk-to-boundary propagator can be used to calculate the three-point function involving a conserved $U(1)$ current as well as two scalar operators of dimension Δ, for instance. The relevant part in the supergravity action takes the form

$$S[\phi, A_m] = \frac{1}{2\kappa_{d+1}^2} \int d^{d+1}z \sqrt{g}\, (g^{mn} D_m\phi D_n\phi + m^2\phi^2),$$

$$D_m\phi = \partial_m\phi - iA_m\phi. \tag{6.61}$$

The cubic vertices present in this action give rise to the following expression for the three-point function considered,

$$\langle J_\mu(\vec{z})\mathcal{O}(\vec{x})\mathcal{O}(\vec{y})\rangle = -\int \frac{d^{d+1}w}{w_0^{d+1}} G_{m\mu}(w,\vec{z})w_0^2 K_\Delta(w,\vec{x}) \frac{\overset{\leftrightarrow}{\partial}}{\partial w_m} K_\Delta(w,\vec{x}) \tag{6.62}$$

where K_Δ is the bulk-to-boundary propagator for a scalar operator of dimension Δ as in (5.71) and

$$\frac{\overset{\leftrightarrow}{\partial}}{\partial w_m} \equiv -\frac{\overset{\leftarrow}{\partial}}{\partial w_m} + \frac{\overset{\rightarrow}{\partial}}{\partial w_m}. \tag{6.63}$$

Exercise 6.2.1   Calculate the integral in (6.62) using an inversion, and show that the result is

$$\langle J_\mu(\vec{z})\mathcal{O}(\vec{x})\mathcal{O}(\vec{y})\rangle = -\xi(d-2)\frac{1}{(\vec{x}-\vec{y})^{2\Delta}}(Z^2)^{(d-2)/2}Z_\mu, \tag{6.64}$$

with Z_μ defined in (3.65) in chapter 3 and

$$\xi = \frac{(\Delta-d/2)\Gamma(d/2)\Gamma(\Delta)}{\pi^{d/2}(d-2)\Gamma(\Delta-d/2)}. \tag{6.65}$$

Exercise 6.2.2   Show that (6.64) satisfies the Ward identity

$$\frac{\partial}{\partial z_\mu}\langle J_\mu(\vec{z})\mathcal{O}(\vec{x})\mathcal{O}(\vec{y})\rangle = \left(\delta^d(\vec{x}-\vec{z})+\delta^d(\vec{y}-\vec{z})\right)\langle\mathcal{O}(\vec{x})\mathcal{O}(\vec{y})\rangle, \tag{6.66}$$

with $\langle\mathcal{O}(\vec{x})\mathcal{O}(\vec{y})\rangle$ the holographic two-point function (5.89).

For the graviton, which is dual to the energy-momentum tensor, the result for the bulk-to-boundary propagator analogous to (6.58) is obtained from

$$g^m{}_n(z) = \int d^d\vec{x}\,\sqrt{g}\,G^m{}_{n\mu\nu}(z,\vec{x})g^{\mu\nu}(\vec{x}) \tag{6.67}$$

and reads

$$G^m{}_{n\mu\nu}(z-\vec{x}) = \frac{d+1}{d-1}\frac{\Gamma(d)}{\pi^{d/2}\Gamma(d/2)}\frac{z_0^d}{(z-\vec{x})^{2d}}I^{m\rho}(z-\vec{x})I_n{}^\sigma P_{\rho\sigma,\mu\nu}, \tag{6.68}$$

where I is the inversion tensor defined in (3.62) and P is the projection operator onto traceless symmetric tensors defined in (3.12).

6.3 The conformal anomaly

As a second example of astonishing agreement between computations in AdS gravity and in $\mathcal{N}=4$ Super Yang–Mills theory, we compute the conformal anomaly or trace anomaly using both approaches. As introduced in 3.2.6, the conformal anomaly arises from the failure of the energy-momentum tensor to remain traceless under quantum corrections in a classically conformal field theory.

6.3.1 The conformal anomaly on the field theory side

Recall from equation (3.121) that the operator insertion of the energy-momentum tensor can be obtained from

$$\langle T_{\mu\nu}(x)\rangle = -\frac{2}{\sqrt{g}}\frac{\delta W}{\delta g^{\mu\nu}(x)}. \tag{6.69}$$

In this definition, $g^{\mu\nu}$ is a classical background field: it does not propagate, but is a source for $T_{\mu\nu}$. Generically, $\langle T_\mu{}^\mu\rangle \neq 0$ at the quantum level, since under a conformal transformation, counterterms needed for regularisation give finite contributions to the trace. This applies even to theories which are conformally invariant at the classical level.

As explained in section 3.2.6, the conformal anomaly of a four-dimensional quantum field theory is of the generic form

$$\langle T_\mu{}^\mu(x) \rangle = \frac{c}{16\pi^2} \, C^{\mu\sigma\rho\nu} \, C_{\mu\sigma\rho\nu} - \frac{a}{16\pi^2} \, E \tag{6.70}$$

where C is the Weyl tensor and E is the Euler density. The coefficients c and a are model-dependent. Many explicit calculation methods, for instance dimensional regularisation or the heat-kernel approach, can be used to calculate the coefficients within field theory at weak coupling. To lowest order in perturbation theory, they are determined by the number of scalar, fermionic and vector fields present in the field theory considered, and are found to take the form

$$c = \frac{1}{120} \left(N_s + 6N_f + 12N_v \right), \qquad a = \frac{1}{360} \left(N_s + 11N_f + 62N_v \right), \tag{6.71}$$

with N_s, N_f and N_v the number of scalars, Dirac fermions and vectors, respectively. For a theory with $\mathcal{N} = 1$ supersymmetry with N_Φ chiral multiplets and N_V vector multiplets, we may reexpress these numbers as

$$c = \frac{1}{24} \left(N_\Phi + 3N_V \right), \qquad a = \frac{1}{48} \left(N_\Phi + 9N_V \right). \tag{6.72}$$

For $\mathcal{N} = 4$ $SU(N)$ Super Yang–Mills theory with $N_\Phi = 3(N^2 - 1)$, $N_V = (N^2 - 1)$ this implies

$$c = a = \frac{1}{4} \, (N^2 - 1). \tag{6.73}$$

It is a very special property of $\mathcal{N} = 4$ Super Yang–Mills theory that c and a coincide, $c = a$. This is not the case in generic quantum field theories. In the large N limit, we have to leading order

$$c = a = \frac{1}{4} \, N^2. \tag{6.74}$$

In total, the agreement of c and a implies that the conformal anomaly for $\mathcal{N} = 4$ theory at large N takes the form

$$\langle T_\mu^\mu(x) \rangle = \frac{c}{8\pi^2} \left(R^{\mu\nu} R_{\mu\nu} - \frac{1}{3} R^2 \right) \overset{N \to \infty}{\to} \frac{N^2}{32\pi^2} \left(R^{\mu\nu} R_{\mu\nu} - \frac{1}{3} R^2 \right). \tag{6.75}$$

It can be shown that this result is one-loop exact in $\mathcal{N} = 4$ theory, which implies that it is independent of λ to all orders in perturbation theory. It is therefore ideally suited for a test of the AdS/CFT correspondence, since the perturbative result (6.75) can be compared directly to the strong coupling result obtained by mapping to AdS space.

6.3.2 The conformal anomaly on the gravity side

The gravity counterpart, i.e. the conformal anomaly from AdS space, is computed from the action of $(d + 1)$-dimensional AdS gravity. We begin the analysis in general d dimensions and specify to $d = 4$ at the end for comparison with the previous section 6.3.1. The gravity action is

$$S = -\frac{1}{16\pi G} \left[\int \mathrm{d}^{d+1}x \, \sqrt{g} \left(R + \frac{d(d-1)}{L^2} \right) + 2 \int \mathrm{d}^d x \, \sqrt{\gamma} K \right], \tag{6.76}$$

including a Gibbons–Hawking boundary term as introduced in (2.155). Recall from section 5.5 that the metric for AdS_{d+1} is given by

$$ds^2 = L^2 \left(\frac{d\rho^2}{4\rho^2} + \frac{1}{\rho} \delta_{\mu\nu} dx^\mu dx^\nu \right) \tag{6.77}$$

as long as the $\rho = 0$ boundary remains flat. If we allow for a curved boundary spacetime with metric $g^{(0)}_{\mu\nu}(x)$ this generalises to

$$ds^2 = L^2 \left(\frac{d\rho^2}{4\rho^2} + \frac{1}{\rho} g_{\mu\nu}(\rho, x) \, dx^\mu \, dx^\nu \right), \quad \lim_{\rho \to 0} g_{\mu\nu}(\rho, x) = g^{(0)}_{\mu\nu}(x). \tag{6.78}$$

The coordinate singularity of the metric at $\rho \to 0$ can be avoided by means of a cutoff at $\rho = \epsilon$. The integration region in the action is then restricted to $\rho \geq \epsilon$.

On the field theory side, a Weyl transformation of the metric gives the trace of the energy-momentum tensor. Therefore, we need to translate a Weyl transformation in the boundary theory into a transformation in the bulk, i.e. in $(d + 1)$-dimensional AdS space. The task is to find a $(d + 1)$-dimensional diffeomorphism which reduces to a Weyl transformation on the boundary. The desired diffeomorphism is known as the *Penrose–Brown–Henneaux transformation* or *PBH transformation* [4, 5, 6]

$$\rho = \rho' \left(1 - 2\sigma(x') \right), \quad x^\mu = x'^\mu + a^\mu(x', \rho'). \tag{6.79}$$

We have to ensure that the form of the metric (6.78) is covariant under this transformation, i.e. that $g'_{\rho\rho} = g_{\rho\rho}$ and $g'_{\rho\mu} = g_{\rho\mu} = 0$. This imposes the constraints

$$\partial_\rho a^\mu = \frac{L^2}{2} g^{\mu\nu} \partial_\nu \sigma \tag{6.80}$$

on the functions a^μ and σ of (6.79). It follows that

$$a^\mu(x, \rho) = \frac{L^2}{2} \int_0^\rho d\hat{\rho} \, g^{\mu\nu}(x, \hat{\rho}) \, \partial_\nu \sigma(x). \tag{6.81}$$

Under this diffeomorphism, the d-dimensional part $g_{\mu\nu}(\rho, x)$ of the metric transforms as

$$g_{\mu\nu} \mapsto g_{\mu\nu} + 2\sigma \left(1 - \rho \frac{\partial}{\partial \rho} \right) g_{\mu\nu} + \nabla_\mu a_\nu + \nabla_\nu a_\mu \tag{6.82}$$

such that at the boundary where $\rho \to 0$, we have $a_\mu \to 0$ and $\rho \frac{\partial}{\partial \rho} g_{\mu\nu} \to 0$ and therefore we recover the Weyl transformation $\delta g_{\mu\nu}(x) = 2\sigma(x) g^{(0)}_{\mu\nu}(x)$. Applying the PBH transformation to the action (6.76)) using the metric (6.78) gives the expected boundary value of the trace of the energy-momentum tensor

$$\delta S = \frac{1}{2} \int d^d x \sqrt{g^{(0)}} \langle T^{\mu\nu} \rangle \delta g^{(0)}_{\mu\nu}, \quad \delta g^{(0)}_{\mu\nu} = 2\sigma \, g^{(0)}_{\mu\nu}. \tag{6.83}$$

Our aim is to analyse the boundary behaviour of the action (6.76). This requires further information on the structure of the metric $g_{\mu\nu}(\rho, x)$. This information is provided by the Fefferman–Graham theorem introduced in section 5.5. If a metric of the form (6.78) satisfies the Einstein equations and if d is even, then $g_{\mu\nu}(\rho, x)$ can be expanded as

$$g_{\mu\nu}(x, \rho) = g^{(0)}_{\mu\nu}(x) + \rho \, g^{(2)}_{\mu\nu}(x) + \rho^2 \, g^{(4)}_{\mu\nu}(x) + \rho^{d/2} \ln(\rho) \, h^{(d)}_{\mu\nu} + \cdots. \tag{6.84}$$

The coefficients $g_{\mu\nu}^{(n)}(x)$ are built out of the curvature for the boundary metric $g_{\mu\nu}^{(0)}(x)$. As explained in section 5.5, they are calculated by inserting the expansion into the vacuum Einstein equation. For example, the linear coefficient $g_{\mu\nu}^{(2)}(x)$ was determined in exercise 5.5.2.

Using holographic renormalisation as introduced in section 5.5.2, the holographic conformal anomaly may now be calculated as follows. Inserting the metric (6.78) with the expansion (6.84) into the action (6.76) gives rise to a boundary expansion of the gravity action, which takes the form

$$S = -\frac{1}{16\pi G} \int d^{d+1}x \sqrt{\det g^{(0)}} \left(\epsilon^{-d/2} a_{(0)} + \epsilon^{-d/2+1} a_{(2)} + \cdots - \ln\epsilon \, a_{(d)} \right) + S_{\text{finite}},$$

$$(6.85)$$

where the action contribution S_{finite} summarises the finite contributions. The remaining terms characterise the divergences for $\epsilon \to 0$. The coefficients $a_{(n)}$ are determined in terms of $g_{\mu\nu}^{(0)}$. We recall from (5.122) that they read

$$a_{(0)} = \frac{2(d-1)}{L}, \quad a_{(2)} = \frac{L}{2(d-1)}R, \quad a_{(4)} = \frac{L^3}{2(d-2)^2}(R^{\mu\nu}R_{\mu\nu} - \frac{1}{d-1}R^2). \quad (6.86)$$

The simplest form of regularisation is the minimal subtraction scheme which amounts to adding the counterterms

$$S_{\text{ct}} = \frac{1}{16\pi G} \int d^{d+1}x \sqrt{\det g^{(0)}} \left(\epsilon^{-d/2} a_{(0)} + \epsilon^{-d/2+1} a_{(2)} + \cdots - \ln\epsilon \, a_{(d)} \right). \quad (6.87)$$

This counterterm ensures finiteness of the AdS action, but it spoils invariance under the PBH transformation. This may be seen as follows. Close to the boundary, the PBH transformation amounts to

$$\delta(S + S_{\text{ct}}) = 2 \int d^d x \, \sigma(x) \left(\epsilon \frac{\delta}{\delta\epsilon} - g^{(0)\mu\nu} \frac{\delta}{\delta g^{(0)\mu\nu}} \right) (S + S_{\text{ct}}). \quad (6.88)$$

While S is diffeomorphism invariant and thus invariant under PBH transformations, applying the PBH transformation as in (6.88) to the counterterms (6.87) gives rise to finite contributions to the trace of the energy-momentum tensor even in the limit $\epsilon \to 0$. This may be seen using the results of the two following exercises. To compare with $\mathcal{N} = 4$ Super Yang–Mills theory, we are restricted to $d = 4$ boundary dimensions from now on.

Exercise 6.3.1   Show that for a function $\sigma(x)$ corresponding to a conformal transformation, i.e. $\sigma(x) = (\text{constant} - 2b \cdot x)$ as discussed in 3.2.1, the terms involving $\sqrt{\det g^{(0)}}\epsilon^{-2} a_{(0)}$ and $\sqrt{\det g^{(0)}}\epsilon^{-1} a_{(2)}$ in S_{ct} are invariant under the residual PBH transformation (6.88).

Exercise 6.3.2   For the remaining term involving $\sqrt{\det g^{(0)}}\ln\epsilon \, a_{(4)}$, show for the transformation (6.88) that

$$\delta S_{\text{ct}} = 2 \int d^4 x \, \sigma(x) \left(\epsilon \frac{\delta}{\delta\epsilon} - g^{(0)\mu\nu} \frac{\delta}{\delta g^{(0)\mu\nu}} \right) S_{\text{ct}} = 2 \int d^4 x \, \sigma(x) \epsilon \frac{\delta}{\delta\epsilon} S_{\text{ct}}$$

$$= -\frac{L^3}{64\pi G} \int d^4 x \sqrt{\det g^{(0)}} \left(R^{\mu\nu} R_{\mu\nu} - \frac{R^2}{3} \right). \quad (6.89)$$

Hint: Use the conformal transformations

$$\delta E = 4\sigma E - G^{\mu\nu}\nabla_\mu\nabla_\nu\sigma, \qquad \delta R = 2\sigma R - \nabla^2\sigma \quad (6.90)$$

for the Euler density and the Ricci scalar, with $G_{\mu\nu}$ the Einstein tensor.

In the limit $\epsilon \to 0$, the PBH transformation reduces to a Weyl variation of the metric at the boundary. Therefore, when applied to (6.85) and (6.87), in the limit $\epsilon \to 0$ the PBH transformation (6.88) gives rise to the energy-momentum trace

$$-2g^{(0)\mu\nu} \frac{\delta}{\delta g^{(0)\mu\nu}} (S + S_{ct}) = g^{(0)\mu\nu} \langle T_{\mu\nu} \rangle = \frac{L^3}{64\pi G} \left(R^{\mu\nu} R_{\mu\nu} - \frac{R^2}{3} \right)$$

$$= \frac{N^2}{32\pi^2} \left(R^{\mu\nu} R_{\mu\nu} - \frac{R^2}{3} \right) \qquad (6.91)$$

with the five-dimensional Newton constant

$$G = \frac{G_{10}}{\text{vol}(S^5)} = \frac{\pi L^3}{2N^2}, \qquad (6.92)$$

with G_{10} as given by (4.76) with $2\kappa_{10}^2 = 16\pi G_{10}$. Note that the S^5 factor enters in (6.92) and is essential for fixing the anomaly coefficient to the value expected for $\mathcal{N} = 4$ Super Yang–Mills theory. The gravity result (6.91) for the anomaly coincides with the $N \to \infty$ limit of the field theory result (6.75), which provides a second test for the AdS/CFT correspondence in addition to the agreement found for the three-point function.

6.4 Further reading

A review of tests of the AdS/CFT correspondence for correlation functions and the conformal anomaly may be found in [7].

The AdS/CFT three-point function for scalar and vector fields was computed in [2]. The non-renormalisation theorems for two- and three-point functions of one-half BPS operators in $\mathcal{N} = 4$ Super Yang–Mills theory were shown in [2]. The exact matching of the perturbative field theory and AdS/CFT results for the three-point function of one-half BPS operators was obtained in [1]. Four-point functions were studied using the AdS/CFT correspondence in [8, 9, 10, 11, 12, 13]. The analysis of the Kaluza–Klein reduction for cubic couplings of [1] was extended to quartic couplings and four-point functions in [14]. Extremal correlators were investigated in [3] and next-to-extremal correlators in [15]. The graviton propagator in AdS/CFT was studied in [16]. The three point-function of the energy-momentum tensor was calculated using AdS/CFT in [17].

Within quantum field theory on a curved space background, the conformal anomaly was found in 1973 by Capper and Duff [18], see also [19]. The holographic conformal anomaly was calculated in [20, 21]. An analysis of this anomaly using holographic renormalisation may be found in [22]. The Penrose–Brown–Henneaux transformation goes back to the work of Penrose [4] and Brown and Henneaux [5]. In the holographic context, it was studied in in [6, 23]. For the case of AdS_3, Brown and Henneaux [5] anticipated the conformal anomaly found in the AdS/CFT correspondence.

References

[1] Lee, Sangmin, Minwalla, Shiraz, Rangamani, Mukund, and Seiberg, Nathan. 1998. Three-point functions of chiral operators in $D = 4$, $\mathcal{N} = 4$ SYM at large N. *Adv. Theor. Math. Phys.*, **2**, 697–718.

[2] Freedman, Daniel Z., Mathur, Samir D., Matusis, Alec, and Rastelli, Leonardo. 1999. Correlation functions in the $\mathrm{CFT}_d/\mathrm{AdS}_{(d+1)}$ correspondence. *Nucl. Phys.*, **B546**, 96–118.

[3] D'Hoker, Eric, Freedman, Daniel Z., Mathur, Samir D., Matusis, Alec, and Rastelli, Leonardo. 1999. Extremal correlators in the AdS/CFT correspondence. ArXiv:hep-th/9908160.

[4] Penrose, R., and Rindler, W. 1986. *Spinors and Space–Time*. Vol. 2: *Spinor and Twistor Methods in Space–Time Geometry*, Chapter 9. Cambridge University Press.

[5] Brown, J. David, and Henneaux, M. 1986. Central charges in the canonical realization of asymptotic symmetries: an example from three-dimensional gravity. *Commun. Math. Phys.*, **104**, 207–226.

[6] Imbimbo, C., Schwimmer, A., Theisen, S., and Yankielowicz, S. 2000. Diffeomorphisms and holographic anomalies. *Class. Quantum. Grav.*, **17**, 1129–1138.

[7] D'Hoker, Eric, and Freedman, Daniel Z. 2002. Supersymmetric gauge theories and the AdS/CFT correspondence. TASI 2001 School Proceedings. ArXiv:hep-th/0201253. 3–158.

[8] Freedman, Daniel Z., Mathur, Samir D., Matusis, Alec, and Rastelli, Leonardo. 1999. Comments on four-point functions in the CFT/AdS correspondence. *Phys. Lett.*, **B452**, 61–68.

[9] Liu, Hong, and Tseytlin, Arkady A. 1999. On four-point functions in the CFT/AdS correspondence. *Phys. Rev.*, **D59**, 086002.

[10] D'Hoker, Eric, and Freedman, Daniel Z. 1999. Gauge boson exchange in AdS_{d+1}. *Nucl. Phys.*, **B544**, 612–632.

[11] D'Hoker, Eric, and Freedman, Daniel Z. 1999. General scalar exchange in $\mathrm{AdS}_{(d+1)}$. *Nucl. Phys.*, **B550**, 261–288.

[12] D'Hoker, Eric, Freedman, Daniel Z., Mathur, Samir D., Matusis, Alec, and Rastelli, Leonardo. 1999. Graviton exchange and complete four-point functions in the AdS/CFT correspondence. *Nucl. Phys.*, **B562**, 353–394.

[13] D'Hoker, Eric, Freedman, Daniel Z., and Rastelli, Leonardo. 1999. AdS/CFT four-point functions: how to succeed at z integrals without really trying. *Nucl. Phys.*, **B562**, 395–411.

[14] Arutyunov, G., and Frolov, S. 2000. Four-point functions of lowest weight CPO's in $\mathcal{N} = 4$ SYM in supergravity approximation. *Phys. Rev.*, **D62**, 064016.

[15] Erdmenger, J., and Pérez-Victoria, M. 2000. Nonrenormalization of next-to-extremal correlators in $\mathcal{N} = 4$ SYM and the AdS/CFT correspondence. *Phys. Rev.*, **D62**, 045008.

[16] Liu, Hong, and Tseytlin, Arkady A. 1998. $D = 4$ Super Yang-Mills, $D = 5$ gauged supergravity, and $D = 4$ conformal supergravity. *Nucl. Phys.*, **B533**, 88–108.

[17] Arutyunov, G., and Frolov, S. 1999. Three-point Green function of the stress energy tensor in the AdS/CFT correspondence. *Phys. Rev.*, **D60**, 026004.

[18] Capper, D. M., and Duff, M. J. 1974. Trace anomalies in dimensional regularization. *Nuovo Cimento*, **A23**, 173–183.

[19] Christensen, S. M., and Duff, M. J. 1978. Axial and conformal anomalies for arbitrary spin in gravity and supergravity. *Phys. Lett.*, **B76**, 571.

[20] Henningson, M., and Skenderis, K. 1998. The holographic Weyl anomaly. *J. High Energy Phys.*, **9807**, 023.

[21] Henningson, M., and Skenderis, K. 2000. Holography and the Weyl anomaly. *Fortschr. Phys.*, **48**, 125–128.

[22] de Haro, Sebastian, Solodukhin, Sergey N., and Skenderis, Kostas. 2001. Holographic reconstruction of space-time and renormalization in the AdS/CFT correspondence. *Commun. Math. Phys.*, **217**, 595–622.

[23] Schwimmer, A., and Theisen, S. 2000. Diffeomorphisms, anomalies and the Fefferman-Graham ambiguity. *J. High Energy Phys.*, **0008**, 032.

Solving interacting quantum (field) theories exactly for all values of the coupling constant, and not just for very small coupling constant where perturbation theory is applicable, is a long-standing open problem of theoretical physics. By exactly solving we mean diagonalising the corresponding Hamiltonian, such that both eigenstates and eigenvalues are known explicitly. This is an extraordinarily difficult task: for instance, for QCD this would mean finding the complete mass spectrum from first principles from the QCD Lagrangian using analytical methods, as well as the associated eigenstates. This is certainly impossible at present for such a complicated theory.

The same problem occurs also for quantum mechanics. For most quantum systems, the complete set of eigenstates is not known. However, within quantum mechanics there are some cases where an exact solution is possible: the harmonic oscillator and the hydrogen atom, for instance. These systems should be viewed as toy models since they are ideal approximations to systems realised in nature. For instance, to describe real oscillators in solids, higher order interaction terms have to be added.

In quantum field theory there are also exactly solvable toy models: however, they are defined in low spacetime dimensions, for instance in $d = 1 + 1$. An example is the Thirring model which is *integrable* in the sense that it has an infinite number of conserved quantities. Are there any interacting exactly solvable quantum field theories also in 3+1 dimensions? The surprising answer is yes: there is a large amount of evidence that $\mathcal{N} = 4$ Super Yang–Mills theory has an integrable structure, at least in the planar (large N) limit. The evidence for such an integrable structure at strong coupling has been found using the AdS/CFT correspondence. Since $\mathcal{N} = 4$ Super Yang–Mills is a conformal field theory, in the planar limit the theory is characterised completely by the scaling dimensions Δ of the composite local operators built from products of elementary fields. In this chapter we provide evidence that, ultimately, the methods presented will lead to a precise determination of the scaling dimension $\Delta_{\mathcal{O}}$ for any gauge invariant local observable $\mathcal{O}$ as a function of the coupling λ.

In this sense, $\mathcal{N} = 4$ Super Yang–Mills theory in 3+1 dimensions may be viewed as the harmonic oscillator of the twenty-first century. We may push this analogy even further by stating that any gauge theory in 3+1 dimensions may be obtained by adding further interaction terms to $\mathcal{N} = 4$ Super Yang–Mills theory, or by removing them. In the same way, we may add anharmonic terms to the potential of the harmonic oscillator.

Moreover, there are also integrable structures in classical string theory on $AdS_5 \times S^5$, as we will discuss. These are related to the conjectured integrability of $\mathcal{N} = 4$ Super Yang–Mills theory, as may be seen in particular by taking a specific limit, the *BMN limit*. In this limit, field theory operators and string theory configurations can be mapped directly to each other. This provides evidence for the strong form of the AdS/CFT correspondence as defined in table 5.1.

We begin this chapter by discussing integrable structures perturbatively in $\mathcal{N} = 4$ theory at one and higher order loops. Similar structures also occur at strong coupling. We discuss them using string theory and the AdS/CFT correspondence. In addition to local composite operators, we also consider scattering amplitudes. These have a new emergent symmetry at both weak and strong coupling, the *dual superconformal symmetry*, discussed in section 7.4.3. Moreover, we describe how the scattering amplitudes may be mapped to lightlike Wilson loops in section 7.4.2. This map was first found using the AdS/CFT correspondence for strongly coupled $\mathcal{N} = 4$ theory, and subsequently was also established directly at weak coupling. This provides an unexpected and powerful use of the AdS/CFT correspondence. Finally, we describe how the standard and dual superconformal symmetries present in $\mathcal{N} = 4$ Super Yang–Mills theory may be combined to obtain a *Yangian structure* with appealing mathematical properties.

Note that in the integrability literature, it is customary to define $g_{\text{YM}}^2 = 4\pi g_s$. For consistency, we also adopt this definition in this chapter. Note, however, that this is in contrast to (5.9) which we use in the remainder of the book.

7.1 Integrable structures on the gauge theory side

Integrable structures are found both perturbatively within the gauge theory and at strong coupling by considering string theory on AdS space. In this section we discuss how the integrable structure in the gauge theory appears by considering anomalous dimensions of certain single-trace operators in $\mathcal{N} = 4$ super Yang–Mills in the large N limit. It turns out that this computation can be rephrased in terms of spin chain models which are integrable.

Anomalous dimensions of operators $\mathcal{O}$ can be determined by two-point functions of the operators. For example, the two-point function of conveniently normalised operators $\mathcal{O}$ satisfies in perturbation theory

$$\langle \mathcal{O}(x)\bar{\mathcal{O}}(y) \rangle \sim \frac{1}{(x-y)^{2\Delta_0}} \left(1 - \gamma \ln \mu^2 (x-y)^2 + \cdots \right) \tag{7.1}$$

where for regularity, the dimensionful scale μ has to be introduced, for example by using an appropriate regularisation scheme.

We will determine these anomalous dimensions γ to one-loop order for local gauge invariant operators in the $\mathcal{N} = 4$ Super Yang–Mills theory. In the large N limit in which we will work from now on, the dimension of the multi-trace operators is equal to the sum of the dimensions of the corresponding single-trace operators. Therefore it is sufficient to compute the anomalous dimensions of the single-trace operators. In this section we further restrict ourselves to single-trace operators involving only the scalar fields without any covariant derivatives acting on them. The six real scalar fields can be arranged into three complex scalar fields W, Z, Y defined by

$$W = \frac{1}{\sqrt{2}} \left(\phi^1 + i\phi^2 \right), \qquad Z = \frac{1}{\sqrt{2}} \left(\phi^3 + i\phi^4 \right), \qquad Y = \frac{1}{\sqrt{2}} \left(\phi^5 + i\phi^6 \right). \tag{7.2}$$

An important example of a single-trace operator is $\mathcal{O}_l \sim \mathrm{Tr}(Z^l)$ with $l \geq 2$ which is normalised by

$$\mathcal{O}_l = \frac{(2\pi)^l}{\sqrt{l}N^{l/2}}\mathrm{Tr}Z^l, \tag{7.3}$$

such that at tree level

$$\langle \mathcal{O}(x)\bar{\mathcal{O}}(y)\rangle_{\mathrm{tree}} = \frac{1}{(x-y)^{2l}}. \tag{7.4}$$

As discussed in chapter 6, the anomalous dimension for $\mathcal{O}_l$ vanishes to all orders in perturbation theory and thus the scaling dimension satisfies $\Delta_0 = l$. This is due to the fact that $\mathcal{O}_l$ is a chiral primary operator as discussed in chapter 6, and thus the scaling dimension is not renormalised for the interacting theory.

In the following we generalise the single-trace operator (7.3) to

$$\mathcal{O}_{i_1 i_2 \ldots i_l}(x) = \frac{(2\pi)^l}{\sqrt{C_{i_1 i_2 \ldots i_l}}N^{l/2}}\mathrm{Tr}\left(\phi^{i_1}(x)\phi^{i_2}(x)\ldots\phi^{i_l}(x)\right), \tag{7.5}$$

where $C_{i_1 i_2 \ldots i_l}$ is a symmetry factor. $C_{i_1 i_2 \ldots i_l}$ takes the maximal value n if the l-tuple of indices $(i_1, i_2, \ldots, i_l)$ is invariant under shifting each position in the tuple by l/n, For example, if all the indices are the same, i.e. $i_1 = i_2 = \ldots = i_l$, $(i_1, i_2, \ldots, i_l)$ are invariant under shifts by 1, and thus $n = l$. Thus we obtain the normalisation as in (7.3)

The tree level contribution for the correlation function of the operator $\mathcal{O}_{i_1 i_2 \ldots i_l}$ and its complex conjugate $\bar{\mathcal{O}}_{i_1 i_2 \ldots i_l}$ inserted at x and y respectively, is given by

$$\langle \mathcal{O}_{i_1 i_2 \ldots i_l}(x)\bar{\mathcal{O}}^{j_1 j_2 \ldots j_l}(y)\rangle_{\mathrm{tree}} = \frac{1}{C_{i_1, i_2, \ldots, i_l}}\left(\delta_{i_1}^{j_1}\delta_{i_2}^{j_2}\ldots\delta_{i_l}^{j_l} + \mathrm{cyc.\ perm.}\right) \cdot \frac{1}{(x-y)^{2l}}. \tag{7.6}$$

Here, cyc. perm. refers to cyclic permutations of the $l-1$ possible cyclic shifts of the indices j_n. All other contractions give rise to non-planar diagrams as discussed in chapter 6. Let us now determine the same correlator at one-loop level. In order to determine the corresponding contribution, we have to consider several types of Feynman diagrams, in particular those that contain the scalar vertex. Since again only planar Feynman diagrams contribute, the scalar vertex has to be contracted with two neighbouring fields in the incoming and outgoing operators as shown in figure 7.1.

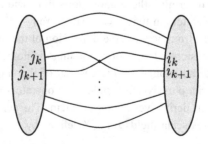

Figure 7.1   One-loop contribution from a quartic scalar vertex.

Performing the corresponding Wick contractions, we obtain, defining

$$\mathcal{K} \equiv \frac{ig_{YM}^2}{4} \int d^4z \sum_{I,j} \left(\text{Tr}(\phi_i\phi_i\phi_j\phi_j)(z) - \text{Tr}(\phi_i\phi_j\phi_i\phi_j)(z) \right) \tag{7.7}$$

for the scalar vertex,

$$\langle (\phi_{i_k}\phi_{i_{k+1}})^a{}_b(x) \cdot \mathcal{K} \cdot (\phi^{j_{k+1}}\phi^{j_k})^{b'}{}_{a'}(y) \rangle$$

$$= \frac{iN}{(4\pi^2)^2}\delta^a_{a'}\delta^{b'}_b \frac{Ng_{YM}^2}{64\pi^4} \left(2\delta^{j_k}_{i_k}\delta^{j_{k+1}}_{i_{k+1}} + 2\delta_{i_k i_{k+1}}\delta^{j_k j_{k+1}} \right.$$

$$\left. -4\delta^{j_{k+1}}_{i_k}\delta^{j_k}_{i_{k+1}} \right) \cdot \int \frac{d^4z}{(z-x)^4(z-y)^4}. \tag{7.8}$$

This integral is logarithmically divergent for $z \to x$ and $z \to y$. We regularise the integral by a Wick rotation $d^4z \to id^4z_E$ and by introducing a UV cutoff Λ. In particular, we restrict the integral to regions with $|z_E - x| \geq \Lambda^{-1}$ and $|z_E - y| \geq \Lambda^{-1}$. Then the integral may be approximated by

$$i \int \frac{d^4z_E}{(z_E-x)^4(z_E-y)^4} \simeq \frac{2i}{(x-y)^4} \int_{\Lambda^{-1}}^{|x-y|} \frac{d\xi \, d\Omega_\xi}{\xi} = \frac{2\pi^2 i}{(x-y)^4} \cdot \ln\left(\Lambda^2(x-y)^2\right). \tag{7.9}$$

This gives

$$\langle \mathcal{O}_{i_1 i_2 \ldots i_l}(x)\bar{\mathcal{O}}^{j_1 j_2 \ldots j_l}(y) \rangle_{\text{1-loop}}$$

$$= \frac{\lambda}{16\pi^2} \frac{\ln\left(\Lambda^2(x-y)^2\right)}{(x-y)^{2l}} \sum_{k=1}^{l} \left(2P_{k,k+1} - K_{k,k+1} - 1 \right) \cdot \frac{1}{\sqrt{C_{i_1\ldots i_l}C_{j_1\ldots j_l}}}\delta^{j_1}_{i_1}\ldots\delta^{j_l}_{i_l}. \tag{7.10}$$

Here $P_{k,k+1}$ is the exchange operator which as its name indicates exchanges the indices of the site k and $k+1$, i.e.

$$P_{k,k+1}\delta^{j_1}_{i_1}\ldots\delta^{j_l}_{i_l}\delta^{j_{l+1}}_{i_{l+1}}\ldots\delta^{j_l}_{i_l} = \delta^{j_1}_{i_1}\ldots\delta^{j_{k+1}}_{i_k}\delta^{j_k}_{i_{k+1}}\ldots\delta^{j_l}_{i_l}. \tag{7.11}$$

$K_{k,k+1}$ is the trace operator contracting the indices of fields at sites k and $k+1$,

$$K_{k,k+1}\delta^{j_1}_{i_1}\ldots\delta^{j_k}_{i_k}\delta^{j_{k+1}}_{i_{k+1}}\ldots\delta^{j_l}_{i_l} = \delta^{j_1}_{i_1}\ldots\delta_{i_k i_{k+1}}\delta^{j_k j_{k+1}}\ldots\delta^{j_l}_{i_l}. \tag{7.12}$$

Due to periodicity we have $P_{l,l+1} = P_{1,l}$ and $K_{l,l+1} = K_{1,l}$. Equation (7.10) corresponds to a special class of one-loop contributions, namely to the insertion of the scalar vertex. In addition, there are gluon exchange contributions between neighbouring scalars as shown in figure 7.2, as well as self-energy diagrams of the form shown in figure 7.3. The two diagrams in figure 7.3 lead to terms in which all incoming indices are sequentially contracted with the outgoing indices, i.e. to terms which do not change the index structure. Surprisingly, this is also true for the diagram of figure 7.2, since the R-charge is conserved and gluons do not carry any R-charges.

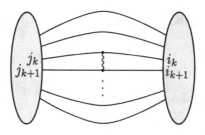

Figure 7.2   Gluon exchange contribution.

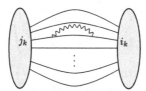
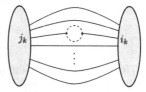

Figure 7.3   Self-energy contributions involving a gluon (left) and a fermion loop (right).

Finally, adding the one-loop contribution for the correlator to the tree level result we obtain

$$
\begin{aligned}
&\left\langle \mathcal{O}_{i_1 i_2 \ldots i_l}(x) \bar{\mathcal{O}}^{j_1 j_2 \ldots j_l}(y) \right\rangle \\
&= \frac{1}{(x-y)^{2l}} \left(1 - \ln(\mu^2(x-y)^2) \mathcal{D}^{\text{1-loop}} \right) \delta_{i_1}^{j_1} \delta_{i_2}^{j_2} \ldots \delta_{i_l}^{j_l} + \text{cyc. perm.},
\end{aligned}
\tag{7.13}
$$

where the operator $\mathcal{D}^{\text{1-loop}}$ is given by

$$
\mathcal{D}^{\text{1-loop}} = \frac{\lambda}{16\pi^2} \sum_{k=1}^{l} \left(1 - C - 2P_{k,k+1} + K_{k,k+1} \right).
\tag{7.14}
$$

The one-loop anomalous dimensions are determined by diagonalising $\mathcal{D}^{\text{1-loop}}$. The constant C in equation (7.14) can be fixed without an explicit calculation: for the chiral primary $\mathcal{O}_l$, the anomalous dimension is zero. Furthermore $P_{k,k+1}\mathcal{O}_l = \mathcal{O}_l$ and $K_{k,k+1}\mathcal{O}_l = 0$, and therefore we find $C = -1$ and $\mathcal{D}^{\text{1-loop}}$ reads

$$
\mathcal{D}^{\text{1-loop}} = \frac{\lambda}{8\pi^2} \sum_{k=1}^{l} \left(1 - P_{k,k+1} + \frac{1}{2} K_{k,k+1} \right).
\tag{7.15}
$$

So far we have presented a very general discussion of the one-loop contribution to the dilatation operator. Note that the dilatation operator is not diagonal, but rather only block diagonal. One such block consists of all correlation functions with L fields inserted, where M fields are for example $W = \frac{1}{\sqrt{2}}(\phi^1 + i\phi^2)$ while the other $L - M$ fields are $Z = \frac{1}{\sqrt{2}}(\phi^3 + i\phi^4)$. No other fields besides Z and W appear in the correlation function, in particular not the complex conjugates of Z and W. This set of fields is known as the $\mathfrak{su}(2)$ sector. No field of the $\mathfrak{su}(2)$ sector mixes with other fields, i.e. the sector is closed.

The full set of scalar operators ϕ^i is referred to as the $\mathfrak{so}(6)$ sector. In contrast to the $\mathfrak{su}(2)$ sector, the $\mathfrak{so}(6)$ sector is not closed since at two or more loops, operators in the $\mathfrak{so}(6)$ sector mix with operators not in this sector. In fact, the smallest closed sector containing the $\mathfrak{so}(6)$ sector is already the full $\mathfrak{psu}(2,2|4)$. This implies that it is non-trivial to find closed sectors. A non-trivial example of a closed sector contains all three complex scalars, Z, W and $Y = \frac{1}{\sqrt{2}}(\phi^5 + i\phi^6)$, as well as two fermions since the combined field ZWY mixes with them. This sector is referred to as $\mathfrak{su}(2|3)$. Here, we restrict our attention to the $\mathfrak{su}(2)$ sector for simplicity.

7.1.1 $\mathfrak{su}(2)$ sector and Heisenberg spin chain

As discussed in the preceding paragraph, we consider only single-trace operators with two different scalar fields, denoted by Z and W. For the $\mathfrak{su}(2)$ sector, $\mathcal{D}^{\text{1-loop}}$ is given by

$$\mathcal{D}^{\text{1-loop}}_{\mathfrak{su}(2)} = \frac{\lambda}{8\pi^2} \sum_{k=1}^{l} (1 - P_{k,k+1}). \tag{7.16}$$

In particular there is no contribution from $K_{k,k+1}$, since the single-trace operators involve Z and W fields, but not their conjugates. Since the fields Z and W transform as a doublet of $\mathfrak{su}(2)$ we may view Z as spin up, $\uparrow$, and W as spin down, $\downarrow$. Due to the identification of the sites $l + 1$ and 1, such that $P_{l,l+1} = P_{1,l}$, the single-trace operator may therefore be identified with a *spin chain* with periodic boundary condition.

We introduce spin operators $\vec{S}_i$ which satisfy

$$\vec{S}_i \cdot \vec{S}_{i+1} = \frac{1}{2} \left(S_i^+ S_{i+1}^- + S_i^- S_{i+1}^+ \right) + S_i^z S_i^z, \tag{7.17}$$

where $S_i^\pm$ are the standard ladder operators, $S_i^\pm = S_i^x \pm i S_i^y$. These operators act on the spin chain as follows,

$$S_i^z | \ldots \uparrow \ldots \rangle = \tfrac{1}{2} | \ldots \uparrow \ldots \rangle, \qquad S_i^z | \ldots \downarrow \ldots \rangle = -\tfrac{1}{2} | \ldots \downarrow \ldots \rangle, \tag{7.18}$$

$$S_i^- | \ldots \uparrow \ldots \rangle = | \ldots \downarrow \ldots \rangle, \qquad S_i^+ | \ldots \uparrow \ldots \rangle = 0, \tag{7.19}$$

$$S_i^+ | \ldots \downarrow \ldots \rangle = | \ldots \uparrow \ldots \rangle, \qquad S_i^- | \ldots \downarrow \ldots \rangle = 0, \tag{7.20}$$

where we show only the spin at the ith site explicitly. In terms of spin operators the operator $\mathcal{D}^{\text{1-loop}}_{\mathfrak{su}(2)}$ can be rewritten as

$$\mathcal{D}^{\text{1-loop}}_{\mathfrak{su}(2)} = \frac{\lambda}{8\pi^2} \sum_{i=1}^{l} \left(\frac{1}{2} - 2\, \vec{S}_i \cdot \vec{S}_{i+1} \right). \tag{7.21}$$

Remarkably, $\mathcal{D}^{\text{1-loop}}_{\mathfrak{su}(2)}$ is the Hamiltonian of the Heisenberg $XXX_{\frac{1}{2}}$ spin chain with l lattice sites. $\mathcal{D}^{\text{1-loop}}_{\mathfrak{su}(2)}$ commutes with the total spin

$$\vec{S} = \sum_{i=1}^{l} \vec{S}_i \cdot \vec{S}_{i+1}. \tag{7.22}$$

Consequently, $\mathcal{D}^{\text{1-loop}}_{\mathfrak{su}(2)}$ and S^z, the z component of the total spin, can be diagonalised simultaneously.

In order to diagonalise $\mathcal{D}^{\text{1-loop}}_{\mathfrak{su}(2)}$ we use the coordinate *Bethe ansatz*, which was originally invented by Bethe in 1931 to diagonalise the Hamiltonian of the Heisenberg spin chain. To use this analogy, it is important to drop the cyclicity constraint which arises from the trace structure of the single-trace operators. In particular, we consider a non-cyclic spin chain with length l, for which we still impose periodic boundary conditions. In other words, we still identify the $l + 1$st spin with the first spin, which amounts to an invariance of the kth spin under shifts of the form $k \mapsto k + l$. However, we do not impose invariance under $k \mapsto k + 1$, which corresponds to cyclicity, unless explicitly mentioned.

The spin chain is ferromagnetic, i.e. the energy favoured ground state has all spins aligned, for example

$$|\Omega\rangle = |\uparrow\uparrow \ldots \uparrow\uparrow\rangle. \tag{7.23}$$

Indeed $|\Omega\rangle$ is an eigenstate of $\mathcal{D}^{\text{1-loop}}_{\mathfrak{su}(2)}$ with eigenvalue zero and has maximal total spin $s^z = l/2$. This state of the spin chain corresponds to the single-trace operator $\mathcal{O}_l \sim \text{Tr}(Z^l)$ which is a chiral primary operator. Therefore we have again confirmed that there are no one-loop contributions to the anomalous dimension of this operator.

Moreover, we consider *magnons*, which are impurities in this ground state with spin down. For example, a typical *single magnon* state at site k is of the form $|\uparrow \ldots \uparrow\overset{k}{\downarrow}\uparrow \ldots \uparrow\rangle$, which is an eigenstate with respect to the z component of the total spin. The operator $\mathcal{D}^{\text{1-loop}}_{\mathfrak{su}(2)}$ acts on such a single magnon state as

$$
\mathcal{D}^{\text{1-loop}}_{\mathfrak{su}(2)}|\uparrow \ldots \uparrow\overset{k}{\downarrow}\uparrow \ldots \uparrow\rangle
$$
$$
= \frac{\lambda}{8\pi^2}\left(2|\uparrow \ldots \uparrow\overset{k}{\downarrow}\uparrow \ldots \uparrow\rangle - |\uparrow \ldots \overset{k-1}{\downarrow}\uparrow\uparrow \ldots \uparrow\rangle - |\uparrow \ldots \uparrow\uparrow\overset{k+1}{\downarrow} \ldots \uparrow\rangle\right). \tag{7.24}
$$

We see that the single magnon state is not an eigenstate of $\mathcal{D}^{\text{1-loop}}_{\mathfrak{su}(2)}$. However, a linear combination of single magnon states of the form

$$|p\rangle \equiv \frac{1}{\sqrt{l}}\sum_{k=1}^{l} e^{ipk}|\uparrow\uparrow \ldots \overset{k}{\downarrow} \ldots \uparrow\uparrow\rangle \tag{7.25}$$

is an eigenstate of $\mathcal{D}^{\text{1-loop}}_{\mathfrak{su}(2)}$ with eigenvalue

$$\mathcal{D}^{\text{1-loop}}_{\mathfrak{su}(2)}|p\rangle = E(p)\,|p\rangle, \qquad E(p) = \frac{\lambda}{2\pi^2}\sin^2\frac{p}{2}. \tag{7.26}$$

Note that p (which labels the eigenstate $|p\rangle$) and hence its eigenvalue $E(p)$ have to be quantised, since the state $|p\rangle$ has to be invariant under shifts $i \mapsto i + l$. Thus p is given by $p = 2\pi n/l$, where n is an integer. For $n \neq 0$, the state $|p\rangle$ has eigenvalue $l/2 - 1$ with respect to the z component of the total spin, while for $n = 0$, the state $|0\rangle$ is symmetric and has $s^z = l/2$, as well as $E(p) = 0$.

Note that for $p \neq 0$, the eigenstate $|p\rangle$ is not invariant under cyclic shifts $i \mapsto i+1$. Since we also have to impose this cyclicity, the only allowed state is $|0\rangle$ whose one-loop dimension vanishes. This state is symmetric and corresponds to a chiral primary composed of a string of Z fields plus one W field.

We move on to consider the *two-magnon state*, which is given by

$$|p_1, p_2\rangle = \sum_{k_1 < k_2} e^{ip_1 k_1 + ip_2 k_2} | \ldots \overset{k_1}{\downarrow} \ldots \overset{k_2}{\downarrow} \ldots \rangle + e^{i\phi} \sum_{k_1 > k_2} e^{ip_1 k_1 + ip_2 k_2} | \ldots \overset{k_2}{\downarrow} \ldots \overset{k_1}{\downarrow} \ldots \rangle,$$

$$(7.27)$$

where we assume that $p_1 > p_2$. Note that in general the state $|p_1, p_2\rangle$ is not an eigenstate of $\mathcal{D}^{\text{1-loop}}_{\mathfrak{su}(2)}$ unless we choose the phase $e^{i\phi}$ to be

$$e^{i\phi} = -\frac{e^{ip_1+ip_2} - 2e^{ip_2} + 1}{e^{ip_1+ip_2} - 2e^{ip_1} + 1}.$$

$$(7.28)$$

These phases can be viewed as scattering matrices of the two magnons.

The eigenvalue of the state $|p_1, p_2\rangle$ with (7.28) is simply given by the sum of the eigenvalues of the individual magnons,

$$\mathcal{D}^{\text{1-loop}}_{\mathfrak{su}(2)} |p_1, p_2\rangle = (E(p_1) + E(p_2)) |p_1, p_2\rangle,$$

$$(7.29)$$

where $E(p)$ is given by (7.26). So far we have not implemented the periodicity constraint. In technical terms, we have considered an infinite spin chain. Let us consider now a finite spin chain of length l. Looking at (7.27) we see that the state $|p_1, p_2\rangle$ is invariant (up to constant phase factors, such as $e^{i\phi}$) under shifting all spins from k_1 to $k_1 + l$ provided that $e^{ip_1 l} = e^{-i\phi}$. Similarly, the state is invariant under shifts of all spins from k_2 to $k_2 + l$ provided that $e^{ip_2 l} = e^{i\phi}$. Thus we conclude that p_1 and p_2 have to be quantised since

$$e^{i(p_1+p_2)l} = 1 \quad \Rightarrow \quad p_1 + p_2 = \frac{2\pi m}{l},$$

$$(7.30)$$

where m is an integer. However, implementing the cyclicity constraint $k_i \rightarrow k_i + 1$, we conclude that $m = 0$ and thus $p \equiv p_1 = -p_2$. Therefore using (7.28) we know that

$$e^{ipl} = e^{-i\phi} = -\frac{1 - e^{ip}}{1 - e^{-ip}} = e^{ip}$$

$$(7.31)$$

where we have used (7.28). The last equality is a mathematical identity. Thus we conclude that

$$e^{ip(l-1)} = 1 \quad \Rightarrow \quad p = \frac{2\pi n}{l-1}.$$

$$(7.32)$$

Consequently, after implementing all constraints, the eigenvalue of the state $|p_1, p_2\rangle$ with respect to $\mathcal{D}^{\text{1-loop}}_{\mathfrak{su}(2)}$, i.e. the anomalous conformal dimension, is given by

$$\Delta^{\text{1-loop}} = \frac{\lambda}{\pi^2} \sin^2 \left(\frac{\pi n}{l-1} \right).$$

$$(7.33)$$

Again, for $n = 0$ we have a state which is symmetric. It has maximal z component of the total spin, $s^z = l/2$ and thus is a chiral primary. Note also that the one-loop anomalous dimension, i.e. the eigenvalue of $\mathcal{D}^{\text{1-loop}}_{\mathfrak{su}(2)}$ vanishes as expected.

Unlike the one-magnon case, there are also states with $n \neq 0$ satisfying the constraints. These states are not symmetric, and have $s^z = l/2 - 2$ as well as a non-vanishing eigenvalue of $\mathcal{D}^{\text{1-loop}}_{\mathfrak{su}(2)}$. Therefore these states correspond to single-trace operators which are not chiral primaries and thus their conformal dimension is not protected. To be precise, these states correspond to operators of the form

$$\mathcal{O} \sim \sum_{k=0}^{l-2} \cos\left(\pi n \frac{2k+1}{l-2}\right) \text{Tr}\left(WZ^k WZ^{l-2-k}\right). \tag{7.34}$$

We can generalise the construction to states with $M > 2$ magnons,

$$|p_1, p_2, \ldots, p_M\rangle = \sum_{k_1 < k_2 \cdots < k_M} e^{ip_1 k_1 + ip_2 k_2 + \cdots + ip_M k_M} |\ldots \overset{k_1}{\downarrow} \ldots \overset{k_2}{\downarrow} \ldots \overset{k_M}{\downarrow} \ldots\rangle + \cdots \tag{7.35}$$

with $p_1 > p_2 > \cdots > p_M$ and where the last set of dots refers to the other possible orderings for the magnons, with appropriate phase factors. It can be shown that the phase factors are products of the corresponding phase factors for two magnons. Since we may interpret the phase factors as scattering matrices, the scattering matrix of M magnons factors into scattering matrices for two magnons. This is evidence for the fact that the system studied here is integrable.

7.1.2 Integrability beyond one-loop order

Above we considered the dilatation operator to one-loop order. For the $\mathfrak{su}(2)$ sector, we mapped this operator to a spin chain with nearest neighbour interactions. This map is applicable not only to the $\mathfrak{su}(2)$ sector, but also to any other closed sector as well as to higher loop contributions to the dilatation operator. For example, the two-loop dilatation operator in the $\mathfrak{su}(2)$ sector contains a permutation of the next-to-nearest neighbours. This is a general feature of the planar limit: at nth loop order, the spin chain interacts at most with the nth nearest neighbour. Thus the dilatation operator in perturbation theory can be rewritten in terms of a spin chain with the range of interaction given by the order of the perturbation theory. It turns out that all these spin chains are integrable. This provides strong evidence for the fact that $\mathcal{N} = 4$ Super Yang–Mills theory is integrable to all orders in perturbation theory.

7.2 Integrability on the gravity (string theory) side

We now turn to possible integrable structures on the gravity side. Here, we again work in the large N limit, so that we may restrict ourselves to the strong form of the conjectured duality involving classical type IIB string theory on $AdS_5 \times S^5$ as defined in table 5.1.

Since this background has a non-trivial Ramond-Ramond flux, we cannot use the construction of superstring theory reviewed in chapter 4. Thus we first have to introduce an alternative formalism for superstrings, the *Green–Schwarz formalism*. We first discuss this formalism in flat space and then move on to study type IIB string theory on $AdS_5 \times S^5$ in section 7.2.2 below.

7.2.1 Green–Schwarz formalism in flat spactime

In bosonic string theory as discussed in chapter 4, it is obvious that the bosonic fields denoted by $X^M(\tau, \sigma)$ are objects defined in the target space, since these fields are labelled by an index M. However, if we wish to supersymmetrise the action, we have to introduce spinors. This can be done in two different ways which are referred to as Ramond–Neveu–Schwarz formalism or Green–Schwarz formalism.

In the Ramond–Neveu–Schwarz (RNS) formalism, worldsheet supersymmetry is manifest since the fermionic degrees of freedom are present in two component spinors on the worldsheet. We reviewed this RNS formalism in chapter 4. In contrast, in the Green–Schwarz (GS) formalism, target space supersymmetry is preserved and thus the fermionic degrees of freedom are grouped into $2^{D/2}$ component spinors, where D is the dimension of target space, i.e. $D = 10$. Note that the corresponding actions for the string in the two formalisms are different and also have different symmetry.

Remarkably, these two different formalisms are equivalent – at least for flat spacetime in the light-cone gauge – and this equivalence is believed to hold also in other gauges.

The important difference between these two approaches presents itself in curved target spaces, i.e. when the superstring is coupled to gravity. In curved backgrounds Ramond–Ramond fields, originating from the combination of the fermionic creation operator on the string vacuum in the RNS formalism, are realised by non-local spin operators and it is not clear how one should couple them to the worldsheet metric. This forces one to abandon the RNS description in favour of the target space supersymmetry formalism by Green and Schwarz.

The action in the Green–Schwarz formalism in flat spacetime reads

$$\mathcal{S}_1 = -\frac{1}{2\pi\alpha'} \int d^2\sigma \sqrt{h} h^{\alpha\beta} \Pi_\alpha^M \Pi_{\beta M}, \tag{7.36}$$

where Π_α^M is given by

$$\Pi_\alpha^M = \partial_\alpha X^M - i\overline{\Theta}^A \Gamma^M \partial_\alpha \Theta^A \tag{7.37}$$

and $\alpha \in \{0, 1\}$. M is a target spacetime index, i.e. $M = 0, \ldots, 9$, and A is a spinor index. Spinors in ten dimensions have $2^{10/2} = 32$ complex dimensions in general. The Majorana–Weyl condition reduces this number to sixteen real components. The Majorana–Weyl condition implies that for any spinor Ψ, $\overline{\Psi} = \Psi^\dagger \Gamma^0$ and thus $\overline{\Psi}\Theta = \overline{\Theta}\Psi$. Defining $\tilde{\Gamma}_* = \Gamma_0 \ldots \Gamma_9$ in ten dimensions, by construction this anticommutes with Γ_M, i.e. $\{\tilde{\Gamma}_*, \Gamma_M\} = 0$. We may now distinguish two spinors of different chirality, with $\tilde{\Gamma}_*\Theta = \pm\Theta$, known as Weyl spinors. As for the RNS formalism we can calculate the variation with respect to the worldsheet metric $h_{\alpha\beta}$. This yields the Virasoro constraint

$$\Pi_\alpha^M \Pi_{\beta M} - \frac{1}{2} h_{\alpha\beta} h^{\gamma\delta} \Pi_\gamma^M \Pi_{\delta M} = 0. \tag{7.38}$$

Since our theory has sixteen real spinors but only eight bosonic degrees of freedom in light-cone gauge, the theory cannot be target spacetime supersymmetric unless there is a peculiar symmetry reducing the sixteen spinors to the eight spinors. The eight bosonic degrees of freedom are $X^M(\tau, \sigma)$ transverse to the light-cone coordinates, i.e. $M \neq +$ and $M \neq -$. The required local fermionic symmetry is indeed present here which is known as κ-symmetry. To obtain this symmetry, it is necessary to add the following term to the action,

$$
S_2 = \frac{1}{\pi \alpha'} \int d^2\sigma \, \epsilon^{\alpha\beta} \left((\overline{\Theta}^1 \Gamma^M \partial_\alpha \Theta^1)(\overline{\Theta}^2 \Gamma_M \partial_\beta \Theta^2) \right.
$$
$$
\left. - i \partial_\alpha X^M \left(\overline{\Theta}^1 \Gamma_M \partial_\beta \Theta^1 - \overline{\Theta}^2 \Gamma_M \partial_\beta \Theta^2 \right) \right), \qquad (7.39)
$$

which is known as the Wess–Zumino term. Therefore the combined action $S = S_1 + S_2$ is supersymmetric and invariant under κ-symmetry. The $\mathcal{N} = 2$ supersymmetry transformations act as

$$
\delta_\epsilon \Theta^A = \epsilon^A \qquad \text{and} \qquad \delta_\epsilon X^M = i \overline{\epsilon}^A \Gamma^M \Theta^A. \qquad (7.40)
$$

Exercise 7.2.1 Show that S_1 and S_2 are independently invariant under the supersymmetry transformations (7.40). For the case of the Wess–Zumino term, we have to use the fact that Θ is a Majorana–Weyl spinor. If Θ^1 and Θ^2 have the same chirality then we describe type IIB string theory in Green–Schwarz formalism, while in the case that Θ^1 and Θ^2 have opposite chirality, this is a type IIA string theory.

The κ-symmetry transformation reads

$$
\delta_\kappa \Theta^A = 2i\Gamma^M \Pi_{\alpha M} \kappa^{A\alpha}, \qquad \delta_\kappa X^M = i\overline{\Theta}^A \Gamma^M \delta_\kappa \Theta^A, \qquad (7.41)
$$

where κ^A is self-dual for $A = 1$ and anti-self-dual for $A = 2$ with respect to the projection operator

$$
P_\pm^{\alpha\beta} = \frac{1}{2}\left(h^{\alpha\beta} \pm \frac{1}{\sqrt{h}} \epsilon^{\alpha\beta} \right), \qquad (7.42)
$$

i.e.

$$
\kappa^{1\alpha} = P_-^{\alpha\beta} \kappa_\beta^1 \quad \text{and} \quad \kappa^{2\alpha} = P_+^{\alpha\beta} \kappa_\beta^2. \qquad (7.43)
$$

Exercise 7.2.2 Check that Π_α^M transforms under a κ-symmetry transformation as

$$
\delta_\kappa \Pi_\alpha^M = 2i\partial_\alpha \overline{\Theta}^A \Gamma^M \delta_\kappa \Theta^A. \qquad (7.44)
$$

Exercise 7.2.3 Show that the action S is invariant under the κ-symmetry transformation. How does $h^{\alpha\beta}$ transform under the κ-symmetry transformation?

A special virtue of the Green–Schwarz formalism is that it may be easily coupled to gravity by considering strings propagating in curved backgrounds. For a general curved background, the S is also invariant under global target space isometry transformations. However, it is very tedious to construct a κ-symmetric action S in the Green–Schwarz formalism for generic curved backgrounds. In most cases the action is only known in the first few orders of Θ. For higher order terms in Θ it is necessary to construct them by hand.

7.2.2 Green–Schwarz formalism on $AdS_5 \times S^5$

In a few explicit examples the target spacetime may be written as a coset. It turns out that the Green–Schwarz superstring is very simple in the coset construction. The resulting theory is highly non-linear and has all the symmetries discussed above by construction. As we will also see, it is manifestly classically integrable. We already saw in chapter 2, box 2.2, that Minkowski space and Anti-de Sitter space are coset spaces. In both cosets we start with the isometry group of the underlying spacetime and divide by the little group H. In the case of AdS_5, the isometry group is $SO(4,2)$ while the little group of a generic point p, i.e. the transformations which leave the point unchanged, is $SO(4,1)$. Therefore AdS_5 may be represented by

$$AdS_5 = \frac{SO(4,2)}{SO(4,1)}. \qquad (7.45)$$

Using the same arguments, the sphere S^5 may be represented by

$$S^5 = \frac{SO(6)}{SO(5)}. \qquad (7.46)$$

To describe Green–Schwarz strings in $AdS_5 \times S^5$, we therefore have to consider the coset space

$$G/H = \frac{SO(4,2) \times SO(6)}{SO(4,1) \times SO(5)}. \qquad (7.47)$$

Obtaining a manifestly supersymmetric theory is ensured by starting with the supergroup $PSU(2,2|4)$ instead of its bosonic subgroup $SO(4,2) \times SO(6)$. Therefore we consider the coset

$$G/H = \frac{PSU(2,2|4)}{SO(4,1) \times SO(5)}. \qquad (7.48)$$

The supergroups may be represented by the superalgebras. Since $PSU(2,2|4)$ does not have a realisation in terms of supermatrices, we use $SU(2,2|4)$ instead which may be represented by 8×8 matrices. We construct these below.

The superalgebra $\mathfrak{psu}(2,2|4)$ is a real form of the superalgebra $\mathfrak{gl}(4|4)$ whose matrices are given by

$$M = \begin{pmatrix} A & \Theta \\ \eta & B \end{pmatrix}, \qquad (7.49)$$

where A and B are Grassmann even, while Θ and η are Grassmann odd. In terms of the matrix M, the supertrace Str acting on such a matrix M is defined by [1]

$$\mathrm{Str}(M) = \mathrm{Tr}(A) - \mathrm{Tr}(B). \qquad (7.50)$$

By definition the superalgebra $\mathfrak{sl}(4|4)$ contains those matrices M which in addition satisfy $\mathrm{Str}(M) = 0$. In order to define the superalgebra $\mathfrak{su}(2,2|4)$ we have to consider the real

---

[1] The supertrace used here should not be confused with the symmetrised trace.

form of $\mathfrak{sl}(4|4)$ which is specified by the relation

$$M^\dagger H + HM = 0, \qquad H = \begin{pmatrix} \mathbb{1}_2 & 0 & 0 \\ 0 & -\mathbb{1}_2 & 0 \\ 0 & 0 & \mathbb{1}_4 \end{pmatrix}. \qquad (7.51)$$

Note that in particular the unit matrix $\mathbb{1}_8$ is part of the algebra $\mathfrak{su}(2,2|4)$. Since $\mathbb{1}_8$ commutes with all other generators of $\mathfrak{su}(2,2|4)$, it may be projected out, which yields $\mathfrak{psu}(2,2|4)$.

The superalgebra $\mathfrak{psu}(2,2|4)$ admits an invertible map Ω from $\mathfrak{psu}(2,2|4)$ onto itself, preserving the (anti-)commutation relations of the algebra and satisfying $\Omega^4 = 1$. In technical terms, Ω is a fourth-order *outer automorphism*. A convenient realisation is given by

$$\Omega(M) = -\mathcal{K}M^{\text{st}}\mathcal{K}^{-1}, \qquad (7.52)$$

where $\mathcal{K} = \text{diag}(i\sigma_2, i\sigma_2, i\sigma_2, i\sigma_2)$ involving the Pauli matrix σ_2. Moreover, M^{st} is the supertransposition of the supermatrix (7.49),

$$M^{\text{st}} = \begin{pmatrix} A^{\text{t}} & -\eta^{\text{t}} \\ \Theta^{\text{t}} & B^{\text{t}} \end{pmatrix}. \qquad (7.53)$$

Using the outer automorphism Ω, we may define four subspaces $\mathfrak{a}^{(k)}$ satisfying

$$\Omega\left(\mathfrak{a}^{(k)}\right) = i^k \mathfrak{a}^{(k)} \qquad (7.54)$$

and an associated natural $\mathbb{Z}_4$ grading. We may decompose any matrix M in $\mathfrak{psu}(2,2|4)$ into elements $M^{(i)}$ according to their $\mathbb{Z}_4$ gradings,

$$M = M^{(0)} + M^{(1)} + M^{(2)} + M^{(3)}. \qquad (7.55)$$

We apply this decomposition to the representative g of the coset, to which we can associate a current A of $\mathfrak{psu}(2,2|4)$ by

$$A = -g^{-1}dg = A^{(0)} + A^{(1)} + A^{(2)} + A^{(3)}. \qquad (7.56)$$

Using (7.56), the Lagrangian density for the superstring on $AdS_5 \times S^5$, i.e. the sum of its kinetic term and the topological Wess–Zumino terms, is given by the coset spacetime representation

$$\mathcal{L} = \mathcal{L}_{\text{kin}} + \mathcal{L}_{\text{WZ}} = -\frac{\sqrt{\lambda}}{4\pi} \text{Str} \left(h^{\alpha\beta}\sqrt{-h}A_\alpha^{(2)}A_\beta^{(2)} + \kappa\epsilon^{\alpha\beta}A_\alpha^{(1)}A_\beta^{(3)} \right), \qquad (7.57)$$

where $\kappa = \pm 1$.

By construction, the Lagrangian (7.57) depends on the coset elements $PSU(2,2|4)/(SO(4,1) \times SO(5))$ and thus encodes the isometries of $AdS_5 \times S^5$. Moreover, it is supersymmetric. The restriction of (7.57) to those terms which contain only bosonic variables coincides with the Polyakov action on $AdS_5 \times S^5$.

Box 7.1	Lax pairs

The notion of classical integrability may be illustrated by considering a system of partial differential equations in two dimensions, with coordinates σ and τ, of the form

$$\partial_\sigma \psi(\tau,\sigma,z) = L_\sigma(\tau,\sigma,z)\psi,$$
$$\partial_\tau \psi(\tau,\sigma,z) = L_\tau(\tau,\sigma,z)\psi, \tag{7.58}$$

where ψ is an n-dimensional vector and L_σ, L_τ are $n \times n$ matrices which are referred to as a *Lax pair*. The additional parameter z is referred to as a spectral parameter. The system (7.58) has a well-defined solution provided that

$$\partial_\sigma L_\tau - \partial_\tau L_\sigma + [L_\sigma, L_\tau] = 0. \tag{7.59}$$

In other words, L_σ and L_τ satisfy an integrability condition. Introducing the *Lax connection* by

$$L = L_\sigma d\sigma + L_\tau d\tau, \tag{7.60}$$

we may rewrite (7.59) as $dL + L \wedge L = 0$, which means that the curvature of the Lax connection vanishes.

7.2.3 Classical integrability of the sigma model

We now discuss the integrability of the superstring on $AdS_5 \times S^5$. To establish classical integrability, we have to use a *Lax pair* as reviewed in box 7.1. Given such a Lax pair, which we assume to be 2π periodic in σ and to have vanishing curvature, i.e. satisfying (7.59), the theory automatically possesses an infinite number of integrals of motion. These integrals of motion are encoded in the monodromy matrix $T(z)$ associated with the σ-component of the Lax connection,

$$T(z) = \mathcal{P}\exp \int_0^{2\pi} d\sigma \, L_\sigma(\tau,\sigma,z), \tag{7.61}$$

where $\mathcal{P}$ denotes path ordering. Taking the derivative with respect to τ and using (7.59), we obtain

$$\partial_\tau T(z) = \int_0^{2\pi} d\sigma \, \partial_\sigma \left(\left(\mathcal{P}\exp \int_\sigma^{2\pi} d\sigma' L_{\sigma'} \right) L_\tau(\tau,\sigma,z) \left(\mathcal{P}\exp \int_0^\sigma d\sigma' L_{\sigma'} \right) \right). \tag{7.62}$$

Note that the integrand is a derivative with respect to σ. Performing the integral and using that the Lax connection is 2π-periodic in σ, i.e. $L_\tau(\tau,\sigma = 0,z) = L_\tau(\tau,\sigma = 2\pi,z)$, we obtain

$$\partial_\tau T(z) = [L_\tau(\tau,\sigma = 0,z), T(z)]. \tag{7.63}$$

Consequently, the trace as well as all eigenvalues of $T(z)$ are independent of τ. Hence we may express $T(z)$ as

$$T(z) = \sum_{n=0}^{\infty} z^n Q_n, \qquad Q_0 = \mathbb{1}. \tag{7.64}$$

The infinite set of matrices Q_n encodes the integrals of motion. For the classical superstring on $AdS_5 \times S^5$ as considered in section 7.2.2, we have to consider the Lax connection [1]

$$
\begin{aligned}
L_\alpha(\tau, \sigma, z) = {} & l_0(z) A_\alpha^{(0)}(\tau, \sigma) + l_1(z) A_\alpha^{(2)}(\tau, \sigma) \\
& + l_2(z) \sqrt{-h} h_{\alpha\beta}(\tau, \sigma) \epsilon^{\beta\gamma} A_\gamma^{(2)}(\tau, \sigma) \\
& + l_3(z) A_\alpha^{(1)}(\tau, \sigma) + l_4(z) A_\alpha^{(3)}(\tau, \sigma).
\end{aligned}
\tag{7.65}
$$

Here, the l_i $(i = 0, \ldots, 4)$ are functions which depend only on the spectral parameter z. So far, they are undetermined. These functions are constrained by curvature conditions for L_α and also for $A_\alpha^{(i)}$. These conditions are satisfied by [2]

$$
\begin{aligned}
& l_0(z) = 1, \qquad l_1(z) = \frac{1}{2}\left(z^2 + \frac{1}{z^2}\right), \qquad l_2(z) = \frac{1}{2\kappa}\left(z^2 - \frac{1}{z^2}\right), \\
& l_3(z) = z, \qquad l_4(z) = \frac{1}{z}.
\end{aligned}
\tag{7.66}
$$

Note that we have to impose $\kappa = \pm 1$. The equations of motion of the coset σ-model Lagrangian (7.57) may be represented by a Lax connection (7.65). The existence of such a Lax connection guarantees that the classical string theory on $AdS_5 \times S^5$ is integrable.

7.3 BMN limit and classical string configurations

Above we found integrable structures in perturbative $\mathcal{N} = 4$ Super Yang–Mills theory, as well as in type IIB string theory on $AdS_5 \times S^5$ which is dual to the strong coupling regime of $\mathcal{N} = 4$ Super Yang–Mills theory. These two regimes do not overlap and, consequently, results obtained in the two regimes cannot be compared to each other. To make a comparison between perturbative $\mathcal{N} = 4$ Super Yang–Mills theory and the dual string theory approach possible, Berenstein, Maldacena and Nastase [3] proposed considering a new regime in which the string moves along the equator of S^5 at very high speed. The corresponding large angular momentum J of S^5 is related to the spin chain of length L with M impurities by $J = L - M$. It turns out that the effective coupling constant for this setup is given by

$$\lambda' = \frac{\lambda}{J^2}. \tag{7.67}$$

We are interested in a small effective coupling constant $\lambda' \ll 1$. This may be obtained in two ways: either λ and J are large such that λ' is small, or λ is small and J is large. While in the first case the correct description is in terms of strings rotating along the equator of S^5, the result in the second limit may be obtained by spin chains.

7.3.1 Strings in plane wave background

The plane wave background is the geometry seen by a particle moving at or close to the speed of light. This may be seen by considering the $AdS_5 \times S^5$ geometry written in the form

$$ds^2 = L^2 \left(-dt^2 \cosh^2 \rho + d\rho^2 + \sinh^2 \rho \, d\Omega_3^2 + d\psi^2 \cos^2 \theta + d\theta^2 + \sin^2 \theta \, d\Omega_3'^2 \right). \tag{7.68}$$

Let us consider a massless particle moving around the equator of S^5 given by $\theta = 0$, placed at $\rho = 0$ in the ρ direction. Its motion is determined by $\psi = \psi(t)$. In light-cone coordinates,

$$x^{\pm} = \frac{1}{2}(t \pm \psi), \tag{7.69}$$

a particle moving close to the speed of light, for which $\psi(t)$ increases with time, satisfies $x^- \ll 1$. Introducing finite coordinates $(\tilde{x}^{\pm}, r, y)$ by

$$\tilde{x}^+ = \frac{x^+}{\mu}, \qquad \tilde{x}^- = \mu L^2 x^-, \qquad r = L\rho, \qquad y = L\theta, \tag{7.70}$$

where μ is a parameter of dimension mass to keep the dimension correct, an ultrafast particle has to satisfy $L \to \infty$. In this limit with $\tilde{x}^{\pm}$, r, y finite, the $AdS_5 \times S^5$ metric reads

$$ds_{\text{pw}}^2 = -4d\tilde{x}^+ d\tilde{x}^- - \mu^2 (y^2 + r^2)(d\tilde{x}^+)^2 + d\vec{y}^2 + d\vec{r}^2, \tag{7.71}$$

where $(d\vec{y})^2 = dy^2 + y^2 d\Omega_3^2$ and $(d\vec{r})^2 = dr^2 + r^2 d\Omega_3'^2$. This is the plane wave geometry. Note that in the limit $\mu \to \infty$, we recover ten-dimensional Minkowski spacetime. Moreover, the apparent $SO(8)$ symmetry of the transverse coordinates $\vec{y}, \vec{r}$ is broken down to $SO(4) \times SO(4)$ due to the non-vanishing self-dual five-form flux

$$F_{+1234} = F_{+5678} = 4\mu. \tag{7.72}$$

Using the light-cone variables $\tilde{x}^{\pm}$, the energy $E = i\partial_t$ and the angular momentum $J = -i\partial_\psi$ are given by

$$H_{\text{lc}} \equiv 2\tilde{p}^- = i\partial_{\tilde{x}^+} = i\mu \left(\partial_t + \partial_\psi \right) = \mu(E - J), \tag{7.73}$$

$$2\tilde{p}^+ = i\partial_{\tilde{x}^-} = \frac{i}{\mu L^2} \left(\partial_t - \partial_\psi \right) = \frac{E + J}{\mu L^2}. \tag{7.74}$$

Taking again the $L \to \infty$ limit, we observe that the momentum $\tilde{p}^+$ vanishes unless J scales as $L^2 = \sqrt{\lambda}\alpha'$. In particular, we are interested in situations with finite H_{lc}, i.e. finite light-cone momentum $\tilde{p}$, with the component $\tilde{p}^+$ also finite. This requires $E \approx J$, with both E and J scaling as $L^2 = \sqrt{\lambda}\alpha'$.

Let us translate this particular limit to the gauge theory side. The energy E in global coordinates is identified with the scaling dimension Δ of a composite local gauge invariant Super Yang–Mills operator, while the angular momentum J corresponds to the charge of a $U(1)$ subgroup of the $SO(6) \simeq SU(4)$ R-symmetry group of this composite operator. The limit $L \to \infty$ with $E \approx J \sim L^2$ as discussed above translates to

$$N \to \infty, \qquad J \sim \sqrt{N}, \qquad g_{\text{YM}} \text{ kept fixed}. \tag{7.75}$$

Because $E \approx J$, only operators with conformal dimension of order of the $U(1)$ R-charge, i.e. with $\Delta \approx J$, will be present in this limit. These correspond to finite light-cone energy states on the string theory side. The string coupling constant is given by $4\pi g_s = g_{\text{YM}}^2$.

7.3.2 Type IIB string theory in plane wave background

To quantise the superstring in the plane wave background, we have to use the Green–Schwarz formalism introduced in section 7.2.1. In order to simplify the discussion, we work in light-cone coordinates and consider only the bosonic sector. The action for the bosonic sector reads

$$
\begin{aligned}
\mathcal{S}_{\text{B}} &= \frac{1}{4\pi\alpha'} \int d^2\sigma \, \sqrt{-h} h^{\alpha\beta} G_{MN} \partial_\alpha X^M \partial_\beta X^N, \\
&= \frac{1}{4\pi\alpha'} \int d^2\sigma \, \sqrt{-h} h^{\alpha\beta} \left(-2\partial_\alpha X^+ \partial_\beta X^- + \partial_\alpha X^I \partial_\beta X^I - \mu^2 (X^I)^2 \partial_\alpha X^I \partial_\beta X^I \right),
\end{aligned}
\tag{7.76}
$$

where $I = 1, \ldots, 8$ labels the transverse directions $\vec{r}$ and $\vec{y}$. Fixing the diffeomorphism and Weyl symmetries by choosing

$$
\sqrt{-h} h^{\alpha\beta} = \eta^{\alpha\beta}, \quad \eta_{\tau\tau} = -1, \quad \eta_{\sigma\sigma} = 1,
\tag{7.77}
$$

and by imposing

$$
X^+ = \alpha' \tilde{p}^+ \tau \quad \text{with} \quad \tilde{p}^+ > 0,
\tag{7.78}
$$

we may express $X^-(\tau, \sigma)$ in terms of the X^I. Moreover, the action for the bosonic sector becomes quadratic in the X^I,

$$
\mathcal{S}_{\text{B}} = \frac{1}{4\pi\alpha'} \int d\tau \int_0^{2\pi\alpha'\tilde{p}^+} d^2\sigma \left(\partial_\tau X^I \partial_\tau X^I - \partial_\sigma X^I \partial_\sigma X^I - \mu^2 X_I^2 \right).
\tag{7.79}
$$

This is an action for eight scalars X^I with mass μ. The fermionic sector in the light-cone gauge reduces to eight non-interacting massive fermions in the same manner. This is expected since the background is supersymmetric.

Exercise 7.3.1 Show that the equations of motion obtained from (7.79) are given by

$$
(\partial_\tau^2 - \partial_\sigma^2 - \mu^2) X^I(\tau, \sigma) = 0.
\tag{7.80}
$$

Moreover, show that the general solution of this equation may be written as

$$
\begin{aligned}
X^I = \cos(\mu\tau) \frac{x_0^I}{\mu} + \sin(\mu\tau) \frac{p_0^I}{\mu} \\
+ \sum_{n \neq 0} \frac{i}{\sqrt{2\omega_n}} \left(\alpha_n^I e^{-i(\omega_n \tau - k_n \sigma)} + \tilde{\alpha}_n^I e^{-i(\omega_n \tau + k_n \sigma)} \right),
\end{aligned}
\tag{7.81}
$$

with

$$
\omega_n = \text{sign}(n) \sqrt{k_n^2 + \mu^2} \quad \text{and} \quad k_n = \frac{n}{\alpha' \tilde{p}^+},
\tag{7.82}
$$

subject to the usual boundary condition for closed strings,

$$X^M(\tau, \sigma + 2\pi\alpha'\tilde{p}^+) = X^M(\tau, \sigma). \tag{7.83}$$

The action (7.79) may be quantised similarly to the bosonic string in flat spacetime as discussed in chapter 4. Imposing the usual commutation relations as in (4.24), the associated Hamiltonian takes the form

$$H_{\text{lc}} = \frac{1}{p^+} \int d\sigma \left((P^I)^2 + (\partial_\sigma X^I)^2 + \mu^2 (X^I)^2 + \text{fermions} \right). \tag{7.84}$$

Introducing the zero modes

$$\alpha_0^I = \frac{1}{\sqrt{2\mu}} (p_0^I + i\mu x_0^I), \tag{7.85}$$

the Hamiltonian may be written in terms of oscillator modes as

$$H_{\text{lc}} = \mu \, (\alpha_0^{\dagger I} \alpha_0^I + \text{fermions})$$
$$+ \frac{1}{\alpha'\tilde{p}^+} \sum_{n=1}^{\infty} \sqrt{n^2 + (\alpha'\tilde{p}^+\mu)^2} \left(\alpha_{-n}^I \alpha_n^I + \tilde{\alpha}_{-n}^I \tilde{\alpha}_n^I + \text{fermions} \right). \tag{7.86}$$

As in chapter 4, the creation and annihilation operators define physical states which are subject to the Virasoro constraint (4.11). The resulting spectrum is given by

$$E_{\text{lc}} = \mu \, N_0 + \mu \sum_{n=1}^{\infty} (N_n + \tilde{N}_n) \sqrt{1 + \frac{n^2}{(\alpha'\tilde{p}^+\mu)^2}}, \tag{7.87}$$

where $N_n = \alpha_{-n}^I \alpha_n^I +$ fermions and $\tilde{N}_n = \tilde{\alpha}_{-n}^I \tilde{\alpha}_n^I +$ fermions.

7.3.3 BMN limit: a particular example

In order to show the BMN limit at work, we consider a string in the plane wave background for which only the modes $N_n = \tilde{N}_n = 1$ for some $n \in \mathbb{N}$ are excited. This string has energy

$$E_{\text{lc}} = 2\mu\sqrt{1 + \frac{n^2}{(\alpha'\tilde{p}^+\mu)^2}} = 2\mu\sqrt{1 + \frac{\lambda n^2}{J^2}} = 2\mu\sqrt{1 + \lambda'n^2}, \tag{7.88}$$

where we take the BMN limit $E \simeq J$ and use $L^4 = \lambda\alpha'^2$, such that

$$\frac{1}{(\alpha'\tilde{p}^+\mu)^2} \simeq \frac{\lambda}{J^2} \equiv \lambda'. \tag{7.89}$$

According to the BMN conjecture, this string configuration corresponds to a local operator of dimension Δ, where

$$\Delta - J = \frac{E_{\text{lc}}}{\mu} = 2\sqrt{1 + \lambda'n^2} = 2 + \lambda'n^2 + \mathcal{O}\left(\frac{\lambda^2 n^4}{J^4}\right). \tag{7.90}$$

Note that in the BMN limit $N \to \infty, J \to \infty$ with $\lambda' = \lambda/J^2$ fixed, this prediction matches precisely with the conformal dimension of the two-magnon operator (7.34),

$$\Delta = l + \frac{\lambda}{\pi^2} \sin^2 \left(\frac{\pi n}{l-1} \right) = l + \frac{\lambda n^2}{(l-1)^2} + \mathcal{O}(l^{-4}), \qquad (7.91)$$

provided that we identify $J = l - 2$.

In addition to the one-loop example, we note also that at two loops on the gauge theory side, there is perfect agreement between the perturbative field theory and the string theory results. At three loops, however, there is a discrepancy which is due to a problem of ordering of limits. On the string theory side, the limit $\lambda \to \infty$ is taken with J^2/λ fixed. On the other hand, on the field theory side, the perturbative calculation is performed for $\lambda \ll 1$, subsequently taking $J \to \infty$ while keeping only terms which scale as λ/J^2. The discrepancy observed indicates that these limits do not commute.

7.4 Dual superconformal symmetry

In addition to the superconformal symmetry, $\mathcal{N} = 4$ supersymmetric Yang–Mills theory has a further rather unexpected symmetry, which however cannot be seen at the level of the Lagrangian. This *dual superconformal symmetry* was discovered at the level of scattering amplitudes, which turn out to be related to certain lightlike Wilson loops. This was first discovered using AdS/CFT techniques at strong coupling, and later confirmed in perturbative calculations. This result is an important example of the impact and power of the AdS/CFT approach.

These rather unexpected relations are discussed in the subsequent sections. First we introduce the concept of scattering amplitudes and their IR divergences. In particular we focus on $\mathcal{N} = 4$ Super Yang–Mills theory and show that scattering amplitudes exhibit a dual conformal symmetry which may be extended even to a dual superconformal theory. Then we study the Wilson loop/scattering amplitude correspondence in detail, providing us with more evidence for the dual superconformal symmetry. Finally we show that the superconformal symmetry and dual superconformal symmetry can be unified in a Yangian structure.

7.4.1 Scattering amplitudes and IR divergences in field theory

Let us first consider a pure Yang–Mills theory with gauge group $SU(N)$ whose generators are denoted by T_a. Its action is given by (1.185) which, however, we normalise canonically in the following (see box 1.2). In this theory we consider n gluons scattering with each other. The corresponding scattering amplitude $\mathcal{A}_{n \, \mu_1 \ldots \mu_n}$ is given in terms of gauge fields A_{μ_1} as

$$\mathcal{A}_{n \, \mu_1 \ldots \mu_n} = \langle A_{\mu_1}(\vec{p}_1) \ldots A_{\mu_n}(\vec{p}_n) \rangle, \qquad (7.92)$$

where we suppress the indices $\mu_1 \ldots \mu_n$ in the following to simplify the notation. The scattering amplitude $\mathcal{A}_n$ depends on the momenta $\vec{p}_i$, helicities h_i and colour index a_i (where $i = 1, \ldots, n$ labels the different gluons) of the gluons and may be expanded perturbatively,

$$\mathcal{A}_n(\vec{p}_i, h_i, a_i) = g^{n-2} \sum_l g^{2l} \mathcal{A}_n^{(l)}(\vec{p}_i, h_i, a_i) \tag{7.93}$$

where $\mathcal{A}_n^{(l)}(\vec{p}_i, h_i, a_i)$ is the scattering amplitude at l loop order which furthermore may be decomposed as follows,

$$\mathcal{A}_n^{(l)} = N^l \sum_{\rho \in S_n/\mathbb{Z}_n} \text{Tr}(T_{a_{\rho(1)}} \ldots T_{a_{\rho(n)}}) A_n^{(l)}(p_{\rho(1)} \ldots p_{\rho(n)}, N^{-1}) + \text{multi-traces}. \tag{7.94}$$

Here the sum extends over all non-cyclic permutations ρ of the n-tuple $(1, \ldots, n)$. The coefficients $A_n^{(l)}(p_{\rho(1)} \ldots p_{\rho(n)}, N^{-1})$ are referred to as *colour-ordered* or *partial amplitudes*. In the limit of large number of colours, $N \to \infty$, in which we are interested, the multi-trace terms left unspecified in the equation above drop out. The same is true for all N-dependent terms in $A_n^{(l)}(p_{\rho(1)} \ldots p_{\rho(n)}, N^{-1})$, reducing them to planar partial amplitudes $A_n^{(l)}(p_{\rho(1)} \ldots p_{\rho(n)})$. These planar partial amplitudes contain all the kinematic information and they are gauge invariant since we factored out all the generators of the gauge group $SU(N)$ [4]. From now on, we restrict our discussion to partial amplitudes and just refer to them as amplitudes.

The physical on-shell degrees of freedom of Yang–Mills theory, i.e. the gluons with helicity $h = \pm 1$, are most easily described in terms of the spinor helicity formulation. The basic idea behind this formulation is the following: a four-momentum p_i^μ may be written as a 2×2 matrix of the form

$$p_i^{\alpha\dot\alpha} = \sigma_\mu^{\alpha\dot\alpha} p_i^\mu. \tag{7.95}$$

Consider the massless case in which the square of the four-momentum vanishes, $p_\mu p^\mu = 0$. Since $p_\mu p^\mu$ is the determinant of the matrix $p_i^{\alpha\dot\alpha}$ we can rewrite $p_i^{\alpha\dot\alpha}$ as the product of two commuting spinors,

$$p_i^{\alpha\dot\alpha} = \lambda_i^\alpha \tilde\lambda_i^{\dot\alpha}. \tag{7.96}$$

With the help of these commuting spinors, we can define the polarisation vectors by

$$\epsilon_{+,i}^{\alpha\dot\alpha} = \frac{\tilde\lambda_i^{\dot\alpha} \mu_i^\alpha}{\langle \lambda_i \mu_i \rangle}, \qquad \epsilon_{-,i}^{\alpha\dot\alpha} = \frac{\lambda_i^\alpha \tilde\mu_i^{\dot\alpha}}{[\tilde\lambda_i \tilde\mu_i]}, \tag{7.97}$$

where l_i^μ is an auxiliary lightlike momentum with associated commuting spinors μ_i^α and $\tilde\mu_i^{\dot\alpha}$, i.e. $l_i^{\alpha\dot\alpha} = \mu_i^\alpha \tilde\mu_i^{\dot\alpha}$. Here, we have introduced the notation for the spinor scalar products,

$$\langle \lambda\mu \rangle = \lambda^\alpha \mu_\alpha = \lambda^\alpha \mu^\beta \epsilon_{\beta\alpha}, \qquad [\lambda\mu] = \lambda_{\dot\alpha} \mu^{\dot\alpha} = \lambda_{\dot\alpha} \mu_{\dot\beta} \epsilon^{\dot\alpha\dot\beta}. \tag{7.98}$$

We use these polarisation vectors to contract the indices of partial amplitude $A_n^{(l)}{}_{\mu_1 \ldots \mu_n}$ $(p_1 \ldots p_n)$, where we have reinstated the indices $\mu_1 \ldots \mu_n$. For example, consider an n-point ordered amplitude with two negative helicities and $n - 2$ positive helicities. Let us assume that the first and second gluons of the ordered amplitude have negative helicity. Denoting this amplitude by $A(1^-, 2^-)$, we have

$$A_n^{\text{tree}}(1^-, 2^-) = \epsilon_{-,1}^{\mu_1}\epsilon_{-,2}^{\mu_2}\epsilon_{+,3}^{\mu_3}\cdots\epsilon_{+,n}^{\mu_n}A_{n\,\mu_1\ldots\mu_n}^{tree}(p_{\rho(1)}\cdots p_{\rho(n)}). \qquad (7.99)$$

Let us analyse the partial amplitudes in more detail. Due to conservation of momentum, these amplitudes always contain a δ function $\delta^4(p) \equiv \delta^4(p_1 + \cdots + p_n)$. After factoring out this δ function, the remaining partial amplitude is a single Lorentz invariant rational function which contains local poles of the form $(p_k + p_{k+1} + \cdots + p_l)^{-2}$. At tree level, these amplitudes may be written in a simple form. First of all, amplitudes with only positive helicity gluons or with just one negative helicity gluon have to vanish. The same is true for amplitudes with only negative helicity gluons or with just one positive helicity gluon. Thus the simplest non-trivial amplitudes, referred to as maximally helicity violating (MHV) amplitudes, have $n - 2$ gluons with positive helicity and two gluons with negative helicity. Next-to-MHV amplitudes (for $n \geq 5$) contain three gluons with negative helicity and $n - 3$ gluons with positive helicity. MHV amplitudes are very important since they can be written in a surprisingly simple manner. In particular, the four-gluon scattering amplitude reads

$$A_4^{\text{tree,MHV}}(i^-, j^-) = i\frac{\langle ij\rangle^4}{\langle 12\rangle\langle 23\rangle\langle 34\rangle\langle 41\rangle}\delta^4(p). \qquad (7.100)$$

It is straightforward to generalise this to an n-point MHV amplitude

$$A_n^{\text{tree,MHV}}(i^-, j^-) = i\frac{\langle ij\rangle^4}{\langle 12\rangle\langle 23\rangle\ldots\langle n1\rangle}\delta^4(p). \qquad (7.101)$$

Supersymmetric tree level amplitudes

So far we have considered partial gluon scattering amplitudes in pure Yang–Mills theory which contain only two gluon states. The story can be extended in a simple manner to supersymmetric theories, in particular to $\mathcal{N} = 4$ Super Yang–Mills theory in which we are interested. In addition to the two gluon states G^+ and G^- with helicity $h = \pm 1$, $\mathcal{N} = 4$ Super Yang–Mills theory has four fermion states (gluinos) Γ_a and $\bar{\Gamma}^a$ with helicities $1/2$ and $-1/2$, respectively, and six scalars of helicity zero $S_{ab} = -S_{ba}$. Here $a, b, c, d = 1, \ldots, 4$ are indices of the (anti-)fundamental representation of the R-symmetry group $SU(4)$. These particles can scatter into each other in many different combinations, which results in a large variety of amplitudes. These various scattering amplitudes are related to each other through supersymmetric Ward identities.

To discuss the symmetry properties of the scattering amplitudes for $\mathcal{N} = 4$ Super Yang–Mills theory, it is desirable to find a way to present all scattering amplitudes in the $\mathcal{N} = 4$ theory as one simple and compact object with manifest supersymmetry. In fact, such an object exists. It is given by the superfield

$$\Phi(p, \eta) = G^+(p) + \eta^a\Gamma_a(p) + \frac{1}{2}\eta^a\eta^bS_{ab}(p) + \frac{1}{3!}\eta^a\eta^b\eta^c\epsilon_{abcd}\bar{\Gamma}^d(p)$$

$$+ \frac{1}{4!}\eta^a\eta^b\eta^c\eta^d\epsilon_{abcd}G^-(p), \qquad (7.102)$$

where p is the four-momentum and η is a four-component Grassmann variable. The on-shell supersymmetry generators are given by

$$q_\alpha^b = \lambda_\alpha \eta^b, \qquad \bar{q}_{b\dot\alpha} = \tilde\lambda_{\dot\alpha} \frac{\partial}{\partial \eta^b}, \tag{7.103}$$

such that under infinitesimal supersymmetry transformations we have

$$\delta_\epsilon \Phi(p, \eta^a) = \left(\epsilon_b^\alpha q_\alpha^b + \bar{\epsilon}^{b\dot\alpha} \bar{q}_{b\dot\alpha} \right) \Phi(p, \eta^a) \equiv \left(\epsilon_b \eta^b + \bar{\epsilon}^b \frac{\partial}{\partial \eta^b} \right) \Phi(p, \eta^a). \tag{7.104}$$

Next we construct a superamplitude which gives a compact description of the scattering amplitudes of n particles in the $\mathcal{N} = 4$ theory, $A_n(\Phi(1), \Phi(2), \ldots, \Phi(n))$. This n-particle superamplitude contains all correlation functions involving gluons, fermions and scalars as can be seen by expanding it in the Grassmann variable η,

$$A_n(\Phi(1), \Phi(2), \ldots, \Phi(n)) = A_n(+, +, \ldots, +) + (\eta_1)^4 (\eta_2)^4 A_n(-, -, +, \ldots, +)$$
$$+ (\eta_1)^4 (\eta_2)^3 \eta_3 A_n(-, \bar\Gamma, \Gamma, +, \ldots, +) + \cdots . \tag{7.105}$$

Here, the $\pm$ signs in the amplitude A_n symbolise positive and negative helicity gluons $G^\pm$, while Γ and $\bar\Gamma$ symbolise fermions. To improve readability, we have suppressed the $\mathfrak{su}(4)$ indices in the expression above.

In $\mathcal{N} = 4$ theory, this superamplitude takes a remarkably simple form

$$A_n(\lambda, \tilde\lambda, \eta) = i\delta^{(4)}(p)\delta^{(8)}(q)\mathcal{P}_n(\lambda, \tilde\lambda, \eta), \tag{7.106}$$

with the function $\mathcal{P}_n$ satisfying

$$\bar{q}_{a\dot\alpha} \mathcal{P}_n(\lambda, \tilde\lambda, \eta) = 0, \tag{7.107}$$

where the supersymmetry generators acting on scattering amplitudes are just given by the sum of the single-particle supersymmetry generators, i.e. for $\bar{q}_{a\dot\alpha}$ we have

$$\bar{q}_a^{\dot\alpha} = \sum_{i=1}^n \tilde\lambda_i^{\dot\alpha} \frac{\partial}{\partial \eta_i^a}. \tag{7.108}$$

For notational simplicity, we do not distinguish in the following between symmetry generators acting on scattering amplitudes or single-particle symmetry generators, since the only difference is the sum over $i = 1, \ldots, n$.

Exercise 7.4.1   Check that the amplitude given by (7.106) satisfies

$$q_\alpha^A A_n = \bar{q}_{A\dot\alpha} A_n = p_{\alpha\dot\alpha} A_n = 0, \tag{7.109}$$

which implies that this amplitude is supersymmetric since it is annihilated by the supersymmetry generators. Note that $p_{\alpha\dot\alpha} = \sigma_{\alpha\dot\alpha}^\mu p_\mu$.

Moreover, the amplitude is also covariant under conformal transformations, such as the dilatation operator given by

$$d = \frac{1}{2} \sum_{i=1}^n \left(\lambda_i^\alpha \frac{\partial}{\partial \lambda_i^\alpha} + \tilde\lambda_i^{\dot\alpha} \frac{\partial}{\partial \tilde\lambda_i^{\dot\alpha}} \right) \tag{7.110}$$

as well as the remaining (super)conformal generators and the R-symmetry generator. In particular, the R-symmetry imposes restrictions on the form of the amplitude since due to the $SU(4)$ R-symmetry, $\mathcal{P}_n$ may be expanded into terms of Grassmann degree $0, 4, 8, \ldots$, i.e.

$$\mathcal{P}_n(\lambda, \tilde{\lambda}, \eta) = \mathcal{P}_n^{(0)} + \mathcal{P}_n^{(4)} + \mathcal{P}_n^{(8)} + \cdots + \mathcal{P}_n^{(4n-16)}. \tag{7.111}$$

The lowest order term $\mathcal{P}_n^{(0)}$ corresponds to MHV amplitudes, $\mathcal{P}_n^{(4)}$ to next-to-MHV amplitudes, and so on. Let us consider the example of four-point amplitudes, $n = 4$. In this case, the expansion is trivial and contains only $\mathcal{P}_4^{(0)}$. Thus the amplitude of MHV type, with two gluon fields of helicity $h = +1$ and two fields of helicity $h = -1$ is given by

$$A_4^{\text{tree}}(\lambda, \tilde{\lambda}, \eta) = i\delta^{(4)}(p)\delta^{(8)}(q)\frac{1}{\langle 1\,2\rangle\langle 2\,3\rangle\langle 3\,4\rangle\langle 4\,1\rangle}. \tag{7.112}$$

This is of the same structure as the non-supersymmetric result (7.101). To discuss loop contributions, we therefore return to the non-supersymmetric case.

Beyond tree level: the BDS conjecture

We move on to consider amplitudes beyond tree level. In general, loop contributions lead to IR divergences of the amplitudes. Unlike UV divergences, IR divergences cannot be renormalised away. Nevertheless, they cancel in IR-safe quantities, such as cross sections of colour-singlet states or anomalous dimensions.

To analyse the IR divergences, we use dimensional regularisation, i.e. we consider $d = 4 - 2\epsilon$ dimensions. This requires the introduction of a renormalisation scale μ which breaks scale invariance. The four-gluon amplitude has external momenta $p_1, \ldots, p_4$, where the index indicates the colour ordering. We take the particles labelled by 1, 3 to be incoming, and those labelled by 2, 4 to be outgoing. All four momenta are taken to be pointing inwards, such that conservation of momentum implies $\sum_i p_i = 0$. In general, the one-loop contribution to the four-gluon amplitude contains the divergent integral

$$I_4^{(1)} = \int \mathrm{d}^d k \, \frac{1}{k^2(k-p_1)^2(k-p_1-p_2)^2(k+p_3)^2}, \tag{7.113}$$

which using conservation of momentum may be shown to have the correct symmetry properties under exchanges of the four momenta. Generically, as observed in this integral, there are two types of divergences: *soft* divergences for $p^\mu \sim 0$, which correspond to exchanging soft gluons between the external gluons, and *collinear* divergences for $p^\mu \sim k_i^\mu$, for which the internal momentum is proportional to one of the external momenta. The IR divergences lead to poles in $1/\epsilon$, i.e. taking both divergence types together we have a pole of order $1/\epsilon^2$ at one loop. At lth order in the perturbative expansion, we have a pole of order $1/\epsilon^{2l}$, i.e. the amplitude has a divergent contribution

$$A_4^{(l)} \simeq 1/\epsilon^{2l}. \tag{7.114}$$

In this context, let us now consider the MHV amplitudes (7.101). These planar partial amplitudes can be determined for any loop order l. Using the three-loop result, as well as resummation and exponentiation of IR singularities, Bern, Dixon and Smirnov (BDS) conjectured an all-loop expression for the partial amplitudes $A_n(k_{\rho(1)} \ldots k_{\rho(n)})$. For four-point amplitudes with $n = 4$, their expression reads

$$A_4 = A_4^{\text{tree}} \cdot \left(A_{\text{div},s}\right)^2 \cdot \left(A_{\text{div},t}\right)^2 \cdot \exp\left(\frac{f(\lambda)}{8}\ln^2\left(\frac{s}{t}\right)^2 + \text{constant}\right). \tag{7.115}$$

In this expression, s and t are the Mandelstam variables

$$s = -(k_1 + k_2)^2, \qquad t = -(k_2 + k_3)^2 \tag{7.116}$$

and λ is the 't Hooft coupling. The divergent contributions in (7.115) are given by

$$A_{\text{div},s} = \exp\left(-\frac{1}{8\epsilon^2}f^{(-2)}\left(\frac{\lambda\mu^{2\epsilon}}{(-s)^\epsilon}\right) - \frac{1}{4\epsilon}g^{(-1)}\left(\frac{\lambda\mu^{2\epsilon}}{(-s)^\epsilon}\right)\right), \tag{7.117}$$

with a similar expression for $A_{\text{div},t}$. Here, μ is an IR renormalisation scale. The functions $f^{(-2)}$ and $g^{(-1)}$ are related to the *cusp anomalous dimension* $f(\lambda)$ and the *collinear anomalous dimension* $g(\lambda)$ by

$$f(\lambda) = \left(\lambda\frac{\partial}{\partial\lambda}\right)^2 f^{(-2)}, \qquad g(\lambda) = \lambda\frac{\partial}{\partial\lambda}g^{(-1)}(\lambda). \tag{7.118}$$

To provide evidence for this conjecture, we calculate $f(\lambda)$ and $g(\lambda)$ on the gravity side below.

7.4.2 Relation between Wilson loops and scattering amplitudes

At strong coupling, the amplitude A_4 may be calculated using the AdS/CFT correspondence, where the external gluons correspond to open strings ending on D-branes. In fact, on the gravity side the scattering amplitude can be determined by the minimal area of a fundamental string ending on a curve with lightlike segments [5]. Since such a fundamental string is also used to calculate the expectation value of the Wilson loop on the gauge theory side, an interesting relation between amplitudes and Wilson loops is revealed at strong coupling: colour ordered amplitudes may be calculated by Wilson loops with lightlike segments.

In order to calculate the colour ordered amplitude with the AdS/CFT correspondence, we need an appropriate regulator of IR divergences on the gravity side. A candidate for such a regulator is a D-brane, as we now explain. More precisely, we start with the AdS_5 metric written in Poincaré coordinates

$$ds^2 = \frac{L^2}{z^2}\left(\eta_{\mu\nu}dx^\mu dx^\nu + dz^2\right) \tag{7.119}$$

and place a D3-brane at some fixed large value of $z = z_{\text{IR}}$ which also extends along the field theory directions given by the coordinates x^μ. We now study open strings on these Dp-branes and scatter them. These open strings correspond precisely to the external gluons. In order to make the colour ordering in this regularisation prescription manifest, we should consider not only one D3-brane but N of them.

The proper momentum $p_{(\text{pr})}^\mu$ of the strings corresponding to external gluons is given by $p_{(\text{pr})}^\mu = p^\mu z_{\text{IR}}/L$, where p^μ is the momentum in the dual field theory since it is conjugate to the field theory coordinates x^μ. We keep the momentum p^μ fixed as we remove the IR cut-off, $z_{IR} \to \infty$. Due to the metric warp factor z^2, the proper momentum p_{pr}^μ is very large, such that this approach corresponds to string scattering at fixed angle and with very large momentum. In particular, we are interested in the regime of string scattering where all kinematic invariants (such as the Mandelstam variables s, t and u) are much larger than the IR cut-off given by z_{IR}^{-2}.

In flat space, the scattering amplitude of strings with very large momentum is dominated by a saddle point of the classical string action, see for example [6]. By analogy we therefore expect to calculate scattering amplitudes at strong coupling by evaluating the classical string action on AdS spacetime subject to special boundary conditions encoding the colour ordering and the kinematical invariants. A hand-waving argument is as follows: the insertions of open strings in scattering amplitudes can be described by *vertex operators*[2] of momentum $p_{(pr)i}^\mu$ placed on the regulator D3-brane. To calculate the scattering amplitude of the open strings at string tree level we have to consider worldsheets with the topology of a disc, with vertex operator insertions of fixed large momentum $p_{(pr)i}^\mu$ on the boundary of the disc. At large momenta, we expect that the scattering amplitude A_n can be approximated by

$$A_n \sim e^{iS_{\min}}, \tag{7.120}$$

where $S_{\min}$ is the value of the worldsheet string action at the saddle point.

Fixing the momenta $p_{(pr)i}^\mu$ of the open strings thus corresponds to imposing Neumann boundary conditions on the IR regulating D-brane. Finding the saddle point $S_{\min}$ and thus computing the amplitude via (7.120) explicitly in the AdS case is difficult because of these Neumann boundary conditions.

A solution to this problem is to perform a T-duality transformation in all of the field theory directions x^0, x^1, x^2, x^3. Note, however, that we do not assume that the coordinates x^μ are compact. We should view the T-duality transformation as a technical trick to find solutions with the correct boundary conditions, which are Dirichlet boundary conditions in this case. Performing the T-duality, the Neumann boundary conditions are indeed replaced by Dirichlet boundary conditions, as we now explain.

T-dualising the geometry by applying the Buscher rules (4.101) to the metric (7.119) we obtain

$$ds^2 = \frac{L^2}{\tilde{z}^2}\left(\eta_{\mu\nu}dy^\mu dy^\nu + d\tilde{z}^2\right), \qquad \tilde{z} = \frac{L^2}{z}. \tag{7.121}$$

Note that this is again an Anti-de Sitter space, which we refer to as the T-dual AdS space. However, the T-duality transformation maps the boundary of the string worldsheet from z_{IR} in the interior of the AdS space to $\tilde{z}_{IR} = L^2/z_{IR}$, which is located at the boundary $\tilde{z} \to 0$ of the dual AdS space if the regulator is taken to zero, i.e. for $z_{IR} \to \infty$. Moreover, it maps the Neumann to Dirichlet boundary conditions and the momenta p_i to winding numbers. This last relation is realised as follows. The string zero mode with momentum p_i, as described by a local vertex operator, is replaced by a winding mode, which in the present setting

---

[2] For a very brief discussion of vertex operators see page 155.

means that the difference between the two endpoints of the string satisfies

$$\Delta y^\mu = 2p_i^\mu. \tag{7.122}$$

This implies that each vertex operator is replaced by a line segment connecting two points whose coordinate difference is a multiple of the vertex operator momentum p_i.

Thus we are ready to state the recipe for calculating colour ordered scattering amplitudes within the AdS/CFT correspondence. In the T-dual background, find the minimum of the worldsheet string action $S_{\min}$, taking the worldsheet to end on the polygon at $\tilde{z} = L^2/z_{\mathrm{IR}}$. This polygon is obtained as follows.

- For every gluon with momentum p_i^μ draw a lightlike interval of length $\Delta y^\mu = 2p_i^\mu$.
- The intervals are assembled according to the corresponding colour ordering of the partial amplitude A_n.

The colour ordered amplitude A_n is then given to leading order in λ by

$$A_n = e^{iS_{\min}} \tag{7.123}$$

for the minimum of worldsheet action determined by the recipe given.

Scattering of four gluons

As an example we consider the scattering of four gluons with momenta $p_1, \ldots, p_4$. We take the gluons labelled by 1, 3 to be incoming, and those labelled by 2, 4 to be outgoing. All four momenta p_i are taken to be pointing inwards. This is precisely the situation considered when we discussed the all-loop BDS conjecture. The kinematical information is stored in the Mandelstam variables s and t given by (7.116). For simplicity, we consider the case where the Mandelstam variables satisfy $s = t$.

According to the prescription given above, to calculate the four-gluon scattering amplitude we need to find the minimal surface ending on the lightlike polygon given by figure 7.4.

We use AdS Poincaré coordinates $(\tilde{z}, y^\mu)$, $\mu = 0, 1, 2, 3$, whose metric reads (7.121). Using translation symmetry, we set $y^3 = 0$. Moreover, we parametrise the string worldvolume by y^1, y^2, i.e.

$$\tilde{z} = \tilde{z}(y^1, y^2), \qquad y^0 = y^0(y^1, y^2). \tag{7.124}$$

The Nambu–Goto action with this embedding is given by

$$S = \frac{1}{2\pi\alpha'} \int dy_1 dy_2 \sqrt{-\det P[g]} \tag{7.125}$$

$$= \frac{iL^2}{2\pi\alpha'} \int dy_1 dy_2 \frac{1}{\tilde{z}^2} \sqrt{1 + (\partial_i \tilde{z})^2 - (\partial_i y^0)^2 - (\partial_1 \tilde{z} \partial_2 y^0 - \partial_2 \tilde{z} \partial_1 y^0)^2},$$

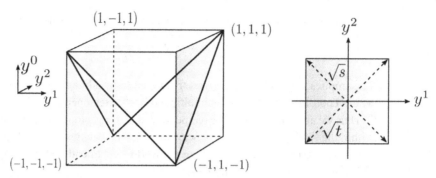

Figure 7.4 Polygon for calculating the four-gluon scattering amplitude at strong coupling (left), which we have to insert at the boundary of AdS space at $\tilde{z} = 0$. The right-hand side shows the projection to the (y^1, y^2)-plane, which is a square when imposing $s = t$ for the Mandelstam variables.

with boundary conditions

$$\tilde{z}(\pm 1, y^2) = \tilde{z}(y^1, \pm 1) = 0, \tag{7.126}$$

$$y^0(\pm 1, y^2) = \pm y^2, \tag{7.127}$$

$$y^0(y^1, \pm 1) = \pm y^1. \tag{7.128}$$

These boundary conditions are visualised in figure 7.4. The solution to the associated equations of motion subject to the boundary conditions given is

$$y^0(y^1, y^2) = y^1 y^2, \quad \tilde{z}(y^1, y^2) = \sqrt{(1 - (y^1)^2)(1 - (y^2)^2)}. \tag{7.129}$$

This solution has the following induced metric on the worldsheet,

$$ds^2 = \frac{dy_1^2}{(1 - y_1^2)^2} + \frac{dy_2^2}{(1 - y_2^2)^2} = du_1^2 + du_2^2, \tag{7.130}$$

where we have introduced new coordinates u^1 and u^2 given by $\tanh u^i = y^i$. Note that we embedded a spacelike surface into AdS spacetime with Lorentzian signature.

For later purposes, in particular to reinsert the dependence on the Mandelstam variables, we have to scale the solution by a factor a, such that the solution in the coordinates (u^1, u^2) reads

$$y^0(u^1, u^2) = a \tanh u_1 \ \tanh u_2, \quad \tilde{z}(u^1, u^2) = \frac{a}{\cosh u_1 \ \cosh u_2}. \tag{7.131}$$

The parameter a is related to the Mandelstam variables $s = t$ by

$$a^2 = -\frac{\pi^2}{2} s. \tag{7.132}$$

Note that in our convention, the Mandelstam variables are negative for spacelike momentum transfer and thus a is real.

Similar arguments apply to the case $s \neq t$. Starting from the square depicted on the right-hand side of figure 7.4, in order to achieve $s \neq t$ we have to deform this square in the (y^1, y^2) plane into a parallelogram.

Exercise 7.4.2 Using the isometries of the dual AdS_5 spacetime (7.121), i.e. $SO(4,2)$ transformations, generalise the solution (7.129) or (7.131) to cases where $s \neq t$.

Inserting (7.131) into the string action, we obtain an infinite result, as expected, which requires regularisation. In particular, the on-shell Lagrangian is constant. The appearance of divergences should not be surprising, since on the field theory side we also needed regularisation. For the BDS conjecture discussed above we used dimensional regularisation, setting $d = 4 - 2\epsilon$. To find the corresponding metric of the gravity background, let us first look at a hypothetical Dp-brane with $p = 3 - 2\epsilon$ spatial dimensions. In string frame, its metric is

$$\mathrm{d}s^2 = H^{-1/2}(r)\mathrm{d}x_{4-2\epsilon}^2 + H^{1/2}(r)\left(\mathrm{d}r^2 + r^2\mathrm{d}\Omega_{5+2\epsilon}^2\right), \tag{7.133}$$

with

$$H(r) = 1 + \frac{\lambda_{4-2\epsilon}c_{4-2\epsilon}\alpha'^2}{r^{4+2\epsilon}}, \tag{7.134}$$

where $\lambda_{4-2\epsilon} = Ng_{\mathrm{YM},4-2\epsilon}^2$ is the coupling constant in $d = 4 - 2\epsilon$ dimensions. This is related to the 't Hooft coupling constant λ in $d = 4$ dimensions by

$$\lambda_{4-2\epsilon} = \frac{\lambda\mu^{2\epsilon}}{(4\pi e^{-\gamma})^\epsilon}, \tag{7.135}$$

where γ is the Euler–Mascheroni constant defined as the negative derivative of the Gamma function $\Gamma(x)$ at $x = 1$, i.e. $\gamma = -\Gamma'(1)$. Moreover, in (7.134) the constant $c_{4-2\epsilon}$ is given by $c_{4-2\epsilon} = 2^{4\epsilon}\pi^{3\epsilon}\Gamma(2 + \epsilon)$. Thus taking the limit $\epsilon \to 0$ in (7.134) we obtain the usual geometry of N D3-branes, as expected.

In order to obtain the metric of AdS space in $d = 5 - 2\epsilon$ dimensions, we just have to take the near-horizon limit of (7.134) by dropping the constant 1 in $H(r)$. However, we are interested in the dual AdS space. Performing T-dualities along the four dimensions of the field theory by applying the Buscher rules (4.101), the metric of the dual AdS in $d = 5 - 2\epsilon$ dimensions reads

$$\mathrm{d}s^2 = \frac{\sqrt{\lambda_{4-2\epsilon}c_{4-2\epsilon}}\alpha'}{\tilde{z}^{2+\epsilon}}\left(\eta_{\mu\nu}\mathrm{d}y^\mu\mathrm{d}y^\nu + \mathrm{d}\tilde{z}^2\right). \tag{7.136}$$

Thus, the only differences are the modified prefactor $\sqrt{\lambda_{4-2\epsilon}c_{4-2\epsilon}}\alpha'$ which reduces to L^2 in the limit $\epsilon \to 0$ as well as the exponent of $\tilde{z}$, which is now $2 + \epsilon$ instead of 2.

Now we have to perform the same analysis as beforehand in this dual AdS space. This gives

$$S = \frac{i\sqrt{\lambda_{4-2\epsilon}c_{4-2\epsilon}}}{2\pi}\int \mathrm{d}y_1\mathrm{d}y_2 \frac{1}{\tilde{z}^{2+\epsilon}}\sqrt{1 + (\partial_i\tilde{z})^2 - (\partial_i y^0)^2 - (\partial_1\tilde{z}\partial_2 y^0 - \partial_2\tilde{z}\partial_1 y^0)^2}. \tag{7.137}$$

Note the differences between (7.137) and (7.125), in particular the modified exponent in the denominator of the Lagrangian. As a result of this modification, we have to find the solutions to the equations of motion for this ϵ-deformed Lagrangian. However these solutions are not known. Since we are interested in the divergence structure of the amplitude up to finite terms, it is sufficient to insert the original solution (7.129) into the ϵ-deformed action (7.137).

Performing this calculation explicitly using the more convenient coordinates u_i instead, we obtain the following on-shell action

$$S_{\min} = \frac{i\sqrt{\lambda_{4-2\epsilon}c_{4-2\epsilon}}}{2\pi\, a^\epsilon} \int du_1 du_2 \, (\cosh u_1 \, \cosh u_2)^\epsilon \, (1 + \mathcal{O}(\epsilon)), \qquad (7.138)$$

where a is given by (7.132). The integrand in (7.138) contains terms summarised by $\mathcal{O}(\epsilon)$ which are higher order in ϵ. The integral is finite for negative ϵ. To show the general structure, let us ignore the terms summarised by $\mathcal{O}(\epsilon)$ and compute the resulting integral explicitly,

$$S_{\min} = \frac{i\sqrt{\lambda_{4-2\epsilon}c_{4-2\epsilon}}}{2\pi\, a^\epsilon} \frac{\pi\,\Gamma\left(-\frac{\epsilon}{2}\right)^2}{\Gamma\left(\frac{1-\epsilon}{2}\right)^2}. \qquad (7.139)$$

From this expression, we can read off the most divergent part, which is of order ϵ^{-2}. This divergent part coincides with the most divergent part of the BDS conjecture for colour ordered scattering amplitude A_4. To see this explicitly, we calculate A_4 as the exponential of $iS_{\min}$ and expand it in $1/\epsilon$,

$$A_4 = e^{iS_{\min}} = \exp\left(-\frac{1}{\epsilon^2}\frac{1}{2\pi}\sqrt{\frac{\lambda\mu^{2\epsilon}}{(-s)^\epsilon}} + \mathcal{O}(1/\epsilon)\right). \qquad (7.140)$$

This is in agreement with the BDS conjecture (7.115) provided that we identify

$$f(\lambda) = \frac{\sqrt{\lambda}}{\pi} \qquad (7.141)$$

for the cusp anomalous dimension of (7.118), which leads to the expected divergent parts in (7.117).

To check the BDS conjecture further, we have to generalise to the case $s \neq t$ since according to the BDS conjecture (7.115), $f(\lambda)$ should also appear in front of the term proportional to $\log(s/t)$. A detailed calculation not explicitly shown here shows that for $s \neq t$ the minimal value $S_{\min}$ is given by

$$S_{\min} = \frac{i\sqrt{\lambda_{4-2\epsilon}c_{4-2\epsilon}}}{2\pi\, a^\epsilon} \left(\frac{\pi\,\Gamma\left(-\frac{\epsilon}{2}\right)^2}{\Gamma\left(\frac{1-\epsilon}{2}\right)^2} {}_2F_1\left(\frac{1}{2}, -\frac{\epsilon}{2}, \frac{1-\epsilon}{2}; b^2\right) + \frac{1}{2}\right), \qquad (7.142)$$

where a and b are related to the Mandelstam variables s and t by

$$\frac{s}{t} = \frac{(1+b)^2}{(1-b)^2}, \qquad -(2\pi)^2 s = \frac{8a^2}{(1-b^2)}, \qquad -(2\pi)^2 t = \frac{8a^2}{(1+b^2)}. \qquad (7.143)$$

Carefully expanding (7.142) in powers of $1/\epsilon$ up to finite terms, we find for the four-gluon scattering amplitude A_4,

$$A_4 = e^{iS_{\min}} = \exp\left(iS_{\mathrm{div}} + \frac{\sqrt{\lambda}}{8\pi}\ln^2\left(\frac{s}{t}\right) + \text{constant}\right), \qquad (7.144)$$

where

$$S_{\mathrm{div}} = 2S_{\mathrm{div},s} + 2S_{\mathrm{div},t} \qquad (7.145)$$

with

$$iS_{\text{div},s} = -\frac{1}{\epsilon^2}\frac{1}{2\pi}\sqrt{\frac{\lambda\mu^{2\epsilon}}{(-s)^\epsilon}} - \frac{1}{\epsilon}\frac{1}{4\pi}(1-\ln 2)\sqrt{\frac{\lambda\mu^{2\epsilon}}{(-s)^\epsilon}}, \qquad (7.146)$$

and the same expression for $S_{\text{div},s}$ with s and t exchanged. This verifies the BDS conjecture (7.115) for the case of four gluons, provided we identify

$$f(\lambda) = \frac{\sqrt{\lambda}}{\pi}, \qquad g(\lambda) = \frac{\sqrt{\lambda}}{2\pi}(1-\ln 2) \qquad (7.147)$$

for the cusp and collinear anomalous dimensions.

In summary, the structure of IR divergences in the gravity calculation of the colour ordered amplitude for four gluons agrees in all details with the BDS conjecture motivated from general field theory reasoning. This is further astonishing evidence for the AdS/CFT correspondence.

With the recipe given above, we may also determine the IR divergences for a scattering amplitude with $n > 4$ external gluons from the gravity side. Note that on the field theory side, there are discrepancies between the BDS conjecture and the actual field theory calculation, for instance for the six-gluon amplitude at two loops [7]. The difference is referred to as a *remainder function* and depends only on cross ratios. It is common understanding that the BDS conjecture for $n > 5$ gluons receives some non-trivial corrections. It may be possible to obtain these from the result on the gravity side.

Moreover, the construction of the colour ordered scattering amplitudes on the gravity side tells us two more surprising facts. First of all, the construction – finding a minimal surface of the Nambu–Goto action and exponentiating the regularised on-shell action – in the dual AdS spacetime reminds us of the same calculation of Wilson loop expectation values on the gravity side. In fact, we can identify – at least at strong coupling – colour ordered amplitudes with Wilson loops whose contour is given by lightlike straight lines. It turns out that this is also true at weak coupling. Second, another interesting feature which we have not appreciated so far, is that when T-dualising the AdS space we obtain another AdS space. While the isometries of the original AdS space induce conformal transformations on the conformal boundary, the isometries of the new AdS space give rise to *dual conformal transformations*. The dual AdS space appears in the construction of scattering amplitudes on the gravity side and thus the dual conformal transformations have to act on scattering amplitudes.

These examples show the powerful nature of AdS/CFT dualities: hidden symmetry relations as discussed in the previous paragraph become obvious when realising them geometrically on the gravity side.

7.4.3 Dual superconformal symmetry and Yangians

At least at tree level, the scattering amplitudes considered are invariant under the superconformal symmetry. As noted at the end of the preceding section, the scattering amplitudes are also covariant under a dual conformal symmetry, which can be extended to a dual superconformal symmetry.

Let us describe briefly the dual superconformal symmetry. Just as for the superconformal symmetry, the transformations of the dual superconformal symmetry act on the on-shell superspace variables $\{\lambda, \tilde{\lambda}, \eta\}$ which are introduced to formulate scattering amplitudes in $\mathcal{N} = 4$ Super Yang–Mills theory. As examples of the corresponding dual superconformal generators, we state the expressions for the dual supersymmetry generators and the dual dilatation,

$$P_{\alpha\dot{\alpha}} = \sum_i \frac{\partial}{\partial x_i^{\alpha\dot{\alpha}}}, \tag{7.148}$$

$$Q_{\alpha,A} = \sum_i \frac{\partial}{\partial \theta_i^{\alpha A}}, \tag{7.149}$$

$$\bar{Q}_{\dot{\alpha}}^A = \sum_i \left[\theta_i^{\alpha A} \frac{\partial}{\partial x_i^{\alpha\dot{\alpha}}} + \eta_i^A \frac{\partial}{\partial \tilde{\lambda}_i^{\dot{\alpha}}} \right], \tag{7.150}$$

$$D = \sum_i \left[-x_i^{\alpha\dot{\alpha}} \frac{\partial}{\partial x_i^{\alpha\dot{\alpha}}} - \frac{1}{2} \theta_i^{\alpha A} \frac{\partial}{\partial \theta_i^{\alpha A}} - \frac{1}{2} \lambda_i^\alpha \frac{\partial}{\partial \lambda_i^\alpha} - \frac{1}{2} \tilde{\lambda}_i^{\dot{\alpha}} \frac{\partial}{\partial \tilde{\lambda}_i^{\dot{\alpha}}} \right]. \tag{7.151}$$

Here, the dual supersapce coordinates $(x, \theta, \bar{\theta})$ are defined as follows. For n momenta labelled by i, the dual space coordinates $x_i^{\alpha\dot{\alpha}}$ are given by $\lambda_i^\alpha \tilde{\lambda}_i^{\dot{\alpha}} = (x_i - x_{i+1})^{\alpha\dot{\alpha}}$, with $x_{n+1} = x_1$ due to momentum conservation. A similar definition holds for θ and $\bar{\theta}$. The dual generators $\{P, Q, \bar{Q}, K, M, \bar{M}, R, D, S, \bar{S}\}$ may be written in a form in which they annihilate the scattering amplitudes A_n.

The combination of superconformal and dual superconformal symmetry leads to a *Yangian* symmetry, which is defined in box 7.2.

To combine both superconformal and dual superconformal symmetries, it is useful to reconstruct the dual superconformal generators to act only on the on-shell superspace

---

Box 7.2   **Yangian symmetry**

Consider a finite-dimensional simple Lie algebra g with generators T_a and structure constants f_{ab}^c satisfying

$$[T_a, T_b] = i f_{ab}^c T_c. \tag{7.152}$$

The *Yangian* of this Lie algebra g, denoted by $Y(g)$, is a deformation of the universal enveloping algebra defined in appendix B. In addition to the generators T_a which are referred to as level zero generators, we also introduce level one generators $\hat{T}_a$ satisfying

$$[T_a, \hat{T}_b] = i f_{ab}^c \hat{T}_c. \tag{7.153}$$

Equations (7.152) and (7.153) imply two different Jacobi identities: one just involving level zero generators and one involving two level zero and one level one operator. Moreover, there is a third Jacobi identity involving two level one generators. This identity, usually referred to as the Serre relation, is quantum deformed and reads

$$\left[[T_a, \hat{T}_b], \hat{T}_c \right] + \left[[T_b, \hat{T}_c], \hat{T}_a \right] + \left[[T_c, \hat{T}_a], \hat{T}_b \right] - f_{ag}^d f_{bh}^e f_{ci}^f f^{ghi} T_{\{d} T_e T_{f\}} = 0. \tag{7.154}$$

This Yangian structure is present in $\mathcal{N} = 4$ Super Yang–Mills theory. The Yangian of $\mathcal{N} = 4$ theory is based on the Lie superalgebra $\mathfrak{psu}(2, 2|4)$ and is referred to as $Y(\mathfrak{psu}(2, 2|4))$.

variables $(\lambda_i, \tilde{\lambda}_i, \eta_i)$. In this case, the P, Q generators are trivial, while the generators $\{\bar{Q}, M, \bar{M}, R, D, \bar{S}\}$ coincide with those of the superconformal algebra. The non-trivial dual generators which are not part of the superconformal generators are the dual K and S. The Yangian $Y(\mathfrak{psu}(2,2|4))$ is generated by all superconformal generators $\{J_a^{\mathrm{sc}}\}$ together with the dual S, or alternatively by J_a^{sc} and the dual K.

7.5 Further reading

Integrability is reviewed for instance in [8, 9], and scattering amplitudes are reviewed in in [10, 9, 11]. The one-loop calculation of the anomalous dimension of composite operators and its relation to spin chains is reviewed in [12].

The BMN limit was proposed in [3]. Spin chains and the Bethe ansatz for $\mathcal{N} = 4$ Super Yang–Mills theory were introduced in [13]. A string theory/gauge theory comparison to two loops was performed in [14]. The dilatation operator for $\mathcal{N} = 4$ Super Yang–Mills theory is discussed in [15], with further results in [16]. The three-loop anomalous dimension was calculated in [17, 18] and the discrepancy observed was traced back to a limit ordering problem in [19].

Classical integrability of the string sigma model on $AdS_5 \times S^5$ is considered in [1] and also in [2].

The Bern–Dixon–Smirnov conjecture was proposed in [20]. The high-energy behaviour of string scattering amplitudes was determined by Gross and Mende in [6]. At strong coupling, the map of the four-gluon amplitude to a lightlike Wilson loop was found in [5]. The value of the cusp anomalous dimension $f(\lambda)$ found from this approach coincides with expectations from integrability in field theory [21]. Discrepancies between the BDS conjecture for the six-gluon amplitude and the two-loop result were found in [7].

The dual superconformal symmetry of amplitudes in $\mathcal{N} = 4$ supersymmetric Yang–Mills theory was discovered in [22, 23]. Their Yangian symmetry is discussed in [24, 25, 26].

References

[1] Bena, Iosif, Polchinski, Joseph, and Roiban, Radu. 2004. Hidden symmetries of the $AdS_5 \times S^5$ superstring. *Phys. Rev.*, **D69**, 046002.

[2] Arutyunov, Gleb, and Frolov, Sergey. 2009. Foundations of the $AdS_5 x S^5$ superstring. Part I. *J. Phys.*, **A42**, 254003.

[3] Berenstein, David Eliecer, Maldacena, Juan Martin, and Nastase, Horatiu Stefan. 2002. Strings in flat space and pp waves from $\mathcal{N} = 4$ Super Yang-Mills. *J. High Energy Phys.*, **0204**, 013.

[4] Mangano, Michelangelo and Parke, S. 1991. Multiparton amplitudes in gauge theories. *Phys. Rep.*, **200**, 301.

[5] Alday, Luis F., and Maldacena, Juan. 2008. Gluon scattering amplitudes at strong coupling. *J. High Energy Phys.*, **0706**, 064.

[6] Gross, David J., and Mende, Paul F. 1987. The high-energy behavior of string scattering amplitudes. *Phys. Lett.*, **B197**, 129.

[7] Drummond, J. M., Henn, J., Korchemsky, G. P., and Sokatchev, E. 2008. The hexagon Wilson loop and the BDS ansatz for the six-gluon amplitude. *Phys. Lett.*, **B662**, 456–460.

[8] Plefka, Jan. 2005. Spinning strings and integrable spin chains in the AdS/CFT correspondence. *Living Rev. Relativity*, **8**, 9.

[9] Beisert, Niklas, Ahn, Changrim, Alday, Luis F., Bajnok, Zoltan, Drummond, James M., *et al.* 2012. Review of AdS/CFT integrability: an overview. *Lett. Math. Phys.*, **99**, 3–32.

[10] Alday, Luis F., and Roiban, Radu. 2008. Scattering amplitudes, Wilson loops and the string/gauge theory correspondence. *Phys. Rep.*, **468**, 153–211.

[11] Elvang, Henriette, and Huang, Yu-tin. 2015. *Scattering Amplitudes in Gauge Theory and Gravity*. Cambridge University Press.

[12] Minahan, J. 2012. Spin chains in $\mathcal{N} = 4$ Super Yang–Mills. *Lett. Math. Phys.*, **99**, 33–58.

[13] Minahan, J. A., and Zarembo, K. 2003. The Bethe ansatz for $N = 4$ Super Yang-Mills. *J. High Energy Phys.*, **0303**, 013.

[14] Kazakov, V. A., Marshakov, A., Minahan, J. A., and Zarembo, K. 2004. Classical/quantum integrability in AdS/CFT. *J. High Energy Phys.*, **0405**, 024.

[15] Beisert, N., Kristjansen, C., and Staudacher, M. 2003. The dilatation operator of conformal $\mathcal{N} = 4$ Super Yang-Mills theory. *Nucl. Phys.*, **B664**, 131–184.

[16] Beisert, Niklas. 2004. The dilatation operator of $\mathcal{N} = 4$ Super Yang-Mills theory and integrability. *Phys. Rep.*, **405**, 1–202.

[17] Beisert, Niklas. 2004. The $su(2|3)$ dynamic spin chain. *Nucl. Phys.*, **B682**, 487–520.

[18] Eden, B., Jarczak, C., and Sokatchev, E. 2005. A Three-loop test of the dilatation operator in $\mathcal{N} = 4$ SYM. *Nucl. Phys.*, **B712**, 157–195.

[19] Beisert, N., Dippel, V., and Staudacher, M. 2004. A novel long range spin chain and planar $\mathcal{N} = 4$ super Yang-Mills. *J. High Energy Phys.*, **0407**, 075.

[20] Bern, Zvi, Dixon, Lance J., and Smirnov, Vladimir A. 2005. Iteration of planar amplitudes in maximally supersymmetric Yang-Mills theory at three loops and beyond. *Phys. Rev.*, **D72**, 085001.

[21] Beisert, Niklas, Eden, Burkhard, and Staudacher, Matthias. 2007. Transcendentality and crossing. *J. Stat. Mech.*, **0701**, P01021.

[22] Drummond, J. M., Henn, J., Korchemsky, G. P., and Sokatchev, E. 2010. Dual superconformal symmetry of scattering amplitudes in $\mathcal{N} = 4$ Super Yang-Mills theory. *Nucl. Phys.*, **B828**, 317–374.

[23] Drummond, J. M., Henn, J., Korchemsky, G. P., and Sokatchev, E. 2013. Generalized unitarity for $\mathcal{N} = 4$ super-amplitudes. *Nucl. Phys.*, **B869**, 452–492.

[24] Dolan, Louise, Nappi, Chiara R., and Witten, Edward. 2004. Yangian symmetry in $D = 4$ superconformal Yang-Mills theory. ArXiv:hep-th/0401243.

[25] Drummond, James M., Henn, Johannes M., and Plefka, Jan. 2009. Yangian symmetry of scattering amplitudes in $\mathcal{N} = 4$ Super Yang-Mills theory. *J. High Energy Phys.*, **0905**, 046.

[26] Drummond, J. M., and Ferro, L. 2010. Yangians, Grassmannians and T-duality. *J. High Energy Phys.*, **1007**, 027.

Further examples of the AdS/CFT correspondence

In this chapter we study further examples of the AdS/CFT correspondence relating gravity theories and conformal field theories. These examples involve branes placed in other, more involved backgrounds than flat ten-dimensional space, as well as branes other than D3-branes.

8.1 D3-branes at singularities

The prototype example of the AdS/CFT correspondence involving a field theory in 3+1 dimensions is obtained from considering a stack of D3-branes in (9+1)-dimensional flat space (see chapter 5). The rotational symmetry in the six perpendicular directions is $SO(6)$. This reflects itself both in the $SO(6) \sim SU(4)$ R-symmetry of the $\mathcal{N} = 4$ Super Yang–Mills theory on the field theory side and in the isometries of S^5 on the gravity side.

We may now ask whether it is possible to obtain examples of the AdS/CFT correspondence where some of the supersymmetry charges are broken. There is actually a natural realisation of this by considering D3-branes placed at the tip of a suitable singular space. The singularity is essential in this construction; a smooth curved space is not sufficient since locally it still looks flat in a suitable coordinate system.

On the gravity side, considering D3-branes at the tip of a suitable singular space leads to a geometry $AdS_5 \times X$, with X a suitable manifold, examples of which we discuss below. We consider conical singularities where the radial direction of the Anti-de Sitter space and X form a cone with base X, with metric

$$ds^2 = dr^2 + r^2 ds_X^2. \tag{8.1}$$

The point $r = 0$ is singular unless X is a round sphere. Since the AdS_5 space and its $SO(4,2)$ symmetry are preserved in this construction, the dual field theory is conformal.

8.1.1 Orbifold

The simplest example of a suitable singular space is the *orbifold*. This is a manifold M/Γ quotiented by a subgroup of its isometries. Let us consider the example $M = \mathbb{C}^2/\mathbb{Z}_2$ for the case of ten-dimensional space with D3-branes in the $x^0, \ldots, x^3$ directions. For four of the coordinates perpendicular to the D3-branes, the $\mathbb{Z}_2$ orbifold projection acts as

$$x^i \mapsto -x^i, \quad i \in \{6, 7, 8, 9\}. \tag{8.2}$$

The other coordinates, both those along the D3-brane worldvolume and the two remaining perpendicular coordinates, are inert under the orbifold action. The space perpendicular to the D3-branes is thus $\mathbb{C}^2/\mathbb{Z}_2 \times \mathbb{C}$.

The projection which identifies x^i with $-x^i$ in the 6, 7, 8, 9 directions defines the singular point $x^i = 0$, which is invariant under the orbifold action. The original space present before carrying out the projection is referred to as the *underlying space*. A D3-brane located at an arbitrary point in transverse space is not invariant under the orbifold action. For an invariant configuration, a D3-brane located at the point $x^i = y^i$ in the 6, 7, 8, 9 directions must have an image located at $x^i = -y^i$ in the underlying space. To describe the open string spectrum, we thus have to consider four different kinds of strings: those with both ends attached to either the brane or its image, and those linking the two branes with different orientation. This means that the Chan–Paton factor associated with the string endpoints is a 2×2 matrix λ_{CP} of schematic form

$$\lambda_{CP} = \begin{pmatrix} D - D & D - D' \\ D' - D & D' - D' \end{pmatrix}, \tag{8.3}$$

where the entries stand for the different possibilites for strings stretched between the brane D and its image D'. The $\mathbb{Z}_2$ orbifold acts both on the string states and on the Chan–Paton matrix. The action on the latter is given by

$$\lambda_{CP} \mapsto \gamma(g)\lambda_{CP}\gamma(g)^{-1}, \tag{8.4}$$

with γ a representation of $\mathbb{Z}_2$. An appropriate choice of representation which interchanges the brane with its image is

$$\gamma(g) = \sigma^1, \tag{8.5}$$

with σ^1 the first Pauli matrix. This choice corresponds to the *regular* representation of the orbifold group, for which the dimension is equal to the order of the group. This representation is reducible. By analysing the orbifold action on both the Chan–Paton matrix and the string states, a $U(1) \times U(1)$ gauge group is obtained, under which the matter fields transform in bifundamental representations. Such a product gauge group is referred to as a *quiver gauge group*. This example may be generalised to other *regular D-branes*, i.e. D-branes whose Chan–Paton factors transform under the regular representation of the orbifold group. If we place a stack of N D3-branes at the orbifold singular point, we obtain a $U(N) \times U(N)$ quiver gauge theory.

Taking the near-horizon limit, the $U(1) \times U(1)$ degrees of freedom in the $U(N) \times U(N)$ quiver gauge theory become non-dynamical, such that the product gauge group is $SU(N) \times SU(N)$. The matter field content of the associated field theory is best determined by considering the chiral $\mathcal{N} = 1$ superspace multiplets which contain complex scalars in their lowest component, which may be thought of as being constructed from the four real scalars x^i introduced above in (8.2) together with the two $x^j, j \in \{4, 5\}$, with additional structure arising from the orbifold projection. It turns out that there are two chiral multiplets $\Phi_j, j = 1, 2$, one of which is in the adjoint representation of the first of the two $SU(N)$ product groups, and the other is in the adjoint of the second $SU(N)$. Moreover, there are chiral multiplets $A_k, B_l, k, l = 1, 2$, in the bifundamental representations $(\mathbf{N}, \overline{\mathbf{N}})$ and $(\overline{\mathbf{N}}, \mathbf{N})$

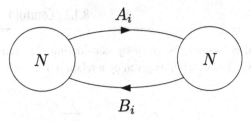

Figure 8.1 Quiver diagram for D3-branes at the $\mathbb{C}^2/\mathbb{Z}_2 \times \mathbb{C}$ orbifold. The nodes correspond to the two gauge groups and the arrows to the bifundamental fields A_i, B_i. Moreover, Φ_1 transforms in the adjoint of the first gauge group and Φ_2 in the adjoint of the second.

of $SU(N) \times SU(N)$, respectively. The superpotential for these fields respecting $\mathcal{N} = 2$ supersymmetry is given by

$$W = g\,\mathrm{Tr}\,\Phi_1(A_1 B_1 - A_2 B_2) + g\,\mathrm{Tr}\,\Phi_2(B_1 A_1 - B_2 A_2). \tag{8.6}$$

This simple example of a *quiver gauge theory* may be visualised as in figure 8.1.

Let us now consider the supergravity side for the $\mathbb{Z}_2$ example considered above, with N D3-branes at the orbifold singularity. In the near-horizon limit of N D3-branes, the $\mathbb{Z}_2$ orbifold projection leaves the AdS_5 space unchanged, and acts on the original S^5 as follows. With S^5 described by

$$\sum_{i=4}^{9} x_i^2 = 1, \tag{8.7}$$

the projection (8.2) implies that points opposite to each other on S^5 are identified with each other, including the signs given above. In this way, we obtain the space $AdS_5 \times S^5/\mathbb{Z}_2$. The orbifold action fixes a plane given by $x_i = 0$ for $i \in \{6, 7, 8, 9\}$ in the original $\mathbb{R}^6$ space. This plane intersects S^5 in a great circle S^1.

Exercise 8.1.1   Show that the $\mathbb{C}^2/\Gamma$ orbifold with a discrete subgroup $\Gamma \subset SU(2)$ preserves one half of the supersymmetry charges of flat space, while the $\mathbb{C}^3/\Gamma$ orbifold with $\Gamma \subset SU(3)$ preserves one quarter of the flat space supersymmetry charges.

Exercise 8.1.2   Show that for D3-branes at a $\mathbb{C}^2/\mathbb{Z}_n$ orbifold, the field theory has a $U(N_1) \times \cdots \times U(N_n)$ quiver symmetry. Moreover, draw the corresponding quiver diagram and show that it corresponds to the Dynkin diagram of the group A_{n-1} (i.e. $SU(n)$) of which $\mathbb{Z}_n$ is a subgroup.

A further interesting brane configuration at an orbifold singularity is given by M2-branes, as introduced in section 4.4.3, placed at a $\mathbb{C}^4/\mathbb{Z}_k$ singularity, where $\mathbb{C}^4$ involves all eight real dimensions perpendicular to the M2-brane in eleven-dimensional spacetime. We will consider this below in section 8.2.

8.1.2 Conifold

A further example is given by considering D3-branes placed at the tip, or *apex*, of a *conifold*. The conifold is given by a relation in complex space $\mathbb{C}^4$,

$$\sum_{n=1}^{4} z_n^2 = 0 \,. \tag{8.8}$$

This describes a cone since for any z_n satisfying (8.8), λz_n also satisfies (8.8) for any $\lambda \in \mathbb{C}\backslash\{0\}$. This cone has an apex at $z_n = 0$, where the surface given by (8.8) is no longer smooth. The conifold has symmetry $SO(4) \times U(1)$, which is isomorphic to $SU(2) \times SU(2) \times U(1)$. It is a standard example of a Calabi–Yau manifold, with the required three-form given by

$$\Omega = \frac{dz_2 \wedge dz_3 \wedge dz_4}{z_1}, \tag{8.9}$$

which is charged under the $U(1)$ R-symmetry. Let us identify the topological nature of the base $T^{1,1}$ of the cone (8.8). This is obtained by intersecting (8.8) with the unit sphere

$$\sum_{i=1}^{4} |z_i|^2 = 1, \tag{8.10}$$

omitting the singularity at the origin of the conifold. The group $SO(4)$ acts transitively on the intersection. Any point on the intersection is invariant only under a $U(1) \subset SO(4)$, which implies $T^{1,1} = SO(4)/U(1)$, which is isomorphic to $X_5 = (SU(2) \times SU(2))/U(1)$.

Equation (8.8) may be rewritten as

$$\det_{i,j} z_{ij} = 0 \tag{8.11}$$

using $z_{ij} = \sum_n \sigma_{ij}^n z_n$, where σ^n are the Pauli matrices for $n = 1, 2, 3$ and σ^4 is i times the unit matrix. We may solve (8.11) in terms of unconstrained variables by writing

$$z_{ij} = a_i b_j \tag{8.12}$$

with complex scalars $a_i, b_j, i, j \in \{1, 2\}$. The a_i, b_j have an additional $SU(2) \times SU(2)$ global symmetry, which is quotiented by the $U(1)$ symmetry generated by

$$a_i \mapsto e^{i\alpha} a_i, \qquad b_j \mapsto e^{-i\alpha} b_j, \tag{8.13}$$

such that the global symmetry is $(SU(2) \times SU(2))/U(1)$.

The a_i, b_j are the starting point for constructing the associated field theory. Calabi–Yau manifolds, of which the conifold is an example, preserve one quarter of the original supersymmetry. This follows from the relevant Killing spinor equations. Consequently, here we have $\mathcal{N} = 1$ supersymmetry and associate chiral $\mathcal{N} = 1$ superfields $A_i, B_j, i, j = 1, 2$, to the complex scalars a_i, b_j. In addition to the global symmetry discussed above, we have a $U(1)_R$ symmetry. Similarly to the orbifold case, the gauge symmetry of the field theory is $SU(N) \times SU(N)$ for N D3-branes at the tip of the conifold, with the A_i transforming in the $(\mathbf{N}, \overline{\mathbf{N}})$ representation and the B_j in the $(\overline{\mathbf{N}}, \mathbf{N})$ representation. This is again an example of a *quiver gauge theory*. Moreover, cancellation of the anomaly in the

$U(1)$ R-symmetry requires that the A_i and B_j each have R-charge $1/2$. The gauge theory then also involves a marginal superpotential which is uniquely fixed by the symmetries up to an overall factor,

$$W = \epsilon^{ij} \epsilon^{kl} \text{Tr} A_i B_k A_j B_l. \tag{8.14}$$

It is interesting to note that the field theory with this superpotential is associated with an IR fixed point of the renormalisation group flow obtained from the orbifold field theory (8.6) by a relevant perturbation. In fact, by adding a relevant perturbation of the form

$$W_{\text{pert}} = \frac{m}{2} (\text{Tr} \, \Phi_1^2 - \text{Tr} \, \Phi_2^2) \tag{8.15}$$

to (8.6), and integrating out the Φ_i, we obtain

$$W = -\frac{g^2}{m} \left(\text{Tr}(A_1 B_1 A_2 B_2) - \text{Tr}(B_1 A_1 B_2 A_2) \right), \tag{8.16}$$

which, up to the prefactor, coincides with the conifold superpotential (8.14).

Let us now look at the gravity side of the correspondence for D3-branes at the apex of the conifold. Equation (8.11) describes a cone whose base is a coset space $T^{1,1} = (SU(2) \times SU(2))/U(1)$. To obtain the metric of the space $T^{1,1}$, we parametrise the z_{ij} of (8.11) by

$$\begin{aligned}
z_{11} &= r^{3/2} \, e^{i/2 \, (\psi + \phi_1 + \phi_2)} \, \sin(\theta_1/2) \sin(\theta_2/2), \\
z_{12} &= r^{3/2} \, e^{i/2 \, (\psi - \phi_1 + \phi_2)} \, \cos(\theta_1/2) \sin(\theta_2/2), \\
z_{21} &= r^{3/2} \, e^{i/2 \, (\psi + \phi_1 - \phi_2)} \, \sin(\theta_1/2) \cos(\theta_2/2), \\
z_{22} &= r^{3/2} \, e^{i/2 \, (\psi - \phi_1 - \phi_2)} \, \cos(\theta_1/2) \cos(\theta_2/2),
\end{aligned} \tag{8.17}$$

with Euler angles ψ, ϕ_i, θ_i, $i = 1, 2$. The Einstein metric of the space $T^{1,1}$ is then given by [1]

$$ds_{T^{1,1}}^2 = \frac{1}{9} \left(d\psi + \sum_{i=1}^{2} \cos\theta_i d\phi_i \right)^2 + \frac{1}{6} \sum_{i=1}^{2} \left(d\theta_i^2 + \sin^2\theta_i d\phi_i^2 \right) \tag{8.18}$$

with $0 \leq \psi \leq 4\pi$, $0 \leq \theta_i \leq \pi$ and $0 \leq \phi_i \leq 2\pi$. From this metric we can read off that $T^{1,1}$ is a S^1 bundle over $S^2 \times S^2$ and thus has symmetry group $SU(2) \times SU(2) \times U(1)$. Moreover, $T^{1,1}$ is topologically equivalent to $S^2 \times S^3$: the two-cycle corresponding to S^2 is given by $\psi = 0$, $\theta_1 = \theta_2$, $\phi_1 = -\phi_2$, and the three-cycle corresponding to S^3 is given by $\theta_1 = \phi_1 = 0$.

The conifold is the cone over $T^{1,1}$; the radii of both the S^2 and S^3 spheres shrink to zero at its origin. Taking the near-horizon limit of N D3-branes placed at the apex of the cone over $T^{1,1}$, we obtain the geometry $AdS_5 \times T^{1,1}$. In exact analogy to the flat space case, we may now conjecture that type IIB supergravity on $AdS_5 \times T^{1,1}$ is dual to the $\mathcal{N} = 1$ superconformal quantum gauge theory involving the A_i, B_j and the superpotential (8.14) as discussed above.

Further aspects of the structure of the field theory may be inferred from the string theory construction [2]. In type IIB theory on $AdS_5 \times T^{1,1}$, there are two complex moduli. The first is the standard axion dilaton $\tau = C_{(0)} + ie^{-\phi}$. The second is obtained from integrating $C_{(2)}$ and $B_{(2)}$ over S^2 in $T^{1,1}$. The two integrals form the real and imaginary parts of the second modulus. Both moduli are related to the two complex coupling constants $4\pi/g_i^2 + \theta_i$, $i = 1, 2$, of the $SU(N) \times SU(N)$ gauge theory in the following way. S-duality, or more precisely its non-trivial centre, the negative unit matrix, reverses the sign of $C_{(2)}$ and $B_{(2)}$ while leaving the axion-dilaton invariant. Therefore the sum of the two gauge couplings $4\pi/g_i^2$ gives the dilaton $e^{-\phi}$, while the sum of the two theta parameters θ_i gives the axion $C_{(0)}$. The same argument implies that the difference of the real gauge couplings gives the integral over $B_{(2)}$, while the difference of the theta parameters gives the integral over $C_{(2)}$. In particular, we find for the gauge couplings

$$\frac{4\pi}{g_1^2} + \frac{4\pi}{g_2^2} = e^{-\phi}, \tag{8.19}$$

$$\frac{4\pi}{g_1^2} - \frac{4\pi}{g_2^2} = e^{-\phi}\left(-1 + \frac{1}{2\pi^2\alpha'}\int_{S^2} B_2\right). \tag{8.20}$$

Since the integral over $B_{(2)}$ corresponds to an axion, it is periodic. The second complex modulus, as introduced at the beginning of this paragraph, thus describes a torus. In the example considered above, the gauge couplings are constants which do not run. However, in chapter 9 we will consider non-conformal examples where both ϕ and $B_{(2)}$ depend on the radial variable r, which leads to a running of the gauge couplings.

8.2 M2-branes: AdS$_4$/CFT$_3$

In addition to D3-branes, the AdS/CFT correspondence can also be established for other types of branes. A very important example is provided by the branes in eleven-dimensional supergravity, which is expected to be the low-energy limit of M-theory. As discussed in chapter 4, supergravity in eleven dimensions supports M2-branes and M5-branes. In this chapter we study an explicit realisation of the AdS/CFT correspondence based on M2-branes. This correspondence relates a (2+1)-dimensional superconformal field theory (denoted by CFT$_3$) to a gravity theory on AdS_4, and is therefore referred to as the AdS$_4$/CFT$_3$ correspondence.

It is more difficult to work out the dual CFT for M2-branes compared with the examples for AdS/CFT duality which we have studied so far. In type IIB string theory, we had a dimensionless coupling constant, g_sN, controlling the interaction strength of the fundamental strings. Whereas for $g_sN \gg 1$ we viewed the branes as gravitational sources curving the surrounding spacetime, for $g_sN \ll 1$ the D-branes were just hyperplanes where open strings can end. These two views of branes allowed us to motivate the AdS/CFT correspondence and in particular to determine the dual CFT description.

In M-theory we do not have the possibility of choosing the value of the coupling, since M-theory is already the strong coupling regime of type IIA string theory. Therefore it is quite difficult to find the dual CFT description of the near-horizon limit of M2-branes. In recent years there has been significant progress in the understanding of the interactions of coincident M2-branes, despite the fact that a fundamental perturbative description is not available. In particular, the low-energy effective action for M2-branes at a $\mathbb{C}^4/\mathbb{Z}_k$ singularity has been established, which is referred to as *ABJM theory* [3], after its authors Aharony, Bergman, Jafferis and Maldacena.

8.2.1 Gravity dual for M2-branes

The starting point is given by the M2-brane solution (4.127) of eleven-dimensional supergravity which was introduced in section 4.4.3. Taking the near-horizon limit of (4.127), i.e. $r \ll L$, we may approximate the function $H(r)$ by L^6/r^6 and thus the metric reduces to

$$ds^2 = L^2 \left(\frac{1}{4} ds^2_{AdS_4} + ds^2_{S^7} \right). \tag{8.21}$$

$ds^2_{AdS_4}$ and $ds^2_{S^7}$ are the metrics of AdS_4 and S^7 with unit radius. Note that due to the different prefactors in (8.21), the radii of S^7 and of AdS_4 are not equal as was the case for D3-branes. Here, the radius of curvature of S^7 is twice the radius of curvature of AdS_4, as we may see from (8.21).

Exercise 8.2.1 Starting from the M2-brane solution (4.127), take the near-horizon limit $r \ll L$ and calculate the metric and the four-form $F_{(4)}$ in this limit. Show that the result is

$$ds^2 = \frac{r^4}{L^4} \left(-dt^2 + dx^2 + dy^2 \right) + \frac{L^2}{r^2} \left(dr^2 + r^2 d\Omega_7^2 \right), \tag{8.22}$$

$$F_{(4)} = dt \wedge dx \wedge dy \wedge dH^{-1}(r) = 6 \frac{r^5}{L^6} dt \wedge dx \wedge dy \wedge dr. \tag{8.23}$$

Exercise 8.2.2 Use the coordinate transformation $z = \frac{L^3}{2r^2}$ and compute the metric as well as the four-form $F_{(4)}$ in the coordinates (z, t, x, y, Ω_7). Show that the result is

$$ds^2 = \frac{L^2}{4z^2} \left(-dt^2 + dx^2 + dy^2 + dz^2 \right) + L^2 d\Omega_7^2, \tag{8.24}$$

$$F_{(4)} = -\frac{3}{8} \frac{L^3}{z^4} dt \wedge dx \wedge dy \wedge dz. \tag{8.25}$$

The bosonic symmetry subgroup for this supergravity solution is given by $SO(3,2) \times SO(8)$. The solution preserves sixteen Poincaré supercharges. Moreover, in the near-horizon limit where the AdS_4 factor is present, there is an enhancement of the supersymmetry by sixteen conformal supercharges.

For the strongly coupled theory of N M2-branes, the AdS/CFT correspondence predicts an interesting feature: the number of degrees of freedom scales as $N^{3/2}$ [4]. In contrast, for N D3-branes the number of degrees of freedom scales as N^2 as expected for a gauge theory. The peculiar scaling $N^{3/2}$ can be understood in terms of a gauge theory in which not all degrees of freedom are dynamical. We will consider the counting of degrees of freedom in more detail in chapter 11.

An important geometrical configuration is given by N M2-branes placed at a $\mathbb{C}^4/\mathbb{Z}_k$ orbifold singularity, as introduced in section 8.1.1. Here, the orbifold acts on the four complex coordinates z_n of $\mathbb{C}^4$ as

$$z_n \mapsto \exp\left(\frac{2\pi i}{k}\right) z_n. \tag{8.26}$$

The orbifold thus breaks the $SO(8)$ symmetry and preserves an $SU(4) \times U(1)$ symmetry. For the orbifold geometry, the near-horizon limit gives the geometry $AdS_4 \times S^7/\mathbb{Z}_k$. Let us state this conjectured duality for M2-branes in analogy to the D3-brane duality given on page 180. The subsequent sections below will then explain the ingredients of this conjectured duality. The conjecture is as follows.

$\mathcal{N} = 6$ superconformal Chern–Simons matter theory in 2+1 dimensions with gauge group $U(N) \times U(N)$ and Chern–Simons levels k and $-k$,

referred to as *ABJM theory*,

is dynamically equivalent to

M-theory on $AdS_4 \times S^7/\mathbb{Z}_k$
with N units of R-R four-form flux $F_{(4)}$ through AdS_4.

The 't Hooft coupling is given by $\lambda = \frac{N}{k}$ and is related to the AdS_4 radius L and the eleven-dimensional Planck length ℓ_p by

$$\frac{L^3}{\ell_p^3} = 4\pi\sqrt{2kN} = 4\pi k\sqrt{2\lambda}. \tag{8.27}$$

In the limit of large 't Hooft coupling, the M-theory side of the correspondence reduces to eleven-dimensional supergravity on $AdS_4 \times S^7/\mathbb{Z}_k$.

8.2.2 Dual field theory

For the supergravity solution given by (8.24), we expect the dual field theory to be a superconformal field theory in 2+1 dimensions which has a global $SO(8)$ symmetry. As noted in box 8.2, in 2+1 dimensions, a theory with $\mathcal{N}$ supersymmetries has an $SO(\mathcal{N})$ R-symmetry. Therefore the dual field theory is a theory with $\mathcal{N} = 8$ supersymmetry in 2+1 dimensions.

For a conformal theory in 2+1 dimensions, the natural candidate is a *Chern–Simons theory* as introduced in box 8.1 since the Yang–Mills theory in 2+1 dimensions has a dimensionful coupling. We note, however, that the action (8.28) breaks parity, while the supergravity solution introduced in the preceding section preserves parity. In order to have a parity-even field theory, we need a product gauge group and two gauge fields with opposite Chern–Simons levels. It will turn out that such a field theory corresponds precisely to the field theory associated with N M2-branes located at the apex $\mathbb{C}^4/\mathbb{Z}_k$ orbifold. In this

The action of Chern–Simons theory is given by

$$S_{CS} = \frac{k}{4\pi} \int d^3x \, \epsilon^{\mu\nu\rho} \text{Tr} \left(A_\mu \partial_\nu A_\rho - i\frac{2}{3} A_\mu A_\nu A_\rho \right), \tag{8.28}$$

where k is the *Chern–Simons level*. A_μ may be taken to be a $U(N)$ gauge field transforming as

$$A_\mu(x) \mapsto A'_\mu(x) = g(x) \left(A_\mu + i\partial_\mu \right) g^{-1}(x), \qquad g(x) \in U(N). \tag{8.29}$$

Exercise 8.2.3   Show that the action (8.28) is invariant under an *infinitesimal* gauge transformation $g(x) = 1 + i\alpha^a(x)T_a$.

Nevertheless, under a *finite* gauge transformation, (8.28) transforms as

$$S \mapsto S' = S + 2\pi k \mathcal{I}, \tag{8.30}$$

where

$$\mathcal{I} = -\frac{1}{24\pi^2} \int d^3x \, \epsilon^{\mu\nu\rho} \text{Tr} \left((\partial_\mu g^{-1})g (\partial_\nu g^{-1})g (\partial_\rho g^{-1})g \right). \tag{8.31}$$

$\mathcal{I}$ takes only integer values, such that $\exp(iS) = \exp(iS')$ under large gauge transformations provided that $k \in \mathbb{Z}$.

case, the global symmetry is broken to $SU(4) \times U(1)$, which corresponds to $\mathcal{N} = 6$ supersymmetry in 2+1 dimensions.

Based on these considerations, we now construct the field theory Lagrangian involved in the M2-brane duality, which is referred to as *ABJM field theory*. This is a $U(N) \times U(N)$ gauge theory with a Chern–Simons term for each gauge group factor. The two Chern–Simons terms have equal but opposite levels, k and $-k$, which we denote by $U(N)_k \times U(N)_{-k}$. The starting point for constructing the ABJM field theory is the $\mathcal{N} = 2$ supersymmetric completion of Chern–Simons theory. In addition to the gauge field, the vector multiplet of $\mathcal{N} = 2$ supersymmetry in 2+1 dimensions contains a real scalar field σ, two real (Majorana) gauginos, and an auxiliary real scalar field D, all in the adjoint representation of the gauge group. Combining the gauginos into one complex fermionic field χ, the action for each factor of the gauge group is given by

$$S_{CS}^{\mathcal{N}=2} = \frac{k}{4\pi} \int d^3x \, \text{Tr} \left(\epsilon^{\mu\nu\rho} \left(A_\mu \partial_\nu A_\rho - i\frac{2}{3} A_\mu A_\nu A_\rho \right) + i\bar{\chi}\chi - 2D\sigma \right). \tag{8.32}$$

The Lagrangian of the full ABJM theory is conveniently written in $\mathcal{N} = 2$ superspace in 2+1 dimensions, as introduced in box 8.2. The ABJM theory includes the following fields.

- Two $\mathcal{N} = 2$ vector superfields V_i with field content as given in box 8.2. There is one of these fields for each gauge group, hence $i = 1, 2$ labels the $U(N)$ factor.
- Two $\mathcal{N} = 2$ chiral superfields Φ_i, each of which is in the adjoint representation.

Box 8.2 $\mathcal{N} = 2$ **algebra and superspace in** $2 + 1$ **dimensions**

In 2+1 dimensions, the $\mathcal{N}$ supersymmetric algebra contains $\mathcal{N}$ Majorana spinors and has $SO(\mathcal{N})$ R-symmetry. Consequently, the (2+1)-dimensional $\mathcal{N} = 2$ supersymmetry algebra includes two Majorana spinors, which we combine into a single complex spinor. The $\mathcal{N} = 2$ algebra in 2+1 dimensions reads

$$\left\{ Q_\alpha, \bar{Q}_\beta \right\} = 2\gamma^\mu_{\alpha\beta} P_\mu, \tag{8.33}$$

where $\gamma^0 = \sigma_2$, $\gamma^1 = i\sigma_1$, and $\gamma^2 = i\sigma_3$, with σ_1, σ_2, and σ_3 the usual Pauli matrices, and $\alpha, \beta = 1, 2$ the spinor index. The (2+1)-dimensional $\mathcal{N} = 2$ supersymmetry algebra (8.33) is obtained from dimensional reduction of the (3+1)-dimensional $\mathcal{N} = 1$ supersymmetry algebra discussed in chapter 3. Q_α is precisely the complex spinor charge of the (3+1)-dimensional $\mathcal{N} = 1$ supersymmetry algebra. Equation (8.33) gives rise to a superspace with complex spinors θ_α, $\bar{\theta}^\alpha$. The superspace covariant derivatives are then

$$D_\alpha = \frac{\partial}{\partial\theta^\alpha} + \left(\gamma^\mu\bar{\theta}\right)_\alpha \frac{\partial}{\partial x^\mu}, \qquad \bar{D}_\alpha = -\frac{\partial}{\partial\bar{\theta}^\alpha} - (\theta\gamma^\mu)_\alpha \frac{\partial}{\partial x^\mu}. \tag{8.34}$$

Chiral superfields Φ obey $\bar{D}_\alpha \Phi = 0$. In this superspace we may define the following superfields:

- an $\mathcal{N} = 2$ vector superfield which includes a vector potential A_μ, a real scalar field σ, two real (Majorana) gauginos, and an auxiliary real scalar field D, all in the adjoint representation of the gauge group;
- an $\mathcal{N} = 2$ chiral superfield which includes two real (Majorana) fermions, two real scalars, and a complex auxiliary scalar F.

- Four $\mathcal{N} = 2$ chiral superfields, A_1, A_2, B_1 and B_2, where A_1 and A_2 are in the bifundamental $(\mathbf{N}, \overline{\mathbf{N}})$ representation and B_1 and B_2 are in the anti-bifundamental $(\overline{\mathbf{N}}, \mathbf{N})$ representation.

We divide the action into three pieces,

$$\mathcal{S}_{\text{ABJM}} = \mathcal{S}_{\text{CS}} + \mathcal{S}_{\text{bifund}} + \mathcal{S}_{\text{pot}}. \tag{8.35}$$

The three action contributions are given in $\mathcal{N} = 2$ superspace as follows. The Chern–Simons contribution is the action (8.32) for each gauge group factor. To write this in $\mathcal{N} = 2$ superspace, it is necessary to introduce an auxiliary integration parameter t. For the two gauge fields, for each of which the component action is given by (8.32), we then have

$$\mathcal{S}_{\text{CS}} = -i\frac{k}{4\pi} \int d^3x \, d^4\theta \int_0^1 dt \, \text{Tr} \left(V_1 \bar{D}^\alpha \left(e^{tV_1} D_\alpha e^{-tV_1} \right) \right.$$
$$\left. - V_2 \bar{D}^\alpha \left(e^{tV_2} D_\alpha e^{-tV_2} \right) \right). \tag{8.36}$$

In addition, the remaining two action contributions to (8.35) are given in $\mathcal{N} = 2$ superspace by

$$\mathcal{S}_{\text{bifund}} = -\int d^3x \, d^4\theta \, \text{Tr} \left(\bar{A}_a e^{-V_1} A_a e^{V_2} + \bar{B}_a e^{-V_2} B_a e^{V_1} \right), \tag{8.37}$$

$$\mathcal{S}_{\text{pot}} = \int d^3x \, d^2\theta \, W + \text{c.c.}, \tag{8.38}$$

with the superpotential

$$W = -\frac{k}{8\pi}\text{Tr}\left(\Phi_1^2 - \Phi_2^2\right) + \text{Tr}\left(B_a\Phi_1 A_a\right) + \text{Tr}\left(A_a\Phi_2 B_a\right). \tag{8.39}$$

In S_{bifund} and the superpotential, summation over $a = 1, 2$ is implicit. All traces are taken in the fundamental representation. Without the superpotential the action has $\mathcal{N} = 2$ supersymmetry. The chiral superfields Φ_i combine with the corresponding V_i to form $\mathcal{N} = 4$ vector multiplets. However, the Chern–Simons terms only preserve $\mathcal{N} = 3$ supersymmetry, since the supersymmetry partner of (8.36) is an additional superpotential contribution involving $\text{Tr}\left(\Phi_1^2 - \Phi_2^2\right)$ as given in (8.39), which breaks $\mathcal{N} = 4$ to $\mathcal{N} = 3$. The form of the superpotential (8.39) is completely fixed by $\mathcal{N} = 3$ supersymmetry. The theory has an $SO(3)_R \cong SU(2)_R$ R-symmetry.

The fields Φ_i do not have kinetic terms, hence at low energy they can be integrated out, which means we may use their equation of motion to eliminate them from the original action. This leads to a supersymmetry enhancement to $\mathcal{N} = 6$ supersymmetry, as we now show. In fact, integrating out the Φ_i as described, the superpotential becomes

$$W_{\text{ABJM}} = \frac{2\pi}{k}\epsilon^{ab}\epsilon^{\dot{a}\dot{b}}\,\text{Tr}\left(A_a B_{\dot{a}} A_b B_{\dot{b}}\right), \tag{8.40}$$

which clearly exhibits an $SU(2)$ symmetry acting on A_a and a separate $SU(2)$ symmetry acting on $B_{\dot{a}}$. We denote this symmetry as $SU(2)_A \times SU(2)_B$. The R-symmetry of the theory, $SO(3)_R \cong SU(2)_R$, does not commute with the $SU(2)_A \times SU(2)_B$: under the $SU(2)_R$ symmetry, (A_1, B_1^*) and (A_2, B_2^*) are each a doublet. We thus conclude that the full symmetry is $SU(4)$, under which (A_1, A_2, B_1^*, B_2^*) transforms in the representation **4**. The supercharges also transform under this $SU(4)$, hence the full R-symmetry is $SU(4)_R \equiv SO(6)_R$, and hence the theory is in fact $\mathcal{N} = 6$ supersymmetric. An important property of the model is thus that at low energies the supersymmetry is enhanced.

The theory additionally has a $U(1)_b$ *baryon number* symmetry under which $A_i \mapsto e^{i\alpha}A_i$ and $B_i \mapsto e^{-i\alpha}B_i$. Remarkably, due to the product gauge group, the theory also has a parity symmetry in spite of the Chern–Simons terms present. The parity transformation involves inverting one spatial coordinate (say $x^1 \to -x^1$), exchanging the two gauge groups, and performing charge conjugation on all of the fields. Finally, the moduli space of the theory is $\mathbb{C}^4/\mathbb{Z}_k$, where the $\mathbb{Z}_k$ acts as $(A_1, A_2, B_1^*, B_2^*) \mapsto e^{2\pi i/k}(A_1, A_2, B_1^*, B_2^*)$, where here A_a and B_a denote only the scalar component of the corresponding superfields.

8.2.3 * Brane construction for the ABJM theory

Let us consider the brane construction leading to the $\mathcal{N} = 6$ Chern–Simons matter theory with gauge group $U(N)_k \times U(N)_{-k}$ as described above [3], making use of the string theory concepts introduced in chapter 4. The starting point is the type IIB brane configuration given in table 8.1. The x^6 direction is a circle, and the NS5- and NS5'-branes are separated in the x^6 direction. The N D3-branes, which are extended in the x^6 direction, break on the NS5-branes. The k D5-branes and the NS5'-brane are at the same position in x^6.

	0	1	2	3	4	5	6	7	8	9
Table 8.1 Type IIB brane construction leading to ABJM theory										
NS5	•	•	•	•	•	•	–	–	–	–
NS5′	•	•	•	•	•	•	–	–	–	–
N D3	•	•	•	–	–	–	•	–	–	–
k D5	•	•	•	•	•	–	–	–	–	•

The D3-branes, together with the NS5- and NS5′-branes, give rise to an $\mathcal{N} = 4$ supersymmetric $U(N) \times U(N)$ Yang–Mills theory in 2+1 dimensions. The bosonic part of the $\mathcal{N} = 4$ vector multiplet in each $U(N)$ gauge group consists of the (2+1)-dimensional components of the D3-brane worldvolume gauge field together with the three real scalars describing each D3-brane's position in the (x^3, x^4, x^5) directions. Each $\mathcal{N} = 4$ vector multiplet consists of an $\mathcal{N} = 2$ vector multiplet V_i and an $\mathcal{N} = 2$ chiral multiplet Φ_i. The real scalars are the two real scalars in Φ_i plus the real scalar σ_i in V_i, which thus form a vector representation of $SO(3)_R$. Similarly, the auxiliary fields D and F form a vector of the R-symmetry.

The theory also has (anti-)bifundamental $\mathcal{N} = 2$ chiral multiplets, coming from strings stretched between the two stacks of D3-branes. These are the fields A_a and B_a of the last subsection, with $a = 1, 2$. The k D5-branes coincident with the NS5′-branes introduce massless D3/D5 strings, and break the supersymmetry to $\mathcal{N} = 2$. The field theory thus has k massless $\mathcal{N} = 2$ chiral multiplets in the fundamental and k massless $\mathcal{N} = 2$ chiral multiplets in the anti-fundamental of *each* $U(N)$ factor.

This construction gives rise to Chern–Simons theory in the following way. If the same mass is given to both the fundamental and anti-fundamental fields, then there is a parity anomaly which corresponds precisely to Chern–Simons terms being present at low energies. This requires real masses of equal sign. The deformation of the brane construction that produces such masses is to bind the k D5-branes to the NS5′-brane, producing a $(1, k)$5-brane. A bound state of this type was introduced in section 4.3.2.

To preserve $\mathcal{N} = 2$ supersymmetry, the $(1, k)$5-brane must be tilted at an angle θ in the $(5, 9)$ plane. This rotation is denoted by $[5, 9]_\theta$. The angle θ depends on the complex axion-dilaton $\tau = C_{(0)} + i\exp(-\phi)$ as $\theta = \arg(\tau) - \arg(k + \tau)$, where $\exp(\phi) = g_s$. In what follows, we set $\exp(\phi) = g_s = 1$ and $C_{(0)} = 0$, which implies $\tau = i$. Such a deformation actually gives the fundamental and anti-fundamental fields infinite mass. Integrating out these fields then gives rise to Chern–Simons terms with levels k and $-k$ for the two $U(N)$ gauge groups. Moreover, we may enhance the supersymmetry to $\mathcal{N} = 3$ if we additionally rotate the $(1, k)$5-brane by the same angle θ in the $(3, 7)$ and $(4, 8)$ planes. We thus arrive at the brane construction of table 8.2.

The field theory associated with this setup is an $\mathcal{N} = 3$ $U(N)_k \times U(N)_{-k}$ Yang–Mills theory with Chern–Simons terms and four massless bifundamental matter multiplets (A_a, B_b). As we saw above in section 8.2.2, when integrating out the Φ_i fields, at low energies this theory flows to the $\mathcal{N} = 6$ superconformal $U(N)_k \times U(N)_{-k}$ Chern–Simons

| Table 8.2 | ABJM brane construction in IIB theory | | | | | | | | |

	0	1	2	3	4	5	6	7	8	9
NS5	•	•	•	•	•	•	–	–	–	–
$(1,k)5$	•	•	•	$[3,7]_\theta$	$[4,8]_\theta$	$[5,9]_\theta$	–	–	–	–
N D3	•	•	•	–	–	–	•	–	–	–

theory with the same bifundamental matter content. The easiest way to see this happen in the brane setup is to T-dualise to type IIA theory, which is described equivalently by M-theory with the eleventh dimension compactified on a small circle. Changing from the type IIA to the M-theory description with compacitified eleventh dimension is referred to as 'lifting to M-theory'.

Performing a T-duality along the x^6 direction, the N D3-branes become N D2-branes, and subsequently M2-branes when lifting to M-theory, whereas the $NS5$- and $(1,k)5$-branes are mapped to a non-trivial geometry through this procedure. The spacetime is now $\mathbb{R}^{1,2} \times X_8$, where the M2-branes are extended along $\mathbb{R}^{1,2}$ and the space X_8 preserves 3/16 of the 32 supersymmetries of M-theory. We thus expect the M2-branes' worldvolume theory to have $\mathcal{N} = 3$ supersymmetry. However, the space X_8 has a singularity which locally is $\mathbb{C}^4/\mathbb{Z}_k$. In the low-energy limit, we retain only this singular contribution to the X_8 space. $\mathbb{C}^4/\mathbb{Z}_k$ preserves 12 supersymmetries, or 3/8 of the 32 supersymmetries of M-theory. Twelve real supercharges is of course the correct amount for a (2+1)-dimensional $\mathcal{N} = 6$ supersymmetric theory. This corresponds to the enhancement of supersymmetry that we saw in the field theory.

Recall also that the moduli space of the $\mathcal{N} = 6$ Chern–Simons matter theory is precisely $\mathbb{C}^4/\mathbb{Z}_k$. Furthermore, $\mathbb{C}^4 \cong \mathbb{R}^8$ has an $SO(8)$ isometry, of which only $SU(4) \times U(1)$ remains after the $\mathbb{Z}_k$ orbifold. These symmetries match the $SU(4)_R \times U(1)_b$ symmetry of the $\mathcal{N} = 6$ Chern–Simons theory. The central conclusion is, therefore, that the $\mathcal{N} = 6$ superconformal $U(N)_k \times U(N)_{-k}$ Chern–Simons matter theory of section 8.2.2 describes the low-energy dynamics of N coincident M2-branes at the $\mathbb{C}^4/\mathbb{Z}_k$ singularity.

Moreover, recalling that in the field theory, the $\mathbb{Z}_k$ acts on the bifundamentals as $(A_1, A_2, B_1^*, B_2^*) \mapsto e^{2\pi i/k}(A_1, A_2, B_1^*, B_2^*)$, and also that they transform as a **4** of $SU(4)_R$, we may identify (z^1, z^2, z^3, z^4) with (A_1, A_2, B_1^*, B_2^*), where here A_a and B_a represent the bosonic components of the corresponding superfields. The $U(1)_b$ symmetry of the field theory thus appears as a phase shift $z^i \mapsto e^{i\alpha} z^i$ which turns out to be equivalent to shifts in the eleventh compactified dimension.

8.2.4 A special limit

A special limit of the ABJM construction arises when $k^5 \gg N$ with k the Chern–Simons level. The radius of the compactified eleventh dimension is of the order

---

Box 8.3   **Projective spaces**

Generally, the projective space of a vector space V is the set of lines passing through the origin of V. The simplest example is the real projective space $\mathbb{RP}^2$. This has three equivalent definitions.

(1) The set of all lines in $\mathbb{R}^3$ passing through the origin at $(0, 0, 0)$.
(2) The set of points on S^2 with antipodal points identified.
(3) The set of equivalence classes of $\mathbb{R}^3 \setminus (0, 0, 0)$, with two points P given by (x_1, x_2, x_3) and P' given by (x_1', x_2', x_3') being equivalent if and only if there is a real number λ such that $(x, y, z) = (\lambda x', \lambda y', \lambda z')$.

This is generalised straightforwardly to $\mathbb{RP}^n$ in arbitrary dimensions. Similarly, this is generalised to complex numbers: the complex projective space $\mathbb{CP}^n$ is given by the set of equivalence classes of $\mathbb{C}^{n+1} \setminus \{0\}$ with two points P given by $(x_1, \ldots, x_{n+1})$ and P' given by $(x_1', \ldots, x_{n+1}')$ being equivalent if and only if there is a complex number λ such that $x_i = \lambda x_i'$ for all i.

$L/(k\ell_p) \propto (Nk)^{1/6}/k$ in Planck units. For $k^5 \gg N$, this becomes small. This means that M-theory can be replaced by type IIA theory. Let us explain this limit in some detail.

We note that the sphere S^7 may be written as an S^1 fibre over the *projective space* $\mathbb{CP}^3$, which is defined in box 8.3. The S^7 metric may be written as

$$ds_{S^7}^2 = L^2(d\phi' + \omega)^2 + L^2 ds_{\mathbb{CP}^3}^2, \tag{8.41}$$

where in terms of the complex coordinates z_n, $n = 1, \ldots, 4$ on the $\mathbb{C}^4$ perpendicular to the M2-branes we have

$$ds_{\mathbb{CP}^3}^2 = \frac{1}{r^2} \sum_n dz_n d\bar{z}_n - \frac{1}{r^4} \left| \sum_n z_n d\bar{z}^n \right|^2, \qquad r^2 \equiv \sum_{n=1}^{4} |z_n|^2, \tag{8.42}$$

$$d\phi' + \omega \equiv \frac{i}{2r^2}(z_n d\bar{z}_n - \bar{z}_n dz_n), \qquad J = d\omega = id\left(\frac{z_n}{r}\right) \wedge d\left(\frac{\bar{z}_n}{r}\right), \tag{8.43}$$

where ϕ' is periodic with period 2π and J corresponds to the Kähler form on $\mathbb{CP}^3$. To perform the $\mathbb{Z}_k$ orbifold quotient as in (8.26), we write $\phi' = \phi/k$, and the metric becomes

$$ds_{S^7/\mathbb{Z}_k}^2 = \frac{L^2}{k^2}(d\phi + k\omega)^2 + L^2 ds_{\mathbb{CP}^3}^2. \tag{8.44}$$

We read off from (8.44) that in Planck units ℓ_p, the radius of S^1 is given by $L/(k\ell_p)$. This implies that for $k^5 \gg N$ the radius of the circle S^1 in the eleventh dimension becomes very small, and M-theory as given in the duality stated on page 280 may be replaced by

---

type IIA string theory on $AdS_4 \times \mathbb{CP}^3$, with N units of R-R four-form flux $F_{(4)}$ through AdS_4 and k units of R-R two-form flux $F_{(2)}$ through a $\mathbb{CP}^1 \subset \mathbb{CP}^3$.

Since the string coupling is related to N and k via $g_s \propto (N/k^5)^{1/4}$, it becomes small in the limit $k^5 \gg N$.

8.3 Gravity duals of conformal field theories: further examples

8.3.1 M5-branes: AdS$_7$/CFT$_6$

The second type of branes present in M-theory and eleven-dimensional supergravity are M5-branes, which are the magnetic dual of the M2-branes as introduced in section 4.4.3. In the near-horizon limit $r \ll L$ we may approximate $H(r) = L^3/r^3$ and, consequently, the corresponding solution of eleven-dimensional supergravity as given in (4.129) reads

$$\mathrm{d}s^2 = \frac{r}{L}\eta_{\mu\nu}\mathrm{d}x^\mu \mathrm{d}x^\nu + \frac{L^2}{r^2}(\mathrm{d}r^2 + r^2 \mathrm{d}\Omega_4^2). \tag{8.45}$$

With the coordinate transformation

$$z \equiv \frac{2L^{3/2}}{r^{1/2}} \tag{8.46}$$

this metric becomes

$$\mathrm{d}s^2 = \frac{4L^2}{z^2}\left(\eta_{\mu\nu}\mathrm{d}x^\mu \mathrm{d}x^\nu + \mathrm{d}z^2\right) + L^2 \mathrm{d}\Omega_4^2, \tag{8.47}$$

which corresponds to $AdS_7 \times S^4$. The radius of S^4 is L while the radius of AdS_7 is $2L$. Moreover, there are N units of four-form flux on S^4. The bosonic subgroup of the symmetries of this supergravity solution is $SO(6,2) \times SO(5)$.

The dual field theory is a six-dimensional conformal field theory with R-symmetry group $SO(5)$. The theory preserves sixteen Poincaré supercharges which may be grouped into two left-handed supersymmetry generators transforming in the spinorial representation $\mathbf{4}_l$ of $SO(5)$. The theory is therefore known as $\mathcal{N} = (2,0)$ theory. The supersymmetry algebra has one irreducible massless representation, a tensor multiplet which consists of a two-form, five real scalars and the associated fermions. A Lagrangian formulation for the theory corresponding to N M5-branes is not known. Nevertheless, on the gravity side it is possible to consider the Kaluza–Klein reduction on S^4 to obtain the spectrum of the dual chiral primary operators, in analogy to the reduction performed in chapter 5 for S^5. Since here the Kaluza–Klein reduction involves only particles of spin less than two in small supersymmetry representations, the dual operators are protected against quantum corrections.

Note that for N coincident M5-branes, the degrees of freedom scale as N^3. So far it has not been possible to reproduce this result within field theory.

8.3.2 D1/D5 system: AdS$_3$/CFT$_2$

A further example of AdS/CFT correspondence is the D1/D5-brane system which gives rise to an AdS$_3$/CFT$_2$ duality. This allows us to make full use of the infinite-dimensional conformal symmetry of the two-dimensional CFT involved.

The brane setup for this duality is shown in table 8.3. We consider IIB string theory on $\mathbb{R}^{4,1} \times S^1 \times M^4$, where M^4 is an internal compact manifold which may be taken to be the four-torus T^4. We wrap N_5 D5-branes on $S^1 \times M^4$, and N_1 D1-branes on S^1. This setup preserves eight of the original thirty-two supercharges. When the length scale associated with M^4 is small compared to S^1, the low-energy dynamics of this system is described by a theory on the $(1 + 1)$-dimensional intersection. Acccording to standard weak coupling open string quantisation, this is a $U(N_1) \times U(N_5)$ supersymmetric gauge theory, which flows to a non-trivial CFT in the IR. This theory has $(4, 4)$ supersymmetry in 1+1 dimensions, with four left-handed and four right-handed supercharges. The central charge of this theory can be calculated using standard supersymmetry and CFT techniques. The result is $c = 6N_1N_5$.

On the gravity side, the solution of type IIB supergravity which corresponds to the D1/D5-brane system is given by

$$\mathrm{d}s^2 = (H_1H_5)^{-1/2}(\mathrm{d}t^2 + \mathrm{d}x_5^2) + (H_1H_5)^{1/2}\mathrm{d}x^i\mathrm{d}x^i + (H_1/H_5)^{1/2}\mathrm{d}s_{M^4}^2, \tag{8.48}$$

$$H_{(3)} = 2Q_5\mathrm{dVol}(S^3) + 2Q_1e^{-2\phi} *_6 \mathrm{dVol}(S^3), \tag{8.49}$$

$$e^{-2\phi} = H_5/H_1. \tag{8.50}$$

Here $\mathrm{dVol}(S^3)$ is the volume form on the unit three-sphere and $*_6$ is the Hodge dual in six dimensions. $H_{(3)}$ is the three-form in the type IIB supergravity action. The coordinates wrapped by the D1-branes are (t, x^5). The four non-compact directions are denoted by x^i, $i = 1, \ldots, 4$. H_1 and H_5 are harmonic functions of the radial coordinate r given by $r^2 = \sum_i(x^i)^2$,

$$H_1(r) = 1 + \frac{Q_1}{r^2}, \qquad Q_1 = \frac{(2\pi)^4 g_s N_1 \alpha'^3}{V_4}, \tag{8.51}$$

$$H_5(r) = 1 + \frac{Q_5}{r^2}, \qquad Q_5 = g_s N_5 \alpha'. \tag{8.52}$$

Q_1 and Q_5 provide the length scale $L^2 = (Q_1Q_5)^{1/2}$. In the usual Maldacena limit $\alpha' \to 0$ with $u = r/\alpha'$ fixed, we may drop the 1 in $H_1(r)$, $H_5(r)$ and get

$$\mathrm{d}s^2 = \frac{r^2}{L^2}(\mathrm{d}t^2 + \mathrm{d}x_5^2) + \frac{L^2}{r^2}\mathrm{d}r^2 + L^2\mathrm{d}\Omega_3^2 + (Q_1/Q_5)^{1/2}\mathrm{d}s_{M_4}^2, \tag{8.53}$$

Table 8.3	D1/D5-brane configuration									
	0	1	2	3	4	5	6	7	8	9
N_1 D1	•	–	–	–	–	•	–	–	–	–
N_5 D5	•	–	–	–	–	•	•	•	•	•

$$H_{(3)} = 2Q_5(\text{dVol}_3 + i *_6 \text{dVol}_3), \tag{8.54}$$
$$e^{2\phi} = Q_1/Q_5. \tag{8.55}$$

The metric obtained corresponds to the space $AdS_3 \times S^3 \times M^4$.

We can also calculate the conformal anomaly for this solution using the methods of holographic renormalisation introduced in section 5.5. In fact, for AdS_3 this result was obtained even before the AdS/CFT correspondence [5] and was found to be $c = 3L/2G$. In the present configuration, this gives $c = 6N_1N_5$ which agrees precisely with the field theory result.

8.4 Towards non-conformal field theories

8.4.1 Duality for Dp-branes with $p \neq 3$

The examples of gauge/gravity duality we have considered so far involve brane systems whose near-horizon limit gives rise to an Anti-de Sitter space. Consequently, the dual field theory is conformal. Here we turn to the near-horizon limit of Dp-branes with $p \neq 3$. In this case, in the string frame the near-horizon limit no longer involves an Anti-de Sitter space. The dual field theory is thus no longer conformal, which is obvious from the fact that in dimensions other than $3 + 1$, the gauge coupling is dimensionful and runs with the energy scale.

We embed N coincident Dp-branes into flat (9+1)-dimensional space along the directions $0, 1, \ldots, p$ as shown in table 8.4.

These Dp-branes break half of the thirty-two supercharges preserved by the ten-dimensional space. The sixteen preserved supercharges correspond to the Poincaré supercharges of the dual field theory. Since the symmetry in the $9 - p$ transversal directions corresponds to the R-charge of this theory, it has an $SO(9 - p)$ global symmetry. In addition, the field theory is again expected to be an $SU(N)$ Yang–Mills theory which we may derive by compactifying $\mathcal{N} = 1$ Super Yang–Mills theory in ten dimensions to $p + 1$ spacetime dimensions. The resulting theory has a gauge field, $9 - p$ scalars and fermions all transforming in the adjoint representation of the gauge group.

Also in this case we may motivate the correspondence by considering the N Dp-branes from the two different perspectives, i.e. from both the open and the closed string perspectives. From the open string point of view, the dynamics is governed by the

Table 8.4 Embedding of N coincident Dp-branes in flat ten-dimensional spacetime

	0	1	...	p	$p+1$	...	8	9
N Dp	•	•	•	•	–	–	–	–

DBI action describing the open strings attached to the Dp-brane, as well as by type II supergravity in ten dimensions.

Exercise 8.4.1   By expanding the DBI action for a Dp-brane,

$$S_{\text{DBI}} = -\tau_p \int d^{p+1}\xi \, e^{-\phi} \sqrt{-\det(\eta_{ab} + 2\pi\alpha' F_{ab})}, \tag{8.56}$$

considering a flat embedding in Minkowski space with $B = 0$, and

$$\tau_p = (2\pi)^{-p}\alpha'^{-\frac{p+1}{2}}, \qquad e^{\phi} = g_{\text{s}}, \tag{8.57}$$

and using the methods of chapter 4, show that the leading term

$$S_{\text{YM}} = -\frac{1}{4g_{\text{YM}}^2} \int d^{p+1}\xi F^2 \tag{8.58}$$

has the Yang–Mills coupling

$$g_{\text{YM}}^2 = (2\pi)^{p-2} g_{\text{s}} \alpha'^{\frac{p-3}{2}}. \tag{8.59}$$

Note that the Yang–Mills coupling is dimensionful as expected. Therefore we consider the effective dimensionless 't Hooft coupling constant

$$\lambda_{\text{eff}} = g_{\text{YM}}^2 N u^{p-3} = \lambda u^{p-3}, \tag{8.60}$$

where u is an energy scale in the field theory. Alternatively, u may correspond to a vacuum expectation value of one of the scalars. The next step is to take the decoupling limit as in the case of D3-branes (5.10),

$$\alpha' \to 0, \qquad u = r/\alpha' = \text{fixed}. \tag{8.61}$$

The open string perspective requires the effective string coupling λ_{eff} to be small, $\lambda_{\text{eff}} \ll 1$. This is equivalent to

$$\begin{aligned} u &\gg \lambda^{1/(3-p)} \quad &\text{for } p < 3, \\ u &\ll \lambda^{1/(p-3)} \quad &\text{for } p > 3. \end{aligned} \tag{8.62}$$

For $p \leq 3$ the limit (8.61) implies directly a decoupling of the Yang–Mills theory from the bulk gravity theory since the ten-dimensional Newton constant goes to zero. For $p > 3$, (8.59) implies $g_{\text{s}} \to \infty$. The analysis of this case requires a duality transformation, as we will discuss below.

Now consider the closed string perspective, i.e. the Dp-branes as heavy objects in type II supergravity which curve the space around them. To obtain the explicit form of the metric, we have to solve the equations of motion of type II supergravity. The relevant part of the action of section 4.2.3 reads, in string frame for $p \neq 3$,

$$S_{\text{II}} = \frac{1}{2\kappa_{10}^2} \int d^{10}x \sqrt{-g} \left[e^{-2\phi} \left(R + 4 \, (\partial\phi)^2 - \frac{1}{2}|H_{(3)}|^2 \right) - \frac{1}{2}|F_{(p+2)}|^2 \right], \tag{8.63}$$

where R is the Ricci scalar, ϕ is the dilaton, $H_{(3)}$ is the NS-NS three-form field strength, and $F_{(p+2)}$ is the R-R $(p+2)$-form field strength. The general asymptotically flat solution describing N coincident Dp-branes was given in chapter 4, equations (4.115)–(4.118). For $p < 3$ these solutions admit a decoupling limit [6]

$$g_s \to 0, \qquad \alpha' \to 0, \qquad g_{YM}^2 N = \text{fixed}, \qquad u \equiv \frac{r}{\alpha'} = \text{fixed}. \qquad (8.64)$$

In that limit, we may approximate H_p as given by (4.120) by

$$H_p = \frac{L_p^{7-p}}{r^{7-p}} = \frac{(4\pi)^{\frac{5-p}{2}} \Gamma(\frac{7-p}{2}) g_s N \alpha'^{\frac{7-p}{2}}}{r^{7-p}} \equiv \frac{1}{\alpha'} \left(\frac{u_p}{u} \right)^{7-p} \qquad (8.65)$$

where L_p is given by (4.122). Equation (8.65) implicitly defines u_p. Inserting H_p into the solution (4.115)–(4.118) gives the near-horizon geometry of the Dp-branes,

$$ds^2/\alpha' = \left(\frac{u}{u_p} \right)^{(7-p)/2} \eta_{\mu\nu} dx^\mu dx^\nu + \left(\frac{u_p}{u} \right)^{(7-p)/2} \left(du^2 + u^2 d\Omega_{8-p}^2 \right), \qquad (8.66)$$

$$e^\phi = g_s \alpha'^{\frac{p-3}{2}} \left(\frac{u}{u_p} \right)^{(7-p)(p-3)/4}, \qquad (8.67)$$

$$C_{(p+1)} = \alpha'^2 \left(\frac{u}{u_p} \right)^{7-p} dx^0 \wedge \cdots \wedge dx^p. \qquad (8.68)$$

The conjectured duality then states that this type II supergravity solution is dual to the worldvolume Yang–Mills theory in $d = p + 1$ dimensions. From the field theory point of view, the radial coordinate u corresponds to an energy scale. The UV limit of the field theory corresponds to $u \to \infty$. For $p < 3$ the effective coupling vanishes in this limit and the theory becomes free. For $p > 3$, the coupling increases in this limit and a dual description is required, which we now introduce. This is related to the fact that for $p > 3$, Super Yang–Mills theories are non-renormalisable and new degrees of freedom appear at short distances.

Exercise 8.4.2 Show that the near-horizon Dp-brane metric is conformal to $AdS_{p+2} \times S^{8-p}$. For this purpose, perform a Weyl transformation to the *dual frame*

$$ds_{\text{dual}}^2 = (Ne^\phi)^{\frac{2}{p-7}} ds^2. \qquad (8.69)$$

If we also change coordinates from u to U, where

$$U^2 = \left(\frac{5-p}{2} \right)^2 u_p^{p-7} u^{5-p}, \qquad (8.70)$$

then the metric becomes

$$ds_{\text{dual}}^2/\alpha' = A_p \left(\frac{dU^2}{U^2} + U^2 \eta_{\mu\nu} dx^\mu dx^\nu + \left(\frac{5-p}{2} \right)^2 d\Omega_{8-p}^2 \right), \qquad (8.71)$$

with

$$A_p \equiv \alpha'^{\frac{p-3}{p-7}} (Ng_s)^{\frac{2}{p-7}} u_p^2 \left(\frac{2}{5-p} \right)^2. \qquad (8.72)$$

In the dual frame, the metric (8.71) is obviously that of $AdS_{p+2} \times S^{8-p}$. We also note that in the dual frame, the dilaton reads

$$e^\phi = B_p U^{(p-3)(7-p)/2(5-p)}, \qquad B_p \equiv g_s \alpha'^{\frac{p-3}{2}} U_p^{\frac{(7-p)(p-3)}{2(5-p)}} \left(\frac{2}{5-p}\right)^{\frac{(7-p)(p-3)}{2(5-p)}}. \tag{8.73}$$

This means that the dilaton runs, which corresponds to a running coupling in the dual theory. Nevertheless, since e^ϕ has a power-law scaling, we may define a generalised conformal symmetry under which the gauge couping also transforms. The theory will be invariant under this generalised conformal symmetry.

In the dual frame, we may see that for $p = 4$, there is also a decoupling limit which corresponds to $N \to \infty$ for fixed λ. In this limit $e^\phi \ll 1$. The case $p = 5$ is singular since U becomes a constant. For $p = 6$, the gravity theory does not decouple from the Yang–Mills theory. This is best seen by lifting the D6-brane in IIA theory to M-theory: the decoupling limit requires keeping $g_{YM}^2 \propto g_s \alpha'^{3/2}$ fixed. However, in this case the eleven-dimensional Planck length $l_p = g_s^{1/3} \alpha'^{1/2}$ also remains fixed, and gravity does not decouple.

8.4.2 D4-branes and pure Yang–Mills theory

A first step towards finding gravity duals of QCD-like theories is to construct a gravity dual of pure Yang–Mills theory without supersymmetry. This was achieved by Witten very early on in the AdS/CFT correspondence [7]. The starting point is to consider N D4-branes in type IIA string theory, for which the existence of a decoupling limit was found in the preceding section. Due to supersymmetry, the action describing open strings attached to these D4-branes involves gauge, fermionic and bosonic degrees of freedom. To break supersymmetry, and to obtain a theory which is effectively $(3 + 1)$-dimensional at low energies, one of the spatial dimensions wrapped by the branes is compactified on a circle of radius M_{KK}^{-1}. Then, anti-periodic boundary conditions are imposed on this circle for the fermionic degrees of freedom present. These anti-periodic boundary conditions break supersymmetry and generate a mass for the fermions. By quantum effects, the scalars present will then acquire masses at first order in perturbation theory. Both fermions and scalars decouple from the gauge field in the low-energy limit, such that we are left with pure Yang–Mills theory in 3+1 dimensions. For completeness, we note that periodic boundary conditions would preserve supersymmetry, and we would recover $\mathcal{N} = 4$ Super Yang–Mills theory in four dimensions.

For N D4-branes embedded as shown in table 8.5, the metric (8.66) gives, after a rescaling,

$$ds^2 = \left(\frac{u}{L}\right)^{3/2} \left(\eta_{\mu\nu} dx^\mu dx^\nu + f(u) dx_4^2\right) + \left(\frac{L}{u}\right)^{3/2} \left(\frac{du^2}{f(u)} + u^2 d\Omega_4^2\right),$$
$$f(u) \equiv 1 - \left(\frac{u_{KK}}{u}\right)^3. \tag{8.74}$$

Moreover, there is a dilaton given by $e^\phi = g_s(u/L)^{3/4}$, and $L^3 = \pi g_s N(\alpha')^{3/2}$. The new additional factor $f(u)$ in (8.74) ensures that the coordinate x^4 is compactified on a circle S^1

Table 8.5 Embedding of N coincident D4-branes compactified on S^1

	0	1	2	3	4	5	6	7	8	9
N D4	•	•	•	•	∘	–	–	–	–	–

Table 8.6 Low-energy field theory degrees of freedom for N D4-branes compactified on S^1

Field	$U(N)$	$SO(3,1)$	$SO(5)$
A_μ	adjoint	4	1
a_4	1	1	1
ϕ	1	1	5

with period given by

$$\delta x_4 = \frac{4\pi}{3} \frac{L^{3/2}}{u_{KK}^{1/2}} \equiv \frac{2\pi}{M_{KK}}. \tag{8.75}$$

This compactification is necessary in order to make the space smooth, such that there is no conical deficit angle in the $u - x_4$ plane. At $u = u_{KK}$, the radius of S^1 shrinks to zero. The point $u = u_{KK}$ is the tip of a cigar-shaped subspace spanned by x_4 and the holographic coordinate u. Consequently, the coordinate u is restricted to the range $[u_{KK}, \infty]$. The scale set by $M_{KK} \sim u_{KK}^{-1}$ represents the *mass gap* of the pure Yang–Mills theory. Note though that the quantum field theory obtained in this way coincides with pure $SU(N)$ Yang–Mills theory only at very low energies. At energies larger than the scale M_{KK}, the theory becomes five-dimensional again, while remaining strongly coupled. It is thus very different from standard four-dimensional $SU(N)$ Yang–Mills theory, which is confining at low energies, but becomes *asymptotically free* at high energies, i.e. its coupling goes to zero in this limit.

For scales lower than M_{kk}, the field theory dual to (8.74) is a four-dimensional $U(N)$ gauge theory in the large N limit. Anti-periodic boundary conditions on the S^1 for the fermions ensure that these become massive with mass of order M_{KK}. Consequently, they decouple from the low-energy theory, and supersymmetry is completely broken. The remaining massless degrees of freedom are the gauge field A_μ, $\mu = 0, 1, 2, 3$, and the scalar fields A_4, the component of the gauge field in the compactified direction, as well as ϕ^i, $i = 5, 6, \ldots, 9$. All of these fields are in the adjoint representation of $U(N)$. Since the scalar fields are no longer protected by supersymmetry, A_4 and the ϕ^i will receive quantum corrections, making them massive with mass of order M. However, the trace parts of of A_4 and the ϕ^i, denoted by a_4 and ϕ, remain massless since they are protected by the $U(1)$ shift symmetry $a_4 \mapsto a_4 + \alpha \mathbf{1}_N$, $\phi^i \mapsto \phi^i + \alpha^i \mathbf{1}_n$. However, since they couple to the other massless modes only through irrelevant operators, they are not expected to play a role in the low-energy theory. The field content of the low-energy theory is summarised in table 8.6.

8.5 Further reading

Branes at singularities and quiver theories are discussed extensively in [8]. A first example of AdS/CFT correspondence with branes at singularities was given in [9]. From a mathematical point of view, conifolds and also the $T^{1,1}$ space are examined in [1]. The AdS/CFT dual for branes at the conifold is given in [2], see also [10]. A review of D-branes at conifolds is given in [11]. The ABJM theory and its gravity dual were introduced in [3]. Reviews may be found for instance in [12] and in [13] and [14]. The entropy of near-extremal black p-branes and its scaling with N was found in [4]. The relation between AdS$_3$ and CFT$_2$ was studied even before the AdS/CFT correspondence in [5]. A review of AdS$_3$/CFT$_2$ is given in [15].

Original references for the Dp-brane duality with $p \neq 3$ are [6, 16, 17, 18, 19]. For a detailed discussion of where the supergravity solution is reliable, see [6].

The duality for the field theory defined on either $\mathbb{R} \times S^3$ or $S^1 \times \mathbb{R}^3$ was proposed in [7], as well as for D4-branes with the x^4 direction compactified.

References

[1] Candelas, Philip, and de la Ossa, Xenia C. 1990. Comments on conifolds. *Nucl. Phys.*, **B342**, 246–268.

[2] Klebanov, Igor R., and Witten, Edward. 1998. Superconformal field theory on three-branes at a Calabi-Yau singularity. *Nucl. Phys.*, **B536**, 199–218.

[3] Aharony, Ofer, Bergman, Oren, Jafferis, Daniel Louis, and Maldacena, Juan. 2008. $\mathcal{N} = 6$ superconformal Chern-Simons-matter theories, M2-branes and their gravity duals. *J. High Energy Phys.*, **0810**, 091.

[4] Klebanov, Igor R., and Tseytlin, Arkady A. 1996. Entropy of near extremal black p-branes. *Nucl. Phys.*, **B475**, 164–178.

[5] Brown, J. David, and Henneaux, M. 1986. Central charges in the canonical realization of asymptotic symmetries: An example from three-dimensional gravity. *Commun. Math. Phys.*, **104**, 207–226.

[6] Itzhaki, Nissan, Maldacena, Juan Martin, Sonnenschein, Jacob, and Yankielowicz, Shimon. 1998. Supergravity and the large N limit of theories with sixteen supercharges. *Phys. Rev.*, **D58**, 046004.

[7] Witten, Edward. 1998. Anti-de Sitter space, thermal phase transition, and confinement in gauge theories. *Adv. Theor. Math. Phys.*, **2**, 505–532.

[8] Douglas, Michael R., and Moore, Gregory W. 1996. D-branes, quivers, and ALE instantons. ArXiv:hep-th/9603167.

[9] Kachru, Shamit, and Silverstein, Eva. 1998. 4d conformal theories and strings on orbifolds. *Phys. Rev. Lett.*, **80**, 4855–4858.

[10] Gubser, Steven S., and Klebanov, Igor R. 1998. Baryons and domain walls in an $\mathcal{N} = 1$ superconformal gauge theory. *Phys. Rev.*, **D58**, 125025.

[11] Herzog, Christopher P., Klebanov, Igor R., and Ouyang, Peter. 2002. D-branes on the conifold and $\mathcal{N} = 1$ gauge/gravity dualities. ArXiv:hep-th/0205100.

[12] Klebanov, Igor R., and Torri, Giuseppe. 2010. M2-branes and AdS/CFT. *Int. J. Mod. Phys.*, **A25**, 332–350.

[13] Klose, Thomas. 2012. Review of AdS/CFT integrability, Chapter IV.3: $\mathcal{N} = 6$ Chern-Simons and strings on $AdS_4 \times \mathbb{CP}^3$. *Lett. Math. Phys.*, **99**, 401–423.

[14] Ammon, Martin, Erdmenger, Johanna, Meyer, Rene, O'Bannon, Andy, and Wrase, Timm. 2009. Adding flavor to AdS_4/CFT_3. *J. High Energy Phys.*, **0911**, 125.

[15] Kraus, Per. 2008. Lectures on black holes and the AdS_3/CFT_2 correspondence. In *Supersymmetric Mechanics*, Vol. 3, pp. 193–247. Lecture Notes in Physics, Vol. 755. Springer.

[16] Wiseman, Toby, and Withers, Benjamin. 2008. Holographic renormalization for coincident Dp-branes. *J. High Energy Phys.*, **0810**, 037.

[17] Kanitscheider, Ingmar, Skenderis, Kostas, and Taylor, Marika. 2008. Precision holography for non-conformal branes. *J. High Energy Phys.*, **0809**, 094.

[18] Kanitscheider, Ingmar, and Skenderis, Kostas. 2009. Universal hydrodynamics of non-conformal branes. *J. High Energy Phys.*, **0904**, 062.

[19] Benincasa, Paolo. 2009. A note on holographic renormalization of probe D-Branes. ArXiv:0903.4356.

9

Holographic renormalisation group flows

In the preceding chapters we have studied examples of very non-trivial tests of the AdS/CFT correspondence. At the same time we have seen that the AdS/CFT correspondence is a new useful approach for studying strongly coupled field theories by mapping them to a weakly coupled gravity theory. This raises the question whether a similar procedure may be used to study less symmetric strongly coupled field theories, thus generalising the AdS/CFT correspondence to *gauge/gravity duality*. The prototype example where such a procedure is desirable is Quantum Chromodynamics (QCD), the theory of quarks and gluons, which is strongly coupled at low energies. Although a holographic description of QCD itself is not yet available, decisive progress has been achieved in many respects. We will discuss the achievements and open questions in this direction in chapter 13. Here we begin the discussion of generalisations of the AdS/CFT correspondence in a more modest, though well-controlled and simpler, way by considering the gravity duals of $\mathcal{N} = 4$ Super Yang–Mills theory deformed by relevant and marginal operators. These deformations break part of supersymmetry, and relevant operators also break conformal symmetry.

9.1 Renormalisation group flows in quantum field theory

9.1.1 Perturbing UV fixed points

The term *interpolating flows* refers to renormalisation group flows which connect an unstable UV fixed point to an IR fixed point at which the field theory is conformal again. A flow of this type is obtained for instance by perturbing the theory at a UV fixed point by a relevant or marginal operator. Marginal operators typically lead to a line of fixed points, while relevant operators generate a genuine RG flow, which may end at an IR fixed point. A further issue is whether the theory flows to a *confining* theory in the IR, as we will discuss in more detail in chapter 13. A field theory example of an interpolating flow will be given in section 9.1.3 below.

9.1.2 The C-theorem

A very important theorem for renormalisation group flows connecting two conformal field theories was proved by Zamolodchikov for field theories in two dimensions in 1986. This theorem makes a statement about interpolating RG flows relating a UV to an IR fixed point.

The theorem states that in two dimensions there is a function, named C, which decreases monotonically along the flow from the UV to the IR. At the fixed points, this function reduces to the central charge, which is proportional to the coefficient of the conformal anomaly.

Let us consider the statement and proof of the C-theorem in two-dimensional quantum field theory as found by Zamolodchikov. The theorem has a beautiful field theory proof in two dimensions, which relies on conservation of the energy-momentum tensor – i.e. on translational invariance – on rotational invariance and on reflection positivity, the analogue of unitarity in Euclidean quantum field theory. At the same time, it makes a statement of deep significance within physics, since the function C may be interpreted as an entropy function which counts degrees of freedom. Let us begin by stating the theorem.

Theorem   There exists a function $C(g^i)$ of the coupling constants which is non-increasing along RG flows, and is stationary only at the fixed points where conformal invariance is recovered. Moreover, at the fixed points it takes the value of the central charge c of the corresponding conformally invariant theory.

Proof   For the proof, we use complex coordinates $z = x_1 + ix_2$, $\bar{z} = x_1 - ix_2$ in two-dimensional Euclidean field theory, as they are conveniently used in two-dimensional conformal field theory. We consider a general point along the flow where conformal symmetry is broken. The energy-momentum tensor has components $T \equiv T_{zz}, \bar{T} \equiv T_{\bar{z}\bar{z}}$ and the trace $\Theta \equiv T_z{}^z + T_{\bar{z}}{}^{\bar{z}} = 4T_{z\bar{z}}$. These three components, T, Θ and $\bar{T}$, have spins $s = 2$, 0, -2 under rotations $z \mapsto z \exp(i\phi)$. In generalisation of the results of two-dimensional conformal field theory, we construct their two-point functions by writing the most general expressions compatible with translational and rotational symmetry, scaling dimensions and spin. These are given by

$$\langle T(z,\bar{z})T(0,0)\rangle = \frac{F(z\bar{z}\mu^2)}{z^4}, \tag{9.1}$$

$$\langle \Theta(z,\bar{z})T(0,0)\rangle = \langle T(z,\bar{z})\Theta(0,0)\rangle = \frac{G(z\bar{z}\mu^2)}{z^3\bar{z}}, \tag{9.2}$$

$$\langle \Theta(z,\bar{z})\Theta(0,0)\rangle = \frac{H(z\bar{z}\mu^2)}{z^2\bar{z}^2}, \tag{9.3}$$

where μ is a mass scale and F, G and H are non-trivial scalar functions of the dimensionless variable $z\bar{z}\mu$. Moreover, conservation of the energy-momentum tensor $\partial^\mu T_{\mu\nu} = 0$ implies, in complex variables,

$$\partial_{\bar{z}}T + \frac{1}{4}\partial_z\Theta = 0. \tag{9.4}$$

Taking the correlation function of the left-hand side of (9.4) with $T(0,0)$ and $\Theta(0,0)$, respectively, yields the two equations

$$\dot{F} = \frac{1}{4}(\dot{G} - 3G) = 0, \tag{9.5}$$

$$\dot{G} - G + \frac{1}{4}(\dot{H} - 2H) = 0, \tag{9.6}$$

where the derivative is defined by $\dot{F} \equiv z\bar{z}F'(z\bar{z}\mu)$, and similarly for G, H.

We now define a new function C by

$$C \equiv 2F - G - \frac{3}{8}H. \tag{9.7}$$

For its derivative we find

$$\dot{C} = -\frac{3}{4}H. \tag{9.8}$$

Now *reflection positivity* requires that states $|\Theta <= \Theta|0\rangle$ satisfy $\langle\Theta|\Theta\rangle \geq 0$, which implies $H \geq 0$. This is equivalent to the statement that $\langle\Theta|\Theta\rangle$ corresponds to a probability density and therefore must be non-negative. Reflection positivity is the Euclidean equivalent of unitarity in a field theory with Minkowski signature. Equation (9.8) implies that C is a non-increasing function of $z\bar{z}$, and is stationary only when $H = 0$.

We now translate C into a function of the couplings. Within quantum field theory we have $C = C(g^i(z\bar{z}\mu^2), z\bar{z}\mu^2)$. Since C is a dimensionless physical RG invariant quantity and thus independent of μ, we have

$$\mu\frac{\mathrm{d}}{\mathrm{d}\mu}C = 0 \Rightarrow \mu\frac{\partial}{\partial\mu}C + \beta^i\frac{\partial}{\partial g^i}C = 0. \tag{9.9}$$

This implies

$$-\beta^i\partial_i C(g^i) \leq 0. \tag{9.10}$$

In addition, at the fixed points the function C as given by (9.7) takes the value of the central charge. This proves the theorem. $\qquad\qquad\square$

Since Zamolodchikov's proof of the C-theorem in two dimensions in 1986, it has remained as an open question in quantum field theory whether there is also a proof for this theorem in more than two dimensions. In four spacetime dimensions, it is the coefficient a of the Euler density contribution to the trace of the energy-momentum tensor, as defined in chapter 3, which is expected to appear in the function C. Like the Ricci scalar in two dimensions, the Euler term in four dimensions leads to a topological density. For the anomaly coefficient c of the Weyl tensor squared anomaly contribution however, there are counterexamples which mean that it is ruled out as a candidate C function.

Very recently, an approach to proving the C-theorem in four dimensions was proposed [1, 2], showing that under certain assumptions,

$$a_{\mathrm{UV}} - a_{\mathrm{IR}} = \frac{f^4}{\pi}\int\limits_{s'>0}\mathrm{d}s'\frac{\sigma(s')}{s'^2}, \tag{9.11}$$

with f a positive scalar decay constant and $\sigma(s)$ the positive definite cross section for the scattering of two massless scalars, related to their four-point function. One of the assumptions made is that an equation of motion may be imposed for the massless scalar, the dilaton which acts as source for the trace of the energy-momentum tensor.

9.1.3 Deformations of $\mathcal{N} = 4$ Super Yang–Mills theory

Within gauge/gravity duality, the simplest examples of RG flows to consider are based on a UV fixed point, at which the field theory is $\mathcal{N} = 4$ Super Yang–Mills theory, and to add

$SU(4)$ representation	Dynkin labels	Operator	Dimension
20'	$[0,2,0]$	$\mathrm{Tr}(\phi^{(i}\phi^{j)})$ traces	$\Delta = 2$
50	$[0,3,0]$	$\mathrm{Tr}(\phi^{(i}\phi^j\phi^{k)})$ traces	$\Delta = 3$
10$_c$	$[2,2,0]$	$\mathrm{Tr}\lambda_a\lambda_b +$ SUSY completion	$\Delta = 3$
105	$[0,4,0]$	$\mathrm{Tr}(\phi^{(i}\phi^j\phi^k\phi^{l)})$ traces	$\Delta = 4$
45$_c$	$[2,3,0]$	$\mathrm{Tr}\lambda_a\lambda_b\phi^i +$ SUSY completion	$\Delta = 4$
1	$[0,0,0]$	Lagrangian	$\Delta = 4$

Table 9.1 Relevant and marginal deformations of $\mathcal{N} = 4$ theory

marginal or relevant operators to this theory which generate a flow. Let us first study this on the field theory side.

For deformations which preserve $\mathcal{N} = 1$ supersymmetry, it is convenient to use $\mathcal{N} = 1$ superspace language. Relevant deformations of $\mathcal{N} = 4$ Super Yang–Mills theory are obtained by adding a mass term for the three chiral multiplets to the superpotential in the Lagrangian, of the form

$$W_m = m^{ij}\mathrm{Tr}(\Phi_i\Phi_j), \qquad (9.12)$$

where m_{ij} is a 3×3 mass matrix. Moreover, marginal deformations are obtained by adding a superpotential term cubic in the chiral multiplets,

$$W_h = h^{ijk}\mathrm{Tr}(\Phi_i\Phi_j\Phi_k). \qquad (9.13)$$

In component fields, relevant and marginal operators are obtained from chiral primary fields and their descendants as listed in table 9.1.

In table 9.1, the representations **20'**, **50** and **105** correspond to chiral primaries, whereas the **10$_c$** and **45$_c$** are obtained by acting twice with the supersymmetry generator Q_α^a on the **20'** and **50**, respectively. The SUSY completions for both these descendant operators involve terms of the form $[\phi^i, \phi^j]\phi^k$ and $[\phi^i, \phi^j][\phi^k, \phi^l]$, respectively.

Marginal deformations of $\mathcal{N} = 4$ Super Yang–Mills theory

An example of a marginal deformation of $\mathcal{N} = 4$ Super Yang–Mills theory as given by (9.13) amounts to changing the superpotential of the theory by adding a phase,

$$\mathrm{Tr}(\Phi_1\Phi_2\Phi_3 - \Phi_1\Phi_3\Phi_2) \mapsto \mathrm{Tr}(e^{i\pi\beta}\Phi_1\Phi_2\Phi_3 - e^{-i\pi\beta}\Phi_1\Phi_3\Phi_2). \qquad (9.14)$$

With this deformation, the theory preserves $\mathcal{N} = 1$ supersymmetry and a global $U(1) \times U(1)$ symmetry in addition to the $U(1)_R$ symmetry. This superpotential defines the β-deformed $\mathcal{N} = 4$ theory.

More generally, relevant deformations which do not necessarily preserve $\mathcal{N} = 1$ supersymmetry correspond to adding dimension two mass terms and dimension three interaction terms for the six scalars of $\mathcal{N} = 4$ theory to the $\mathcal{N} = 4$ Lagrangian,

$$\mathcal{L} = \mathcal{L}_{\mathcal{N}=4} + \frac{m_{ij}}{2}\,\mathrm{Tr}\,(\phi^i\phi^j) + \frac{M_{ab}}{2}\,\mathrm{Tr}\,(\lambda^a\lambda^b) + b_{ijk}\,\mathrm{Tr}\,(\phi^i\phi^j\phi^k). \qquad (9.15)$$

Relevant deformations of $\mathcal{N} = 4$ Super Yang–Mills theory

Let us consider an example of an interpolating RG flow triggered by a relevant deformation of $\mathcal{N} = 4$ theory at its UV fixed point. This flow is best described in $\mathcal{N} = 1$ superfield language. The starting point is the Lagrangian of $\mathcal{N} = 4$ Super Yang–Mills theory, to which as a special case of (9.12), (9.13) a superpotential of the form

$$W_{LS} \equiv h \operatorname{Tr}(\Phi_3 [\Phi_1, \Phi_2]) + \frac{m}{2} \operatorname{Tr}(\Phi_3^2) \tag{9.16}$$

is added. Due to the scaling dimensions $[h] = 0$ and $[m] = 1$, the former term is marginal and the mass term is relevant. This deformation leads to a reduced R-symmetry $SU(2) \times U(1)$ of the original $\mathcal{N} = 4$ Lagrangian. The $\Phi_{1,2}$ fields are an $SU(2)$ doublet. The $U(1)$ charges of the chiral superfields $\Phi_{1,2,3}$ are $(1/2, 1/2, -1)$.

A necessary condition for an IR fixed point is that all beta functions vanish. An expression for the gauge β function $\beta(g)$ is well known for $\mathcal{N} = 1$ theories to all orders in perturbation theory. It is given by the *NSVZ β function* introduced in (3.223) of chapter 3. Here, we are dealing with the gauge group $G = SU(N)$ and all the fields transform in the adjoint representation. Therefore, $C(\mathbf{R}) = C(\mathbf{adj}(G)) = N$ and

$$\beta(g) \sim 2N (\gamma_1 + \gamma_2 + \gamma_3). \tag{9.17}$$

The β functions for the matter fields are simple due to non-renormalisation theorems in SUSY theories: the running of the parameters h, m in (9.16) is governed by

$$\beta_h = \gamma_1 + \gamma_2 + \gamma_3, \qquad \beta_m = 1 - 2\gamma_3. \tag{9.18}$$

To find a non-trivial fixed point, we look for solutions to the condition $\beta(g) = \beta_h = \beta_m = 0$. Requiring $SU(2)$ symmetry, which implies that $\gamma_1 = \gamma_2$ such that the chiral superfields Φ_1 and Φ_2 continue to form a doublet, this condition has a unique solution

$$\gamma_1 = \gamma_2 = -\frac{\gamma_3}{2} = -\frac{1}{4}. \tag{9.19}$$

The IR fixed point theory corresponding to these values of the anomalous dimensions has $\mathcal{N} = 1$ superconformal symmetry $SU(2, 2|1)$, with an additional global $SU(2)$ symmetry. The RG flow from $\mathcal{N} = 4$ theory in the UV to the $\mathcal{N} = 1$ IR fixed point is referred to as *Leigh–Strassler flow* [3].

The non-trivial scaling dimensions of the superfields at the IR fixed point are given by $\Delta_i = 1 + \gamma_i$. Short superconformal multiplets $\mathcal{O}$ may be constructed from gauge invariant combinations of Φ_1, Φ_2 and the field strength superfield W_α, which contains $F_{\mu\nu}$. The possible multiplets and the conformal dimension of their lowest component are listed in table 9.2. The $\Phi_{i=1,2}$ form an $SU(2)$ doublet, and T^a denote the associated $SU(2)$ generators. The first three multiplets in table 9.2 are chiral, the fourth contains the $SU(2)$ current and the last is the supercurrent which has the $U(1)_R$ current, the supersymmetry currents and the energy-momentum tensor among its components.

The conformal anomaly coefficients a and c at both the UV and the IR fixed point may be calculated in the component formalism, noting that $\mathcal{N} = 1$ supersymmetry, which relates

$\mathcal{O}$	$\mathrm{Tr}(\Phi_i\,\Phi_j)$	$\mathrm{Tr}(W_\alpha\,\Phi_i)$	$\mathrm{Tr}(W^\alpha\,W_\alpha)$	$\mathrm{Tr}(\bar{\Phi}^{\dagger i}\,(T^a)_i{}^j\,\Phi_j)$	$\mathrm{Tr}(W_\alpha\,\bar{W}_{\dot{\beta}})$
Δ	3/2	9/4	3	2	3

Table 9.2 Short multiplets at the $\mathcal{N}=1$ IR fixed point

the trace of the energy-momentum tensor to the divergence of the R-symmetry current, implies

$$\partial^\mu \langle R_\mu \rangle = -\frac{a-c}{24\pi^2} R_{\mu\nu\rho\sigma} \tilde{R}^{\mu\nu\rho\sigma} + \frac{5a-3c}{9\pi^2} \mathcal{A}_{\mu\nu} {}^*\mathcal{A}^{\mu\nu}, \tag{9.20}$$

where

$$\mathcal{A}_{\mu\nu} \equiv \partial_\mu \mathcal{A}_\nu - \partial_\nu \mathcal{A}_\mu, \tag{9.21}$$

with $\mathcal{A}_\mu$ a source for the R-symmetry current R_μ. By calculating three-point functions involving R_μ to one loop by functionally varying (9.20) with respect to $\mathcal{A}_\nu$, $\mathcal{A}_\lambda$, we obtain the UV values for the anomaly coefficients, a_{UV} and c_{UV}. These satisfy $a_{\mathrm{UV}} - c_{\mathrm{UV}} = 0$ since the theory at the UV fixed point is $\mathcal{N}=4$ Super Yang–Mills theory. In the IR, a similar one-loop calculation requires considering the current

$$S_\mu = R_\mu + \frac{2}{3} \sum_i (\gamma_{\mathrm{IR}}^i - \gamma^i) K_\mu^i, \tag{9.22}$$

with the anomalous dimensions γ_{IR}^i given by (9.19). In (9.22), K_μ^i is the *Konishi current* as given in box 9.1. The coefficients of the two terms in the current S_μ of (9.22) are

---

Box 9.1   **Konishi current**

In a general $(3+1)$-dimensional $\mathcal{N}=1$ supersymmetric Yang–Mills theory with chiral superfields Φ as introduced in chapter 3, the Konishi current is given by

$$K_\mu = \sigma_\mu^{\alpha\dot{\alpha}} [D_\alpha, \bar{D}_{\dot{\alpha}}] K, \qquad K = \bar{\Phi}\Phi, \tag{9.23}$$

where care has to be taken of the correct gauge representation of the chiral multiplets Φ. K is referred to as the Kähler potential. For the three chiral superfields Φ^i of $\mathcal{N}=4$ Super Yang–Mills theory and its deformations we have

$$K = \mathrm{Tr}\,(\bar{\Phi}^i e^{-V} \Phi^i e^V). \tag{9.24}$$

The Konishi current has an anomaly which is one-loop exact [4]. For the lowest-order component of the superfield, the one-loop exact anomaly reads

$$\langle \partial^\mu K_\mu \rangle = \frac{N_\mathrm{f}}{16\pi^2} \mathrm{Tr}\,(F_{\mu\nu}^a \tilde{F}^{a\mu\nu}), \tag{9.25}$$

with $\tilde{F}$ the Hodge dual of F and N_f the number of flavours.

chosen to ensure that S_μ is free of dynamical anomalies, i.e. its divergence is free of $F\tilde{F}$ terms. Due to 't Hooft anomaly matching as described in box 1.4 in chapter 1, the IR gravitational anomaly coefficients may be obtained from the calculation of three-point functions involving S_μ in the UV where $\gamma^i = 0$. The IR anomaly coefficients also satisfy $a_{\text{IR}} - c_{\text{IR}} = 0$. Putting all the results together, we obtain

$$\frac{a_{\text{IR}}}{a_{\text{UV}}} = \frac{c_{\text{IR}}}{c_{\text{UV}}} = \frac{27}{32} \tag{9.26}$$

in agreement with the four-dimensional C-theorem, since $a_{\text{IR}} < a_{\text{UV}}$.

9.2 Holographic renormalisation group flows

9.2.1 Domain wall flows

Our aim is now to describe the construction of supergravity backgrounds which can be conjectured to be dual to renormalisation group flows within quantum field theory, and in particular to the interpolating flows described above. A promising candidate is provided by the *domain wall flows* which interpolate between stationary points of a potential on the gravity side. At the stationary points, the potential reduces to the cosmological constant of an Anti-de Sitter space and the metric becomes an AdS metric. However, at different stationary points, the AdS radius L may differ. We will discuss the construction of these flows and subsequently present some non-trivial tests of the proposed conjecture.

To find a gravity analogue of field theory RG equations, we begin by considering a toy model of a supergravity dual to an RG flow. The model we consider is five-dimensional gravity with a single scalar field, with a general potential. This model may be part of an action obtained by dimensionally reducing ten- or eleven-dimensional supergravity. However, we may also view it more generally as a genuinely five-dimensional model within gauge/gravity duality. This takes us beyond the AdS/CFT correspondence as discussed in the previous chapters. In this case though, in general it will not be possible to identify the Lagrangian of the dual field field theory.

The idea is to obtain an RG equation as a *gradient flow* equivalent to the supergravity equations of motion. This will provide the first-order differential equation necessary for formulating an RG equation. The model we consider is

$$S = \int \mathrm{d}^{d+1}x \sqrt{-g} \left(\frac{R}{16\pi G} - \frac{1}{2} \partial_m\phi \, \partial^m\phi - V(\phi) \right). \tag{9.27}$$

In the following, we write $\mathrm{d}^{d+1}x = \mathrm{d}^d x \, \mathrm{d}r$. G is the Newton constant in $d + 1$ dimensions.

We choose the potential $V(\phi)$ such that it has one or more stationary points with $V'(\phi) = 0$. The equations of motion for ϕ and g_{mn} read

$$\frac{1}{\sqrt{-g}} \partial_m \left(\sqrt{-g} \, g^{mn} \, \partial_n\phi \right) - V'(\phi) = 0, \tag{9.28}$$

as well as

$$R_{mn} - \frac{R}{2}\, g_{mn} = 8\pi G \left(\partial_m \phi\, \partial_n \phi\ -\ \frac{1}{2}\, g_{mn}\, \partial_l \phi\, \partial^l \phi\ -\ g_{mn} V(\phi) \right) \equiv 8\pi G\, T_{mn}. \tag{9.29}$$

At the stationary points ϕ_i, there is a trivial solution of the scalar equation of motion with constant $\phi(r) = \phi_i$. Here, the Einstein equation reduces to $R_{mn} - \frac{R}{2} g_{mn} = -8\pi G\, g_{mn} V(\phi_i)$. This is identical to the Einstein equation of AdS space $G_{mn} + \Lambda g_{mn} = 0$, if we identify

$$\Lambda_i = 8\pi G V(\phi_i) = -\frac{d(d-1)}{L_i^2}. \tag{9.30}$$

In other words, constant scalar fields with AdS_{d+1} geometry of scale L_i are critical solutions which correspond to conformal theories at RG fixed points on the field theory side.

A more general ansatz than (9.27) for solving the equations of motion involves a metric with *warp factor* $A(r)$,

$$ds^2 = e^{2A(r)}\, \eta_{\mu\nu}\, dx^\mu\, dx^\nu + dr^2, \qquad \phi = \phi(r). \tag{9.31}$$

This is known as the *domain wall ansatz*. For a linear function $A(r) = r/L$, we recover the AdS metric. Together with a constant scalar we recover the dual of a conformal field theory, as expected at an RG fixed point. Here we will consider solutions which have linear $A(r)$ and constant ϕ near the boundary at $r \to \infty$ and in the deep interior for $r \to -\infty$. This is conjectured to be dual to an RG flow from a UV fixed point to an IR fixed point. It is natural to identify the radial coordinate r with the field theory RG scale via $\mu = \mu_0 \exp\left(\frac{r}{L}\right)$. This choice guarantees that in the UV at the AdS boundary, we have $\mu \to \infty$ for $r \to \infty$, while in the deep interior we have $\mu \to 0$ for $r \to -\infty$. The exact identification of the RG scale is scheme dependent and we will see more generally that a particular choice of coordinates on the supergravity side corresponds to a particular choice of renormalisation scheme on the field theory side.

Calculating the Riemann tensor for the metric (9.31), we obtain the Einstein equations

$$G^\mu{}_\nu = (d-1)\, \delta^\mu{}_\nu \left(A'' + \frac{d}{2}\, (A')^2 \right) = 8\pi G\, T^\mu{}_\nu,$$
$$G^r{}_r = \frac{d(d-1)}{2}\, (A')^2 = 8\pi G\, T^r{}_r \tag{9.32}$$

with μ, ν labelling the d boundary directions and r the radial direction. By considering the difference $G^t{}_t - G^r{}_r$, we extract a bound on the second derivative of the warp factor from (9.32),

$$A'' = \frac{8\pi G}{d-1}\, \left(T^t{}_t\ -\ T^r{}_r \right) = -\frac{8\pi G}{d-1}\, (\phi')^2 \quad \Rightarrow \quad A'' \leq 0. \tag{9.33}$$

This is consistent with the null energy condition $T_{mn} \zeta^m \zeta^n \geq 0$ for ζ^m a null vector as introduced in chapter 2, which here translates into $T^r{}_r - T^t{}_t \geq 0$. Combining the Einstein equation with the equation of motion for ϕ gives rise to

$$\phi'' + d\, A'\, \phi' = \frac{dV(\phi)}{d\phi},$$
$$(\phi')^2 - 2\, V(\phi) = \frac{1}{8\pi G} d(d-1)\, (A')^2. \tag{9.34}$$

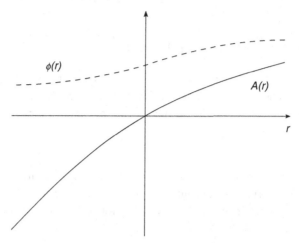

Figure 9.1　Functions $A(r)$ and $\phi(r)$ for an interpolating holographic RG flow.

These equations of motion can be simplified by introducing an auxiliary function, the *superpotential* $W(\phi)$,

$$V(\phi) = \frac{1}{2}\left(\frac{dW}{dr}\right)^2 - \frac{d}{d-1}\,W^2. \tag{9.35}$$

$W(\phi)$ is referred to as superpotential since within supergravity, it corresponds to the contribution of the scalar field to the F-term potential as defined in section 3.3.3.

It may be shown that any solution to the first-order *gradient flow* equations

$$\sqrt{8\pi G}\frac{d\phi}{dr} = \frac{dW}{d\phi}, \quad A' = -\frac{\sqrt{8\pi G}}{d-1}W \tag{9.36}$$

is also a solution to the equations of motion (9.34).

Our goal is to find a solution of (9.34) which interpolates between two stationary points. In AdS/CFT language, this means that we are looking for a domain wall solution interpolating between an AdS space of radius L_{UV} for $r \to +\infty$ and another AdS space of radius L_{IR} for $r \to -\infty$, where (9.33) implies that $L_{\mathrm{UV}} \geq L_{\mathrm{IR}}$. At the same time, the scalar ϕ is expected to flow from a constant ϕ_{UV} in the UV to a constant ϕ_{IR} in the IR, with $\phi_{\mathrm{IR}} \leq \phi_{\mathrm{UV}}$. A domain wall solution of this type is expected to be dual to a field theory RG flow between two fixed points. The expected behaviour of $A(r)$ and $\phi(r)$ is shown in figure 9.1. We will study an example of this type in section 9.2.3 below. First, however, we will introduce a holographic version of the C-theorem for the domain wall flows introduced here.

9.2.2　Holographic C-theorem

A very important aspect of the holographic interpolating flows is that they allow for a holographic proof of the *C-theorem* introduced in section 9.1.2 in all even dimensions. Assuming the validity of the AdS/CFT conjecture, the holographic C-theorem as described here shows that such a theorem holds for those higher dimensional quantum field theories in even dimensions which have a gravity dual.

For definiteness, let us consider the case of four dimensions. The starting point for constructing a holographic C-function is the conformal anomaly for $\mathcal{N} = 4$ Super Yang–Mills theory in the large N limit, which as discussed in chapter 6 is given in four spacetime dimensions by

$$\langle T_\mu{}^\mu \rangle = \frac{c}{8\pi^2}\left(R^{\mu\nu}R_{\mu\nu} - \frac{1}{3}R^2\right), \qquad c = \frac{N^2}{4} \tag{9.37}$$

on the field theory side, and by

$$\langle T_\mu{}^\mu \rangle = \frac{L^3}{64\pi G_5}\left(R^{\mu\nu}R_{\mu\nu} - \frac{1}{3}R^2\right), \qquad G_5 = \frac{G_{10}}{\text{Vol}(S^5)} = \frac{\pi L^3}{2N^2} \tag{9.38}$$

on the gravity side, with G_5 and G_{10} the Newton constants in five and ten dimensions, respectively. The field theory and gravity results coincide.

Let us now consider a holographic interpolating flow with metric

$$ds^2 = e^{2A(r)}\eta_{\mu\nu}dx^\mu dx^\nu + dr^2, \tag{9.39}$$

where the radial coordinate may be interpreted as an energy scale. At fixed points, this metric has to coincide with the AdS metric which implies

$$A(r)\Big|_{\text{FP}} = \frac{r}{L}, \qquad A'(r) = \frac{1}{L}. \tag{9.40}$$

This suggests a generalised expression for the trace anomaly which is obtained by replacing the AdS radius L by $1/A'(r)$ in (9.38),

$$\langle T_\mu{}^\mu \rangle = C(r)\frac{1}{64\pi}\left(R^{\mu\nu}R_{\mu\nu} - \frac{1}{3}R^2\right), \qquad C(r) = \frac{\pi}{G_5 A'(r)^3}. \tag{9.41}$$

From the Einstein equation of motion for the interpolating flow we have, using $A'' \le 0$ as derived in (9.33),

$$C'(r) = -3\frac{1}{G_5}\frac{A''(r)}{A'(r)^4} \ge 0. \tag{9.42}$$

This is in agreement with the C-theorem since C decreases monotonically when moving to the IR at $r \to -\infty$. Moreover, at the fixed points, $C(r)$ takes the form

$$C_i = \frac{L_i^3}{G_5}, \tag{9.43}$$

which corresponds to the conformal anomaly as given by (9.38).

The holographic C-theorem may also be written in a form which makes obvious its identification with the field theory C-theorem. To see this, we consider the gravity action for several scalars given by

$$S = \int d^5x\,\sqrt{-g}\left(\frac{1}{16\pi G_5}R - \frac{1}{2}\mathcal{G}_{IJ}\partial^m\phi^I\partial_m\phi^J - V(\phi^I)\right). \tag{9.44}$$

Here, $\mathcal{G}_{IJ}$ must be a positive definite metric on the space of scalar fields in order to obtain a unitary theory of scalars. This is in agreement with the null energy condition again, which in this case implies

$$T^r{}_r - T^t{}_t = \mathcal{G}_{IJ}\partial^m\phi^I\partial_m\phi^J \geq 0. \tag{9.45}$$

The flow equations in this case read

$$\sqrt{8\pi G_5}\,\frac{d\phi^I}{dr} = \mathcal{G}^{IJ}\frac{\partial W}{\partial\phi^J}, \quad \frac{dA}{dr} = -\frac{\sqrt{8\pi G_5}}{3}W(\phi), \tag{9.46}$$

which implies

$$A''(r) = -\frac{8\pi G_5}{3}\mathcal{G}_{IJ}\frac{d\phi^I}{dr}\frac{d\phi^J}{dr}. \tag{9.47}$$

Using this we may write the C-theorem for a function of the scalars ϕ^i, which have the interpretation of sources or generalised couplings on the field theory side,

$$C = \frac{\pi}{G_5 A'(r)^3} = -\frac{27\pi}{G_5(8\pi G_5)^{3/2}}\frac{1}{W^3}. \tag{9.48}$$

This function is obtained from (9.41) using the second equation in (9.46). We now define a gravity β function in analogy to the field theory β function. There is an arbitrariness in this choice which corresponds to fixing a renormalisation scheme on the field theory side. A change of coordinates will correspond to a change to a different renormalisation scheme. We choose

$$\beta^I \equiv \frac{d\phi^I}{dr}. \tag{9.49}$$

Using this we have

$$-\beta^I\partial_I C = -\frac{d\phi^I}{dr}\partial_I C = -27\cdot 8\,\frac{1}{W^4}\mathcal{G}_{IJ}\frac{d\phi^I}{dr}\frac{d\phi^J}{dr} \leq 0, \tag{9.50}$$

which coincides with the field theory result (9.10). We thus confirm that we have a function which is decreasing along RG flows towards the IR. Moreover, at the fixed points, C coincides with the conformal anomaly coefficients given by (9.38).

The essential positivity condition for the proof of the C-theorem is provided by the null energy condition in the curved five-dimensional space. A similar argument will be possible in general $d+1$ odd bulk dimensions, which allows for a proof of the holographic C-theorem for field theory RG flows dual to interpolating flows in any even dimension. In odd boundary dimensions, the standard gravitational anomalies considered here are absent.

9.2.3 Holographic interpolating flows: example

Here we present a conjectured holographic dual to the interpolating RG flow of section 9.1.3. On the field theory side, this flow, which is referred to as Leigh–Strassler flow, is obtained by perturbing $\mathcal{N} = 4$ theory at the UV fixed point by relevant operators, and flows to an $\mathcal{N} = 1$ supersymmetric theory at an IR fixed point.

The gravity dual of this flow is referred to as *FGPW flow* after Freedman, Gubser, Pilch and Warner [5]. These authors obtained this flow by considering $\mathcal{N} = 8$ gauged

Table 9.3 Graviton multiplet of $\mathcal{N} = 8, D = 5$ supergravity

Field	$g_{\mu\nu}$	Ψ_μ^a	A_μ^A	$B_{\mu\nu}$	χ^{abc}	ϕ^I
$SO(6)$ representation	1	8	15	12	48	42

supergravity in five dimensions with gauge group $SO(6)$, which is expected to be a *consistent truncation* of type IIB supergravity on $AdS_5 \times S^5$ to five dimensions. While the ten-dimensional theory has contributions from Kaluza–Klein towers of modes dual to operators with increasing values of dimension Δ, $\mathcal{N} = 8, D = 5$ supergravity just contains the five-dimensional graviton multiplet. This is given in table 9.3.

The theory thus contains forty-two scalars. These enter a complicated potential of which it is very hard to determine the extrema. To reduce the numbers of scalars, it is necessary to make use of symmetries. One possibility [6, 5] is to consider an $SU(2)$ subgroup of the original $SO(6)$ symmetry. The forty-two scalars may then be organised into singlets ϕ and non-trivial representations χ of $SU(2)$. Since the potential V is invariant under $SU(2)$, i.e. a singlet, Schur's lemma of group theory implies that the original potential V takes the form

$$V(\phi, \chi) = V_0(\phi) + V_2(\phi)\chi^2 + \mathcal{O}(\chi^3). \tag{9.51}$$

Note that there cannot be any term linear in χ present in this expansion, since it is impossible to form an $SU(2)$ singlet with only one χ. Due to (9.51), any stationary point $\bar\phi$ of $V_0(\phi)$ corresponds to a stationary point $(\bar\phi, \chi = 0)$ of $V(\phi, \chi)$. This significantly reduces the number of scalars to be considered.

Considering $SU(2)$ singlets as described allows the construction of a gravity flow preserving the same symmetries as the field theory example (9.16). This is generated by perturbing the UV fixed point theory by two fields, ϕ_2 a field with $\Delta = 2$ in the $\mathbf{20'}$ of $SO(6)$ in the full theory, and ϕ_3 a field with $\Delta = 3$ in the $\mathbf{10} + \overline{\mathbf{10}}$ representation. These may be identified as being dual to the relevant deformations on the field theory side as given in table 9.1 and in (9.16). The gravity potential for the scalars ϕ_2 and ϕ_3 is obtained by an involved supergravity analysis [5] using the reduction to $SU(2)$ singlets as described above. Writing $\rho = e^{\frac{\phi_2}{\sqrt{6}}}$, the superpotential is found to be

$$W(\phi_2, \phi_3) = \frac{1}{4L\rho^2}\left[\cosh(2\phi_3)(\rho^6 - 2) - 3\rho^6 - 2\right]. \tag{9.52}$$

Exercise 9.2.1 Determine the stationary points of this potential.

One critical point is given by $\phi_2 = \phi_3 = 0$ and corresponds to the original case with $SO(6)$ symmetry dual to $\mathcal{N} = 4$ Super Yang–Mills theory. This is a maximum of the potential at which $W = -\frac{3}{2L}$. Moreover, there are three unstable stationary points which may be shown to be non-supersymmetric using a Killing spinor analysis in supergravity. Finally, there are two further equivalent stationary points related by a $\mathbb{Z}_2$ symmetry. These are saddle points of the potential at $\phi_2 = \frac{1}{\sqrt{6}}\ln 2, \phi_3 = \pm\frac{1}{2}\ln 3$, for which $W = -\frac{2^{2/3}}{L}$. The supergravity solution at these points preserves an $SU(2) \times U(1)$ symmetry in addition to the $SU(2)$ considered to generate the solution.

There is a possible supersymmetric domain wall flow between the maximum at $\phi_2 = \phi_3 = 0$ and any of the two saddle points. This flow may be considered as a gravity dual of the field theory Leigh–Strassler flow of section 9.1.3. The UV fixed point is dual to $\mathcal{N} = 4$ Super Yang–Mills theory and the symmetry at any of the saddle points is $SU(2) \times SU(2) \times U(1)$, which corresponds to the symmetries at the IR fixed point of the Leigh–Strassler flow. Unfortunately, the equations of motion of this flow have so far only been solved numerically and an analytical solution is not yet known. However, the field-operator map for the IR fixed point has been established. There is a one-to-one correspondence between the operators at the Leigh–Strassler fixed point, given in table 9.2 in section 9.1.3, and the operators at the IR fixed point of the domain wall flow discussed here. This map provides substantial evidence for the conjectured duality.

We now illustrate that this interpolating flow is in agreement with the holographic C-theorem of section 9.2.2. For $d = 4$, we have

$$\frac{c_{\text{IR}}}{c_{\text{UV}}} = \frac{W_{\text{UV}}{}^3}{W_{\text{IR}}{}^3}. \tag{9.53}$$

Let us insert the values for the FGPW interpolating flow. At the UV fixed point, we have the gravity dual of $\mathcal{N} = 4$ theory, at which the superpotential (9.52) takes the value $W_{\text{UV}} = -3/(2L)$. In the IR, as discussed above, the superpotential (9.52) takes the value $W_{\text{IR}} = -\frac{2^{2/3}}{L}$. Therefore we have

$$\frac{c_{\text{IR}}}{c_{\text{UV}}} = \frac{27}{32}. \tag{9.54}$$

This agrees with the field-theory result for the Leigh–Strassler flow, providing further evidence for the conjectured duality. Moreover, this is also in agreement with the C-theorem since the coefficient is smaller in the IR than in the UV.

To conclude this section, let us comment on the counting of supersymmetries. On the field theory side, the Leigh–Strassler flow preserves $\mathcal{N} = 1$ supersymmetry with four supercharges. At the IR fixed point, the theory is conformal and there are four further superconformal charges, so in total there are eight supercharges. On the supergravity side, we have five-dimensional minimal supergravity with again eight real supercharges. This is referred to as $\mathcal{N} = 2$ supergravity. Note that the $SU(2) \times SU(2) \times U(1)$ symmetry is local within supergravity and global in the boundary field theory. It is a generic feature of the AdS/CFT correspondence that additional local symmetries in the bulk become global at the boundary.

9.3 *Supersymmetric flows within IIB Supergravity in $D = 10$

In addition to the domain wall flows within five-dimensional gauged supergravity, there are also a number of appealing examples of holographic RG flows within ten-dimensional type

IIB supergravity. We present prominent examples here, first for a marginal deformation and subsequently for relevant deformations.

In contrast to the example studied in section 9.2.3, for the case of relevant deformations these examples do not flow to an IR conformal fixed point. Some of the IR properties of these flows are very similar to those of QCD, which we will discuss in chapter 13. In particular, these flows display *confinement* in the IR, which means that there are no free colour charges. A criterion for confinement is that the Wilson loop has an area law behaviour. A further important QCD-like property is *spontaneous chiral symmetry breaking*. In the case of the examples considered here, this corresponds to a spontaneous breaking of the $U(1)$ R-symmetry to a discrete subgroup.

9.3.1 Marginal deformation

We begin with the example of a gravity dual to the marginal deformation (9.14), the β-deformation. Recall from (9.14) that the dual field theory has a global $U(1) \times U(1)$ symmetry in addition to the $U(1)_R$ symmetry. Geometrically, this additional global symmetry corresponds to a two-torus with parameter

$$\tau \equiv B + i\sqrt{g}, \tag{9.55}$$

where $\sqrt{g}$ is the volume of the two-torus and B is the two-form in the torus directions. An eight-dimensional supergravity theory obtained by compactifying on this two-torus has a $SL(2, \mathbb{R}) \times SL(2, \mathbb{R})$ symmetry. Each of the $SL(2, \mathbb{R})$ symmetries acts as

$$\tau \to \tau_\beta = \frac{\tau}{1 + \beta\tau} \tag{9.56}$$

on the torus parameter. This transformation generates a new solution of the supergravity theory compactified on the torus. It corresponds to a T-duality, an angular shift and a second T-duality and is known as *TsT transformation*. This type of transformation also plays a role when constructing non-relativistic examples of gauge/gravity duality, as may be of interest for condensed matter physics.

For the geometry $AdS_5 \times S^5$ in ten dimensions, the TsT transformation preserves the AdS_5 part of the $AdS_5 \times S^5$ geometry and thus conformal symmetry of the dual field theory, as appropriate for a marginal deformation. On the other hand, S^5 is deformed. We write the S^5 metric in terms of radial and toroidal variables (ρ_i, ϕ_i), $i = 1, 2, 3$,

$$ds^2_{S_5} = L^2 \sum_{i=1}^{3} \left(d\rho_i d\rho_i + \rho_i^2 d\phi_i^2\right), \qquad \sum_{i=1}^{3} \rho_i^2 = 1. \tag{9.57}$$

In these coordinates, the τ-parameter of the two-torus is given by

$$\tau = i\sqrt{g} = iL^2(\rho_1^2\rho_2^2 + \rho_2^2\rho_3^2 + \rho_3^2\rho_1^2)^{1/2}. \tag{9.58}$$

The metric of the β-deformed S^5 is obtained by applying the TsT transformation (9.56) to (9.57) with (9.58). It takes the form

$$ds^2 = L^2 \sum_i (d\rho_i^2 + \mathcal{G}\rho_i^2 d\phi_i^2) + L^2 \hat{\beta}^2 \mathcal{G} \rho_1^2 \rho_2^2 \rho_3^2 \left(\sum_i d\phi_i \right)^2, \tag{9.59}$$

$$\mathcal{G}^{-1} = 1 + \hat{\beta}^2(\rho_1^2 \rho_2^2 + \rho_2^2 \rho_3^2 + \rho_1^2 \rho_3^2), \qquad \hat{\beta} = L^2 \beta, \tag{9.60}$$

$$e^{2\phi} = e^{2\phi_0}\mathcal{G} \tag{9.61}$$

in string frame when setting $\alpha' = 1$. Moreover, in the supergravity solution there are non-trivial contributions to the NS-NS B field, as well as to the R-R forms $C_{(2)}$ and $C_{(4)}$. The factor $\mathcal{G}$ in (9.60) clearly displays the effect of the TsT transformation.

9.3.2 Wrapped and fractional branes

We now turn to relevant deformations. String theory and supergravity provide a number of ways of breaking supersymmetry and conformal symmetry in a controlled way. An example of a running of couplings is obtained by considering non-trivial geometries obtained by deforming the $T^{1,1}$ geometry discussed in section 8.1.2. Turning on a non-trivial dilaton or $B_{(2)}$ field which depends on the radial direction will lead to a running of the gauge coupling given by (8.19) and (8.20) . More generally, two examples of brane constructions which give rise to holographic RG flows involve *wrapped* or *fractional* branes. We discuss RG flows generated in both of these configurations in the subsequent sections. Here we begin by outlining the concepts of wrapped and fractional branes.

For an example of *wrapped branes*, consider D5-branes with worldvolume $\mathbb{R}^{3,1} \times S^2$. At energies lower than the inverse radius of S^2, the theory whose action is given by the DBI action for the D5-branes is effectively four dimensional. Wrapped branes which preserve $\mathcal{N} = 1$ supersymmetry correspond to D5-branes wrapped on a non-vanishing two-cycle, for instance on the S^2 inside $T^{1,1}$.

Fractional branes correspond to configurations of branes wrapped on cycles collapsed to a singular point. These branes cannot move away from the singularity. From the open string perspective, this may be seen as follows. Unlike the regular branes at singularities discussed for the orbifold case in section 8.1.1, fractional branes do not have images in the covering space. Thus a symmetry invariant configuration is possible only at the origin in those spatial directions in which the identifications take place. In the $\mathbb{Z}_2$ orbifold example discussed in section 8.1.1, this means that the Chan–Paton matrix is just a real number and the representation matrix $\gamma(g)$ may be chosen to be one of the two one-dimensional *irreducible* representations of $\mathbb{Z}_2$, $\gamma(g) = +1$ or $\gamma(g) = -1$ with g generating $\mathbb{Z}_2$. The two different representations correspond to two equivalent types of fractional branes which may be present in different numbers, N_+ and N_-. When $N_+ = N_- = N$, we recover the regular case of section 8.1.1. In general, a fractional brane is a D-brane whose Chan–Paton factors transform in an irreducible representation of the orbifold group. The term *fractional* refers to the fact that these branes carry fractional D-brane charge. From the closed string (supergravity) perspective, fractional D3-branes correspond to D5-branes wrapped on a collapsed two-cycle. Returning to the example of D5-branes wrapping the S^2 factor in the

$T^{1,1}$ conifold, these become fractional branes at the tip of the conifold, where they carry fractional D3-brane charge.

To conclude this section, let us state the mechanism which generates non-trivial RG flows both for wrapped and for fractional branes. As may be seen by factorising the DBI action for a D5-brane in one integral over $\mathbb{R}^{3,1}$ and one over S^2, the gauge coupling for the effective D3-branes is given by

$$\frac{1}{g_{YM}^2} = \frac{e^{-\phi}}{2(2\pi)^3\alpha'} \int_{S^2} d\Omega_2 \sqrt{-\det\left([G+B]_{ij}\right)}, \tag{9.62}$$

where the indices i, j run over the S^2 directions. For wrapped branes, the flux of the B field may be zero, however a non-trivial contribution arises from the volume of the two-cycle wrapped by the D5-branes. For fractional branes, $\int_{S^2} B \neq 0$ even in the collapsed case. In both cases, we find that the gauge coupling runs and the theory is non-conformal.

9.3.3 Flow from fractional branes

An example of a holographic RG flow based on fractional branes is given by the supergravity solution of Klebanov and Strassler [7]. This is based on N D3-branes and M fractional D3-branes placed at a conifold singularity. This leads to an $\mathcal{N} = 1$ supersymmetric field theory with product gauge group $SU(N) \times SU(N+M)$. Both of the gauge couplings associated with the two gauge group factors run.

The starting point is the conformal theory with gauge group $SU(N) \times SU(N)$ obtained from D3-branes placed at a conifold singularity, as discussed in section 8.1.2. The rank of one of the $SU(N)$ factors of the $SU(N) \times SU(N)$ gauge group may be changed by adding M fractional D3-branes [8]. As discussed in section 9.3.2 above, these are D5-branes with two directions wrapped on the collapsed S^2 at the tip of the $T^{1,1}$ conifold. These wrapped D5-branes carry fractional D3-brane charge. Placing N D3-branes and M fractional D3-branes at the tip of the conifold gives rise to an $SU(N+M) \times SU(N)$ gauge group. The matter fields in this theory are two chiral superfields A_1, A_2 in the $(\mathbf{N+M}, \overline{\mathbf{N}})$ representation and two fields B_1, B_2 in the $(\overline{\mathbf{N+M}}, \mathbf{N})$ representation. The superpotential of the model is given by

$$W = \lambda_1 \operatorname{Tr}\, (A_i B_j A_k B_\ell)\epsilon^{ik}\epsilon^{j\ell}. \tag{9.63}$$

The Lagrangian has $SU(2) \times SU(2) \times U(1)$ global symmetry.

The M fractional D3-branes, which correspond to M D5-branes wrapped over the S^2 of $T^{1,1}$, source magnetic R-R three-form flux through the S^3 of $T^{1,1}$ in addition to the five-form flux through $T^{1,1}$,

$$\frac{1}{4\pi^2\alpha' g_s} \int_{S^3} F_{(3)} = M, \qquad \frac{1}{(2\pi)^4\alpha'^2 g_s} \int_{T^{1,1}} F_{(5)} = N, \tag{9.64}$$

at a cut-off scale r_0. This induces conformal symmetry breaking since now the supergravity equations of motion imply that $B_{(2)}$ acquires a radial dependence,

$$\frac{1}{2(2\pi)^3\alpha'} \int_{S^2} B_{(2)} = Me^\phi \ln(r/r_0). \tag{9.65}$$

According to the relation between couplings (8.20), which was discussed for the conformal $SU(N) \times SU(N)$ theory in section 8.1.2, this implies a logarithmic running of the gauge couplings in the dual $SU(N + M) \times SU(N)$ theory and the gauge coupling β functions are non-zero. This may be determined from the NSVZ β function introduced in (3.223) to be

$$\frac{d}{d \log(\Lambda/\mu)} \frac{8\pi^2}{g_1^2} \sim 3(N + M) - 2N(1 - \gamma), \tag{9.66}$$

$$\frac{d}{d \log(\Lambda/\mu)} \frac{8\pi^2}{g_2^2} \sim 3N - 2(N + M)(1 - \gamma), \tag{9.67}$$

where γ is the anomalous dimension of operators $\text{Tr}A_i B_j$, which leads to

$$\frac{8\pi^2}{g_1^2} - \frac{8\pi^2}{g_2^2} \sim M \ln(\Lambda/\mu)[3 + 2(1 - \gamma)] \tag{9.68}$$

in agreement with the supergravity result. The dilaton is constant to linear order in M, which implies $\gamma = -\frac{1}{2} + \mathcal{O}[(M/N)^2]$.

The non-trivial behaviour of $B_{(2)}$ as given by (9.65) also determines the behaviour of the supergravity metric and five-form, as may be seen by going beyond linear order in M. Since $F_{(5)} = dC_{(4)} + B_{(2)} \wedge F_{(3)}$, the five-form has a radial dependence, which is obtained as

$$F_{(5)} = \mathcal{F}_{(5)} + *\mathcal{F}_{(5)}, \qquad \mathcal{F}_{(5)} = \mathcal{K}(r)\text{Vol}(T^{1,1}), \tag{9.69}$$

$$\mathcal{K}(r) = N + a g_s M^2 \ln(r/r_0), \tag{9.70}$$

with a a constant of order one. The five-form flux vanishes at an IR scale $\tilde{r}$ where $\mathcal{K}(\tilde{r}) = 0$. This hints at the fact that the rank of the gauge groups $SU(N)$ and $SU(N + M)$ is decreased along the RG flow towards the IR, which is denoted as an *RG cascade*. We will discuss this in detail below from the field theory perspective.

To conclude the supergravity analysis, we now consider the metric. Starting from the ansatz

$$ds^2 = H^{-1/2}(r)\eta_{\mu\nu}dx^\mu dx^\nu + H^{1/2}(r)(dr^2 + r^2 ds^2_{T^{1,1}}), \tag{9.71}$$

the Einstein equation implies that

$$H(r) = \frac{4\pi g_s}{r^4}[\mathcal{K}(r) + a g_s M^2/4], \quad \mathcal{K}(r) = a g_s M^2 \ln(r/\tilde{r}). \tag{9.72}$$

This solution has a naked singularity at $r = r_s$, where $H(r_s) = 0$. With

$$H(r) = \frac{L^4}{r^4} \ln(r/r_s), \qquad L^2 \sim g_s M, \tag{9.73}$$

we may write the metric in the form

$$ds^2 = \frac{r^2}{L^2 \sqrt{\ln(r/r_s)}}\eta_{\mu\nu}dx^\mu dx^\nu + \frac{L^2 \sqrt{\ln(r/r_s)}}{r^2}dr^2 + L^2 \sqrt{\ln(r/r_s)}ds^2_{T^{1,1}}. \tag{9.74}$$

As mentioned above, this metric is dual to a gauge theory where the ranks of the two product gauge groups change along the flow. In the UV where $r \to \infty$, the five-form flux diverges, generating larger and larger N in the UV. On the other hand, the flow towards the IR must stop before N becomes negative, i.e. before $\mathcal{K}(r)$ becomes negative. The $T^{1,1}$ radius at $r = \tilde{r}$ is of order $\sqrt{g_s M}$, which corresponds to a gauge group $SU(M)$. At this scale, the metric (9.74) must be modified to obtain regular behaviour in the IR. We discuss the mechanism which ensures this below.

Duality cascade

Let us now consider the field theory RG flow. This requires the essential concept of *Seiberg duality*, which is introduced in box 9.2.

Let us apply Seiberg duality to the Klebanov–Strassler RG flow. From the β functions as given by (9.66), (9.67) we see that $1/g_1{}^2$ grows while $1/g_2{}^2$ decreases. Moreover, there is a scale at which the $SU(N + M)$ coupling g_1 diverges. To continue past this point, we perform a Seiberg duality transformation on the $SU(N + M)$ gauge group factor, which has $2N$ flavours in the fundamental representation. Under the duality transformation, this becomes an $SU(2N - [N + M]) = SU(N - M)$ group with $2N$ flavours and additional massive meson bilinears. The mesons may be integrated out using the equations of motion. This gives rise to a theory with a similar field content to the original one, except that the gauge group has now become $SU(N) \times SU(N - M)$. It turns out that the gauge couplings g_1 and g_2 have interchanged their behaviour: it is now g_2 which is growing until it diverges at a given, lower, energy scale, at which another Seiberg duality is performed. This then changes the gauge group to $SU(N - M) \times SU(N - 2M)$. This procedure then repeats itself again. The RG flow thus is a *duality cascade*.

Box 9.2   **Seiberg duality**

Seiberg duality relates two $\mathcal{N} = 1$ supersymmetric non-Abelian gauge theories with different Lagrangian. The first of the two theories involved, named SQCD, has gauge group $SU(N_c)$ and N_f flavours of both fundamental and anti-fundamental chiral multiplets, referred to as quarks and antiquarks. Its dual theory has gauge group $SU(N_f - N_c)$ and N_f flavours. The duality holds for $N_f > N_c + 1$. Seiberg argued in a series of seminal papers that these theories are equivalent since they flow to the same theory in the IR and are thus in the same universality class. Moreover, in the IR the two theories describe the same physics, just as QCD and an effective field theory of pions (quark–antiquark bound states) are expected to describe the same physics below the confinement scale.

The dual theory contains a gauge invariant meson chiral superfield M which transforms as a bifundamental under the flavour symmetries. As a generalisation of electromagnetic duality, Seiberg duality maps chromoelectric fields (gluons) to chromomagnetic fields (gluons of the dual gauge group), and particles with chromoelectric charge (quarks) to non-Abelian 't Hooft–Polyakov monopoles. The Higgs phase is dual to the confinement phase. Evidence for Seiberg duality is provided by the fact that the moduli spaces of both theories coincide, and that the global symmetries agree. Moreover, the anomaly coefficients in both theories agree in accordance with *'t Hooft anomaly matching* which states that coefficients of chiral anomalies are not renormalised.

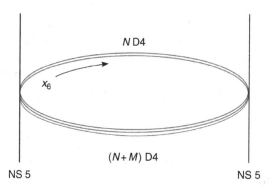

Figure 9.2   Type IIA picture of the Klebanov–Strassler flow.

IIA picture

The duality cascade has a nice geometrical representation which is obtained from T-dualising the type IIB setup of D3-branes and fractional D3-branes at the conifold tip to type IIA theory. Performing a T-duality transformation along the x_6 direction, we obtain the brane configuration displayed in figure 9.2. This is given by an NS5- and an $\overline{\mathrm{NS5}}$-brane which are linked by a stack of $(N + M)$ D4-branes on one side and a stack of N D4-branes on the other. Since the numbers of D4-branes on the two sides differ, the NS5-branes feel a force and begin to move. When they cross each other, a Seiberg duality transformation takes place.

Deformation of the conifold – chiral symmetry breaking

An important question is how the duality cascade ends in the IR. On the gravity side, this will provide the resolution of the naked singularity present in the metric (9.74). As was found by Klebanov and Strassler, in the IR, the singularity in the metric (9.74) is removed by a blow-up of S^3 in $T^{1,1}$. This leads to a deformation of the conifold (8.11),

$$\sum_{i=1}^{4} z_i^2 = -2\det_{i,j} z_{ij} = \epsilon^2 . \tag{9.75}$$

On the field theory side, the deformed conifold at the end of the duality cascade is obtained as follows. After a series of Seiberg dualities, the product gauge group finally reaches $SU(M+1) \times SU(1)$. The factor $SU(1)$ just corresponds to the identity and may be omitted. The theory has fields C_i and D_j in the $\mathbf{M+1}$ and $\overline{\mathbf{M+1}}$ representations, $i,j = 1, 2$, and with superpotential $W = \lambda C_i D_j C_k D_l \epsilon^{ik} \epsilon^{jl}$. Define $N_{ij} = C_i D_j$, which is gauge invariant. The expectation values of N_{ij} specify the position of the probe brane; in the classical theory, we have $\det_{i,j} N_{ij} = 0$, indicating the probe is moving on the original, singular conifold. At low energy the theory can be written in terms of these invariants and develops the non-perturbative superpotential first written down by Affleck, Dine and Seiberg [9]

$$W_{\mathrm{L}} = \lambda N_{ij} N_{kl} \epsilon^{ik} \epsilon^{jl} + (M - 1) \left[\frac{2\Lambda^{3M+1}}{N_{ij} N_{kl} \epsilon^{ik} \epsilon^{jl}} \right]^{\frac{1}{M-1}} . \tag{9.76}$$

The equations for a supersymmetric vacuum are

$$0 = \left(\lambda - \left[\frac{2\Lambda^{3M+1}}{(N_{ij} N_{kl} \epsilon^{ik} \epsilon^{jl})^M} \right]^{\frac{1}{M-1}} \right) N_{ij}. \tag{9.77}$$

The apparent solution $N_{ij} = 0$ for all i, j actually gives infinity on the right-hand side. The only solutions are then

$$(N_{ij} N_{kl} \epsilon^{ik} \epsilon^{jl})^M = \frac{2\Lambda^{3M+1}}{\lambda^{M-1}}. \tag{9.78}$$

There are thus M branches which spontaneously break the R-symmetry present. We have

$$W = M\lambda \langle N_{ij} N_{kl} \epsilon^{ik} \epsilon^{jl} \rangle \propto M \left[2\lambda \Lambda^{3M+1} \right]^{1/M} \tag{9.79}$$

for the new ground state, which corresponds to the deformed conifold (9.75).

The spontaneous R-symmetry breaking is an example of *spontaneous chiral symmetry breaking*, a concept of central importance in QCD. Moreover, the remaining $SU(M + 1)$ Yang–Mills theory in the IR is confining. The regular ten-dimensional supergravity solution based on fractional branes thus provides a gravity dual for an $\mathcal{N} = 1$ supersymmetric theory which has properties similar to those of QCD.

9.3.4 Maldacena–Núñez flow from wrapped branes

An alternative, though related construction leading to a flow to $\mathcal{N} = 1$ pure gauge theory in the IR has been proposed by Maldacena and Núñez [10], based on earlier work of Chamseddine and Volkov [8]. In contrast to the Klebanov–Strassler approach, where *fractional* branes lead to the running of the coupling (8.20), in this case the running is provided by *wrapped* branes as introduced in section 9.3.2. String theory provides powerful techniques for studying D-branes wrapped on non-trivial cycles within Calabi–Yau manifolds. Some of these come into play for constructing the gravity flow considered here, as we now sketch.

Consider a geometry of a Calabi–Yau threefold with a non-trivial supersymmetric two-cycle which is topologically equivalent to S^2. We wrap a D5-brane on this cycle. This brane also extends in the four flat directions perpendicular to the Calabi–Yau threefold. The world-volume of the brane is thus of the form $\mathbb{R}^{3,1} \times S^2$, and at energies lower than the inverse radius of S^2 the theory living on the worldvolume is effectively four dimensional. This construction preserves four supercharges. It is displayed in table 9.4, where the symbol o denotes a compactified direction. The Calabi–Yau manifold is placed in the $4, 5, \ldots, 9$ directions.

Table 9.4 Embedding of D5-branes									
	0	1	2	3	4	5	6	7	8	9
N D5	•	•	•	•	∘	∘	–	–	–	–

The volume of S^2 is a geometrical modulus which is a parameter of the associated string theory. By expanding the six-dimensional DBI action for the D5-brane, we find that the four-dimensional gauge theory has an effective coupling given by

$$\frac{1}{g_{\text{YM}}^2} = \frac{e^{-\phi}}{2(2\pi)^3\alpha'} \int_{S^2} d\Omega_2 \sqrt{G}. \tag{9.80}$$

In contrast to the fractional brane case (9.62), a B field is not necessary here. Configurations of D5-branes wrapped in a supersymmetric way on a non-vanishing two-cycle with $\text{Vol}(S^2) \neq 0$ thus lead a running of the coupling and broken conformal symmetry.

Similarly, for the theta angle in the complex gauge coupling $\tau = \vartheta/2\pi + i4\pi/g_{\text{YM}}^2$ we have

$$\Theta = -\frac{e^{-\phi}}{2\pi\alpha'^2} \int_{S^2} C_{(2)}. \tag{9.81}$$

The results (9.80) and (9.81) are analogous to the discussion of complex moduli leading to (8.19) and (8.20) for the conifold.

Wrapping a brane on a generic cycle breaks supersymmetry. It turns out that the condition for a cycle to be supersymmetric is equivalent to imposing a *twist* on the brane theory. Let us explain this by considering the case of N IIB D5-branes wrapped on a cycle with the topology of a two-sphere. Using S-duality, we can think equivalently in terms of NS5-branes. In four dimensions, supersymmetry is preserved if and only if there exists a covariantly constant spinor,

$$(\partial_\mu + \omega_\mu)\epsilon = 0, \tag{9.82}$$

where $\omega_\mu \equiv \frac{1}{4}\omega_{\mu ab}\Gamma^{ab}$ and $\omega_{\mu ab}$ is the spin connection. Equation (9.82) does not have a solution for the example considered here, since a non-trivial cycle does not admit covariantly constant spinors. However, if there is an R-symmetry present, we may introduce an external gauge field which couples to the corresponding current. In the case considered here, we have an $SO(4) \simeq SU(2) \times SU(2)$ R-symmetry. We may therefore redefine the covariant derivative to include a gauge connection A_μ in a $U(1)$ subgroup of the R-symmetry group,

$$D_\mu = \partial_\mu + \omega_\mu - A_\mu. \tag{9.83}$$

This is referred to as a *twist* of the original theory. Mathematically, since the cycle is non-trivially fibred within the Calabi–Yau space, A_μ is the connection on the non-trivial normal bundle in the Calabi–Yau directions perpendicular to the brane. Supersymmetry may be preserved by taking $\omega_\mu = A_\mu$, such that (9.83) gives rise to the spinor Killing equation

$$D_\mu\epsilon = \partial_\mu\epsilon = 0, \qquad (9.84)$$

which admits constant spinors as solutions. The term *twist* refers to the fact that when proceding from (9.83) to (9.84), the spin of the fields present is changed. In our case, the fields in the $\mathbb{R}^{3,1}$ part of the worldvolume of the D5-branes remains unchanged, so that we still have an ordinary field theory in these directions. The number of surviving supersymmetries depends on the way the $U(1)$ gauge connection A_μ is embedded in $SO(4)$. The subgroup of $SO(4)$ left unbroken by the twist provides the R-symmetries of the four-dimensional theory. The twist also determines the field content of the four-dimensional theory. In the case considered, we obtain the $\mathcal{N} = 1$ vector multiplet of pure $\mathcal{N} = 1$ Super Yang–Mills theory with R-symmetry group $U(1)_R$.

Let us now consider the supergravity solution for the NS5-branes in the wrapped geometry. The $SO(4)$ symmetry in the four directions transverse to the NS5-branes is isomorphic to the product of two $SU(2)$ groups, denoted $SU(2)^+ \times SU(2)^-$. The $U(1)$ responsible for the twist is taken to be a subgroup of $SU(2)^+$. The gravity solution may be found using seven-dimensional gauged supergravity. We may consistently truncate this theory to the sector invariant under $SU(2)^-$. Only one scalar field, the dilaton, survives in this truncation. Moreover, we expect the metric warp factor and the $U(1)_R$ gauge field to be non-trivial. To obtain a solution which is regular in the IR, it is necessary also to turn on the non-Abelian part of the $SU(2)^+$ gauge connection. An ansatz consistent with these requirements is given by

$$\mathrm{d}s_7^2 = e^{2f}(\mathrm{d}x_4^2 + N\alpha'\mathrm{d}\rho^2) + e^{2g}(\mathrm{d}\theta^2 + \sin^2\theta\,\mathrm{d}\varphi^2),$$
$$A = \frac{1}{2}\left[\sigma^3\cos\theta\,\mathrm{d}\varphi + \frac{a}{2}(\sigma^1 + i\sigma^2)(\mathrm{d}\theta - i\sin\theta\,\mathrm{d}\varphi) + \text{c.c.}\right], \qquad (9.85)$$

where the functions f, g and a depend only on ρ. Substituting the ansatz (9.85) into the seven-dimensional supergravity Lagrangian and integrating over S^2, we obtain the effective Lagrangian

$$\mathcal{L} = \frac{3}{16}e^{4Y}\left[16(Y')^2 - 2(h')^2 - \frac{1}{2}e^{-2h}|a'|^2 + 2e^{-2h} - \frac{1}{4}e^{-4h}(|a|^2 - 1)^2 + 4\right], \quad (9.86)$$

where $h = f - g$ and $4Y = 2h - 2\phi + \log(16/3)$, with ϕ the dilaton. The associated superpotential is

$$W = -\frac{3}{8}e^{-2h}\sqrt{(1 + 4e^{2h})^2 + 2(-1 + 4e^{2h})|a|^2 + |a|^4}. \qquad (9.87)$$

The supersymmetric solution to the associated equations of motion is given by [8, 11, 10]

$$e^{2h} = \rho\coth 2\rho - \frac{\rho^2}{\sinh^2 2\rho} - \frac{1}{4}, \qquad a = \frac{2\rho}{\sinh 2\rho}, \qquad e^{2\phi} = \frac{2e^h}{\sinh 2\rho}. \qquad (9.88)$$

This solution may be lifted to ten dimensions, for which it is convenient to use Euler angles on the three-sphere. Defining an $SU(2)$ group element by

$$g = \exp\left(\frac{i\psi\sigma^3}{2}\right)\exp\left(\frac{i\tilde\theta\sigma^1}{2}\right)\exp\left(\frac{i\tilde\varphi\sigma^3}{2}\right), \qquad (9.89)$$

and left-invariant one-forms by

$$\frac{i}{2}w^a \sigma^a = \mathrm{d}gg^{-1},\tag{9.90}$$

we have

$$w^1 + iw^2 = e^{-i\psi}(\mathrm{d}\tilde{\theta} + i\sin\tilde{\theta}\,\mathrm{d}\tilde{\varphi}),\quad w^3 = \mathrm{d}\psi + \cos\tilde{\theta}\,\mathrm{d}\tilde{\varphi},$$

with $\psi \in [0, 4\pi]$.

In string frame, the ten-dimensional solution for the wrapped NS5-branes is, with $A = \frac{1}{2}A^a\sigma^a$,

$$\mathrm{d}s^2 = \mathrm{d}x_4^2 + N\alpha'\left[\mathrm{d}\rho^2 + e^{2h(\rho)}(\mathrm{d}\theta^2 + \sin^2\theta\,\mathrm{d}\varphi^2) + \frac{1}{4}\sum_a (w^a - A^a)^2\right],$$

$$H_{(3)} = \frac{N\alpha'}{4}\left[-(w^1 - A^1) \wedge (w^2 - A^2) \wedge (w^3 - A^3) + \sum_a F^a \wedge (w^a - A^a)\right],\tag{9.91}$$

$$e^{2\phi} = e^{2\phi_0}\frac{2e^{h(\rho)}}{\sinh 2\rho}.$$

This is the Maldacena–Núñez solution. It has the important property of being regular in the IR.

Exercise 9.3.1   Show that in the IR, (9.91) asymptotes to $\mathbb{R}^7 \times S^3$ and is thus regular. This may be done using the result that in the IR, $a \to 1$, moreover A is pure gauge and can be reabsorbed by a coordinate transformation on S^3. Also, since $e^{2h} \to \rho^2$, the original S^2 is now contractible and combines with ρ to give $\mathbb{R}^3$.

In the IR, the square of the radius of the three-sphere is of order $N\alpha'$ and the supergravity approximation is valid when $N \gg 1$. The string coupling vanishes for large ρ and reaches its maximum value, e^{ϕ_0}, at $\rho = 0$. For $e^{\phi_0} \ll 1$, the string coupling is small everywhere.

From the gauge theory point of view, we would like to decouple the Kaluza–Klein modes in order to obtain pure Super Yang–Mills theory. The ratio between the gauge theory and Kaluza–Klein scales is of order $e^{-\phi_0}N$, so a decoupling requires $e^{\phi_0} \to \infty$. In order to be able to take this limit, we have to use the S-dual D5-brane solution

$$\mathrm{d}s^2 = e^{\phi_{D5}}\left[\mathrm{d}x_4^2 + N\alpha'\left[\mathrm{d}\rho^2 + e^{2h(\rho)}(\mathrm{d}\theta^2 + \sin^2\theta\,\mathrm{d}\phi^2) + \frac{1}{4}\sum_a (w^a - A^a)^2\right]\right],$$

$$F_{(3)} = \frac{N\alpha'}{4}\left[-(w^1 - A^1) \wedge (w^2 - A^2) \wedge (w^3 - A^3) + \sum_a F^a \wedge (w^a - A^a)\right],\tag{9.92}$$

$$e^{2\phi_{D5}} = e^{2\phi_{D5,0}}\frac{\sinh 2\rho}{2e^{h(\rho)}}.$$

This solution is very similar to the Klebanov–Strassler solution of section 9.3.3 since it involves a deformed conifold in the IR, as we discuss in more detail below. The squared radius of the IR three-sphere is $e^{\phi_{D5,0}}N\alpha'$ and the smallest value of the string coupling is $e^{\phi_{D5,0}} = e^{-\phi_0}$, reached for $\rho = 0$. The string coupling grows with ρ and eventually diverges in the UV.

Field theory properties

The field theory dual to the supergravity solution given has a running gauge coupling. This is obtained by inserting the supergravity solution (9.91) into the expression (9.80) for the gauge coupling. For large ρ, the result is

$$\frac{1}{g_{\text{YM}}^2} = \frac{N^2}{16\pi^2} \left(e^{2h(\rho)} + (a(\rho-1)^2 - 1)^2 \right) = \frac{N}{4\pi^2} \rho \tanh \rho. \tag{9.93}$$

The evaluation of (9.80) makes use of the fact that for large ρ, the angular part of the metric (9.91) corresponds to the space $T^{1,1}$ introduced in section 8.1.2, up to a rescaling. Consequently, the S^2 cycle in $T^{1,1}$ provides the appropriate integration region in (9.80). To derive the field theory β function from (9.93), we have to identify ρ with the field theory energy scale. This requires us to discuss chiral symmetry breaking first.

In the UV, the $U(1)_R$ symmetry is spontaneously broken to $\mathbb{Z}_{2N}$. In the Maldacena–Núñez solution, the $U(1)_R$ symmetry acts as a shift of the angle ψ. The metric is invariant under such a shift, however this is not the case for the R-R two-form $C_{(2)} \sim -\frac{N\alpha'}{2} \psi \sin\theta \, d\theta \wedge d\varphi$. In particular, the theta angle given by (9.81) takes the form

$$\Theta = \frac{e^{-\phi}}{2\pi\alpha'^2} \frac{N}{2\pi} \psi. \tag{9.94}$$

This is invariant under shifts

$$\psi \mapsto \psi + \frac{2\pi k}{N}, \tag{9.95}$$

which corresponds to a $\mathbb{Z}_{2N}$ symmetry in the UV at large ρ, since ψ has period 4π.

In the full solution, however, this symmetry is broken further to $\mathbb{Z}_2$, corresponding to $\psi = 0$ and $\psi = 2\pi$, for which S^2 has minimal volume. This is in agreement with the fact that in the dual gauge theory, $\mathbb{Z}_{2N}$ is broken to $\mathbb{Z}_2$ by the gaugino condensate $\langle \mathbb{Tr} \, \lambda\lambda \rangle = \Lambda^3$, which is a protected operator and related to a dynamical scale Λ. It is natural to identify the gaugino condensate with the gravity function $a(\rho)$ by virtue of

$$\mu^3 a(\rho) = \Lambda^3, \tag{9.96}$$

with μ the renormalisation scale and $a(\rho)$ given by (9.88).

The result (9.88) also provides the relation between ρ and μ required for calculating the β function. From (9.93) and (9.96) this is found to be

$$\beta(g_{\text{YM}}) \equiv \mu \frac{\partial}{\partial\mu} g_{\text{YM}}$$

$$= -\frac{3N g_{\text{YM}}^3}{16\pi^2} \left(1 - \frac{g_{\text{YM}}^2 N}{8\pi^2} + \frac{2\exp\left(-16\pi^2/(g_{\text{YM}}^2 N)\right)}{1 - \exp\left(-16\pi^2/(g_{\text{YM}}^2 N)\right)} \right)^{-1}. \tag{9.97}$$

Note that when neglecting the term involving exponentials, this corresponds to the NSVZ β function (3.223) with the appropriate group theory factors, while the term involving exponentials corresponds to non-perturbative corrections.

Finally, we note that the dual field theory is confining in the IR. This may be seen by calculating the Wilson loop in the D5-brane geometry (9.92). As explained in section 5.6.1, the gravity dual of the Wilson loop is obtained from a fundamental string with endpoints on the boundary at $\rho = \infty$. Here, it is found that the Wilson loop has an area law consistent with confinement. The string will minimise its energy by reaching $\rho = 0$ where the metric components $\sqrt{g_{xx} g_{tt}}$ have a minimum. The relevant contribution to the energy between two external sources is then due to a string placed at $\rho = 0$ and stretched in the x direction. We will return to a discussion of confinement in chapter 13, noting that the supergravity solution based on wrapped branes as presented in this section is regular in the IR and provides a further example of a dual to an $\mathcal{N} = 1$ field theory with confinement and chiral symmetry breaking.

9.3.5 Polchinski–Strassler flow

Our final example of an RG flow obtained from supergravity in ten dimensions also displays confinement. On the gravity side, it is closer in nature to the RG flows discussed in section 9.2.3 than the two previous examples, since it is obtained by adding relevant operators to $\mathcal{N} = 4$ Super Yang–Mills theory at the UV fixed point.

In the Polchinski–Strassler flow, the $\mathcal{N} = 4$ supersymmetry of the unperturbed $SU(N)$ gauge theory is broken down to $\mathcal{N} = 2$ or to $\mathcal{N} = 1$ by turning on mass terms for all the three chiral superfields Φ_i in (9.12). The case of turning on equal masses for two of the three $\mathcal{N} = 1$ chiral multiplets leads to a theory named $\mathcal{N} = 2^*$ theory. Likewise, turning on equal masses for all three chiral multiplets leads to $\mathcal{N} = 1^*$ theory. As we discuss below, for the dual type IIB supergravity solution this corresponds to turning on two-form potentials leading to a non-trivial three-form flux $G_{(3)} = F_{(3)} - \tau H_{(3)}$. Here $F_{(3)} = dC_{(2)}$ and $H_{(3)} = dB_{(2)}$ denote the R-R and the NS-NS three-form field strengths, respectively, and $\tau = C_{(0)} + ie^{-\phi}$ denotes the type IIB axion-dilaton.

$\mathcal{N} = 4$ Super Yang–Mills theory has Weyl fermions λ_α transforming as a **4** of the $SO(6)$ R-symmetry. We add a mass term

$$m^{\alpha\beta} \lambda_\alpha \lambda_\beta + \text{h.c.}, \tag{9.98}$$

which we assume to be diagonal, $m^{\alpha\beta} = m_\alpha \delta^{\alpha\beta}$. When one of the masses, say m_4, vanishes, the theory has an $\mathcal{N} = 1$ supersymmetric completion, giving rise to the $\mathcal{N} = 1$ superpotential contribution

$$\Delta W = \frac{1}{g_{\text{YM}}^2} (m_1 \operatorname{Tr} \Phi_1^2 + m_2 \operatorname{Tr} \Phi_2^2 + m_3 \operatorname{Tr} \Phi_3^2). \tag{9.99}$$

The fermion λ_4 is then the gluino. As given in table 9.1, the fermion bilinear transforms as the $(\mathbf{4} \otimes \mathbf{4})_{\text{sym}} = \mathbf{10_c}$ of $SO(6)$, and the mass matrix transforms as the $\overline{\mathbf{10_c}}$.

To find the corresponding supergravity solution, we note that the $\mathbf{10_c}$ and $\overline{\mathbf{10_c}}$ representations may also be written as imaginary self-dual antisymmetric three-tensors,

$$*_6 T_{mnp} \equiv \frac{1}{3!} \epsilon_{mnp}{}^{qrs} T_{qrs} = \pm i T_{mnp}, \tag{9.100}$$

with $+$ for the $\mathbf{10_c}$ and $-$ for the $\overline{\mathbf{10_c}}$ representation. $*_6$ denotes the Hodge dual in six-dimensional flat space. The indices run from 4 to 9. To relate the tensor T to the fermion masses, it is convenient to adopt complex coordinates z^i,

$$z^1 = \frac{x^4 + ix^7}{\sqrt{2}}, \quad z^2 = \frac{x^5 + ix^8}{\sqrt{2}}, \quad z^3 = \frac{x^6 + ix^9}{\sqrt{2}}. \tag{9.101}$$

Under a rotation $z^i \to e^{i\phi_i} z^i$, the spinors in the **4** transform as

$$\begin{aligned}
\lambda_1 &\to e^{i(\phi_1 - \phi_2 - \phi_3)/2}\lambda_1, \quad \lambda_2 \to e^{i(-\phi_1 + \phi_2 - \phi_3)/2}\lambda_2, \\
\lambda_3 &\to e^{i(-\phi_1 - \phi_2 + \phi_3)/2}\lambda_3, \quad \lambda_4 \to e^{i(\phi_1 + \phi_2 + \phi_3)/2}\lambda_4.
\end{aligned} \tag{9.102}$$

This implies that a diagonal mass term transforms in the same way as the form

$$T_3 = m_1 dz^1 \wedge d\bar{z}^2 \wedge d\bar{z}^3 + m_2 d\bar{z}^1 \wedge dz^2 \wedge d\bar{z}^3 + m_3 d\bar{z}^1 \wedge d\bar{z}^2 \wedge dz^3 + m_4 dz^1 \wedge dz^2 \wedge dz^3. \tag{9.103}$$

In the $\mathcal{N} = 1$ supersymmetric case, the non-zero components are

$$T_{1\bar{2}\bar{3}} = m_1, \quad T_{\bar{1}2\bar{3}} = m_2, \quad T_{\bar{1}\bar{2}3} = m_3, \tag{9.104}$$

and in the equal mass case

$$T_{\bar{i}jk} = T_{i\bar{j}k} = T_{\bar{i}j\bar{k}} = m\epsilon_{ijk}. \tag{9.105}$$

Both cases correspond to an anti-self-dual form satisfying $*^6 T = -iT$. This condition ensures that T_3 forms a $\overline{\mathbf{10}}_{\mathbf{c}}$ representation of the $SO(6)$ isometry group of S^5, and hence transforms in the same way as the fermion mass matrix.

Let us now consider the IIB supergravity solution which corresponds to the added mass terms. It is natural to assume that the tensor T_{ijk} leads to a non-trivial contribution to the IIB supergravity field $G_{(3)} \equiv F_{(3)} - \tau H_{(3)}$, with τ the unperturbed complex gauge coupling of $\mathcal{N} = 4$ Super Yang–Mills theory. The tensor field $G_{(3)}$ with the necessary asymptotic behaviour to be dual to the mass perturbation is given by

$$G_{(3)} = e^{-\phi} \frac{1}{\sqrt{2}} d(HS_2), \tag{9.106}$$

where the two-form potential S_2 is constructed from the components of T_3 by virtue of

$$S_2 = \frac{1}{2} T_{ijk} x^i dx^j \wedge dx^k, \tag{9.107}$$

and $H_{(r)} = L^4/r^4$. The unperturbed background is given by the $AdS_5 \times S^5$ metric with the five-form field strength F_5 and constant axion-dilaton. At linear order in the masses m_p which parametrise the mass deformation, the supergravity solution is given by the unperturbed metric of $AdS_5 \times S^5$, the non-trivial $G_{(3)}$ as given by (9.106) and an induced six-form potential

$$C_{(6)} = \frac{2}{3} B_{(2)} \wedge C_{(4)}. \tag{9.108}$$

Beyond the linear approximation, at quadratic order in the masses m the corrections also backreact on the metric, the four-form potential C_4, and the complex axion-dilaton τ. At this order, the deformed ten-dimensional near-horizon metric reads

$$ds^2 = (H^{-1/2} + h_0)\eta_{\mu\nu}dx^\mu dx^\nu + \left[(5H^{1/2} + p)I_{ij} + (H^{1/2} + q)\frac{x^i x^j}{r^2} + wW_{ij}\right]dx^i dx^j, \tag{9.109}$$

where $H(r) = L^4/r^4$ and the tensors I_{ij} and W_{ij} are given by

$$I_{ij} = \frac{1}{5}\left(\delta_{ij} - \frac{x^i x^j}{r^2}\right),$$

$$W_{ij} = \frac{1}{|T_3|^2}\mathrm{Re}(T_{ipk}\bar{T}_{jpl})\frac{x^k x^l}{r^2} - I_{ij}, \qquad |T_3|^2 = \frac{1}{3!}T_{ijk}\bar{T}_{ijk}. \tag{9.110}$$

The indices i, j run from 4 to 9. The functions h_0, w, p, q read

$$w = -M^2 L^2 H, \qquad\qquad p = -\frac{3M^2 L^2}{8}H,$$

$$q = \frac{M^2 L^2}{72}H, \qquad\qquad h_0 = \frac{7M^2 L^2}{72}, \tag{9.111}$$

where $M^2 = m_1^2 + m_2^2 + m_3^2 = |T_3|^2$. The dilaton solution obtained from the equations of motion can be factorised into a purely radial and a purely angular part according to $\phi = \varphi Y_+$, given by

$$\varphi = \frac{M^2 L^2}{6}Z^{1/2},$$

$$Y_+ = \frac{3}{M^2 r^2}\left(m_2 m_3(x_4^2 - x_7^2) + m_1 m_3(x_5^2 - x_8^2) + m_1 m_2(x_6^2 - x_9^2)\right).$$

It is essential to note that the metric (9.109) has a curvature singularity at the origin, where the Ricci scalar is given by

$$R = M^2 \frac{5}{2}\frac{L^2}{r^2}. \tag{9.112}$$

Polarisation of D3-branes

The IR metric singularity may be cured by the *Myers dielectric effect*. As was found by Myers [12], a stack of Dp-branes couples to higher r-form potentials ($r > p + 1$) due to the non-commutativity of its matrix valued positions. This coupling has an interpretation as a polarisation of the Dp-brane, with its worldvolume becoming higher-dimensional. The D3-branes considered here may polarise either into D5-branes or into NS5-branes. To lowest non-trivial order in the masses, in the presence of potentials $B_{(2)}$ and C_2 which generate the non-vanishing three-form flux $G_{(3)}$, the effective potential for the positions of the matrix valued coordinates x^i is minimised if

$$[x^i, x^j] = i4\sqrt{2}\pi\alpha'\zeta\,\mathrm{Im}\,T_{ijk}\,x^k. \tag{9.113}$$

Using the expressions for T_{ijk} for the real coordinates $x^4, x^5, \ldots, x^9$, we find the concrete form of the polarisations. Let us discuss the $\mathcal{N} = 2$ supersymmetric case, where $m_1 = 0$, $m_2 = m_3 = m$, explicitly. The only non-vanishing independent components of T are given by

$$T_{456} = iT_{789} = iT_{567} = T_{489} = \frac{m}{\sqrt{2}}. \tag{9.114}$$

Inserting the non-vanishing imaginary parts into the equation for the embedding matrices x^i gives rise to two $su(2)$ Lie algebras in the x^5, x^6, x^7 and x^7, x^8, x^9 directions. That means the D3-branes polarise into two S^2, having in common the x^7 direction. The equations for the two-spheres read

$$(x^5)^2 + (x^6)^2 + (x^7)^2 = r_0^2, \qquad (x^7)^2 + (x^8)^2 + (x^9)^2 = r_0^2. \qquad (9.115)$$

In the $\mathcal{N} = 1^*$ case, a similar polarisation mechanism takes place, involving a total of four superposed two-spheres. It is expected that in the supergravity background which is exact to all orders in the mass perturbation, the polarisation shells will remove the IR singularity. However, the all-order solution has not yet been found.

Field theory dual

The $\mathcal{N} = 1^*$ theory is a confining gauge theory which has a mass gap. This may be shown by a Wilson loop computation. Moreover, the theory has a rich vacuum structure which arises from its superpotential

$$W = \int d^2\theta \left(2\sqrt{2}\, \mathrm{Tr}(\Phi_1[\Phi_2, \Phi_3]) + m \sum_{i=1}^{3}(\Phi_i)^2 \right). \qquad (9.116)$$

The associated F-term equation reads

$$[\Phi_i, \Phi_j] = -\frac{m}{\sqrt{2}}\epsilon_{ijk}\Phi_k. \qquad (9.117)$$

Its solutions are N-dimensional, generally reducible, representations of $SU(2)$. For a generic vacuum, the matrices Φ will have a block diagonal structure, where the blocks represent irreducible $SU(2)$ representations of different dimension n_i, including dimension 1, such that $\sum_i n_i = N$. There are two particularly interesting vacuum solutions. One of them is the Higgs vacuum, corresponding to the N-dimensional irreducible representation of $SU(2)$. In this case the gauge group is completely broken and there is a mass gap already at the classical level. The other vacuum is characterised by vanishing vacuum expectation values for the scalar fields, $\langle \Phi \rangle = 0$, i.e. N copies of the trivial representation. $SU(N)$ is unbroken and the theory is expected to confine and to have N distinct vacua parametrised by the gaugino condensate $\langle \lambda\lambda \rangle$, similarly to the wrapped and fractionalised brane scenarios.

9.4 Further reading

Zamolodchikov's proof of the C-theorem in two dimensions is given in [13]. A pedagogical description of this theorem may be found in [14]. Recent advances to the C-theorem in four dimensions are found in [1, 2]. For a discussion of this approach, see also [15].

The interpolating field theory flow of section 9.1.3 was constructed by Leigh and Strassler in [3]. The holographic interpolating flow dual to the Leigh–Strassler flow was constructed by Freedman, Gubser, Pilch and Warner in [5].

The gravity dual of the marginally β-deformed theory is given in [16].

The Klebanov–Strassler flow from fractional branes was given in [7] This is based on earlier results on flows from fractional branes in [17, 18, 19]. In particular, in [18] there is a discussion of how N D3-branes and M fractional D3-branes lead to an $SU(N) \times SU(N + M)$ gauge group. For the T-duality transformation to the type IIA picture of the Klebanov–Strassler flows, see [20, 21, 22]. The Affleck–Dine–Seiberg superpotential is given in [9].

The Maldacena–Núñez flow was constructed in [10], based on earlier results of Chamseddine and Volkov [8, 11].

Introductory reviews of the flows based on wrapped and fractional branes may be found in [23, 24, 25]. Fractional D-branes and their gauge duals are discussed in [26].

The Polchinski–Strassler flow is given in [27]. A second-order solution was given in [28]. The description here is based on [29]. A further discussion of this flow and its non-supersymmetric version may be found in [30].

References

[1] Komargodski, Zohar, and Schwimmer, Adam. 2011. On renormalization group flows in four dimensions. *J. High Energy Phys.*, **1112**, 099.

[2] Komargodski, Zohar. 2012. The constraints of conformal symmetry on RG flows. *J. High Energy Phys.*, **1207**, 069.

[3] Leigh, Robert G., and Strassler, Matthew J. 1995. Exactly marginal operators and duality in four-dimensional $\mathcal{N} = 1$ supersymmetric gauge theory. *Nucl. Phys.*, **B447**, 95–136.

[4] Konishi, K. 1984. Anomalous supersymmetry transformation of some composite operators in SQCD. *Phys. Lett.*, **B135**, 439.

[5] Freedman, D. Z., Gubser, S. S., Pilch, K., and Warner, N. P. 1999. Renormalization group flows from holography supersymmetry and a c-theorem. *Adv. Theor. Math. Phys.*, **3**, 363–417.

[6] Khavaev, Alexei, Pilch, Krzysztof, and Warner, Nicholas P. 2000. New vacua of gauged $\mathcal{N} = 8$ supergravity in five-dimensions. *Phys. Lett.*, **B487**, 14–21.

[7] Klebanov, Igor R., and Strassler, Matthew J. 2000. Supergravity and a confining gauge theory: Duality cascades and χ SB resolution of naked singularities. *J. High Energy Phys.*, **0008**, 052.

[8] Chamseddine, Ali H., and Volkov, Mikhail S. 1997. Non-Abelian BPS monopoles in $\mathcal{N} = 4$ gauged supergravity. *Phys. Rev. Lett.*, **79**, 3343–3346.

[9] Affleck, Ian, Dine, Michael, and Seiberg, Nathan. 1984. Dynamical supersymmetry breaking in supersymmetric QCD. *Nucl. Phys.*, **B241**, 493–534.

[10] Maldacena, Juan Martin, and Núñez, Carlos. 2001. Towards the large N limit of pure $\mathcal{N} = 1$ Super Yang-Mills. *Phys. Rev. Lett.*, **86**, 588–591.

[11] Chamseddine, Ali H., and Volkov, Mikhail S. 1998. Non-Abelian solitons in $\mathcal{N} = 4$ gauged supergravity and leading order string theory. *Phys. Rev.*, **D57**, 6242–6254.

[12] Myers, Robert C. 1999. Dielectric branes. *J. High Energy Phys.*, **9912**, 022.

[13] Zamolodchikov, A.B. 1986. Irreversibility of the flux of the renormalization group in a 2D field theory. *JETP Lett.*, **43**, 730–732.

[14] Cardy, John L. 1996. *Scaling and Renormalization in Statistical Physics*. Cambridge University Press.

[15] Jack, I., and Osborn, H. 2010. Constraints on RG flow for four dimensional quantum field theories. *Nucl. Phys.*, **B883**, 425–500.

[16] Lunin, Oleg, and Maldacena, Juan Martin. 2005. Deforming field theories with $U(1) \times U(1)$ global symmetry and their gravity duals. *J. High Energy Phys.*, **0505**, 033.

[17] Klebanov, Igor R., and Witten, Edward. 1998. Superconformal field theory on three-branes at a Calabi-Yau singularity. *Nucl. Phys.*, **B536**, 199–218.

[18] Gubser, Steven S., and Klebanov, Igor R. 1998. Baryons and domain walls in an $\mathcal{N} = 1$ superconformal gauge theory. *Phys. Rev.*, **D58**, 125025.

[19] Klebanov, Igor R., and Tseytlin, Arkady A. 2000. Gravity duals of supersymmetric $SU(N) \times SU(N + M)$ gauge theories. *Nucl. Phys.*, **B578**, 123–138.

[20] Aharony, Ofer, and Hanany, Amihay. 1997. Branes, superpotentials and superconformal fixed points. *Nucl. Phys.*, **B504**, 239–271.

[21] Dasgupta, Keshav, and Mukhi, Sunil. 1999. Brane constructions, conifolds and M-theory. *Nucl. Phys.*, **B551**, 204–228.

[22] Uranga, Angel M. 1999. Brane configurations for branes at conifolds. *J. High Energy Phys.*, **9901**, 022.

[23] Bertolini, M. 2003. Four lectures on the gauge/gravity correspondence. Lectures given at SISSA/ISAS Trieste. ArXiv:hep-th/0303160.

[24] Bigazzi, F., Cotrone, A.L., Petrini, M., and Zaffaroni, A. 2002. Supergravity duals of supersymmetric four-dimensional gauge theories. *Riv. Nuovo Cimento*, **25N12**, 1–70.

[25] Imeroni, Emiliano. 2003. The gauge/string correspondence towards realistic gauge theories. ArXiv:hep-th/0312070.

[26] Bertolini, M., Di Vecchia, P., Frau, M., Lerda, A., Marotta, R., and Pesando, I. 2001. *J. High Energy Phys.*, **0102**, 014.

[27] Polchinski, Joseph, and Strassler, Matthew J. 2000. The string dual of a confining four-dimensional gauge theory. ArXiv:hep-th/0003136.

[28] Freedman, Daniel Z., and Minahan, Joseph A. 2001. Finite temperature effects in the supergravity dual of the $\mathcal{N} = 1^*$ gauge theory. *J. High Energy Phys.*, **0101**, 036.

[29] Apreda, Riccardo, Erdmenger, Johanna, Lüst, Dieter, and Sieg, Christoph. 2007. Adding flavour to the Polchinski-Strassler background. *J. High Energy Phys.*, **0701**, 079.

[30] Taylor, Marika. 2001. Anomalies, counterterms and the $\mathcal{N} = 0$ Polchinski-Strassler solutions. ArXiv:hep-th/0103162.

10

Duality with D-branes in supergravity

So far we have studied examples of the AdS/CFT correspondence which are motivated by the near-horizon limit of a stack of D-branes placed either in flat space or in a more involved geometry such as the conifold. In this chapter we will consider examples where additional D-branes are placed in the supergravity solution after the near-horizon limit has been taken. This approach has several motivations. One possible application is to wrap branes on non-trivial cycles in the geometry resulting from the near-horizon limit. Such branes correspond to soliton-like states in the dual conformal field theory. These states are non-perturbative from the point of view of the $1/N$ expansion. Consequently, they allow information about the stringy nature of the correspondence to be uncovered even in its weakest form, where λ and N are large. The soliton-like field theory states include the pointlike *baryon vertex*, one-dimensional *flux tubes* and higher dimensional *domain walls*.

Here, however, we will focus on the second important application of embedding additional D-branes into the near-horizon geometry, the *flavour branes*. Adding additional D-branes to the supergravity solution in the near-horizon limit gives rise to a modification of the original AdS/CFT correspondence which involves field theory degrees of freedom that transform in the *fundamental* representation of the gauge group. This is in contrast to the fields of $\mathcal{N} = 4$ Super Yang–Mills theory which transform in the *adjoint* representation of the gauge group. This is particularly useful for describing strongly coupled quantum field theories which are similar to QCD, since the quark fields in QCD transform in the fundamental representation. From an anti-fundamental and a fundamental field, a gauge invariant bilinear or *meson* operator may be formed. The key idea is then to conjecture that the meson operators are dual to the fluctuations of flavour branes embedded in the dual supergravity background. As in the original form of AdS/CFT correspondence, it is essential that the field theory meson operators and the flavour brane fluctuations are in the same representation of the underlying symmetry group.

10.1 Branes as flavour degrees of freedom

To introduce fundamental flavour fields, we return to the example of D3-branes in flat space and add a stack of N_f Dp-branes, the *flavour* branes. In addition to 3−3 strings beginning and ending on the stack of D3-branes, which give rise to the degrees of freedom of $\mathcal{N} = 4$ super Yang–Mills theory, there are now other types of open strings present: 3−p strings beginning on the stack of D3-branes and ending on the stack of Dp-branes, as well as p−3 and $p-p$ strings. The latter begin and end on the stack of Dp-branes. We will see below that in the Maldacena limit, the $p-p$ strings decouple from the $p-3$, $3-p$ and $3-3$ strings. On the other hand, the endpoints of $3-p$ and $p-3$ strings correspond to pointlike excitations in the $\mathbf{N}$ or $\bar{\mathbf{N}}$ of $SU(N)$ on the worldvolume of the D3-branes and thus transform in the fundamental representation of $SU(N)$.

Similar brane constructions are also possible in type IIA theory, for instance with a background of D4-branes as discussed in section 8.4. As an example, in chapter 13 we will encounter the Sakai–Sugimoto model, which involves D4, D8 and $\overline{\text{D8}}$ branes.

10.1.1 D3/Dp-brane systems

As described above, let us add a stack of N_f Dp-branes to the stack of N D3-branes which are extended along the spacetime directions x^0, x^1, x^2 and x^3. For simplicity, we refer to the 0, 1, 2 and 3 directions, respectively.

In type IIB string theory, there are Dp-branes for any odd $p \leq 7$. Nevertheless, we restrict our attention to D3-, D5- and D7-branes. Instanton-like D(-1)-branes do not introduce flavour degrees of freedom on the D3-brane worldvolume. Although D1-branes are present in type IIB string theory, we cannot use them as flavour branes since the 1-1 strings, i.e. the open strings which begin and end on the stack of D1-branes, will be dynamical and therefore do not decouple in the Maldacena limit. This implies that for D1-branes there would be unwanted additional degrees of freedom, namely gauge bosons due to $1-1$ strings, in addition to the $\mathcal{N} = 4$ vector multiplet and the desired flavour degrees of freedom given by $3-1$ and $1-3$ strings.

To see whether the gauge theory arising from $p-p$ strings is dynamical, we have to compare the 't Hooft coupling λ_{Dp} of the flavour Dp-brane to that of the colour D3-branes. For the D3-branes we have

$$\lambda_{D3} \equiv g_{D3}^2 N = 2\pi g_s N, \tag{10.1}$$

whereas for the flavour Dp-brane

$$\lambda_{Dp} \equiv g_{Dp}^2 N_f = (2\pi)^{p-2} \alpha'^{(p-3)/2} g_s N_f, \tag{10.2}$$

where g_{D3} and g_{Dp} are the Yang–Mills couplings in the appropriate dimension for the D3- and Dp-branes, respectively, see (4.110). Their quotient reads

$$\frac{\lambda_{Dp}}{\lambda_{D3}} = \frac{N_f}{N}(2\pi)^{p-3}\alpha'^{(p-3)/2}. \tag{10.3}$$

For $p > 3$ the quotient vanishes in the Maldacena limit $\alpha' \to 0$, i.e. the $p-p$ strings are non-dynamical in this limit. For $p < 3$ it clearly diverges, i.e. we have to take gauge bosons into account which arise from $p-p$ strings. For $p = 3$ the quotient is finite.

We require the flavour Dp-branes to satisfy the following conditions.

(1) The D3/Dp brane intersection is supersymmetric, i.e. the number of Neumann–Dirichlet directions is 0, 4 or 8. This ensures stability.
(2) The flavour Dp-branes extend into the timelike direction, i.e. x^0.

Table 10.1 All possible supersymmetric flavour branes in D3-brane background									
	0	1	2	3	4	5	6	7	8	9
N D3	•	•	•	•	–	–	–	–	–	–
N_f D7	•	•	•	•	•	•	•	•	–	–
N_f D7	•	•	–	–	•	•	•	•	•	•
N_f D5	•	•	•	–	•	•	•	–	–	–
N_f D5	•	–	–	–	•	•	•	•	•	–
N_f D3	•	•	–	–	•	•	–	–	–	–

(3) The branes extend into at least one of the six extra dimensions perpendicular to the D3-branes, denoted by $x^4, x^5, \ldots, x^8$ or x^9. This condition is necessary since the flavour branes should extend in the radial direction r of AdS_5, with r defined by $r^2 = \sum_{i=4}^{9}(x^i)^2$. Otherwise, the flavour degrees would only be present for one particular energy scale.

In table 10.1, we list the possible flavour D-brane embeddings which satisfy these conditions. Directions which are filled by the D-branes considered are marked by •, while directions perpendicular to these branes are marked by −.

For all the D3/Dp intersections given in table 10.1, it is possible to write down the Lagrangian of the corresponding field theory explicitly. Generically, the presence of the flavour branes breaks part of the supersymmetry. For the examples of the D3/D7 and D3/D5 intersections, we work out the field theory Lagrangian as well as the duality conjecture in detail below. For all intersections, let us note that if a D3/Dp intersection has four Neumann–Dirichlet (ND) directions, then the corresponding flavour fields as obtained from $3-p$ and $p-3$ strings give rise to non-chiral flavour multiplets in the corresponding field theory. This is due to the fact that the fields are arranged in supersymmetry hypermultiplets. On the other hand, with eight ND directions the flavour fields are chiral.

10.1.2   D3/D7-brane system

The first example which we consider in detail is the D3/D7-brane system, which has important generalisations and widespread applications. The D3-branes are extended along 0123, whereas the N_f D7-branes are located in the 01234567 directions. This configuration preserves one quarter of the total amount of supersymmetry in type IIB string theory, corresponding to eight real Poincaré supercharges. It has an $SO(4) \times SO(2)$ isometry in the directions transverse to the D3-branes. The $SO(4)$ group rotates the x^4, x^5, x^6, x^7 directions filled by the D7-branes, while the $SO(2)$ group acts on the x^8, x^9 which are perpendicular to the D7-branes. Separating the D7-branes from the D3-branes in the 89 plane, by placing the D7-branes at $x^8 = l_q$ and $x^9 = 0$, we explicitly break the $SO(2)$ rotation symmetry. We will confirm that these geometrical symmetries are also present in the dual field theory. The dual field theory is (3+1)-dimensional $\mathcal{N} = 4$ Super Yang–Mills theory coupled to flavour fields preserving $\mathcal{N} = 2$ supersymmetry. The Lagrangian of this field theory can conveniently be written down in $\mathcal{N} = 1$ superspace formalism. As reviewed in section 3.3.6, the $\mathcal{N} = 4$ vector multiplet decomposes into the vector multiplet W_α and the three chiral superfields Φ_1, Φ_2, Φ_3 under $\mathcal{N} = 1$ supersymmetry. The vector multiplet W_α and one of the three chiral superfields (e.g. Φ_3 without loss of generality) can be grouped into an $\mathcal{N} = 2$ vector multiplet. The remaining two chiral multiplets Φ_1 and Φ_2 form an $\mathcal{N} = 2$ hypermultiplet. Moreover, the flavour fields are given in terms of the $\mathcal{N} = 1$ chiral multiplets $Q^r, \tilde{Q}_r$ ($r = 1, ..., N_f$). The Lagrangian is thus given by

$$\mathcal{L} = \int d^4\theta \left(\text{Tr}(\bar{\Phi}_I e^V \Phi_I e^{-V}) + Q_r^\dagger e^V Q^r + \tilde{Q}_r^\dagger e^{-V} \tilde{Q}^r \right)$$
$$+ \text{Im}\left(\tau \int d^2\theta \, \text{Tr}(W^\alpha W_\alpha) \right) + \int d^2\theta \, W + \text{c.c.}, \qquad (10.4)$$

where the superpotential W is

$$W = \text{Tr}(\varepsilon_{IJK} \Phi_I \Phi_J \Phi_K) + \tilde{Q}_r (m_q + \Phi_3) Q^r, \qquad (10.5)$$

and $\tau = \vartheta/(2\pi) + 4\pi i/g^2$ is the complex gauge coupling. m_q is the mass of the hypermultiplet of flavour fields.

For massless flavour fields, i.e. for $m_q = 0$, the Lagrangian is classically invariant under conformal transformations $SO(4, 2)$.[1] Moreover, if we assign the quantum numbers listed in table 10.2 to the components of the $\mathcal{N} = 1$ superfields, the theory is invariant under the following global symmetries: the R-symmetries $SU(2)_R$ and $U(1)_R$ as well as $SU(2)_\Phi$. The global symmetry $SU(2)_\Phi$ rotates the scalars Φ_1 and Φ_2 in the adjoint hypermultiplet. Note that the mass term in the Lagrangian breaks the $U(1)_R$ symmetry explicitly. If all N_f flavour fields have the same mass m_q, the field theory is invariant under a global $U(N_f)$ flavour group. The baryonic $U(1)_B$ symmetry is a subgroup of the $U(N_f)$ flavour group. The fundamental superfields Q^r ($\tilde{Q}_r$) are charged $+1$ (-1) under $U(1)_B$, while the adjoint fields are inert.

The mass m_q in (10.5) is related to the separation distance l_q between D3-branes and D7-branes by the relation

$$m_q = \frac{l_q}{2\pi\alpha'}. \qquad (10.6)$$

To see this, consider the energy of a non-excited string stretched between the D3-branes and the D7-brane probe. Its energy is given precisely by the right-hand side of (10.6). This string may be identified with a quark in the dual field theory. Consequently, the quark mass is given by the mass of this string.

The field theory symmetries described above may be mapped to symmetries of the D3/D7-brane intersection and hence also to the dual gravity description. The $U(N_f)$ flavour symmetry and therefore also the baryonic $U(1)_B$ symmetry, which are both global on the

---

[1] However, note that the scale invariance is broken at the quantum level since the β function is proportional to N_f/N and therefore non-vanishing. In the limit $N \to \infty$ with N_f being fixed, which we will use in later chapters, the β function is approximately zero, i.e. we can treat the theory as being scale invariant also at the quantum level.

Table 10.2	Fields of the D3/D7 low-energy effective field theory and their quantum numbers under global symmetries						
$\mathcal{N}=2$	Components	Spin	$SU(2)_\Phi \times SU(2)_R$	$U(1)_R$	Δ	$U(N_f)$	$U(1)_B$
(Φ_1, Φ_2)	X^4, X^5, X^6, X^7	0	$(\frac{1}{2}, \frac{1}{2})$	0	1	1	0
hypermultiplet	λ_1, λ_2	$\frac{1}{2}$	$(\frac{1}{2}, 0)$	-1	$\frac{3}{2}$	1	0
(Φ_3, W_α)	$X_V^A = (X^8, X^9)$	0	$(0, 0)$	$+2$	1	1	0
vector	λ_3, λ_4	$\frac{1}{2}$	$(0, \frac{1}{2})$	$+1$	$\frac{3}{2}$	1	0
multiplet	A_μ	1	$(0, 0)$	0	1	1	0
$(Q, \tilde{Q})$	$q^m = (q, \bar{\tilde{q}})$	0	$(0, \frac{1}{2})$	0	1	N_f	$+1$
fundamental							
hypermultiplet	$\psi_i = (\psi, \tilde{\psi}^\dagger)$	$\frac{1}{2}$	$(0, 0)$	∓ 1	$\frac{3}{2}$	N_f	$+1$

field theory side, are realised by a local gauge symmetry, which in the case of N_f D7-branes is $U(N_f)$. The $U(1)_R$ symmetry corresponds to the $SO(2)$ symmetry of rotations in the 89 plane. Evidence for the matching of these symmetries is the fact that both symmetries are only present for massless flavour fields, i.e. if the D3- and D7-branes are not separated in the transverse 89 plane. The $SO(4)$ rotational invariance in the 4567 subspace can be decomposed into two $SU(2)$ groups, denoted $SU(2)_L$ and $SU(2)_R$. The $SU(2)_R$ symmetry of the brane intersection is mapped to the $\mathcal{N} = 2$ R-symmetry $SU(2)_R$ on the field theory side. Finally, the global $SU(2)_L$ symmetry of the brane intersection is identified with $SU(2)_\Phi$.

In table 10.2, we summarise the component fields and their quantum numbers for the symmetries present, using the following nomenclature. In the first column, we write $\mathcal{N} = 2$ multiplets in terms of $\mathcal{N} = 1$ superfields. In the second column, the scalar component fields $X^4, \ldots, X^9$ correspond to coordinates in the $x^4, \ldots, x^9$ directions. Strictly speaking, they are related to the six scalar fields ϕ^i of $\mathcal{N} = 4$ Super Yang–Mills theory as given in table 3.6 by $\phi^i = X^{i+3}/(2\pi\alpha')$. Here we follow the notation introduced in chapter 5 on page 185 and refer to the $\mathcal{N} = 4$ scalar fields as X^i. The fermions ψ, $\tilde{\psi}$ are Dirac spinors like those in (1.141).

For the field theory given, gauge invariant composite operators may now be constructed which transform in suitable representations of the $SU(2)_\Phi \times SU(2)_R \times U(1)_R$ symmetry group. These are then expected to be dual to the supergravity fluctuations which transform in the same representations. We continue by discussing the supergravity fluctuations in the D3/D7-brane setup, after which we will construct the holographic dictionary matching field theory operators and supergravity fields. For concreteness, already at this stage let us give an example of a mesonic operator. Such an example is provided by the scalar field theory operator

$$\mathcal{M}^A = \bar{\psi}_i \sigma^A{}_{ij} \psi_j + \bar{q}^m X_V^A q^m, \qquad i, m = 1, 2, \tag{10.7}$$

constructed from the fields given in table 10.2 with $\sigma^A \equiv (\sigma^1, \sigma^2)$ a doublet of Pauli matrices, a singlet under both $SU(2)_\Phi$ and $SU(2)_R$ and has charge $+2$ under the

$U(1)_R$ symmetry. The conformal dimension is $\Delta = 3$. This operator may be viewed as a supersymmetric generalisation of a meson in QCD, which is made of a quark bilinear combining two fermionic quark fields.

10.2 AdS/CFT correspondence with probe branes

10.2.1 Probe limit

On the supergravity side, the action describing the combined system of D3-branes in IIB supergravity plus the additional Dp-branes is given by

$$S = S_{\text{IIB}} + S_{\text{D}p}, \tag{10.8}$$

where $S_{\text{D}p}$ is the sum of the DBI and CS actions as given in section 4.4.1. In general, the presence of $S_{\text{D}p}$ gives rise to source terms in the equations of motion of type IIB supergravity, such that $AdS_5 \times S^5$ is no longer a solution. This is referred to as *backreaction* of the Dp-branes on the D3-brane geometry. In particular, the backreaction may cause a running of the dilaton. This is the case for example when embedding D7-branes for which the equation of motion for the dilaton reads schematically

$$\nabla^2 \phi = \frac{\partial}{\partial \phi} \left(\mathcal{L}_{\text{IIB}} - 2\kappa_{10}^2 N_f \mu_7 \mathcal{L}_{\text{D7}} \right), \tag{10.9}$$

where $2\kappa_{10}^2 = (2\pi)^7 g_s^2 \alpha'^4$ and $\mu_p = (2\pi)^{-p} g_s^{-1} \alpha'^{-(p+1)/2}$ and $\mathcal{L}$ refers to the respective Lagrangians.

The simplest way to analyse the D3/Dp system on the gravity side is to work in the limit where the Dp-branes are treated as *probes*. The term *probe brane* refers to the fact that only a very small number N_f of D7-branes is added, while the number N of D3-branes is taken to infinity, such that $N_f/N \to 0$ in the near-horizon limit. Usually, $N_f = 1$ or $N_f = 2$. On the gravity side, this limit implies that we neglect the backreaction of the Dp-branes on the near-horizon geometry of the D3-branes. This implies that in the limit $N_f/N \to 0$, the terms involving $\mathcal{L}_{\text{D7}}$ in (10.9) are neglected. On the field theory side, this corresponds to the *quenched approximation* often used in lattice gauge theory, in which quark loops are neglected.

In the case of D7-branes, the backreaction leads to a positive β function for the field theory gauge coupling, with $\beta \propto N_f/N$ according to (10.9). The gauge coupling will diverge at a finite value of the renormalisation scale, giving rise to a Landau pole. Therefore for D7-branes, only the low-energy physics can be described. This applies also to the probe limit. Nevertheless, as we will see, there are important applications even in this limit.

10.2.2 Probe D7-branes

Let us consider a single probe D7-brane embedded into the near-horizon limit of D3-branes, i.e. into $AdS_5 \times S^5$. Due to the presence of the D7-brane, there are new degrees of

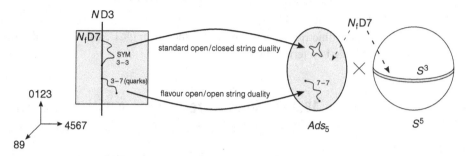

Figure 10.1 Schematic representation of AdS/CFT with added flavour for a probe of N_f D7-branes embedded in a background of N D3-branes, with $N_f \ll N$. In addition to the original AdS/CFT duality, open string degrees of freedom representing quarks are conjectured to be mapped to open strings beginning and ending on the D7-brane probe, which asymptotically near the boundary wraps $AdS_5 \times S^3$ inside $AdS_5 \times S^5$.

freedom, whose low-energy dynamics is described by the Dirac–Born–Infeld and Chern–Simons actions for the D7-brane probe as considered in section 4.4.1. These new degrees of freedom correspond to open string fluctuations on the D7-brane.

The new additional duality is conjectured to map mesonic operators in the field theory and D7-brane fluctuations on the gravity side on top of the original AdS/CFT correspondence between $\mathcal{N} = 4$ Super Yang–Mills theory and the near-horizon geometry of D3-branes. The new additional duality is an open–open string duality, as opposed to the original AdS/CFT correspondence which is an open–closed string duality. The duality states that in addition to the original AdS/CFT duality, gauge invariant bilinear field theory operators involving fundamental fields are mapped to fluctuations of the D7-brane probe inside $AdS_5 \times S^5$, as shown in figure 10.1.

Let us determine the D7-brane probe embedding explicitly. The D7-brane probe dynamics is described by the action (4.104) in general. In the present situation, the relevant bosonic terms in this action are

$$S_{D7} = -\tau_7 \int d^8\xi \, e^{-\phi} \sqrt{-\det\left(P[g]_{ab} + 2\pi\alpha' F_{ab}\right)} + \frac{(2\pi\alpha')^2}{2}\mu_7 \int P[C_{(4)}] \wedge F \wedge F,$$

$$(10.10)$$

where $\mu_7 = [(2\pi)^7 g_s \alpha'^4]^{-1}$ is the D7-brane tension and $\tau_7 = \mu_7 g_s$. P denotes the pullback of a bulk field to the worldvolume of the brane. F_{ab} is the worldvolume field strength tensor associated with a $U(1)$ gauge field $A = A_b d\xi^b$ on the D7-brane. In the probe limit, the ten-dimensional background metric is the one of $AdS_5 \times S^5$, and the R-R form $C_{(4)}$ and the dilaton are those given in chapter 5. The D7-brane action also contains a fermionic term S^f_{D7} which will be discussed separately below.

In studies of probe brane embeddings, it is customary to use the notation $w_1, w_2, \ldots, w_6$ for the six coordinates perpendicular to the N D3-branes. These six coordinates coincide with the six coordinates $x_4, x_5, \ldots, x_9$ used on page 327. We embed the D7-brane probe in such a way that it extends in the $x_0, \ldots, x_3$ directions as well as in the $w_1, \ldots, w_4$ directions. We work in the *static gauge* where the D-brane worldvolume coordinates ξ^a are identified with the spacetime coordinates $x_0, \ldots, x_3$ and $w_1, \ldots, w_4$. We also assume that w_5 and w_6

are functions of ρ with $\rho^2 = w_1^2 + \cdots + w_4^2$ only, in order to preserve Poincaré invariance and $SO(4)$ symmetry. Moreover, we write the $AdS_5 \times S^5$ metric in the form

$$
\begin{aligned}
ds^2 &= g_{MN} dx^M dx^N \\
&= \frac{r^2}{L^2} \eta_{\mu\nu} dx^\mu dx^\nu + \frac{L^2}{r^2} (d\rho^2 + \rho^2 d\Omega_3^2 + dw_5^2 + dw_6^2),
\end{aligned}
\tag{10.11}
$$

with $r^2 = \rho^2 + w_5^2 + w_6^2$, and (ρ, Ω_3) spherical coordinates in the 4567 space. Then, the induced metric obtained from the pullback of the metric to the worldvolume of the D7-brane is given by

$$
ds_{\text{ind}}^2 = \frac{\rho^2 + w_5^2 + w_6^2}{L^2} \eta_{\mu\nu} dx^\mu dx^\nu + \frac{L^2}{\rho^2 + w_5^2 + w_6^2} d\rho^2 + \frac{L^2 \rho^2}{\rho^2 + w_5^2 + w_6^2} d\Omega_3{}^2,
\tag{10.12}
$$

with $w_5 = w_5(\rho)$ and $w_6 = w_6(\rho)$. The action (10.10), for which F_{ab} may be consistently set to zero on its worldvolume, simplifies to

$$
S_{\text{D7}} = -\mu_7 \text{Vol}(\mathbb{R}^{3,1}) \text{Vol}(S^3) \int d\rho \; \rho^3 \sqrt{1 + \dot{w}_5^2 + \dot{w}_6^2},
\tag{10.13}
$$

where dots indicate a ρ derivative, for example

$$
\dot{w}_5 \equiv \frac{dw_5}{d\rho}.
\tag{10.14}
$$

The ground state configuration of the D7-brane then corresponds to the solution of the equation of motion

$$
\frac{d}{d\rho} \left(\frac{\rho^3}{\sqrt{1 + \dot{w}_5^2 + \dot{w}_6^2}} \dot{w} \right) = 0,
\tag{10.15}
$$

where w denotes either w_5 or w_6. Clearly these equations of motion are solved by w_5, w_6 being any arbitrary constant. In this case, the embedded D7-brane probe is flat. The choice of the position in the w_5, w_6 plane corresponds to choosing the quark mass in the gauge theory action. We may use the symmetry in the w_5, w_6 plane to identify $w_5(\rho) = l_q$ with $m_q = l_q/(2\pi\alpha')$ and $w_6(\rho) = 0$. The fact that w_5, w_6 are constant at all values of the radial coordinate ρ, which corresponds to the holographic renormalisation scale, may be interpreted as non-renormalisation of the mass in the dual field theory. The non-renormalisation of the mass is an expected characteristic of supersymmetric gauge theories.

In general, the equations of motion (10.15) have asymptotic ($\rho \to \infty$) solutions of the form

$$
w = l_q + \frac{c}{\rho^2} + \cdots
\tag{10.16}
$$

with non-zero c for the subleading term. Naively, we might read off from this equation that the dimension of the dual operator is two. However, this is not the case as we now explain. In principle, for the standard AdS/CFT procedure we have to use the asymptotic expansion (5.49). However, here the kinetic term in (10.13) is not canonically normalised. Due to the

ρ^3 factor, it still encompasses information about the eight-dimensional theory. Together with the fact that $\rho \sim 1/z$, this leads to the modified asymptotic behaviour

$$w(\rho) \sim l_q \rho^{\Delta - d + \alpha} + c\rho^{-\Delta + \alpha}, \tag{10.17}$$

with some number α, as compared to (5.49). The operator dimension Δ is then determined by the difference of the two exponents in (10.17), which gives

$$(\Delta - d + \alpha) - (-\Delta + \alpha) = 2\Delta - d. \tag{10.18}$$

For (10.16), we have $2\Delta - d = 2$. Since $d = 4$, we obtain $\Delta = 3$ and the dual field theory operator is of dimension three. This operator may be obtained from the field theory action (10.4) by

$$\mathcal{O}_q = -\partial_{m_q} \mathcal{L}_{\mathcal{N}=2} = \tilde{\psi}\psi + \tilde{q}(m_q + \Phi_3)\tilde{q}^\dagger + q^\dagger(m_q + \Phi_3)q + \text{h.c.} \tag{10.19}$$

in terms of the component fields of the supermultiplets involved in (10.4). Moreover, the precise relation between the vacuum expectation value of $\mathcal{O}_q$ and c may be worked out by using the relation

$$\langle \mathcal{O}_q \rangle = \frac{\delta \mathcal{H}}{\delta m_q}, \tag{10.20}$$

where the Hamiltonian $\mathcal{H}$ is the Legendre transform of (10.13) with respect to ρ.

Exercise 10.2.1 Starting from the action (10.13), and using (10.20) as well as $\text{Vol}(S^3) = \pi^3$, show that

$$\langle \mathcal{O}_q \rangle = -2\pi^3 \alpha' \mu_7 c. \tag{10.21}$$

It is important to note that for supersymmetric configurations such as the one considered here, c must be zero. In fact, $\langle \mathcal{O}_q \rangle$ contains the vacuum expectation value of an F-term,

$$\langle \tilde{\psi}\psi \rangle \sim \left\langle \int d^2\theta \, \tilde{Q}Q \right\rangle, \tag{10.22}$$

which breaks supersymmetry. This is also reflected in the supergravity solution. The solutions to the supergravity equations of motion with c non-zero are not regular in AdS space and are therefore excluded.

We therefore consider the regular supersymmetric embeddings of the D7-brane probe for which the quark mass m_q may be non-zero, but the condensate $\langle \tilde{\psi}\psi \rangle$ vanishes. For massive embeddings, the D7-brane probe is separated by l_q from the stack of D3-branes in either the w^5 or w^6 directions, where the indices refer to the coordinates given in (10.11). This corresponds to giving a mass $m_q = l_q/(2\pi\alpha')$ to the hypermultiplet $(Q, \tilde{Q})$ in the fundamental representation. In this case the radius of S^3 becomes a function of the radial coordinate r in AdS_5, as is seen from the induced metric (10.12), which with $w_5 = l_q$, $w_6 = 0$ becomes

$$ds^2 = \frac{\rho^2 + l_q^2}{L^2} \eta_{\mu\nu} dx^\mu dx^\nu + \frac{L^2}{\rho^2 + l_q^2} d\rho^2 + \frac{L^2 \rho^2}{\rho^2 + l_q^2} d\Omega_3^2. \tag{10.23}$$

We see that for $\rho = 0$, the radius of S^3 in (10.23) shrinks to zero. Since ρ is related to the radial coordinate r by $r^2 = \rho^2 + l_q^2$, this implies that there exists a minimal value $r_{\min} = l_q$ beyond which the D7-brane probe cannot extend further into the interior of the AdS space. In the opposite limit, for $\rho \to \infty$ the induced metric asymptotes to the metric of $AdS_5 \times S^3$.

10.3 D7-brane fluctuations and mesons in $\mathcal{N} = 2$ theory

We now consider the computation of meson masses in the framework of gauge/gravity duality. We will see that these masses are determined by the energy eigenvalues of the D7-brane fluctuations. In the present context, mesons are bound states which correspond to gauge invariant operators involving quark–antiquark pairs.

10.3.1 Scalar field fluctuations (spin 0)

As a first example, we discuss the fluctuation modes and meson masses for the scalar fields of the D7-brane. The directions transverse to the D7-brane are chosen to be w^5 and w^6, and the embedding is chosen to be

$$w_5 = 0 + \delta w_5, \qquad w_6 = l_q + \delta w_6, \tag{10.24}$$

where δw_5 and δw_6 are the transverse scalar fluctuations of the D7-brane. To calculate the spectra of the worldvolume fields it is sufficient to work to quadratic order in the fluctuations in the action, so as to obtain linearised equations of motion for the fluctuations. For the scalars, the relevant quadratic part of the Lagrangian density is

$$\mathcal{L} \simeq -\mu_7 \sqrt{-\det P[g]^{(0)}} \left(1 + \frac{1}{2} \frac{L^2}{r^2} P[g]^{(0)ab} \partial_a \varphi \partial_b \varphi \right). \tag{10.25}$$

Here, φ is used to denote either (real) fluctuation, $\delta w_{5,6}$, and $P[g]^{(0)}_{ab}$ is the induced metric on the D7-brane worldvolume to zeroth order in the fluctuations, as given by (10.23). In spherical coordinates with $r^2 = \rho^2 + l_q^2$, the equation of motion becomes

$$\partial_a \left(\frac{\rho^3 \sqrt{\det \tilde{g}}}{\rho^2 + l_q^2} P[g]^{(0)ab} \partial_b \varphi \right) = 0. \tag{10.26}$$

$\tilde{g}$ is the metric on the unit sphere spanned by Ω_3. The equation of motion may be expanded as

$$\frac{L^4}{(\rho^2 + l_q^2)^2} \partial^\mu \partial_\mu \varphi + \frac{1}{\rho^3} \partial_\rho (\rho^3 \partial_\rho \varphi) + \frac{1}{\rho^2} \nabla^i \nabla_i \varphi = 0, \tag{10.27}$$

where ∇_i is the covariant derivative on the three-sphere. Using separation of variables, an ansatz for the modes may be written as

$$\varphi = \phi(\rho) e^{ik \cdot x} \mathcal{Y}^\ell(\Omega^3), \tag{10.28}$$

where $\mathcal{Y}^\ell(\Omega^3)$ are the scalar spherical harmonics on Ω^3, which satisfy

$$\nabla^i \nabla_i \mathcal{Y}^\ell = -\ell(\ell+2)\mathcal{Y}^\ell. \tag{10.29}$$

The meson masses are defined by $M^2 = -k^2$ for the wavevector k introduced in (10.28). Then equation (10.27) gives rise to an equation for $\phi(\rho)$ that, with the redefinitions

$$\varrho = \frac{\rho}{l_q}, \qquad \bar{M}^2 = -\frac{k^2 L^4}{l_q^2}, \tag{10.30}$$

becomes

$$\partial_\varrho^2 \phi + \frac{3}{\varrho}\partial_\varrho \phi + \left(\frac{\bar{M}^2}{(1+\varrho^2)^2} - \frac{\ell(\ell+2)}{\varrho^2}\right)\phi = 0. \tag{10.31}$$

This equation may be solved in terms of a hypergeometric function. Imposing regularity in the interior of *AdS*, the solution is

$$\phi(\rho) = \frac{\rho^\ell}{(\rho^2+L^2)^{n+\ell+1}} F\left(-(n+\ell+1), -n; \ell+2; -\rho^2/L^2\right) \tag{10.32}$$

with

$$\bar{M}^2 = 4(n+\ell+1)(n+\ell+2). \tag{10.33}$$

Using this, and $M^2 = -k^2 = \bar{M}^2 l_q^2/L^4$, the four-dimensional mass spectrum of scalar mesons is given by

$$M_s(n,\ell) = \frac{2l_q}{L^2}\sqrt{(n+\ell+1)(n+\ell+2)}. \tag{10.34}$$

Normalisability of the modes results in a discrete spectrum with a mass scale set by l_q, the position of the D7-brane. Note that the prefactor in (10.34) may be rewritten as a function of the quark mass and the 't Hooft coupling using

$$\frac{l_q}{L^2} = \sqrt{2}\pi \frac{m_q}{\sqrt{\lambda}}. \tag{10.35}$$

Since λ is large, $M_s(n,l)$ is smaller than m_q by a factor of $1/\sqrt{\lambda}$. This implies that the mesons described are tightly bound.

10.3.2 Fermionic fluctuations (spin $\frac{1}{2}$)

We now turn to the spectrum of fermionic fluctuations of the D7-brane probe [1]. These fluctuations are dual to *mesino* operators which are the fermionic superpartners of the mesons. Typical mesino operators with conformal dimension $\Delta = \frac{5}{2}$ and $\Delta = \frac{9}{2}$ are $\mathcal{F} \sim \bar{\psi}q$ and $\mathcal{G} \sim \bar{\psi}\lambda\psi$, where ψ (q) is a quark (squark) and λ is an adjoint fermion.

The dual fluctuations have spin $\frac{1}{2}$ and are described by the *fermionic* part of the D7-brane action, that is the supersymmetric completion of the Dirac–Born–Infeld action. This action is given by [2]

$$S_{\text{D7}}^{\text{f}} = \frac{\tau_7}{2}\int d^8\xi \sqrt{-\det g}\, \bar{\Psi}\mathcal{P}_-\Gamma^a\left(D_a + \frac{1}{8}\frac{i}{2\cdot 5!}F_{NPQRS}\Gamma^{NPQRS}\Gamma_a\right)\Psi. \tag{10.36}$$

Here ξ^a are the worldvolume coordinates ($a = 0, ..., 7$) which are identified with the space-time coordinates $x^0, x^1, ..., x^7$. This identification is referred to as the static gauge. The field Ψ is the ten-dimensional positive chirality Majorana–Weyl spinor of type IIB string theory and Γ_a is the pullback of the ten-dimensional Gamma matrix Γ_M ($M, N, ... = 0, ..., 9$), $\Gamma_a = \Gamma_M \partial_a x^M$. The integration volume is given by the world-volume of the D7-brane which as before wraps a submanifold of $AdS_5 \times S^5$ which asymptotes to $AdS_5 \times S^3$. The spinor $\Psi = \Psi(x^M, \Omega_3)$ depends on the coordinates x^M of AdS_5 and on the three angles Ω_3 of the three-sphere S^3. The operator $\mathcal{P}_-$ is a κ-symmetry projector ensuring supersymmetry of the action. The action $\mathcal{S}_{D7} = \mathcal{S}_{D7}^{b} + \mathcal{S}_{D7}^{f}$ with $\mathcal{S}_{D7}^{b}$ and $\mathcal{S}_{D7}^{f}$ given by (10.10) and (10.36) is invariant under supersymmetries corresponding to bulk Killing spinors.

We evaluate the five-form F_{NPQRS} as well as the curved spacetime covariant derivative D_M on $AdS_5 \times S^5$. This gives a Dirac-type equation which will then be transformed into a second-order differential equation. The fluctuations are assumed to be of the form

$$\Psi(x, \rho, \Omega_3) = \psi_{\ell,\pm}(\rho) e^{ik_\mu x^\mu} \chi_\ell^\pm(\Omega_3), \tag{10.37}$$

where $\psi_{\ell,\pm}(\rho)$ and $\chi_\ell^\pm(\Omega_3)$ are spinors on AdS_5 and S^5, respectively. Here, the $\pm$ signs refer to the eigenvalues of the spinor spherical harmonics on S^3, given by $\pm(\ell + \frac{3}{2})$. As in the scalar case, the mesino masses are obtained from $M^2 = -k^2$. Solving the equation of motion for the fluctuations is somewhat involved. For the fluctuations $\psi_{\ell,+}$, the result for the meson masses is, using the rescaling (10.30),

$$\bar{M}_{\mathcal{G}}^2 = 4(n + \ell + 2)(n + \ell + 3) \tag{10.38}$$

which corresponds to the spectrum of the operators $\mathcal{G}_\alpha^\ell$. The spectrum of $\mathcal{F}_\alpha^\ell$ is obtained in a similar way by solving the equations of motion for $\psi_{\ell,-}$.

10.3.3 Gauge field fluctuations (spin 1)

The fluctuations of the D7-brane worldvolume gauge field A_M ($M = 0, ..., 7$) give rise to three further mass towers, denoted by $M_{I,\pm}$, M_{II} and M_{III} [3]. These spectra are generated by planewave fluctuations of the components A_i (along S^3), A_μ (along $x^0, ..., x^3$) and A_ρ (along radial direction ρ) of the eight-dimensional worldvolume gauge field $A_M = (A_\mu, A_\rho, A_i)$.

10.3.4 Fluctuation operator matching

So far we have discussed the mass spectra of open string fluctuations on the D7-branes. In order to interpret these spectra as those of meson-like operators, we have to map the fluctuations to the corresponding meson operators in the dual field theory. In the following we construct these operators and assign them to the corresponding open string fluctuations.

Table 10.3 Field content of the $\mathcal{N} = 2$ supermultiplets in D3/D7 theory

	Fluctuation	Degree of freedom	$(j_\Phi, j_R)_q$	Lowest five-dimensional mass	Spectrum		Operator	Δ
Mesons	1 scalar	1	$(\frac{\ell}{2}, \frac{\ell}{2}+1)_0$	$m^2 = -4$	$M_{I,-}(n, \ell+1)$	$(\ell \geq 0)$	$\mathcal{C}^{I\ell}$	2
(bosons)	2 scalars	2	$(\frac{\ell}{2}, \frac{\ell}{2})_2$	$m^2 = -3$	$M_s(n, \ell)$	$(\ell \geq 0)$	$\mathcal{M}_s^{A\ell}$	3
	1 scalar	1	$(\frac{\ell}{2}, \frac{\ell}{2})_0$	$m^2 = -3$	$M_{III}(n, \ell)$	$(\ell \geq 1)$	$\mathcal{J}^{5\ell}$	3
	1 vector	3	$(\frac{\ell}{2}, \frac{\ell}{2})_0$	$m^2 = 0$	$M_{II}(n, \ell)$	$(\ell \geq 0)$	$\mathcal{J}^{\mu\ell}$	3
	1 scalar	1	$(\frac{\ell}{2}, \frac{\ell}{2}-1)_0$	$m^2 = 0$	$M_{I,+}(n, \ell-1)$	$(\ell \geq 2)$	–	4
Mesinos	1 Dirac	4	$(\frac{\ell}{2}, \frac{\ell+1}{2})_1$	$\lvert m \rvert = \frac{1}{2}$	$M_\mathcal{F}(n, \ell)$	$(\ell \geq 0)$	$\mathcal{F}_\alpha^\ell$	$\frac{5}{2}$
(fermions)	1 Dirac	4	$(\frac{\ell}{2}, \frac{\ell-1}{2})_1$	$\lvert m \rvert = \frac{5}{2}$	$M_\mathcal{G}(n, \ell-1)$	$(\ell \geq 1)$	$\mathcal{G}_\alpha^\ell$	$\frac{9}{2}$

The complete set of D7-brane fluctuations fits into a series of massive gauge supermultiplets of the $\mathcal{N} = 2$ supersymmetry algebra. These multiplets contain $16(\ell + 1)$ states with masses

$$M^2 = \frac{4l_q^2}{L^4}(n + \ell + 1)(n + \ell + 2), \qquad n, \ell \geqslant 0. \tag{10.39}$$

Since the supercharges commute with the generators of the global symmetry group $SU(2)_\Phi$, the $SU(2)_\Phi$ quantum number $\frac{\ell}{2}$ is the same for all fluctuations in a supermultiplet.

All D7-brane fluctuations and their quantum numbers are listed in table 10.3, where we set $L = 1$ for simplicity. In this table, $(j_\Phi, j_R)_q$ label representations of $SO(4) \approx SU(2)_\Phi \times SU(2)_R$, and q is the $U(1)_R$ charge. In order to count the number of states in a multiplet we must take into account the degeneracy in the $SU(2)_R$ quantum number, i.e. we count the degrees of freedom of a particular massive fluctuation and multiply it with $(2j_R + 1)$. Then, the number of bosonic components in a multiplet is

$$1 \cdot (2(\tfrac{\ell}{2} + 1) + 1) + (2 + 3 + 1) \cdot (2\tfrac{\ell}{2} + 1) + 1 \cdot (2(\tfrac{\ell}{2} - 1) + 1) = 8(\ell + 1). \tag{10.40}$$

Of course, this agrees with the number of fermionic components,

$$4(2 \cdot \tfrac{\ell+1}{2} + 1) + 4(2 \cdot \tfrac{\ell-1}{2} + 1) = 8(\ell + 1), \tag{10.41}$$

giving $16(\ell + 1)$ states in total.

We now assign operators to the D7-brane fluctuations appearing in table 10.3. Note that the masses are above the Breitenlohner–Freedman bound. Open strings are dual to composite operators with fundamental fields at their ends: scalars $q^m = (q, \tilde{q})^\mathsf{T}$ and spinors $\psi_i = (\psi, \tilde{\psi}^\dagger)^\mathsf{T}$. We will refer to these operators as mesons and their superpartners as *mesinos*. We must ensure that the operators have the same quantum numbers (i.e. spin, global symmetries, etc.) as the corresponding fluctuations. Also, the five-dimensional mass of a fluctuation and the conformal dimension of the dual operator must satisfy the usual mass relation depending on the spin, for example $m^2 = \Delta(\Delta - 4)$ for scalars.

Let us construct gauge invariant operators for the bosonic fluctuations. First, there is a scalar in $(\frac{\ell}{2}, \frac{\ell}{2} + 1)_0$ with five-dimensional mass $m^2 = -4 + \ell^2 \geq m_{\text{BF}}^2$. The lowest fluctuation has negative mass squared, $m^2 = -4$, saturating the Breitenlohner–Freedman bound, $m_{\text{BF}}^2 = -d^2/4 = -4$ in four dimensions. These scalar fluctuations correspond to the $\Delta = \ell + 2$ chiral primaries

$$\mathcal{C}^{I\ell} = \bar{q}^m \sigma_{mn}^I X^\ell q^n. \tag{10.42}$$

Here the Pauli matrices σ_{mn}^I ($I = 1,2,3$) transform in the triplet representation of $SU(2)_R$, while q^m, ψ^i and X^ℓ have the $SO(4)$ quantum numbers $(0, \frac{1}{2})$, $(0,0)$ and $(\frac{\ell}{2}, \frac{\ell}{2})$, respectively. X^ℓ denotes the symmetric, traceless operator insertion $X^{\{i_1} \cdots X^{i_\ell\}}$ of ℓ adjoint scalars X^i ($i = 4,5,6,7$). This operator insertion generates operators with higher angular momentum ℓ.

Then, there are two scalars in $(\frac{\ell}{2}, \frac{\ell}{2})_2$ which are dual to the scalar meson operators

$$\mathcal{M}_s^{A\ell} = \bar{\psi}_i \sigma_{ij}^A X^\ell \psi_j + \bar{q}^m X_V^A X^\ell q^m, \qquad i, m = 1, 2 \tag{10.43}$$

which have conformal dimensions $\Delta = \ell + 3$. Here X_V^A denotes the vector (X^8, X^9) and $\sigma^A = (\sigma^1, \sigma^2)$ is a doublet of Pauli matrices. Both X_V^A and σ^A have charge $+2$ under $U(1)_R$. The operators $\mathcal{M}_s^{A\ell}$ thus transform in $(\frac{\ell}{2}, \frac{\ell}{2})$ of $SO(4)$ and have charge $+2$ under $U(1)_R$.

Next, there is a vector in the $(\frac{\ell}{2}, \frac{\ell}{2})_0$ associated with the $\Delta = \ell + 3$ operator

$$\mathcal{J}^{\mu\ell} = \bar{\psi}_i^\alpha \gamma_{\alpha\beta}^\mu X^\ell \psi_i^\beta + i\bar{q}^m X^\ell D^\mu q^m - i\bar{D}^\mu \bar{q}^m X^\ell q^m, \qquad \mu = 0, 1, 2, 3 \tag{10.44}$$

which we identify as the $U(N_{\text{f}})$ flavour current.

Finally, for $\ell \geq 1$ there is a (pseudo-)scalar in the $(\frac{\ell}{2}, \frac{\ell}{2})_0$ dual to

$$\mathcal{J}^{5\ell-1} = \bar{\psi}_i^\alpha \gamma_{\alpha\beta}^5 X^{\ell-1} \psi_i^\beta + \cdots, \tag{10.45}$$

as well as a scalar in $(\frac{\ell}{2}, \frac{\ell}{2} + 1)_0$ ($\ell \geq 2$) which corresponds to a higher descendant of $\mathcal{C}^{I\ell}$. These operators do not appear in the lowest ($\ell = 0$) multiplet.

We now turn to the fermionic fluctuations dual to mesino operators. The spin-$\frac{1}{2}$ operators dual to the fluctuations $\psi_{\ell, \pm}$ in (10.37) are denoted by $\mathcal{G}_\alpha^\ell$ and $\mathcal{F}_\alpha^\ell$. The mass dimension relation for spin-$\frac{1}{2}$ fields, $|m|L = \Delta - 2$, determines the conformal dimensions of these operators,

$$\Delta_\mathcal{G} = \tfrac{9}{2} + \ell, \qquad \Delta_\mathcal{F} = \tfrac{5}{2} + \ell, \qquad \ell \geq 0. \tag{10.46}$$

We have to ensure that the operators $\mathcal{G}_\alpha^\ell$ and $\mathcal{F}_\alpha^\ell$ have the same $SO(4)$ and $U(1)_R$ quantum numbers as the fluctuations. For instance, the spinorial spherical harmonics on S^3 transform in the $(\frac{\ell+1}{2}, \frac{\ell}{2})$ and $(\frac{\ell}{2}, \frac{\ell+1}{2})$ of $SO(4) = SU(2)_\Phi \times SU(2)_R$, while the $U(1)_R$ charge is $+1$. These properties uniquely fix the structure of $\mathcal{G}_\alpha^\ell$ and $\mathcal{F}_\alpha^\ell$ as

$$\mathcal{F}_\alpha^\ell = \bar{q} X^\ell \tilde{\psi}_\alpha^\dagger + \tilde{\psi}_\alpha X^\ell q, \tag{10.47}$$

$$\mathcal{G}_\alpha^{\ell-1} = \bar{\psi}_i \sigma_{ij}^B \lambda_{\alpha C} X^{\ell-1} \psi_j + \bar{q}^m X_V^B \lambda_{\alpha C} X^{\ell-1} q^m, \qquad A, B, C = 1, 2 \tag{10.48}$$

which have the conformal dimensions $\Delta = \frac{5}{2} + \ell$ ($\ell \geq 0$) and $\Delta = \frac{7}{2} + \ell$ ($\ell \geq 1$), respectively. As their bosonic partners, mesinos have fundamental fields at their ends. The spinors $\lambda_{\alpha A}$ ($A = 1, 2$) have the $SO(4)$ quantum numbers $(\frac{1}{2}, 0)$ and belong to the adjoint hypermultiplets (Φ_1, Φ_2).

10.4 *D3/D5-brane system

We now turn to another useful probe brane configuration and consider the case of a D5-brane probe embedded in the D3-brane near-horizon geometry. In accordance with table 10.1, this is a codimension one intersection: there is one dimension in which the D3-branes extend, but not the D5-brane. This implies that the fundamental flavour fields live in 2+1 dimensions. The associated field theory is thus a *defect field theory* in which (2+1)-dimensional matter fields are coupled to a gauge theory in 3+1 dimensions. This field theory is supersymmetric with eight real supercharges. In contrast to the field theory associated with the D3/D7 intersection, the field theory considered here is conformal to all orders in perturbation theory even at finite N; its β function vanishes.

It is instructive to consider the construction of the quantum field theory associated with the D3/D5 brane configuration and its supergravity dual in some detail. This provides a further example of the AdS/CFT correspondence at work. In particular, the D3/D5 brane system will prove useful for applications of gauge/gravity duality to systems of relevance for condensed matter physics, as discussed in chapter 15. We begin by listing the field content and map the symmetries of the brane intersection with the symmetries of the field theory. Then, we construct the mesonic operators and determine their dual supergravity modes. For simplicity, we restrict ourselves to one D5-brane, $N_{\rm f} = 1$.

The D3-branes extend along the 0123 directions, whereas the D5-branes wrap the 012456 directions. The brane intersection preserves eight of the thirty-two real supercharges. Hence the dual field theory is (3+1)-dimensional $\mathcal{N} = 4$ Super Yang–Mills coupled to defect flavour fields preserving (2+1)-dimensional $\mathcal{N} = 4$ supersymmetry. Coupling the defect fields to the fields in $(3 + 1)$ dimensions requires decomposing the (3+1)-dimensional $\mathcal{N} = 4$ multiplet into two (2+1)-dimensional $\mathcal{N} = 4$ multiplets, a vector multiplet and a hypermultiplet. The bosonic content of the (3+1)-dimensional $\mathcal{N} = 4$ multiplet is the vector A_μ and six scalars $X^4, X^5, \ldots, X^9$. The bosonic content of the (2+1)-dimensional vector multiplet is the (2+1)-dimensional vector field A_k and the three scalars $X_V = (X^7, X^8, X^9)$. The bosonic content of the (2+1)-dimensional hypermultiplet is the scalar A_3 and the three scalars $X_H = (X^4, X^5, X^6)$. The flavour fields form a (2+1)-dimensional hypermultiplet with two fermions (quarks) ψ and two complex scalars (squarks) q.

The classical Lagrangian preserves (2+1)-dimensional $SO(3,2)$ conformal symmetry for massless flavour degrees, but breaks the $SO(6)_R$ R-symmetry down to a subgroup $SU(2)_H \times SU(2)_V$, under which the scalars in X_H transform in the $(1,0)$ representation and the scalars in X_V transform in the $(0,1)$ representation. We use an upper index to denote these representations: X_V^A and X_H^I. The adjoint fermions λ^{im} transform in the $(1/2, 1/2)$ representation. Here, i is the $SU(2)_V$ index and m is the $SU(2)_H$ index. The quarks ψ^i transform in the $(1/2, 0)$ and the squarks q^m transform in the $(0, 1/2)$ representation. Table 10.4 summarises the field content and quantum numbers, including the conformal dimensions of the fields. Here, A_k, X_V^A, A_3, X_H^I and λ^{im} are the adjoint fields of (3+1)-dimensional $\mathcal{N} = 4$ Super Yang–Mills theory decomposed into (2+1)-dimensional $\mathcal{N} = 4$

Table 10.4 Field content of D3/D5 theory					
Mode	Spin	$SU(2)_H$	$SU(2)_V$	$SU(N)$	Δ
A_k	1	0	0	adjoint	1
X_V^A	0	0	1	adjoint	1
A_3	0	0	0	adjoint	1
X_H^I	0	1	0	adjoint	1
λ^{im}	$\frac{1}{2}$	$\frac{1}{2}$	$\frac{1}{2}$	adjoint	$\frac{3}{2}$
q^m	0	$\frac{1}{2}$	0	N	$\frac{1}{2}$
ψ^i	$\frac{1}{2}$	0	$\frac{1}{2}$	N	1

Table 10.5 Meson operators of D3/D5 theory and their quantum numbers				
Operator	Δ	$SU(2)_H$	$SU(2)_V$	Operator in lowest multiplet ($l = 0$)
$\mathcal{J}_l$	$l+2$	$l,\ l \geq 0$	0	$i\bar{q}^m \overset{\leftrightarrow}{D^k} q^m + \bar{\psi}^i \rho^k \psi^i$
$\mathcal{E}_l$	$l+2$	$l,\ l \geq 0$	1	$\bar{\psi}_i \sigma_{ij}^A \psi_j + 2\bar{q}^m X_V^{Aa} T_a q^m$
$\mathcal{C}_l$	$l+1$	$l+1,\ l \geq 0$	0	$\bar{q}^m \sigma_{mn}^I q^n$
$\mathcal{D}_l$	$l+3$	$l-1, l \geq 1$	0	–
$\mathcal{F}_l$	$l+3/2$	$l+1/2, l \geq 0$	1/2	$\bar{\psi}^i q^m + q^{\dagger m} \psi^i$
$\mathcal{G}_l$	$l+5/2$	$l-1/2, l \geq 1$	1/2	–

multiplets. A_k and X_V^A are the bosons in a (2+1)-dimensional vector multiplet, while A_3 and X_H^I are the bosons in a (2+1)-dimensional hypermultiplet. q^m and ψ^i are the (2+1)-dimensional flavour fields, which are in an $\mathcal{N} = 4$ hypermultiplet.

Let us now consider the meson operators in the field theory dual to the D3/D5 intersection which may be arranged into a (2+1)-dimensional massive $\mathcal{N} = 4$ supersymmetric multiplet. The operators and their quantum numbers are summarised in table 10.5. σ^I are Pauli matrices, T_a are the generators of $SU(2)_V$, and ρ_k are the (2+1)-dimensional Γ matrices.

Let us first review the meson multiplets. All operators with the same l are in one multiplet. Note that we have to distinguish two cases: the $l = 0$ multiplet which will be short, and the $l > 0$ multiplets.

We begin with the short multiplet with $l = 0$. According to table 10.5, the operator $\mathcal{C}_0^I = \bar{q}^{\dagger m} \sigma_{mn}^I q^n$, where σ^I are the Pauli matrices of $SU(2)_H$, is the lowest chiral primary in the multiplet since all other operators dual to D5-brane fluctuations have larger conformal dimensions. $\mathcal{C}_0$ transforms in the $(1,0)$ representation of $SU(2)_H \times SU(2)_V$. We can thus construct all operators in the same multiplet as $\mathcal{C}_0$ by applying supersymmetry generators to $\mathcal{C}_0$. The supersymmetry generators form a 2×2 matrix of Majorana spinors η^{im}, which transforms like λ^{im}, i.e. in the representation $(1/2, 1/2)$ of $SU(2)_H \times SU(2)_V$. Applying the supersymmetry generators to $\mathcal{C}_0$ we obtain the fermionic operator $\mathcal{F}_0^{im} = \bar{\psi}^i q^m + q^{\dagger m} \psi^i$ with conformal dimension $\Delta = 3/2$ and $SU(2)_H \times SU(2)_V$ quantum numbers $(1/2, 1/2)$. Applying another supersymmetry generator to $\mathcal{F}_0^{im}$, we obtain either $\mathcal{J}_0$ or $\mathcal{E}_0$, the forms of which appear in table 10.5. Both $\mathcal{J}_0$ and $\mathcal{E}_0$ have conformal dimension $\Delta = 2$ and are singlets under $SU(2)_H$. However, they may be distinguished by their $SU(2)_V$ quantum number: $\mathcal{J}_0$ is a singlet whereas $\mathcal{E}_0$ is a triplet under $SU(2)_V$.

We now move on to the general multiplet dual to the higher l mesonic operators. As in the $l = 0$ case, we construct the multiplet by applying supersymmetry generators to the lowest chiral primary in the multiplet, $\mathcal{C}_l$. The lowest chiral primary is $\mathcal{C}_l^{I_0 I_1 \dots I_l} = C_0^{(I_0} \left(X_H^l \right)^{I_1 \dots I_l)}$, where (X_H^l) stands for the traceless symmetric product of l copies of the field X_H^I. $\mathcal{C}_l$ has conformal dimension $\Delta = l + 1$ and is in the $(l + 1, 0)$ representation of $SU(2)_H \times SU(2)_V$. Applying a supersymmetry generator to $\mathcal{C}_l$, we find the fermionic operator $\mathcal{F}_l$ with conformal dimension $\Delta = l + 3/2$. $\mathcal{F}_l$ is in the $(l + 1/2, 1/2)$ representation of $SU(2)_H \times SU(2)_V$. Explicitly, $\mathcal{F}_l$ is of the form

$$\mathcal{F}_l^{I_1 \dots I_l \, im} = \bar{\psi}^i \left(X_H^l \right)^{I_1 \dots I_l} q^m + q^{\dagger m} \left(X_H^l \right)^{I_1 \dots I_l} \psi^i. \tag{10.49}$$

Applying another supersymmetry generator to $\mathcal{F}_l$, we obtain $\mathcal{J}_l$ or $\mathcal{E}_l$, which have the same conformal dimension $\Delta = l + 2$, but differ in the $SU(2)_H \times SU(2)_V$ representation. $\mathcal{J}_l$ transforms in the $(l, 0)$ representation whereas $\mathcal{E}_l$ has quantum numbers $(l, 1)$. To obtain the precise form of $\mathcal{J}_l$ or $\mathcal{E}_l$, we insert the operator X_H^l into the operator $\mathcal{J}_0$ or $\mathcal{E}_0$, respectively.

In contrast to the $l = 0$ multiplet, other operators also appear in the multiplet for $l \geq 1$, which we construct by applying three or four supersymmetry generators to $\mathcal{C}_l$: a fermionic operator $\mathcal{G}_l$ and a bosonic operator $\mathcal{D}_l$. $\mathcal{G}_l$ has conformal dimension $\Delta = l + 5/2$ and $SU(2)_H \times SU(2)_V$ quantum numbers $(l - 1/2, 1/2)$. Explicitly, $\mathcal{G}_l$ has the form

$$\mathcal{G}_l^{I_1 \dots I_{l-1} \, im} = \bar{\psi}^j \left(X_H^{l-1} \right)^{I_1 \dots I_{l-1}} \lambda^{im} \psi_j + q^{\dagger n} \left(X_H^{l-1} \right)^{I_1 \dots I_{l-1}} \lambda^{im} X_{H,I} \sigma_{np}^I q^p. \tag{10.50}$$

Finally, $\mathcal{D}_l$ has conformal dimension $\Delta = l + 3$ and $SU(2)_H \times SU(2)_V$ quantum numbers $(l - 1, 0)$. This completes the spectrum as given in table 10.5. On the dual gravity side, the spectrum of D5-brane probe fluctuations coincides precisely with the representations given in this table.

10.5 Further reading

The D3/D7 system system introduced in section 10.1.2 was proposed in [4] as a way to add flavour to the AdS/CFT correspondence. The meson spectrum for the D3/D7 intersection of section 10.3 was worked out in [3], with the fermionic operators given in [1]. The mesino spectrum of section 10.3.2 was analysed in detail in [1], based on the fermionic contribution to the DBI action as given in [2].

Moreover, flavour degrees of freedom in a UV finite theory based on F-theory were introduced in [5]. A review of mesons in the AdS/CFT correspondence may be found in [6]. A review including the backreaction for N_f/N finite may be found in [7], which also provides a guide to further references. The dual of the Landau pole for D7-branes was studied in [8].

The dictionary for the D3/D5 intersection of section 10.4 was worked out in [9]. Conformal invariance of the associated defect field theory at the quantum level was shown in [10]. The dictionary for the D3/D3 intersection was worked out in [11].

Probe brane configurations leading to chiral flavour are considered in [12, 13, 14].

References

[1] Kirsch, Ingo. 2006. Spectroscopy of fermionic operators in AdS/CFT. *J. High Energy Phys.*, **0609**, 052.

[2] Martucci, Luca, Rosseel, Jan, Van den Bleeken, Dieter, and Van Proeyen, Antoine. 2005. Dirac actions for D-branes on backgrounds with fluxes. *Class.Quantum Grav.*, **22**, 2745–2764.

[3] Kruczenski, Martin, Mateos, David, Myers, Robert C., and Winters, David J. 2003. Meson spectroscopy in AdS/CFT with flavor. *J. High Energy Phys.*, **0307**, 049.

[4] Karch, Andreas, and Katz, Emanuel. 2002. Adding flavour to AdS/CFT. *J. High Energy Phys.*, **0206**, 043.

[5] Aharony, Ofer, Fayyazuddin, Ansar, and Maldacena, Juan Martin. 1998. The Large N limit of $N = 2$, $N = 1$ field theories from three-branes in F theory. *J. High Energy Phys.*, **9807**, 013.

[6] Erdmenger, Johanna, Evans, Nick, Kirsch, Ingo, and Threlfall, Ed. 2008. Mesons in gauge/gravity duals – a review. *Eur. Phys. J.*, **A35**, 81–133.

[7] Núñez, Carlos, Paredes, Angel, and Ramallo, Alfonso V. 2010. Unquenched flavor in the gauge/gravity correspondence. *Adv. High Energy Phys.*, **2010**, 196714.

[8] Kirsch, Ingo, and Vaman, Diana. 2005. The D3/D7 background and flavor dependence of Regge trajectories. *Phys. Rev.*, **D72**, 026007.

[9] DeWolfe, Oliver, Freedman, Daniel Z., and Ooguri, Hirosi. 2002. Holography and defect conformal field theories. *Phys. Rev.*, **D66**, 025009.

[10] Erdmenger, Johanna, Guralnik, Zachary, and Kirsch, Ingo. 2002. Four-dimensional superconformal theories with interacting boundaries or defects. *Phys. Rev.*, **D66**, 025020.

[11] Constable, Neil R., Erdmenger, Johanna, Guralnik, Zachary, and Kirsch, Ingo. 2003. Intersecting D-3 branes and holography. *Phys. Rev.*, **D68**, 106007.

[12] Harvey, Jeffrey A., and Royston, Andrew B. 2008. Localized modes at a D-brane–O-plane intersection and heterotic Alice Atrings. *J. High Energy Phys.*, **0804**, 018.

[13] Buchbinder, Evgeny I., Gomis, Jaume, and Passerini, Filippo. 2007. Holographic gauge theories in background fields and surface operators. *J. High Energy Phys.*, **0712**, 101.

[14] Harvey, Jeffrey A., and Royston, Andrew B. 2008. Gauge/gravity duality with a chiral $N = (0,8)$ string defect. *J. High Energy Phys.*, **0808**, 006.

11

Finite temperature and density

The gravity dual of a quantum field theory at finite temperature is readily obtained by considering a *black brane* or a black hole in an asymptotically Anti-de Sitter space. Black holes and black branes are thermal objects themselves which radiate and are therefore associated with a temperature, the Hawking temperature T_{H}. We will see that the Hawking temperature T_{H} equals the temperature T on the field theory side.

We begin this chapter with a summary of finite temperature quantum field theory, which highlights its differences compared with vacuum quantum field theory which we considered before. In particular, we explain the essence of the real and imaginary time formalisms. We then move on to describe black hole thermodynamics and show how a gravity dual description of finite temperature field theory naturally arises. Finally we describe how to obtain a holographic description of finite density and chemical potentials.

11.1 Finite temperature field theory

We consider a field theory whose dynamics is specified by the Hamiltonian $\hat{H}$ in the Heisenberg picture. Moreover, this field theory has one or more global $U(1)$ symmetries associated with conserved currents $J^{\mu a}$ labelled by a. The commuting Noether charges associated with these currents are denoted by $\hat{Q}_a$. Their expectation values $N_a = \langle \hat{Q}_a \rangle$ are referred to as the *particle number* from now on. In this language, the index a refers to different particle species.

11.1.1 Canonical ensemble

One of the essential features of the canonical ensemble of statistical mechanics is that the particle number is fixed. The Hamilton operator $\hat{H}$ may be time dependent. At finite temperature there are thermal fluctuations in addition to the quantum mechanical fluctuations. This implies that the quantum mechanical system may be found in different states $|n\rangle$ with energy E_n. The probability of finding the system in the state $|n\rangle$ with energy E_n is proportional to $\exp(-\beta E_n)$ with $\beta \equiv 1/T$ the inverse temperature. Here, we set the Boltzmann constant k_{B} to one, i.e. $k_{\mathrm{B}} = 1$. The partition function in the canonical ensemble is then given by

$$Z_{\mathrm{can}} = \sum_n \exp(-\beta E_n) = \operatorname{tr} \exp(-\beta \hat{H}), \qquad (11.1)$$

where tr refers to the trace in the Hilbert space of the quantum-mechanical system. In equilibrium, the information of the system is given by the density matrix $\hat{\rho}_{can}$,

$$\hat{\rho}_{can} = \frac{\exp(-\beta\hat{H})}{Z_{can}} = \frac{\exp(-\beta\hat{H})}{\operatorname{tr}\exp(-\beta\hat{H})}. \tag{11.2}$$

In the canonical ensemble, the expectation value for an operator $\mathcal{O}$ is then given by

$$\langle\mathcal{O}\rangle_{can} = \operatorname{tr}(\mathcal{O}\hat{\rho}_{can}). \tag{11.3}$$

For example, the energy average $\langle\hat{H}\rangle_{can} \equiv U$ is given by

$$U = \langle\hat{H}\rangle_{can} = \operatorname{tr}(\hat{\rho}_{can}\hat{H}) = -\partial_\beta \ln Z_{can}. \tag{11.4}$$

The entropy S is defined by

$$S = \langle-\ln\hat{\rho}\rangle_{can} = -\operatorname{tr}(\hat{\rho}_{can}\ln\hat{\rho}_{can}) = \beta U + \ln Z_{can}. \tag{11.5}$$

We have

$$F = U - ST = -T\ln Z_{can}, \tag{11.6}$$

where F is the free energy, which is the thermodynamic potential of the canonical ensemble. In the canonical ensemble, we work at given temperature T, given volume V and at constant particle number N_a, where the label a refers to different species of particles. Therefore, F is a function of T, V, N_a,

$$F = F(T, V, N_a). \tag{11.7}$$

If either F or Z_{can} is known, then all the other thermodynamical variables can be calculated via

$$S = -\left(\frac{\partial F}{\partial T}\right)_{V,N_a}, \qquad p = -\left(\frac{\partial F}{\partial V}\right)_{T,N_a}, \qquad \mu_a = \left(\frac{\partial F}{\partial N_a}\right)_{V,T}. \tag{11.8}$$

Here, S is the entropy, p the pressure and μ_a the chemical potential associated with the particles counted by N_a. The variables written next to the bracket are kept fixed. The variation of the free energy is given by

$$dF = -p\,dV - S\,dT + \mu_a\,dN_a. \tag{11.9}$$

When considering a quantum field theory, the question arises of how to calculate the partition function Z_{can}. We discuss this below. Since in quantum field theory, the number of particles is not fixed in general, we first have to introduce a system in which the particle number is also allowed to fluctuate, i.e. we introduce the grand canonical ensemble.

11.1.2 Grand canonical ensemble

In the canonical ensemble, the temperature T, the volume V and the particle number N_a are kept fixed. Now we allow the particle number N_a to fluctuate and thus consider the grand canonical ensemble, for which the density operator is given by

$$\hat{\rho}_{grand} = \frac{\exp\left(-\beta(\hat{H} - \mu_a\hat{Q}_a)\right)}{Z_{grand}}, \tag{11.10}$$

where the μ_a are the chemical potentials associated with the charges $\hat{Q}_a$, and Z_{grand} is the partition function of the grand canonical ensemble,

$$Z_{\text{grand}} = \text{tr} \exp\left(-\beta(\hat{H} - \mu_a \hat{Q}_a)\right). \tag{11.11}$$

The grand canonical potential Ω, which depends only on T, V and the chemical potentials μ_a, is given by

$$\Omega(T, V, \mu_a) = \langle \hat{H} \rangle_{\text{grand}} - ST - \mu_a \langle \hat{Q}_a \rangle_{\text{grand}}. \tag{11.12}$$

In the grand canonical ensemble, the expectation value for any operator $\mathcal{O}$ is given by

$$\langle \mathcal{O} \rangle_{\text{grand}} = \text{tr}(\hat{\rho}_{\text{grand}} \mathcal{O}). \tag{11.13}$$

For instance, $\langle \hat{H} \rangle_{\text{grand}} \equiv U$ is the internal energy and $\langle \hat{Q}_a \rangle_{\text{grand}} \equiv N_a$ is the average particle number for particles of species a. The entropy S is defined as

$$S = -\langle \ln \hat{\rho} \rangle_{\text{grand}} = -\text{tr}(\hat{\rho} \ln \hat{\rho}) = \beta U - \mu_a N_a + \ln Z_{\text{grand}}. \tag{11.14}$$

This implies

$$\Omega(T, V, \mu_a) = -T \ln Z_{\text{grand}}. \tag{11.15}$$

By a Legendre transformation, we obtain the free energy from the grand canonical potential,

$$\Omega(T, V, \mu_a) = F(T, V, N_a(\mu_a)) - \mu_a \langle \hat{Q}_a \rangle. \tag{11.16}$$

11.1.3 Quantum field theory at finite temperature

We now aim to calculate ensemble averages of the form

$$\langle \mathcal{O} \rangle_\beta = \text{tr}\left(\frac{\exp(-\beta \mathcal{H})}{\text{tr} \exp(-\beta \mathcal{H})} \mathcal{O}\right), \tag{11.17}$$

where $\mathcal{H} = \hat{H}$ for the canonical ensemble, or $\mathcal{H} = \hat{H} - \mu_a \hat{Q}_a$ for the grand canonical ensemble. Formally, the operator $\exp(-\beta \hat{H})$ is identical to the time evolution operator $\exp(i\hat{H}t)$ if we identify $t = i\beta$. Since $\beta = 1/T$ is real, we have to consider imaginary times t.

For simplicity, let us consider a scalar field, $\hat{\phi}(x)$ in the Heisenberg picture, whose dynamics is described by the time-independent Hamilton operator $\hat{H}$. The time evolution of $\hat{\phi}(x) = \hat{\phi}(t, \vec{x})$ is given by

$$\hat{\phi}(t, \vec{x}) = \exp(i\hat{H}t)\, \phi(0, \vec{x}) \exp(-i\hat{H}t), \tag{11.18}$$

where we allow for complex times $t \in \mathbb{C}$ in this definition. We now consider thermal Green's functions defined by

$$G^{\mathcal{C}}(x_1, \ldots, x_n) = \langle T_{\mathcal{C}} \hat{\phi}(x_1) \hat{\phi}(x_2) \ldots \hat{\phi}(x_n) \rangle_\beta, \tag{11.19}$$

where $\langle\ldots\rangle_\beta$ denotes the thermal average defined by (11.17). Note that $G^{\mathcal{C}}(x_1,\ldots,x_n)$ is defined for complex times t_n. This raises the question how to define time ordering for complex times $t_1, t_2, \ldots, t_n$, since in general these cannot be ordered. For complex times, the 'time ordering' $T_{\mathcal{C}}$ is only defined along a curve $\mathcal{C}$ in the complex plane. We restrict our attention to curves $\mathcal{C}$ which may be written in parameter form $t = \gamma(\tau)$, with τ real and monotonically decreasing. We introduce a step function $\Theta_{\mathcal{C}}(t - t')$ and a delta function $\delta_{\mathcal{C}}(t - t')$ by virtue of

$$\Theta_{\mathcal{C}}(t - t') = \Theta(\tau - \tau'), \qquad \delta_{\mathcal{C}}(t - t') = \left(\frac{\partial\gamma}{\partial\tau}\right)^{-1}\delta(\tau - \tau'). \qquad (11.20)$$

Then, we may define the time ordering $T_{\mathcal{C}}$ by

$$T_{\mathcal{C}}\hat{\phi}(x)\hat{\phi}(x') = \Theta_{\mathcal{C}}(t - t')\hat{\phi}(x)\hat{\phi}(x') + \Theta_{\mathcal{C}}(t' - t)\hat{\phi}(x')\hat{\phi}(x). \qquad (11.21)$$

This ensures that the fields whose argument τ is small appear on the right. Moreover, if we define a functional derivative by

$$\frac{\delta J(x')}{\delta J(x)} = \delta_{\mathcal{C}}(t - t')\delta(\vec{x} - \vec{x}'), \qquad (11.22)$$

we may define a generating functional $Z[J]$ for the thermal Green function (11.19) such that

$$G^{\mathcal{C}}(x_1,\ldots,x_n) = \frac{1}{Z[0]}\left(\frac{1}{i}\right)^n \frac{\delta^n Z[J]}{\delta J(x_1)\ldots\delta J(x_n)}\bigg|_{J=0}. \qquad (11.23)$$

Here, $Z[J]$ is given by

$$Z[J] = Z[0]\left\langle T_{\mathcal{C}}\exp\left[i\int_{\mathcal{C}} \mathrm{d}^d x\, J(x)\hat{\phi}(x)\right]\right\rangle_\beta, \qquad (11.24)$$

with $\langle\ldots\rangle_\beta$ again denoting the thermal average.

So far we have not considered which curves $\mathcal{C}$ are actually allowed in the above argument. Requiring all thermal Green's functions to be analytic with respect to their time arguments implies that

$$-\beta \leq \mathrm{Im}\,(t - t') \leq \beta, \qquad (11.25)$$

if $\Theta_{\mathcal{C}}(t - t') = 0$ for $\mathrm{Im}\,(t - t') \geq 0$. This is equivalent to the statement that any point on the curve $\mathcal{C}$ has to have a monotonically decreasing or constant imaginary part.

In the following, we consider two different curves $\mathcal{C}$. The first one is $t = -i\tau$ with $\tau \in [0, \beta]$, as shown in the left graph in figure 11.1. This leads to the *imaginary time formalism*. However, this formalism is inappropriate for studying transport in finite temperature systems. In this case it is desirable to consider a curve $\mathcal{C}$ which covers a large time interval $[t_i, t_f]$ along the real axis. A possible curve is shown as the right graph in figure 11.1. The corresponding formalism is known as the *real time* or *Schwinger–Keldysh formalism*.

Before discussing both formalisms in some detail, we first have to introduce the generalisation of the path integral to a curve $\mathcal{C}$. We work in the Heisenberg picture, in which

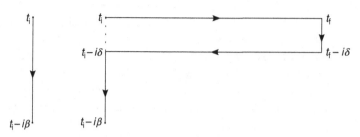

Figure 11.1   Curve C for the imaginary time formalism (left) and for the real time formalism (right).

the operator $\hat{\phi}(x)$ is time dependent. We introduce a state vector $|\phi(\vec{x}); t\rangle$ as an eigenstate of the field operator $\hat{\phi}(x)$ at a fixed time t,

$$\hat{\phi}(x)|\phi(\vec{x}); t\rangle = \phi(\vec{x})|\phi(\vec{x}); t\rangle, \tag{11.26}$$

with eigenvalue $\phi(\vec{x})$. Using the time evolution of the Heisenberg field $\hat{\phi}(x)$ given by (11.18), the eigenstate has the time evolution

$$|\phi(\vec{x}); t\rangle = e^{i\hat{H}t}|\phi(\vec{x}); 0\rangle, \tag{11.27}$$

which is defined to hold also for complex times t. Assuming that the Heisenberg states $|\phi(\vec{x}); t\rangle$ form a complete set at any time t, we then have for the generating functional

$$Z[J] =$$
$$\int \mathcal{D}\phi'(\vec{x}) \langle \phi'(\vec{x}); t_i| e^{-\beta\hat{H}} T_C \exp\left[i \int_C \mathrm{d}^d x \, J(x)\hat{\phi}(x)\right] |\phi'(\vec{x}); t_i\rangle, \tag{11.28}$$

where in general

$$\langle \phi'(\vec{x}); t_i| e^{-\beta\hat{H}} = \langle \phi'(\vec{x}); t_i - i\beta|. \tag{11.29}$$

Moreover, we have

$$\langle \phi'(\vec{x}); t_f| T_C f[\hat{\phi}]|\phi'(\vec{x}); t_i\rangle = N \int \mathcal{D}\phi \, f[\phi] \exp\left[i \int_C \mathrm{d}^d x \, \mathcal{L}\right], \tag{11.30}$$

where we integrate over the field $\phi(x)$ with boundary conditions

$$\phi(t_f, \vec{x}) = \phi'(\vec{x}), \qquad \phi(t_i, \vec{x}) = \phi'(\vec{x}). \tag{11.31}$$

We thus obtain for the generating functional

$$Z[J] = N \int \mathcal{D}\phi \exp\left[i \int_C \mathrm{d}^d x \, (\mathcal{L} + J(x)\phi(x))\right], \tag{11.32}$$

where we integrate over fields for which $\phi(t_i - i\beta, \vec{x}) = \phi(t_i, \vec{x})$. The last step in the calculation leading to (11.32) is valid only if there are no couplings involving derivatives of the field in the field theory Lagrangian.

For fermions, we may derive a formula analogous to (11.32). In the fermionic case, the fields have to satisfy anti-periodic boundary conditions, i.e.

$$\psi(t_i - i\beta, \vec{x}) = -\psi(t_i, \vec{x}). \tag{11.33}$$

This may be seen as follows. Consider two bosonic or fermionic operators $\mathcal{O}_1$ and $\mathcal{O}_2$. For the time ordered correlation function of two operators we obtain

$$
\begin{aligned}
\langle \mathcal{O}_1(t)\mathcal{O}_2(t' + i\beta)\rangle_\beta &= \text{tr}\left[\exp(-\beta\hat{H})\mathcal{O}_1(t)\mathcal{O}_2(t' + i\beta)\right] \\
&= \text{tr}\left[\mathcal{O}_2(t' + i\beta)\exp(-\beta\hat{H})\mathcal{O}_1(t)\right] \\
&= \text{tr}\left[\exp(-\beta\hat{H})\mathcal{O}_2(t')\exp(\beta\hat{H})\exp(-\beta\hat{H})\mathcal{O}_1(t)\right] \\
&= \langle \mathcal{O}_2(t')\mathcal{O}_1(t)\rangle_\beta \\
&= \pm\langle \mathcal{O}_1(t)\mathcal{O}_2(t')\rangle_\beta.
\end{aligned}
\tag{11.34}
$$

We see that the thermal Green function has to be periodic for bosons and anti-periodic for fermions.

11.1.4 Imaginary time formalism

The imaginary time formalism is straightforward to derive by considering a curve $\mathcal{C}$ from 0 to $-i\beta$ along the imaginary time axis. We introduce a new Euclidean time $\tau = it$ and compactify it on a circle $\tau \in [0, \beta]$ with $\beta = 1/T$. Bosonic and fermionic fields satisfy periodic or anti-periodic boundary conditions, respectively, i.e.

$$\mathcal{O}(t, \vec{x}) = \pm\mathcal{O}(t - i\beta, \vec{x}). \tag{11.35}$$

This implies that after Fourier transforming to four-dimensional momentum space, the frequencies ω are quantised,

$$\omega_n = \frac{2n\pi}{\beta} \qquad \text{for bosons,} \tag{11.36}$$

$$\omega_n = \frac{(2n+1)\pi}{\beta} \qquad \text{for fermions.} \tag{11.37}$$

The frequencies ω_n are the *Matsubara frequencies*. It is straightforward to establish the Feynman rules for the calculation of thermal Green's functions once the Feynman rules for the vacuum theory at $T = 0$ are known in momentum space. The translation prescriptions are given in table 11.1. These allow the calculation of any Euclidean Green's function G_E for quantum field theories at finite temperature, for example by using perturbation theory.

Note that if the Euclidean correlation functions G_E are known exactly, we may obtain the retarded Green function G_R by a simple analytic continuation. G_R is defined in exercise 1.3.3 and in section 12.1. We have

$$G_R(\omega, \vec{k}) = G_E(-i(\omega + i\epsilon), \vec{k}), \tag{11.38}$$

or

$$G_E(\omega_E, \vec{k}) = G_R(i\omega_E, \vec{k}) \quad \text{for } \omega_E > 0. \tag{11.39}$$

Table 11.1 Feynman rules at $T = 0$ and rules derived at $T \neq 0$		
Vacuum theory at $T = 0$		Imaginary time formalism for $T \neq 0$
$G(k_1, \ldots, k_n)$	$\mapsto$	$(-i)^n G_E(k_1, \ldots, k_n)$
$\int \frac{\mathrm{d}^d k}{(2\pi)^d} f(\omega, \vec{k})$	$\mapsto$	$-i\beta \sum_n \int \frac{\mathrm{d}^{d-1} k}{(2\pi)^{d-1}} f(\omega_n, \vec{k})$
$(2\pi)^d \delta^{(d)}(k)$	$\mapsto$	$-i\beta (2\pi)^{d-1} \delta_{n,0} \delta(\vec{k})$

However, in most cases we know the Euclidean correlation function only numerically or only for the Matsubara frequencies, so the analytical continuation becomes difficult. Therefore it is important to have techniques available which allow us to calculate real time correlation functions such as the retarded Green function directly. Such techniques will be introduced in the next section.

11.1.5 Real time formalism

The real time or Schwinger–Keldysh formalism is particularly useful for describing transport processes, since it allows deviations from equilibrium. We now consider a curve $\mathcal{C}$ which first runs along the real time axis from t_i to t_f, then moves into the lower imaginary half-plane to $t_f - i\delta$, then returns parallel to the real axis to $t_i - i\delta$ and finally runs to $t_i - i\beta$, as depicted in figure 11.1. Moreover, we identifiy the starting point with the endpoint and impose periodic boundary conditions for bosons and anti-periodic boundary conditions for fermions. This curve $\mathcal{C}$ is parametrised by a free parameter σ which takes values in the interval $\sigma \in [0; \beta]$. We will see that $\sigma = \beta/2$ is special both on the field theory side and on the gravity side.

The action $\mathcal{S}$ may be split into four parts corresponding to the four different parts of the curve $\mathcal{C}$,

$$
\begin{aligned}
\mathcal{S} &= \int_{\mathcal{C}} \mathrm{d}t \int \mathrm{d}^{d-1}x \, \mathcal{L}\left(\phi(t, \vec{x})\right) \\
&= \int_{t_i}^{t_f} \mathrm{d}t \int \mathrm{d}^{d-1}x \, \mathcal{L}\left(\phi(t, \vec{x})\right) - i \int_0^{\sigma} \mathrm{d}\tau \int \mathrm{d}^{d-1}x \, \mathcal{L}\left(\phi(t_f - i\tau, \vec{x})\right) \\
&\quad - \int_{t_i}^{t_f} \mathrm{d}t \int \mathrm{d}^{d-1}x \, \mathcal{L}\left(\phi(t - i\sigma, \vec{x})\right) - i \int_{\sigma}^{\beta} \mathrm{d}\tau \int \mathrm{d}^{d-1}x \, \mathcal{L}\left(\phi(t_i - i\tau, \vec{x})\right), \quad (11.40)
\end{aligned}
$$

where we have dropped the dependence of the Lagrangian on $\partial_\mu \phi$ for simplicity. From now on we write

$$
\phi_1(t, \vec{x}) \equiv \phi(t, \vec{x}), \qquad \phi_2(t, \vec{x}) \equiv \phi(t - i\sigma, \vec{x}), \qquad (11.41)
$$

with the sources J_1, J_2 for ϕ_1, ϕ_2 given by

$$J_1(t, \vec{x}) \equiv J(t, \vec{x}), \qquad J_2(t, \vec{x}) \equiv J(t - i\sigma, \vec{x}). \tag{11.42}$$

The generating functional then reads

$$Z[J_1, J_2] = \int \mathcal{D}\phi \exp\left(iS + i\int_{t_i}^{t_f} dt \int d^{d-1}x \, (\phi_1(t, \vec{x})J_1(t, \vec{x}) - \phi_2(t, \vec{x})J_2(t, \vec{x}))\right). \tag{11.43}$$

Since the fields ϕ_1 and ϕ_2 are independent, the variations with respect to J_1 and J_2 may be taken independently of each other. Varying Z with respect to both sources, we obtain the Schwinger–Keldysh propagator

$$i\, G_{ab}(x - y) \equiv i\begin{pmatrix} G_{11}(x-y) & -G_{12}(x-y) \\ -G_{21}(x-y) & G_{22}(x-y) \end{pmatrix} = \frac{1}{i^2}\frac{\delta^2 \ln Z[J_1, J_2]}{\delta J^a(x)\delta J^b(y)}. \tag{11.44}$$

Since in the operator formalism we have to path order along the contour $\mathcal{C}$, we have to time order along $t_f - i\sigma$ to $t_i - i\sigma$. Therefore we obtain

$$\begin{aligned} iG_{11}(x) &= \langle T\phi_1(x)\phi_1(0)\rangle, & iG_{12}(x) &= \langle \phi_2(0)\phi_1(x)\rangle, \\ iG_{21}(x) &= \langle \phi_2(x)\phi_1(0)\rangle, & iG_{22}(x) &= \langle \bar{T}\phi_2(x)\phi_2(0)\rangle, \end{aligned} \tag{11.45}$$

where T denotes time ordering and $\bar{T}$ denotes reversed time ordering. The components $G_{ab}(\vec{x}, t)$ of the Schwinger–Keldysh propagator may be related to the *retarded* and *advanced* Green functions which are defined by

$$G_R(x - y) = -i\Theta(x^0 - y^0)\langle[\phi(x), \phi(y)]\rangle, \tag{11.46}$$

$$G_A(x - y) = -i\Theta(y^0 - x^0)\langle[\phi(y), \phi(x)]\rangle, \tag{11.47}$$

see also Exercise 1.3.3. For fermionic operators, anticommutators have to be used in these expressions. Transforming to momentum space by using

$$G(k) = \int d^d x \, e^{-ikx} G(x), \tag{11.48}$$

we find that $G_A(k) = G_R^*(k)$. The matrix elements $G_{ab}(k)$ are related to the retarded Green functions $G_R(k)$ by

$$G_{11}(k) = \mathrm{Re}\, G_R(k) + i\coth\left(\frac{\omega}{2T}\right)\mathrm{Im}\, G_R(k), \tag{11.49}$$

$$G_{12}(k) = \frac{2i\exp(-(\beta - \sigma)\omega)}{1 - \exp(-\beta\omega)}\mathrm{Im}\, G_R(k), \tag{11.50}$$

$$G_{21}(k) = \frac{2i\exp(-\sigma\omega)}{1 - \exp(-\beta\omega)}\mathrm{Im}\, G_R(\omega), \tag{11.51}$$

$$G_{22}(k) = -\mathrm{Re}\, G_R(k) + i\coth\left(\frac{\omega}{2T}\right)\mathrm{Im}\, G_R(k), \tag{11.52}$$

where $\omega = k^0$. In particular, for $\sigma = \beta/2$ we obtain a symmetric matrix G_{ab}, i.e.

$$G_{12} = G_{21}. \tag{11.53}$$

The case $\sigma = \beta/2$ turns out to be the natural formulation for black holes which we will study as gravity duals of finite-temperature field theories.

11.2 Gravity dual thermodynamics

11.2.1 Thermodynamics on $\mathbb{R}^3$

We now turn to the gravity dual of the thermodynamics of $\mathcal{N} = 4$ Super Yang–Mills theory. For finite temperature, this theory is considered on flat space with spatial dimensions $\mathbb{R}^3$, while the time direction is compactified on a circle. Note that this time compactification breaks supersymmetry completely, due to anti-periodic boundary conditions imposed for the fermions. On the gravity side, this field theory thermodynamics is identified with thermodynamics of *black D3-branes*. These correspond to non-extremal D3-brane solutions as discussed in (4.125) in chapter 4. Their background metric is given by

$$ds^2 = H(r)^{-1/2} \left(-f(r)dt^2 + d\vec{x}^2 \right) + H(r)^{1/2} \left(\frac{dr^2}{f(r)} + r^2 d\Omega_5^2 \right), \qquad (11.54)$$

with the *blackening factor* $f(r)$

$$f(r) = 1 - \left(\frac{r_{\rm h}}{r} \right)^4 \qquad (11.55)$$

and

$$H(r) = 1 + \frac{L^4}{r^4}. \qquad (11.56)$$

At $r = r_{\rm h}$, there is an event horizon similar to the black hole horizon introduced in section 2.4. As we discuss below, this horizon may be related to the Hawking temperature of the black brane. In contrast to a black hole, for a black brane the spatial directions $\vec{x}$ are not compactified. Taking the near-horizon limit $r/L \ll 1$ and using the coordinate $z \equiv L^2/r$, we obtain

$$ds^2 = \frac{L^2}{z^2} \left(-\left(1 - \frac{z^4}{z_{\rm h}^4} \right) dt^2 + d\vec{x}^2 + \frac{1}{1 - \frac{z^4}{z_{\rm h}^4}} dz^2 \right) + L^2 d\Omega_5^2, \qquad (11.57)$$

where $z_{\rm h} = L^2/r_{\rm h}$. Introducing Euclidean time $\tau \equiv it$ we find

$$ds^2 = \frac{L^2}{z^2} \left(\left(1 - \frac{z^4}{z_{\rm h}^4} \right) d\tau^2 + d\vec{x}^2 + \frac{1}{1 - \frac{z^4}{z_{\rm h}^4}} dz^2 \right) + L^2 d\Omega_5^2. \qquad (11.58)$$

Recall from chapter 4 that while ordinary D3-branes are *extremal*, i.e. they satisfy the BPS condition $M = Q$ and are thus supersymmetric, the metric for black D3 branes at finite temperature is *non-extremal*, and temperature mildly breaks the supersymmetry condition.

To study the physics associated with this metric, we restrict our attention to the five-dimensional deformed AdS space and neglect the five-sphere S^5. First, we observe that

$$g_{tt} \to 0 \quad \text{and} \quad g_{zz} \to \infty \quad \text{for } z \to z_{\rm h}. \qquad (11.59)$$

Let us introduce a further radial variable ρ given by

$$z = z_{\mathrm{h}}\left(1 - \frac{\rho^2}{L^2}\right). \tag{11.60}$$

ρ is a measure for the distance from the horizon at z_{h}. Then, to lowest order in ρ the Euclidean metric becomes

$$ds^2 \simeq \frac{4\rho^2}{z_{\mathrm{h}}^2}\mathrm{d}\tau^2 + \frac{L^2}{z_{\mathrm{h}}^2}\mathrm{d}\vec{x}^2 + \mathrm{d}\rho^2. \tag{11.61}$$

We now show that regularity at the horizon is obtained only if τ is periodic. The period given by $\beta = \frac{1}{T}$ is identified with the inverse temperature, with $k_{\mathrm{B}} \equiv 1$ for the Boltzmann constant. Consider the behaviour of the metric near the horizon z_{h}. In this region, the metric in the (τ, ρ) plane becomes, rescaling τ to $\phi = 2\tau/z_{\mathrm{h}}$,

$$ds^2 = \mathrm{d}\rho^2 + \rho^2\mathrm{d}\phi^2. \tag{11.62}$$

This metric corresponds to a plane in polar coordinates if we impose periodicity, $\phi \sim \phi + 2\pi$, to avoid a conical singularity at $\rho = 0$. For τ this implies periodicity with a period $\Delta\tau = \pi \cdot z_{\mathrm{h}}$. From quantum field theory at finite temperature, we know that $\Delta\tau = \beta = 1/T$. Thus we conclude

$$z_{\mathrm{h}} = \frac{1}{\pi T}, \tag{11.63}$$

where T is the temperature of the field theory.

Therefore we may identify the Anti-de Sitter black brane as the gravity dual of the strongly coupled $\mathcal{N} = 4$ Super Yang–Mills plasma at finite T. Let us now calculate thermodynamical quantities. First we compute the entropy of the field theory from the Bekenstein–Hawking entropy of the associated black brane. We start from the expression

$$S_{\mathrm{BH}} = \frac{A}{4G}, \tag{11.64}$$

for the Bekenstein–Hawking entropy, where A is the horizon area and G is Newton's constant. The area of the horizon is given by

$$A = \int \mathrm{d}^3 x \sqrt{g_{3\mathrm{d}}|_{z=z_{\mathrm{h}}}}\,\mathrm{Vol}(S^5). \tag{11.65}$$

The determinant $g_{3d} = g_{11}g_{22}g_{33}$ gives $g_{3d} = L^6/z^6$ and therefore with $\mathrm{Vol}(S^5) = \pi^3 L^5$ we have

$$A = \pi^6 L^8 T^3 \mathrm{Vol}(\mathbb{R}^3), \tag{11.66}$$

where $\mathrm{Vol}(\mathbb{R}^3)$ is the three-dimensional spatial volume of the field theory. Using the result (6.92) for the ten-dimensional Newton constant, we obtain the Bekenstein–Hawking entropy of the five-dimensional Schwarzschild Anti-de Sitter black brane [1]

$$S_{\mathrm{BH}} = \frac{\pi^2}{2}N^2 T^3 \mathrm{Vol}(\mathbb{R}^3). \tag{11.67}$$

This is identified with the entropy of the strongly coupled $\mathcal{N} = 4$ Super Yang–Mills plasma, $S = S_{\mathrm{BH}}$. Note that using gauge/gravity duality, we have calculated the entropy of a strongly coupled plasma in a thermal field theory by a simple gravity calculation. It is impossible to perform the same calculation directly in the strongly coupled field theory because this would require summing an infinite number of Feynman diagrams. The result obtained above is even more surprising if we compare it to the entropy of free, i.e. non-interacting, $\mathcal{N} = 4$ Super Yang–Mills theory. In that case the entropy reads [2]

$$S_{\text{free}} = \frac{2\pi^2}{3} N^2 T^3 \text{Vol}(\mathbb{R}^3). \tag{11.68}$$

Introducing an interpolating function $a(\lambda)$ depending on the 't Hooft coupling, we see that in both cases the entropy takes the form

$$S = \frac{2\pi^2}{3} N^2 T^3 \text{Vol}(\mathbb{R}^3) \cdot a(\lambda), \tag{11.69}$$

which is fixed by dimensional analysis. The factor of T^3 has to be present since the entropy scales as a mass to the third power and T is the only dimensionful quantity present. The large N limit explains the factor N^2. Assuming the validity of gauge/gravity duality, we have determined two limits of $a(\lambda)$,

$$\lim_{\lambda \to 0} a(\lambda) = 1, \qquad \lim_{\lambda \to \infty} a(\lambda) = \frac{3}{4}. \tag{11.70}$$

To justify the entropy relation (11.64), which was the starting point of our calculation, we may calculate the partition function, or equivalently the free energy density, using the gravitational on-shell action. The five-dimensional Euclidean action giving rise to the Schwarzschild AdS black brane is given by

$$S_E[g] = \frac{1}{-2\kappa_5^2} \int d^5x \sqrt{g} \left(R + \frac{12}{L^2} \right) - \frac{1}{\kappa_5^2} \int d^4x \sqrt{\gamma}\, K, \tag{11.71}$$

where $\kappa_5^2 = 8\pi G_5$. The second term is the Gibbons–Hawking boundary term introduced in (2.155), which is required for a well-posed variational principle. $\gamma^{\mu\nu}$ is the induced metric at the boundary $z \to 0$,

$$\gamma_{\mu\nu} dx^\mu dx^\nu = \frac{L^2}{z^2} \left(f(z)d\tau^2 + d\vec{x}^2 \right), \qquad f(z) = 1 - \left(\frac{z}{z_h} \right)^4, \tag{11.72}$$

and the outward pointing unit normal vector n_m at the boundary, which is needed to evaluate the extrinsic curvature $K_{mn} = \nabla_m n_n$, is given by

$$n_m dx^m = \frac{L}{z\sqrt{f(z)}} dz. \tag{11.73}$$

Since the on-shell action is divergent due to the infinite volume of AdS space, we introduce a cut-off at $z = \varepsilon \ll 1$. For the regularised on-shell action we obtain, applying the methods of holographic renormalisation introduced in section 5.5,

$$S_{\text{reg}} = -\frac{L^3}{T\kappa_5^2} \text{Vol}(\mathbb{R}^3) \left(\frac{3}{\epsilon^4} - \frac{1}{z_h^4} + \mathcal{O}(\varepsilon) \right). \tag{11.74}$$

The divergent term scales as ϵ^{-4}. This corresponds precisely to the infinite volume of AdS_5 space. This divergence may be removed by adding the counterterm

$$\mathcal{S}_{\text{ct}} = \frac{3}{L\kappa_5^2} \int\limits_{z=\varepsilon} \mathrm{d}^4x\sqrt{\gamma} = \frac{L^3}{T\kappa_5^2}\text{Vol}(\mathbb{R}^3)\left(\frac{3}{\epsilon^4} - \frac{3}{2z_{\text{h}}^4} + \mathcal{O}(\epsilon)\right) \tag{11.75}$$

to $\mathcal{S}_{\text{reg}}$. The renormalised on-shell action is given by

$$\mathcal{S}_{\text{ren}} = \lim_{\varepsilon\to 0}(\mathcal{S}_{\text{reg}} + \mathcal{S}_{\text{ct}}) = -\frac{L^3}{T\kappa_5^2}\frac{1}{2z_{\text{h}}^4}\text{Vol}(\mathbb{R}^3)$$

$$= -\frac{\pi^2}{8}N^2T^3\,\text{Vol}(\mathbb{R}^3). \tag{11.76}$$

From this it is straightforward to obtain the free energy which is given by

$$F = -T\ln Z = -\frac{\pi^2}{8}N^2T^4\,\text{Vol}(\mathbb{R}^3). \tag{11.77}$$

This is consistent with the entropy S given by (11.67) since we have

$$S = -\frac{\partial F}{\partial T} = \frac{\pi^2}{2}N^2T^3\text{Vol}(\mathbb{R}^3). \tag{11.78}$$

The average energy $U = F + TS$ reads

$$U = F + TS = \frac{3\pi^2}{8}N^2T^4\text{Vol}(\mathbb{R}^3). \tag{11.79}$$

This energy may also be calculated in a different way by determining the expectation value of the energy-momentum tensor, $\langle T_{\mu\nu}\rangle$, in the Super Yang–Mills plasma, using the method of holographic renormalisation of section 5.5. For this purpose, we have to determine the Fefferman–Graham form of the metric of black D3-branes. Since

$$\mathrm{d}s^2 = \frac{L^2}{z^2}\left[\left(1 - \frac{z^4}{z_{\text{h}}^4}\right)\mathrm{d}\tau^2 + \mathrm{d}\vec{x}^2 + \frac{\mathrm{d}z^2}{1 - \frac{z^4}{z_{\text{h}}^4}}\right], \tag{11.80}$$

a convenient coordinate transformation is

$$z = \frac{\tilde{z}}{\sqrt{1 + \frac{\tilde{z}^4}{4z_{\text{h}}^4}}}. \tag{11.81}$$

The metric written in the $\tilde{z}$ coordinate is indeed of Fefferman–Graham form,

$$\mathrm{d}s^2 = \frac{L^2}{\tilde{z}^2}\left[\frac{\left(1 - \frac{\tilde{z}^4}{4z_{\text{h}}^4}\right)^2}{1 + \frac{\tilde{z}^4}{4z_{\text{h}}^4}}\mathrm{d}\tau^2 + \left(1 + \frac{\tilde{z}^4}{4z_{\text{h}}^4}\right)\mathrm{d}\vec{x}^2 + \mathrm{d}\tilde{z}^2\right]. \tag{11.82}$$

Expanding the factor multiplying $\mathrm{d}\tau^2$ for small $\tilde{z}$, gives $1 - 3\tilde{z}^4/4z_{\text{h}}^4$. We now make use of the result (5.126) for the energy-momentum tensor with $\rho = z^2$, noting that the Einstein tensor vanishes for the boundary metric considered here. The overall normalisation is given by (6.92). Calculating (5.126) for the metric (11.82) gives

$$\langle T_{\mu\nu} \rangle = \frac{\pi^2}{8} N^2 T^4 \text{diag}(-3, 1, 1, 1). \tag{11.83}$$

We observe that $\langle T_{\mu\nu} \rangle$ is symmetric and traceless, in agreement with the fact that we started with a conformal field theory. Note also that $\langle T_{tt} \rangle = -\epsilon$ with ϵ the energy density, while $\langle T_{xx} \rangle = p$, with p the pressure. Therefore for the free energy of (11.77) we have $F = -\langle T_{xx} \rangle \cdot \text{Vol}(\mathbb{R}^3)$.

11.2.2 Thermodynamics on S^{d-1}

Let us now consider a conformal field theory defined on the spacetime manifold $\mathbb{R} \times S^{d-1}$, i.e. the time direction is not compactified but the spatial dimensions are compactified on S^{d-1}. An example is given by $\mathcal{N} = 4$ Super Yang–Mills theory on $\mathbb{R} \times S^3$.

At finite temperature, the time direction is compactified too, and the spacetime manifold becomes $S^1 \times S^{d-1}$. There are now two dimensionful quantities, $\beta = 1/T$ and $\beta' = 1/l$, where l is the radius of S^{d-1}, and the physics will depend on the quotient β/β'.

What is the gravity dual of this field theory? We may write down two different metrics which both have an $S^1 \times S^{d-1}$ boundary.

- $(d+1)$-dimensional global AdS space with metric given by

$$ds^2 = \left(1 + \frac{r^2}{L^2}\right) d\tau^2 + \left(1 + \frac{r^2}{L^2}\right)^{-1} dr^2 + r^2 d\Omega_{d-1}^2, \tag{11.84}$$

where we have introduced periodic Euclidean time $\tau = it$. Global AdS space with the time direction compactified is referred to as *thermal AdS* space.

- The $(d+1)$-dimensional AdS–Schwarzschild black hole with Euclidean metric

$$ds^2 = f(r) d\tau^2 + \frac{1}{f(r)} dr^2 + r^2 d\Omega_{d-1}^2, \tag{11.85}$$

$$f(r) \equiv 1 - \frac{\mu}{r^{d-2}} + \frac{r^2}{L^2}, \tag{11.86}$$

where μ is related to the black hole mass as discussed in section 2.4.4.

While the associated temperature of the first spacetime is given by the inverse of the compactified Euclidean time direction, we may determine the temperature of the second by relating it to the horizon radius $r_{\rm h}$, which is given by the larger root of $f(r_{\rm h}) = 0$, i.e. of

$$1 - \frac{\mu}{r_{\rm h}^{d-2}} + \frac{r_{\rm h}^2}{L^2} = 0. \tag{11.87}$$

Since the metric near the horizon at $r \simeq r_{\rm h}$ reads

$$ds^2 = f'(r_{\rm h})(r - r_{\rm h}) d\tau^2 + \frac{1}{f'(r_{\rm h})(r - r_{\rm h})} dr^2 + r_{\rm h}^2 d\Omega_{d-1}^2, \tag{11.88}$$

the temperature of the black hole is given by

$$T = \frac{|f'(r_h)|}{4\pi} = \frac{d\,r_h^2 + (d-2)L^2}{4\pi L^2 r_h}. \tag{11.89}$$

Viewing T as a function of r_h, we find that there exists a minimal temperature $T_{\min}$ which is given by

$$T_{\min} = \frac{1}{2\pi L}\sqrt{d(d-2)}. \tag{11.90}$$

For $T < T_{\min}$, there is no solution of the second type of spacetime metric introduced above, i.e. of AdS–Schwarzschild form, so thermal AdS is the only possible solution. For $T \geq T_{\min}$, both solutions (11.84) and (11.85) are accessible. Therefore we have to calculate the free energy in order to determine which of the two possible solutions is thermodynamically favoured. The free energy is given by

$$F^{(i)} = T S^{(i)}_{\text{on-shell}}, \tag{11.91}$$

where (i) labels the two possible spacetimes. It turns out that

$$\Delta F = F^{(2)} - F^{(1)} = \frac{r_h^{d-2}}{2\kappa_5^2}\,\text{Vol}(S^{d-1})\left(1 - \frac{r_h^2}{L^2}\right). \tag{11.92}$$

For $r_h < L$, the difference ΔF of the free energies is positive, i.e. $F^{(1)} < F^{(2)}$, and therefore thermal AdS is preferred. For $r_h > L$, the Schwarzschild black hole is preferred. For $r_h = L$, there is a phase transition, the *Hawking–Page phase transition* [3], with transition temperature

$$T = \frac{1}{2\pi L}(d-1). \tag{11.93}$$

To conclude, we see that large black holes are stable, while small ones decay into thermal AdS for $T \leq (d-1)/(2\pi L)$ and into large black holes for $T > (d-1)/(2\pi L)$. The relevance of this phase transition within gauge/gravity duality will be discussed in chapter 14.

11.2.3 Holographic Green's functions

In section 11.1.5 we introduced real time thermal Green's functions on the field theory side. These functions, as well as the Schwinger–Keldysh formalism, are naturally implemented on the gravity side of the correspondence [4], as we now explain. The first step is to transfer the Euclidean approach to calculated holographically the two-point functions of section 5.4.3 to the case of Lorentzian signature and to impose boundary conditions compatible with causality.

We consider the case of $3 + 1$ boundary dimensions for simplicity. It is convenient to introduce a new variable $u = r_h^2/r^2$, such that the metric (11.54) becomes, with $H(r) = L^4/r^4$,

$$ds^2 = \frac{(\pi TL)^2}{u}\left(-f(u)dt^2 + d\vec{x}^2\right) + \frac{L^2}{4u^2 f(u)}du^2 + L^2 d\Omega_5^2, \tag{11.94}$$

with $f(u) = 1 - u^2$. The boundary of this asymptotically AdS space is at $u = 0$, while at the horizon $u = 1$. Let us consider a massive scalar field ϕ with equation of motion $(\Box - m^2)\phi = 0$ dual to the scalar operator $\mathcal{O}$. Fourier transforming in the boundary directions, writing the Fourier transform as $\phi_k(u)$ for notational consistency with chapter 5, we have

$$\phi(x, u) = \int \frac{d^4 k}{(2\pi)^4} e^{ik \cdot x} \phi(u, k). \tag{11.95}$$

The equation of motion for the modes $\phi_k(u)$ reads

$$0 = 4u^3 \partial_u \left(\frac{f}{u} \partial_u \phi(u, k) \right) + \frac{u}{(\pi T)^2 f} \left(\omega^2 - |\vec{k}|^2 f \right) \phi(u, k) - m^2 L^2 \phi(u, k). \tag{11.96}$$

Near the boundary, the solutions have the asymptotic behaviour

$$\phi(u, k) \sim \phi_{(0)}(k) u^{(d-\Delta)/2} (1 + \mathcal{O}(u)) + \phi_{(+)}(k) u^{\Delta/2} (1 + \mathcal{O}(u)). \tag{11.97}$$

Here, as in chapter 5, Δ is the larger root of $\Delta(\Delta - d) = m^2 L^2$ with $d = 4$. $\phi_{(0)}$ and $\phi_{(+)}$ correspond to the leading and subleading terms as given in (5.49).

Next we need to impose the correct boundary conditions for obtaining the retarded Green function. One condition is given by fixing $\phi_{(0)}(k)$ at the boundary $u = 0$. The second condition needs to be imposed at the horizon. Near the horizon at $u = 1$, the solutions to (11.96) scale as

$$\phi_k(u) \sim (1 - u)^\kappa, \tag{11.98}$$

with κ a coefficient which we now discuss. It turns out that the behaviour of the solutions depends crucially on whether we consider Euclidean or Lorentzian signature. In Euclidean signature, κ is real: $\kappa = \pm\omega/(4\pi T)$. Thus only the solution with $+\omega/(4\pi T)$ is regular at the horizon, assuming $\omega > 0$.

In Lorentzian signature as relevant for the real time formalism, however, we have

$$\kappa = \pm \frac{i\omega}{4\pi T} \tag{11.99}$$

in (11.98). Both solutions are regular. Let us consider the boundary condition necessary to fix the solution uniquely. The solution with $-$ sign in (11.99) corresponds to an *infalling* boundary condition while the solution with $+$ sign corresponds to an *outgoing* boundary condition.

Exercise 11.2.1   Identify the infalling and outgoing solutions in (11.99) by restoring the time dependence from the Fourier decomposition and considering $e^{-i\omega t}(1 - u)^{\pm i\omega/(4\pi T)}$. Introducing the variable $\tilde{r} \equiv (\ln(1 - u))/(4\pi T)$, show that the solution with $\kappa = -i\omega/(4\pi T)$ in (11.99) behaves as $\sim e^{-i\omega(t+\tilde{r})}$, which corresponds to a wave moving towards the horizon. Perform a similar analysis for the other solution.

The infalling boundary condition will be associated in a natural way with a retarded Green's function: for infalling boundary conditions at the horizon, only those boundary sources located in the past may influence the bulk physics, which ensures causality.

For calculation of the retarded Green's function, we have, therefore, the following recipe [15].

(i) Linearise the equations of motion for ϕ and solve them in Fourier space. Split the solution $\phi(u, k)$ into a function $\phi_{(0)}(k)$ and a function $\phi_k(u)$,

$$\phi(u, k) = \phi_{(0)}(k) \cdot \phi_k(u). \tag{11.100}$$

At the boundary, $\phi_k(u) = 1$. At the horizon $\phi_k(u)$ has to satisfy the infalling wave boundary condition, i.e. $\phi_k(u) \sim (1 - u)^{-i\omega/(4\pi T)}$.

(ii) The action, evaluated for the solution of the form (11.100), reduces to a surface integral

$$S = \int \frac{d^4 k}{(2\pi)^4} \phi_{(0)}(-k) \, \mathcal{F}(k, u) \, \phi_{(0)}(k) \Big|_{u=0}^{u=u_{\mathrm{h}}}. \tag{11.101}$$

Remember that the conformal boundary in the u-coordinates is located at $u = 0$.

(iii) The retarded Green function is then given by

$$G_{\mathrm{R}}(k) = -2 \, \mathcal{F}(k, u = 0). \tag{11.102}$$

This recipe is analogous to the Euclidean calculation of the two-point function in section 5.4.3. The essential feature added for the Lorentzian case considered here is the infalling boundary condition at the horizon which ensures causality. Note, however, that unlike in the Euclidean case, at the present stage the functional derivative taking (11.101) to (11.102) is not yet defined such as to reflect unambiguously the causality structure. To achieve this, an implementation of the Schwinger–Keldysh formalism introduced in section 11.1.3 is required also on the gravity side [4]. For a further elucidation of the causality structure we now move on to consider Kruskal coordinates as introduced in (2.144), which cover the entire Penrose diagram for the Anti-de Sitter black brane. This allows also for a derivation of the Schwinger–Keldysh formalism on the gravity side as follows. Near the horizon, the metric (11.94) becomes a Schwarzschild metric of the form

$$ds^2 \sim 2(\pi TL)^2 \left(-\left(1 - \frac{2M}{\rho}\right) dt^2 + \left(1 - \frac{2M}{\rho}\right)^{-1} d\rho^2 \right) + \cdots, \tag{11.103}$$

where $T = 1/(8\pi M)$ and the radial variable u is replaced by $u = 2M/\rho$. We may thus introduce the same Kruskal coordinates U and V as for a Schwarzschild black hole,

$$U = -4Me^{-(t-r_*)/4M}, \qquad V = 4Me^{(t+r_*)/4M}, \tag{11.104}$$

with $r_* = \rho - 2M + 2M \ln |(\rho/2M) - 1|$ and $\rho \simeq 2M$ in the near-horizon limit. In Kruskal coordinates, time is defined as $t_{\mathrm{K}} = U + V$ while our radial coordinate corresponds to $x_{\mathrm{K}} = V - U$. In these coordinates, we obtain the full Penrose diagram for the black hole in asymptotically AdS space, as shown in figure 11.2. Above, we have considered the R quadrant only for which $U < 0$ and $V > 0$. To derive the Schwinger–Keldysh formalism, the L quadrant is important too, as discussed below. Finally, the past and future singularities lie in the quadrants P and F.

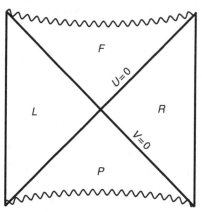

Figure 11.2   Penrose diagram of a black hole in asymptotically AdS space in Kruskal coordinates.

We note that in Kruskal coordinates, we have both infalling and outgoing modes as well as positive and negative frequency modes $\pm\omega$ with $\omega > 0$,

$$e^{-i\omega U} = e^{-i\omega(t_K - x_K)/2}, \quad e^{-i\omega V} = e^{-i\omega(t_K + x_K)/2}, \tag{11.105}$$

$$e^{i\omega U} = e^{i\omega(t_K - x_K)/2}, \quad e^{i\omega V} = e^{i\omega(t_K + x_K)/2}. \tag{11.106}$$

In the R quadrant, modes depending on V are infalling and modes depending on U are outgoing, and vice versa in the L quadrant. For each of the two quadrants, we may use a coordinate system as given by (11.54) and solve the wave equation separately in both quadrants. We obtain a set of mode functions for each quadrant,

$$v_{k,R,\pm} = \begin{cases} e^{ik\cdot x} w_{\pm k}(r) & \text{in } R \\ 0 & \text{in } L \end{cases} \qquad v_{k,L,\pm} = \begin{cases} 0 & \text{in } R \\ e^{ik\cdot x} w_{\pm k}(r) & \text{in } L. \end{cases} \tag{11.107}$$

These modes may be expanded in the Kruskal modes (11.106). However, they contain both positive- and negative-frequency parts. To separate modes with different signs, linear combinations are used that mix modes of the two quadrants,

$$\begin{aligned}
v_{1,k} &= v_{k,R,+} + e^{-\omega/2T} v_{k,L,+}, & \text{outgoing, positive-frequency}, \\
v_{2,k} &= v_{k,R,+} + e^{\omega/2T} v_{k,L,+}, & \text{outgoing, negative-frequency}, \\
v_{3,k} &= v_{k,R,-} + e^{\omega/2T} v_{k,L,-}, & \text{incoming, negative-frequency}, \\
v_{4,k} &= v_{k,R,-} + e^{-\omega/2T} v_{k,L,-}, & \text{incoming, positive-frequency}.
\end{aligned} \tag{11.108}$$

To obtain the Schwinger–Keldysh propagator, we require that positive-frequency modes should be infalling at the horizon in the R quadrant, while negative-frequency modes should be outgoing at the horizon in the R quadrant. Moreover, for a Penrose diagram with two boundaries, we have to fix the solutions to the equation of motion (11.96) on both of them, i.e. ϕ will be equal to ϕ_1 on the boundary of the R quadrant and equal to ϕ_2 on the boundary of the L quadrant. The infalling and outoing boundary conditions described above select v_2 and v_4 as the only modes that can appear in the expansion of a real scalar field, such that

$$\phi(x,r) = \sum_k \alpha_k v_{2,k} + \beta_k v_{4,k}. \tag{11.109}$$

The boundary conditions given specify the field uniquely. By requiring that (11.109) approaches ϕ_1 and ϕ_2 on the two boundaries, we can solve for the coefficients α_k and β_k and obtain

$$\phi(k,r)|_R = \big((n+1)w_k^*(r_R) - nw_k(r_R)\big)\phi_1(k)$$
$$+ \sqrt{n(n+1)}\,\big(w_k(r_R) - w_k^*(r_R)\big)\phi_2(k), \tag{11.110}$$

$$\phi(k,r)|_L = \sqrt{n(n+1)}\,\big(w_k^*(r_L) - w_k(r_L)\big)\phi_1(k)$$
$$+ \big((n+1)w_k(r_L) - nw_k^*(r_L)\big)\phi_2(k), \tag{11.111}$$

where $n \equiv (\exp(\omega/T) - 1)^{-1}$ is the occupation number for bosons. r_L and r_R refer to the radial variable in the two copies of the coordinate system describing the L and R quadrants. The $w_k(r)$ are normalised such that $w_k(r_B) = 1$ at the boundary.

With these considerations we may now apply the standard AdS/CFT procedure for calculating Green's functions. The classical boundary action is

$$S_{\text{bdy}} = \frac{1}{2}\int_R \frac{d^4k}{(2\pi)^4}\,\sqrt{-g}\,g^{rr}\phi(-k,r)\partial_r\phi(k,r)$$
$$- \frac{1}{2}\int_L \frac{d^4k}{(2\pi)^4}\,\sqrt{-g}\,g^{rr}\phi(-k,r)\partial_r\phi(k,r). \tag{11.112}$$

According to the recipe given above on page 359, the retarded Green function is related to w_k by

$$G_R(k) = -2\sqrt{-g}\,g^{rr}w_k(r)\partial_r w_k^*(r)|_{r_B}\,,$$
$$G_A(k) = -2\sqrt{-g}\,g^{rr}w_k^*(r)\partial_r w_k(r)|_{r_B}. \tag{11.113}$$

The advanced Green function $G_A(k)$ may be calculated from $G_A(k) = G_R(k)^*$. Using the normalisation of the w_k, the radial derivative of $\phi(k,r)$ evaluated close to the R or L boundary is then

$$-2\sqrt{-g}g^{rr}\partial_r\phi|_R = [(1+n)G_R - nG_A]\phi_1$$
$$+ \sqrt{n(1+n)}(G_A - G_R)\phi_2, \tag{11.114}$$
$$-2\sqrt{-g}g^{rr}\partial_r\phi|_L = [(1+n)G_A - nG_R]\phi_2$$
$$+ \sqrt{n(1+n)}(G_R - G_A)\phi_1. \tag{11.115}$$

In terms of the boundary values $\phi_1(k)$ and $\phi_2(k)$, the action becomes

$$\mathcal{S}_{\text{bdy}} = -\frac{1}{2} \int \frac{d^4 k}{(2\pi)^4} \Big[\phi_1(-k) \left((1+n)G_R(k) - nG_A(k) \right) \phi_1(k)$$
$$- \phi_2(-k) \left((1+n)G_A(k) - nG_R(k) \right) \phi_2(k)$$
$$+ \phi_1(-k) \sqrt{n(1+n)}(G_A(k) - G_R(k))\phi_2(k)$$
$$+ \phi_2(-k) \sqrt{n(1+n)}(G_A(k) - G_R(k))\phi_1(k) \Big]. \qquad (11.116)$$

This expression now allows us to take functional derivatives which preserve the causality structure.[1] Indeed, taking functional derivatives of $\mathcal{S}$ with respect to $\phi_1(k)$ and $\phi_2(k)$ yields precisely the Schwinger–Keldysh propagators (11.49)–(11.52) derived within field theory with $\sigma = \beta/2$. Consequently, the method presented above provides a well-defined approach for holographic calculation of advanced and retarded Green's functions, including taking a well-defined functional derivative. Holographic retarded Green's functions will play a central role in applications of gauge/gravity duality as discussed below in part III.

11.3 Finite density and chemical potential

In addition to finite temperature, within gauge/gravity duality it is also straightforward to describe a finite density $\rho^a = N^a/V$ and the associated chemical potential μ_a as given by (11.14). We introduce this here for the case of vanishing temperature. In part III of this book, where we discuss applications, we will also consider examples with both finite temperature and finite density present. The way a chemical potential is introduced into gauge/gravity duality is very similar to how it appears in quantum field theory. Consequently, we begin with the quantum field theory case.

11.3.1 Quantum field theory at finite density

Consider the Lagrangian of a quantum field theory with a $U(1)$ gauge symmetry and a scalar and a Dirac fermion charged under this symmetry,

$$\mathcal{L} = -(D_\mu \phi)^* D^\mu \phi + i\bar{\psi}\gamma^\mu D_\mu \psi - \frac{1}{4g^2} F^{\mu\nu} F_{\mu\nu}, \qquad (11.117)$$

with the covariant derivative $D_\mu = \partial_\mu + iA_\mu$. Let us consider a non-vanishing background field $\tilde{A}_0 = \mu$ for the time component of A_μ, such that $A_0 = \tilde{A}_0 + \delta A_0$. This generates a potential of the form

$$V = -\mu^2 \phi^* \phi - \mu\, \psi^\dagger \psi, \qquad (11.118)$$

where μ is the chemical potential and $\psi^\dagger \psi = \hat{N}$ gives the number density operator as in the grand canonical potential (11.12). Moreover, $-\mu^2$ in (11.118) is the mass square of

---

[1] Comparing (11.116) to (5.83), we have performed the integral over the delta distribution. This is a shorthand notation frequently used in the gauge/gravity duality context. We will use this in the subsequent chapters too, however, the reader should bear in mind that to evaluate functional derivatives such as in (5.84) unambiguously, working with the non-local form of the action is essential.

the scalar field. Note that the scalar field has a negative square mass, which leads to an upside-down potential and potentially to instabilities.

11.3.2 Finite density and chemical potential: gravity side

Inspired by the above field theory considerations, it is natural to propose that on the gravity side of the correspondence, finite density and chemical potential are obtained by allowing for a non-trivial profile for the time component of the gauge field in the radial direction of the gravity theory, $A = A_t(r)\mathrm{d}t$. As we discuss below by considering an example in four dimensions, the field-operator dictionary implies that near the boundary at $r \to \infty$, this profile behaves as

$$A_t(r) \sim \mu + \frac{d}{r^2},\tag{11.119}$$

with μ the chemical potential and d proportional to the density.

This structure is readily observed when considering the example of a D7-brane probe as discussed in chapter 10, with action given by (10.10). For a single D7-brane probe leading to a $U(1)$ symmetry, A_t is dual to the quark charge density, i.e. the time component of the conserved $U(1)$ current. For the field theory given by the D3/D7 system, this charge density operator is given by, using the notation of chapter 10,

$$J^t = \psi^\dagger\psi + \tilde{\psi}\tilde{\psi}^\dagger + i\left(q^\dagger D_t q - q(D_t q)^\dagger\right) + i\left(\tilde{q}(D_t\tilde{q})^\dagger - (D_t\tilde{q})\tilde{q}^\dagger\right),\tag{11.120}$$

where D_t is the time component of the covariant derivative in the $SU(N)$ gauge theory. This operator is normalised so that when acting on a particular state, it gives rise precisely to the quark density. According to the field-operator dictionary, its source is given by $A_t(\infty)$, such that $\mathcal{L}'_{\mathcal{N}=2} = \mathcal{L}_{\mathcal{N}=2} + A_t(\infty)J^t$. Comparing to the grand potential (11.12), we have $A_t(\infty) = \mu$ with μ the chemical potential. Similarly, the expectation value of J^t as given by (11.120) coincides with the quark density, $n_q = \langle J^t \rangle$.

Let us establish and solve the equations of motion obtained from the DBI action for a D7-brane with non-trivial profile for A_t, setting the temperature to zero for simplicity. For N_{f} coincident probe D7-branes, the action is

$$S_{\mathrm{D7}} = -N_{\mathrm{f}}\mu_7 \int \mathrm{d}^8\xi \sqrt{-\det(P[g]_{ab} + (2\pi\alpha')F_{ab})},\tag{11.121}$$

in the notation of chapter 10. The contribution from the Chern–Simons term vanishes. With $F_{\rho t} = \dot{A}_t(\rho)$ and using the embedding scalar $w(\rho) \equiv w_5(\rho)$, the zero temperature D7-brane Lagrangian is

$$\mathcal{L} = -\mathcal{N}_{\mathrm{D7}}\rho^3\sqrt{1 + \dot{w}^2 - (2\pi\alpha')^2\dot{A}_t^2},\tag{11.122}$$

where $\mathcal{N}_{\mathrm{D7}} = N_{\mathrm{f}}\mu_7\mathrm{Vol}(S^3)$, with $\mathrm{Vol}(S^3) = 2\pi^2$ the volume of the unit S^3. In terms of field theory quantities, we have $\mathcal{N}_{\mathrm{D7}} = 2\lambda N_{\mathrm{f}}N/(2\pi)^4$, with $\lambda = 2\pi g_{\mathrm{s}}N = g_{\mathrm{YM}}^2 N$. We have also rescaled all coordinates by $1/L$ to make them dimensionless, such that an overall factor L^8 enters $\mathcal{N}_{\mathrm{D7}}$. For simplicity, we divide both sides of (11.121) by the volume of $\mathbb{R}^{3,1}$ and work with the action density $S_{\mathrm{D7}} = \int \mathrm{d}\rho\,\mathcal{L}$.

We see that only derivatives of $w(\rho)$ and $A_t(\rho)$ appear in the Lagrangian, such that there are two conserved charges

$$\frac{\delta\mathcal{L}}{\delta\dot{w}} = -\mathcal{N}_{D7}\rho^3 \frac{\dot{w}}{\sqrt{1+\dot{w}^2-(2\pi\alpha')^2\dot{A}_t^2}} \equiv -c, \tag{11.123}$$

$$\frac{\delta\mathcal{L}}{\delta\dot{A}_t} = \mathcal{N}_{D7}\rho^3 \frac{(2\pi\alpha')^2\dot{A}_t}{\sqrt{1+\dot{w}^2-(2\pi\alpha')^2\dot{A}_t^2}} \equiv d. \tag{11.124}$$

We will see below that c is related to the quark condensate as discussed in chapter 10, while d is related to the quark density. The square of their ratio gives

$$\dot{A}_t^2 = \frac{d^2}{(2\pi\alpha')^4 c^2}\dot{w}^2. \tag{11.125}$$

We solve algebraically for $\dot{w}(\rho)$ and $\dot{A}_t(\rho)$ in terms of the integration constants c and d,

$$\dot{w} = \frac{c}{\sqrt{\mathcal{N}_{D7}^2\rho^6 + \frac{d^2}{(2\pi\alpha')^2} - c^2}}, \qquad \dot{A}_t = \frac{d/(2\pi\alpha')^2}{\sqrt{\mathcal{N}_{D7}^2\rho^6 + \frac{d^2}{(2\pi\alpha')^2} - c^2}}. \tag{11.126}$$

These may be integrated using incomplete β functions. The result depends on the sign of $d^2/(2\pi\alpha')^2 - c^2$. When $c = d = 0$, we obtain the solution with $\dot{w}(\rho) = 0$ and $\dot{A}_t(\rho) = 0$, so $w(\rho)$ and $A_t(\rho)$ are constants, as discussed in chapter 10 where we had $A_t(\rho) = 0$. Solutions with $d^2/(2\pi\alpha')^2 - c^2$ positive correspond to D7-brane embeddings which bend to reach the D3-branes. When $\frac{d^2}{(2\pi\alpha')^2} - c^2$ is negative, the D7-branes bend and turn away from the D3-branes.

The action evaluated on these solutions is

$$\mathcal{S}_{D7} = -\mathcal{N}_{D7}\int^\Lambda d\rho\rho^3 \sqrt{\frac{\mathcal{N}_{D7}^2\rho^6}{\mathcal{N}_{D7}^2\rho^6 + \frac{d^2}{(2\pi\alpha')^2} - c^2}}. \tag{11.127}$$

The lower endpoint of the integration depends on the sign of $\frac{d^2}{(2\pi\alpha')^2} - c^2$. Moreover, the integral diverges if we integrate to $\rho = \infty$. We therefore regulate the integral with a cut-off at $\rho = \Lambda$. For $d = c = 0$, i.e. for straight D7-branes with $A_t(r) = 0$, the divergent term takes the form

$$\mathcal{S}_{ct} = \mathcal{N}_{D7}\int_0^\Lambda d\rho\rho^3 = \frac{1}{4}\mathcal{N}_{D7}\Lambda^4. \tag{11.128}$$

We obtain the renormalised on-shell action $\mathcal{S}_{\text{ren}}$ by adding the relevant counterterm,

$$\mathcal{S}_{\text{ren}} = \lim_{\Lambda\to\infty}(\mathcal{S}_{D7} + \mathcal{S}_{ct}). \tag{11.129}$$

The grand canonical potential Ω is given by

$$\Omega = -\mathcal{S}_{\text{ren}}. \tag{11.130}$$

In analogy to exercise 10.2.1, the conserved charges c and d determine $\langle\mathcal{O}\rangle$ of (10.19) and $\langle J^t\rangle$ of (11.120) as follows,

$$\langle\mathcal{O}\rangle = \frac{\delta\Omega}{\delta m_q} = -(2\pi\alpha')\frac{\delta\mathcal{S}_{\text{ren}}}{\delta w(\infty)}, \qquad \langle J^t\rangle = -\frac{\delta\Omega}{\delta\mu} = \frac{\delta\mathcal{S}_{\text{ren}}}{\delta A_t(\infty)}, \tag{11.131}$$

where in each case one field is varied while holding the other fixed. We then have

$$\delta S_{\text{D7}} = \int \text{d}\rho \left(\frac{\delta \mathcal{L}}{\delta A_t(\rho)} \partial_\rho \delta A_t(\rho) + \frac{\delta \mathcal{L}}{\delta \dot{w}(\rho)} \partial_\rho \delta w(\rho) \right) = d\delta A_t(\infty) - c\delta w(\infty), \quad (11.132)$$

where we impose that $\delta A_t(\rho)$ and $\delta w(\rho)$ are always zero at the lower endpoint of the ρ integration. If we vary $A_t(\rho)$ while holding $w(\rho)$ fixed, $\delta w(\rho) = 0$, we find

$$\langle J^t \rangle = -(2\pi\alpha')^2 \mu_7 d, \qquad \langle \mathcal{O} \rangle = -2\pi\alpha'^3 \mu_7 c. \quad (11.133)$$

This implies that the subleading term in (11.119) is indeed the charge density since $J^\mu = (\rho, \vec{J})$ with $J^t = \rho$.

11.4 Further reading

Reviews on finite temperature quantum field theory include [6, 7]. The original references for the Schwinger–Keldysh formalism of quantum field theory are [8, 9].

The black hole was proposed as the gravity dual of a finite temperature field theory in [10]. The temperature and entropy of black D3-branes were calculated in [1], whereas the corresponding result for free $\mathcal{N} = 4$ Super Yang–Mills theory was obtained in [2]. The Hawking–Page transition was found in [3]. Within the AdS/CFT correspondence, Witten interpreted the Hawking–Page transition as the gravity dual of a deconfinement phase transition in [10].

The study of holographic two-point functions in Minkowski space was initiated in [5, 11]. The gravity dual of the Schwinger–Keldysh formalism was established in [4].

For the D7-brane probe, the chemical potential and finite density are discussed in [12, 13, 14, 15].

References

[1] Gubser, S. S., Klebanov, Igor R., and Peet, A. W. 1996. Entropy and temperature of black 3-branes. *Phys. Rev.*, **D54**, 3915–3919.

[2] Burgess, C. P., Constable, N. R., and Myers, Robert C. 1999. The free energy of $N = 4$ super Yang-Mills and the AdS/CFT correspondence. *J. High Energy Phys.*, **9908**, 017.

[3] Hawking, S. W., and Page, Don N. 1983. Thermodynamics of black holes in anti-de Sitter space. *Commun. Math. Phys.*, **87**, 577.

[4] Herzog, C. P., and Son, D. T. 2003. Schwinger-Keldysh propagators from AdS/CFT correspondence. *J. High Energy Phys.*, **0303**, 046.

[5] Son, Dam T., and Starinets, Andrei O. 2002. Minkowski space correlators in AdS/CFT correspondence: recipe and applications. *J. High Energy Phys.*, **0209**, 042.

[6] Kapusta, J. I., and Gale, Charles. 2006. *Finite-Temperature Field Theory: Principles and Applications*, 2nd edition. Cambridge University Press.

[7] Das, Ashok K. 1997. *Finite Temperature Field Theory*. World Scientific, Singapore.

[8] Schwinger, Julian S. 1961. Brownian motion of a quantum oscillator. *J. Math. Phys.*, **2**, 407–432.

[9] Keldysh, L.V. 1964. Diagram technique for nonequilibrium processes. *Zh. Eksp. Teor. Fiz.*, **47**, 1515–1527.

[10] Witten, Edward. 1998. Anti-de Sitter space, thermal phase transition, and confinement in gauge theories. *Adv. Theor. Math. Phys.*, **2**, 505–532.

[11] Policastro, Giuseppe, Son, Dam T., and Starinets, Andrei O. 2002. From AdS/CFT correspondence to hydrodynamics. *J. High Energy Phys.*, **0209**, 043.

[12] Karch, Andreas, and O'Bannon, Andy. 2007. Holographic thermodynamics at finite baryon density: some exact results. *J. High Energy Phys.*, **0711**, 074.

[13] Kobayashi, Shinpei, Mateos, David, Matsuura, Shunji, Myers, Robert C., and Thomson, Rowan M. 2007. Holographic phase transitions at finite baryon density. *J. High Energy Phys.*, **0702**, 016.

[14] Nakamura, Shin, Seo, Yunseok, Sin, Sang-Jin, and Yogendran, K. P. 2008. Baryon-charge chemical potential in AdS/CFT. *Prog. Theor. Phys.*, **120**, 51–76.

[15] Ghoroku, Kazuo, Ishihara, Masafumi, and Nakamura, Akihiro. 2007. D3/D7 holographic gauge theory and chemical potential. *Phys. Rev.*, **D76**, 124006.

PART III

APPLICATIONS

12

Linear response and hydrodynamics

A very successful and important application of gauge/gravity duality has emerged in the context of hydrodynamics. In generalisation of the dynamics of fluids, the term *hydrodynamics* generically refers to an effective field theory describing long-range, low-energy fluctuations about equilibrium.

Recently, experimental evidence has accumulated that the *quark–gluon plasma* observed in heavy-ion collision experiments is best described by a strongly coupled relativistic fluid, rather than by a gas of weakly interacting particles. Strongly coupled fluids are intrinsically difficult to describe by standard methods. This explains the success of applying gauge/gravity duality to this area of physics. In particular, gauge/gravity duality has made predictions of universal values of certain transport coefficients in strongly coupled fluids. The most famous example of this is the ratio of shear viscosity over entropy density, which takes a very small value. Beyond these results, gauge/gravity duality has provided a fresh look at relativistic hydrodynamics, for which many new non-trivial properties have been uncovered using the *fluid/gravity correspondence*.

We will describe these results in some detail. The starting point is to introduce linear response theory and Green's functions which respect the causal structure. Then we move on to an introduction to relativistic hydrodynamics. We consider the energy-momentum tensor and a conserved current and their dissipative contributions in an expansion in derivatives of fluctuations. We define the associated first-order transport coefficients and subsequently relate them to the retarded Green's function by virtue of appropriate Green–Kubo relations. This provides a link between macroscopic hydrodynamic properties and microscopic physics as described by the Green's functions. Using gauge/gravity duality methods to evaluate the relevant Green's functions, we compute the charge diffusion constant and the shear viscosity. This leads to the well-known universal result for the ratio of shear viscosity over entropy density: gauge/gravity duality gives the famous result $\eta/s = 1/(4\pi)$. This is in agreement with experimental results obtained at the RHIC accelerator which give a range η/s of $1/(4\pi)$ to $2.5/(4\pi)$ in units where $\hbar = k_B = 1$. Two very important properties of the gauge/gravity duality result are that it has a very small value, and that it is *universal*. Perturbative calculations within their domain of validity, i.e. for small coupling, give a much larger result.

Finally, we show how the calculation of transport coefficients may be approached systematically using the general formalism of fluid/gravity correspondence. Using this framework, we show how new transport coefficients arise which have not been considered so far in the context of relativistic hydrodynamics as applied to the quark–gluon plasma. In particular, we consider a coefficient associated with vorticity which is related to the axial anomaly within quantum field theory.

12.1 Linear response

The idea of linear response theory is to consider small space- and time-dependent perturbations about the equilibrium state of a physical system. In this section we introduce linear response in quantum field theory. In addition to transport processes, linear response is also essential for obtaining *spectral functions*.

The basic object in linear response theory is the retarded Green's function introduced in (11.46) in chapter 11. The retarded Green's function relates linear fluctuations of sources to the corresponding expectation values. This allows transport coefficients to be calculated from two-point correlation functions.

12.1.1 Linear response: field theory

We begin by discussing the linear response formalism within field theory. Consider the response of a system to the presence of external fields φ_I coupled to a set of operators $\mathcal{O}^I(x)$. These fields may have arbitrary Lorentz index structure. They modify the unperturbed Hamiltonian of the system considered by a term of the form

$$\delta\mathcal{H} = -\int \mathrm{d}^d x\, \varphi_I(t,\vec{x})\, \mathcal{O}^I(t,\vec{x}) . \tag{12.1}$$

According to time-dependent perturbation theory, these external fields generate a change in the expectation values of the operators, which to linear order is given by

$$\delta\langle\mathcal{O}^I(x)\rangle = \int \mathrm{d}^d y\, G_R^{IJ}(x,y)\, \varphi_J(y) + \mathcal{O}(\varphi^2), \tag{12.2}$$

$$G_R^{IJ}(x,x') = -i\Theta(t-t')\left\langle\left\{\hat{\mathcal{O}}^I(x),\hat{\mathcal{O}}^J(x')\right\}_\pm\right\rangle, \tag{12.3}$$

where $G_R^{IJ}(x,y)$ is the retarded Green's function (11.46) introduced in chapter 11. The bracket $\{\cdot,\cdot\}_\pm$ denotes a commutator in the case of bosonic fields and an anticommutator in the case of fermionic fields. The retarded Green's function is non-vanishing only in the forward light-cone and therefore provides a causality structure: $\delta\langle\mathcal{O}^I(t,x)\rangle$ is influenced only by sources $\phi^I(t',\vec{x}')$ with $t' < t$. In Fourier space we have

$$\delta\langle\mathcal{O}^I(k)\rangle = G_R^{IJ}(k)\,\varphi_J(k) + \mathcal{O}(\varphi^2), \quad G_R^{IJ}(k) = \int \mathrm{d}^d x\, e^{-ik\cdot x}\, G_R^{IJ}(x,0). \tag{12.4}$$

The retarded Green's function encodes important physical information. For example, the *spectral function* $\mathcal{R}^{IJ}$ is defined by

$$G_R^{IJ}(\omega,\vec{q}) = \int \frac{\mathrm{d}\omega'}{2\pi}\, \frac{\mathcal{R}^{IJ}(\omega',\vec{q})}{\omega' - \omega + i\epsilon}, \qquad \epsilon \to 0^+. \tag{12.5}$$

The spectral function $\mathcal{R}^{IJ}$ counts the states propagating with energy ω'. Equation (12.5) can be inverted to give

$$\mathcal{R}^{IJ}(\omega,\vec{q}) \equiv i\left(G_R^{IJ}(\omega,\vec{q}) - \left(G_R^{IJ}(\omega,\vec{q})\right)^\dagger\right). \tag{12.6}$$

Box 12.1   **Linear response for AC conductivity**

As an illustration, let us write Ohm's law in linear response formalism. Ohm's law states that for a spatially constant electric field $\vec{E}(\omega)$, which may oscillate in time with frequency ω, the spatial part of the charge current response, $J^i(\omega)$, is determined by

$$\langle J^i(\omega) \rangle = \sigma^{ij}(\omega) E_j(\omega). \tag{12.8}$$

$\sigma^{ij}(\omega)$ is the conductivity tensor for alternating currents. In the language of the previous paragraphs, we consider an external vector potential $A_\mu(x)$ (playing the role of $\phi^J(x)$) and a conserved current $J^\mu(x)$ which corresponds to the operator $\mathcal{O}_J$. In the gauge $A_t = 0$, the electric field E_k is given by $E_k = -\partial_t A_k$. Fourier decomposing $A_k \sim e^{-i\omega t}$, we obtain for the electric field $E_k = i\omega A_k$. Comparing Ohm's law (12.8) to (12.4) which expresses the linear response in terms of the retarded Green function, we obtain a simple expression for the conductivity tensor of alternating currents,

$$\sigma^{ij}(\omega) = \frac{G_R^{ij}(\omega, \vec{0})}{i\omega}. \tag{12.9}$$

Here, $G_R^{ij}(\omega, \vec{0})$ are the components of the retarded correlator of currents in Fourier space. Note that the spatial momentum $\vec{q}$ is set to zero.

The spectral function is thus given by the anti-Hermitian part of the retarded Green's function. Note that in the case of only one source (or more sources which do not couple to each other), the spectral function $\mathcal{R}^{IJ}$ may be written in the simple form

$$\mathcal{R}^{IJ}(\omega, \vec{q}) = -2 \operatorname{Im} G_R^{IJ}(\omega, \vec{q}). \tag{12.7}$$

Example: R-current in $\mathcal{N} = 4$ Super Yang–Mills theory

As an example, let us consider the retarded Green's functions for the energy-momentum tensor and for the conserved R-symmetry current in $\mathcal{N} = 4$ Super Yang–Mills theory [1]. According to (12.3), these are given by

$$C_{\mu\nu}(x - y) = -i\Theta(x^0 - y^0)\langle [J_\mu(x), J_\nu(y)] \rangle, \tag{12.10}$$

$$G_{\mu\nu,\rho\sigma}(x - y) = -i\Theta(x^0 - y^0)\langle [T_{\mu\nu}(x), T_{\rho\sigma}(y)] \rangle. \tag{12.11}$$

Fourier transforming, we obtain

$$C_{\mu\nu}(x - y) = \int \frac{\mathrm{d}^4 k}{(2\pi)^4} e^{ik(x-y)} C_{\mu\nu}(k), \tag{12.12}$$

and similarly for $G_{\mu\nu,\rho\sigma}$. $C_{\mu\nu}(k)$ is symmetric in its indices, $C_{\mu\nu}(k) = C_{\nu\mu}(k)$. $G_{\mu\nu,\rho\sigma}(k)$ is symmetric under the exchange of the two pairs of indices. It satisfies the symmetry and tracelessness properties of the energy-momentum tensor in each of the two pairs. Moreover, conservation of J_μ and $T_{\mu\nu}$ imposes the Ward identities

$$k^\mu C_{\mu\nu}(k) = 0, \qquad k^\mu G_{\mu\nu,\rho\sigma}(k) = 0. \tag{12.13}$$

These properties constrain the possible form of the retarded Green's function. Let us take the conserved current as an example. The Ward identity (12.13) implies that $C_{\mu\nu}(k)$ is proportional to the projector onto the space of conserved vectors,

$$\mathcal{P}_{\mu\nu} \equiv \eta_{\mu\nu} - \frac{k_\mu k_\nu}{k^2}, \qquad k^\mu \mathcal{P}_{\mu\nu} = 0. \tag{12.14}$$

$C_{\mu\nu}$ thus takes the form

$$C_{\mu\nu}(k) = \mathcal{P}_{\mu\nu} \Pi(k^2) \tag{12.15}$$

and is determined up to the scalar function $\Pi(k^2)$. For the special case of rotationally invariant systems, it is convenient to separate the projector into longitudinal and transversal parts,

$$\mathcal{P}_{\mu\nu} = \mathcal{P}_{\mu\nu}^{\mathrm{T}} + \mathcal{P}_{\mu\nu}^{\mathrm{L}}, \tag{12.16}$$

$$\mathcal{P}_{tt}^{\mathrm{T}} = 0, \quad \mathcal{P}_{ti}^{\mathrm{T}} = 0, \quad \mathcal{P}_{ij}^{\mathrm{T}} = \delta_{ij} - \frac{k_i k_j}{k^2}, \tag{12.17}$$

$$\mathcal{P}_{\mu\nu}^{\mathrm{L}} = \mathcal{P}_{\mu\nu} - \mathcal{P}_{\mu\nu}^{\mathrm{T}}. \tag{12.18}$$

We have $k^\mu \mathcal{P}_{\mu\nu}^{\mathrm{T}} = k^\mu \mathcal{P}_{\mu\nu}^{\mathrm{L}} = 0$. The current retarded Green's function then becomes, with $k^\mu = (\omega, \vec{q})$,

$$C_{\mu\nu}(k) = \mathcal{P}_{\mu\nu}^{\mathrm{T}} \Pi^{\mathrm{T}}(\omega, |\vec{q}|^2) + \mathcal{P}_{\mu\nu}^{\mathrm{L}} \Pi^{\mathrm{L}}(\omega, |\vec{q}|^2). \tag{12.19}$$

For a rotationally invariant system, we may take $k^\mu = (\omega, 0, 0, q)$ with momentum in the 3-direction without loss of generality. Using (12.17), (12.18) we then have

$$C_{11}(k) = C_{22}(k) = \Pi^{\mathrm{T}}(\omega, q) \tag{12.20}$$

for the transversal part of the Green function, and

$$C_{tt}(k) = \frac{q^2}{\omega^2 - q^2} \Pi^{\mathrm{L}}(\omega, q), \qquad C_{t3}(k) = -\frac{\omega q}{\omega^2 - q^2} \Pi^{\mathrm{L}}(\omega, q),$$

$$C_{33}(k) = \frac{\omega^2}{\omega^2 - q^2} \Pi^{\mathrm{L}}(\omega, q) \tag{12.21}$$

for the longitudinal part. For $q \to 0$, we have $\Pi^{\mathrm{T}} = \Pi^{\mathrm{L}} = \Pi$.

We will determine the precise form of Π^{T}, Π^{L} using holography in section 12.1.2. Here, we note that in the long-time, long-wavelength limit, where $\omega/T \ll 1$, $q/T \ll 1$, the Π^{T} and Π^{L} exhibit universal behaviour: Π^{T} is non-singular as function of the frequency, while Π^{L} has a simple pole at

$$\omega = -iDq^2, \tag{12.22}$$

where D is the charge diffusion constant. For the example of $\mathcal{N} = 4$ Super Yang–Mills theory, this is the R-charge diffusion constant. The pole given by (12.22) is obtained by inverting the diffusion equation

$$(\partial_t - D\vec{\nabla}^2)\rho = 0 \tag{12.23}$$

for the charge density $\rho = J_t$. Equation (12.22) is an example of a *hydrodynamic pole*, i.e. ω vanishes for $q \to 0$.

In addition to hydrodynamic poles, the retarded Green's functions may have other singularities located in the lower half of the complex frequency plane. For the example of a simple pole, the Green's function takes the schematic form

$$G_\text{R} \sim \frac{1}{\omega - \Omega + i\Gamma}.$$

(12.24)

The imaginary part Γ is associated to dissipation. The spectral function associated to (12.24) reads

$$R(\omega) \sim \frac{\Gamma}{(\omega - \Omega)^2 + \Gamma^2}.$$

(12.25)

For $\omega \sim \Omega$, the spectral function has a peak of width Γ. At weak coupling where there is a one-to-one correspondence between states in the free and in the interacting systems, the peak may be viewed as a quasiparticle if $\Gamma \ll \Omega$. At strong coupling, it is expected that excitations in the spectral function may also be identified with quasiparticles, however, in this case a one-to-one map between excitations and free theory particles with coinciding quantum numbers may no longer exist.

Let us now consider the energy-momentum tensor two-point function $G_{\mu\nu,\rho\sigma}(k)$ as defined in (12.11). Its index symmetries and conservation allow for contributions of five independent forms in general. A convenient way to write (12.11) is

$$G_{\mu\nu,\rho\sigma}(k) = \mathcal{E}_{\mu\nu,\rho\sigma}\, G_\text{S}(k^2) + \mathcal{P}_{\mu\nu}\mathcal{P}_{\rho\sigma}\, G_\text{B}(k^2),$$

(12.26)

with $\mathcal{P}_{\mu\nu}$ the projector of (12.14) and $\mathcal{E}_{\mu\nu,\rho\sigma}$ the projector onto traceless symmetric tensors given by

$$\mathcal{E}_{\mu\nu,\rho\sigma} = \frac{1}{2}\left(\mathcal{P}_{\mu\rho}\mathcal{P}_{\nu\sigma} + \mathcal{P}_{\mu\sigma}\mathcal{P}_{\nu\rho}\right) - \frac{1}{3}\mathcal{P}_{\mu\nu}\mathcal{P}_{\rho\sigma}.$$

(12.27)

For a scale invariant theory, the second term in (12.26) must vanish, $G_\text{B}(k) = 0$. In what follows, we just consider scale invariant theories. Moreover, for rotationally invariant systems, it is useful to split the projector (12.27) into mutually orthogonal parts constructed from $\mathcal{P}_{\mu\nu}^\text{T}$ and $\mathcal{P}_{\mu\nu}^\text{L}$ as in (12.16). Two projectors with this property are given by

$$\mathcal{S}_{\mu\nu,\rho\sigma} = \frac{1}{2}\left(\mathcal{P}_{\mu\rho}^\text{T}\mathcal{P}_{\nu\sigma}^\text{L} + \mathcal{P}_{\mu\sigma}^\text{T}\mathcal{P}_{\nu\rho}^\text{L} + \mathcal{P}_{\mu\rho}^\text{L}\mathcal{P}_{\nu\sigma}^\text{T} + \mathcal{P}_{\mu\sigma}^\text{L}\mathcal{P}_{\nu\rho}^\text{T}\right),$$

(12.28)

$$\mathcal{Q}_{\mu\nu,\rho\sigma} = \frac{1}{d-1}\left((d-2)\mathcal{P}_{\mu\nu}^\text{L}\mathcal{P}_{\rho\sigma}^\text{L} + \frac{1}{d-2}\mathcal{P}_{\mu\nu}^\text{T}\mathcal{P}_{\rho\sigma}^\text{T} - (\mathcal{P}_{\mu\nu}^\text{T}\mathcal{P}_{\rho\sigma}^\text{L} + \mathcal{P}_{\mu\nu}^\text{L}\mathcal{P}_{\rho\sigma}^\text{T})\right).$$

(12.29)

The projector $\mathcal{L}$ given by

$$\mathcal{L}_{\mu\nu,\rho\sigma} = \mathcal{E}_{\mu\nu,\rho\sigma} - \mathcal{S}_{\mu\nu,\rho\sigma} - \mathcal{Q}_{\mu\nu,\rho\sigma}$$

(12.30)

is orthogonal to both $\mathcal{S}_{\mu\nu,\rho\sigma}$ and $\mathcal{Q}_{\mu\nu,\rho\sigma}$.

Exercise 12.1.1 Confirm that the projectors $\mathcal{S}_{\mu\nu,\rho\sigma}$, $\mathcal{Q}_{\mu\nu,\rho\sigma}$ and $\mathcal{L}_{\mu\nu,\rho\sigma}$ are mutually orthogonal.

For scale invariant theories, the correlator (12.26) may then be written as

$$G_{\mu\nu,\rho\sigma}(k) = \mathcal{S}_{\mu\nu,\rho\sigma}G_1(\omega, |\vec{k}|^2) + \mathcal{Q}_{\mu\nu,\rho\sigma}G_2(\omega, |\vec{k}|^2) + \mathcal{L}_{\mu\nu,\rho\sigma}G_3(\omega, |\vec{k}|^2). \qquad (12.31)$$

Rotational invariance implies that the $\vec{k} \to 0$ limit of the three functions G_I, $I = 1, 2, 3$, in (12.31) must coincide. As examples of components of the correlator for a four-dimensional theory at vanishing temperature, we find for the correlations of transverse momentum density the expressions

$$G_{t1,t1}(k) = \frac{1}{2}\frac{q^2}{\omega^2 - q^2}G_1(\omega, q), \qquad G_{t1,13}(k) = -\frac{1}{2}\frac{\omega q}{\omega^2 - q^2}G_1(\omega, q),$$

$$G_{13,13}(k) = \frac{1}{2}\frac{\omega^2}{\omega^2 - q^2}G_1(\omega, q), \qquad (12.32)$$

choosing $k_\mu = (-\omega, 0, 0, q)$ as before. Moreover, for the energy density correlator we have

$$G_{tt,tt}(k) = \frac{2}{3}\frac{q^4}{(\omega^2 - q^2)^2}G_2(\omega, q), \qquad (12.33)$$

and for the longitudinal momentum density and diagonal stress correlators

$$G_{tt,t3}(k) = -\frac{2}{3}\frac{\omega q^3}{(\omega^2 - q^2)^2}G_2(\omega, q), \qquad G_{tt,11} = \frac{1}{3}\frac{q^2}{q^2 - \omega^2}G_2(\omega, q). \qquad (12.34)$$

For the transverse stress correlator we have

$$G_{12,12}(k) = \frac{1}{2}G_3(\omega, q). \qquad (12.35)$$

Let us consider the lowest-order or hydrodynamic poles of these correlators. Recall that for the example of the current (12.22), this pole is related to charge diffusion. Here, the poles are related to energy diffusion. $G_3(\omega, q)$ is non-singular as function of ω since it does not couple to energy or momentum density fluctuations. On the other hand, $G_1(\omega, q)$ and $G_2(\omega, q)$ do exhibit poles. These are related to *shear modes* and *sound modes* of the energy-momentum tensor. These will be discussed in detail in section 12.2.3 below, where dissipative relativistic hydrodynamics is introduced, and in particular in box 12.2. Here, let us note that $G_1(\omega, q)$ has a simple pole at $\omega = -i\gamma q^2$, with γ the damping constant of the shear mode, and $G_2(\omega, q)$ has simple poles at $\omega = \pm v_s q - i\Gamma_s q^2$, with v_s the speed of sound. Γ_s is the damping constant of the sound mode. Both γ and Γ_s are related to the *shear viscosity*, a very important transport coefficient, as we will see in section 12.2.3.

12.1.2 Linear response: gauge/gravity duality

Let us now turn to linear response within gauge/gravity duality. The calculation of retarded Green's functions from the gravity side as introduced in section 11.2.3 takes a particularly elegant form by combining it with the linear response approach.

Let us consider the example of a scalar operator. While introducing the recipe for holographically calculating the retarded Green's function in section 11.2.3, we noted in (11.97) that near the boundary, the solution to the equation of motion takes the form

$$\phi(u, k) \sim \phi_{(0)}(k)u^{(d-\Delta)/2}(1 + \mathcal{O}(u)) + \phi_{(+)}(k)u^{\Delta/2}(1 + \mathcal{O}(u)). \qquad (12.36)$$

Moreover, we recall from the discussion of holographic renormalisation in section 5.5 that for (12.36), the one-point function in presence of the sources is given by

$$\langle \mathcal{O}(k) \rangle_s = -\lim_{\epsilon \to 0} \left(\frac{L^d}{\epsilon^{\Delta/2}} \frac{1}{\sqrt{\gamma}} \frac{\delta S_{\text{sub}}}{\delta \phi(k, \epsilon)} \right)$$

$$= L^{d-1}(2\Delta - d)\phi_{(+)}(k) + \mathcal{C}(\phi_{(0)}) \qquad (12.37)$$

with a cut-off at $u = \epsilon$ and S_{sub} as defined in section 5.5. γ is the determinant of the induced metric at $u = \epsilon$, and $\mathcal{C}(\phi_{(0)})$ is a local term as in (5.114). Consequently, in the context of linear response, the subleading term in the near-boundary expansion of a fluctuation $\delta\phi(u, k)$ encodes the response of the expectation value $\langle \mathcal{O} \rangle$ to this fluctuation. This is consistent with the fact of (5.49) that the subleading term of the asymptotic expansion encodes information about the vacuum expectation value of the dual operator. From (12.4), which for gauge/gravity duality states that $\delta\langle \mathcal{O}(k) \rangle = G(k)\delta\phi(k)$ with $\delta\phi(k)$ a boundary fluctuation, we find using (12.37) that

$$G_R(\omega, \vec{k}) = L^{d-1}(2\Delta - d)\frac{\phi_{(+)}(\omega, \vec{k})}{\phi_{(0)}(\omega, \vec{k})} \qquad (12.38)$$

for the retarded Green's function, subject to imposing infalling boundary conditions at the horizon as described in section 11.2.3.

A further useful formulation of the result (12.38) is obtained in an approach similar to the Hamiltonian approach of classical mechanics, which considers generalised momenta [2, 3, 4]. This approach proceeds in analogy to the classical mechanics for a particle in one dimension with action

$$S_{\text{one-particle}} = \int_{t_i}^{t_f} dt \, \mathcal{L}(x(t), \dot{x}(t)), \qquad (12.39)$$

where the variation of this action with respect to the initial value of the coordinate gives the generalised momentum $\pi(t)$,

$$\pi(t) \equiv \frac{\partial \mathcal{L}}{\partial \dot{x}}, \qquad \pi(t_i) = \frac{\delta S_{\text{one-particle}}}{\delta x(t_i)}. \qquad (12.40)$$

Generalising this to gauge/gravity duality by identifying the radial coordinate with the time of the mechanics example, we may write

$$\langle \mathcal{O}(k) \rangle = -\frac{\delta S[\phi_{(0)}]}{\delta \phi_{(0)}(k)} = -\lim_{\epsilon \to 0} \left(\epsilon^{(d-\Delta)/2} \pi(\epsilon, k) \right), \qquad (12.41)$$

where the generalised momentum is given by

$$\pi(\epsilon, k) \equiv \frac{\partial \mathcal{L}(\phi(\epsilon, k))}{\partial (\partial_\epsilon \phi(\epsilon, k))} + \frac{\partial \mathcal{L}_{\text{bdy}}(\phi_{(0)})}{\partial \phi_{(0)}}. \qquad (12.42)$$

Here, the first term on the right-hand side is present in analogy to (12.40) and the second involving the boundary fields accounts for contact terms. Moreover, for the Green's function (12.38) we have

$$G_R(k) = \frac{\delta\langle \mathcal{O}(k) \rangle}{\delta\phi_{(0)}(k)} = \lim_{\epsilon \to 0} \frac{\delta\pi(k)}{\delta\phi_{(0)}(k)} \bigg|_{\delta\phi=0}, \qquad (12.43)$$

with $\pi^I(k)$ given by (12.42). This result may straightforwardly be generalised to general fields ϕ^I with sources $\phi_{(0)}(k)^I$.

Example: R-current

As an example, let us calculate the R-current retarded Green's function for $\mathcal{N} = 4$ Super Yang–Mills theory using gauge/gravity duality [1, 5]. The starting point is the five-dimensional Yang–Mills action in the black brane background,

$$\mathcal{S}_{\mathrm{YM}} = -\frac{1}{4g_{\mathrm{YM}}^2} \int \mathrm{d}^5x \sqrt{-g}\, F_{mn}^a F^{amn}, \tag{12.44}$$

with $SU(4)$ gauge fields with field strength

$$F_{mn}^a = \partial_m A_n^a - \partial_n A_m^a + f_{bc}{}^a A_m^b A_n^c. \tag{12.45}$$

The $f_{bc}{}^a$ are the structure constants of the algebra $su(4)$. The coupling constant is given by $g_{\mathrm{YM}}^2 = 16\pi^2 L/N^2$.

For calculation of the two-point function, it is sufficient to consider linearised fluctuations. To study these, it is convenient to use the metric written in the radial coordinate u as in (11.94). Since we do not consider a non-vanishing background gauge field here, the action for these linearised fluctuations is simply

$$\mathcal{S}_{\mathrm{YM}} = -\frac{1}{4g_{\mathrm{YM}}^2} \int \mathrm{d}^5x \sqrt{-g}\, f_{mn}^a f^{amn}, \tag{12.46}$$

with

$$f_{mn}^a = \partial_m a_n^a - \partial_n a_m^a. \tag{12.47}$$

Since interactions between fields with different gauge indices are absent, the Green's functions will be diagonal in the gauge indices, $C_{\mu\nu}^{ab} = \delta^{ab} C_{\mu\nu}$. We may therefore suppress them in the following discussion as far as the fluctuations are concerned. Nevertheless, it is still necessary to identify gauge invariant fields in order to remove unphysical redundancies. The background gauge fields A_m^a are in the adjoint representation of $SU(4)$ and transform as

$$\delta_\Lambda A_m^a = \nabla_m \Lambda^a + f^{abc} A_m^b \Lambda^c. \tag{12.48}$$

It is convenient to choose a gauge in which the radial component of the background gauge field is zero, $A_u^a = 0$. This fixes the gauge freedom only partially. In particular, for the fluctuations, this leaves diagonal $U(1)$ transformations $\delta_\lambda a_n = \nabla_n \lambda$ at the linearised level. Fourier transforming,

$$A_m(u,x) = \int \frac{\mathrm{d}^4k}{(2\pi)^4}\, e^{ik^\mu x_\mu} A_m(u,k), \tag{12.49}$$

and choosing $k^\mu = (\omega, 0, 0, q)$, the remaining gauge transformations are

$$\delta_\lambda a_t = -i\omega\lambda, \qquad \delta_\lambda a_3 = iq\lambda, \qquad \delta a_a = 0, \quad a = 1, 2. \tag{12.50}$$

The gauge invariant fluctuation fields are then given by

$$E_L = qa_t + \omega a_3 \propto f_{3t},$$ (12.51)

$$E_T = \omega a_a \propto f_{at},$$ (12.52)

which are the longitudinal and transversal components of a $U(1)$ electric field. Using the background metric of (11.94), the equations of motion obtained from the Maxwell equation read as follows,

$$E_T'' + \frac{f'}{f}E_T' + \frac{\mathfrak{w}^2 - \mathfrak{q}^2 f}{uf^2}E_T = 0,$$ (12.53)

$$E_L'' + \frac{\mathfrak{w}^2 f'}{f(\mathfrak{w}^2 - \mathfrak{q}^2 f)}E_L' + \frac{\mathfrak{w}^2 - \mathfrak{q}^2 f}{uf^2}E_L = 0,$$ (12.54)

where the prime denotes the derivative with respect to u and we have introduced the dimensionless quantities

$$\mathfrak{w} = \frac{\omega}{2\pi T}, \qquad \mathfrak{q} = \frac{q}{2\pi T}.$$ (12.55)

Looking at the asymptotic behaviour of the solutions near the horizon, we find that they take the form of the infalling and outgoing solutions $\pm i\mathfrak{w}/2$ as in (11.99). According to our previous discussion, we choose the infalling behaviour $-i\mathfrak{w}/2$. Near the boundary at $u = 0$, the solutions take the asymptotic form

$$E_L(u) = E_L^{(0)}(\mathfrak{w}, \mathfrak{q}) + E_L^{(1)}(\mathfrak{w}, \mathfrak{q})u + \cdots,$$ (12.56)

$$E_T(u) = E_T^{(0)}(\mathfrak{w}, \mathfrak{q}) + E_T^{(1)}(\mathfrak{w}, \mathfrak{q})u + \cdots,$$ (12.57)

with $\cdots$ denoting subleading terms. Integrating the Yang–Mills action (12.46) by parts and using the equations of motion, we obtain the boundary action

$$S_{YM} = \lim_{u \to 0} \frac{N^2 T^2}{16} \int \frac{d\omega \, dq}{(2\pi)^2} \left[a_t'(u, k)a_t(u, -k) - f(u)a_3'(u, k)a_3(u, -k) \right],$$ (12.58)

where the prime denotes derivatives with respect to u. Equation (12.58) may be written in terms of the gauge invariant fields E_L, E_T as

$$S_{YM} = \lim_{u \to 0} \frac{N^2 T^2}{16} \int \frac{d\omega dq}{(2\pi)^2} \left[\frac{f(u)}{\mathfrak{q}^2 f(u) - \mathfrak{w}^2} E_L'(u, k)E_L(u, -k) \right.$$

$$\left. - \frac{f(u)}{\mathfrak{w}^2} E_T'(u, k)E_T(u, -k) \right].$$ (12.59)

In addition, S_{YM} contains contact terms which do not contain derivatives of the fields and will lead to local delta function contributions to the correlation functions.

In order to obtain the retarded Green's functions, we now apply the procedure introduced in section 11.2.3. To determine the retarded Green's functions as given by (12.20), (12.21), we have to express the action in terms of the boundary values of the fields as defined in (12.56), (12.57), in order to obtain an expression of the form (11.101). Performing this calculation, we obtain

$$\Pi^T(\omega, q) = -\frac{N^2 T^2}{8} \frac{E_T^{(1)}}{E_T^{(0)}}, \qquad \Pi^L(\omega, q) = -\frac{N^2 T^2}{8} \frac{E_L^{(1)}}{E_L^{(0)}}$$ (12.60)

for the transversal and longitudinal contributions of (12.20), (12.21).

Let us now compute these self-energies explicitly, which requires solving the equations (12.53), (12.54). In general, this is possible only by using numerics. However, for $q = 0$, the two equations for E_T and E_L coincide to give the equation

$$E'' + \frac{f'}{f}E' + \frac{\mathfrak{w}^2}{(1-x)f^2}E = 0 \tag{12.61}$$

for $E = E_T = E_L$, where we have introduced the new variable $x = 1 - u$. We choose infalling boundary conditions at the horizon by making the ansatz

$$E(x) = x^{-i\frac{\mathfrak{w}}{2}}(2-x)^{-\mathfrak{w}/2}F(x), \tag{12.62}$$

where $F(x)$ is regular at the horizon. Inserting this ansatz into (12.61) gives two linearly independent solutions for $F(x)$. The one which preserves the infalling boundary condition, and thus leads to the retarded Green's function, gives

$$E(x) = x^{-i\frac{\mathfrak{w}}{2}}(2-x)^{-\mathfrak{w}/2}(1-x)^{\frac{(1+i)\mathfrak{w}}{2}}$$

$$\times \, _2F_1\left(1 - \frac{(1+i)\mathfrak{w}}{2}, -\frac{(1+i)\mathfrak{w}}{2}; 1 - i\mathfrak{w}; \frac{x}{2(x-1)}\right), \tag{12.63}$$

with $_2F_1$ the Gauss hypergeometric function. Inserting this into (12.60) then gives

$$\Pi(\omega) = \frac{N^2 T^2}{8}\left(i\mathfrak{w} + \mathfrak{w}^2\left[\psi\left(\frac{(1-i)\mathfrak{w}}{2}\right) + \psi\left(-\frac{(1+i)\mathfrak{w}}{2}\right)\right]\right), \tag{12.64}$$

where $\psi(z) = \Gamma'(z)/\Gamma(z)$ is the logarithmic derivative of the Gamma function. This result encodes important information about the physical properties of the theory considered, as we discuss in the next section.

Exercise 12.1.2   Using the definition of the spectral function in (12.6), as well as the following properties of the ψ function,

$$\psi(z^*) = \psi(z)^*, \qquad \psi(z) - \psi(-z) = -\pi\cot\pi z - 1/z, \tag{12.65}$$

show that the spectral function for the current correlator is given by

$$\mathcal{R}(\omega) = \frac{N^2 T^2}{4}\frac{\pi\mathfrak{w}^2\sinh\pi\mathfrak{w}}{\cosh\pi\mathfrak{w} - \cos\pi\mathfrak{w}}. \tag{12.66}$$

Asymptotically, for large frequencies at small temperature $\omega \gg T$, the result (12.66) reduces to

$$\mathcal{R}(\omega) = \frac{N^2\omega^2}{16\pi}. \tag{12.67}$$

This coincides with the expected result for a conformal field theory at vanishing temperature, $\mathcal{R}(\omega) \propto \omega^{2\Delta-d}$, since the dimension of the conserved R-current in four dimensions is $\Delta = 3$. The factor N^2 comes from the number of degrees of freedom, its appearance corresponding to the adjoint representation of $SU(N)$.

12.1.3 Quasinormal modes

Since gauge/gravity duality proposes the identification of planar black holes or black branes in asymptotically AdS spaces with thermal quantum field theories, it is natural to expect that fluctuations of the black hole background will lead to small deviations from equilibrium in the thermal field theory. Within gravity, quasinormal modes are resonant linearised fluctuation modes about a classical background, with specific boundary conditions imposed. A prominent example is fluctuations about a black hole background. Since the fluctuations may fall into the black hole and decay, they are damped and therefore associated with complex eigenfrequencies. Mathematically, this is implemented by infalling boundary conditions at the future horizon. Within gauge/gravity duality, the small deviations from thermal equilibrium induced by the fluctuations give rise to dissipation on the field theory side. This leads to dispersion relations such as (12.22) which correspond to singularities of the retarded Green's function in the complex frequency plane. In fact, there is a precise identification. Using the holographic prescription for calculating causal Green's functions introduced in chapter 11 with its particular realisation for linear response as given in section 12.1.2, we will see that the *quasinormal modes* correspond to poles of the Green's functions.

Let us identify the quasinormal modes for the example considered in section 12.1.2. Consider again the result (12.64). The function $\psi(z) = \Gamma'(z)/\Gamma(z)$ has poles at $z = -n$ for $n \in \mathbb{N}$. This determines the poles of the retarded Green's function as

$$\omega = n(\pm 1 - i), \qquad n = 0, 1, 2, \ldots. \tag{12.68}$$

These poles correspond to the quasinormal modes of the fluctuations about the black hole background with infalling boundary conditions. Their structure is very typical, for this reason we show them in figure 12.1.

12.1.4 Causality and stability

It is inherent in the definition of the retarded Green's function as given by (12.3) that it is causal, i.e. the expectation value (12.2) at time t depends on the source at times t' only for $t' < t$. This fact is also reflected in the pole structure in the complex frequency plane, as

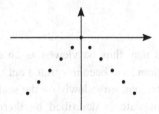

Figure 12.1   Quasinormal modes in the complex frequency plane as given by (12.68). This pole pattern is universal for many systems. The pole at vanishing frequency arises for systems with vanishing chemical potential. Systems with spontaneously broken symmetry may also display poles at vanishing frequency, which correspond to Goldstone modes.

we now discuss by considering the inverse Fourier transform

$$G_R^{IJ}(t, \vec{k}) = \int \frac{d\omega}{2\pi} e^{-i\omega t} G_R^{IJ}(\omega, \vec{k}). \tag{12.69}$$

For $t < 0$ we evaluate this integral by closing the contour in the upper half-plane. Causality implies that this integral vanishes for $t < 0$, and therefore the momentum-space Green's function $G_R^{IJ}(\omega, \vec{k})$ must be analytic in ω for $\text{Im}\,\omega > 0$. For non-analyticities in the upper half of the complex frequency plane, instabilities may occur as is seen by assuming the presence of a pole there at ω_*. This leads to an exponentially growing mode

$$G_R^{IJ}(t, \vec{k}) \sim e^{-i\omega_* t} \sim e^{|\text{Im}\,\omega_*|t}, \tag{12.70}$$

which indicates that the vacuum in which the Green function has been computed is unstable against perturbations.

Moreover, the external sources $\phi_I(t, \vec{x})$ exert work on the system. The spectral function $\mathcal{R}^{IJ}$ measures the time-averaged rate of change of the total energy $\overline{\frac{dW}{dt}}$ to leading order in the external sources, which may be time-dependent. The dissipation of the system is given by the spectral function by virtue of

$$\overline{\frac{dW}{dt}} = \omega \int d^{d-1}\vec{x} \quad d^{d-1}\vec{x}'\, \phi_I(\omega, \vec{x}) \mathcal{R}^{IJ}(\omega, \vec{x} - \vec{x}')\phi_J(\omega, \vec{x}'). \tag{12.71}$$

Stability requires the eigenvalues of the spectral function $\mathcal{R}^{IJ}$, and hence both the diagonal elements of $\mathcal{R}^{IJ}$ and the spectral measure, i.e. the sum of the eigenvalues, to be strictly non-negative. Otherwise, the resulting excitation would experience *negative* energy dissipation into the medium, i.e. the excitation would extract energy from the medium. This is not possible and signals an instability.

12.2 Hydrodynamics

12.2.1 Hydrodynamic approximation

The central idea of hydrodynamics is to consider small fluctuations about thermal equilibrium for which the wavelength λ_{wave} is much larger than the mean free path l_{mfp}, i.e.

$$\lambda_{\text{wave}} \gg l_{\text{mfp}}. \tag{12.72}$$

Hydrodynamics may thus be viewed as an effective field theory whose large k and ω degrees of freedom have been integrated out.

Since perturbations vary slowly on the scale l_{mfp}, the system is locally in equilibrium. The equilibrium state is described by thermodynamical variables. These include the temperature T, the pressure p, the energy density $\varepsilon = U/V$ and the charge density $\rho_a = N_a/V$. Since the equilibrium is local rather than global, these thermodynamical quantities are functions of the space time coordinates x. They vary slowly on the scale l_{mfp}.

Their dynamics is given by conservation laws. For a system with charges and associated conserved currents J_a^μ, these conservation laws are given by

$$\nabla_\mu T^{\mu\nu} = 0, \qquad \nabla_\mu J_a^\mu = 0, \tag{12.73}$$

where $T^{\mu\nu}$ is the conserved energy-momentum tensor.

12.2.2 Ideal fluid

Let us first consider an ideal and isotropic fluid in d-dimensional spacetime with metric $g_{\mu\nu}$. The notion *ideal* refers to the fact that the fluid does not dissipate energy, for instance there is no friction. Moreover, we have an *isotropic* fluid if both the background geometry and the fluid itself are invariant under rotations of the $d-1$ spatial dimensions. In the local rest-frame of the fluid, where the velocity is $u^\mu = (1, 0, 0, 0, \ldots)$, we have

$$\langle T^{tt} \rangle = \varepsilon, \quad \langle T^{ii} \rangle = p, \quad \langle J_a^t \rangle = \rho_a. \tag{12.74}$$

If the fluid has the relativistic velocity u_μ, with $u_\mu u^\mu = -1$, the energy-momentum tensor and conserved current take the form

$$(T^{\mu\nu})_{\text{ideal}} = \varepsilon u^\mu u^\nu + p P^{\mu\nu}, \quad (J_a^\mu)_{\text{ideal}} = \rho_a u^\mu, \tag{12.75}$$

where

$$P^{\mu\nu} = g^{\mu\nu} + u^\mu u^\nu \tag{12.76}$$

is a projector onto the spatial directions for which

$$P^{\mu\nu} u_\nu = 0, \quad P^{\mu\sigma} P_{\sigma\nu} = P^\mu{}_\nu. \tag{12.77}$$

ε and p are not independent, they are related by the equation of state of thermodynamics. Let us now consider a conformal fluid for which the trace of the energy-momentum tensor vanishes. This implies

$$\varepsilon = (d-1)p. \tag{12.78}$$

In addition to $(T^{\mu\nu})_{\text{ideal}}$ and $(J_a^\mu)_{\text{ideal}}$, there is a further conserved quantity, the *entropy current* $(s^\mu)_{\text{ideal}}$,

$$(s^\mu)_{\text{ideal}} = s u^\mu, \tag{12.79}$$

where s is the entropy density, $s = S/V$. As exercise 12.2.1 shows, this current is conserved, i.e. the ideal fluid does not generate entropy.

Exercise 12.2.1 From the conservation equation for the energy-momentum tensor, with the help of the thermodynamical relations

$$\varepsilon + p = sT + \mu_a \rho_a \tag{12.80}$$

and

$$dp = s\,dT + \rho_a d\mu_a, \tag{12.81}$$

show that

$$\nabla_\mu (s^\mu)_{\text{ideal}} = 0. \tag{12.82}$$

12.2.3 Dissipative fluid

As its name tells, the ideal fluid corresponds to an ideal situation and is not realised in nature. Every fluid perturbed by long-wavelength fluctuations is expected to return to local equilibrium. Therefore there are additional contributions to $T^{\mu\nu}$ and J_a^μ which correspond to dissipation. For long-wavelength fluctuations, this may be described by allowing T and u^μ to be slowly varying functions of the boundary coordinates and by considering a *derivative expansion*. The derivatives of the slowly-varying functions are expected to be small. The coefficients in this expansion are the *transport coefficients*.

The zeroth order in this expansion corresponds to the ideal fluid. To describe dissipation or entropy production, we have to proceed to higher orders in the derivative expansion. To first order, we write for the energy-momentum tensor and the conserved current

$$T^{\mu\nu} = \varepsilon u^\mu u^\nu + p P^{\mu\nu} - \sigma^{\mu\nu} + \mathcal{O}(\partial^2), \tag{12.83}$$

$$J_a^\mu = \rho_a u^\mu + \Upsilon_a^\mu. \tag{12.84}$$

$\sigma^{\mu\nu}$ and Υ_a^μ are of first order in the derivative expansion. There is an arbitrariness in choosing these non-equilibrium terms which arises from the freedom of redefining the local temperature field $T(x)$, the local velocity $u_\mu(x)$ and the local chemical potential $\mu_a(x)$ by gradients of the hydrodynamic variables. Fixing this freedom corresponds to a choice of *frame*. For this purpose, here we impose

$$u_\mu \sigma^{\mu\nu} = 0, \qquad u_\mu \Upsilon_a^\mu = 0. \tag{12.85}$$

These relations define the *Landau frame* and the *Eckart frame*. In the Landau frame, we have $u_\mu u_\nu T^{\mu\nu} = \epsilon$ in the local rest-frame of the fluid, which implies $T^{00} = \epsilon$ to all orders in the gradient expansion, such that there is no energy flow in the local rest-frame. For the current, in the Eckart frame we have $u_\mu J_a^\mu = \rho_a$ and thus $J_a^t = \rho_a$ for the charge density to all orders in the local rest-frame, such that there is no charge flow in the local rest-frame.

To find expressions for $\sigma^{\mu\nu}$ and Υ_a^μ in terms of the thermodynamic variables and the four-velocity of the fluid, we consider the entropy current again whose derivative no longer vanishes. Using (12.79) and (12.85), we have

$$(s^\mu)_{\text{dissipative}} = s u^\mu - \frac{\mu^a}{T} \Upsilon_a^\mu, \tag{12.86}$$

$$\nabla_\mu (s^\mu)_{\text{dissipative}} = -\Upsilon_a^\mu \nabla_\mu \frac{\mu^a}{T} + \frac{\sigma^{\mu\nu}}{T} \nabla_\mu u_\nu, \tag{12.87}$$

where we have used $T\nabla_\mu (s u^\mu) = \mu^a \nabla_\mu \Upsilon_a^\mu + u_\nu \nabla_\mu \sigma^{\mu\nu}$, which follows from conservation of the energy-momentum tensor. $(s^\mu)_{\text{dissipative}}$ is the entropy current to first order in the derivative expansion. Due to the second law of thermodynamics, $\nabla_\mu (s^\mu)_{\text{dissipative}}$ in (12.87) must be positive, which in flat space is achieved by writing

$$\sigma^{\mu\nu} = P^{\mu\alpha} P^{\nu\beta} \left[\eta \left(\partial_\alpha u_\beta + \partial_\beta u_\alpha - \frac{2}{3} \delta_{\alpha\beta} \partial_\lambda u^\lambda \right) + \zeta \, \delta_{\alpha\beta} \partial_\lambda u^\lambda \right], \qquad (12.88)$$

$$\Upsilon_a^\mu = -\kappa_{ab} P^{\mu\nu} \partial_\nu \left(\frac{\mu^b}{T} \right), \qquad (12.89)$$

with positive semi-definite coefficients η, ζ and κ_{ab}. $P^{\mu\nu}$ is the projector onto directions perpendicular to u^μ as defined in (12.76).

η and ζ in (12.88) are referred to as as the *shear viscosity* and *bulk viscosity*, respectively, since η is the coefficient of the symmetric traceless contribution and ζ is the coefficient of the trace part. In conformal field theories, tracelessness of the energy-momentum tensor implies that the bulk viscosity vanishes, $\zeta_{CFT} = 0$. The coefficients κ are related to the charge diffusion constant D_{ab}. This may be seen for any given model by using its specific relation between the chemical potential and the charge density. Then, for a charged fluid in flat space with a conserved $U(1)$ current $\partial_\mu J^\mu = 0$ we have

$$J^\mu = \rho \, u^\mu - D P^{\mu\nu} \partial_\nu \rho. \qquad (12.90)$$

In the rest-frame of the fluid, this gives rise to Fick's law of diffusion $\vec{j} = -D\vec{\nabla}\rho$. The relation between the diffusion constants D in (12.90) and κ in (12.89) (i.e. κ_{ab} with $a, b = 1$) is given by the Einstein relation

$$\frac{\kappa}{T} = D \frac{\partial \rho}{\partial \mu}. \qquad (12.91)$$

Note that there is a different transport coefficient for the response to ∇T which we consider in chapter 15. Of course, in addition to (12.88) and (12.89) there are also contributions at second and higher orders in the derivative expansion, which we do not discuss explicitly here.

Exercise 12.2.2   **Shear viscosity at weak coupling** In weakly coupled theories, the viscosity η is governed by the mean free path $l_{\mathrm{mfp}} \sim (n\sigma v)^{-1}$, where n denotes the density, σ the cross section for interactions and v a typical velocity. Consider ϕ^4 theory at finite temperature, for which a perturbative calculation gives

$$n \sim T^3, \qquad \sigma \sim \left(\frac{g}{T} \right)^2, \qquad v \sim 1. \qquad (12.92)$$

The viscosity is obtained from kinetic theory by multiplying l_{mfp} with the energy density ε for which the Stefan–Boltzmann law in four dimensions, $\varepsilon \sim T^4$, is assumed.

Show that for ϕ^4 theory at weak coupling,

$$l_{\mathrm{mfp}} \sim \frac{1}{g^2 T}, \qquad \eta \sim \varepsilon \, l_{\mathrm{mfp}} \sim \frac{T^3}{g^2}. \qquad (12.93)$$

This implies that the mean free path is large at weak coupling.

The entropy density scales in the same way with temperature, $s \sim T^3$. Consequently, at weak coupling (12.93) implies that the quotient

$$\frac{\eta}{s} \sim \frac{1}{g^2} \qquad (12.94)$$

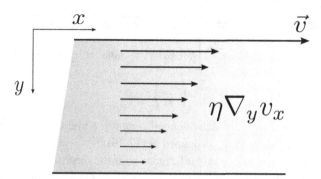

Figure 12.2 Non-relativistic shear viscosity: η measures the velocity gradient $\nabla_y v_x$ for a fluid between two plates, of which the upper one moves in the x-direction.

depends on g only and becomes large at weak coupling, $g \ll 1$. Note that (12.94) diverges for the limit of a free theory for which $g \to 0$. This divergence signals that the hydrodynamic limit and the limit $g \to 0$ do not commute. For a free theory, the mean free path l_{mfp} becomes infinite, such that the hydrodynamic approximation $\lambda_{\mathrm{wave}} \gg l_{\mathrm{mfp}}$ is no longer valid. On the other hand, note that for large coupling $g \sim 1$, (12.94) gives $\eta/s \sim 1$. However, in this coupling regime, the perturbative expansion is expected to break down.

To explain the physical significance of the shear viscosity, let us briefly consider the non-relativistic case. Consider a fluid between two plates, one of which moves with a velocity v_x parallel to the other. The shear viscosity measures the velocity gradient as shown in figure 12.2.

12.3 Transport coefficients from linear response

12.3.1 General remarks

In sections 12.1 and 12.2, we introduced the linear response formalism and hydrodynamics, respectively. We now make connections between these concepts and explain how transport coefficients within hydrodynamics can be calculated using linear response: the transport coefficients are related to the retarded Green's function.

The first-order dissipative corrections to $T^{\mu\nu}$ and J_a^μ given by (12.83) and (12.84) together with the results (12.88) and (12.89) give the linear response of $T^{\mu\nu}$ and J_a^μ to a metric or gauge field fluctuation, respectively. The relations between the macroscopic transport coefficients and the microscopic Green's functions are referred to as *Green–Kubo relations*, for which we now consider two examples.

12.3.2 Green–Kubo relation for charge diffusion

We begin with the $U(1)$ current at finite chemical potential μ and charge density ρ. Let us look at fluctuations of the background gauge field which sources the conserved current J^μ

of the form

$$A_t = \mu + \delta a_t, \tag{12.95}$$

i.e. the time component A_t fluctuates. The dissipative relations (12.84), (12.89) imply

$$\delta J^\mu = \frac{\kappa}{T} P^{\mu\nu} \nabla_\nu \delta a_t = D \frac{\partial \rho}{\partial \mu} P^{\mu\nu} \nabla_\nu \delta a_t, \tag{12.96}$$

where we used (12.91). In the fluid rest-frame $u^\mu = (1,0,0,0)$ this becomes

$$\delta J^i = D \frac{\partial \rho}{\partial \mu} \partial^i \delta a_t. \tag{12.97}$$

Fourier transforming, and choosing $k^\mu = (\omega, 0, 0, q)$, we obtain

$$\delta J^3(\omega, q) = i D \frac{\partial \rho}{\partial \mu} q \, \delta a_t(\omega, q). \tag{12.98}$$

This hydrodynamic result may now be compared to the linear response result (12.4), which for the current considered here is given by

$$\delta J^3(\omega, q) = G_R^{3t} \, \delta a_t(\omega, q). \tag{12.99}$$

Comparing to (12.98), we obtain the Green–Kubo relation

$$iqD \frac{\partial \rho}{\partial \mu} = G_R^{3t}(\omega, q), \tag{12.100}$$

which implies

$$\frac{\kappa}{T} = D \frac{\partial \rho}{\partial \mu} = -i \lim_{q \to 0} \frac{1}{q} G_R^{3t}(\omega, q). \tag{12.101}$$

12.3.3 Green–Kubo relation for the shear viscosity

Similarly, we obtain the Green–Kubo relation for the shear viscosity by considering appropriate metric fluctuations. Let us concentrate on the particular case when metric perturbations are time dependent but homogeneous in space, i.e.

$$g_{ij}(t, \vec{x}) = \delta_{ij} + h_{ij}(t), \qquad h_{ii} = 0, \tag{12.102}$$
$$g_{tt}(t, \vec{x}) = -1, \qquad g_{ti}(t, \vec{x}) = 0. \tag{12.103}$$

The velocity vector hence depends on time only, $u^i = u^i(t)$. Consider the case where the fluid remains at rest at all times, $u^\mu = (1, 0, 0, 0)$.

In curved spacetime, equation (12.88) for the $\mathcal{O}(\partial)$ contributions to $T^{\mu\nu}$ generalises to

$$\sigma^{\mu\nu} = P^{\mu\alpha} P^{\nu\beta} \left[\eta \left(\nabla_\alpha u_\beta + \nabla_\beta u_\alpha \right) + \left(\zeta - \frac{2\eta}{3} \right) g_{\alpha\beta} \nabla_\lambda u^\lambda \right]. \tag{12.104}$$

In the situation considered above, this simplifies to

$$\sigma_{xy} = 2\eta \, \Gamma^t{}_{xy} = \eta \, \partial_t h_{xy}. \tag{12.105}$$

Box 12.2   **Shear modes and sound modes**

To find the poles in the energy-momentum tensor correlator, it is helpful to make use of the normal modes, which organise themselves into two types, the *shear* and *sound* modes. Choosing the momentum $k = (\omega, 0, 0, q)$, with q in the 3-direction, shear modes correspond to fluctuations of pairs of components T^{0a} and T^{3a}, where $a = 1, 2$,

$$T^{3a} = -\eta \, \partial_3 u^a = -\frac{\eta}{\epsilon + p} \, \partial_3 T^{ta},$$

$$\partial_t T^{ta} - \frac{\eta}{\epsilon + p} \, \partial_3^2 T^{ta} = 0. \tag{12.108}$$

For plane waves $h \sim e^{-i\omega t + iqx_3}$, we have

$$\omega = -i\gamma q^2, \qquad \gamma = \frac{\eta}{\epsilon + p}. \tag{12.109}$$

Sound waves, on the other hand, are longitudinal fluctuations of T^{00}, T^{03}, T^{33} with speed $v_s = \sqrt{\frac{dp}{d\epsilon}}$ and frequency

$$\omega = v_s q - i\Gamma_s q^2, \qquad \Gamma_s = \frac{1}{2(\epsilon + p)} \left(\frac{4\eta}{3} + \zeta \right). \tag{12.110}$$

By comparison with linear response theory, we find the zero spatial momentum, low-frequency limit of the retarded Green function of T^{xy},

$$G_R^{xy,xy}(\omega, \vec{0}) = \int dt \, d^3x \, e^{-i\omega t} \, \theta(t) \left\langle \left[T^{xy}(t, \vec{x}), \, T^{xy}(0, \vec{0}) \right] \right\rangle$$

$$= -i\eta \, \omega + \, \mathcal{O}(\omega^2). \tag{12.106}$$

The associated Green–Kubo relation is

$$\eta = - \lim_{\omega \to 0} \frac{1}{\omega} \, \text{Im} \, G_R^{xy,xy}(\omega, \vec{0}). \tag{12.107}$$

12.3.4 AdS/CFT calculation of the diffusion constant

Using the Green–Kubo relations established within field theory above, we now calculate the transport coefficient by evaluating the retarded Green's function using gauge/gravity duality. We consider again the AdS–Schwarzschild geometry dual to $\mathcal{N} = 4$ theory at finite temperature.

For the diffusion constant, we combine the field theory results (12.100) and (12.21) to obtain

$$q D \frac{\partial \rho}{\partial \mu} = i \frac{\omega q}{\omega^2 - q^2} \Pi^L(\omega, q). \tag{12.111}$$

Within gauge/gravity duality, the dependence of the density on the chemical potential is

$$\frac{\partial \rho}{\partial \mu} = \frac{N^2 T^2}{8}.$$ (12.112)

We now use the holographic linear response formalism developed in section 12.1. We evaluate $\Pi^L(\omega, q)$ holographically using (12.60) in the hydrodynamic limit of small frequencies and small momenta. In this limit, the expansion coefficients near the boundary in (12.56) are given by

$$E_L^{(0)} = \frac{1}{\mathfrak{w}}(\mathfrak{w} + i\mathfrak{q}^2),$$ (12.113)

$$E_L^{(1)} = \frac{i}{\mathfrak{w}}(\mathfrak{w}^2 - \mathfrak{q}^2).$$ (12.114)

We thus obtain using (12.60)

$$\Pi^L = -\frac{N^2 T^2}{8} \frac{E_L^{(1)}}{E_L^{(0)}} = -\frac{N^2 T^2}{8} \frac{i}{2\pi T} \frac{\omega^2 - q^2}{\omega + iq^2/(2\pi T)}.$$ (12.115)

Using (12.111), (12.112) and (12.101) we then obtain the diffusion constant

$$D = \frac{1}{2\pi T}.$$ (12.116)

This constant determines the diffusion of R-charge in the dual field theory.

12.3.5 AdS/CFT calculation of the shear viscosity

To obtain the shear viscosity at strong coupling, we compute holographically the retarded Green's function $G_R^{xy,xy}$ in (12.106) associated with the correlator $\langle T^{xy} T^{xy} \rangle$. This requires us to examine the propagation of the dual graviton mode h_{xy} in AdS spacetime. For this purpose, let us start from the Einstein–Hilbert action in five dimensions, as given in (6.76). The part of the Einstein–Hilbert action quadratic in h_{xy} is given by

$$S_{\text{quad}}[h_{xy}] = \frac{N^2}{8\pi^2 L^3} \int d^4x \, dr \sqrt{-g} \left(-\frac{1}{2} g^{\mu\nu} \partial_\mu h_{xy} \partial_\nu h_{xy} \right),$$ (12.117)

where we consider only those terms giving rise to non-local contributions to the retarded Green's function, i.e. we neglect local contact terms. Equation (12.117) gives rise to the linearised equation of motion $\partial_\mu(\sqrt{-g}g^{\mu\nu}\partial_\nu h_{xy}) = 0$. Since this has precisely the form of the equation of motion for a scalar field, we replace h_{xy} by ϕ in the subsequent. The background metric is again the AdS–Schwarzschild metric written in the form

$$ds^2 = \frac{(\pi TL)^2}{u} \left(-f(u) \, dt^2 + d\vec{x}^2 \right) + \frac{L^2}{4 u^2 f(u)} \, du^2 + L^2 \, d\Omega_5^2,$$ (12.118)

where $u = r_h^2/r^2$ and $f(u) = 1 - u^2$. The boundary is located at $u = 0$, the horizon at $u = 1$.

We now apply the procedure given in section 11.2.3 for holographically calculating retarded Green's functions. We perform a Fourier transformation of the boundary coordinates and impose the boundary condition $\phi(u = 0, p) = \phi_{(0)}(p)$ at the boundary of the asymptotically AdS space. In agreement with the holographic prescription for calculating bulk-to-boundary propagators, we write

$$\phi(u, p) = w_p(u)\, \phi_{(0)}(p) \tag{12.119}$$

where the *zero mode function* w_p satisfies

$$w_p'' - \frac{1 + u^2}{u f(u)} w_p' + \frac{\mathfrak{w}^2}{u f^2(u)} w_p - \frac{\mathfrak{q}^2}{u f(u)} w_p = 0, \tag{12.120}$$

where $p = (\omega, \vec{q})$ and with abbreviations $\mathfrak{w} = \omega/(2\pi T)$ and $\mathfrak{q} = q/(2\pi T)$. Near the boundary $u = 0$, the two solutions to this equation behave as $w_1 \sim 1$ and $w_2 \sim u$. Near the horizon, the solutions take the asymptotic form $w_p \sim (1 - u)^{-i\mathfrak{w}/2}$ and $w_p^* \sim (1 - u)^{i\mathfrak{w}/2}$. As seen in section 11.2.3, w_p corresponds to a plane wave moving towards the horizon, i.e. an infalling wave, and w_p^* corresponds to a plane wave moving away from the horizon, i.e. an outgoing wave.

According to section 11.2.3, the retarded Green's function is obtained from the on-shell action by virtue of

$$G_R(\omega, \vec{q}) = -2 \lim_{u \to 0} \mathcal{F}(\omega, \vec{q}, u), \tag{12.121}$$

where $\mathcal{F}$ is defined in equation (11.101). In the case considered here, the on-shell action obtained from (12.117) reads

$$S = \frac{\pi^2 N^2 T^4}{8} \int du \frac{f(u)}{u} \phi(u, x) \partial_u \phi(u, x)|_{u \to 0}. \tag{12.122}$$

Exercise 12.3.1   Starting from (12.122), calculate the retarded Green's function (12.121) for the energy-momentum tensor component T^{xy} in the limit $|\vec{q}| \to 0$. Hint: $\mathcal{F}$ is given by

$$\mathcal{F}(\omega, u) = \frac{\pi^2 N^2 T^4}{8} \frac{1}{u} w_p^*(u) \partial_u w_p(u), \tag{12.123}$$

and the result is

$$G_R(\omega) = -\frac{\pi N^2 T^3}{8} i\omega. \tag{12.124}$$

According to the Green–Kubo relation (12.107), we thus obtain

$$\eta = \frac{\pi}{8} N^2 T^3. \tag{12.125}$$

To obtain the ratio of the shear viscosity η over the entropy density s, we recall from chapter 11 that the entropy is given by the Bekenstein–Hawking formula $S = A/(4\pi G)$, with A the area of the black hole horizon. For the entropy density of the strongly coupled dual field theory, we thus have, see (11.67),

$$s = \frac{S}{\text{Vol}(\mathbb{R}^3)} = \frac{\pi^2}{2} N^2 T^3. \tag{12.126}$$

Combining (12.125) and (12.126), we obtain the famous result

$$\frac{\eta}{s} = \frac{1}{4\pi}. \tag{12.127}$$

Reinstating the units, we have $\eta/s = 1/4\pi \cdot \hbar/k_B$.

This is an extremely important result, which is remarkable from a number of perspectives. First of all, the result is *universal*. For all gravity duals involving the Einstein–Hilbert action, the same result (12.127) is obtained, irrespective of whether other fields such as gauge or scalar fields are added to the gravity action, or whether conformal symmetry or supersymmetry is broken or not. Moreover, the result is independent of the spacetime dimension. Also for gravity duals of non-commutative field theories, the same result is obtained.

It was conjectured that since (12.127) is obtained for large coupling, it actually provides a lower bound on the shear viscosity over entropy density ratio, making strongly coupled holographic fluids the most perfect fluids, next to ideal fluids for which the shear viscosity vanishes. In fact, experimental results at the RHIC and LHC accelerators show that the value of η/s measured for the quark–gluon plasma is in good agreement with (12.127), of the order $\eta/s = 1/(4\pi)$ to $\eta/s = 2.5/(4\pi)$ in units where $\hbar = k_{\mathrm{B}} = 1$. Other liquids, such as water or liquid helium, have a value of η/s which is larger by orders of magnitude. Only some cold atomic gases have a similarly small value of η/s. The measurement of η/s at RHIC and its good agreement with (12.127) was the first example of a successful measurement of an observable calculated using gauge/gravity duality.

Within gauge/gravity duality, there are a few cases where the bound (12.127) is violated, leading to results smaller than $1/(4\pi)$. This happens for gravity duals of particular $Sp(N)$ theories, or if terms of higher order in the curvature are present in the gravity action, or in some cases where the system has an anisotropy in space.

We will return to the physics of the quark–gluon plasma in chapter 14.

12.4 Fluid/gravity correspondence

12.4.1 General method

The calculation of transport coefficients as presented above for the example of the shear viscosity may be systematised in an elegant way which we now describe [6]. Ten-dimensional type IIB supergravity has many consistent truncations on AdS_5 space. More generally, by starting from an appropriate ten- or eleven-dimensional supergravity action, we find consistent truncations to theories on AdS_{d+1}, with Einstein equations

$$R_{mn} - \frac{1}{2}Rg_{mn} + \Lambda g_{mn} = 0, \qquad \Lambda = -\frac{d(d-1)}{2}\frac{1}{L^2}. \tag{12.128}$$

Assuming the gauge/gravity conjecture holds, this implies that there is a class of dual quantum field theories for which (12.128) describes the universal decoupled dynamics of the energy-momentum tensor in these field theories. Moreover, in the long-wavelength regime (12.72), we expect that the Einstein equation (12.128) gives rise to the hydrodynamic equations of d-dimensional effective field theories. This relation is referred to as the *fluid/gravity correspondence* [7].

The starting point for this approach is the planar Schwarzschild–AdS_{d+1} black hole or black brane with metric

$$ds^2 = -r^2 f(br)dt^2 + \frac{dr^2}{r^2 f(br)} + r^2 dx_i dx^i, \tag{12.129}$$

$$f(r) = 1 - \frac{1}{r^d}. \tag{12.130}$$

This is a one-parameter family of solutions parametrised by $b \equiv 1/r_h$, which is related to the Hawking temperature by

$$T = \frac{d}{4\pi b}. \tag{12.131}$$

Moreover, a family of solutions with d parameters is generated by boosting the solution along the spatial directions x^i, given by

$$ds^2 = \frac{dr^2}{r^2 f(br)} + r^2(-f(br)u_\mu u_\nu + P_{\mu\nu})dx^\mu dx^\nu, \tag{12.132}$$

$$u^t = \frac{1}{\sqrt{1 - \beta^2}}, \qquad u^i = \frac{\beta^i}{\sqrt{1 - \beta^2}}, \tag{12.133}$$

with $P^{\mu\nu} = \eta^{\mu\nu} + u^\mu u^\nu$ the projector onto spatial directions and velocities β^i with $\beta^i \beta_i \equiv \beta^2$. The constant parameters T and u^μ of this solution are precisely the parameters of d-dimensional relativistic hydrodynamics, the temperature and relativistic fluid velocity.

For regularity at the future horizon, we introduce infalling Eddington–Finkelstein coordinates, as introduced in (2.143) in chapter 2, to replace the Schwarzschild coordinates. Then, (12.132) becomes

$$ds^2 = -2u_\mu dx^\mu dr - r^2 f(br) u_\mu u_\nu dx^\mu dx^\nu + r^2 P_{\mu\nu} dx^\mu dx^\nu. \tag{12.134}$$

The relativistic d-dimensional vector x^μ now stands for (v, x^i) with v the ingoing Eddington–Finkelstein variable $v = t + r^*$ introduced in chapter 2.

The equilibrium solution (12.132) leads to a conserved boundary energy-momentum tensor, i.e. to an ideal fluid. To describe dissipation, the system has to be perturbed away from global equilibrium. This is achieved in a natural way by allowing the parameters b and β^i to be slowly varying functions of the boundary coordinates v and x^i. Note that (12.134) with arbitrary functions $b(v, \vec{x})$ and $\beta^i(v, \vec{x})$ does not satisfy the Einstein equations. However, assuming that both these functions have only long-wavelength fluctuations as in the hydrodynamic regime, a new solution to the gravity equations of motion may then be constructed order by order in the derivative expansion introduced in section 12.2.3. The Eddington–Finkelstein coordinates provide a clear picture of the procedure to be followed. Boundary domains smaller than the fluctuation wavelength, in which there is local thermal equilibrium, extend to 'tubes' in the bulk along radial null geodesics, as shown in figure 12.3. These tubes may be patched together to obtain solutions of Einstein's equations in the bulk. This patching may be done order by order in boundary derivatives, just as in the hydrodynamic expansion.

The starting point for obtaining the derivative expansion is the metric

$$ds^2 = r^2 g_{\mu\nu}(r, u^\lambda, b)dx^\mu dx^\nu - 2S(r, u^\lambda, b)u_\mu dx^\mu dr. \tag{12.135}$$

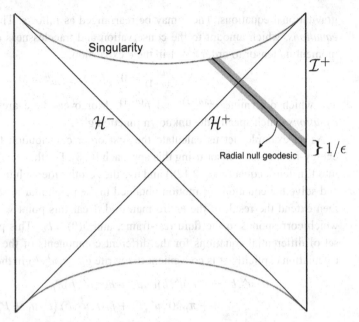

Figure 12.3 Causal structure of AdS space in Eddington–Finkelstein coordinates. The grey shaded region is the domain of validity for the hydrodynamic expansion of the radial null geodesic.

Note that we have gauge fixed the metric since the rr-component vanishes and the $r\mu$-component is proportional to u_μ with proportionality factor $-S(r, u^\lambda, b)$.

In the metric (12.135), the velocities u^i and the parameter b, or equivalently u^μ and T, are allowed to be slowly varying functions of the d-dimensional coordinates $x = (v, x^i)$. The original metric (12.134), which we denote by $g^{(0)}(b, u^i)$, is then no longer a solution to Einstein's equations in general: it appears as the zeroth order contribution in the derivative expansion.[1] The goal is now to solve Einstein's equations (12.128) perturbatively. For this we write

$$u^i = u^{i,(0)} + \epsilon u^{i,(1)} + \mathcal{O}(\epsilon^2), \qquad b = b^{(0)} + \epsilon b^{(1)} + \mathcal{O}(\epsilon^2), \tag{12.136}$$

where $u^{i,(m)}(x)$ and $b^{(m)}(x)$ are mth order coefficients in the derivative expansion. The parameter ϵ counts the order in this expansion. Moreover, we consider an ansatz for the metric of the form

$$g = g^{(0)}(u^i, b) + \epsilon g^{(1)}(u^i, b) + \epsilon^2 g^{(2)}(u^i, b) + \mathcal{O}(\epsilon^3), \tag{12.137}$$

where $g^{(0)}$ is the metric given by (12.135).

We then insert this ansatz into Einstein's equations. Assume that we have solved the resulting equations to order ϵ^{n-1}. Then at order ϵ^n, we obtain a set of differential equations for the components of $g^{(n)}$, with the differential operator given in terms of $g^{(0)}(u^{i,(0)}, b^{(0)})$, and with a source term involving derivatives of $u^{i,(0)}$ and $b^{(0)}$. This gives $(d+1)(d+2)/2$

---

[1] Note that here, $g^{(0)}_{\mu\nu}$ denotes the zeroth order contribution in the derivative expansion, and is not to be confused with the zeroth order term in an expansion around the boundary of AdS space, as we considered in previous chapters.

gravitational equations. These may be rearranged as follows. There are $d + 1$ *constraint equations*, which amount to the conservation and tracelessness of the boundary energy-momentum tensor to order $n - 1$ in the ϵ expansion,

$$\nabla_\mu T^{\mu\nu}_{(n-1)} = 0, \qquad T_{(n-1)\mu}{}^\mu = 0 \qquad (12.138)$$

and which determine $u^{i,(n-1)}$ and $b^{(n-1)}$. Moreover, there are $d(d + 1)/2$ *dynamical equations* which specify the unknown function $g^{(n)}$.

As an example, let us calculate the first order contribution to the metric and to the energy-momentum tensor using this approach [6, 8]. For this we insert the ansatz (12.135) into Einstein's equations (12.128) and use the zeroth order solution (12.134). We establish and solve the equations of motion obtained in the neighbourhood of the point $x^\mu = 0$ and then extend the result to the entire manifold. Near this point we may set $u^\mu(0) = (1, \vec{0})$, which corresponds to the fluid rest-frame, and $b(0) = b_0$. This procedure gives rise to a set of differential equations for the different components of the metric. To perform this calculation explicitly, it is convenient to rewrite $g_{\mu\nu}(r, u^\lambda, b)$ in the metric (12.135) as

$$g_{\mu\nu}(r, u^\lambda, b) = k(r, u^\lambda, b)u_\mu u_\nu + h(r, u^\lambda, b)P_{\mu\nu}$$
$$+ \pi_{\mu\nu}(r, u^\lambda, b) + j_\sigma(r, u^\lambda, b)\left(P^\sigma_\mu u_\nu + P^\sigma_\nu u_\mu\right). \qquad (12.139)$$

The functions S, k, h, j_μ and $\pi_{\mu\nu}$ are determined order by order in the derivative expansion.

As an example, we restrict our attention to the traceless symmetric contribution $\pi_{\mu\nu}$. This is relevant for obtaining the energy-momentum tensor. The lowest order contribution $\pi^{(0)}_{\mu\nu}$ vanishes, as the comparison with (12.134) shows. To nth order in the derivative expansion, the Einstein equations give, in the fluid rest-frame,

$$\partial_r \left(r(1 - b_0^d r^d)\partial_r \pi^{(n)}_{ij}(r) \right) = \mathcal{P}^{(n)}_{ij}(r). \qquad (12.140)$$

Equation (12.140) can be integrated to give an explicit expression for $\pi^{(n)}_{ij}$,

$$\pi^{(n)}_{ij} = -\frac{1}{b_0} \int\limits_{b_0 r}^{\infty} dx \, \frac{\int_1^x dx' \mathcal{P}^{(n)}_{ij} x'/(b_0)}{x(1 - x^d)}. \qquad (12.141)$$

The limits of the inner integration ensure regularity. This integral may be evaluated in closed form. Similar expressions are obtained for the remaining metric functions in (12.139).

Let us consider the first order term $\pi^{(1)}_{ij}$ explicitly, for which $\mathcal{P}^{(1)}_{ij}(r)$ is obtained as

$$\mathcal{P}^{(1)}_{ij}(r) = (d - 1)b_0^2 (rb_0)^{d-2}\tilde{\sigma}_{ij}, \qquad (12.142)$$

$$\tilde{\sigma}_{ij} = 2\partial_{(i}u_{j)}{}^{(0)} - \frac{2}{d - 1}\delta_{ij}\partial_k u^{k,(0)}. \qquad (12.143)$$

Equation (12.143) is of the same form as the first term in (12.88). To obtain the contribution of $\pi^{(1)}_{ij}$ to the energy-momentum tensor, we have to perform a near-boundary expansion. By expanding (12.141) in $1/r$ for $n = 1$, the contribution to $\pi^{(1)}_{ij}$ which leads to a finite contribution to $T_{\mu\nu}$ is found to be

$$\pi^{(1)}_{\mu\nu} = -\frac{1}{r^d b_0^{d-1}}\tilde{\sigma}_{\mu\nu}, \qquad (12.144)$$

where we have reinstated full covariance in the indices.

From this result we may now calculate the contribution to the energy-momentum tensor. For this we note that for asymptotically AdS spaces, the boundary energy-momentum tensor is obtained from

$$T_{\mu\nu} = \lim_{r \to \infty} \frac{r^{d-2}}{8\pi G} \left[K_{\mu\nu} - K g_{\mu\nu} - (d-1)g_{\mu\nu} - \frac{1}{d-2} G_{\mu\nu} \right], \qquad (12.145)$$

with $K^{\mu\nu}$ the extrinsic curvature given by

$$K_{\mu\nu} = g_\mu{}^\rho \nabla_\rho n_\nu. \qquad (12.146)$$

n^ν is the outward pointing normal vector at the boundary. For a traceless energy-momentum tensor, we have $K \equiv g^{\mu\nu} K_{\mu\nu} = -d$. The result (12.145) is obtained by using the methods of holographic renormalisation as described in section 5.5.

Inserting the lowest order solution (12.134) into (12.145) gives rise to the ideal fluid contribution to the energy-momentum tensor as in (12.75),

$$T_{\mu\nu}^{(0)} = \frac{1}{16\pi G} \frac{1}{b_0^d}(d\, u_\mu u_\nu + \eta_{\mu\nu}). \qquad (12.147)$$

To first order in the derivative expansion, (12.145) gives

$$16\pi G T_{\mu\nu}^{(1)} = \pi_{\mu\nu}^{(1)}, \qquad (12.148)$$

with $\pi_{\mu\nu}^{(1)}$ as in (12.144). Finally we thus obtain

$$T_{\mu\nu}^{(1)} = -\frac{1}{16\pi G} \frac{1}{b_0^{d-1}} \tilde\sigma_{\mu\nu}, \qquad (12.149)$$

which for the traceless case agrees with (12.83) together (12.89). Using the relation (12.131) between b_0 and T, we reproduce the result (12.127) for the shear viscosity.

12.4.2 Anomalous flows

The method of fluid/gravity introduced in the previous section is easily generalised to cases with further fields present. A key example is the charged fluid which requires the presence of an additional gauge field on the gravity side. As discussed in chapter 11, charged fluids have a chemical potential μ. This is obtained from a non-trivial profile for the time component of the gauge field on the gravity side, $A_t = A_t(r)$. Here we consider a consistent truncation of IIB supergravity on $AdS_5 \times S^5$ with such a gauge field profile. The consistent truncation we choose is dual to a subsector of the $\mathcal{N} = 4$ Super Yang–Mills theory in which a single conserved $U(1)$ current is excited. This $U(1)$ is the diagonal $U(1)$ of the maximal Abelian subgroup of the $SU(4)$ R-symmetry and is dual to a $U(1)$ bulk gauge field. The gravity action obtained from this consistent truncation is given by

$$S = \int d^5 x \sqrt{-g} \left(\frac{1}{16\pi G_5}(R+12) - \frac{1}{4g_{YM}^2} F^2 + \frac{1}{3\sqrt{3}} A \wedge F \wedge F \right). \qquad (12.150)$$

The equations of motion obtained from this gravity action are solved by charged black holes, i.e. by *Reissner–Nordström* black holes in AdS_5. In the context of hydrodynamics the corresponding metric is written as

$$ds^2 = -r^2 f(r) u_\mu u_\nu dx^\mu dx^\nu + r^2 P_{\mu\nu} dx^\mu dx^\nu - 2u_\mu dx^\mu dr, \qquad (12.151)$$

using Eddington–Finkelstein coordinates, with the vector u^μ satisfying $u^\mu u_\mu = -1$ and the projector $P_{\mu\nu} = \eta_{\mu\nu} + u_\mu u_\nu$. The function $f(r)$ in (12.151) is given by

$$f(r) = 1 - M\left(\frac{r_h}{r}\right)^4 + Q^2\left(\frac{r_h}{r}\right)^6. \tag{12.152}$$

M and Q are associated with the mass and charge of a *black brane solution*, Note that the temperature of the black brane is given by

$$T = \frac{r_+}{2\pi}\left(2 - \frac{r_-^2}{r_+^2} - \frac{r_-^4}{r_+^4}\right), \tag{12.153}$$

with r_+ the larger of the two positive roots of $f(r_h) = 0$ and r_- the smaller one. The horizon is located at r_+ and the boundary of the asymptotically AdS space at $r \to \infty$. In addition to the metric, there is a non-trivial gauge field of the form

$$A_\mu = \sqrt{\frac{3L^2 g_{YM}^2}{16\pi G_5}} \times \frac{Q}{r_+^2}\left(1 - \frac{r_+^2}{r^2}\right)u_\mu, \qquad A_r = 0. \tag{12.154}$$

Similarly to section 11.3, the chemical potential is given by

$$\mu = A_t(r_+) - A_t(\infty) = \sqrt{\frac{3L^2 g_{YM}^2}{16\pi G_5}}\frac{Q}{r_+^2}. \tag{12.155}$$

For the hydrodynamic expansion, we proceed in the spirit of the fluid/gravity approach and allow u^μ, M and Q to be slowly varying functions of the space-time coordinates. This allows us to calculate hydrodynamic coefficients in a derivative expansion. The key point of this approach is that the Chern–Simons term present in the action naturally leads to a *vorticity* in the dual fluid. In fact, in addition to the hydrodynamic expansion of the energy-momentum tensor, the application of fluid/gravity duality to the charged solution considered here leads to first order contributions in the derivative expansion of the $U(1)$ current of the form

$$\Upsilon_\mu = -\kappa T P_\mu{}^\alpha \partial_\alpha\left(\frac{\mu}{T}\right) + \xi\,\omega_\mu, \tag{12.156}$$

$$\omega_\mu = \epsilon_\mu{}^{\rho\sigma\tau} u_\rho \partial_\sigma u_\tau. \tag{12.157}$$

The second term ω_μ corresponds to a parity-breaking vorticity. It reduces to $\vec{\nabla} \times \vec{v}$, the curl of the velocity, in the local rest-frame, which means that there is a current directed along the vorticity. The fluid/gravity approach allows determination of the transport coefficients κ and ξ and in particular yields

$$\xi \neq 0, \tag{12.158}$$

which signals the presence of a vorticity in (12.157).

Applying fluid/gravity duality to charged solutions thus naturally leads to a vorticity contribution to the derivative expansion of the current. This is referred to as the *chiral vortical effect*. For instance, for a volume of rotating quark matter, quarks with opposite helicities will move in opposite directions. Experiments are being proposed to observe this effect in the quark–gluon plasma. Note that the chiral vortical effect requires a relativistic fluid with a quantum anomaly, so it is not expected to be observed in non-relativistic or classical fluids.

The chiral vortical effect may also be obtained from a purely field theoretical analysis. It is a consequence of the presence of both an entropy current with non-vanishing divergence indicating dissipation and a $U(1)$ current which has an axial anomaly in the presence of a background electromagnetic field, i.e.

$$\partial_\mu J^\mu = -\frac{C}{8}\epsilon^{\mu\nu\sigma\rho}F_{\mu\nu}F_{\sigma\rho}. \tag{12.159}$$

In the presence of such a background field, we have

$$\partial_\mu T^{\mu\nu} = F^{\nu\lambda}J_\lambda, \qquad \partial^\mu J_\mu = CE^\mu B_\mu, \tag{12.160}$$

with

$$E^\mu = F^{\mu\nu}u_\nu, \qquad B^\mu = \frac{1}{2}\epsilon^{\mu\nu\alpha\beta}u_\nu F_{\alpha\beta} \tag{12.161}$$

the electric and magnetic fields in the fluid rest-frame. Using the equation of state $\epsilon + p = Ts + \mu\rho$, it can be shown that the condition $\partial_\mu s^\mu \geq 0$ for the entropy current necessarily requires a vorticity term in the hydrodynamic expansion of the current. To see this let us recall the derivative expansion to first order for the energy-momentum tensor and a $U(1)$ current, which from (12.84) and (12.83) is given by

$$T^{\mu\nu} = (\epsilon + p)u^\mu u^\nu + pg^{\mu\nu} - \sigma^{\mu\nu}, \tag{12.162}$$

$$J^\mu = \rho u^\mu + \Upsilon^\mu, \tag{12.163}$$

where $\pi^{\mu\nu}$ and Υ^μ represent the first order terms as given by (12.88), (12.89). In the presence of the background field, Υ^μ has an additional contribution involving E^μ, such that we have

$$\Upsilon^\mu = -\kappa T P^{\mu\nu}\partial_\nu\left(\frac{\mu}{T}\right) + \kappa E^\mu. \tag{12.164}$$

In addition, according to (12.79), the entropy current reads to first order

$$s^\mu = su^\mu - \frac{\mu}{T}\Upsilon^\mu. \tag{12.165}$$

Equation (12.160) and $\epsilon + p = Ts + \mu\rho$ then imply

$$\partial_\mu\left(su^\mu - \frac{\mu}{T}v^\mu\right) = \frac{1}{T}\partial_\mu u_\nu\sigma^{\mu\nu} - \Upsilon^\mu\left(\partial_\mu\left(\frac{\mu}{T}\right) - \frac{E_\mu}{T}\right) - C\frac{\mu}{T}E \cdot B. \tag{12.166}$$

When the current is non-anomalous with $C = 0$, then the explicit expressions for $\sigma^{\mu\nu}$ and v^μ, (12.88) and (12.89), imply that the right-hand side of (12.166) is manifestly positive, which corresponds to entropy production. However, when $C \neq 0$ the right-hand side is no longer necessarily positive. This implies that the currents (12.164) and (12.165) have to be modified to

$$\Upsilon'^\mu = \Upsilon^\mu + \xi\omega^\mu + \xi_B B^\mu, \tag{12.167}$$

$$s'^\mu = s^\mu + D\omega^\mu + D_B B^\mu, \tag{12.168}$$

with ω^μ the vorticity term of (12.157) and B^μ the magnetic field (12.161). The coefficients ξ, ξ_B, D and D_B are functions of T and μ. Requiring $\partial_\mu s'^\mu \geq 0$ implies that

$$\xi = C\left(\mu^2 - \frac{2}{3}\frac{\rho\mu^3}{\epsilon+p}\right), \qquad \xi_B = C\left(\mu - \frac{1}{2}\frac{\rho\mu^2}{\epsilon+p}\right). \qquad (12.169)$$

12.5 Further reading

Reviews of gauge/gravity duality and hydrodynamics are found in [9, 10, 11]. The Hamiltonian approach to holographic renormalisation and correlation functions was developed in [2, 3]. The shear viscosity was calculated holographically in [12]. The shear viscosity over entropy ratio and its universality were calculated and discussed in [13, 14]. Contributions of higher order in the inverse coupling were calculated in [15]. Violations of the bound for theories with higher curvature terms were found in [16, 17, 18], violations in $Sp(N)$ theories at order $1/N^2$ in [16] and violations in anisotropic systems to leading order in N and λ in [19]. Note that in the anisotropic systems discussed in [20, 21], η/s is temperature dependent, but satisfies the bound.

Quasinormal modes and linear response are discussed within gauge/gravity duality in [22, 1]. A review of quasinormal modes within gauge/gravity duality and their general relativity origin is provided in [23]. Holographic spectral functions are discussed in [5], also for flavour branes. Holographic hydrodynamics for R-charged black holes is discussed in [24], where, in particular, the relation (12.112) between charge density and chemical potential is obtained.

The fluid/gravity correspondence was introduced in [7]. A complementary way of obtaining second order hydrodynamic coefficients within gauge/gravity duality using Weyl invariance was introduced in [6]. Fluid/gravity duality in general d dimensions is discussed in [8]. Methods for deriving the boundary energy-momentum tensor from the bulk metric are given in [25, 26].

Anomalous hydrodynamics is discussed within gauge/gravity duality in [27, 28], and within field theory in [29]. A cousin of the chiral vortical effect is the chiral magnetic effect, discussed within field theory in [30] and within gauge/gravity duality for instance in [31].

References

[1] Kovtun, Pavel K., and Starinets, Andrei O. 2005. Quasinormal modes and holography. *Phys. Rev.*, **D72**, 086009.

[2] de Boer, Jan, Verlinde, Erik P., and Verlinde, Herman L. 2000. On the holographic renormalization group. *J. High Energy Phys.*, **0008**, 003.

[3] Papadimitriou, Ioannis, and Skenderis, Kostas. 2004. Correlation functions in holographic RG flows. *J. High Energy Phys.*, **0410**, 075.

[4] McGreevy, John. 2010. Holographic duality with a view toward many-body physics. *Adv. High Energy Phys.*, **2010**, 723105.

[5] Myers, Robert C., Starinets, Andrei O., and Thomson, Rowan M. 2007. Holographic spectral functions and diffusion constants for fundamental matter. *J. High Energy Phys.*, **0711**, 091.

[6] Baier, Rudolf, Romatschke, Paul, Son, Dam Thanh, Starinets, Andrei O., and Stephanov, Mikhail A. 2008. Relativistic viscous hydrodynamics, conformal invariance, and holography. *J. High Energy Phys.*, **0804**, 100.

[7] Bhattacharyya, Sayantani, Hubeny, Veronika E., Minwalla, Shiraz, and Rangamani, Mukund. 2008. Nonlinear fluid dynamics from gravity. *J. High Energy Phys.*, **0802**, 045.

[8] Haack, Michael, and Yarom, Amos. 2008. Nonlinear viscous hydrodynamics in various dimensions using AdS/CFT. *J. High Energy Phys.*, **0810**, 063.

[9] Rangamani, Mukund. 2009. Gravity and hydrodynamics: lectures on the fluid-gravity correspondence. *Class.Quantum Grav.*, **26**, 224003.

[10] Son, Dam T., and Starinets, Andrei O. 2007. Viscosity, black holes, and quantum field theory. *Ann. Rev. Nucl. Part. Sci.*, **57**, 95–118.

[11] Kovtun, Pavel. 2012. Lectures on hydrodynamic fluctuations in relativistic theories. *J. Phys.*, **A45**, 473001.

[12] Policastro, G., Son, D. T., and Starinets, A. O. 2001. The shear viscosity of strongly coupled $N = 4$ supersymmetric Yang-Mills plasma. *Phys. Rev. Lett.*, **87**, 081601.

[13] Kovtun, P., Son, D. T., and Starinets, A. O. 2005. Viscosity in strongly interacting quantum field theories from black hole physics. *Phys. Rev. Lett.*, **94**, 111601.

[14] Buchel, Alex, and Liu, James T. 2004. Universality of the shear viscosity in supergravity. *Phys. Rev. Lett.*, **93**, 090602.

[15] Buchel, Alex, Liu, James T., and Starinets, Andrei O. 2005. Coupling constant dependence of the shear viscosity in $N = 4$ supersymmetric Yang-Mills theory. *Nucl. Phys.*, **B707**, 56–68.

[16] Kats, Yevgeny, and Petrov, Pavel. 2009. Effect of curvature squared corrections in AdS on the viscosity of the dual gauge theory. *J. High Energy Phys.*, **0901**, 044.

[17] Brigante, Mauro, Liu, Hong, Myers, Robert C., Shenker, Stephen, and Yaida, Sho. 2008. Viscosity bound violation in higher derivative gravity. *Phys. Rev.*, **D77**, 126006.

[18] Buchel, Alex, Myers, Robert C., and Sinha, Aninda. 2009. Beyond eta/s = 1/4pi. *J. High Energy Phys.*, **0903**, 084.

[19] Rebhan, Anton, and Steineder, Dominik. 2012. Violation of the holographic viscosity bound in a strongly coupled anisotropic plasma. *Phys. Rev. Lett.*, **108**, 021601.

[20] Erdmenger, Johanna, Kerner, Patrick, and Zeller, Hansjorg. 2011. Non-universal shear viscosity from Einstein gravity. *Phys. Lett.*, **B699**, 301–304.

[21] Erdmenger, Johanna, Kerner, Patrick, and Zeller, Hansjorg. 2012. Transport in anisotropic superfluids: a holographic description. *J. High Energy Phys.*, **1201**, 059.

[22] Horowitz, Gary T., and Hubeny, Veronika E. 2000. Quasinormal modes of AdS black holes and the approach to thermal equilibrium. *Phys. Rev.*, **D62**, 024027.

[23] Berti, Emanuele, Cardoso, Vitor, and Starinets, Andrei O. 2009. Quasinormal modes of black holes and black branes. *Class.Quantum Grav.*, **26**, 163001.

[24] Son, Dam T., and Starinets, Andrei O. 2006. Hydrodynamics of R-charged black holes. *J. High Energy Phys.*, **0603**, 052.

[25] Balasubramanian, Vijay, and Kraus, Per. 1999. A stress tensor for anti-de Sitter gravity. *Commun. Math. Phys.*, **208**, 413–428.

[26] de Haro, Sebastian, Solodukhin, Sergey N., and Skenderis, Kostas. 2001. Holographic reconstruction of space-time and renormalization in the AdS/CFT correspondence. *Commun. Math. Phys.*, **217**, 595–622.

[27] Erdmenger, Johanna, Haack, Michael, Kaminski, Matthias, and Yarom, Amos. 2009. Fluid dynamics of R-charged black holes. *J. High Energy Phys.*, **0901**, 055.

[28] Banerjee, Nabamita, Bhattacharya, Jyotirmoy, Bhattacharyya, Sayantani, Dutta, Suvankar, Loganayagam, R., and Surowka, Piotr. 2011. Hydrodynamics from charged black branes. *J. High Energy Phys.*, **1101**, 094.

[29] Son, Dam T., and Surowka, Piotr. 2009. Hydrodynamics with triangle anomalies. *Phys. Rev. Lett.*, **103**, 191601.

[30] Fukushima, Kenji, Kharzeev, Dmitri E., and Warringa, Harmen J. 2008. The chiral magnetic effect. *Phys. Rev.*, **D78**, 074033.

[31] Kalaydzhyan, Tigran, and Kirsch, Ingo. 2011. Fluid/gravity model for the chiral magnetic effect. *Phys. Rev. Lett.*, **106**, 211601.

13 QCD and holography: confinement and chiral symmetry breaking

In most of the examples given in part II of this book, we studied the correspondence between quantum gauge theories and gravity for systems with supersymmetry. These theories provide a large symmetry and the field-operator map is readily established by matching representations. We learned in particular that gauge/gravity duality is an important tool for studying strongly coupled gauge theories to which it is sometimes difficult to apply standard methods. Thus the question arises to what extent gauge/gravity duality may also be applied to non-supersymmetric strongly coupled gauge theories which are realised in nature, such as QCD (Quantum Chromodynamics), the theory of the strong interaction in the standard model of elementary particles. In addition to the absence of supersymmetry, a further central feature of QCD is the presence of quarks, i.e. of flavour degrees of freedom which transform in the fundamental representation of the gauge group. In this chapter we will see that while a gravity dual of the precise form of the QCD Lagrangian and its RG flow is beyond reach, central features of low-energy QCD, such as confinement and chiral symmetry breaking, may be realised within gauge/gravity duality, and masses of bound states such as mesons may be calculated.

13.1 Review of QCD

Quantum Chromodynamics (QCD), the theory of strong interactions in the standard model of elementary particles, may be formulated as a non-Abelian gauge theory with gauge group $SU(3)$. Moreover, there are matter degrees of freedom, referred to as quarks, which come in six *flavours*. The masses of the lighter and heavier quark flavours differ by two orders of magnitude, a fact not explained until the present day since the quark masses enter the standard model as free parameters. At very low energies, essentially only three quark flavours contribute, traditionally denoted as the *up, down* and *strange* quarks.

13.1.1 Non-Abelian gauge theory with fundamental matter

Let us consider non-Abelian gauge theories with N_f additional Dirac fermions ψ^i, described by the Lagrangian

$$\mathcal{L} = -\frac{1}{4} F^a_{\mu\nu} F^{a\mu\nu} + \bar{\psi}_j \left(i\gamma^\mu D_\mu \delta^{jk} - M^{jk} \right) \psi_k. \tag{13.1}$$

$F^a_{\mu\nu}$ transforms in the adjoint representation of the gauge group $SU(N)$, which is referred to as *colour*. The quanta of the gauge field are the *gluons*. The *quark* fields are given by Dirac spinors ψ_j transforming in the fundamental representation of the gauge group. For simplicity, we suppress their spinorial indices. M_{jk} is the quark mass matrix, which may be diagonalised as $M = m_q \mathbb{1}$, where we have chosen all quark masses to be equal. The Dirac spinors also transform in the fundamental representation of a global *flavour symmetry* $U(N_f)$ with $j = 1, \ldots, N_f$. In the quantised theory, if the quark masses are taken to be equal, the one-loop β function for the gauge coupling is given by

$$\beta(g) = -\frac{g^3}{48\pi^2}\left(11N - 2N_f\right). \tag{13.2}$$

For $11N > 2N_f$, the one-loop β function is negative. This implies that there is a UV fixed point for $\mu \to \infty$, with μ the renormalisation scale, at which the theory is *asymptotically free*. Asymptotic freedom is a key feature of non-Abelian gauge theories which distinguishes them from QED, for instance.

On the other hand, for low energies, the coupling in non-Abelian gauge theories becomes large and perturbation theory breaks down. Other approaches to describe the dynamics become necessary. A well-established approach to low-energy non-Abelian gauge theories is *lattice gauge theory*, as described in box 13.1.

Below, we will explore how gauge/gravity duality can be used to study low-energy non-Abelian gauge theories. As we shall see, some of the gauge/gravity duality results may be compared directly to lattice gauge theory results, while there are other examples where gauge/gravity duality is readily used while lattice gauge theory is hard to apply. For instance, this is the case for situations in which it is necessary to consider Minkowski signature. Of course, at the present stage gauge/gravity is not directly applicable to QCD as it appears in the standard model, but only to related gauge theories. However, in many cases a comparison is nevertheless meaningful, as the examples below will show. In particular, the application of gauge/gravity duality in its present form to low-energy non-supersymmetric non-Abelian gauge theories requires the planar large N limit. However, this limit is also established as a useful tool in standard quantum field theoretical approaches to QCD.

Box 13.1   **Lattice gauge theory**

In the lattice gauge theory approach to strongly coupled quantum field theories such as QCD at low energies, a Wick rotation to Euclidean signature is performed. Space and Euclidean time are discretised to a finite set of points in each direction. For a finite number of points, path integrals of fields in Euclidean signature become finite. The factor $\exp(-S)$ in the Euclidean partition function

$$Z = \int \mathcal{D}\phi\, e^{-S} \tag{13.3}$$

is used as a probability distribution, to which a stochastic approach such as the Monte Carlo method is applied.

13.1.2 Confinement

At low energies, non-Abelian gauge theories display the property of *confinement*. Denoting the quantum number associated with the non-Abelian gauge symmetry as *colour*, it is observed that colour charged particles cannot be isolated and that coloured bound states do not exist.

In particular, QCD shows colour confinement. It is observed experimentally that quarks are confined by the strong interaction to form either pairs (mesons, $\bar{\psi}\psi$) or triplets (baryons, $\psi\psi\psi$), which are colourless. So far, an analytical proof of confinement has not been found.

Let us recall the main difference between Abelian and non-Abelian gauge theories from chapter 1: in non-Abelian theories, the gluons carry charge and self-interact, which is not the case for photons in QED with Abelian gauge group $U(1)$.

Intuitively, we may think of the quarks in a bound state as being connected by gluon flux tubes, as shown in figure 13.1. These flux tubes have the properties of strings. This leads to a potential of the form $V(R) \sim \kappa \cdot R$ for a quark and an antiquark separated by a distance R. Once the distance reaches a critical value corresponding to the energy of a quark–antiquark pair, the string breaks. This results in two new strings each connecting a quark–antiquark pair.

Moreover, the flux tube geometry implies that mesons have a characteristic relation between mass and angular momentum,

$$M_J^2 = \frac{1}{\alpha'}J - M_0^2. \tag{13.4}$$

Here, α' is proportional to the inverse string tension of the flux tube. Equation (13.4) is referred to as a *Regge trajectory*. In fact, it is observed experimentally that QCD mesons and baryons are organised in Regge trajectories, with the inverse string tension of the order of $\alpha' \sim (1\,\text{GeV})^{-2}$. For this reason, string theory was originally proposed in the late 1960s as a theory of the strong interaction. This was subsequently abandoned since string theory in four dimensions is non-critical and predicts massless particles which are not observed. Moreover, it includes a graviton in a natural way, which makes string theory a natural candidate for a quantum theory of gravity, as discussed in chapter 4. Of course, modern string theory as discussed in chapter 4 has $\alpha' \sim l_s^2$ with the string length l_s of order of the the Planck length $l_p = \sqrt{\hbar G/c^3} \sim 1.6 \times 10^{-35}$ m.

Exercise 13.1.1 Show that the Regge slope may be obtained from string theory in flat spacetime as introduced in chapter 4. In particular, using the results of section 4.1.2 in

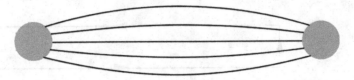

Figure 13.1   Schematic representation of flux tubes between a quark–antiquark pair.

$D = 26$ dimensions, show that for strings, the Regge slope is given by

$$J \leq 1 + \alpha' M^2. \tag{13.5}$$

This calculation requires the use of the string mass M found in section 4.1.2, which is given by $M^2 = (N-1)/\alpha'$ in $D = 26$ dimensions. Moreover, the angular momentum J is given by the eigenvalue of the spin operator $\mathcal{J}^{IJ}$, where for simplicity, we choose a rotation in the x^2, x^3 directions. From the Lorentz generators of the little group, the spin operator is given by

$$\mathcal{J}^{23} = -i \sum_{n=1}^{\infty} \frac{1}{n} (\alpha_{-n}^2 \alpha_n^3 - \alpha_{-n}^3 \alpha_n^2) \tag{13.6}$$

in terms of the string mode operators.

13.1.3 Wilson loops

A criterion for determining whether a gauge theory displays confinement is obtained from the Wilson loop. The Wilson loop was introduced in chapter 1 and discussed in chapter 5 for $\mathcal{N} = 4$ Super Yang–Mills theory. In the present context the Wilson loop is given by

$$\mathcal{W}[\mathcal{C}] = \frac{1}{N} \mathrm{Tr}\, \mathcal{P} \exp \left(ig \int_{\mathcal{C}} \mathrm{d}x^\mu A_\mu^a T_a \right), \tag{13.7}$$

where Tr stands for the trace over the gauge group in the fundamental representation, the generators of the gauge group satisfy $[T_a, T_b] = if_{ab}{}^c T_c$ and $\mathcal{P}$ denotes path ordering. As discussed in chapter 1, the ordering is performed in the following way. We choose a parametrisation for the closed loop $\mathcal{C}$ over which the integration in (13.7) is performed to be of the general form $x^\mu = x^\mu(\tau)$ with parameter τ. Then, any product of gauge fields is ordered such that gauge fields with larger τ appear to the left of those with smaller τ. These products of gauge fields are present in the Wilson loop since the exponential involved can be expanded in a power series at least for small g.

In Euclidean signature, the expectation value of the Wilson loop is then given by

$$\langle \mathcal{W}(\mathcal{C}) \rangle = \int \mathcal{D}A \, \exp\left(-\mathcal{S}_{\mathrm{YM}}[A]\right) \mathcal{W}(\mathcal{C}), \tag{13.8}$$

where $\mathcal{S}_{\mathrm{YM}}$ is the Yang–Mills action. Let us consider a rectangular Wilson loop as shown in figure 13.2, where T is the time direction and R the separation of the quarks.

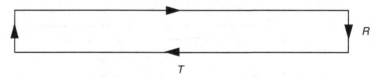

Figure 13.2 Closed contour with $T \gg R$ used to evaluate the rectangular Wilson loop.

The quark–antiquark potential is obtained from the Wilson loop by virtue of (1.214), $\langle \mathcal{W}(\mathcal{C}) \rangle \sim e^{-TV(R)}$. The criterion for confinement is now that this potential scales linearly with the distance between the quarks, i.e.

$$V(R) \sim \kappa \cdot R, \tag{13.9}$$

where κ is the string tension of the flux tube. This corresponds to the Wilson loop being of the form

$$\langle W(\mathcal{C}) \rangle \sim \exp\left(-\kappa \operatorname{Area}(\mathcal{C})\right). \tag{13.10}$$

For confining theories, the Wilson loop thus follows an area law.

Let us consider some examples. As obtained in exercise 1.7.4, for an Abelian gauge theory such as quantum electrodynamics (QED), the path integral is Gaussian and no path ordering is needed. The perturbative calculation of exercise 1.7.4 yields

$$V_{\text{QED}}(R) = \frac{e^2}{4\pi R} + \text{self-energy}, \tag{13.11}$$

which corresponds to the standard Coulomb potential for the electron charge e, with an additional divergent term reflecting the divergent self-energy of pointlike electric charges. Performing a similar perturbative calculation for QCD, we obtain

$$V_{\text{perturbative QCD}}(R) = \frac{g_{\text{YM}}^2}{3\pi R} + \text{self-energy}. \tag{13.12}$$

In this case we again obtain a Coulomb behaviour and no confinement. Since confinement is observed in QCD, this tells us that the perturbative expansion of QCD breaks down at low energies where confinement occurs, in agreement with the β function of QCD being negative.

Below in section 13.2, we will consider Wilson loops within gauge/gravity duality. We will discuss how to calculate Wilson loops holographically, and how to define gravity backgrounds which display confinement. First, however, we turn to a further key property of low-energy QCD, chiral symmetry breaking.

13.1.4 Chiral symmetry breaking

Chiral symmetry and its spontaneous breaking belong to the key features of low-energy QCD. This is the key mechanism for the generation of light particle masses. For instance, the pions are pseudo-Goldstone bosons of the spontaneously broken chiral symmetry.

Chiral symmetry is a feature of the Lagrangian of massless QCD, given by

$$\mathcal{L}_{\text{QCD}}|_{m=0} = -\frac{1}{4} F^a_{\mu\nu} F^{a\mu\nu} + i\bar{\psi}_L \slashed{D} \psi_L + i\bar{\psi}_R \slashed{D} \psi_R. \tag{13.13}$$

ψ_L and ψ_R are the chiral projections of the Dirac spinors ψ of (13.1), which in the massless case we can rewrite as

$$\psi = \begin{pmatrix} \psi_L \\ \psi_R \end{pmatrix}. \tag{13.14}$$

In the massless case, the left-handed and right-handed fields have separate invariances under flavour symmetry. For the case of three flavours u, d, s, i.e. $N_f = 3$, we have

$$\psi_L \mapsto \exp(-i\theta_L \cdot \lambda)\psi_L, \qquad \psi_R \mapsto \exp(-i\theta_R \cdot \lambda)\psi_R, \tag{13.15}$$

where we have used $T_a = \lambda_a$, $a = 1, \ldots, 8$. The λ^a are the $SU(3)$ Gell-Mann matrices. These transformations can also be expressed as vector and axial-vector transformations,

$$\psi \mapsto \exp(-i\theta_V \cdot \lambda)\psi, \qquad \psi \mapsto \exp(-i\theta_A \cdot \lambda\gamma_5)\psi, \tag{13.16}$$

with $\theta_V = (\theta_L + \theta_R)/2$, $\theta_A = (\theta_L - \theta_R)/2$. The Lagrangian (13.13) is thus invariant under $SU(3)_L \times SU(3)_R$ or $SU(3)_V \times SU(3)_A$.

Given the symmetry transformations of the spinors as described above, we might have expected a $U(3)_V \times U(3)_A$ global symmetry, equivalent to $SU(3)_V \times SU(3)_A \times U(1)_V \times U(1)_A$. However, it turns out that in QCD, the $U(1)_A$ symmetry is anomalous, and thus not present in the quantised theory. The divergence of the associated axial current receives non-trivial quantum contributions through the triangle quark loop graph, $\langle \partial_\mu J_5^\mu \rangle \neq 0$. Let us consider this for general N and N_f. The only exception to the anomalous $U(1)_A$ symmetry breaking arises when $N_f \ll N$. In this case, the triangle graph gives rise to

$$\langle \partial_\mu J_5^\mu \rangle = \frac{1}{16\pi^2} \frac{N_f}{N} \tilde{F}F. \tag{13.17}$$

The triangle graph becomes suppressed in the $1/N$ expansion, and the coefficient in (13.17) vanishes in the limit $N \to \infty$ for fixed N_f. The $U(1)_A$ symmetry is thus not anomalous at large N, provided that N_f is finite. On the other hand, the vector $U(1)_V$ symmetry which corresponds to baryon number conservation is non-anomalous and is preserved for any N and N_f.

This chiral symmetry may be broken *explicitly* if a mass term is present in the Lagrangian,

$$\mathcal{L}_{m_q} = -m_q \bar{\psi}\psi, \tag{13.18}$$

with ψ a Dirac spinor.

On the other hand, there is also a *spontaneous* breaking of chiral symmetry in QCD. The dynamics of the strong force generates a vacuum expectation value for the operator

$$\langle \bar{\psi}\psi \rangle = \langle \bar{\psi}_L \psi_R \rangle + \text{h.c.} \neq 0. \tag{13.19}$$

In both symmetry breaking cases, the flavour symmetry is broken down to a single vector $SU(3)_V$ factor,

$$SU(3) \times SU(3) \to SU(3)_V. \tag{13.20}$$

Goldstone's theorem ensures that for the case of spontaneous symmetry breaking, eight massless Goldstone bosons are expected, one for each generator for which the associated symmetry is broken. In QCD these correspond to quark bound states, the pions and kaons $\pi^\pm, \pi^0, K^\pm, K^0, \bar{K}^0$ and the eta-particle η. In the large N limit where the $U(1)_A$ symmetry is restored, the eta-prime particle η' joins these particles as a Goldstone boson.

A low-energy effective action for the Goldstone modes, which are lighter than all other QCD bound states, can be obtained [1] for instance by writing the Goldstone fields, π^a, as part of a field

$$U = e^{i\pi^a(x)T_a/f_\pi}, \qquad (13.21)$$

where f_π is the pion decay constant. U transforms under the underlying chiral symmetries as $L^\dagger U R$ and its vacuum expectation value, which involves the 3×3 unit matrix, breaks this symmetry to the diagonal. The effective Lagrangian, known as *chiral perturbation theory*, can be constructed as a derivative expansion with leading term

$$\mathcal{L} = -\frac{1}{2}\partial^\mu \pi^a \partial_\mu \pi^a + \cdots = -f_\pi^2 \operatorname{Tr} \partial^\mu U^\dagger \partial_\mu U. \qquad (13.22)$$

If a small explicit symmetry breaking by a quark mass term is present, the Goldstone bosons acquire mass to become pseudo-Goldstone bosons. Since the 3×3 mass matrix M with diagonal entries m_q transforms under the (now spurious) chiral symmetries as $L^\dagger m_q R$, we may add a term of the form

$$\Delta\mathcal{L} = v^3 \operatorname{Tr}\left(M^\dagger U^\dagger + M U\right) \qquad (13.23)$$

to the low-energy action, where v^3 is a coefficient of dimension three that measures the size of the quark condensate and must be fitted phenomenologically. This term generates a mass for the Goldstone bosons with $M_\pi^2 \sim m_q$.

We will see below how this symmetry breaking is realised in gravity duals. In the first examples, we will make use of the large N limit of the AdS/CFT correspondence and realise the breaking of the $U(1)_A$ symmetry, under which ψ_L and ψ_R transform as

$$\psi_L \mapsto e^{i\alpha}\psi_L, \qquad \psi_R \mapsto e^{-i\alpha}\psi_R. \qquad (13.24)$$

The associated Goldstone boson has the quantum number of the η' particle, although in some respects its behaviour is similar to the pions. We will also describe a model that can realise the full non-Abelian chiral symmetry breaking pattern as seen in QCD.

13.2 Gauge/gravity duality description of confinement

The general idea for describing confinement in gauge/gravity duality is to consider a domain wall ansatz similar to those discussed in chapter 9 when describing holographic RG flows. While the RG flows of chpater 9 asymptote to an IR Anti-de Sitter space dual to an IR RG fixed point, a way to obtain confinement is to consider domain wall solutions in which $A(r)$ as in (9.31) diverges at a finite $r = r_0$. The fact that such a background is dual to a confining theory is established by calculating the Wilson loop holographically and demonstrating that it has an area law.

13.2.1 Wilson Loops and their dual description

The holographic calculation of Wilson loops was introduced in section 5.6 for $\mathcal{N} = 4$ Super Yang–Mills theory and supergravity in $AdS_5 \times S^5$. The field theory Wilson loop corresponds to the minimal surface in Anti-de Sitter space which has the field theory loop as its boundary [2]. This calculation can be extended to more general gravity backgrounds [3].

Let us briefly recapitulate the main result of section 5.6 for the holographic Wilson loop written in a form suitable for the present applied context. Here, the starting point is the metric of AdS_5 written in the form

$$ds^2 = \frac{L^2}{z^2}\left(\eta_{\mu\nu}dx^\mu dx^\nu + dz^2\right). \tag{13.25}$$

The expectation value of the Wilson loop $\mathcal{C}$ is obtained holographically by calculating the minimal surface which ends on the loop $\mathcal{C}$ and determining its area. This is given by the Nambu–Goto action for a string sweeping out a two-dimensional surface in AdS_5,

$$S_{\text{NG}} = \frac{1}{2\pi\alpha'}\int d\tau d\sigma \sqrt{-\det_{\alpha,\beta}(g_{mn}\partial_\alpha X^m \partial_\beta X^n)}, \tag{13.26}$$

where $X^m(\tau,\sigma)$ are the embeddings of the string worldvolume into AdS_5. The expectation value of the Wilson loop is then given by

$$\langle \mathcal{W}(\mathcal{C})\rangle = \exp(-S_{\text{NG,min}} - S_{\text{ct}}), \tag{13.27}$$

where S_{ct} denotes the counterterms necessary for regularisation. The divergent counterterms in the Nambu–Goto action correspond to the self-energy of pointlike charges in the field theory.

Naively we might expect that the area law for the Wilson loop, which signals confinement, is built in automatically into the holographic calculation outlined above since the Nambu–Goto action measures the area of a surface. However, this is not the case as may be seen by calculating the holographic Wilson loop for $\mathcal{N} = 4$ Super Yang–Mills theory. In this case we obtain

$$\langle \mathcal{W}(\mathcal{C})\rangle = \exp\left(C\sqrt{\lambda}\frac{T}{R}\right), \tag{13.28}$$

where T is the timelike extension of the Wilson loop as shown in figure 13.2, with the constant C given by $C = 4\pi^2\sqrt{2}/(\Gamma(\frac{1}{4}))^2$, as calculated in (5.144). For the quark–antiquark potential, (13.28) implies

$$V(R) = -C\sqrt{\lambda}\frac{1}{R}. \tag{13.29}$$

This $1/R$ behaviour may also be derived from dimensional analysis. Since a conformal field theory does not have any length scales, the only dimensionful scale is R and the potential has to be proportional to $1/R$, i.e. of Coulomb form. Consequently, a conformal field theory such as $\mathcal{N} = 4$ Super Yang–Mills theory is not confining.

In the next section, we investigate an example where the Wilson loop shows area law behaviour as expected for a confining theory.

13.2.2 Confinement in gauge/gravity duals

Geometrically, the gravity dual of a flux tube connecting a quark and an antiquark corresponds to a string dipping into Anti-de Sitter space. The more the quarks are separated, the further the string reaches into the interior of the AdS space. This is in analogy to the gravity dual of the Wilson loop. It is energetically favourable for the string to dip into AdS space. This is seen from the metric of AdS space written in the form (13.25). Strings with their global turning point deep in the interior have a smaller worldsheet area due to the $1/z^2$ factor in the metric (13.25). This implies that the quark–antiquark potential has the conformal behaviour $V(R) \sim 1/R$.

Confinement, i.e. linear behaviour of the quark–antiquark potential $V(R) \sim \kappa R$, requires an alteration of the AdS metric. In fact, confinement may be achieved by placing some obstruction in the interior of AdS space. This may be a hard wall, a gravitational potential, a brane, or similar. In this case, the string connecting the quark–antiquark pair will behave as in the AdS case above for small separations. For large separations, however, when the string reaches further into the interior of the modified AdS space such as to reach the obstruction, it extends along this barrier in a direction orthogonal to the radial direction. Therefore in this case, the energy scales with the quark separation, as is expected for a confining theory. The difference between conformal behaviour in AdS space and confinement in the obstructed geometry is displayed in figure 13.3.

In the subsequent sections, we discuss examples for realising such barriers in the dual geometry.

Renormalisation group flows to confining theories

One of the first models of this type was proposed by Girardello, Petrini, Porrati and Zaffaroni [4, 5] and is referred to as the GPPZ flow. These authors considered a holographic RG flow which does not run to an IR conformal fixed point, but rather has a singularity in both the warp factor $A(r)$ and the dilaton $\phi(r)$ at a finite value of the radial coordinate,

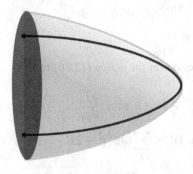
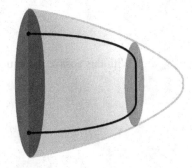

Conformal AdS   Infrared wall

Figure 13.3 A string dipping into AdS space (left) and into the hard wall geometry (right). The hard wall geometry leads to a potential $V(r) \sim \kappa \cdot r$ for the dual quark–antiquark pair located at the boundary.

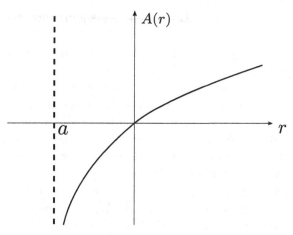

Figure 13.4 Typical behaviour of the warp factor $A(r)$ in a confining geometry. $A(r)$ is linear in the UV where the metric asymptotes to the AdS metric. On the other hand, $A(r)$ diverges in the IR at a finite $r = a$.

$r = a$. The behaviour of $A(r)$ is displayed in figure 13.4. We first discuss the general features of this flow. Then we move on to calculating holographically the Wilson loop for this geometry. It displays an area law, which is a criterion for confinement, as discussed in section 13.1.3 above.

The starting point of this class of models is, similarly to the interpolating RG flows, a domain wall ansatz for solving the equations of motion of $\mathcal{N} = 8$ gauged supergravity in five dimensions with gauge group $SU(4)$. As before in section 9.2.2, the action reads, for several scalars ϕ^I,

$$S = \int \mathrm{d}^5x\sqrt{-g}\left(\frac{1}{16\pi G}R - \frac{1}{2}\partial^m\phi^I\partial_m\phi_I - V(\phi^I)\right), \tag{13.30}$$

where compared to (9.44) we have set the metric on the space of scalars to $\mathcal{G}^{IJ} = \delta^{IJ}$. For the domain wall ansatz we take

$$\begin{aligned}
\mathrm{d}s^2 &= e^{2A(r)}\eta_{\mu\nu}dx^\mu dx^\nu + \mathrm{d}r^2, \\
\phi^I &= \phi^I(r).
\end{aligned} \tag{13.31}$$

With this ansatz, the equations of motion for the action (13.30) read

$$\partial_r{}^2\phi^I + 4\partial_rA\,\partial_r\phi^I = \frac{\partial V}{\partial\phi_I}, \tag{13.32}$$

$$\frac{3}{2\pi G}(\partial_rA)^2 = \sum_I(\partial\phi^I)^2 - 2V. \tag{13.33}$$

The key point in view of confinement is that there are solutions which have a singularity at a finite value of the radial coordinate, $r = a$. As an ansatz for such a solution, consider a potential $V(\phi^I)$ such that there are a dilaton and a warp factor with logarithmic divergences, such that in a vicinity of the singularity at $r = a$ we have

$$\phi^I = -\kappa^I \ln |r - a| + \text{constant}, \tag{13.34}$$

$$A(r) = \frac{1}{4} \ln |r - a| + \text{constant}, \tag{13.35}$$

with constants κ^I.

Near the singularity, the potential and its derivative are small compared to the other terms present in the equations of motion (13.32), (13.33), and may be ignored. This may be achieved for instance if the potential is a polynomial. Then inserting (13.34) and (13.35) into (13.32), (13.33) we obtain the condition

$$\sum_I (\kappa^I)^2 = \frac{1}{8\pi G} \frac{3}{4}. \tag{13.36}$$

For other potentials which are not small compared to the derivative terms, other model-dependent (in)equalities have to hold for the κ^I. For the dilaton, κ in (13.34) must be negative, such that the string coupling remains small.

Using (13.31) and (13.35), we find that near the singularity, the metric is independent of the κ^I and takes the form

$$ds^2 = dr^2 + |r - a|^{1/2} \eta_{\mu\nu} dx^\mu dx^\nu \quad \text{for } r \sim a. \tag{13.37}$$

For simplicity, we set $a = 0$ from now on. In the UV, near the boundary at $r \to \infty$, we know that $A(r) = r/L$ and the metric asymptotes to AdS_5 again.

The singularity acts as an IR boundary inside the deformed AdS space. On the field theory side, the presence of this boundary corresponds to confinement, as we now show by calculating the Wilson loop. For this, we embed a string into the deformed AdS space. The spacetime coordinates x^μ, r and worldsheet coordinates τ, σ are identified as follows: $x^0 = \tau$, $x^1 = x = \sigma$, $x^2 = x^3 = 0$. The embedding into the background (13.31) is specified by a non-trivial function $r = r(x)$ with boundary condition $r(0) = r(R) \to \infty$. The Nambu–Goto action for the string is then given by

$$\mathcal{S}_{\text{NG}} = \int_0^R d\sigma \int_0^T d\tau \, \tau_{\text{F1}}(r) \sqrt{-\det P[g]}, \tag{13.38}$$

where

$$\det P[g] = \det_{\alpha, \beta}(g_{mn} \partial_\alpha X^m \partial_\beta X^n)$$
$$= \det \begin{pmatrix} -e^{2A(r)} & 0 \\ 0 & (\partial_x r)^2 + e^{2A(r)} \end{pmatrix}. \tag{13.39}$$

$\tau_{\text{F1}}(r)$ is the effective tension of the fundamental string in five dimensions, which is obtained from a dimensional reduction from ten to five dimensions. We will consider an explicit example below.

Introducing a new radial variable u and the function $f(u)$ by

$$\frac{\partial u}{\partial r} = \tau_{\text{F1}}(r) e^{A(r)}, \qquad f(u) = \tau_{\text{F1}}^2(u) e^{4A(u)}, \tag{13.40}$$

we obtain

$$S_{NG} = T \cdot \int_0^R dx \sqrt{(\partial_x u)^2 + f(u)} \tag{13.41}$$

for the Nambu–Goto action and, using (13.27) with $\langle \mathcal{W}(\mathcal{C}) \rangle \sim e^{-TV(R)}$,

$$V(R) = \int_0^R dx \sqrt{(\partial_x u)^2 + f(u)} \tag{13.42}$$

for the quark–antiquark potential.

For confinement where $V(R) \propto R$, the function $f(u)$ has to have a finite minimum. This may occur in two cases: (1) $f(u)$ diverges both in the UV and in the IR, i.e. at the boundary where $u \to \infty$ and near the singularity at $u = 0$; (2) $f(0) \neq 0$ is finite, $f(u)$ increases monotonically and diverges for $u \to \infty$.

Using these Wilson loop methods, let us now consider an example of a supergravity background of the form (13.30) where confinement occurs. To determine the necessary non-trivial scalars in (13.30), let us first consider the relevant operators to be switched on on the field theory side. Within field theory, the starting point is $\mathcal{N} = 4$ Super Yang–Mills theory in the UV, which is perturbed by three mass terms for the three $\mathcal{N} = 1$ chiral multiplets within $\mathcal{N} = 4$ theory, of the form

$$W_M = \int d^2\theta \, m_{ij} \text{Tr} \Phi_i \Phi_j \tag{13.43}$$

in $\mathcal{N} = 1$ superspace, with $i, j \in \{1, 2, 3\}$. This corresponds to a relevant deformation which flows to pure $\mathcal{N} = 1$ Super–Yang Mills theory in the IR, involving just the $\mathcal{N} = 1$ vector supermultiplet V. This theory is confining. The supersymmetric mass term m_{ij} transforms in the representation **6** of $SU(3)$. The corresponding supergravity mode appears in the decomposition $\mathbf{10} \to \mathbf{1+6+3}$ of $SU(4)$ under $SU(3)$. If the matrix is taken to be a multiple of the identity matrix, $m_{ij} = m\mathbb{1}_{ij}$, then group theory implies that the mass operator can be chosen consistently to be the only operator perturbing the $\mathcal{N} = 4$ theory. This case corresponds to giving equal masses to all three chiral multiplets.

The gravity dual of this flow, given by an action of type (13.30), involves the dilaton ϕ as well as the scalar dual to the mass deformation. We denote this scalar by $\tilde{m}$. This scalar is dimensionless if written in units of $\sqrt{8\pi G}$. A careful analysis of the five-dimensional gauged supergravity theory as outlined in chapter 9, with the corresponding deformation in the **6** of $SU(3)$, shows that this scalar has a potential of the form

$$V(\tilde{m}) = -\frac{1}{8\pi G \cdot L^2} \frac{3}{4} \left(\cosh^2(2\tilde{m}) + 7 \right). \tag{13.44}$$

For $\tilde{m} \ll 1$, $8\pi GV(\tilde{m})$ asymptotes to $-6/L^2$, the value of the cosmological constant of the $SU(4)$ symmetric AdS vacuum. For $\tilde{m} \gg 1$, it takes the form $V(\tilde{m}) \sim \exp(4\tilde{m})$.

The solution of the equations of motion for the Lagrangian (13.30) with the potential (13.44) shows a singular behaviour of the type (13.35). Denoting by κ and κ_0 the constants associated with $\tilde{m}$ and the dilaton respectively, defined in (13.34), the condition (13.36) becomes

$$\kappa^2 + \kappa_0^2 = \frac{1}{8\pi G}\frac{3}{4}. \tag{13.45}$$

We note that the asymptotic behaviour of the potential $V(\tilde{m}) \sim \exp{(4\tilde{m})}$ ensures that near the singularity, its contribution to the equations of motion is negligible.

We proceed by calculating the Wilson loop for this model. This requires the five-dimensional fundamental string tension of (13.40). In the present case, the reduction from ten to five dimensions gives [6]

$$\tau_{\mathrm{F}1}(r) \simeq r^{-\kappa-\kappa_0}, \tag{13.46}$$

where we have set $8\pi G = 1$ and $2\pi\alpha' = 1$ for simplicity. Then, $f(u)$ is given by

$$f(u) \sim u^4 \qquad\qquad \text{for } u \text{ large (UV)}, \tag{13.47}$$

$$f(u) \sim u^{\frac{1-2(\kappa+\kappa_0)}{5/4-(\kappa+\kappa_0)}} \qquad\qquad \text{for } u \to 0 \quad \text{(IR)}. \tag{13.48}$$

In the IR, $f(u)$ diverges to $+\infty$ for all allowed values of κ and κ_0. Since it is a continuous function, the combination of IR and UV behaviour implies that $f(u)$ must have a minimum at a value $u = u^*$. The Wilson loop potential may then be approximated by

$$V(R) = \int\limits_0^R \mathrm{d}x \sqrt{(\partial_x u)^2 + f(u)} \approx \sqrt{f(u^*)}R. \tag{13.49}$$

This displays the linear behaviour required for confinement. This result is almost independent of the UV and IR behaviour of the solution, except for the fact that $f(u)$ must diverge for both $u \to \infty$ and $u \to 0$ in order to ensure the minimum at $u = u^*$. Moreover, the minimum must occur for a finite $f(u^*)$.

Dilaton flows

For studying gauge/gravity duals of chiral symmetry breaking by a quark condensate, which we will discuss in the sections below, it is essential to consider confining gravity models in which supersymmetry is completely broken. This is due to the fact that a quark condensate breaks supersymmetry and thus cannot be present in a supersymmetric gravity background. Moreover, for embedding probe branes we wish to consider full ten-dimensional gravity backgrounds rather than five-dimensional reductions such as the GPPZ flow.

A class of models which lead to non-supersymmetric confining theories in the IR are the *dilaton flows*. In these flows, the dilaton has a non-trivial profile in the radial coordinate of the deformed AdS space.

An example of a dilaton flow is the *Constable–Myers* flow [7]. In a coordinate system which is convenient for embedding probe branes, as we will do below, this gravity solution is given by

$$ds^2 = H^{-1/2}(w) \left(\frac{w^4 + b^4}{w^4 - b^4} \right)^{\delta/4} \eta_{\mu\nu} dx^\mu dx^\nu$$

$$+ H^{1/2}(w) \left(\frac{w^4 + b^4}{w^4 - b^4} \right)^{(2-\delta)/4} \frac{w^4 - b^4}{w^4} \sum_{i=1}^{6} dw_i^2, \qquad (13.50)$$

where b is the scale of the geometry that determines the size of the deformation. Equation (13.50) is given in Einstein frame. We use the coordinates introduced in chapter 10 for D7-brane probe embeddings, with $w^2 = \sum_{i=1}^{6} w_i^2$. The parameter δ is given by $\delta = L^4/(2b^4)$ with L the AdS radius. Moreover,

$$H(w) = \left(\frac{w^4 + b^4}{w^4 - b^4} \right)^\delta - 1 \qquad (13.51)$$

and the dilaton and four-form are given by

$$e^{2\phi} = e^{2\phi_0} \left(\frac{w^4 + b^4}{w^4 - b^4} \right)^\Delta, \quad C_{(4)} = -\frac{1}{4} H^{-1} dt \wedge dx \wedge dy \wedge dz, \qquad (13.52)$$

with Δ an additional parameter satisfying $\Delta^2 + \delta^2 = 10$, and $e^{\phi_0} = g_s$. By expanding this geometry for large w, and performing a rescaling, it may be seen that it asymptotes to $AdS_5 \times S^5$ near the boundary.

The field theory dual is therefore $\mathcal{N} = 4$ Super Yang–Mills theory in the far UV. In the IR, the deformation parameter b sets the scale for conformal symmetry breaking. It determines the scale similar to Λ_{QCD} in the gauge theory,

$$\Lambda_b = \frac{b}{2\pi\alpha'}. \qquad (13.53)$$

This may be seen as follows. The running of the dilaton in the gravity theory corresponds to a running of the coupling in the gauge theory. Indeed the dilaton and the geometry diverge at the scale Λ_b, consistent with the interpretation of this scale as Λ_{QCD}.

Let us discuss the field theory implications of introducing b^4 in (13.50). The $SO(6)$ symmetry of the geometry is unbroken, so the equivalent deformation in the gauge theory does not break the R-symmetry. By dimensional analysis, the deformation by b^4 in (13.50) corresponds to an operator of dimension four. There is a natural dimension four R-chargeless operator in the field theory, which is $\text{Tr}(F^2)$. The Constable–Myers geometry thus describes $\mathcal{N} = 4$ Super Yang–Mills theory plus an additional vacuum expectation value for $\text{Tr}(F^2)$ which corresponds to a non-supersymmetric vacuum. Since $\text{Tr}(F^2)$ is the F-term of a composite operator involving the product of two chiral superfields, $\text{Tr}(W^\alpha W_\alpha)$, hence a vacuum expectation value for the operator breaks supersymmetry. Note that $\text{Tr}(F^2)$ is not a modulus of $\mathcal{N} = 4$ theory; there is a potential for $\text{Tr}(F^2)$ which has a minimum when $\text{Tr}(F^2)$ vanishes. Therefore choosing $\text{Tr}(F^2)$ to be non-zero gives rise to an instability. Nevertheless, this remains a simple model for broken supersymmetry from which useful information may be extracted, as we discuss in detail below in section 13.3.1.

On the gravity side, singularities such as those appearing in (13.50) are usually equipped with an additional source, for instance with additional D3-branes, which shield the singularity to ensure a regular background. In the non-supersymmetric case considered here, the identification of branes shielding the singularity is less clear. However, the Wilson loop in this geometry has an area law and the field theory is therefore confining. We will therefore assume that a shielding mechanism exists, which is also expected to take care of the field theory instability mentioned above, and consider the Constable–Myers background as a simple model for a confining $SU(N)$ theory at large N.

13.3 Chiral symmetry breaking from D7-brane probes

In order to obtain chiral symmetry breaking by a quark condensate in a fashion similar to QCD, it is necessary to break supersymmetry completely. This is due to the fact that the operator $\tilde{\psi}\psi$ is the F-term of a composite chiral superfield $\tilde{Q}Q$. Since for a supersymmetric vacuum state, F-terms have to be absent, a non-vanishing vacuum expectation value $\langle\tilde{\psi}\psi\rangle$ will therefore break supersymmetry.

To study quarks in a confining gauge theory holographically, in order to model some of the essential features of QCD, an appropriate approach is therefore to embed probe branes, as introduced in chapter 10, into a non-supersymmetric confining background. As we will see below, this indeed leads to a geometrical picture of chiral symmetry breaking. Moreover, considering the fluctuations of these probe branes, as studied for the supersymmetric case in chapter 10, allows us to identify Goldstone bosons associated with a spontaneously broken global flavour symmetry.

13.3.1 D7-brane probes in non-supersymmetric dilaton background

The simplest approach to holographic chiral symmetry breaking is to consider the embedding of a D7-brane probe into a confining ten-dimensional non-supersymmetric gravity background. As an example, let us consider embedding a D7-brane probe into the Constable–Myers background. The D7-brane is embedded in analogy to the discussion in section 10.2.2, as shown in table 13.1. We use again the static gauge with worldvolume coordinates identified with $x_{0,1,2,3}$ and $w_{1,2,3,4}$. Transverse fluctuations are parametrised by w_5 and w_6. Moreover, as in (10.11) it is convenient to define a coordinate $\rho = \sum_{i=1}^{4} w_i^2$, such that $w^2 = \rho^2 + w_5{}^2 + w_6{}^2$.

As in chapter 10, the brane probe introduces $\mathcal{N} = 2$ quark hypermultiplets with the usual superpotential coupling to the $\mathcal{N} = 4$ fields given by $\tilde{Q}\Phi Q$. There is a $U(1)_R$ symmetry under which Q and $\tilde{Q}$ both have charge -1 and Φ has charge $+2$. For the component

Table 13.1 D7-brane probe embedded in the Constable–Myers background

Dimensions	0	1	2	3	4	5	6	7	8	9
Coordinates D7	x_0 •	x_1 •	x_2 •	x_3 •	w_1 •	w_2 •	w_3 •	w_4 •	w_5 –	w_6 –

fields, the quantum numbers are given in table 10.2. These imply that a condensate involving $\langle \tilde{\psi}\psi \rangle$ will break the $U(1)_R$ symmetry. This symmetry breaking is analogous to the $U(1)_A$ symmetry breaking introduced in section 13.1.4 for large N QCD, for which the condensate is $\langle \bar{\psi}\psi \rangle$. We will therefore take the spontaneous breaking of $U(1)_R$ by $\langle \tilde{\psi}\psi \rangle$ as a model for the spontaneous breaking of $U(1)_A$ symmetry by $\langle \bar{\psi}\psi \rangle$. Geometrically, the $U(1)_R$ symmetry corresponds to rotations in the $w_5 - w_6$ plane. With the supersymmetry breaking induced by b^4 in the geometry (13.50), the scalar quarks in the hypermultiplets are expected to become massive due to quantum corrections, so that we may expect that there are no more contributions from scalars to the condensate involving $\tilde{\psi}\psi$, unlike in the supersymmetric case (10.19).

The Constable–Myers background is convenient for embedding a D7-brane probe since it preserves $SO(6)$ symmetry, which also implies that the $U(1)_R$ symmetry discussed above is present before embedding the D7-brane probe, even if supersymmetry is broken. The embedding functions determining the minimum energy configuration of the D7-brane probe are functions of ρ only, i.e. essentially of the energy scale. It will turn out that the D7-brane probes giving rise to chiral symmetry breaking are embedded in a regular way, avoiding the region of large curvature near the naked singularity at b. The flavour physics is therefore unaffected by the instability mentioned above below (13.53).

To embed the D7-brane probe, we follow the procedure introduced in section 10.1.2 and consider the Dirac–Born–Infeld action for the D7-brane probe. In the configuration considered here, the only non-trivial contributions arise from the DBI action given by

$$S_{D7} = -\tau_7 \int d^8\xi \, e^{-\phi} \sqrt{-\det(P[g]_{ab} + 2\pi\alpha' F_{ab})}. \tag{13.54}$$

Below we first consider solutions to the equations of motion for which the field strength tensor vanishes, $F_{ab} = 0$.

Converting the Constable–Myers background (13.50) to string frame and inserting it into (13.54), we obtain

$$S_{D7} = -\frac{2\lambda N}{(2\pi)^4} \int d\rho \, e^{\phi} \mathcal{G}(\rho, w_5, w_6) \Big(1 + g^{ab}g_{55}\partial_a w_5 \partial_b w_5 + g^{ab}g_{66}\partial_a w_6 \partial_b w_6 \Big)^{1/2}, \tag{13.55}$$

in the coordinates of table 13.1 and with $\mathcal{G}(\rho, w_5, w_6) = \sqrt{-\det g_{ab}}$ given by

$$\mathcal{G}(\rho, w_5, w_6) = \rho^3 \frac{((\rho^2 + w_5^2 + w_6^2)^2 + b^4)((\rho^2 + w_5^2 + w_6^2)^2 - b^4)}{(\rho^2 + w_5^2 + w_6^2)^4},$$

where the determinant is taken with respect to the world volume coordinates. In (13.55), we converted the prefactor to field theory quantities as in (11.122). This implies that in the usual 't Hooft limit $N \to \infty$ with $g_{YM}^2 N$ fixed, the flavour contribution to the free energy grows as N as expected.

From this action we derive the corresponding equation of motion. We look for classical solutions of the form $w_6 = w_6(\rho)$, $w_5 = 0$. The equation of motion reads

$$\frac{d}{d\rho} \left(\frac{e^{\phi}\mathcal{G}(\rho, w_6)}{\sqrt{1 + (\partial_\rho w_6)^2}} (\partial_\rho w_6) \right) - \sqrt{1 + (\partial_\rho w_6)^2} \frac{d}{dw_6} \left(e^{\phi}\mathcal{G}(\rho, w_6) \right) = 0. \tag{13.56}$$

The last term in the above is a potential-like term which is evaluated as

$$\frac{d}{dw_6}\left(e^\phi \mathcal{G}(\rho, w_6)\right) = \frac{4b^4 \rho^3 w_6}{(\rho^2 + w_6^2)^5}\left(\frac{(\rho^2 + w_6^2)^2 + b^4}{(\rho^2 + w_6^2)^2 - b^4}\right)^{\Delta/2}(2b^4 - \Delta(\rho^2 + w_6^2)^2). \quad (13.57)$$

Equation (13.56) has to be solved numerically. We find solutions with the asymptotic behaviour

$$w_6 \sim l_q + c/\rho^2. \quad (13.58)$$

The identification of these constants with field theory operators requires a coordinate transformation, since the scalar kinetic term is not of the usual canonical AdS form. According to the analysis of section 10.2, the leading term l_q is related to the quark mass m_q and the subleading term c to the quark condensate $\langle \tilde{\psi}\psi \rangle$. In this section, we use the dimensionless mass parameter $m = l_q/b$.

Due to the singularity in the background, we have to impose a regularity constraint on the brane embedding, which amounts to a boundary condition for the equation of motion determining the embedding. Brane embeddings reaching the singularity are excluded since they enter a region of strong curvature where the supergravity approximation is no longer valid. In addition, embeddings which intersect the circles of constant energy twice cannot be interpreted as RG flows and thus are unphysical. A boundary condition which selects the physical embeddings requires that the first derivative of the embedding functions vanishes at $\rho = 0$. In the picture of the RG flow induced by a finite quark mass as discussed in chapter 10, this amounts to requiring S^3 to shrink to zero at this point.

We now calculate the embedding functions for the D7-brane probe by solving the equations of motion obtained from the DBI action (13.55). The numerical result is displayed in figure 13.5. For each of these embeddings we fix two boundary conditions, as required for solving a second order differential equation. First, for regularity we require the first derivative of the embedding to vanish at $\rho = 0$. Secondly, the absolute value of the embedding function w at the boundary $\rho \to \infty$ fixes the value of the quark mass in units of the scale b. The condensate $c \sim \langle \tilde{\psi}\psi \rangle$ in units of b may then be read off from the asymptotic behaviour of the embedding at $\rho \to \infty$, where the embedding takes the asymptotic form (13.58).

We see an interesting screening effect in figure 13.5. The regular solutions are repelled by the singularity, such that they remain outside the region of strong curvature near the singularity. The repulsion implies that the brane probes are bent, rather than just being straight lines as in the supersymmetric case. This bending can be related to spontaneous chiral symmetry breaking by a quark condensate. In fact, as is seen from figure 13.5, there is a regular embedding with non-zero condensate, as determined by the subleading term of the embedding coordinate w_6, even for $m \to 0$. The fact that there is a non-zero c for vanishing m corresponds exactly to spontaneous chiral symmetry breaking by a quark condensate. Notice also the finite distance on the w-axis between the singularity and the embedding with $m \to 0$. This ensures that the embedding stays in the weakly curved region of space time, where the large N, large λ approximation remains valid. Moreover, as seen

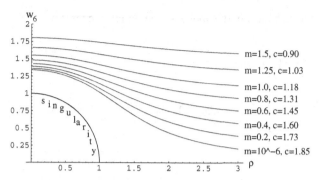

Figure 13.5   Regular D7-brane embeddings in the Constable–Myers background. From reference [8].

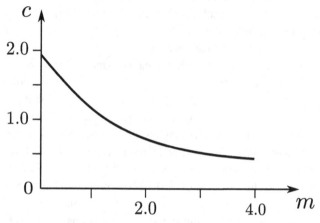

Figure 13.6   Condensate parameter c versus quark mass m for the regular solutions of the equation of motion in the Constable–Myers background. c and m are given in units set by the length scale b.

from figure 13.6, at large m we have $c \sim 1/m$. This behaviour is expected from field theory, as we will see in the discussion of (13.59) below.

Goldstone boson

Since there is spontaneous symmetry breaking for $m \to 0$, we expect a Goldstone boson in the meson spectrum. This Goldstone boson is exactly massless for vanishing quark mass. At small explicit symmetry breaking by a small non-zero quark mass, it turns into a *pseudo-Goldstone boson* of small mass.

Clearly, fluctuations in the angular direction in the $w_5 - w_6$ plane (i.e. along the vacuum manifold) will generate the required massless states. Solving the equation of motion for D7-brane probe fluctuations in the two directions transverse to the probe ($\delta w_5 = f(r) \sin(k \cdot x)$, $\delta w_6 = h(r) \sin(k \cdot x)$) around the D7-brane probe embedding shown in figure 13.5, the meson masses are given by $M^2 = -k^2$. There are indeed two distinct mesons (see figure 13.6): one is massive for every m, and corresponds to fluctuations in

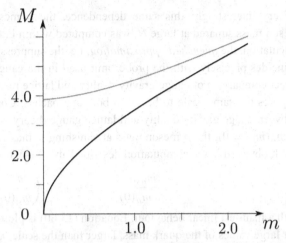

Figure 13.7 Masses of the lowest-lying meson masses for fluctuations about the D7-brane embedding in radial and angular direction, as a function of the quark mass, with scale set by b. The angular fluctuation mode gives rise to a (pseudo-)Goldstone mode.

the radial transverse direction, the other, corresponding to the $U(1)$ symmetric fluctuation, is massless for $m = 0$ and is thus a Goldstone boson. This is similar to the pion in QCD, which is a pseudo-Goldstone boson of flavour symmetry and the lightest bound state in the QCD spectrum. In gauge/gravity duality models involving a D7-brane as considered here, the Goldstone mode is associated with spontaneous breaking of the $U(1)_A$ symmetry which is non-anomalous in the $N \to \infty$ limit. In this sense this Goldstone mode behaves as an η' particle, which becomes a $U(1)_A$ Goldstone boson for $N \to \infty$.

A further important property of the model presented here is the small quark mass behaviour of the meson mass, which is proportional to the square-root of m. It is thus in agreement with the *Gell-Mann–Oakes–Renner relation* of chiral QCD. This relation states that

$$M^2_{\text{meson}} f_\pi^2 = 2\langle \bar\psi \psi \rangle m + \mathcal{O}(m^2), \tag{13.59}$$

where f_π is the meson decay constant. As for the supersymmetric mesons of chapter 10, the mesons found here scale as $M \sim m_q/\sqrt{\lambda}$ and are thus tightly bound at large 't Hooft coupling λ.

Vector mesons

The vector mesons in the model are described by the gauge fields in the DBI action for the D7-branes. Again, solutions of the form $A^\mu = g(\rho) \sin(k \cdot x)\epsilon^\mu$ provide the masses of the ρ meson and its radially excited states. By considering different D7-brane embeddings characterised by different quark mass parameters m in the near-boundary expansion, we obtain the mass of the ρ meson as a function of the pion mass squared in this model, in dimensionless units fixed by the choice of the supergravity scale b. A numerical calculation reveals linear behaviour of $m_\rho(m_\pi^2)$.

Very interestingly, this same dependence, the ρ meson mass as function of the π meson mass squared at large N, was computed within lattice gauge theory. For the lattice calculation, the *quenched approximation*, i.e. the suppression of quark loops, is used. This coincides precisely with the probe limit used in the gauge/gravity duality calculation. A direct comparison of gauge/gravity duality and lattice results is possible. This comparison requires the same scale to be set in both approaches. Consequently, a direct comparison between gauge/gravity duality and lattice gauge theory is best performed by normalising to $m_\rho(m_\pi = 0)$, the ρ meson mass at vanishing π mass. Then, the gauge/gravity duality result obtained in the computation described above is written as

$$\frac{m_\rho(m_\pi)}{m_\rho(0)} = 1 + 0.307 \left(\frac{m_\pi}{m_\rho(0)} \right)^2 , \tag{13.60}$$

with a manifest linear behaviour. Equation (13.60) is plotted as a black line in figure 13.8. For large values of the quark mass, larger than the scale b, we expect $m_\rho \propto m_\pi$ due to the onset of supersymmetry. However, the comparison with lattice gauge theory described is performed at scales much smaller than b, where spontaneous chiral symmetry breaking is present and supersymmetry is broken.

For the lattice computation, it is necessary to work at finite N and then to perform an extrapolation to $N \to \infty$. It is possible to quantify the systematic error of this approach,

$$\frac{m_\rho(m_\pi)}{m_\rho(0)} = 1 + 0.360(64) \left(\frac{m_\pi}{m_\rho(0)} \right)^2 + \cdots , \tag{13.61}$$

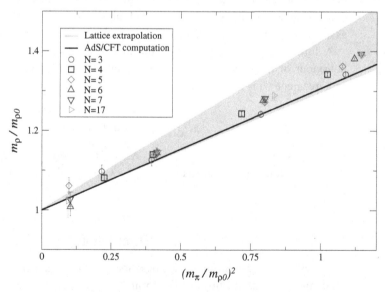

Figure 13.8   Rho and pion meson masses: comparison of gauge/gravity duality and lattice gauge theory. The black line corresponds to the gauge/gravity duality result (13.60) for m_ρ versus m_π^2 at large N in the Constable–Myers background. The dots correspond to recent lattice results of Del Debbio *et al.* [9] for given values of N. The grey zone corresponds to the systematic error of the extrapolation of the lattice results to $N \to \infty$ as given in (13.61). We are grateful to Biagio Lucini for providing this figure.

where the dots stand for higher order terms. As shown in figure 13.8, the gauge/gravity duality and lattice results agree to impressive accuracy.

It is still an unanswered question whether gauge/gravity duality manages to capture the dynamics of the strongly coupled gauge theory, or whether a correct description of the symmetry properties suffices to obtain realistic values for the meson masses. It may also be the case that when considering ratios, dynamical effects – large N effects in particular – cancel out. All these questions are difficult to answer while gauge/gravity duality is available only in the large N, large λ limit. In this limit, the dynamics inside the mesons may not be resolved: since their mass scales with $1/\sqrt{\lambda}$, they are tightly bound.

To summarise, the D7-brane embeddings in confining gravity backgrounds geometrically display spontaneous breaking of the $U(1)_A$ chiral symmetry, as well as the associated Goldstone boson. It is not possible, however, to extend this model in a simple way – by considering a probe of several D7-branes for instance – to describe the spontaneous breaking of the non-Abelian $SU(N_{\mathrm{f}}) \times SU(N_{\mathrm{f}})$ chiral symmetry. This symmetry would already be broken explicitly at the classical level since a product group flavour symmetry requires embeddings of different stacks of D7-branes, one stack in the $x^0, \ldots, x^7$ directions and one in the $x^0, \ldots, x^5$, x^8, x^9 directions. Such a configuration, however, leads to couplings between the quark fields and the adjoint scalar which is a superpartner of the gluons. In the superpotential, a term of the form $\tilde{Q}\Phi Q$ is present which breaks the non-Abelian chiral symmetry explicitly. Nevertheless, the D7-brane embedding in confining backgrounds as discussed above provides an interesting toy model for chiral symmetry breaking since the four-dimensional field theory Lagrangian is known explicitly. The bare quark mass may be chosen as parameter, leading to control over the pseudo-Goldstone behaviour.

13.3.2 Chiral symmetry breaking in the presence of a magnetic field

In the example studied in section 13.3.1, the chiral symmetry breaking is generated holographically by a region of strong curvature which provides a repulsive potential for the D7-brane probe. An alternative geometry which is regular everywhere and generates chiral symmetry breaking for a D7-brane probe is obtained by adding a B-field to the original D3-brane geometry [10].

We begin with the metric of $AdS_5 \times S^5$ written in the form (10.11) of chapter 10 and embed a D7-brane probe. We introduce polar coordinates also in the two directions w_5, w_6 perpendicular to the D7-brane, i.e. $\mathrm{d}w_5^2 + \mathrm{d}w_6^2 = \mathrm{d}l + l^2\mathrm{d}\varphi^2$, such that

$$\mathrm{d}s^2 = \frac{\rho^2 + l^2}{L^2}\eta_{\mu\nu}\mathrm{d}x^\mu\mathrm{d}x^\nu + \frac{L^2}{\rho^2 + l^2}\left(\mathrm{d}\rho^2 + \rho^2\mathrm{d}\Omega_3^2 + \mathrm{d}l^2 + l^2\mathrm{d}\phi^2\right), \qquad (13.62)$$

$$\mathrm{d}\Omega_3^2 = \mathrm{d}\psi^2 + \cos^2\psi\,\mathrm{d}\beta^2 + \sin^2\psi\,\mathrm{d}\gamma^2, \qquad (13.63)$$

where we have also chosen an appropriate parametrisation for S^3. A consistent ansatz for the D7-embedding is $\phi = $ constant, $l = l(\rho)$.

We now introduce a magnetic field by considering a non-trivial $B_{(2)}$-form, given by

$$B_{(2)} = H \mathrm{d}x^2 \wedge \mathrm{d}x^3. \tag{13.64}$$

Since this field is pure gauge, i.e. $\mathrm{d}B = 0$, the background is still a solution to the supergravity equations of motion. In the presence of (13.64), the D7-probe brane action will have a Chern–Simons contribution in addition to the contribution from the DBI action. Consequently, the D7-brane probe action reads

$$\mathcal{S}_{\mathrm{D7}} = -\tau_7 \int \mathrm{d}^8\xi\, e^{-\phi} \sqrt{-\det(P[g]_{ab} + P[B]_{ab} + 2\pi\alpha' F_{ab})} + 2\pi\alpha'\mu_7 \int F_{(2)} \wedge C_{(6)}, \tag{13.65}$$

where we have written the Wess–Zumino term to first order in α'. All other possible contributions of Chern–Simons type vanish. $P[g]_{ab}$ and $P[B]_{ab}$ are the metric and B-field induced on the worldvolume of the D7-brane and F_{ab} is the worldvolume $U(1)$ gauge field. Expanding the DBI action to first order in F_{ab}, the equation of motion for F_{ab} leads to a constraint for $C_{(6)}$. $C_{(6)}$ itself is determined from the action contribution

$$\mathcal{S}_{C_{(6)}} = \mu_7 \int B_{(2)} \wedge C_{(6)} - \frac{1}{4\tilde{\kappa}_{10}^2} \int \mathrm{d}^{10}x \sqrt{-g}\, |\mathrm{d}C_{(6)}|^2, \tag{13.66}$$

which combines a contribution from the D-brane action (13.65) with a contribution of the ten-dimensional gravity action. The equation of motion for $C_{(6)}$ gives

$$\mathrm{d}C_{l01\rho\psi\beta\gamma} = \frac{\mu_7\tilde{\kappa}_{10}^2}{\pi} H \frac{\rho^3 L^4}{(\rho^2 + l^2)^2} \theta(l - l(\rho)) \sin\psi \cos\psi, \tag{13.67}$$

which is compatible with the constraint following from (13.65). Note that $2\tilde{\kappa}_{10}^2 = \tau_7^{-1} = (2\pi)^7\alpha'^4$, such that $\mu_7\tilde{\kappa}_{10}^2 = 1/g_s$. Note that supersymmetry is broken at the level of the probe brane. However, the super gravity background itself is supersymmetric and hence stable.

Let us now find the embedding $l(\rho)$ for the D7-brane. For this purpose we start from the explicit form for the DBI action to leading order in α', which for the geometry considered here is given by

$$\mathcal{S}_{\mathrm{DBI}} = -\mu_7 \int \mathrm{d}^4x\, \mathrm{d}\rho\, \mathrm{d}\Omega_3\, \rho^3 \sin\psi \cos\psi \sqrt{1 + l'^2} \sqrt{1 + \frac{L^2 H^2}{(\rho^2 + l^2)^2}}. \tag{13.68}$$

The embedding has the standard near-boundary leading asymptotic behaviour

$$l(\rho) = m + \frac{c}{\rho^2} \tag{13.69}$$

with non-zero c. In general, the equations of motion have to be solved numerically. However, for a small magnetic field and large masses, which implies $l(\rho) \simeq m + \eta(\rho)$ with η small, an analytic expansion is possible. In this case, the equation of motion for $\eta(\rho)$ may be solved analytically. To leading order, the result implies

$$\langle \tilde{\psi}\psi \rangle \propto -c = -\frac{L^4}{4m} H^2. \tag{13.70}$$

Numerical calculations reveal that $c \neq 0$ also for $m \to 0$, which corresponds to spontaneous chiral symmetry breaking.

The meson spectrum in the presence of the B-field is obtained from the fluctuations as before. As a characteristic feature, it displays a level splitting in agreement with the Zeeman effect.

13.4 Non-Abelian chiral symmetries: the Sakai–Sugimoto model

A more realistic non-Abelian $U(N_f) \times U(N_f)$ chiral symmetry beyond the $U(1)_A$ symmetry considered in the previous section is realised in the model established by Sakai and Sugimoto [11, 12]. In their model (table 13.2), which is based on type IIA string theory, the degrees of freedom in the adjoint representation of the gauge group are provided by N D4-branes with one direction wrapped on a circle, as introduced in section 8.4.2. Quarks are included by adding D8-brane and anti-D8-brane (or $\overline{\text{D8}}$-brane for short) probes. The latter are D8-branes with opposite charge. The probe brane configuration of N_f D8-branes and N_f anti-D8-branes fills the whole space except the single direction compactified on a circle S^1.

As discussed in section 8.4.2, the D4–D4 strings provide the gauge field degrees of freedom. The D4–D8 (D4–$\overline{\text{D8}}$) strings generate chiral (anti-chiral) quark fields in the gauge theory [11]. The two $U(N_f)$ gauge symmetries on the worldvolumes of the N_f D8-branes and $\overline{\text{D8}}$-branes are dual to the the chiral non-Abelian flavour symmetries $U(N_f) \times U(N_f)$ on the field theory side.

In the UV, where the radius of the compactified direction is large, the model describes a five-dimensional theory. In the IR, where the compactification radius is small, this flows to a four-dimensional theory. However, as we will see, the energy scale associated with compactification is of the same order as the confinement scale equivalent to Λ_{QCD} which potentially leaves some unwanted excitations in the spectrum. Nevertheless, this model comes surprisingly close to QCD, as we now describe. The key feature of the model is spontaneous breaking of the non-Abelian chiral symmetry. For N_f D8-branes and $\overline{\text{D8}}$-branes each, there is a $U(N_f)_L \times U(N_f)_R$ gauge symmetry on the gravity side, which corresponds to a global chiral symmetry on the field theory side. As discussed in section 13.1.4, this global symmetry is also spontaneously broken in QCD. On the gravity side, spontaneous breaking of this chiral symmetry is realised as follows. The D8-branes and $\overline{\text{D8}}$-branes will prefer to join into a single curved D8-brane as shown in figure 13.9. There is only one surviving $U(N_f)$ symmetry, corresponding to the chiral symmetries being broken to the vector $U(N_f)_V$ as discussed in section 13.1.4.

Table 13.2 The Sakai–Sugimoto configuration

	0	1	2	3	4	5	6	7	8	9
N D4	•	•	•	•	•	–	–	–	–	–
N_f D8 – $\overline{\text{D8}}$	•	•	•	•	–	•	•	•	•	•

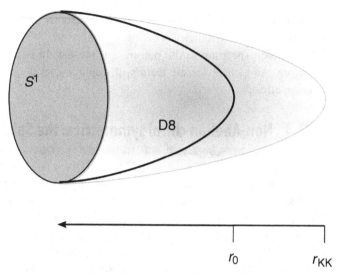

Figure 13.9 Chiral symmetry breaking in the Sakai–Sugimoto model. In the UV, a D8-brane and a $\overline{\text{D8}}$-brane are located at antipodal points of S^1. The S^1 shrinks to zero at $r = r_{\text{KK}}$. The D8–$\overline{\text{D8}}$ pair joins at $r = r_0$, which leads to chiral symmetry breaking.

13.4.1 Gravitational background for the D4–D8–$\overline{\text{D8}}$ system

The gravity background into which the D8-brane and $\overline{\text{D8}}$-brane probe are embedded within type IIA theory is obtained by taking the near-horizon limit of the geometry of a large N D4-brane stack wrapped on a circle S^1. This model was introduced in section 8.4.2. Let us recapitulate this solution in a notation convenient for the present context, in particular with the replacement $u \to r$ to ease comparison with the other flavour models discussed. We have

$$ds^2 = \left(\frac{r}{L}\right)^{3/2}\left(\eta_{\mu\nu}dx^\mu dx^\nu + f(r)dx_4^2\right) + \left(\frac{L}{r}\right)^{3/2}\left(\frac{dr^2}{f(r)} + r^2 d\Omega_4^2\right),$$
$$f(r) \equiv 1 - \left(\frac{r_{\text{KK}}}{r}\right)^3. \tag{13.71}$$

Here, r is the holographic radial direction, $d\Omega_4^2$ is the metric of a four-sphere, and $L^3 = \pi g_s N \alpha'^{3/2}$. There is also a non-zero four-form flux, as well as a dilaton given by

$$e^\phi = g_s\left(\frac{r}{L}\right)^{3/4}. \tag{13.72}$$

As discussed in section (8.4.2), the coordinate x_4 in (13.71) is periodic, with the period given by

$$x_4 \equiv x_4 + 2\pi/M_{\text{KK}}, \qquad M_{\text{KK}} = \frac{3}{2}\frac{r_{\text{KK}}^{1/2}}{L^{3/2}}, \tag{13.73}$$

giving rise to S^1 which is wrapped by the D4-branes. M_{KK} is the Kaluza–Klein mass which sets the mass scale. There is a horizon at $r = r_{\text{KK}}$, at which the radius of S^1 shrinks to zero. The point $r = r_{\text{KK}}$ is the tip of the cigar-shaped subspace spanned by x_4 and the holographic coordinate r, as shown in figure 13.9. Therefore the coordinate r is restricted to the range $[r_{KK}, \infty]$. This scale gives rise to a mass gap. As discussed above in section 13.2.2, the fact that the geometry is restricted by a lower bound on r in the deep interior implies that the dual field theory is confining. A further crucial feature of the model is that the compactified dimension x_4 breaks supersymmetry completely, by giving Kaluza–Klein masses to the adjoint fermions of the dual gauge theory and at higher loop order also to the adjoint scalars. This leaves only gauge bosons in the spectrum of the low-energy theory, as the latter are protected by gauge symmetry.

For embedding probe branes it is convenient to change coordinates from r to ζ, where

$$1 + \zeta^2 = \left(\frac{r}{r_{KK}} \right)^3. \tag{13.74}$$

Then the geometry becomes

$$ds^2 = \left(\frac{r_{KK}}{L} \right)^{3/2} \left(\sqrt{1 + \zeta^2}\, \eta_{\mu\nu} dx^\mu dx^\nu + \frac{\zeta^2}{\sqrt{1 + \zeta^2}}\, dx_4^2 \right)$$
$$+ \left(\frac{L}{r_{KK}} \right)^{3/2} r_{KK}^2 \left(\frac{4}{9} (1 + \zeta^2)^{-5/6}\, d\zeta^2 + (1 + \zeta^2)^{\frac{1}{6}}\, d\Omega_4^2 \right). \tag{13.75}$$

13.4.2 Probe D8-branes

Finding the full backreacted geometry including the D8-branes is challenging in the non-supersymmetric configuration considered here. We therefore work in the probe limit $N_f \ll N$ again, for simplicity restricted to $N_f = 1$. On the field theory side, this corresponds to the quenched approximation, i.e. to the suppression of quark loops in the Feynman diagrams.

The embeddings of a probe D8-brane in the above background are determined by the equations of motion obtained from the DBI action. They form a family of curves in the (ζ, x_4)-plane which we parametrise as $x_4(\zeta)$. The relevant part of the Dirac–Born–Infeld action for the embedding is

$$S_{DBI} = -\tau_8 \int_{D8} d^8\xi\, e^{-\phi} \sqrt{-\det (P[g]_{ab})}. \tag{13.76}$$

This gives

$$S_{DBI} = -\mu_8 \text{Vol}(S^4) \int d^4x\, d\zeta\, \frac{2}{3} r_{KK}^5 \left(\frac{L}{r_{KK}} \right)^{3/2} (1 + \zeta^2)^{2/3}$$
$$\times \sqrt{1 + \frac{9}{4r_{KK}^2} \left(\frac{r_{KK}}{L} \right)^3 \zeta^2 (1 + \zeta^2)^{1/3} x_4'(\zeta)^2}. \tag{13.77}$$

The extremal configurations $x_4(\zeta)$ for the D8-branes satisfy

$$x_4'(\zeta) = \frac{2}{3} \left(\frac{L}{r_{KK}} \right)^{3/2} \frac{J}{\sqrt{r_{KK}^6 \zeta^4 (1 + \zeta^2)^2 - J^2 r_{KK}^{-2} \zeta^2 (1 + \zeta^2)^{1/3}}}. \tag{13.78}$$

Here $J = r_{KK}^4 \zeta_0 (1 + \zeta_0^2)^{5/6}$ is chosen in such a way that $x_4'(\zeta)$ becomes infinite at $\zeta = \zeta_0$, which corresponds to $r = r_0$ as shown in figure 13.9. This point is the point of closest approach of the D8-brane to the horizon at $r = r_{KK}$. There is a one-parameter family of embeddings for which a particular value of ζ_0 specifies one particular curve. Note the curve for $\zeta_0 = 0$ consists of two lines at

$$x_4 = \pm \frac{\pi}{3} \frac{L^{3/2}}{r_{KK}^{1/2}}. \tag{13.79}$$

The large ζ UV asymptotic behaviour of the solutions takes the form

$$x_4 = \alpha - \frac{\beta}{\zeta^3} \tag{13.80}$$

with α, β free parameters. This asymptotic behaviour is relevant for identifying the fluctuation mode dual to the pion.

13.4.3 Pion

Let us restrict the discussion to the $\beta = 0$ solution in (13.80). In this case, $r_0 = r_{KK}$, such that the D8-branes and $\overline{\text{D8}}$-branes lie at antipodal points on the circle until they join at the horizon at $r = r_{KK}$, where the radius of the circle shrinks to zero. This configuration is interpreted as the theory with massless quarks: chiral symmetry breaking at the same scale as the glueball mass gap. The spontaneous symmetry breaking implies the existence of a Goldstone boson.

For spontaneously broken chiral symmetry, there has to be a vacuum manifold with different points corresponding to the different possible phases of the quark condensate. In the Sakai–Sugimoto model, the phase of the quark condensate is identified with the value of the D8-brane gauge field A_ζ. We consider $N_f = 1$ for simplicity. To identify the vacuum manifold, we have to determine background solutions for $A_\zeta(\zeta)$ independent of the x coordinate, which correspond to different global choices of the phase π. A_ζ is described by the DBI action including a $U(1)$ gauge field, which at low energy has a Lagrangian density on the D8-brane worldvolume given by

$$\mathcal{L} = -\tau_8 e^{-\phi} \sqrt{-\det\left(P[g]_{ab}\right)} \left(1 - \frac{(2\pi\alpha')^2}{4} F^{ab} F_{ab}\right). \tag{13.81}$$

For the massless D8-brane embedding, we take

$$x_4(\zeta) = \pm \frac{\delta x_4}{4} = \pm \frac{\pi}{3} \frac{L^{3/2}}{r_{KK}^{1/2}}. \tag{13.82}$$

Working on the upper branch of the D8-brane, for which

$$x_4(\zeta) = +\frac{\pi}{3} L^{3/2} / r_{KK}^{1/2}, \tag{13.83}$$

the quadratic part of the action then takes the form

$$\begin{aligned}
\mathcal{S} = &-\tau_8 \frac{\text{Vol}(S^4)}{2} (2\pi\alpha')^2 \int_0^\infty d\zeta \int d^4x \left(e^{-\phi} \sqrt{-g} g^{\zeta\zeta} g^{11}\right) \\
&\times \left(-(\partial_t A_\zeta)^2 + (\partial_1 A_\zeta)^2 + (\partial_2 A_\zeta)^2 + (\partial_3 A_\zeta)^2\right)
\end{aligned} \tag{13.84}$$

for states of zero spin on S^4.

We see that F^{ab} and hence the action vanishes if A_ζ is the only non-zero field and if it is only a function of ζ. Any function of ζ is allowed, which is an artefact of gauge freedom in the model. To implement a gauge choice $\nabla_a A^a(\zeta) = \kappa(\zeta)$, a convenient trick is to add a gauge fixing term as in (1.202),

$$\delta\mathcal{L} = \frac{1}{\xi}\, e^{-\phi} \sqrt{-\det(P[g]_{ab})}\, \left(\nabla_a A^a - \kappa(\zeta)\right)^2, \tag{13.85}$$

where ξ is a gauge parameter and $\kappa(\zeta)$ is any arbitrary function. A convenient choice for the gauge field is

$$A_\zeta(\zeta) = \frac{\mathcal{C}}{1+\zeta^2}. \tag{13.86}$$

The solution contains the arbitrary multiplicative factor $\mathcal{C}$, which is not fixed by the equations of motion since the action is only quadratic in A_ζ.

The pion field should correspond to spacetime-dependent fluctuations around the vacuum manifold, i.e. to fluctuations depending on the coordinates x^μ. We now show that the leading term in the pion effective action is obtained by considering fluctuations of the form $A_\zeta(\zeta,x) \equiv A_\zeta(\zeta)\pi(x)$ about the background (13.86) [13, 14]. In fact, choosing $\mathcal{C}$ to be

$$\mathcal{C} = \frac{1}{\sqrt{\mathcal{N}_A}}, \qquad \mathcal{N}_A = \frac{\lambda^{2/3} N (L M_{KK})^7}{\pi^4 2^{5/3} 3^6}, \tag{13.87}$$

where $\mathcal{N}_A$ is the overall normalisation of the action (13.84), we have

$$A_\zeta(\zeta,x) = \frac{1}{\sqrt{\mathcal{N}_A}}\, \frac{1}{1+\zeta^2}\, \pi(x). \tag{13.88}$$

Substituting this into the action (13.84), we find a canonically normalised kinetic term for a massless field,

$$S = -\frac{1}{2} \int d^4 x\, \eta^{\mu\nu} \partial_\mu \pi(x) \partial_\nu \pi(x). \tag{13.89}$$

In analogy to (13.22), this is the pion, i.e. the Goldstone mode of chiral symmetry breaking. Note that interchanging the D8-branes and $\overline{\text{D8}}$-branes corresponds to interchanging left-handed and right-handed quarks and is therefore equivalent to parity transformations in the model. The pion state considered has negative parity eigenvalue and is hence a pseudo-scalar, as is the pion in QCD.

13.4.4 Meson spectrum

Fluctuations of the D8-branes about the embeddings discussed above correspond to mesons of the gauge theory. We look for solutions to the linearised field equations obtained from the DBI action. We consider fluctuations of the form $f(r)e^{ikx}$. Even and odd functions $f(r)$ describe even and odd parity states.

Fluctuations of the vector field in the DBI action generate vector and axial-vector mesons. In addition, there is a scalar field corresponding to fluctuations of the embedding. Restricting these fluctuations to the trivial harmonic of the four-sphere on the D8-brane

Table 13.3 Meson masses in the Sakai–Sugimoto model

m_ρ	$0.67 M_{KK}$	m_{a_1}	$1.58 M_{KK}$
m_ρ^*	$1.89 M_{KK}$	$m_{a_1^*}$	$2.11 M_{KK}$
m_ρ^{**}	$2.21 M_{KK}$		

transverse to the x directions, we obtain QCD-like states. It is important to keep in mind that there are additional states with higher harmonics carrying R-charge, indicating that there are light non-degenerate superpartners of the QCD fields in the field theory. Moreover, there are fermionic fields in the DBI action dual to *mesinos*. Finally, there are also Kaluza–Klein modes of the glueballs and gluino balls from the gauge sector. The typical scale for the masses of all of these bound states is given by M_{KK} as given by (13.73). Note that as in previous examples, the mesons are tightly bound in the limit of large 't Hooft coupling λ, and hence are rather different from QCD mesons. The values of the masses for states that can be mapped to QCD are given in table 13.3. Comparing ratios of these masses to the experimentally measured values leads to agreement within about 10-20%.

13.5 AdS/QCD correspondence

13.5.1 A simple model

So far we have considered gravity duals which involve ten-dimensional gravity theories. These were obtained from deformations of ten-dimensional supergravity actions which arise as low-energy limits of string theory. In some cases, however, it is instructive to consider simpler models, which consist of gravity models in five dimensions. These are conjectured to be dual to confining gauge theories. This conjectured duality is inspired by the AdS/CFT correspondence, but less motivated by string theory arguments. Moreover, it is not possible to determine the field content of the dual field theory explicitly. Nevertheless this approach, which is referred to as *AdS/QCD*, gives rise to impressive agreement with experimental results for both ratios of meson masses and structure constants, with an error of the order of 10%. Calculations are generally simpler in these *bottom-up* models than in the *top-down* ten-dimensional models considered above.

Let us consider a simple model of this type. It consists of AdS space in five dimensions, with metric

$$ds^2 = \frac{1}{z^2} \left(dz^2 + \eta_{\mu\nu} dx^\mu dx^\nu \right), \tag{13.90}$$

where we set $L = 1$. As before, the radial coordinate z is interpreted as the holographic energy scale of the theory. To break the $SO(4,2)$ symmetry of this metric, which corresponds to a dual conformal field theory, a rather simple and drastic approach is followed by imposing a cut-off or *hard wall* at $z = z_0$, such that the gravity theory is defined only for $z \leq z_0$. This procedure also ensures confinement, since the scale z_0 corresponds to a mass gap, which is a generic feature of any confining gauge theory.

The aim is now to construct a gravity action dual to a low-energy effective theory which encodes the pion pseudo-scalar meson as well as the vector ρ and axial-vector a_1 meson [15, 16]. By adapting the standard definition from low-energy effective chiral QCD, the quark mass and condensate as well as the pion field are introduced by virtue of the scalar field

$$X(z,x) = X_0(z)e^{2i\pi^a(x)T_a}. \tag{13.91}$$

Here, $X_0(z)$ is a bulk scalar field that contains the quark mass and condensate in its asymptotic boundary expansion. We consider N_f flavours. The T_a are the generators of $SU(N_f)$ in the fundamental representation. π^a describes the pion fields. X may be viewed as the gravity analogue of the field U of (13.21). According to the AdS/CFT prescription for the asymptotic boundary behaviour of the bulk fields, a scalar which describes a quark bilinear operator of dimension $\Delta = 3$ must have mass squared $m^2 = \Delta(\Delta - 4) = -3$ in AdS space. Near the boundary, the bulk scalar field X_0 then has the form

$$X_0(z) \sim \frac{1}{2}Mz + \frac{1}{2}\Sigma z^3, \tag{13.92}$$

where M is the quark mass matrix and $\Sigma^{ij} = \langle \bar{\psi}^i \psi^j \rangle$ is the quark condensate. These are taken to be diagonal in flavour space, $M = m_q \mathbb{1}$, $\Sigma = \sigma \mathbb{1}$.

In addition, the model describes vector and axial-vector states by virtue of two additional massless gauge fields dual to the operators $\bar{\psi}_L \gamma^\mu \psi_L$ and $\bar{\psi}_R \gamma^\mu \psi_R$. Since for vector operators, the mass conformal dimension relation is $m^2 = (\Delta - 1)(\Delta - 3)$, we have $m^2 = 0$ on the gravity side.

Assembling all the ingredients described, the action for the model considered is

$$S = -\int_0^{z_0} d^5x \sqrt{-g} \, \text{Tr} \left\{ |DX|^2 + 3|X|^2 + \frac{1}{4g_5^2}(F_L^2 + F_R^2) \right\}, \tag{13.93}$$

where X transforms on the left under $SU(N_f)_L$ and on the right under $SU(N_f)_R$, i.e.

$$D_\mu X = \partial_\mu X - iA_\mu^L X + iX A_\mu^R. \tag{13.94}$$

The integral is cut off at the hard wall at $z = z_0$. The mass m_q, the condensate σ, the coupling g_5 and the position of the hard wall z_0 are parameters of this model which have to be chosen.

For this model, we calculate the two-point function using the AdS/CFT correspondence. We write the calculation in such a way as to facilitate comparison with low-energy QCD. In the boundary expansion about $z = 0$, the vector field

$$V_\mu^a(x,z) = (A_{L\mu}^a(x,z) + A_{R\mu}^a(x,z))/2 \tag{13.95}$$

is the source for the four-dimensional vector current $J_\mu^a(x)$. The field $V_\mu(x,z) = V_\mu^a(x,z)T_a$ satisfies the linearised equation of motion

$$\partial_\mu \left(\frac{1}{g_5^2} \sqrt{-g} g^{\mu\rho} g^{\nu\sigma} (\partial_\rho V_\sigma - \partial_\sigma V_\rho) \right) = 0. \tag{13.96}$$

We look for solutions of the form $V_\mu(x,z) = V_{0\mu}(x)v(x,z)$ with $\lim_{z \to 0} v(x,z) = 1$, such that $V_{0\mu}(x)$ is of dimension one as required for $J_\mu(x)$ to be a conserved current. Solving the equation of motion (13.96) gives the asymptotic result

$$v(q,z) \sim 1 + \frac{q^2 z^2}{4} \ln\left(q^2 z^2 \right), \qquad \text{for } z \to 0. \tag{13.97}$$

Substituting the solution back into the action and differentiating twice with respect to the source V_0^μ gives the vector current correlator

$$\Pi_V(q^2) = \lim_{\epsilon \to 0} \left[\frac{1}{g_5^2 q^2} \frac{1}{z} \partial_z v(q,z) \right]_{z=\epsilon}, \tag{13.98}$$

which gives

$$\Pi_V(q^2) = \frac{1}{2g_5^2} \ln(q^2) \tag{13.99}$$

up to contact terms.

This result may be compared with correlators from QCD, where the current correlator takes the form

$$\int d^4 x e^{iqx} \langle J_\mu^a(x) J_\nu^b(0) \rangle = \delta^{ab} (q_\mu q_\nu - q^2 \eta_{\mu\nu}) \Pi_V(q^2), \tag{13.100}$$

where $J_\mu^a(x) = \bar\psi \gamma_\mu T^a \psi$, with Dirac spinors ψ. For QCD, the leading order contribution to $\Pi_V(q^2)$ is [17]

$$\Pi_V(q^2) = \frac{N}{24\pi^2} \ln(q^2). \tag{13.101}$$

Comparing the gravity result (13.99) to the perturbative QCD result (13.101) determines the five-dimensional coupling g_5 as

$$g_5^2 = \frac{12\pi^2}{N}. \tag{13.102}$$

We note that the gauge/gravity duality expression is fitted to the asymptotic perturbative result in this approach, in spite of the fact that the gravity dual is inherently a description of a strongly coupled gauge theory. The standard argument for this procedure is that perturbative QCD is conformal in the UV, and therefore it is natural to match it to the UV behaviour in AdS space which is also conformal. This procedure captures conformality, but not the asymptotic freedom of QCD in the UV.

Meson masses may be obtained as in the previous sections by considering fluctuations. For instance, for the ρ vector meson, the mass is obtained by solving (13.96) for fluctuations of the form $V = V(z)e^{ip \cdot x}$, $p^2 = -M^2$. A boundary condition at the hard wall has to be chosen, which we take to be $\partial_z V = 0$. From the solutions we may extract the masses of the ρ meson and its excited states. Similarly, the pion mass is obtained

Table 13.4	Meson masses and decay constants in the hard wall AdS/QCD model	
Observable	Measured (MeV)	AdS/QCD (MeV)
m_π	139.6 ± 0.0004	141
m_ρ	775.8 ± 0.5	832
m_{a_1}	1230 ± 40	1220
f_π	92.4 ± 0.35	84.0
$F_\rho^{1/2}$	345 ± 8	353
$F_{a_1}^{1/2}$	433 ± 13	440

from fluctuations of π^a, and the mass of the axial-vector meson a_1 from fluctuations of $A(z) \equiv (A_L - A_R)/2$. The results are given in table 13.4.

Moreover, we may obtain the decay constants f_π, F_ρ and F_a for these mesons. Let us consider the decay constant f_π, which within low-energy chiral QCD is obtained from

$$\langle 0 | J_{5\mu}^a | \pi^b(p) \rangle = i f_\pi p_\mu \delta^{ab}, \tag{13.103}$$

with $J_{5\mu}^a = \bar{\psi} \gamma_5 \gamma_\mu T^a \psi$ the axial current dual to $A_\mu = (A_{L\mu} - A_{R\mu})$. Equation (13.103) implies that Π_A defined in analogy to (13.100) has a pole at $q^2 = 0$ if $m_\pi = 0$, $\Pi_A(q^2) \sim f_\pi^2/q^2$. A holographic calculation thus gives, with $A_\mu(q,z) = a(q,z)A_\mu(q)$,

$$f_\pi^2 = \frac{1}{g_5^2} \lim_{\epsilon \to 0} \left[\frac{\partial_z a(q=0, z)}{z} \right]_{z=\epsilon}. \tag{13.104}$$

The explicit values of the decay constants have to be evaluated numerically. They are listed in table 13.4 together with the masses, in comparison to the values found experimentally. The AdS/QCD results correspond to the best fit to all the observables. It may also be shown that the results obtained satisfy the Gell-Mann–Oakes–Renner relation (13.59).

13.5.2 Soft wall and excited meson states

Both within QCD and experimentally, it is found that the masses of radially excited meson states follow Regge trajectories. This means that the meson masses scale as $M_n^2 \sim n$ (or equivalently $M_n \sim \sqrt{n}$) with the radial quantum number. However, the hard wall AdS/QCD model as presented above does not display this behaviour: for the hard wall model, $M_n \sim n$.

To see this explicitly, we obtain the dependence of M_n on n for the hard wall model by considering the part of the action (13.93) for the gauge field in AdS space describing the ρ mesons, which may be written as

$$S \sim -\frac{1}{4} \int d^5x e^{-\phi(z)} \sqrt{-g} F_{mn} F^{mn}, \tag{13.105}$$

with F the field strength for the field V defined in (13.95), where for simplicity we now consider just one flavour. For constant dilaton ϕ, the equation of motion for a solution of the form $A_y = f(z)e^{ikx}$ with $k^2 = -M^2$ is

$$\left(\partial_z^2 - \frac{1}{z}\partial_z + M^2\right)f(z) = 0. \tag{13.106}$$

Defining $f \equiv \sqrt{z}\psi$, this equation may be brought into Schrödinger form,

$$-\psi'' + V(z)\psi = M^2\psi, \qquad V(z) = \frac{3}{4}\frac{1}{z^2}. \tag{13.107}$$

When introducing a hard wall cut-off at $z = z_0$, the Schrödinger potential in the IR is that of a square well. The mass spectrum therefore grows as $M_n^2 \sim n^2$ (or equivalently $M_n \sim n$), in contradiction with the physically observed Regge behaviour. This may be viewed as a sign that the supergravity approximation breaks down when applying gauge/gravity duality in the weak form of table 5.1 to QCD. If we were able to study gauge/gravity duality with confinement in the strong form, string theory would naturally give rise to Regge behaviour. Nevertheless, even for the weak form of the correspondence, Regge behaviour may be achieved in the following way [18]. We replace the hard wall by a *soft wall*, i.e. we introduce a non-trivial dilaton in (13.105) which grows as z^2. Then, substituting

$$f = e^{B/2}\psi, \qquad B = \phi + \ln z, \tag{13.108}$$

we find

$$-\psi'' + V(z)\psi = M^2\psi, \qquad V = \frac{1}{4}(B')^2 - \frac{1}{2}B''. \tag{13.109}$$

In the IR, the potential V of (13.109) will take the form

$$V = z^2 + \frac{3}{4z^2}. \tag{13.110}$$

The Schrödinger equation with this new potential can be solved analytically and gives rise to the spectrum $M_n^2 = 4(n+1)$. Scaling behaviour of Regge type is therefore accessible in principle in the supergravity regime. However, so far it has not been possible to derive a potential of this form from the low-energy limit of string theory; the z dependence of the dilaton leading to (13.110) is an ad hoc assumption.

13.6 Further reading

An excellent overview of low-energy QCD is given in [19].

Confining flows in five-dimensional gauged supergravity were found in [4, 5]. Holographic confinement criteria from the Wilson loop were given in [3].

An example of a non-supersymmetric dilaton flow is the gravity solution found by Constable and Myers [7]. Other examples include [20]. The shielding of singularities in supergravity backgrounds is discussed in [21]. Chiral symmetry breaking obtained from embedding a D7-brane probe in the Constable–Myers background was obtained in [8], giving rise to a Goldstone boson similar to the η' in $N \to \infty$ limit. The anomaly of the $U(1)_A$ symmetry and its absence in the $N \to \infty$ limit is discussed in [22, 23]. At finite N, stringy corrections will give the η' meson a non-zero mass in the gravity picture, similarly to instantons in the field theory dual [24, 25]. The Gell-Mann–Oakes–Renner relation of effective field theory was derived in [26]. Holographic vector mesons for D7-branes were considered in [27]. A review of mesons in gauge/gravity duality is given in [14].

Lattice gauge theory computations of the ρ meson mass as function of the pion mass squared were performed by Lucini, Bali and collaborators in [9, 28, 29]. This work is part towards a long-term project towards understanding the spectrum of large N gauge theory using lattice gauge theory.

Holographic chiral symmetry breaking through a magnetic field was found by Johnson and collaborators [10]. For magnetic fields and finite temperature, see [30, 31].

The Sakai–Sugimoto model was established in [11, 12]. The backreaction in this model was addressed as an expansion in the number of D8-branes in [32].

AdS/QCD models were proposed in [15, 16]. The soft wall model leading to the correct Regge behaviour is given in [18].

Effective five-dimensional models involving a running dilaton, dual to QCD-like theories, have been studied extensively by Kiritsis together with Gürsoy, Mazzanti, Michalogiorgakis and Nitti as well as other collaborators, for a review see [33] and references therein. A further approach to AdS/QCD models, based on the light-front approach, was pursued by Brodsky and de Teramond in a series of papers [34, 35, 36].

It is also possible to consider baryons in gravity duals. This is achieved using instanton configurations, see for instance [37, 12]. Glueball spectra were investigated using gauge/gravity duality for instance in [38, 39, 40].

In our discussion of QCD and holography, we have concentrated on confinement and chiral symmetry breaking. There are many further important aspects of QCD which have been studied sucessfully using gauge/gravity duality. A very prominent aspect is deep inelastic scattering including structure functions and the pomeron, see [41, 42, 43].

When a magnetic field is applied to the QCD vacuum, a new ground state appears [44, 45]. This new ground state is a ρ meson condensate which forms a triangular lattice. Although the magnetic field required is probably too high for this effect to be observed experimentally, it is interesting to note that a similar condensation to a triangular lattice ground state is also observed within gauge/gravity duality [46, 47, 48] .

References

[1] Georgi, H. 1984. *Weak Interactions and Modern Particle Theory*. Benjamin Cummings.

[2] Maldacena, Juan Martin. 1998. Wilson loops in large N field theories. *Phys. Rev. Lett.*, **80**, 4859–4862.

[3] Sonnenschein, J., and Loewy, A. 2000. On the supergravity evaluation of Wilson loop correlators in confining theories. *J. High Energy Phys.*, **0001**, 042.

[4] Girardello, L., Petrini, M., Porrati, M., and Zaffaroni, A. 1998. Novel local CFT and exact results on perturbations of $\mathcal{N} = 4$ superYang Mills from AdS dynamics. *J. High Energy Phys.*, **9812**, 022.

[5] Girardello, L., Petrini, M., Porrati, M., and Zaffaroni, A. 2000. The supergravity dual of $\mathcal{N} = 1$ Super Yang-Mills theory. *Nucl. Phys.*, **B569**, 451–469.

[6] Girardello, L., Petrini, M., Porrati, M., and Zaffaroni, A. 1999. Confinement and condensates without fine tuning in supergravity duals of gauge theories. *J. High Energy Phys.*, **9905**, 026.

[7] Constable, Neil R., and Myers, Robert C. 1999. Exotic scalar states in the AdS/CFT correspondence. *J. High Energy Phys.*, **9911**, 020.

[8] Babington, J., Erdmenger, J., Evans, Nick J., Guralnik, Z., and Kirsch, I. 2004. Chiral symmetry breaking and pions in nonsupersymmetric gauge/gravity duals. *Phys. Rev.*, **D69**, 066007.

[9] Del Debbio, Luigi, Lucini, Biagio, Patella, Agostino, and Pica, Claudio. 2008. Quenched mesonic spectrum at large N. *J. High Energy Phys.*, **0803**, 062.

[10] Filev, Veselin G., Johnson, Clifford V., Rashkov, R. C., and Viswanathan, K. S. 2007. Flavoured large N gauge theory in an external magnetic field. *J. High Energy Phys.*, **0710**, 019.

[11] Sakai, Tadakatsu, and Sugimoto, Shigeki. 2005. Low energy hadron physics in holographic QCD. *Prog. Theor. Phys.*, **113**, 843–882.

[12] Sakai, Tadakatsu, and Sugimoto, Shigeki. 2005. More on a holographic dual of QCD. *Prog. Theor. Phys.*, **114**, 1083–1118.

[13] Evans, Nick, and Threlfall, Ed. 2007. Quark mass in the Sakai-Sugimoto model of chiral symmetry breaking. ArXiv:0706.3285.

[14] Erdmenger, Johanna, Evans, Nick, Kirsch, Ingo, and Threlfall, Ed. 2008. Mesons in gauge/gravity duals – a review. *Eur. Phys. J.*, **A35**, 81–133.

[15] Erlich, Joshua, Katz, Emanuel, Son, Dam T., and Stephanov, Mikhail A. 2005. QCD and a holographic model of hadrons. *Phys. Rev. Lett.*, **95**, 261602.

[16] Da Rold, Leandro, and Pomarol, Alex. 2005. Chiral symmetry breaking from five dimensional spaces. *Nucl. Phys.*, **B721**, 79–97.

[17] Shifman, Mikhail A., Vainshtein, A. I., and Zakharov, Valentin I. 1979. QCD and resonance physics: sum rules. *Nucl. Phys.*, **B147**, 385–447.

[18] Karch, Andreas, Katz, Emanuel, Son, Dam T., and Stephanov, Mikhail A. 2006. Linear confinement and AdS/QCD. *Phys. Rev.*, **D74**, 015005.

[19] Donoghue, J. F., Golowich, E., and Holstein, Barry R. 1992. *Dynamics of the Standard Model. Cambridge Monograph on Particle Physics, Nuclear Physics and Cosmology*, Vol. **2**, Cambridge University Press, 2nd edition, 2014.

[20] Gubser, Steven S. 1999. Dilaton driven confinement. ArXiv:hep-th/9902155.

[21] Johnson, Clifford V., Peet, Amanda W., and Polchinski, Joseph. 2000. Gauge theory and the excision of repulson singularities. *Phys. Rev.*, **D61**, 086001.

[22] Witten, Edward. 1979. Current algebra theorems for the $U(1)$ Goldstone boson. *Nucl. Phys.*, **B156**, 269.

[23] 't Hooft, Gerard. 1986. How instantons solve the $U(1)$ problem. *Phys. Rep.*, **142**, 357–387.

[24] Barbon, Jose L. F., Hoyos-Badajoz, Carlos, Mateos, David, and Myers, Robert C. 2004. The holographic life of the eta-prime. *J. High Energy Phys.*, **0410**, 029.

[25] Armoni, Adi. 2004. Witten-Veneziano from Green-Schwarz. *J. High Energy Phys.*, **0406**, 019.

[26] Gell-Mann, Murray, Oakes, R. J., and Renner, B. 1968. Behaviour of current divergences under $SU(3) \times SU(3)$. *Phys. Rev.*, **175**, 2195–2199.

[27] Evans, Nick J., and Shock, Jonathan P. 2004. Chiral dynamics from AdS space. *Phys. Rev.*, **D70**, 046002.

[28] Bali, Gunnar, and Bursa, Francis. 2007. Meson masses at large N_c. Proceedings of Science, ArXiv:0708.3427.

[29] Bali, Gunnar S., Bursa, Francis, Castagnini, Luca, Collins, Sara, Del Debbio, Luigi, *et al.* 2013. Mesons in large-N QCD. *J. High Energy Phys.*, **1306**, 071.

[30] Albash, Tameem, Filev, Veselin G., Johnson, Clifford V., and Kundu, Arnab. 2008. Finite temperature large N gauge theory with quarks in an external magnetic field. *J. High Energy Phys.*, **0807**, 080.

[31] Erdmenger, Johanna, Meyer, Rene, and Shock, Jonathan P. 2007. AdS/CFT with flavour in electric and magnetic Kalb-Ramond fields. *J. High Energy Phys.*, **0712**, 091.

[32] Burrington, Benjamin A., Kaplunovsky, Vadim S., and Sonnenschein, Jacob. 2008. Localized backreacted flavor branes in holographic QCD. *J. High Energy Phys.*, **0802**, 001.

[33] Gürsoy, Umut, Kiritsis, Elias, Mazzanti, Liuba, Michalogiorgakis, Georgios, and Nitti, Francesco. 2011. Improved holographic QCD. In *From Gravity to Thermal Gange Theories: the AdS/CFT Correspondence.*, pp. 79–146. Lecture Notes in Phys.

[34] Brodsky, S. J., and de Teramond, G. F. 2004. Light-front hadron dynamics and AdS/CFT correspondence. *Phys. Lett.*, **B582**, 211–221.

[35] Brodsky, S. J., and de Teramond, G. F. 2005. Hadronic spectrum of a holographic dual of QCD. *Phys. Rev. Lett.*, **94**, 201601.

[36] Brodsky, S. J., and de Teramond, G. F. 2009. Light-front holography: A first approximation to QCD. *Phys. Rev. Lett.*, **102**, 081601.

[37] Witten, Edward. 1998. Baryons and branes in anti-de Sitter space. *J. High Energy Phys.*, **9807**, 006.

[38] Csaki, Csaba, Ooguri, Hirosi, Oz, Yaron, and Terning, John. 1999. Glueball mass spectrum from supergravity. *J. High Energy Phys.*, **9901**, 017.

[39] de Mello Koch, Robert, Jevicki, Antal, Mihailescu, Mihail, and Nunes, Joao P. 1998. Evaluation of glueball masses from supergravity. *Phys. Rev.*, **D58**, 105009.

[40] Brower, Richard C., Mathur, Samir D., and Tan, Chung-I. 2000. Glueball spectrum for QCD from AdS supergravity duality. *Nucl. Phys.*, **B587**, 249–276.

[41] Polchinski, Joseph, and Strassler, Matthew J. 2002. Hard scattering and gauge/string duality. *Phys. Rev. Lett.*, **88**, 031601.

[42] Polchinski, Joseph, and Strassler, Matthew J. 2003. Deep inelastic scattering and gauge/string duality. *J. High Energy Phys.*, **0305**, 012.

[43] Brower, Richard C., Polchinski, Joseph, Strassler, Matthew J., and Tan, Chung-I. 2007. The pomeron and gauge/string duality. *J. High Energy Phys.*, **0712**, 005.

[44] Chernodub, M. N. 2010. Superconductivity of QCD vacuum in strong magnetic field. *Phys. Rev.*, **D82**, 085011.

[45] Nielsen, N.K., and Olesen, P. 1978. An unstable Yang-Mills field mode. *Nucl. Phys.*, **B144**, 376.

[46] Bu, Yan-Yan, Erdmenger, Johanna, Shock, Jonathan P., and Strydom, Migael. 2013. Magnetic field induced lattice ground states from holography. *J. High Energy Phys.*, **1303**, 165.

[47] Donos, Aristomenis, and Gauntlett, Jerome P. 2013. On the thermodynamics of periodic AdS black branes. *J. High Energy Phys.*, **1310**, 038.

[48] Callebaut, N., Dudal, D., and Verschelde, H. 2013. Holographic rho mesons in an external magnetic field. *J. High Energy Phys.*, **1303**, 033.

QCD and holography: finite temperature and density

14.1 QCD at finite temperature and density

A crucial aspect of the strong interaction as decribed by QCD is the study of its properties at finite temperature and density. In the past decade, significant progress has been achieved towards understanding the phase structure of the strong interaction, both experimentally and theoretically. However, many questions, in particular about the detailed structure of the QCD phase diagram, remain open. To a large extent, this is due to the strong coupling nature of QCD in the relevant energy range.

14.1.1 Phase diagram of QCD

QCD has a very non-trivial phase diagram which is only partially understood both experimentally and theoretically. The picture which is generally believed to emerge is shown schematically in figure 14.1.

As a central feature of this diagram, it is generally expected that there is a deconfinement phase transition from bound states at low temperature and chemical potential to deconfined quarks and gluons at high temperature and chemical potential. The order of this phase transition, which is expected to end in a critical point, denoted by a black dot, is not clear at present. Only at very low μ it is known that there is merely a crossover from confinement to deconfinement when increasing the temperature. Moreover, experiments at the RHIC accelerator in Brookhaven, as well as more recently at the Large Hadron Collider (LHC) at CERN, Geneva, strongly suggest that the quark–gluon plasma observed at high temperatures is still a strongly coupled state of matter for which a hydrodynamical description is appropriate. At low temperatures, when increasing the chemical potential, a phase of dense nuclear matter such as found in neutron stars is reached. At very large chemical potential a *colour-flavour locked* (CFL) superconducting phase is expected, in which colour and flavour degrees of freedom are coupled to each other.

14.1.2 Quark–gluon plasma

The quark–gluon plasma is a new state of matter which has been studied experimentally in heavy ion collisions at the RHIC accelerator in Brookhaven and more recently at the LHC at CERN. Historically, it was assumed that at high temperatures above the deconfinement transition, matter subject to the strong interaction as described by QCD essentially dissociates and forms a gas of $SU(3)$ charges or *plasma*. Over the past decade,

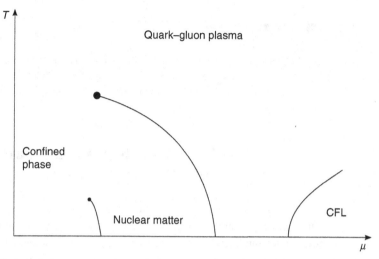

Figure 14.1   Expected QCD phase diagram.

combined experimental and theoretical advances have revealed, however, that this plasma shows a collective behaviour, which implies that it is strongly coupled and well described by hydrodynamics.

The two heavy ions which collide in a suitable detector may be visualised as spheres which, by Lorentz contraction, are transformed to a pancake-like shape. In general they collide non-centrally, which gives rise to an almond shaped region in which the quark–gluon plasma is created: the original non-equilibrium configuration thermalises and relaxes to thermal equilibrium. This equilibrium phase is again short lived, however. Since the pressure in the reaction zone is much larger at its centre than at its boundaries, the reaction zone expands and eventually the quarks and gluons *freeze out* to hadronic particles. An experimental sign of collective behaviour of the quark–gluon plasma is characterised by the *flow*. This refers to the energy and momentum distribution, as well as number distribution of the hadronic particles ejected from the reaction zone. The flow reflects the original almond shaped geometry since it has an anisotropic structure. This may be quantified in the angular distribution

$$\frac{dN}{d\phi} = \frac{v_0}{2\pi} + \frac{v_2}{\pi}\cos(2\phi) + \frac{v_4}{\pi}\cos(4\phi) + \cdots, \tag{14.1}$$

with N the number of hadrons and ϕ the azimuthal angle relative to the reaction plane. The v_i are functions of centrality, transverse momentum and rapidity which specifies longitudinal momentum, and depend on the particle type. v_2, the *elliptic flow*, is the simplest to measure. The fact that it depends on centrality, and thus on the original shape of the reaction zone, is an indication for strong coupling and collective behaviour. This in turn implies that the quark–gluon plasma is appropriately described by hydrodynamics. A non-zero v_2 implies that an original spatial asymmetry is transformed into an anisotropy of the momentum distribution.

An analysis of v_2 within hydrodynamics shows that agreement with the experimental value found at RHIC is good if the thermalisation time for the formation of the quark–gluon plasma is short, of the order of $\tau \sim 0.3$ to 1 fm/c, which is shorter than the time which light needs to cross a proton. Moreover, v_2 decreases with increasing η/s, the shear viscosity over entropy ratio whose calculation within gauge/gravity duality is presented in detail in chapter 12. The measurements of v_2 at RHIC and the level of coherence observed strongly favour a very small value of $\eta/s \leq 0.25$. The value of $\eta/s = 1/4\pi \simeq 0.08$ as obtained from gauge/gravity duality is in excellent agreement with this bound. This non-trivial agreement provided the first major success of applications of gauge/gravity duality. Other fluids such as water or liquid helium have a much larger value of η/s, and perturbative calculations also provide a much larger result.

The quark–gluon plasma displays experimental signatures of a strongly coupled fluid with very low η/s. This behaviour may be tested by further observables that can be measured in heavy-ion collisions. One of these, which is also accessible to calculations in gauge/gravity duality, is *jet quenching*. This refers to the influence of the medium produced in ultrarelativistic heavy-ion collisions on very energetic or *hard* quarks or gluons, which may be viewed as probes. Their interaction with the medium results in energy loss and in changes in the direction of their momentum. The change in momentum direction is referred to as *transverse momentum broadening*. Broadening means that the momentum distribution in a many-particle jet broadens, while 'transverse' here refers to directions perpendicular to the original direction of the hard probe. A particular consequence of jet quenching is that for a pair of back-to-back jets created at the boundary of the reaction zone, the jet immediately leaving the reaction zone is visible in the detector, while the jet travelling in the opposite direction through the medium is suppressed.

Under the assumption that the hard probes generating the jets interact weakly with each other, jet quenching and transverse momentum broadening may be described using the *jet quenching parameter*. This is defined as follows. A hard probe radiates gluons while interacting with the medium. Its transverse momentum broadening leads to a momentum distribution $P(\vec{p}_\perp)$. This is defined as the probability that after travelling a distance l through the medium, the hard probe has acquired a transverse momentum $\vec{p}_\perp$. A convenient normalisation for $P(\vec{p}_\perp)$ is

$$\int \frac{\mathrm{d}^2\vec{p}_\perp}{(2\pi)^2} P(\vec{p}_\perp) = 1. \tag{14.2}$$

The jet quenching parameter $\hat{q}$ is then defined as the mean transverse momentum acquired by the hard probe per unit distance travelled,

$$\hat{q} \equiv \frac{1}{l}\langle |\vec{p}_\perp|^2 \rangle = \frac{1}{l}\int \frac{\mathrm{d}^2\vec{p}_\perp}{(2\pi)^2} P(\vec{p}_\perp)\, |\vec{p}_\perp|^2. \tag{14.3}$$

Here the brackets $\langle\rangle$ stand for the statistical expectation value obtained from the probability distribution $P(\vec{p}_\perp)$. A central field theory result is that $\hat{q}$ as given by (14.3) is related to a correlator of Wilson lines. This is obtained by assuming that the energy loss is dominated by processes involving probe splitting into pairs at weak coupling. On the other hand, the interactions with the plasma may be strong, which leads to modifications of the probe propagators. In this case, for a representation R of $SU(N)$, the distribution $P(\vec{p}_\perp)$ is given by

$$P(\vec{p}_\perp) = \int d^2\vec{x}_\perp \, e^{-i\vec{p}_\perp \cdot \vec{x}_\perp} \mathcal{W}_R(x_\perp), \qquad (14.4)$$

where $\mathcal{W}_R$ is a correlator of Wilson lines,

$$\mathcal{W}_R(x_\perp) = \frac{1}{(\dim R)} \langle \text{Tr} \, W_R^\dagger(0, x_\perp) W_R(0, 0) \rangle, \qquad (14.5)$$

$$W_R(0, x_\perp) = \mathcal{P} \left(\exp \left[i \int_0^{\sqrt{2}l} dx^- \, A_R^+(x^+, x^-, x_\perp) \right] \right). \qquad (14.6)$$

Here, we use light-cone coordinates $x^+ = x+t$, $x^- = t-x$. W_R is the Wilson line along the light cone direction x^-. $\sqrt{2}l$ is the distance along x^- for travelling a distance l along x. The path ordering of the Wilson lines is such that a Schwinger–Keldysh contour is followed, as introduced in section 11.1.3. The first of the two lightlike Wilson lines follows the $\text{Im}\, t = 0$ segment of the Schwinger–Keldysh contour of figure 11.1, while the second follows the $\text{Im}\, t = -\delta$ segment of this contour.

Note that (14.5) is an elegant result within quantum field theory [1]. For strongly coupled plasmas, however, it is not possible to evaluate (14.5) explicitly within field theory. Nevertheless, as we will see below, (14.5) may be evaluated within gauge/gravity duality [2]. For applications of these holographic results to the quark–gluon plasma, it has to be noted that in strongly coupled $\mathcal{N} = 4$ Super Yang–Mills theory, the assumption of probes interacting weakly with each other is no longer valid. Consequently, a hybrid scenario has to be adopted in which the hard probes are assumed to be weakly interacting as in high-energy QCD, while the plasma is strongly coupled. In this approach, the Wilson loop is calculated holographically, and the result is related to energy loss using high-energy QCD.

14.2 Gauge/gravity approach to the quark–gluon plasma

As discussed in the preceding section, there is experimental evidence that the quark–gluon plasma is a strongly coupled fluid. This implies that its observables, such as transport coefficients, are hard to study using conventional methods. Lattice gauge theory is an important tool for studying strongly coupled gauge theories. However, a finite temperature is not easy to introduce, though progress has been achieved recently. Moreover, since lattice gauge theory requires Euclidean signature as discussed in box 13.1, a $U(1)$ baryon chemical potential contribution may generically lead to an imaginary contribution to e^{-S}, such that a probability density interpretation is no longer possible.

On the other hand, as discussed in chapters 11 and 12, Minkowski signature calculations of causal propagators and transport coefficients are naturally performed within gauge/gravity duality. Moreover, finite temperature and density are readily introduced by considering a black hole and a non-trivial radial profile for a gauge field present in the supergravity theory. Therefore it appears natural to use gauge/gravity duality at finite temperature and density to study the phase structure of strongly coupled gauge theories. The simplest example is to consider the five-dimensional AdS–Schwarzschild black brane, which is dual to $SU(N)$ $\mathcal{N} = 4$ Super Yang–Mills theory at finite temperature. While the finite temperature breaks all the supersymmetry and the necessary boundary conditions remove the adjoint fermions and make the adjoint scalars very massive, the field content of this theory is still different from QCD: it is a large N gluon plasma. Nevertheless, in the energy range considered, this theory still has many features in common with QCD, which motivates the comparison of gauge/gravity duality results with the features of the quark–gluon plasma described in the preceding section.

14.2.1 Energy density

A very striking coincidence of a gauge/gravity duality result with the corresponding lattice gauge theory result concerns the energy density of the quark–gluon plasma. Recall that in chapter 11 we explained how to obtain the free energy for the field theory dual to the AdS–Schwarzschild black brane, i.e. for $\mathcal{N} = 4$ Super Yang–Mills theory at finite temperature. In section 11.2.1 we found

$$F_{\text{strong coupling}} = -\frac{\pi^2}{8} N^2 T^4 \text{Vol}(\mathbb{R}^3) \tag{14.7}$$

for the free energy at strong coupling using gauge/gravity duality. This may be compared with the corresponding weak coupling perturbative result for $\mathcal{N} = 4$ theory at finite temperature. Using a heat kernel approach, it was found that to leading order

$$F_{\text{weak coupling}} = -\frac{\pi^2}{6} N^2 T^4 \text{Vol}(\mathbb{R}^3) = \frac{4}{3} F_{\text{strong coupling}}, \tag{14.8}$$

i.e. the strong coupling gauge/gravity result is a factor of three-quarters smaller than the weak coupling result. At leading order, this result is independent of the coupling.

From the free energy we can calculate the entropy $S = -\partial F/\partial T$, which coincides with the Bekenstein–Hawking entropy as discussed in chapter 11, and finally the energy density $\varepsilon = E/\text{Vol}(\mathbb{R}^3)$, obtained from the energy $E = F - TS$ as

$$\varepsilon = \frac{E}{\text{Vol}(\mathbb{R}^3)} = \frac{3\pi^2}{8} N^2 T^4. \tag{14.9}$$

This implies

$$\frac{\varepsilon}{\varepsilon_0} = \frac{3}{4}, \tag{14.10}$$

where the energy density is given by ε at strong coupling and by ε_0 at vanishing coupling. The ratio $\varepsilon/\varepsilon_0$ was also calculated within lattice gauge theory for QCD, for instance for $N = 3$ colours and $N_f = 2$ or $N_f = 3$ flavours, and also extrapolating to large N [3]. Within lattice gauge theory, it is possible to calculate ε as a function of the temperature. When increasing the temperature starting from zero, ε rises rapidly around the deconfinement transition and then reaches a plateau at an almost constant value at $\varepsilon/\varepsilon_0 \simeq 0.8$ to 0.85. This behaviour is shown in figure 14.2, together with the constant value of 0.75

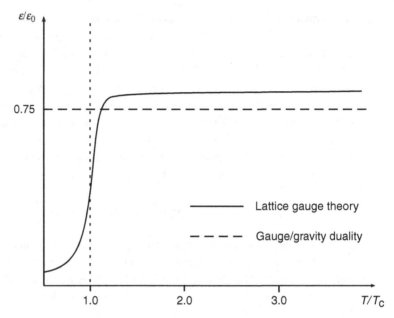

Figure 14.2 Schematic dependence of $\varepsilon/\varepsilon_0$ on the temperature in lattice gauge theory and in gauge/gravity duality. Perturbative calculations give values close to one.

from gauge/gravity duality. This should be compared to perturbative calculations at weak coupling which yield $\varepsilon/\varepsilon_0 = 1 + \mathcal{O}(g^2_{\mathrm{YM}})$.

The gauge/gravity duality result for $\mathcal{N} = 4$ Super Yang–Mills theory at finite T and the lattice gauge theory result for QCD are impressively close to each other. This leads to the expectation that there are *universal* mechanisms at work. *Universality* generally refers to the situation that the macroscopic properties of a physical system are independent of the microscopic degrees of freedom. In the present case this supports the expectation that gauge/gravity duality may provide useful statements about the deconfined phase of QCD.

14.2.2 Jet quenching

As discussed at the end of section 14.1.2, it is possible to evaluate the Wilson line correlator (14.5) using holography. In an approach where the plasma is strongly coupled while the hard probe interaction is weak, this result may be used to evaluate the jet quenching parameter (14.3).

To evaluate the Wilson line correlator holographically, the approach to holographic Wilson loops of section 13.2.1 has to be adapted to the Schwinger–Keldysh ordering of section 14.1.2 discussed below (14.5). This involves some subtleties about constructing the gravity dual of the Im t segment of the Schwinger–Keldysh contour. Here we follow a simpler procedure [4] and discuss the evaluation of (14.5) for lightlike Wilson lines and standard time ordering. We then comment on the extension to the Schwinger–Keldysh case at the end of the calculation.

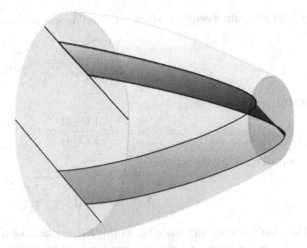

Figure 14.3 Gravity dual worldsheets for two timelike Wilson lines.

We recall that the gravity dual of a Wilson loop with contour $\mathcal{C}$ is given by the exponential of the classical action of an extremised string worldsheet which ends on $\mathcal{C}$. For the case of two long parallel Wilson lines (14.6) separated from each other by a distance $x_\perp$, we have a worldsheet hanging down all the way to the horizon for each of the two lines. At the horizon, both worldsheets join in a smooth fashion. This is shown in figure 14.3.

As discussed in section 13.2.1, the gravity dual of the Wilson loop is obtained from the Nambu–Goto action $\mathcal{S}_{NG}$, parametrising the worldsheet in five-dimensional spacetime with coordinates X^m as $X^m(\sigma, \tau)$. In the case considered here, we obtain

$$\langle \mathcal{W}(\mathcal{C}) \rangle = \exp\left(i\left[\mathcal{S}_{NG,\,reg} + \mathcal{S}_{NG,\,ct}\right]\right), \tag{14.11}$$

$$\mathcal{S}_{NG} = -\frac{1}{2\pi\alpha'} \int d\sigma\, d\tau \sqrt{-\det P[g]_{\alpha\beta}}, \qquad P[g]_{\alpha\beta} = g_{mn}\partial_\alpha X^m \partial_\beta X^n. \tag{14.12}$$

Here, g_{mn} is the metric of the AdS–Schwarzschild black brane (11.54) with horizon at $r = r_h$ and Hawking temperature $T = r_h/(\pi L)^2$. $\mathcal{S}_{NG}$ has to be regularised by introducing a convenient regulator. For the geometry considered here, a convenient counterterm $\mathcal{S}_{NG,\,ct}$ for removing the divergences is given by the Nambu–Goto action for two separate worldsheets hanging down straight to the horizon without joining. For the case that the length of the two lightlike lines $\sqrt{2}l$ is much larger than their separation $x_\perp$, the Nambu–Goto action (14.12) takes the form

$$\mathcal{S}_{NG} = i \frac{r_h^2 \sqrt{2\lambda}l}{\pi L^4} \int d\sigma \sqrt{1 + \frac{r'^2 L^4}{r^4 - r_h^4}}, \qquad r' \equiv \partial_\sigma r, \tag{14.13}$$

where we have used the standard AdS/CFT result $2\lambda = L^4/\alpha'^2$. The function $r(\sigma)$ describes the worldsheet. At $\sigma = \pm x_\perp/2$, it reaches the boundary, $r(\pm x_\perp/2) = \infty$. Moreover, the worldsheet is symmetric under $\sigma \to -\sigma$. The equation of motion obtained from (14.13) is

$$r'^2 = \frac{\gamma^2}{L^4}(r^4 - r_h^4). \tag{14.14}$$

γ is an integration constant which for $\gamma \neq 0$ is specified by

$$\frac{x_\perp}{2} = \frac{L^2}{\gamma} \int\limits_{r_h}^{\infty} \frac{dr}{\sqrt{r^4 - r_h^4}} = \frac{aL^4}{\gamma r_h}, \tag{14.15}$$

where

$$a = \sqrt{\pi} \frac{\Gamma(5/4)}{\Gamma(3/4)} \simeq 1.311. \tag{14.16}$$

The final result for the action is

$$\mathcal{S} = ia\sqrt{2\lambda} Tl \sqrt{1 + \frac{\pi^2 T^2 x_\perp^2}{4a^2}}. \tag{14.17}$$

This result is imaginary since the worldsheet is spacelike if the contour $\mathcal{C}$ is lightlike. An imaginary action ensures that the correlator (14.5) and the probability distribution $P(\vec{p}_\perp)$ are real, as they should be.

A detailed analysis of the Wilson line with the Schwinger–Keldysh ordering as described in section 14.1.2 requires a careful calculation of the worldsheet for the Lorentzian signature gravity dual. The gravity duals of the segments of the Schwinger–Keldysh contour with $\operatorname{Im} t = 0$ and $\operatorname{Im} t = -i\delta$ lie in two superposed copies of the first quadrant of Lorentzian AdS–Schwarzschild space. Both copies join at the horizon, which leads precisely to a worldsheet of the same form as considered in the calculation given above. Thus the calculation of the Wilson line with Schwinger–Keldysh path ordering will give the same result as the calculation for lightlike Wilson lines with standard time ordering of the operators. The jet quenching parameter may thus be evaluated using the action (14.17).

Exercise 14.2.1   Using the action (14.17) and the field theory results for the jet quenching parameter of section 14.1.2, show that for $\sqrt{\lambda} lT \gg 1$, where the calculation is dominated by small values of $x_\perp$, the Wilson loop correlator (14.5) is obtained as

$$\mathcal{W}(x_\perp) = \exp\left(-\frac{\pi^2}{4a}\sqrt{2\lambda} lT^3 x_\perp^2\right). \tag{14.18}$$

Moreover, show that this leads to the probability distribution

$$P(\vec{p}_\perp) = \frac{4a}{\pi\sqrt{2\lambda} lT^3} \exp\left(-\frac{a|\vec{p}_\perp|^2}{\pi^2 \sqrt{2\lambda} lT^3}\right) \tag{14.19}$$

and to the final result

$$\hat{q} = \frac{\pi^{3/2}\Gamma(3/4)}{\Gamma(5/4)}\sqrt{2\lambda} T^3 \tag{14.20}$$

for the jet quenching parameter [2].

The probability distribution (14.19) is consistent with the interpretation that the probability for the hard probe to gain transverse momentum $p_\perp$ is given by diffusion in transverse momentum space with diffusion constant $\hat{q}l$. This interpretation is consistent with the quark–gluon plasma being a strongly coupled fluid.

A rough comparison with experiment is possible by inserting experimental values into the result (14.20) for the jet quenching parameter. Choosing $T = 300$ MeV, as well as $N = 3$, $g_{YM}^2 = \pi$ leading to $\lambda = 3\pi$, we have $\hat{q} = 4.5$ GeV^2/fm, which is of the same order of magnitude as the value measured experimentally. Note that the values chosen here for both N and λ are motivated by QCD and are small compared to $N \to \infty$, λ large as necessary in principle for applying gauge/gravity duality.

To conclude this section, let us discuss briefly the case of a light quark travelling through the plasma. Initially, this corresponds to a probe; however, the energy loss is of the same order of magnitude as the quark mass, such that eventually the light quark thermalises and becomes part of the plasma. An important quantity for describing the energy loss of a light quark in the plasma is the *stopping distance*, which is given by the maximum penetration depth x of a quark of energy E until its thermalisation. The associated energy loss takes the form

$$\frac{dE}{dx} \sim q E^a, \tag{14.21}$$

where for $a = 1/2$, q may be identified with the jet quenching parameter $\hat{q}$. Different theoretical approaches lead to different values for the exponent a in (14.21). In perturbative approaches and also in the hybrid approach discussed above, i.e. for weakly interacting probes in a strongly coupled plasma, $a = 1/2$. At RHIC energies, fits to experimental data require a large q as obtained holographically above, but they also require smaller values of a. Within gauge/gravity duality, models with $a = 1/3$ are obtained by considering string probes falling in the AdS black hole geometry. Models based on holographic three-point functions typically find $a = 1/4$, with a maximal $a = 1/3$. For LHC energy scales, which are much larger than those at RHIC, the hybrid approach gives better agreement with experiment. This implies that the energy loss also depends on the coupling, which is smaller at higher energies.

14.2.3 Confinement–deconfinement transition

Here we discuss examples of realising a confinement–deconfinement phase transition within gauge/gravity duality.

Hawking–Page transition

The field theory dual of the Hawking–Page transition introduced in section 11.2.2 shows some characteristics of a confinement–deconfinement phase transition [5]. In particular in the confined phase at $T < T_{HP}$, all field theory states are singlets of the gauge group. The number of such states is of order one as compared to the rank N of the gauge group. Thus in the confined phase, we find for the free energy that $F/N^2 \to 0$ for $N \to \infty$. On the other hand, in the deconfined phase for $T \geq T_{HP}$ we have liberated charged states, of which there are as many as the number of elements in the group, i.e. of order N^2. This implies that F/N^2 is finite for $N \to \infty$.

Note however that the Wilson loop calculation of the quark–antiquark potential does not provide a confinement criterion for the Hawking–Page transition: due to the finite volume, it is not possible to have an infinite distance between a quark and an antiquark.

Compactified D4-branes

A further gauge/gravity duality realisation of the confinement–deconfinement transition is found by considering the ten-dimensional supergravity description of N D4-branes in type IIA superstring theory compactified on a circle. There are two different solutions for the metric, realised in two different temperature regimes. The transition from one to the other is interpreted as the deconfinement phase transition. In the confined phase, the compactification on the x_4 direction leads to a mass gap and confinement, as explained in section 8.4.2. On the other hand, in the deconfined phase, the time direction is compactified, leading to a black hole dual to thermal field theory.

The Euclidean metric of the confined phase is given by the Euclidean version of the metric (13.71), which is the background metric used for the Sakai–Sugimoto model,

$$ds^2_{\text{conf}} = \left(\frac{r}{L}\right)^{3/2} \left(d\tau^2 + \delta_{ij}dx^i dx^j + f(r)dx_4^2\right) + \left(\frac{L}{r}\right)^{3/2} \left(\frac{dr^2}{f(r)} + r^2 d\Omega_4^2\right) \quad (14.22)$$

using the notation of (13.71) with $\tau = it$. The point at $r = r_{\text{KK}}$ is the tip of the cigar-shaped subspace spanned by x_4 and the holographic coordinate r and

$$f(r) \equiv 1 - \left(\frac{r_{\text{KK}}}{r}\right)^3. \quad (14.23)$$

On the other hand, the subspace spanned by the Euclidean time τ and the coordinate r is cylinder shaped, with the circumference given by the inverse temperature, since we have to identify $\tau \sim \tau + 1/T$. In the deconfined phase the coordinates τ and x_4 interchange their roles, i.e. now the subspace spanned by x_4 and r is cylinder shaped while the subspace spanned by τ and r is cigar shaped. In this case, the metric is

$$ds^2_{\text{deconf}} = \left(\frac{r}{L}\right)^{3/2} \left(\tilde{f}(r)d\tau^2 + \delta_{ij}dx^i dx^j + dx_4^2\right) + \left(\frac{L}{r}\right)^{3/2} \left(\frac{dr^2}{\tilde{f}(r)} + r^2 d\Omega_4^2\right), \quad (14.24)$$

where the temperature is related to the tip of the cigar-shaped t–r space at r_T via

$$T = \frac{3}{4\pi} \frac{r_T^{1/2}}{L^{3/2}}, \quad (14.25)$$

and

$$\tilde{f}(r) \equiv 1 - \left(\frac{r_T}{r}\right)^3. \quad (14.26)$$

The deconfinement phase transition is located at a critical temperature $T = T_c$, where the free energies corresponding to the two phases are identical. This occurs at $r_{\text{KK}} = r_{T_c}$ and thus $T_c = M_{\text{KK}}/(2\pi)$. This critical temperature is independent of the chemical potential. Consequently, the model has a horizontal phase transition line in the T–μ_B plane. This agrees with expectations from QCD at large N [6].

14.3 Holographic flavour at finite temperature and density

In the preceding section we studied the gravity dual of finite temperature $\mathcal{N} = 4$ Super Yang–Mills theory, i.e. of the large N gluon plasma. As in chapter 13 where we considered quark degrees of freedom in the fundamental representation of the gauge group by adding probe branes to confining gravity backgrounds, it is also natural to introduce flavour in the finite temperature context by adding probe branes to the AdS–Schwarzschild black brane background.

14.3.1 D7-brane models

In order to study flavour degrees of freedom at finite temperature, we embed a D7-brane probe into the AdS–Schwarzschild black hole background and study its fluctuations. As we will see, this approach gives rise to a first order phase transition as function of the parameter m_q/T, with m_q the quark mass. This phase transition corresponds to the dissociation of mesons within the plasma.

The approach followed is in exact analogy to the supersymmetric D7-brane embeddings of chapter 10 and of the D7-brane description of chiral symmetry breaking in section 13.3. The dual field theory is the $\mathcal{N} = 2$ gauge theory discussed in chapter 10, now at finite temperature. We start with the AdS–Schwarzschild black brane metric (11.54) in the near-horizon limit,

$$ds^2 = \frac{r^2}{L^2}\left(-f(r)dt^2 + d\vec{x}^2\right) + \frac{L^2}{r^2 f(r)}dr^2 + L^2 d\Omega_5^2, \tag{14.27}$$

$$f(r) = 1 - \left(\frac{r_h}{r}\right)^4. \tag{14.28}$$

Recall that the dual field theory corresponding to this metric is $\mathcal{N} = 4$ Super Yang–Mills theory at finite temperature, in which supersymmetry is broken.

To embed a D7-brane in the AdS black hole background it is useful to recast the metric (14.27) to a form with an explicit flat plane in the six Euclidean directions perpendicular to the D3-branes. To this end, we change variables from r to w, such that

$$\frac{dw}{w} \equiv \frac{r\,dr}{(r^4 - r_h^4)^{1/2}}, \tag{14.29}$$

which is solved by

$$2w^2 = r^2 + \sqrt{r^4 - r_h^4}. \tag{14.30}$$

The metric is then

$$ds^2 = \frac{1}{2}\frac{w^2}{L^2}\left(-\frac{f(w)^2}{\tilde{f}(w)}dt^2 + \tilde{f}(w)d\vec{x}^2\right) + \frac{L^2}{w^2}\sum_{i=1}^{6}dw_i^2, \tag{14.31}$$

$$f(w) = \left(1 - \frac{w_h^4}{w^4}\right), \qquad \tilde{f}(w) = \left(1 + \frac{w_h^4}{w^4}\right). \tag{14.32}$$

The black hole temperature is now given by $T = w_h/(\pi L)^2$. Moreover we have

$$\sum_i dw_i^2 = dw^2 + w^2 d\Omega_5^2 = d\rho^2 + \rho^2 d\Omega_3^2 + dw_5^2 + dw_6^2 \qquad (14.33)$$

as before in chapters 10 and 13, with $\rho^2 = w_1^2 + w_2^2 + w_3^2 + w_4^2$. In the subsequent, we set $L = 1$. For the embedding functions, we consider an ansatz of the form $w_6 = w_6(\rho), w_5 = 0$. The DBI action for the embedding coordinates w_5, w_6 is

$$S_{D7} = -\mu_7 \text{Vol}(\mathbb{R}^{3,1})\text{Vol}(S^3) \int d\rho \, \mathcal{G}(\rho, w_5, w_6) \qquad (14.34)$$

$$\times \left(1 + \frac{g^{ab}}{(\rho^2 + w_5^2 + w_6^2)}\partial_a w_5 \partial_b w_5 + \frac{g^{ab}}{(\rho^2 + w_5^2 + w_6^2)}\partial_a w_6 \partial_b w_6\right)^{1/2},$$

where $\mathcal{G}(\rho, w_5, w_6) = \sqrt{-\det(g_{ab})}$ is the square-root of the determinant with respect to the world volume coordinates, which reads

$$\mathcal{G}(\rho, w_5, w_6) = \rho^3 \frac{((\rho^2 + w_5^2 + w_6^2)^2 + w_h^4)((\rho^2 + w_5^2 + w_6^2)^2 - w_h^4)}{4(\rho^2 + w_5^2 + w_6^2)^4}. \qquad (14.35)$$

With the ansatz $w_5 = 0$ and $w_6 = w_6(\rho)$, the equation of motion becomes

$$\frac{d}{d\rho}\left(\frac{\mathcal{G}(\rho, w_6)}{\sqrt{1 + \left(\frac{dw_6}{d\rho}\right)^2}}\frac{dw_6}{d\rho}\right) - \sqrt{1 + \left(\frac{dw_6}{d\rho}\right)^2}\frac{8w_h^8 \rho^3 w_6}{(\rho^2 + w_6^2)^5} = 0. \qquad (14.36)$$

The solutions of this equation determine the induced metric on the D7-brane, which is given by

$$ds^2 = \frac{1}{2}\left(\tilde{w}^2 + \frac{w_h^4}{\tilde{w}^2}\right)d\vec{x}^2 - \frac{1}{2}\frac{(\tilde{w}^4 - w_h^4)^2}{\tilde{w}^2(\tilde{w}^4 + w_h^4)}dt^2 + \frac{1 + (\partial_\rho w_6)^2}{\tilde{w}^2}d\rho^2 + \frac{\rho^2}{\tilde{w}^2}d\Omega_3^2, \quad (14.37)$$

with $\tilde{w}^2 = \rho^2 + w_6^2(\rho)$. The D7-brane metric becomes $AdS_5 \times S^3$ for $\rho \gg w_h$.

14.3.2 First order phase transition at finite temperature

We now compute the D7-brane solutions explicitly. As in chapter 10, the UV asymptotic (large ρ) solution, where the geometry returns to $AdS_5 \times S^5$, is of the form

$$w_6(\rho) \sim l_q + \frac{c}{\rho^2}. \qquad (14.38)$$

The parameters m and c are related to the quark mass and bilinear quark condensate $\langle \bar{\psi}\psi \rangle$, respectively. A similar analysis as in exercise 10.2.1 yields

$$m_q = \frac{1}{2}\sqrt{\lambda}Tl_q, \qquad \langle \bar{\psi}\psi \rangle = -\frac{1}{8}\sqrt{\lambda}NT^3 c. \qquad (14.39)$$

These parameters provide the boundary conditions for the second order differential equation (14.36). For a given value of m, as defined in the caption of figure 14.4 c is fixed by requiring regularity throughout the space. The equation of motion has to be solved numerically. The numerical solutions obtained using a *shooting technique* are illustrated in

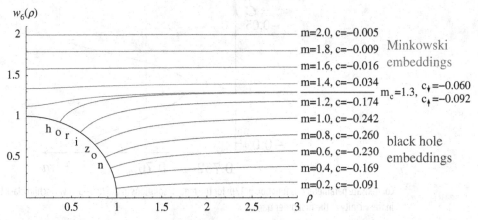

Figure 14.4 Two classes of regular solutions in the AdS black hole background. The quark mass m_q is the parameter m in units of $\Lambda \equiv \frac{w_h}{2\pi\alpha'}$: $m_q = m\Lambda$. We set $\Lambda = w_h = 1$.

figure 14.4 for several choices of m. We choose units such that the horizon is represented as a quarter circle with radius $w_h = 1$.

As can be seen from the figure, there are two qualitatively different D7-brane embeddings. At large quark masses the D7-brane tension is stronger than the attractive force of the black hole. The D7-brane ends at a point outside the horizon, $\rho = 0$, $w_6 \geq w_h$, at which the S^3 wrapped by the D7-brane collapses (see (14.37)). Such a D7-brane solution is referred to as a *Minkowski* embedding. It is similar to the supersymmetric solutions in $AdS_5 \times S^5$. As the mass decreases, there is a critical value of the quark mass parameter at $m = m_c = 1.307$ at which a stable embedding reaches the black hole horizon. The D7-brane ends at the horizon $w = w_h$, at which the S^1 of the black hole geometry collapses. This is a so-called *black hole* embedding. Note that for the critical quark mass, there are two regular embeddings, one with condensate given by $c_\downarrow = -0.060$ for which the embedding is still Minkowski, and one with condensate given by $c_\uparrow = -0.092$, for which the embedding is black hole. This multi-valuedness indicates a first-order phase transition, as we will confirm below. For even smaller values of the quark mass, all stable embeddings are of black hole type.

From a geometrical point of view, the two classes of embeddings differ in their topology. For Minkowski embeddings, the topology is $\mathbb{R}^3 \times B^4 \times S^1$, while it becomes $\mathbb{R}^3 \times S^3 \times B^2$ for black hole embeddings. The change in the embedding topology at m_c points to a phase transition in the dual field theory at this critical value of the quark mass.

The dependence of the condensate on the mass is illustrated in figure 14.4. At $m = 0$ the condensate c is zero (the brane lies flat), so there is no spontaneous chiral symmetry breaking in this gauge theory. As m increases, the condensate c initially increases and then decreases again. At sufficiently large m, the condensate becomes negligible, which is to be expected as the D7-brane ends in the region where the deformation of AdS space is small. Recall that there is no condensate in the Yang–Mills theory with unbroken $\mathcal{N} = 2$ supersymmetry described by D7-branes in undeformed AdS space. Once supersymmetry is broken by the temperature, the presence of a chiral condensate c is generally expected.

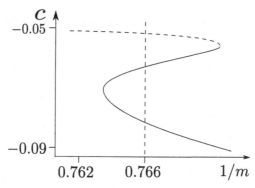

Figure 14.5   Condensate parameter c as function of $1/m$ for the regular solutions in the AdS–Schwarzschild black brane background, in the vicinity of the first order transition.

Zooming in around m_c, we see in figure 14.5 that c is multi-valued around the critical mass m_c. The phase transition at m_c is thus of first order. For a given quark mass in the regime $1.295 \leq m \leq 1.308$, there exist both Minkowski and black hole embeddings. These solutions have the same quark mass m, but different values for the quark condensate c. The exact value of c at the phase transition may be found by a Maxwell construction. The transition happens once for a given m, the black hole solution has lower free energy.

The plot of c versus $1/m$ may also be considered as a plot of the condensate c versus the temperature, since all dimensionful quantities are normalised by the temperature by setting $r_h = 1$. For this we keep the quark mass m fixed and vary the horizon $w_h \sim T$. Then, for small temperatures we recover the Minkowski embeddings, while for high temperatures we have black hole embeddings. Heating up the system from zero temperature, we eventually reach a critical temperature T_c at which further supply of external energy does not increase the temperature of the system. Rather it leads to the formation of a quark condensate. The jump in the quark condensate shows that the phase transition is discontinuous and thus of first order. It is remarkable that this phase transition occurs for a black hole background, which is dual to a deconfined field theory. As we will discuss in the next section, this transition corresponds to meson melting.

14.3.3   Mesons in the AdS black hole background

The physical nature of the phase transition introduced in the previous section is revealed through the behaviour of the mesons in the two phases.

In the Minkowski phase, i.e. when the D7-brane probe does not reach the black hole horizon, the meson spectrum is similar to that in the zero temperature theory. We may study perturbations of the D7-brane about the background embedding of the form $h(\rho)e^{ipx}$, with $p^2 = -M^2$. In the present context, it is convenient to choose with $p = (\omega, 0, 0, 0)$, such that we have $M^2 = \omega^2$. Requiring regularity for $h(\rho)$ determines the allowed meson masses M, which are real numbers.

On the other hand, in the black hole phase where the D7-brane probe terminates on the horizon, the mesons become unstable and decay. In this case, there are no regular mesonic fluctuations with real masses. Instead, the fluctuations have quasinormal modes as discussed in chapter 12, given by complex frequencies ω. These arise from fluctuations of the D7-branes that are purely infalling waves at the horizon. These fluctuation modes experience dissipation. The mass that is extracted from these solutions is complex. This is interpreted as the fact that the mesons are not stable in the thermal plasma, and *melt* into the plasma with a characteristic decay width determined by the imaginary part of the quasinormal eigenfrequency.

Let us consider the fluctuations about a black hole embedding, where we will find complex eigenfrequencies corresponding to melting mesons. Both for the two scalar fluctuation modes perpendicular to the D7-brane and for the vector modes dual to gauge field fluctuations on the brane, a quasinormal mode and spectral function analysis may be performed as in chapter 12. As an example, let us consider the vector modes at zero momentum. We apply the analysis of chapter 12 to gauge field fluctuations governed by the DBI action for the D7-brane probe. To obtain the quasinormal modes, we linearise the equation of motion of the D7-brane around the equilibrium configuration. For simplicity we consider the D7-brane embedding with $m = 0$, which is flat with $w_6(\rho) = 0$ and reaches the black hole horizon.

We consider the fluctuations for the electric field $E_i = i\omega A_i$ at zero momentum. Since we consider black hole embeddings, it is convenient to use the coordinates defined in (14.45) and (14.46). The calculation, given in [7], is analogous to the calculation performed in section 12.1.3, except for the fact that now the DBI action and the induced metric for the black hole embedding have to be used. The DBI action is expanded to second order in the gauge fields using the induced metric. Moreover, the fluctuations of the gauge field on the brane are expanded in Kaluza–Klein modes on S^3 wrapped by the D7-brane, such that

$$A_m = \sum_l \mathcal{Y}^\ell(S^3) A_m^\ell(\rho, x^\mu), \qquad \nabla^2 \mathcal{Y}^\ell = -\ell(\ell+2)\mathcal{Y}^\ell, \qquad (14.40)$$

and $E_i^\ell = i\omega A_i^\ell$. Changing variables to $x \equiv 1 - 2w_{\text{h}}^2/(w^2\tilde{f})$, and taking extra care to ensure a regular behaviour of the two-point function at the boundary as in section 5.4.3, as well as a correct scaling with temperature, we arrive at the equation, with $\mathfrak{w} = \omega/(2\pi T)$,

$$E_m''^{\,\ell} + \frac{f'}{f}E_m'^{\,\ell} + \left(\frac{\mathfrak{w}^2}{(1-x)f^2} - \frac{\ell(\ell+2)}{4(1-x)2f}\right) E_m^{\,\ell} = 0, \qquad (14.41)$$

which is to be compared to (12.61) in section 12.1.2. Equation (14.41) has two linearly independent solutions. We choose the solution which involves a factor $(1-x)^{\frac{(1+i)\mathfrak{w}}{2}}$ and thus satisfies the infalling boundary condition. The spectral function $\mathcal{R} = -2\text{Im}\, G_{\text{R}}$ is obtained from

$$G_{\text{R}} = \frac{\pi^{2\ell}}{2^{\ell+3}} NT^{2l+2} \lim_{w \to \infty} \left(w^{2\ell+3} \frac{\partial_w E_m^{\,\ell}(w)}{E_m^{\,\ell}(w)}\right). \qquad (14.42)$$

Its poles determine the quasinormal modes

$$\mathfrak{w} = \pm \left(n + 1 + \frac{l}{2} \right) (1 \mp i), \qquad n = 0, 1, \ldots, \tag{14.43}$$

similarly to the result (12.68) of chapter 12. Here, there is a symmetry degeneracy of the quasinormal modes since they depend only on the linear combination $n + l/2$.

The main physical characteristic of the phase transition is the melting of the mesons into the background thermal plasma, i.e. the transition from real eigenfrequencies for the fluctuations about Minkowski embeddings to complex eigenfrequencies for fluctuations about black hole embeddings as given in (12.68). Note that since the temperature is $T = r_{\rm h}/(L^2 \pi)$ with $L = 4\pi \lambda \alpha'^2$ and the transition occurs when $m \sim r_{\rm h}$, the temperature scale of the transition is

$$T_{\rm c} \sim \frac{m_q(2\pi \alpha')}{\sqrt{\lambda} \alpha' \pi} \sim \frac{2m_q}{\sqrt{\lambda}}. \tag{14.44}$$

The transition occurs at a temperature of order of the meson mass.

14.3.4 D7-brane embeddings at finite density

As discussed in section 11.3, a chemical potential and finite density for the flavour fields are introduced by considering a non-trivial radial profile for the time component of the gauge field on the brane. Here we discuss the case of both finite temperature and finite density.

In generalisation to the zero temperature case considered in chapter 11, a finite chemical potential and density are obtained by considering a non-trivial gauge field profile $A_t = A_t(\rho)$ on the brane. When studying D7-brane embeddings in the presence of this profile, it is found that all embeddings are of black hole type, i.e. in the IR they reach the black hole horizon. The absence of Minkowski embeddings is consistent with the fact that at finite quark density, only black hole embeddings are physical. This is seen as follows. The finite density is obtained from a non-trivial radial profile for the gauge field, which implies that there is a non-vanishing electric field in the radial direction, $E_\rho = F_{\rho t}$, also near the boundary. Due to Gauss' law, this field must be sourced by an electric charge in the bulk. If the brane is connected to the black hole as in the case of Minkowski embeddings, this charge can be hidden inside the black hole. For large quark mass over temperature ratio, i.e. nearly flat embeddings, a spike forms close to the horizon in order to connect the the brane to the black hole.

To study the black hole embeddings, it is useful to introduce a new set of coordinates in the following way. We take the AdS–Schwarzschild black brane metric (14.31) and write its six-dimensional plane (14.33) as

$$\begin{aligned} dw^2 + w^2 d\Omega_5^2 &= d\rho^2 + \rho^2 d\Omega_3^2 + dl_q^2 + l_q^2 d\phi^2, \\ &= dw^2 + w^2(d\theta^2 + \cos^2\theta\, d\phi^2 + \sin^2\theta\, d\Omega_3^2), \end{aligned} \tag{14.45}$$

where $\rho = w\sin\theta$, $w^2 = \rho^2 + l_q^2$ and $l_q = w\cos\theta$. The coordinates (θ, w) determine the black hole embedding. Moreover, we define $\chi = \cos\theta$. In these variables, the boundary conditions imposed at the black brane horizon read

$$\chi(w = w_h) = \chi_0, \qquad \chi'(w = w_h) = 0. \tag{14.46}$$

The parameter $\chi_0 = \cos\theta_0$ determines the angle under which the brane falls into the horizon. For the massless embedding at $m_q = 0$ we have $\theta_0 = \pi/2$, $\chi_0 = 0$ which corresponds to the equator of S^3.

In coordinate system (14.45), with $\varrho \equiv \rho/\rho_h$, the DBI action for the D7-brane is then given by

$$S_{\text{DBI}} = -N_f\mu_7 \int d^8\xi \frac{\varrho^3}{4} f\tilde{f}(1-\chi^2)\sqrt{1 - \chi^2 + \varrho^2(\partial_\varrho\chi)^2 - 2(2\pi\alpha')^2\frac{\tilde{f}}{f^2}(1-\chi^2)F_{\varrho t}^2}, \tag{14.47}$$

with $E_\varrho = F_{\varrho t} = \partial_\varrho A_t$ the radial electric field. From this action we obtain the equation of motion for the radial electric field which is a constant of motion. Moreover, near the boundary we have the leading asymptotic behaviour

$$A_t \sim \mu - \frac{d}{\varrho^2}. \tag{14.48}$$

As discussed in chapter 11, μ corresponds to the chemical potential and d is related to the density by (11.132), which gives

$$n_{\text{B}} = \frac{\delta S_{\text{DBI}}}{\delta F_{\varrho t}} = N_f\mu_{\text{D7}}(2\pi\alpha')^2 d, \tag{14.49}$$

The index B in n_{B} refers to the fact that the $U(1)$ symmetry considered may be interpreted as a baryon symmetry. Using a Legendre transform, we may eliminate A_t from the action in favour of d,

$$\tilde{S}_{\text{DBI}} = S_{\text{DBI}} - \int d^8\xi \, F_{\varrho t}\frac{\delta S_{\text{DBI}}}{\delta F_{\varrho t}}. \tag{14.50}$$

From the Legendre transformed action, we obtain the embeddings of the brane probes by numerically solving the corresponding equations of motion. The result is shown in figure 14.6, which clearly shows that all embeddings are black hole embeddings.

Exercise 14.3.1 Starting from the DBI action (14.47), perform the Legendre transformation (14.50) explicitly.

Exercise 14.3.2 Using the Legendre transformed action as a starting point, show that the brane develops a spike of strings near the black hole horizon, i.e. for $\chi \simeq 1$ which corresponds to $\theta \simeq 0$, the Legendre transformed action takes the asymptotic form

$$\tilde{S}_{\text{DBI}} \propto \frac{d}{2\pi\alpha'} \int dt \, d\varrho \sqrt{-g_{tt}(g_{\varrho\varrho} + g_{\theta\theta}(\partial_\varrho\theta)^2)}. \tag{14.51}$$

This corresponds to the Nambu–Goto action for a bundle of fundamental strings stretching in the ϱ direction which is free to bend away from $\theta = 0$ on S^5.

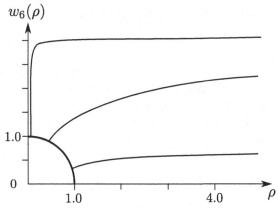

Figure 14.6   Schematic behaviour of D7-brane embeddings at finite density.

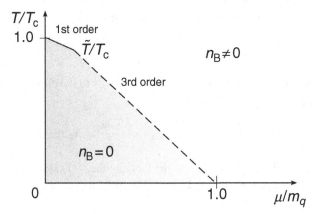

Figure 14.7   Phase diagram for D7-branes in the AdS–Schwarzschild geometry at finite baryon density: The quark chemical potential μ_q divided by the quark mass is plotted versus the temperature T divided by $\bar{M} = 2m_q/\sqrt{\lambda}$.

For Minkowski embeddings, which are not connected to the black hole, the electric charge and field must vanish. However, even for Minkowski embeddings a constant time component of the gauge field is possible. This leads to a finite chemical potential, but zero density.

14.3.5   Phase diagram and spectral functions

As an example we consider here the phase diagram in the (T, μ) plane and discuss the behaviour of the spectral functions in the different phases. The phase diagram is displayed in figure 14.7.

This phase diagram takes the following structure. In the grey shaded area, the density n_B vanishes, while it is non-zero outside this area. For non-zero density, only black hole embeddings are consistent due to charge conservation. At vanishing chemical potential, at $T = T_c$ there is the first order meson melting phase transition between Minkowski embeddings and black hole embeddings discussed in section 14.3.3. This transition persists for small values of the chemical potential. For larger values of the chemical potential, or correspondingly below a further critical temperature $\tilde{T}$, the transition becomes third order [8]. This is due to worldsheet instanton corrections in the DBI action of the probe brane, which turn the Minkowski embeddings into black hole embeddings even at zero density. Then, the transition between different black hole embeddings is smooth. At $T = 0$, there is again a second order phase transition, see section 11.3.2.

Let us consider spectral functions at non-zero density. According to the phase diagram, within the finite density phase, for fixed quark mass there is a temperature dominated region for large temperatures and small chemical potential, and a potential dominated region for small temperatures and large chemical potential. In the two regions, the spectral functions, which have to be evaluated numerically, show qualitatively different behaviour. We consider the spectral functions for the current–current correlator coupling to the gauge field on the D7-brane. To calculate these, the linear response formalism presented in chapter 12 is adapted to the gauge field fluctuations on the D7-brane probe. At vanishing momentum, the spectral function $\mathcal{R}(\omega) = -2\text{Im}\, G_R(\omega)$ is obtained from

$$G_R = \frac{NT^2}{8} \lim_{\rho \to \infty} \left(\rho^3 \frac{\partial_\rho E(\rho)}{E(\rho)} \right) \tag{14.52}$$

where we consider the gauge invariant electric field $E = E_i = i\omega A_i$ for each of the three boundary spatial directions. The gauge potential A_i which enters (14.52) corresponds to linearised fluctuations about the background gauge field with asymptotic behaviour (14.48). The equations of motion for these fluctuations are obtained from the DBI action (14.47). The operator dual to these fluctuations is a vector meson which in some respects is similar to the ρ meson. The result of the spectral function computation is displayed in figure 14.8. In the temperature dominated region, the spectral function, i.e. the imaginary part of the retarded Green function, displays very broad peaks corresponding to unstable vector mesons. This is shown on the left-hand side of the figure. In the potential dominated region, however, the peaks become very narrow and their location coincides exactly with the supersymmetric meson spectrum discussed earlier in chapter 10.

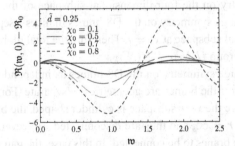
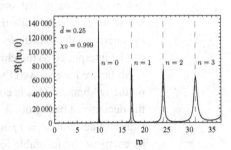

Figure 14.8 Finite density: the spectral function $\mathfrak{R} - \mathfrak{R}_0$ (in units of $NT^2/4$) in the temperature dominated region (left plot) and in the potential dominated region (right plot). $\mathfrak{R}_0$ corresponds to the spectral function at zero temperature given in (12.67). Figures from [9]

14.4 Sakai–Sugimoto model at finite temperature

The Sakai–Sugimoto model provides a compelling geometrical picture for the confinement–deconfinement transition together with the transition between the phases with broken and restored chiral symmetry. For this purpose we consider the embedding of $D8$-brane and $\overline{D8}$-brane probes into the geometry displaying a confinement–deconfinement transition as discussed in section 14.2.3.

The D8- and $\overline{D8}$-branes extend in all dimensions except for the coordinate x_4, whereas the D4-branes extend in the $\tau, x_i, i = 1, \ldots, 4$ directions. The induced metrics on the probe branes in the confined and deconfined backgrounds are

$$ds^2_{\text{D8,conf}} = \left(\frac{r}{L}\right)^{3/2} \left(d\tau^2 + \delta_{ij}dx^i dx^j\right)$$
$$+ \left(\frac{L}{r}\right)^{3/2} \left(\frac{v^2(r)}{f(r)}dr^2 + r^2 d\Omega_4\right), \tag{14.53}$$

$$ds^2_{\text{D8,deconf}} = \left(\frac{r}{L}\right)^{3/2} \left(\tilde{f}(r)d\tau^2 + \delta_{ij}dx^i dx^j\right)$$
$$+ \left(\frac{L}{r}\right)^{3/2} \left(\frac{\tilde{v}^2(r)}{\tilde{f}(r)}dr^2 + r^2 d\Omega_4\right), \tag{14.54}$$

with $f(r)$ and $\tilde{f}(r)$ as in (14.23) and (14.24), and where we have abbreviated

$$v(r) \equiv \sqrt{1 + f^2(r)\left(\frac{r}{L}\right)^3 (\partial_r x_4)^2}, \qquad \tilde{v}(r) \equiv \sqrt{1 + \left(\frac{r}{L}\right)^3 (\partial_r x_4)^2}. \tag{14.55}$$

$x_4(r)$ gives the embedding of the D8-branes in the x_4–r subspace.

The D4/D8–$\overline{D8}$ setup provides the tools to study not only the deconfinement phase transition, but also the chiral phase transition. In the x_4 direction, the D8-branes are separated from the $\overline{D8}$-branes by a distance l. The maximal separation of the branes is $l = \pi/M_{\text{KK}}$, in which case the branes are attached at antipodal points of the circle spanned by x_4. As described in section 13.4, gauge fields on the D8-branes and $\overline{D8}$-branes transforming under a local symmetry group $U(N_{\text{f}})$ induce a global symmetry group $U(N_{\text{f}})$ on the five-dimensional boundary at $r = \infty$. More precisely, a gauge symmetry on the D8-branes induces a global symmetry on the four-dimensional subspace of the holographic boundary at $x_4 = 0$, while the gauge symmetry on the $\overline{D8}$-branes induces a separate global symmetry on the four-dimensional subspace at $x_4 = l$. Therefore the total global symmetry can be interpreted as the chiral group $U(N_{\text{f}})_L \times U(N_{\text{f}})_R$.

So far we have viewed the gauge symmetry on the D8-branes as independent from that on the $\overline{D8}$-branes. This is correct if the branes are geometrically separate. For example, in the deconfined background, where the x_4–r subspace is cylinder shaped, the branes follow straight lines from $r = r_T$ up to $r = \infty$, and thus are disconnected. However, it may also be energetically favorable for the branes to be connected. In this case, the gauge symmetry reduces to joint rotations, given by the vectorial subgroup $U(N_{\text{f}})_{L+R}$. This is exactly the symmetry breaking pattern induced by a chiral condensate. In fact, in the confined phase, where the x_4–r subspace is cigar shaped, the branes *must* connect. In other words, chiral symmetry is always broken in the confined phase. Whether the branes are disconnected in the deconfined phase depends on the separation scale l. For sufficiently large l, they are always disconnected, while for smaller l the connected phase may be favoured for certain temperatures [10]. In other words, in the former case, deconfinement and the chiral phase transition are identical, while in the latter case they differ, and there exists a deconfined but chirally broken phase in the T–μ_B plane [11].

14.5 Holographic predictions for the quark–gluon plasma

Let us conclude this chapter with a summary of the status of the connections between gauge/gravity duality and low-energy QCD. The most striking result from holography is of course the ratio of shear viscosity over entropy density, $\eta/s = \hbar/(4\pi k_B)$. This result is in very good agreement with experimental observations. Moreover, it established the relevance of applications of gauge/gravity duality to the quark–gluon plasma: these provide a successful approach to studying transport properties in strongly coupled systems. The small value for η/s indicates that the quark–gluon plasma is the most strongly coupled fluid known. More recently, this ratio has also been calculated for $SU(3)$ Yang–Mills theory within lattice gauge theory, where the result obtained is $\eta/s = 0.102(56)\hbar/k_B$ at $T = 1.24T_c$, with T_c the deconfinement temperature [12]. This value increases slightly if the temperature is raised.

Moreover, further transport coefficients in the hydrodynamic expansion have been calculated using gauge/gravity duality. One example is the bulk viscosity which is non-zero for non-conformal theories. Within gauge/gravity duality, the bulk viscosity ζ was found to be related to the shear viscosity by $\zeta \sim 4.6\eta(1/3 - c_s^2)$ [13, 14]. This is in contrast to the perturbative result $\zeta \sim 15\eta(1/3 - c_s^2)^2$ [15]. Within lattice gauge theory for the $SU(3)$ gauge group, the bulk viscosity was found to be strongly temperature dependent [16]; it is negligible at very high temperatures, while relevant near the deconfinement transition at T_c. Moreover, for $T = 1.24T_c$, the lattice calculations show that the ratio of bulk over shear viscosity is a factor of six larger than expected from the perturbative result.

A further example of transport coefficients obtained holographically are second order coefficients in the hydrodynamic expansion [17, 18, 19]. These second order coefficients are needed for stable simulations of hydrodynamics, for which the holographic results are routinely used.

In addition, there are anomalous transport coefficients, for instance the chiral vortical effect discussed in section 12.4.2, and its cousin, the chiral magnetic effect, which corresponds to the response to an external magnetic field [20]. In a medium with axial chemical potential μ_5 for the axial symmetry $U(1)_A$, which may be generated by a sphaleron background in QCD, there are currents of the form

$$\vec{J}^b = \frac{N\mu_5}{2\pi^2}\left(a_B^b\,\vec{B} + 2\,a_\omega^b\,\mu\,\vec{\omega}\right), \qquad \vec{J}^e = \frac{N\mu_5}{2\pi^2}\left(a_B^e\,\vec{B} + 2\,a_\omega^e\,\mu\,\vec{\omega}\right). \qquad (14.56)$$

Here, $\vec{J}^b$ and $\vec{J}^e$ are the baryon charge and electric currents, respectively. $\vec{B}$ is a magnetic field and $\vec{\omega} = \nabla \times \vec{v}$ is the vorticity. The coefficients a_B^b, a_ω^b, a_B^e, a_ω^e are obtained from the axial anomaly by calculating the appropriate one-loop triangle diagram. As we saw in section 12.4.2, within gauge/gravity duality, the anomaly arises from a Chern–Simons term on the gravity side, which generates the anomalous transport terms. The chiral and vortical effects lead to predictions for experimental observables. In particular, they predict a baryon number separation of the same sign as the charge separation. It is not clear at present whether in heavy-ion experiments, $\vec{B}$ and $\vec{\omega}$ are large enough to generate observable effects; experiments to address this question are under way.

Another important aspect of quark–gluon plasma physics is thermalisation. When the heavy ions collide, a non-equilibrium state is formed, which then relaxes to thermal equilibrium. There are different models for this within gauge/gravity duality. One possibility is to consider colliding shock waves in the dual gravity theory [21, 22]. Another possibility is to investigate the collapse of a matter shell and black hole formation in asymptotically AdS space [23, 24]. This is a large research area with many studies under way. The gauge/gravity duality results obtained so far imply in particular that for strongly coupled systems, the relaxation time is very short [25].

14.6 Further reading

An extensive review of finite temperature QCD, the quark–gluon plasma and applications of gauge/gravity duality may be found in [4]. A review of lattice gauge theory at finite temperature and density, in particular explaining the calculation of the energy density, is given in [26]. A more recent review of lattice results for large N theories at finite temperature is [27], which contains a wealth of useful references. The black D3-brane result for the energy density was obtained in [28]. At weak coupling, the energy density in $\mathcal{N} = 4$ Super Yang–Mills theory was investigated for instance in [29, 30, 31, 32].

The result (14.5) for the jet quenching parameter was obtained within field theory in [1]. This expression was evaluated using gauge/gravity duality in [2].

Energy loss for a light quark was investigated using falling strings in [33, 34] and using three-point functions in [35].

As a further application of gauge/gravity duality not discussed here, a long D3–D7 string describing a heavy deconfined quark and the energy loss and wake produced by such a string dragged through the plasma was studied in [36, 37, 38, 39].

The lattice gauge theory result for η/s was obtained in [12], and for the bulk viscosity in [16].

The geometric transition between Minkowski and black hole embeddings was discovered in [40] and was shown to be first order in [41]. The meson melting at this first order phase transition was analysed using quasinormal modes in [42]. An extensive analysis of the thermodynamics of $D7$-branes and other brane probes was carried out by Mateos, Myers and collaborators [43, 44, 45]. The evolution of the quasinormal modes at large T into the stable mesons at small T was explicitly followed in [7] using spectral function. Meson spectra at finite density from D7-brane probes were investigated in [9].

The Sakai–Sugimoto model at finite temperature was studied in [10, 46, 47]. Its phase diagram was studied in [48]. Witten interpreted the Hawking–Page transition as the gravity dual of a deconfinement phase transition in [5].

Finite temperature models based on effective five-dimensional theories dual to QCD-like theories in four dimensions, as reviewed in [49], were also studied by Kajantie together with Alanen, Suur-Uski, Tahkokallio, Vuorinen and further collaborators, see for instance [50, 51] and references therein. The results of Kiritsis et al. [49] also allow to obtain the full result displayed in figure 14.2 using AdS/QCD, achieving impressive agreement with the lattice gauge theory result [49].

References

[1] Wiedemann, Urs Achim. 2000. Gluon radiation off hard quarks in a nuclear environment: opacity expansion. *Nucl. Phys.*, **B588**, 303–344.

[2] Liu, Hong, Rajagopal, Krishna, and Wiedemann, Urs Achim. 2006. Calculating the jet quenching parameter from AdS/CFT. *Phys. Rev. Lett.*, **97**, 182301.

[3] Panero, Marco. 2009. Thermodynamics of the QCD plasma and the large N limit *Phys. Rev. Lett.*, **103**, 232001.

[4] Casalderrey-Solana, Jorge, Liu, Hong, Mateos, David, Rajagopal, Krishna, and Wiedemann, Urs Achim. 2014. *Gauge/string duality, hot QCD and heavy ion collisions*. Cambridge University Press.

[5] Witten, Edward. 1998. Anti-de Sitter space, thermal phase transition, and confinement in gauge theories. *Adv. Theor. Math. Phys.*, **2**, 505–532.

[6] McLerran, Larry, and Pisarski, Robert D. 2007. Phases of cold, dense quarks at large N_c. *Nucl. Phys.*, **A796**, 83–100.

[7] Myers, Robert C., Starinets, Andrei O., and Thomson, Rowan M. 2007. Holographic spectral functions and diffusion constants for fundamental matter. *J. High Energy Phys.*, **0711**, 091.

[8] Faulkner, Thomas, and Liu, Hong. 2008. Condensed matter physics of a strongly coupled gauge theory with quarks: some novel features of the phase diagram. ArXiv:0712.4278.

[9] Erdmenger, Johanna, Kaminski, Matthias, and Rust, Felix. 2008. Holographic vector mesons from spectral functions at finite baryon or isospin density. *Phys. Rev.*, **D77**, 046005.

[10] Aharony, Ofer, Sonnenschein, Jacob, and Yankielowicz, Shimon. 2007. A holographic model of deconfinement and chiral symmetry restoration. *Ann. Phys.*, **322**, 1420–1443.

[11] Horigome, Norio, and Tanii, Yoshiaki. 2007. Holographic chiral phase transition with chemical potential. *J. High Energy Phys.*, **0701**, 072.

[12] Meyer, Harvey B. 2007. A calculation of the shear viscosity in $SU(3)$ gluodynamics. *Phys. Rev.*, **D76**, 101701.

[13] Benincasa, Paolo, Buchel, Alex, and Starinets, Andrei O. 2006. Sound waves in strongly coupled non-conformal gauge theory plasma. *Nucl. Phys.*, **B733**, 160–187.

[14] Buchel, Alex. 2008. Bulk viscosity of gauge theory plasma at strong coupling. *Phys. Lett.*, **B663**, 286–289.

[15] Weinberg, Steven. 1971. Entropy generation and the survival of protogalaxies in an expanding universe. *Astrophys. J.*, **168**, 175.

[16] Meyer, Harvey B. 2008. A calculation of the bulk viscosity in $SU(3)$ gluodynamics. *Phys. Rev. Lett.*, **100**, 162001.

[17] Bhattacharyya, Sayantani, Hubeny, Veronika E., Minwalla, Shiraz, and Rangamani, Mukund. 2008. Nonlinear fluid dynamics from gravity. *J. High Energy Phys.*, **0802**, 045.

[18] Baier, Rudolf, Romatschke, Paul, Son, Dam Thanh, Starinets, Andrei O., and Stephanov, Mikhail A. 2008. Relativistic viscous hydrodynamics, conformal invariance, and holography. *J. High Energy Phys.*, **0804**, 100.

[19] Haack, Michael, and Yarom, Amos. 2009. Universality of second order transport coefficients from the gauge-string duality. *Nucl. Phys.*, **B813**, 140–155.

[20] Kharzeev, Dmitri E., and Son, Dam T. 2011. Testing the chiral magnetic and chiral vortical effects in heavy ion collisions. *Phys. Rev. Lett.*, **106**, 062301.

[21] Grumiller, Daniel, and Romatschke, Paul. 2008. On the collision of two shock waves in AdS$_5$. *J. High Energy Phys.*, **0808**, 027.

[22] Chesler, Paul M., and Yaffe, Laurence G. 2009. Horizon formation and far-from-equilibrium isotropization in supersymmetric Yang-Mills plasma. *Phys. Rev. Lett.*, **102**, 211601.

[23] Lin, Shu, and Shuryak, Edward. 2008. Toward the AdS/CFT gravity dual for high energy collisions. 3. Gravitationally collapsing shell and quasiequilibrium. *Phys. Rev.*, **D78**, 125018.

[24] Erdmenger, Johanna, and Lin, Shu. 2012. Thermalization from gauge/gravity duality: evolution of singularities in unequal time correlators. *J. High Energy Phys.*, **1210**, 028.

[25] van der Schee, W. 2013. Holographic thermalization with radial flow. *Phys. Rev.*, **D87**, 061901.

[26] Karsch, Frithjof. 2002. Lattice QCD at high temperature and density. In *Lectures on Quark Matter*, pp. 209–249. Lecture Notes in Physics, Vol. 583. Springer.

[27] Panero, Marco. 2012. Recent results in large-N lattice gauge theories. Proceedings of science. ArXiv:1210.5510.

[28] Gubser, S. S., Klebanov, Igor R., and Peet, A. W. 1996. Entropy and temperature of black 3-branes. *Phys. Rev.*, **D54**, 3915–3919.

[29] Fotopoulos, A., and Taylor, T. R. 1999. Comment on two loop free energy in $\mathcal{N} = 4$ supersymmetric Yang-Mills theory at finite temperature. *Phys. Rev.*, **D59**, 061701.

[30] Nieto, Agustin, and Tytgat, Michel H. G. 1999. Effective field theory approach to $\mathcal{N} = 4$ supersymmetric Yang-Mills at finite temperature. ArXiv:hep-th/9906147.

[31] Burgess, C. P., Constable, N. R., and Myers, Robert C. 1999. The free energy of $N = 4$ superYang-Mills and the AdS/CFT correspondence. *J. High Energy Phys.*, **9908**, 017.

[32] Blaizot, J.-P., Iancu, E., Kraemmer, U., and Rebhan, A. 2007. Hard thermal loops and the entropy of supersymmetric Yang-Mills theories. *J. High Energy Phys.*, **0706**, 035.

[33] Gubser, Steven S. 2008. Momentum fluctuations of heavy quarks in the gauge-string duality. *Nucl. Phys.*, **B790**, 175–199.

[34] Chesler, Paul M., Jensen, Kristan, Karch, Andreas, and Yaffe, Laurence G. 2009. Light quark energy loss in strongly-coupled $\mathcal{N} = 4$ supersymmetric Yang-Mills plasma. *Phys. Rev.*, **D79**, 125015.

[35] Arnold, Peter, and Vaman, Diana. 2010. Jet quenching in hot strongly coupled gauge theories revisited: 3-point correlators with gauge-gravity duality. *J. High Energy Phys.*, **1010**, 099.

[36] Gubser, Steven S. 2006. Drag force in AdS/CFT. *Phys. Rev.*, **D74**, 126005.

[37] Herzog, C. P., Karch, A., Kovtun, P., Kozcaz, C., and Yaffe, L. G. 2006. Energy loss of a heavy quark moving through $\mathcal{N} = 4$ supersymmetric Yang-Mills plasma. *J. High Energy Phys.*, **0607**, 013.

[38] Chesler, Paul M., and Yaffe, Laurence G. 2007. The wake of a quark moving through a strongly-coupled plasma. *Phys. Rev. Lett.*, **99**, 152001.

[39] Gubser, Steven S., Pufu, Silviu S., and Yarom, Amos. 2008. Shock waves from heavy-quark mesons in AdS/CFT. *J. High Energy Phys.*, **0807**, 108.

[40] Babington, J., Erdmenger, J., Evans, Nick J., Guralnik, Z., and Kirsch, I. 2004. Chiral symmetry breaking and pions in nonsupersymmetric gauge / gravity duals. *Phys. Rev.*, **D69**, 066007.

[41] Kruczenski, Martin, Mateos, David, Myers, Robert C., and Winters, David J. 2004. Towards a holographic dual of large N_c QCD. *J. High Energy Phys.*, **0405**, 041.

[42] Hoyos-Badajoz, Carlos, Landsteiner, Karl, and Montero, Sergio. 2007. Holographic meson melting. *J. High Energy Phys.*, **0704**, 031.

[43] Kobayashi, Shinpei, Mateos, David, Matsuura, Shunji, Myers, Robert C., and Thomson, Rowan M. 2007. Holographic phase transitions at finite baryon density. *J. High Energy Phys.*, **0702**, 016.

[44] Mateos, David, Myers, Robert C., and Thomson, Rowan M. 2007. Thermodynamics of the brane. *J. High Energy Phys.*, **0705**, 067.

[45] Mateos, David, Matsuura, Shunji, Myers, Robert C., and Thomson, Rowan M. 2007. Holographic phase transitions at finite chemical potential. *J. High Energy Phys.*, **0711**, 085.

[46] Parnachev, Andrei, and Sahakyan, David A. 2006. Chiral phase transition from string theory. *Phys. Rev. Lett.*, **97**, 111601.

[47] Peeters, Kasper, Sonnenschein, Jacob, and Zamaklar, Marija. 2006. Holographic melting and related properties of mesons in a quark gluon plasma. *Phys. Rev.*, **D74**, 106008.

[48] Bergman, Oren, Lifschytz, Gilad, and Lippert, Matthew. 2007. Holographic nuclear physics. *J. High Energy Phys.*, **0711**, 056.

[49] Gürsoy, Umut, Kiritsis, Elias, Mazzanti, Liuba, Michalogiorgakis, Georgios, and Nitti, Francesco. 2011. Improved Holographic QCD. In *From Gravity to Thermal Gange Theories: the AdS/CFT Correspondence*, pp. 79–146. Lecture Notes in Physics, Vol. 828. Springer.

[50] Alanen, J., Kajantie, K., and Suur-Uski, V. 2009. A gauge/gravity duality model for gauge theory thermodynamics. *Phys. Rev.*, **D80**, 126008.

[51] Kajantie, K., Krssak, Martin, and Vuorinen, Aleksi. 2013. Energy momentum tensor correlators in hot Yang-Mills theory: holography confronts lattice and perturbation theory. *J. High Energy Phys.*, **1305**, 140.

15 Strongly coupled condensed matter systems

Within condensed matter physics, there are many extremely interesting physical systems which are strongly coupled. Although various approaches have been developed within condensed matter physics to deal with strongly coupled systems, there are many important physically relevant examples where a description in terms of theoretical models has not been successful. It thus appears natural to make use of gauge/gravity duality, which is very effective for describing systems at strong coupling, in this context as well. Of course, the microscopic degrees of freedom in a condensed matter system are very different from those described by a non-Abelian gauge theory at large N. For instance, these systems are non-relativistic in general. Nevertheless, the idea is to make use of *universality* again and to consider systems at second order phase transitions or renormalisation group fixed points, where the microscopic details may not be important. A prototype example of this scenario is given by *quantum phase transitions*, i.e. phase transitions at zero temperature which are induced by quantum rather than thermal fluctuations. These transitions appear generically when varying a parameter or coupling, which does not have to be small, such that the use of perturbative methods may not be possible.

In many cases, the study of models relevant to condensed matter physics involves the introduction of a finite charge density in addition to finite temperature. This applies for instance to Fermi surfaces or condensation processes. In the gauge/gravity duality context, this is obtained in a natural way by considering charged black holes of Reissner–Nordström type. Their gravity action involves additional gauge fields. Within this approach, standard thermodynamical quantities such as the free energy and the entropy may be calculated. A further important observable characterising the properties of condensed matter systems is the frequency-dependent conductivity. This is calculated in a straightforward way using gauge/gravity duality techniques.

A very instructive example of a quantum phase transition within gauge/gravity duality is obtained by using a magnetic field as the control parameter. The structure of models with a magnetic field B is different for field theories in 2+1 and in 3+1 dimensions owing to the axial anomaly which may be present in (3+1)-dimensional field theories.

In a number of holographic models, instabilities characterised by violations of the Breitenlohner–Freedman bound occur, leading to new ground states with lower free energy. This includes models with properties of superfluids and superconductors. In some cases, the new ground state is characterised by a spatially modulated condensate.

A further important aspect of condensed matter applications is the study of fermions in strongly coupled systems using gauge/gravity duality. The standard well-understood approach for describing fermions in weakly coupled systems is Landau–Fermi liquid theory. These systems have a Fermi surface, and the low-energy degrees of freedom are quasiparticle excitations around the Fermi surface. However, many systems have been observed in experiments which do not exhibit Landau–Fermi liquid behaviour. Although they have a Fermi surface, their low-energy degrees of freedom do not correspond to weakly coupled quasiparticles. The Fermi surface also contains essential information about the physical properties of strongly coupled systems. For instance for high-T_c superconductors, it reveals the d-wave symmetry structure. Gauge/gravity duality provides means for calculating spectral functions and identifying Fermi surfaces for strongly coupled systems. It has thus proved to be useful – so far, however, the virtue of this approach has been more to uncover universal features than to describe in detail the specific properties of individual physical systems considered in experiments. Within gauge/gravity duality, the simplest approach is to calculate Fermi surfaces for fermionic supergravity fields dual to composite gauge invariant fermionic operators in the dual field theory. Due to the strong coupling, these may be of marginal or non-Fermi liquid type. More recently, progress has been made towards calculating holographically the Fermi surfaces for the elementary fermions present in the dual field theory.

Since condensed matter systems are generically non-relativistic, it is useful to consider extensions of gauge/gravity duality to spaces which have non-relativistic symmetries. Some of these spaces have the additional advantage of naturally providing a zero ground state entropy. Moreover, in addition to the thermodynamic entropy, the quantum mechanical entanglement entropy may also be determined within gauge/gravity duality, with significant consequences for the models considered. Generally, the entanglement entropy provides an order parameter, for instance for topologically ordered states.

15.1 Quantum phase transitions

The prototype example of applications of gauge/gravity duality in the condensed matter context is provided by *quantum phase transitions* which occur at zero temperature, but also influence large areas of the phase diagram at finite temperature. To begin with, let us review the standard definition of these transitions [1].

The starting point is a quantum mechanical system with a Hamiltonian which depends on a coupling $H(g)$. The coupling g can be related for instance to a magnetic field, the pressure or a doping parameter. A *quantum critical point* occurs at a coupling $g = g_c$ where a non-analyticity in the ground-state energy occurs as a function of g. There are two possibilities: either the energy levels of the ground and first excited states cross at $g = g_c$, or they come very close to each other but do not cross. The second possibility is referred to as *avoided level crossing*. It gives rise to a non-analyticity of the ground state energy only in the infinite volume limit.

Let us consider the first case where the energy levels of the ground and first excited states cross at g_c. We denote the energy of fluctuations about the ground state by Δ. Near the level crossing, this energy may also be thought of as the difference between the energy levels of the ground and first excited states. For second order phase transitions, Δ vanishes when approaching g_c, while the coherence length ξ diverges,

$$\Delta \sim |g - g_c|^{z\nu}, \qquad \xi^{-1} \sim |g - g_c|^{-\nu} \tag{15.1}$$

where z, ν are *critical exponents*. Combining these two equations gives

$$\Delta \sim \xi^{-z}. \tag{15.2}$$

z is referred to as the *dynamical scaling exponent*. Note that the energy Δ and the length scale ξ need not be inversely related. The effective theory at the critical point itself, the *quantum critical theory*, is scale invariant. For $z = 1$ the effective theory scales as a relativistic theory, while for $z \neq 1$ the scaling is non-relativistic.

Note that the critical coupling g_c does not have to be small, and therefore perturbative approaches are not applicable in general to the effective theory. In most cases, it is even difficult to construct the effective theory explicitly.

An important property of quantum phase transitions is that they also influence the physics in a large part of the phase diagram at finite temperature, as shown in figure 15.1. For vanishing temperature, there is a second order quantum phase transition at $g = g_c$ from one quantum mechanical ground state of the system to a new one. In figure 15.1, the solid line denotes a second order phase transition driven by thermal fluctuations for $T > 0$, while the dashed line corresponds to a cross-over delimiting the *quantum critical region*. The different regions of this phase diagram may be characterised by their different relevant time scales. The dynamics at finite temperature, $T > 0$, is characterised by the thermal equilibration time τ_{eq}, which sets the time scale on which local thermal equilibrium is reestablished after an external perturbation. In the regime where $k_B T < \Delta$, the system becomes effectively classical and the equilibration time is long, $\tau_{eq} \gg \hbar/k_B T$. On the other

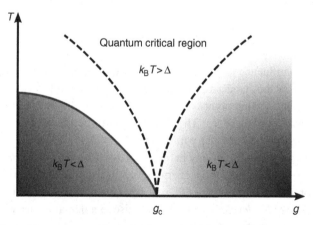

Figure 15.1   Quantum critical region at finite temperature. The dashed lines correspond to a cross-over at $T \sim |g-g_c|^{z\nu}$. For $T > 0$, the solid line denotes a second order thermal phase transition.

hand, for $k_B T > \Delta$ in the case of strong coupling $g_c \gg 1$, a very appealing scenario is present. There is a short equilibration time,

$$\tau_{eq} \sim \frac{\hbar}{k_B T}, \tag{15.3}$$

which is determined by $k_B T$, and is independent of any microscopic model-dependent energy scale. In this quantum critical region of the phase diagram, the dynamics is dominated by the quantum phase transition at g_c even at finite temperature.

Quantum phase transitions have key properties which make them a prime target for applying gauge/gravity duality. They occur at strong coupling, their effective theory is scale invariant at the transition and the dynamics are independent of the microscopic details of the theory considered.

15.2 Charges and finite density

The discussion of quantum phase transitions in the preceding section suggests that we could apply gauge/gravity duality to the strongly coupled effective theory at the transition. As explained above, the detailed structure of this theory is not known in general, so a conceptually simple approach would be, for instance, to replace it by $\mathcal{N} = 4$ Super Yang–Mills theory when considering 3+1 dimensions, or by ABJM theory when considering 2+1 dimensions. In agreement with the concept of *universality*, the physical behaviour at the quantum critical point is expected to be independent of the detailed structure of the microscopic degrees of freedom. However, as we have studied in part II of this book, the gravity duals of $\mathcal{N} = 4$ Super Yang–Mills theory and of ABJM theory are quite involved with a large number of fields. A simpler gravity model is sufficient to describe the physical properties of the quantum critical theory, even if the explicit Lagrangian of such an ad hoc bottom-up model is not known in general.

Let us consider the simplest model on the gravity side. To describe physical behaviour which is prototypical for the condensed matter systems discussed, it is useful to consider a 'bottom-up' model which has the necessary physical properties such as finite temperature and density. Since any physical model posesses an energy-momentum tensor which is sourced by a metric, we are naturally led to introduce gravity. A negative cosmological constant induces the necessary asymptotically AdS geometry. Moreover, on the field theory side, the essential property of a finite charge density is realised by a conserved $U(1)$ symmetry with current J_μ. This current couples to a gauge field A_μ on the gravity side.

We look for charged black hole solutions of the Einstein–Maxwell theory given by the action

$$S = \int d^{d+1}x \sqrt{-g} \left(\frac{1}{2\kappa^2} (R - 2\Lambda) - \frac{1}{4g^2} F^{mn} F_{mn} \right), \tag{15.4}$$

with cosmological constant $\Lambda = -d(d-1)/2L^2$. The Einstein equations of motion involve the energy-momentum tensor of the field strength F_{mn},

$$R_{mn} - \frac{R}{2}g_{mn} - \frac{d(d-1)}{2L^2}g_{mn} = \frac{\kappa^2}{g^2}\left(F_{ml}F_n^l - \frac{1}{4}g_{mn}F_{lr}F^{lr}\right), \tag{15.5}$$

together with the Maxwell equations

$$\nabla_m F^{mn} = 0. \tag{15.6}$$

The simplest solution is of course to set the gauge field to zero, in which case the solution for the metric is the standard AdS metric

$$ds^2 = \frac{L^2}{z^2}\left(-dt^2 + d\vec{x}^2 + dz^2\right). \tag{15.7}$$

Solving (15.6) near the boundary located at $z = 0$, the $U(1)$ gauge field on the gravity side then has to satisfy the asymptotic boundary behaviour

$$A_m(r) = A_m^{(0)} + z^{d-2}A_m^{(1)} + \cdots \text{ as } z \to 0. \tag{15.8}$$

15.2.1 Reissner–Nordström solution

We now look for a solution to the Einstein–Maxwell equations with non-trivial $U(1)$ gauge field on the gravity side. For rotational symmetry in the spatial directions, we require $A_i(z) = 0$. To obtain the important physical property of a finite charge density, the time component of the gauge field is allowed to have a non-trivial profile $A = A_t(z)dt$.

Such a solution to the Einstein and Maxwell equations (15.5) and (15.6) is given by the planar AdS Reissner–Nordström black hole or black brane, whose metric is

$$ds^2 = \frac{L^2}{z^2}\left(-f(z)dt^2 + \frac{dz^2}{f(z)} + d\vec{x}^2\right), \tag{15.9}$$

with

$$f(z) = 1 - M\left(\frac{z}{z_h}\right)^d + Q^2\left(\frac{z}{z_h}\right)^{2(d-1)}, \tag{15.10}$$

$$M = 1 + \frac{z_h^2\mu^2}{\gamma^2}, \qquad Q^2 = \frac{z_h^2\mu^2}{\gamma^2}, \tag{15.11}$$

$$\gamma^2 = \frac{(d-1)L^2g^2}{(d-2)\kappa^2}. \tag{15.12}$$

γ is a dimensionless constant which parametrises the ratio of the couplings g and κ as introduced in (15.4). The horizon is located at $z = z_h$ and has topology $\mathbb{R}^{d-1}$, hence the name *planar* black hole. Its infinite volume will be denoted by $V_{d-1} = \text{Vol}(\mathbb{R}^{d-1})$.

For the non-trivial time component of the gauge field, we have the solution

$$A_t(z) = \mu\left(1 - \left(\frac{z}{z_h}\right)^{d-2}\right). \tag{15.13}$$

This satisfies the boundary condition that $A_t(z)$ has to vanish at the horizon since ∂_t is not well defined as a Killing vector there. Moreover, the μ parameter in the solution (15.13) corresponds to the chemical potential. In agreement with the standard AdS/CFT result for the asymptotic behaviour of gravity fields near the AdS boundary, asymptotically $A_t(z)$ gives the source and the vacuum expectation value of the dual field theory operator. Here, these determine the chemical potential and the density, respectively. By comparison with the field theory analysis in chapter 11, we readily identify the source term μ as the chemical potential. To identify the density, we proceed as follows.

The temperature is again fixed by analytic continuation to the Euclidean regime and is given by

$$T = \frac{1}{4\pi z_h} \left(d - \frac{(d-2) z_h^2 \mu^2}{\gamma^2} \right). \tag{15.14}$$

In order to allow for a variable charge density, we work in the grand canonical ensemble. This means that a boundary term fixing A_t is absent from the gravity action (15.4). [1] In the grand canonical ensemble, by evaluating the Euclidean action on the solution, including the counter terms necessary for regularisation, we find the following Gibbs free energy

$$\Omega = -T \ln Z = -\frac{L^{d-1}}{2\kappa^2 z_h^d} \left(1 + \frac{z_h^2 \mu^2}{\gamma^2} \right) V_{d-1} = \mathcal{F}\left(\frac{T}{\mu}\right) V_{d-1} T^d, \tag{15.15}$$

where the function $\mathcal{F}$ is obtained from solving (15.14) for z_h. V_{d-1} is the spatial volume of the field theory and γ is given by (15.12). This result is in agreement with expectations from thermodynamics as far as the dependence on volume and temperature is concerned. From (15.15), we obtain the charge density

$$\rho = -\frac{1}{V_{d-1}} \frac{\partial \Omega}{\partial \mu} = \frac{L^{d-1} \mu}{\kappa^2 z_h^{d-2} \gamma^2}. \tag{15.16}$$

Moreover, the entropy density is obtained from the area of the black hole horizon as

$$s = \frac{2\pi}{\kappa^2} \left(\frac{L}{z_h}\right)^{d-1}. \tag{15.17}$$

The AdS Reissner–Nordström solution above gives a toy model of a dual quantum phase transition [2], in the sense that the quantum phase transition occurs at $\rho = 0$ where the theory is conformal, with ρ the charge density. For a quantum phase transition at finite control parameters, we have to introduce additional control parameters such as a magnetic field. We will consider this below in sections 15.2.4 and 15.2.5.

Let us also note that while the AdS Reissner–Nordström model is simple on the gravity side, it leads to rather special properties of the dual field theory. Generically, we expect the $U(1)$ symmetry and also the Poincaré symmetry to be spontaneously broken. Later on, we will encounter models in which this is indeed the case.

---

[1] In the canonical ensemble, the charge density is fixed by adding a boundary term $\Delta S_E = 1/g^2 \times \int_{z \to 0} d^d x \sqrt{\gamma} n^a F_{ab} A^b$ to the Euclidean gravity action, with γ the induced metric. Then the thermodynamical relation $F = \Omega + \mu Q$ is satisfied with $Q = \rho V_{d-1}$.

15.2.2 Emergent quantum criticality

Remarkable properties of the field theory dual to the gravity solution discussed in the previous section arise in the low temperature limit of the Reissner–Nordström solution. Let us first consider the case when $T = 0$, which corresponds to an *extremal* black brane. In this case, (15.14) implies

$$\frac{z_{\mathrm{h}}^2 \mu^2}{\gamma^2} = \frac{d}{d-2} \tag{15.18}$$

and the black hole factor $f(z)$ develops a double zero when expanding near the horizon in $\tilde{z} = z_{\mathrm{h}} - z$,

$$f(z) \approx d(d-1)\frac{\tilde{z}^2}{z_{\mathrm{h}}^2} + \cdots . \tag{15.19}$$

We now show that the $(d + 1)$-dimensional AdS Reissner–Nordström metric reduces to the space $AdS_2 \times \mathbb{R}^{d-1}$ near the horizon. To see this, we insert the expansion (15.19) into the metric given by (15.9)–(15.12). The original Reissner–Nordström metric (15.9) asymptotes to

$$\mathrm{d}s^2 = d(d-1)\frac{L^2\tilde{z}^2}{z_{\mathrm{h}}^4}(-\mathrm{d}t^2) + \frac{1}{d(d-1)}\frac{L^2}{\tilde{z}^2}\mathrm{d}\tilde{z}^2 + \frac{L^2}{z_{\mathrm{h}}^2}\mathrm{d}\vec{x}^2. \tag{15.20}$$

Performing a coordinate change $\tilde{z} = \hat{z}\,z_{\mathrm{h}}^2/L^2$, (15.20) becomes

$$\mathrm{d}s^2 = \frac{d(d-1)}{L^2}\hat{z}^2(-\mathrm{d}t^2) + \frac{L^2}{d(d-1)}\frac{1}{\hat{z}^2}\mathrm{d}\hat{z}^2 + \frac{L^2}{z_{\mathrm{h}}^2}\mathrm{d}\vec{x}^2. \tag{15.21}$$

This is the metric of $AdS_2 \times \mathbb{R}^{d-1}$, with the boundary of AdS_2 at $\hat{z} \to \infty$, where the radius of AdS_2 is

$$\tilde{L} = \frac{L}{\sqrt{d(d-1)}}. \tag{15.22}$$

To write the AdS_2 metric (15.21) in a form where its boundary is at $\zeta \to 0$, we perform a further coordinate transformation $\zeta = \tilde{L}^2/\hat{z}$ and obtain the $AdS_2 \times \mathbb{R}^{d-1}$ metric

$$\mathrm{d}s^2 = \frac{\tilde{L}^2}{\zeta^2}(-\mathrm{d}t^2 + \mathrm{d}\zeta^2) + \mathrm{d}\vec{x}^2, \tag{15.23}$$

where we have rescaled $\mathrm{d}\vec{x}$ by a constant factor.

What are the physical consequences of this AdS_2 in the IR? From our experience with the AdS/CFT correspondence, we expect this factor to be dual to a one-dimensional conformal field theory. This one-dimensional CFT can be understood either as conformal quantum mechanics or as the chiral sector of a $(1 + 1)$-dimensional CFT, i.e. for instance just the left-moving sector. The IR dynamics of the dual field theory is governed by this one-dimensional CFT. Since the d-dimensional CFT in the UV is broken by the finite density, this is referred to as *emergent* quantum criticality [3], emerging in the IR.

To investigate whether the Reissner–Nordström solution and in particular its near-horizon geometry $AdS_2 \times \mathbb{R}^{d-1}$ is a candidate for the ground state of the system, we have to consider its thermodynamical properties at $T = 0$. We find that it has a finite entropy density. From (15.17) and (15.16) together with (15.12) we have at $T = 0$ that

$$s = \frac{2\pi (d - 2)}{\sqrt{d(d - 1)}} \frac{\kappa}{gL} \rho. \tag{15.24}$$

In general, it is expected that systems at $T = 0$ have zero entropy, in accordance with the third law of thermodynamics [2]. The finite ground state entropy (15.24) may be a sign of an instability. This is not seen in the simple holographic model (15.4) considered so far.

However, when adding additional matter fields to this model, such as a scalar field, or additional interactions, such as Chern–Simons terms, we find a new ground state with lower free energy. In the case of a scalar field, this field condenses, leading to a superfluid. For the Chern–Simons interaction the new ground state is spatially modulated.

Therefore the following picture arises. The corresponding near-horizon geometry is modified. It is no longer $AdS_2 \times \mathbb{R}^{d-1}$ because of the instabilities discussed above. Nevertheless, the $AdS_2 \times \mathbb{R}^{d-1}$ geometry with its dual emergent one-dimensional CFT may still be present at intermediate energy scales.

All of these systems are discussed in the sections below.

Exercise 15.2.1 Repeat the analysis above at finite temperature T to show that when expanding about the horizon, the geometry becomes a black hole in $AdS_2 \times \mathbb{R}^{d-1}$.

To do this, begin with the metric (15.9) with (15.11). Define a scale z_0 by rewriting M and Q as

$$Q^2 = \frac{d}{d - 2} \left(\frac{z_h}{z_0} \right)^{2(d-1)}, \qquad M = 1 + Q^2. \tag{15.25}$$

Then, redefine the radial coordinate z and its horizon value z_h using

$$z \to \frac{z_0^2}{\zeta} \frac{\epsilon}{d(d - 1)}, \qquad z_h \to \frac{z_0^2}{\zeta_h} \frac{\epsilon}{d(d - 1)}. \tag{15.26}$$

In the limit $\epsilon \to 0$, the metric (15.9) becomes

$$ds^2 = \frac{\tilde{L}^2}{\zeta^2} \left(- \left(1 - \frac{\zeta^2}{\zeta_h^2} \right) dt^2 + \frac{1}{1 - \frac{\zeta^2}{\zeta_h^2}} d\zeta^2 + d\vec{x}^2 \right), \tag{15.27}$$

where $\tilde{L} = L/\sqrt{d(d - 1)}$ and $\vec{x}$ has been rescaled. Equation (15.27) is the metric of a black hole in AdS_2, multiplied by $\mathbb{R}^{d-1}$.

---

[2] Finite entropy is possible for systems with intrinsic order, such as ferromagnets or anti-ferromagnets for instance, or in frustrated systems where the potential energy is much larger than the kinetic energy. Also within gauge/gravity duality, it may be the case that the ground state is degenerate, for instance for a theory with moduli spaces giving rise to additional gapless modes. However, for supersymmetric theories with moduli spaces, the entropy is not extensive in volume, unlike the AdS Reissner–Nordström case which has a genuine thermodynamic entropy even at $T = 0$.

15.2.3 Transport properties

An important physical observable for describing condensed matter systems is the frequency-dependent conductivity, i.e. the description of charge transport. The gauge/gravity duality formalism for transport as presented in chapter 12 may be applied to charge transport in a straightforward way [4, 2]. As an example on the field theory side, we already considered a linear response formulation of Ohm's law in that chapter.

In analogy to the electric current J_x generated by the electric field E_x, there is also a heat current $Q_x = T_{xt} - \mu J_x$ generated by a temperature gradient $(\nabla_x T)/T$. Since the two currents mix, we have a matrix structure for the linear response,

$$
\begin{pmatrix} J_x \\ Q_x \end{pmatrix} = \begin{pmatrix} \sigma & \alpha T \\ \alpha T & \bar{\kappa} T \end{pmatrix} \begin{pmatrix} E_x \\ -(\frac{\nabla_x T}{T}) \end{pmatrix}.
\tag{15.28}
$$

To apply linear response theory, the sources E_x and $\nabla_x T$ need to be related to fluctuations δA_x and δg_{tx} for the gauge field and metric. For the gauge field we have at vanishing momentum

$$
E_j = i\omega \delta A_j,
\tag{15.29}
$$

which leads to Ohm's law as shown in box 12.1. In addition, there are contributions from the temperature gradient. At vanishing chemical potential, the g_{tt} component of the metric depends on the temperature and for a spatially varying temperature we may rescale the metric component g_{tt} in Euclidean signature such that it takes the form $g_{tt} = T_0^2/T(x)^2$, where the temperature $T(x)$ varies about T_0. We thus find

$$
\partial_i g_{tt} = -2\frac{\partial_i T}{T}.
\tag{15.30}
$$

Using the diffeomorphism

$$
\delta g_{\mu\nu} = \partial_\mu \xi_\nu + \partial_\nu \xi_\mu,
\tag{15.31}
$$

we can trade (15.30) for a change in δg^{ti}. We may choose ξ_μ such that $\partial_i(g_{tt} + \delta g_{tt}) = 0$. With $\xi_i = 0$ we find

$$
\delta g_{ti} = \partial_i \xi_t = -\frac{\partial_i T}{i\omega T}.
\tag{15.32}
$$

For $\partial_i T \sim e^{-i\omega t}$ we therefore have

$$
T^{0j} = G_R^{tj,ti}(\omega)\delta g_{ti}(\omega) = -\bar{\kappa}^{ij}(\omega)\partial_i T,
\tag{15.33}
$$

$$
\bar{\kappa}^{ij}(\omega) = \frac{G_R^{ti,tj}(\omega)}{i\omega T}.
\tag{15.34}
$$

The diffeomorphism (15.32) also acts on the vector potential A_μ by virtue of

$$
\delta A_\mu = A_\nu \partial_\mu \xi^\nu + \xi^\nu \partial_\nu A_\mu.
\tag{15.35}
$$

For the choice of diffeomorphism given above, we have

$$
\delta A_i = -A_t \partial_i \xi_t = -\mu \frac{\partial_i T}{i\omega T}.
\tag{15.36}
$$

For the transformation of the action, this gives

$$\delta S = \int d^d x \sqrt{-g}(T^{tx}\delta g_{tx} + J^x \delta A_x)$$

$$= \int d^d x \sqrt{-g}\left(-Q^x \frac{\nabla_x T}{i\omega T} + J^x \frac{E_x}{i\omega}\right), \tag{15.37}$$

with the heat current $Q^x = T^{tx} - \mu J^x$ as expected. We may then rewrite (15.28) as

$$\begin{pmatrix} J_x \\ Q_x \end{pmatrix} = \begin{pmatrix} \sigma & \alpha T \\ \alpha T & \bar{\kappa} T \end{pmatrix} \begin{pmatrix} i\omega(\delta A_x + \mu\delta g_{tx}) \\ i\omega\delta g_{tx} \end{pmatrix}, \tag{15.38}$$

with

$$\sigma(\omega) = -\frac{i}{\omega}G_R^{JJ}, \qquad \alpha(\omega)T = -\frac{i}{\omega}G_R^{QJ}, \qquad \bar{\kappa}(\omega)T = -\frac{i}{\omega}G_R^{QQ}. \tag{15.39}$$

Let us now calculate the conductivities in (15.39) holographically using the gauge/gravity duality linear response formalism given in section 12.1.2, and in particular (12.42) and (12.43), for the case $d = 3$. We use the Reissner–Nordström background (15.9)–(15.12). To apply the holographic linear response, we have to solve the equations of motion for the fluctuations δA, δg in the bulk. Using the equations of motion (15.5) and (15.6), the linearised equations of motion for fluctuations about the AdS Reissner–Nordström background read

$$\delta g'_{tx} + \frac{2}{z}\delta g_{tx} + \frac{4L^2}{\gamma^2}A'_t\delta A_x = 0, \tag{15.40}$$

$$(f\delta A'_x)' + \frac{\omega^2}{f}\delta A_x + \frac{z^2 A'_t}{L^2}\left(\delta g'_{tx} + \frac{2}{z}\delta g_{tx}\right) = 0, \tag{15.41}$$

with (15.9)–(15.12) for the metric, as well as (15.13) for A_t and with the prime denoting derivatives with respect to r. These two equations are easily decoupled to give

$$(f\delta A'_x)' + \frac{\omega^2}{f}\delta A_x - \frac{4\mu^2 z^2}{\gamma^2 z_h^2}\delta A_x = 0. \tag{15.42}$$

Near the boundary at $z \to 0$, the asymptotic solution to this equation takes the form

$$\delta A_x = \delta A_x^{(0)} + \frac{z}{L}\delta A_x^{(+)} + \cdots. \tag{15.43}$$

For the conductivity σ in (15.39), we have

$$\sigma(\omega) = -\frac{1}{g^2 L}\frac{i}{\omega}\frac{\delta A_x^{(+)}}{\delta A_x^{(0)}}. \tag{15.44}$$

Exercise 15.2.2 Calculate π as defined in (12.42) for the example studied in this section. This is done by inserting $A + \delta A$, $g + \delta g$ into the Einstein–Maxwell action (15.4) together with the gravitational Gibbons–Hawking boundary term introduced in section 5.5, with A, g the solutions to the equations of motion and δA, δg the fluctuations.

The result is

$$\pi_{g_{tx}} = \frac{\delta S}{\delta g_{tx}^{(0)}} = -\rho\,\delta A_x^{(0)} + \frac{2L^2}{\kappa^2 z^3}(1 - f^{-1/2})\delta g_{tx}^{(0)},\qquad (15.45)$$

$$\pi_{A_x} = \frac{\delta S}{\delta A_x^{(0)}} = \frac{f\delta A_x'^{\,(0)}}{g^2} - \rho\,\delta g_{tx}^{(0)},\qquad (15.46)$$

with ρ as in (15.16). To obtain $\pi_{g_{tx}}$, a careful integration by parts and use of the Gibbons–Hawking boundary term are necessary.

Moreover, calculate the remaining conductivities in (15.39) and find them to be

$$T\alpha(\omega) = \frac{i\rho}{\omega} - \mu\sigma(\omega),\qquad T\bar{\kappa}(\omega) = \frac{i(\epsilon + p - 2\mu\rho)}{\omega} + \mu^2\sigma(\omega)\qquad (15.47)$$

with the energy density $\epsilon = -2\Omega/V_2$ and Ω as in (15.15). The extra term involving the pressure p in $\bar{\kappa}(\omega)$ is due to translation invariance.

We proceed by calculating $\sigma(\omega)$ from (15.44). The infalling boundary condition at the horizon is imposed by writing

$$A_x(z) = f(z)^{-i\omega/(4\pi T)}a_x(z),\qquad (15.48)$$

and requiring

$$a_x(z) = 1 + k_1(z - z_{\rm h}) + k_2(z - z_{\rm h})^2\qquad (15.49)$$

near the horizon. With this ansatz, (15.42) is solved numerically and the result inserted into (15.44). The result for $\sigma(\omega)$ obtained in this way is shown in figure 15.2.

In figure 15.2 we observe that the real part of the conductivity asymptotes to a constant value for large frequencies. This is due to the fact that the conductivity is dimensionless in 2+1 dimensions. Moreover, due to translation invariance, the conductivity $\mathrm{Re}\,\sigma$ has a contribution proportional to $\delta(\omega)$, i.e. a singular peak at $\omega = 0$. According to the *Kramers–Kronig relation*

$$\mathrm{Im}\,G_{\rm R}(\omega) = -\mathcal{P}\int_{-\infty}^{\infty}\frac{d\omega'}{\pi}\frac{\mathrm{Im}\,G_{\rm R}(\omega')}{\omega' - \omega},\qquad (15.50)$$

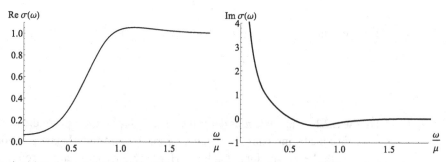

Figure 15.2   Conductivity $\sigma(\omega)$ associated with the $U(1)$ current J_x: real and imaginary parts at finite constant temperature.

where $\mathcal{P}$ denotes the principal value of the integral, the $\delta(\omega)$ contribution to $\mathrm{Re}\,\sigma$ corresponds to a pole in the imaginary part $\mathrm{Im}\,\sigma(\omega)$,

$$\mathrm{Im}\,\sigma(\omega) = \frac{C}{\omega} \quad \Leftrightarrow \quad \mathrm{Re}\,\omega = C\,\delta(\omega) + \text{regular terms}. \tag{15.51}$$

This pole at $\omega \to 0$ in $\mathrm{Im}\,\sigma(\omega)$ is clearly visible in the right panel of figure 15.2.

This delta peak in $\mathrm{Re}\,\sigma(\omega)$ is genuinely different from conductors in solid state physics, where translation invariance is broken by a lattice. In solids, the delta peak is washed out to a broader *Drude peak* at low frequencies. For $\omega > 0$, however, the conductivity shown in figure 15.2 is qualitatively very similar to the conductivity observed experimentally in graphene, for instance. Graphene has features similar to the holographic model described here, since at low energies it can be described by a (2+1)-dimensional relativistic field theory with chemical potential.

Recall from section 12.1 that the real part of the conductivity is related to the spectral function which is a measure of the number of states. In the plot for $\mathrm{Re}\,\sigma(\omega)$ on the left-hand side of figure 15.2, we see that the states with $\omega < \mu$ are depleted, which leads to a drop in the conductivity. This corresponds to the fact the chemical potential μ corresponds to the Fermi energy E_F.

Small ω expansion

The Green's function takes a particular form in the case of the Reissner–Nordström planar black hole solution discussed in the previous section, owing to its asymptotic $AdS_2 \times \mathbb{R}^{d-1}$ behaviour in the IR [3]. The Green function in AdS_2 space can be obtained analytically, and for small ω, the Green function for the full space can be obtained by matching the UV and IR expansions. This implies that the IR behaviour of the full Green's function is determined by the Green's function in AdS_2.

To see this, we consider $T \to 0$. We begin with the IR Green's function for a charged scalar field in AdS_2 space, which is obtained from the action

$$S = -\frac{1}{2} \int \mathrm{d}^2 x \, \sqrt{-g}\left((D^m \phi)^* D_m \phi + m^2 \phi^* \phi\right), \tag{15.52}$$

with the covariant derivative $D_\mu = \partial_\mu - iqA_\mu$ and the background given by the metric (15.23) of section 15.2.2 and an additional gauge field,

$$\mathrm{d}s^2 = \frac{\tilde{L}^2}{\zeta^2}(-\mathrm{d}t^2 + \mathrm{d}\zeta^2), \tag{15.53}$$

$$A_t = \frac{1}{\sqrt{d(d-1)}} \frac{Lg}{\kappa} \frac{1}{\zeta}. \tag{15.54}$$

With $\phi(\zeta, t) = \exp(-i\omega t)\phi(\zeta)$, the wave equation becomes

$$-\partial_\zeta^2 \phi(\zeta) + V(\zeta)\phi(\zeta) = 0, \qquad V(\zeta) = \frac{m^2 \tilde{L}^2}{\zeta^2} - \left(\omega + \frac{q}{\sqrt{d(d-1)}} \frac{Lg}{\kappa} \frac{1}{\zeta}\right)^2. \tag{15.55}$$

Consider the case that $\phi(\zeta, t)$ is obtained from dimensionally reducing a field $\phi(z, t, \vec{k})$ in the $AdS_2 \times \mathbb{R}^{d-1}$ space obtained as the IR limit of the Reissner–Nordström solution (15.9) for $T \to 0$. In this case, $\phi(\zeta, t)$ has the effective mass

$$m_k^2 = k^2 \frac{z_h^2}{L^2} + m^2, \qquad k^2 = |\vec{k}^2|. \tag{15.56}$$

The conformal dimension of the operator $\mathcal{O}_{\vec{k}}(\omega)$ dual to ϕ is obtained by solving (15.55) near the boundary where $\zeta \to 0$, which gives

$$\phi(\zeta, \omega) = A(\omega)\zeta^{\frac{1}{2}-\nu_k}(1 + O(\zeta)) + B(\omega)\zeta^{\frac{1}{2}+\nu_k}(1 + O(\zeta)), \tag{15.57}$$

with

$$\nu_k = \sqrt{m_k^2 \tilde{L}^2 - q^2 \frac{L^2 g^2}{d(d-1)\kappa}}. \tag{15.58}$$

ν_k is related to the conformal dimension δ_k of the dual operator $\mathcal{O}_{\vec{k}}(\omega)$ in the CFT_1 theory by $\delta_k = \nu_k + 1/2$. The wave equation (15.55) may be solved exactly and gives the Green's function

$$\mathcal{G}_k(\omega) = 2\nu_k e^{-i\pi\nu_k} \frac{\Gamma(-2\nu_k)\Gamma(1/2 + \nu_k - iq\frac{\tilde{L}^2 g^2}{\kappa^2})}{\Gamma(2\nu_k)\Gamma(1/2 - \nu_k - iq\frac{\tilde{L}^2 g^2}{\kappa^2})} \omega^{2\nu_k} \tag{15.59}$$

for the associated Green's function. The scale dependence of this exact result is as expected from the holographic linear response prescription (12.38),

$$\mathcal{G}_k(\omega) = K \frac{B(\omega)}{A(\omega)} \sim \omega^{2\nu}, \tag{15.60}$$

with A, B of (15.57). K is a positive normalisation constant which is is obtained from the prefactor of the UV gravity action considered.

Let us now consider the Green function for the full Reissner–Nordström geometry. Its low-frequency limit is not easy to determine because of the double pole of g_{tt} at the horizon. As we now discuss, this low-frequency limit may be obtained by matching the expansion of the full Green's function onto the expansion of the AdS_2 Green's function $\mathcal{G}_k$ discussed above: The solutions in the IR and UV regions of the AdS Reissner–Nordström geometry are expanded in power series which are identified in the region of overlap. The approach followed here is to expand the Green's function for (15.9) in the UV, and in the IR for the rescaled asymptotic metric of $AdS_2 \times \mathbb{R}^{d-1}$ determined by (15.23) and (15.53). The two expansions are then matched at intermediate scales.

This is done as follows [3]. First we divide the radial coordinate into an inner and an outer region, defined using the coordinate ζ of section 15.2.2 by

$$\begin{aligned} \text{inner region:} \quad & z_h - z = (\omega L) \cdot \tilde{L}/\zeta \quad \text{for } \epsilon < \zeta < \infty, \\ \text{outer region:} \quad & (\omega L) \cdot \tilde{L}^2/\epsilon \ < z_h - z. \end{aligned} \tag{15.61}$$

We consider the limit

$$(\omega L) \to 0, \quad \zeta \text{ finite}, \quad \epsilon \to 0, \quad (\omega L)\frac{\tilde{L}}{\epsilon} \to 0. \tag{15.62}$$

For the Green's function in the outer region, we solve the equation of motion for a scalar field $\phi_O(z)$ in the $(d+1)$-dimensional Reissner–Nordström metric (15.53) for small frequencies. This is possible subject to the regularisation provided by splitting the spacetime into the inner and outer regions as given above. We use ζ as coordinate in the inner region and $z_h - z$ as coordinate in the outer region. Then the small ω expansion takes the simple form

$$
\begin{aligned}
\text{inner region :} \quad & \phi_I(\zeta) = \phi_I^{(0)}(\zeta) + \omega\,\phi_I^{(1)}(\zeta) + \cdots, \\
\text{outer region :} \quad & \phi_O(z) = \phi_O^{(0)}(z) + \omega\,\phi_O^{(1)}(z) + \cdots,
\end{aligned}
\tag{15.63}
$$

where here the index in brackets denotes the order in the small ω expansion. The full solution is obtained by matching ϕ_I and ϕ_O in the overlapping region, which is given by $\zeta \to 0$ together with $(z_h - z) \to 0$. In the inner region, the field $\phi_I^{(0)}(\zeta, k)$ takes the asymptotic form

$$
\phi_I^{(0)}(\zeta, k) = \zeta^{1/2 - \nu_k}(1 + O(\zeta)) + \mathcal{G}_k(\omega)\zeta^{1/2 + \nu_k}(1 + O(\zeta)).
\tag{15.64}
$$

Here, $\mathcal{G}_k(\omega)$ is precisely the AdS_2 Green's function of (15.59).

Looking at the equation of motion for $\omega = 0$ in the outer region for $z \to z_h$, it can be seen that it is identical to the inner region equation for $\phi_I^{(0)}$ in the limit $\zeta \to 0$. It is thus convenient to write the solutions $\eta_\pm^{(0)}(z)$ in the outer region in such a way that their boundary behaviour for $z \to z_h$ corresponds to the two linearly independent solutions in (15.64), i.e.

$$
\eta_\pm^{(0)}(z) \sim (z_h - z)^{-(1/2) \pm \nu_k} + \cdots, \qquad z \to z_h.
\tag{15.65}
$$

This ensures matching between the two solutions. The solution in the outer region may then be written as

$$
\phi_O^{(0)}(z) = \eta_+^{(0)}(z) + \mathcal{G}_k(\omega)\eta_-^{(0)}(z).
\tag{15.66}
$$

This may be generalised to the small ω expansion

$$
\eta_{O\pm}(z, k) = \eta_\pm^{(0)}(z, k) + \omega\,\eta_\pm^{(1)}(z, k) + \omega^2\,\eta_\pm^{(2)}(z, k) + \cdots
\tag{15.67}
$$

in a straightforward way, such that perturbatively we have

$$
\phi_O(z, k) = \eta_+(z, k) + \mathcal{G}_k(\omega)\eta_-(z, k)
\tag{15.68}
$$

to all orders in ω.

On the other hand, in the near-boundary expansion for $z \to 0$, the coefficients in the ω expansion of (15.67) take the asymptotic form

$$
\eta_\pm^{(n)}(z) = a_\pm^{(n)} z^{d - \Delta}(1 + \cdots) + b_\pm^{(n)} z^\Delta(1 + \cdots),
\tag{15.69}
$$

with Δ denoting the conformal dimension of the dual operator in $d+1$ dimensions. Combining this with (15.68) and the prescription (12.43) for the Green's function, we obtain the central result [3].

$$
G_R(\omega, k) = K \frac{b_+^{(0)} + \omega b_+^{(1)} + O(\omega^2) + \mathcal{G}_k(\omega)(b_-^{(0)} + \omega b_-^{(1)} + O(\omega^2))}{a_+^{(0)} + \omega a_+^{(1)} + O(\omega^2) + \mathcal{G}_k(\omega)(a_-^{(0)} + \omega a_-^{(1)} + O(\omega^2))}
\tag{15.70}
$$

for the retarded Green's function, with normalisation K as in (15.60). This shows how the low-energy behaviour dual to the AdS_2 region determines the structure of the full Green function for small frequencies.

15.2.4 B-field in 2+1 dimensions

As a further control parameter we may introduce a magnetic field into the boundary field theory, given by $B = F_{xy}^{(0)}$, where $F^{(0)}$ denotes the boundary field strength tensor. Together with the finite charge density, as introduced in section 15.2.1 above, this is realised by considering a bulk $U(1)$ gauge field of the form

$$A = A_t(z)\,\mathrm{d}t + Bx\,\mathrm{d}y, \tag{15.71}$$

with B of mass dimension two. For simplicity, we first consider the case where the boundary field theory has dimension $d = 2 + 1$ [2]. Solving the Einstein–Maxwell equations of motion (15.5), (15.6) with this ansatz gives rise to a *dyonic* black hole of the form

$$f(z) = 1 - \left(1 + \frac{z_h^2\mu^2 + z_h^4 B^2}{\gamma^2}\right)\left(\frac{z}{z_h}\right)^3 + \frac{z_h^2\mu^2 + z_h^4 B^2}{\gamma^2}\left(\frac{z}{z_h}\right)^4, \tag{15.72}$$

$$A = \mu\left(1 - \frac{z}{z_h}\right)\mathrm{d}t + Bx\,\mathrm{d}y. \tag{15.73}$$

The temperature is now given by

$$T = \frac{1}{4\pi z_h}\left(3 - \frac{z_h^2\mu^2}{\gamma^2} - \frac{z_h^4 B^2}{\gamma^2}\right), \tag{15.74}$$

and the grand canonical potential by

$$\Omega = -\frac{L^2}{2\kappa^2 z_h^3}\left(1 + \frac{z_h^2\mu^2}{\gamma^2} - \frac{3z_h^4 B^2}{\gamma^2}\right)V_2. \tag{15.75}$$

The magnetisation density reads

$$m = -\frac{1}{\mathrm{Vol}(\mathbb{R}^2)}\frac{\partial\Omega}{\partial B} = -\frac{2L^2}{\kappa^2}\frac{z_h B}{\gamma^2}. \tag{15.76}$$

Since the field theory is conformal at vanishing temperature, density and magnetic field, it depends only on the dimensionless ratios μ/T and B/T^2 at finite T, μ, B. This implies that the magnetic susceptibility $\chi = \partial^2\Omega/\partial B^2$ is of order $1/T$. This is to be contrasted with weakly coupled systems such as the free electron gas for which the magnetic susceptibility is independent of the temperature.

15.2.5 B-field in 3+1 dimensions

Unlike a gauge theory in 2+1 dimensions, a gauge theory in 3+1 dimensions generically has anomalies of axial $U(1)$ symmetry, of the form

$$\langle\partial^\mu J_{5,\mu}\rangle = \frac{k}{6}\epsilon_{\mu\nu\rho\sigma}F^{\mu\nu}F^{\rho\sigma}, \tag{15.77}$$

with k some coefficient depending on the parameters of the field theory. On the gravity side, this anomaly is generated by a five-dimensional Chern–Simons term in the gravitational action, which is given by

$$S = \int d^5x \sqrt{-g} \left(\frac{1}{2\kappa^2} \left(R + \frac{12}{L^2} \right) - \frac{1}{4g^2} F_{mn} F^{mn} \right) + \frac{1}{6} k \int A \wedge F \wedge F. \quad (15.78)$$

k is the dimensionless Chern–Simons coupling which determines the anomaly coefficient in (15.77). For $k = 2/\sqrt{3}$, the action (15.78) coincides with the bosonic part of minimal supergravity in 4+1 dimensions, which is a consistent truncation of type IIB supergravity or M-theory. Below, k is left as a free parameter. The equations of motion corresponding to (15.78) are given by the Einstein equations (15.5) with $d = 4$, together with the additional equation

$$\frac{1}{g^2} d * F + k F \wedge F = 0. \quad (15.79)$$

The solutions reflecting a finite magnetic field B in the x_3 direction, a finite charge density and finite temperature T are translation invariant in the field theory directions t, x_1, x_2, x_3 and rotation invariant in the x_1, x_2 plane. The Bianchi identities imply that B is independent of the radial variable z.

Let us discuss the resulting metric for particular values of the parameters, as well as the resulting phase diagram [5]. The most important feature of this phase diagram is that a quantum phase transition occurs for a finite critical value of the magnetic field. It is useful to introduce a dimensionless reduced magnetic field

$$\hat{B} = \frac{B}{\rho^{2/3}}, \quad (15.80)$$

with ρ the charge density.

In the case of vanishing temperature and charge density, the metric obtained as solution to (15.79) asymptotes to $AdS_3 \times \mathbb{R}^2$ in the IR, and to AdS_5 in the UV. This solution is referred to as the *magnetic brane*. At finite T, this is replaced by a solution which asymptotes to $BTZ \times \mathbb{R}^2$ in the IR, with BTZ being the AdS_3 black hole defined in chapter 2, and to AdS_5 in the UV.

At finite charge density, there is a charged magnetic brane solution which interpolates between a *Schrödinger* spacetime in the IR and AdS_5 in the UV [5]. For $\hat{B} = \hat{B}_c$, the IR metric is of the general form

$$ds^2 \propto \frac{dz^2}{z^2} + \frac{1}{z^2} dt \, dx_3 - \frac{r^{2k}}{k(2k-1)} dt^2 + dx_1^2 + dx_2^2. \quad (15.81)$$

Metrics of this type are referred to as Schrödinger metrics and have been proposed for obtaining gravity duals of non-relativistic systems.

Regularity implies that the metric component g_{tt} has to be negative to ensure Lorentzian signature, and has to vanish only at the horizon. It turns out that for the charged magnetic brane solution, this condition has two consequences. First, the Chern–Simons level must satisfy $k > 1/2$, and second, it implies the existence of a critical magnetic field above which the charged magnetic brane solution is regular. This critical field is referred to as $\hat{B}_c$, with the reduced field defined as in (15.80). For $\hat{B} < \hat{B}_c$, the solution is given by a deformation of the Reissner–Nordström solution for zero magnetic field. At $\tilde{B}_c$, a quantum phase transition between the two phases occurs. This is supported by the fact that the dimensionless ratio $s/(TB)$ involving the entropy density diverges at this point. Therefore, in a region around $T = 0$ and $\hat{B} = \hat{B}_c$ in the (T, B) plane, we expect quantum critical behaviour. This is indeed the case. A central element for motivating this is to consider the IR metric given by (15.81) for $\hat{B} = \hat{B}_c$. This metric is invariant under the scale transformations

$$z \mapsto \lambda^{-1/2}z, \qquad t \mapsto \lambda^{-k}t, \qquad x_3 \mapsto \lambda^{k-1}x_3, \tag{15.82}$$

with x_1, x_2 invariant. Numerical analysis of the full solution confirms this IR scaling behaviour and gives, defining $\hat{s} = s/B^{3/2}$ [5],

$$\hat{s} \propto \left(\frac{T}{\sqrt{B}}\right)^h, \tag{15.83}$$

with

$$\begin{aligned} h &= \frac{1-k}{k} \quad \text{for} \quad \frac{1}{2} < k \le \frac{3}{4}, \\ h &= \frac{1}{3} \quad\;\; \text{for} \quad \frac{3}{4} < k. \end{aligned} \tag{15.84}$$

On the other hand, for the deformed Reissner–Nordström solution below $\hat{B}_c$, it is found that [5]

$$\hat{s} = \frac{\sqrt{\hat{B}_c - \hat{B}}}{4\sqrt{2}k\hat{B}_c^2}. \tag{15.85}$$

We note that at $T = 0$, the entropy density vanishes for $\hat{B} \ge \hat{B}_c$, while it is finite for $\hat{B} < \hat{B}_c$. The complete phase diagram is visualised in figure 15.3.

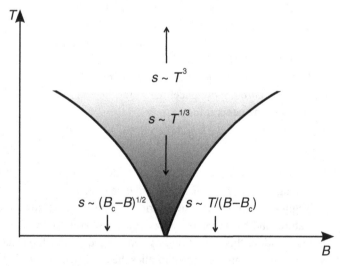

Figure 15.3   Holographic quantum phase transition at finite magnetic field.

We see that the B-field in 3+1 dimensions provides a holographic model with a quantum phase transition which occurs at a finite value $\hat{B}_c$ of the order parameter $\hat{B}$. This model shares the features of quantum phase transitions in condensed matter physics as introduced in section 15.1. In particular, the dynamical scaling exponent defined in (15.2) is given by $z = k(1 - k)$ for the model considered here.

15.3 Holographic superfluids and superconductors

An important aspect of holographic models at finite charge density is that in some cases, as explained below, there is a critical temperature below which a new ground state with lower free energy forms. This new ground state is dual to a condensate. The resistivity calculated from the linear response formalism displays a gap in the condensed phase. These models may therefore be viewed as a holographic dual of a superfluid, in which a global symmety is spontaneously broken and particles can move without energy loss. Some properties of these models are also present in superconductors, in which a local symmetry is spontaneously broken. While these new solutions are forbidden by the *no-hair theorem* for black holes embedded in flat space, this theorem does not apply to black holes in AdS space. This important discovery, which led to gravity duals of strongly coupled superfluids, raises the hope that it may also be used to obtain new information about *high T_c superconductors* which are strongly coupled as well, but whose pairing mechanism remains unknown. While gauge/gravity duality is unlikely to be able to provide an explanation of the pairing mechanism, it can be used to calculate physical properties and observables for strongly coupled superfluids.

15.3.1 High T_c superconductors

Within condensed matter physics, there is a large class of materials which are superconductors with particularly high transition temperatures, the *high T_c superconductors*. These materials cannot be described by the standard BCS theory of superconductivity, in which a small attractive interaction leads to an electron pairing mechanism and the formation of a new ground state with lower free energy. Generically, the high T_c superconductors are expected to be strongly coupled. So far, a complete theoretical explanation of high T_c superconductivity is still lacking. From experimental results, it is known that the phase diagram of high T_c superconductors takes the form given in figure 15.4.

The superconducting region in figure 15.4 is often referred to as the *superconducting dome* due to its shape. In the wedge-shaped region above this dome, non-Fermi liquid behaviour is observed. This is referred to as the *strange metal phase*. Due to the wedge shape of this region, and comparing with figure 15.1, there are suggestions that a quantum critical point is hidden under the superconducting dome. This suggestion implies that the phase above the dome may be a quantum critical region.

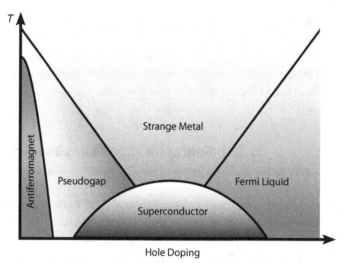

Figure 15.4 Phase diagram of high T_c superconductors.

The fact that high T_c superconductors appear to be strongly coupled systems with a quantum critical region, together with the absence of a theoretical understanding within condensed matter physics, make them a prototype example for applying gauge/gravity duality. Of course, it is difficult to make statements about the precise form of the pairing mechanism which requires a detailed knowledge of microscopic degrees of freedom which are not determined by the analysis of quantum critical points. Nevertheless, gauge/gravity duality can provide a description of universal macroscopic features of strongly coupled systems. As we will see below, there is indeed a condensation mechanism within gauge/gravity duality, which leads to an instability of the normal phase ground state, and to condensation to a new ground state with lower free energy. This new state has the properties of a superfluid or superconductor.

15.3.2 Superfluids and superconductors

The main difference between superfluids and superconductors is that, for the former, a global symmetry is spontaneously broken, while for the latter, the spontaneously broken symmetry is a local gauge symmetry. Superfluidity occurs for instance in helium systems.

Conventional superconductors are described by BCS theory. Here, the $U(1)$ symmetry of electromagnetism is spontaneously broken by the formation of electron pairs, the *Cooper pairs*. The pair formation is due to an attractive force mediated by phonons, i.e. by quanta of lattice vibrations. Near the phase transition, superconductors can be described by an effective field theory, the *Ginzburg–Landau theory* (see box 15.1).

Let us list a few key properties of superconductors. The most important one is the infinite DC conductivity

$$\mathrm{Re}\,\sigma(\omega) = \pi \rho_s \delta(\omega) \tag{15.91}$$

Box 15.1	Ginzburg–Landau theory

Ginzburg–Landau theory is a phenomenological theory of superconductivity based on a complex order parameter ϕ with action

$$\mathcal{S} = -\int d^d x \left(|D\phi|^2 + \alpha |\phi|^2 + \frac{\beta}{2} |\phi|^4 \right), \qquad (15.86)$$

with coefficients α, β. This action, inspired by ϕ^4 theory, is equivalent to the free energy. $D_\mu = \partial_\mu + iqA_\mu$ is the $U(1)$ gauge covariant derivative, α is related to an effective mass squared and β is a coupling coefficient. The equations of motion minimising this action are

$$\alpha\phi + \beta|\phi|^2\phi + D^\mu D_\mu \phi = 0, \qquad (15.87)$$

$$D^\mu J_\mu = 0 \quad \text{with } J_\mu = \text{Re}\, \phi^* D_\mu \phi. \qquad (15.88)$$

For a homogeneous superconductor with spatially non-varying condensate, the first of these equations simplifies to

$$\alpha\phi + \beta|\phi|^2\phi = 0. \qquad (15.89)$$

The trivial solution $\phi = 0$ corresponds to the normal phase. For coefficients $\alpha(T)$, $\beta(T)$ dependent on the temperature, a new ground state is present below a transition temperature $T = T_c$ at which α changes sign while β does not. When α and β have different signs for $T < T_c$, then there is a new ground state given by

$$|\phi|^2 = -\frac{\alpha}{\beta}. \qquad (15.90)$$

This solution minimises the free energy below T_c. The condensate vanishes for $T \to T_c$, which corresponds to a second order phase transition. Above T_c, only $\phi = 0$ is a solution. $|\phi|^2$ may be interpreted as the density of particles – electrons in a conventional superconductor – which have condensed to a superfluid phase. The Ginzburg–Landau model can be derived from the microscopic BCS theory of superconductivity.

near $\omega = 0$, where ρ_s is the superconducting density, which reflects the absence of resistivity. The second key property is the gap in the excitation spectrum at frequencies smaller than the energy given by the order parameter. This gap implies in particular that the superconductor does not absorb any radiation with frequency, i.e. energy, inside the gap.

A feature which distinguishes superfluids and superconductors is the Meissner effect, i.e. the repulsion of magnetic flux by the condensate. The Meissner effect is present in superconductors with gauged symmetry, while in superfluids with global symmetry there is only a remnant of it: a magnetic field inhibits condensation, i.e. T_c is lowered.

15.3.3 Holographic superconductor: s-wave superfluid

Let us now consider gauge/gravity duality models where condensation occurs. The starting point is again our simple holographic model containing a metric which is asymptotically

AdS space, and a $U(1)$ gauge field to describe finite charge density. The simplest way of extending this model is to add a scalar field, which for $U(1)$ gauge invariance has to be complex [6, 7]. As we will see, this scalar field provides the order parameter of a condensation process, which is referred to as an *s-wave* holographic superconductor.

The action corresponding to these ingredients is an Abelian Higgs model with action

$$S = \int d^{d+1}x \sqrt{-g} \left(\frac{1}{2\kappa^2}(R - 2\Lambda) - \frac{1}{4g^2}F_{mn}F^{mn} \right)$$

$$- \int d^{d+1}x \sqrt{-g} \left[|D\Phi|^2 + V(|\Phi|) \right] + S_{\text{bdy}}, \qquad (15.92)$$

with the $U(1)$ covariant derivative $D_\mu\Phi = \partial_\mu\Phi + iA_\mu\Phi$, $|D\Phi|^2 = D_\mu\Phi D^\mu\Phi^*$, the cosmological constant $\Lambda = -\frac{d(d-1)}{2L^2}$, where L is the radius of AdS_{d+1} and d the dimension of the field theory.

We note that $V(|\Phi|)$ is not specified. To determine the specific form of this potential, a string theory embedding of the bottom-up model given by (15.92) is required. Here we continue in the bottom-up philosophy and choose

$$V(|\Phi|) = m^2|\Phi|^2. \qquad (15.93)$$

This choice is motivated by Ginzburg–Landau theory as reviewed in box 15.1. However, in this case, a Φ^4 potential is not necessary, since the AdS space may essentially be thought of as a box which prohibits any runaway behaviour of the scalar field. As in the simplest holographic model without scalar field, finite temperature and finite chemical potential are realised by a charged black hole solution to the equations of motion with $A_t \neq 0$. We now show that for sufficiently large chemical potential, the scalar Φ condenses, i.e. $\Phi \neq 0$. To see this, let us assume that Φ depends only on the radial direction z, i.e. $\Phi = \Phi(z)$. Then for the Reissner–Nordström background, $|D\Phi|^2 + V(|\Phi|)$ reads

$$|D\Phi|^2 + V(|\Phi|) = g^{zz}\partial_z\Phi\partial_z\Phi^* + g^{tt}A_t^2|\Phi|^2 + m^2|\Phi|^2, \qquad (15.94)$$

and the effective mass is given by

$$m_{\text{eff}}^2 = m^2 + g^{tt}A_t^2, \qquad m_{\text{eff}}^2 \leq m^2 \quad \text{since } g^{tt} < 0. \qquad (15.95)$$

Since $g^{tt} \to -\infty$ at the horizon, Φ may be tachyonic, i.e. its effective mass is below the Breitenlohner–Freedman bound for sufficiently large chemical potential.

Exercise 15.3.1 Repeat this calculation for the case of an uncharged scalar in the Reissner–Nordström background. Since this background has an AdS_2 factor in the IR, show that the the scalar may condense since it violates the Breitenlohner–Freedman bound of AdS_2.

As before, the chemical potential is the leading term in the near-boundary expansion of the temporal component of the gauge field, $A_t|_{z=0} = \mu$. The violation of the Breitenlohner–Freedman bound corresponds to an instability: Φ condenses to a new ground state and breaks the $U(1)$ symmetry spontaneously. To analyse this process in detail, we consider the charged scalar case in a model with $(2+1)$-dimensional boundary, with potential given by

$$m^2 = -\frac{2}{L^2} \qquad (15.96)$$

in (15.93) for simplicity. m^2 is negative, but above the Breitenlohner–Freedman bound $m_{BF}^2 = -9/(4L^2)$. Moreover, we choose the decoupling limit $\kappa^2 \ll g^2 L^2$. In this weak gravity (or probe) regime, the gauge and scalar sectors decouple from the gravity sector: both sectors have insufficient energy to curve the spacetime. In this case, the calculations may be performed in a fixed spacetime background,

$$ds^2 = \frac{L^2}{z^2} \left(-f(z)\, dt^2 + dx^2 + dy^2\right) + \frac{L^2}{z^2} \frac{dz^2}{f(z)}, \quad f(z) = 1 - \left(\frac{z}{z_h}\right)^3. \tag{15.97}$$

The equations of motion for A_t and Φ read (for $L = 1$)

$$z^2 \frac{\partial}{\partial z}\left(\frac{f(z)}{z^2}\frac{\partial\Phi}{\partial z}\right) = \left(\frac{m^2}{z^2} - \frac{A_t^2}{f(z)}\right)\Phi(z), \tag{15.98}$$

$$\frac{\partial^2}{\partial z^2}A_t(z) = \frac{2g^2}{z^2 f(z)}\Phi^2(z)A_t(z). \tag{15.99}$$

Near the boundary $z \to 0$, A_t and Φ read

$$A_t(z) \approx \mu - \rho z + \cdots \tag{15.100}$$

$$\Phi(z) \approx \Phi_1 z + \Phi_2 z^2. \tag{15.101}$$

μ and ρ are the chemical potential and the corresponding density. ρ corresponds to the expectation value for the charge density, $\rho = \langle J^0\rangle$, with a source term in the boundary action of the form $S_{bdy} \to S_{bdy} + \mu \int J^0 d^d x$. For Φ, there are two different cases, since the mass of the scalar satisfies the inequality

$$-\frac{d^2}{4} + 1 \geq m^2 \geq -\frac{d^2}{4} = m_{BF}^2. \tag{15.102}$$

As discussed in section 5.3.4, for these values of m, there are two different possibilities for identifying the source and the condensate of the dual field theory operator.

- Φ_1 is the source (which vanishes) and $\langle O_2\rangle \propto \Phi_2$ (i.e. Φ_2 is the condensate). This implies that O_2 is a dimension two operator.
- Φ_2 is the source (which vanishes) and $\langle O_1\rangle \propto \Phi_1$ (i.e. Φ_1 is the condensate). This implies that O_1 is a dimension one operator.

If the vacuum expectation values for $\langle O_1\rangle$ and $\langle O_2\rangle$ are non-zero while the corresponding sources vanish, the $U(1)$ symmetry is broken spontaneously. Non-zero sources would break this symmetry explicitly. As the numerical result in figure 15.5 shows, we indeed have spontaneous symmetry breaking for $T < T_c$. $\langle O_1\rangle$ and $\sqrt{\langle O_2\rangle}$ are order parameters which near T_c scale as

$$\langle O_1\rangle, \sqrt{\langle O_2\rangle} \propto \left(1 - \frac{T}{T_c}\right)^{1/2}. \tag{15.103}$$

This implies that the critical exponent is $1/2$, as is expected for a second order phase transition in mean field theory. This is typical for a Landau–Ginzburg effective description.

We use linear response theory to determine the conductivity for this model. This is done by perturbing the background with a small fluctuation

$$\delta A_x = \delta A_x^{(0)} + \delta A_x^{(1)} z + \mathcal{O}(z^2) \quad \text{for } z \to 0. \tag{15.104}$$

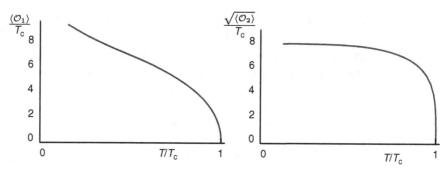

Figure 15.5   Holographic superconductors. Order parameters: left $\langle \mathcal{O}_1 \rangle / T_c$, right $\sqrt{\langle \mathcal{O}_2 \rangle} / T_c$.

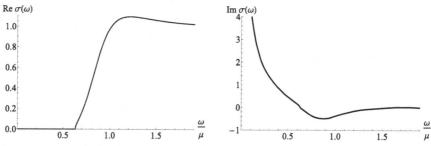

Figure 15.6   Real and imaginary parts of the conductivity $\sigma(\omega)$ at vanishing temperature $T = 0$ for the holographic superfluid with condensate $\langle \mathcal{O}_2 \rangle$. The real part of the conductivity displays a superconducting gap at low frequencies.

$\delta A_x^{(1)}$ is related to $\langle J^x \rangle$, and the electric field is given by

$$E_x = F_{tx} = \partial_t(\delta A_x)|_{z=0} = i\omega \delta A_x^{(0)}, \tag{15.105}$$

where we have assumed an $e^{i\omega t}$ time dependence of the fluctuations. The conductivity $\sigma(\omega) = \frac{\langle J^x \rangle}{E_x}$ is given by $\sigma(\omega) = \delta A_x^{(1)} / i\omega \delta A_x^{(0)}$. For the case of the operator $\mathcal{O}_2$ condensing, the result is shown in figure 15.6.

As shown in figure 15.6, we find a gap in the real frequency-dependent conductivity. This is expected for a superfluid or superconductor: states with energy smaller than the gap energy set by the order parameter cannot be filled. Moreover, there is an infinite DC conductivity at $\omega = 0$. This is expected for a superfluid or superconductor as discussed below (15.91), however here this superfluid delta peak is superposed by the delta peak (15.51) due to translation invariance.

The gauge/gravity duality model introduced here thus displays features expected for a superconductor. Additionally, effects similar to the *Meissner effect* occur in holographic systems, in the sense that a magnetic field reduces the transition temperature. Note, however, that although the $U(1)$ symmetry which is spontaneously broken is local in the gravity description, it corresponds to a global symmetry in the field theory. Generically within gauge/gravity duality, local gauge symmetries in the bulk give rise to global symmetries on the boundary. Therefore, in the discussion above we have found the holographic description of a *superfluid* where a global symmetry is spontaneously broken, whereas a superconductor requires the spontaneous breaking of a local symmetry. Nevertheless, the term *holographic superconductor* is also frequently used for this model, since the conductivity takes the form expected for a superconductor. Moreover, it is expected that when weakly gauging the global symmetry on the field theory side, i.e. by introducing a gauge connection for the $U(1)$ symmetry with a small gauge coupling, the phenomena discussed in this section remain.

15.3.4 Holographic superconductor: p-wave superfluid

A similar condensation mechanism within gauge/gravity duality also arises when considering the gravity side given by the $SU(2)$ *Einstein–Yang–Mills* action

$$S = \int d^4x \sqrt{-g} \left(\frac{1}{2\kappa^2} \left(R + \frac{6}{L^2} \right) - \frac{1}{2g^2} \mathrm{Tr} F^{mn} F_{mn} \right), \qquad (15.106)$$

with A_m in F_{mn} an $SU(2)$ gauge field. The $SU(2)$ symmetry may be interpreted as an isospin symmetry for a two-flavour model. A charge is introduced by considering a non-trivial profile for the temporal component of the gauge field, which asymptotically near the boundary reads, when $d = 4$,

$$A_t(z) = A_t^3(z)\tau^3 = \mu + z\langle J_t^3 \rangle \tau^3 + \mathcal{O}(z^2), \qquad (15.107)$$

with $J_t = J_t^3 \tau^3$ the charge density in the dual field theory. The τ^i are the Pauli matrices, generators of $SU(2)$. This ansatz breaks the $SU(2)$ symmetry to $U(1)$. In this background, there is a current $J_x = J_x^1 \tau^1$ dual to the gauge field condensing, spontaneously breaking the residual $U(1)$ symmetry,

$$A_x = z\langle J_x^1 \rangle \tau^1 + \mathcal{O}(z). \qquad (15.108)$$

For $T < T_c$, the solution with both A_t^3 and A_x^1 turned on has lower free energy than the solution with $A_x^1 = 0$. Moreover, A_x^1 is dual to the current J_x^1. In addition to the spontaneous breaking of the $U(1)$ symmetry, the non-trivial expectation value for this current also breaks rotational symmetry in configuration space.

Exercise 15.3.2   Derive the equations of motion from the action (15.106) . Show that in the presence of the background field (15.107), an effective mass of the form

$$m_{\mathrm{eff}} = m^2 + g^{tt}(A_t(z))^2 \qquad (15.109)$$

arises for the linear perturbations of the charged field about the Reissner–Nordström solution. This effective mass may be below the Breitenlohner–Freedman bound, signalling an instability of the normal state solution and a condensation process.

15.3.5 Spatially modulated phases

A different type of instability occurs when there is a Chern–Simons term present in the gravity action, for Chern–Simons levels above a critical value [8]. We already discussed a gravitational Chern–Simons term in section 12.4.2, where we saw that it leads to an axial anomaly in the dual field theory, as well as in section 15.2.5 above.

The Maxwell–Chern–Simons theory in five spacetime dimensions becomes unstable if a constant charge density is turned on, generating an electric field. This instability is present for non-vanishing momenta, leading to a spatially modulated new ground state. The starting point is again the Chern–Simons action of (15.78),

$$S = \int d^5x \sqrt{-g} \left(\frac{1}{2\kappa^2} \left(R + \frac{12}{L^2} \right) - \frac{1}{4g^2} F_{mn} F^{mn} \right) + \frac{1}{6} k \int A \wedge F \wedge F, \quad (15.110)$$

where

$$k = \frac{2}{\sqrt{3}} \quad (15.111)$$

corresponds to the supersymmetric case.

For Chern–Simons coupling larger than a critical value, there is a critical temperature below which the Reissner–Nordström black hole solution in AdS_5 becomes unstable. In the dual (3+1)-dimensional field theory, this corresponds to a phase transition at finite chemical potential where the charge current develops a position-dependent expectation value of the form

$$\langle \vec{J} \rangle = \text{Re}(\vec{u} e^{ipx}), \quad (15.112)$$

with non-zero momentum p. The constant vector $\vec{u}$ is circularly polarised. For the momentum pointing in the x_3 direction, a possible realisation of circular polarisation is given by the parametrisation

$$\langle \vec{J}_1 \rangle = u_1 \cos(p_3 x_3 - \omega t), \quad \langle \vec{J}_2 \rangle = u_1 \sin(p_3 x_3 - \omega t), \quad \langle \vec{J}_3 \rangle = 0. \quad (15.113)$$

This leads to a helical symmetry, i.e. translation and rotation symmetry are broken, but a combination of both is preserved. This behaviour is generic for a system with axial symmetry broken by the anomaly (15.77).

For the zero temperature extremal black hole solution, the IR dual geometry at finite chemical potential is given by $AdS_2 \times \mathbb{R}^3$, with the AdS_2 metric given by (15.23). The curvature radius of AdS_2 is $\tilde{L} = L/\sqrt{12}$. The charge density generates an electric field which near the horizon is proportional to the volume form of AdS_2,

$$F_{01} = \frac{E}{12r^2}, \qquad E = \pm 2\sqrt{6}. \quad (15.114)$$

Taking gravity to be non-dynamical, gauge field fluctuations near the horizon violate the Breitenlohner–Freedman bound of AdS_2 if

$$-k^2 E^2 < m_{\text{BF}}^2 = -\frac{1}{4\tilde{L}^2}. \quad (15.115)$$

The associated instability then leads to condensation to the new modulated ground state.

Explaining the fact that the effective mass of the gauge field fluctuations is given by $-k^2 E^2$, as implied by (15.115), requires an involved calculation since the condensation only happens for non-zero momenta. However, this may be motivated as follows. Consider the space $\mathbb{R}^{1,1} \times \mathbb{R}^3$ with coordinates $x^0, x^1, y^i, i = 1, 2, 3$. We switch on a constant

E-field in the x^1 direction. The equation of motion is obtained from (15.110) for vanishing curvature. By solving the equation of motion for fluctuations of the form

$$a_\mu(x,y) = \epsilon_\mu e^{-ipx+iqy},   \tag{15.116}$$

we obtain the dispersion relation

$$p_0^2 - p_1^2 = \left(|\vec{q}| \pm \frac{1}{2}kE\right)^2 - \frac{1}{4}k^2E^2.   \tag{15.117}$$

This implies that there are tachyonic modes in $\mathbb{R}^{1,1}$ in the range $0 < |\vec{q}| < |kE|$.

Let us return to the Reissner–Nordström black brane. For the case of dynamical gravity governed by Einstein's equations, there is a critical value for the Chern–Simons level k, above which the instability occurs. This numerical analysis reveals that the supersymmetric value for k given by (15.111) still leads to a stable solution, while being less than 1% away from the critical value. The spatially modulated phase presented here is generic in models with axial symmetry broken by the anomaly (15.77). There are further scenarios where spatially modulated ground states occur, for instance for gravity theories with complex two-forms or in the presence of a magnetic field.

15.4 Fermions

The most straightforward approach to considering fermions within gauge/gravity duality is to consider fermionic contributions to the gravity action, for instance those that occur naturally in supergravity. These are dual to composite gauge invariant operators in the dual field theory. Though these operators describe physical objects which are quite different from gauge variant elementary fermions such as electrons, it is nevertheless instructive to study their thermodynamical properties and to calculate their correlation functions and their conductivity [9, 10]. In particular, *non-Fermi liquid behaviour* is found in the dual strongly coupled field theories, which means that the standard Landau–Fermi approach of describing fermions at weak coupling as described in box 15.2 is not applicable.

The simplest example of a holographic approach is to start again from the Einstein–Maxwell action (15.4) leading to the Reissner–Nordström black hole, written in the form (15.9), and to add the fermionic contribution

$$S_{\text{spinor}} = \int d^{d+1}x\sqrt{-g}\left(i\bar{\Psi}\Gamma^n D_n\Psi - m\bar{\Psi}\Psi\right),   \tag{15.118}$$

where

$$\bar{\psi} = \psi^\dagger\Gamma^t, \qquad D_n = \partial_n + \frac{1}{4}\omega_{nab}\Gamma^{ab} + iqA_n.   \tag{15.120}$$

ω_{nab} is the spin connection for the derivative D_n to be covariant with respect to spacetime symmetry, while the gauge field A_n ensures gauge covariance. When specifying a boundary

Landau-Fermi liquid theory

Fermi liquid theory describes interacting fermions. It is based on the assumption that there is a one-to-one map between the states in an interacting Fermi liquid and those in a non-interacting Fermi gas. The ground state of the Fermi gas is given by the Fermi-Dirac distribution, such that at $T = 0$, all states up to the Fermi energy $\epsilon_F = \mu$ are filled. When an interaction is turned on in the Fermi gas, it is assumed that the ground state deforms adiabatically into the new ground state of the interacting Fermi liquid. The excitations about this ground state are *quasiparticles* with the same quantum numbers such as charge and spin as in the non-interacting case, but with a renormalised effective mass.

A characteristic property of a Fermi liquid is that at low temperatures, the resistivity scales as T^2. Moreover, the retarded Green's functions for the Landau–Fermi liquid have the characteristic scaling behaviour

$$\text{Im } G(a^z\omega, a\tilde{k}) = a^{-\alpha}\text{Im } G(\omega, \tilde{k}), \qquad \tilde{k} = k - k_F,$$

$$\text{with} \qquad z = \alpha = 1. \tag{15.119}$$

Deviations from Fermi liquid behaviour occur for instance in the *Luttinger liquid* in 1+1 dimensions and are generally termed *non-Fermi liquid behaviour*. High T_c superconductors also display non-Fermi liquid behaviour.

field theory in $d = 2 + 1$ dimensions, the relevant components of the spin connection are given by

$$\omega_{ttz} = -\frac{\partial_z g_{tt}}{2\sqrt{-g_{tt}g_{zz}}}, \qquad \omega_{xxz} = -\frac{\partial_z g_{xx}}{2\sqrt{g_{xx}g_{zz}}}, \qquad \omega_{yyz} = -\frac{\partial_z g_{yy}}{2\sqrt{g_{yy}g_{zz}}}. \tag{15.121}$$

The Γ^n are the gamma matrices in general dimensions. A convenient basis for the gamma matrices of the $(3 + 1)$-dimensional bulk theory is given by

$$\Gamma^r = \begin{pmatrix} \mathbb{1}_2 & 0 \\ 0 & -\mathbb{1}_2 \end{pmatrix}, \quad \Gamma^\mu = \begin{pmatrix} 0 & \gamma^\mu \\ \gamma^\mu & 0 \end{pmatrix}, \quad \Psi = \begin{pmatrix} \psi_+ \\ \psi_- \end{pmatrix}, \tag{15.122}$$

where the γ^μ are the gamma matrices of the boundary theory and $\psi_\pm$ are two-component spinors.

By writing

$$\psi_\pm = (-gg^{zz})^{-1/4}e^{-i\omega t + ik_i x^i}\chi_\pm, \tag{15.123}$$

the Dirac equation takes the form

$$\sqrt{\frac{g_{ii}}{g_{zz}}}(\partial_z \pm im\sqrt{g_{zz}})\chi_\pm = \mp iK_\mu\gamma^\mu\chi_\mp, \tag{15.124}$$

where we have introduced

$$K_\mu(z) = (-E(z), k_i), \qquad E(z) = \sqrt{\frac{g_{ii}}{-g_{tt}}}\left(\omega - \mu_q(1 - z)\right). \tag{15.125}$$

Since for $z \to 0$, $E(z) \to \omega + \mu_q$, ω gives the deviation from the Fermi energy μ_q, where $\mu_q = q \cdot \mu$ is the effective chemical potential for a field with charge q.

Our aim is now to calculate the retarded Green's function for the operator dual to (15.123) and to show that it displays a Fermi surface. This requires solving the Dirac equation (15.124) with infalling boundary conditions at the horizon, and identifing the source and the expectation value from the asymptotic behaviour of Ψ near the boundary. A possible prescription ensuring regularity at the horizon is to identify ψ_+ as the source and its canonical momentum with respect to the radial variable, which is related to ψ_-, as the expectation value. On the field theory side, the retarded Green's function corresponds to $G_R \propto \langle \{\mathcal{O}, \mathcal{O}^\dagger\}\rangle$, with the source term given by

$$S_{\text{bdy}} = -i \int d^3x \, (\bar{\chi}_+^{(0)}\mathcal{O} + \bar{\mathcal{O}}\chi_+^{(0)}), \qquad \bar{\mathcal{O}} = \mathcal{O}^\dagger \gamma^0, \tag{15.126}$$

with $\chi_+^{(0)}$ the boundary value of χ_+, as defined in (15.123).

We write the asymptotic behaviour of $\chi_\pm$, as defined in (15.123) as

$$\chi_+ = Az^{-m} + Bz^{m+1}, \qquad \chi_- = Cz^{1-m} + Dz^m, \tag{15.127}$$

where

$$C = \frac{i\gamma^\mu k_\mu}{2m-1}A, \qquad B = \frac{i\gamma^\mu k_\mu}{2m+1}D, \qquad k_\mu = (-(\omega+\mu), k_i). \tag{15.128}$$

We identify D as the expectation value and A as the source. Then for $m > 0$, with the sources as given in (15.126), the retarded Green function $G_R \propto \langle \{\mathcal{O}, \mathcal{O}^\dagger\}\rangle$ can be obtained from

$$G_R = -i\frac{D}{A}\gamma^0. \tag{15.129}$$

It is now convenient to choose a particular basis for the gamma matrices,

$$\gamma^0 = \sigma_2, \quad \gamma^1 = i\sigma_1, \quad \gamma^2 = i\sigma_3, \tag{15.130}$$

in order to simplify the Dirac equation (15.124) further. In addition, without loss of generality we set $k_2 = 0$ and $k_1 = k$. Then, writing

$$\chi_\pm = \begin{pmatrix} y_\pm \\ w_\pm \end{pmatrix}, \tag{15.131}$$

the equations of motion (15.124) decouple to give

$$\sqrt{\frac{g_{ii}}{g_{zz}}}(\partial_z \pm im\sqrt{g_{zz}})y_\pm = \pm i(k + E(r))w_\mp, \tag{15.132}$$

$$\sqrt{\frac{g_{ii}}{g_{zz}}}(\partial_z \pm im\sqrt{g_{zz}})w_\mp = \pm i(k - E(r))y_\pm. \tag{15.133}$$

With

$$\xi_+ = i\frac{y_-}{w_+}, \qquad \xi_- = -i\frac{w_-}{y_+} \tag{15.134}$$

the retarded Green function (15.129) may be written as

$$G_R = \lim_{\epsilon \to 0} \epsilon^{2m} \begin{pmatrix} \xi_+ & 0 \\ 0 & \xi_- \end{pmatrix}\bigg|_{r=\frac{1}{\epsilon}}. \tag{15.135}$$

With (15.134), equations (15.132) and (15.133) may be rewritten as

$$\sqrt{\frac{g_{ii}}{g_{zz}}}\partial_z\xi_\pm = 2im\sqrt{g_{ii}}\xi_\pm \pm (ik \pm iE(z)) \mp (ik \mp iE(z))\xi_\pm^2. \tag{15.136}$$

The infalling boundary condition at the horizon is then

$$\xi_\pm\Big|_{z=z_h} = i. \tag{15.137}$$

This condition allows integratation of (15.136) to the boundary at $z \to 0$ in order to obtain the boundary correlation function. The solutions for the retarded Green's function (15.129) satisfy

$$G_{22}(\omega, k) = G_{11}(\omega, -k), \qquad G_{22}(\omega, k) = -G_{11}(\omega, k)^{-1}. \tag{15.138}$$

In general, the solutions for G_R can only be found numerically. For illustration, we show a plot of the real and imaginary parts of G_{22} for $m = 0$ in figure 15.7. This figure clearly displays a Fermi surface, which is not of Landau–Fermi liquid type. The scaling exponents in (15.119) are [9]

$$\alpha = 1, \qquad z = 2.09 \pm 0.01. \tag{15.139}$$

The emergent IR conformal symmetry dual to the AdS_2 subspace present in the IR geometry, as introduced in section 15.2.2, has important consequences for the fermion spectral functions and the conductivity. Recall that in section 15.2.3, we described how this emergent IR geometry determines the structure of the retarded Green's functions.

Consider a fermionic operator $\mathcal{O}$ of mass m and charge q depending on momentum $\vec{k}$ in a $(2+1)$-dimensional theory. As explained in section 15.2.3, the retarded Green's function of this operator will be determined by the Green function of an operator $\mathcal{O}^{IR}$ in the IR CFT_{0+1}, of scaling dimension δ_k given by

$$\delta_k = \nu_k + \frac{1}{2}, \qquad \nu_k = \frac{1}{\sqrt{6}}\sqrt{m^2L^2 + \frac{3k^2}{\mu^2} - \frac{g^2q^2}{2}}, \tag{15.140}$$

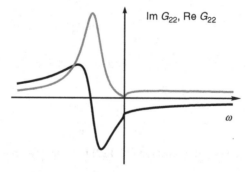

Figure 15.7   Real (black) and imaginary (grey) parts of the fermion spectral function G_{22} as a function of the frequency, for a momentum just below the Fermi surface.

with two-point correlation function given by

$$\mathcal{G}_k(\omega) = c(k)(-i\omega)^{2\nu_k},\tag{15.141}$$

with $c(k)$ complex and analytic in $k \equiv |\vec{k}|$.

For $m^2 L^2 < g^2 q^2/2$, the Dirac equation for the gravity side fermion field dual to the operator $\mathcal{O}$ turns out to have a static normalisable solution at a discrete shell in momentum space. The corresponding momentum is naturally identified with the Fermi momentum k_F. Near this value, and at small frequencies relative to the Fermi energy, the retarded Green's function for the operator $\mathcal{O}$ has the form

$$G_R(k, \omega) \simeq \frac{Z}{\omega - v_F(k - k_F) + \Sigma(\omega)}, \qquad \Sigma(\omega) = a\mathcal{G}_{k_F}(\omega) = ac(k_F)(-i\omega)^{2\nu_{k_F}},$$
$$\tag{15.142}$$

with a a numerical constant and v_F the Fermi velocity. Σ is determined by the correlator (15.141) of the IR operator. $\Sigma(\omega)$ determines the dissipative part of the correlator (15.142), i.e. the imaginary part of the Green function. Note that due to (15.141), in general this imaginary part does not scale as in the Fermi liquid case introduced in box 15.2, where the scaling is given by (15.119). We thus have an example for a *non-Fermi liquid theory*.

Moreover, for the finite temperature generalisation of (15.142), the conductivity is found to depend on the temperature as

$$\sigma_{DC} \propto T^{-2\nu_{k_F}}.\tag{15.143}$$

For $\nu_{k_F} = 1/2$, which corresponds to a *marginal Fermi liquid*, this implies that the resistivity is linear in the temperature. This behaviour is observed for instance for high T_c superconductors in the strange metal phase as shown in figure 15.4. Materials of non-Fermi liquid type are generically hard to describe using conventional methods, and the approach presented here may present a new avenue towards a better understanding of non-Fermi liquids. It is to be noted, however, that so far the parameters q and m which determine ν_k according to (15.140) are not determined by the model presented, and further information is needed to fix them.

15.5 Towards non-relativistic systems and hyperscaling violation

In view of further progress towards condensed matter applications, it is desirable to find gravity duals for non-relativistic systems. Progress in this direction has been achieved by studying the Schrödinger geometries already mentioned above in section 15.2.5, and the related *Lifshitz geometries* [11]. These provide in particular a different scaling behaviour for the time and space directions at the boundary. This is of relevance for describing quantum critical points for instance.

15.5.1 Lifshitz spaces

Consider anisotropic scaling of the form

$$t \to \lambda^z t, \qquad \vec{x} \to \lambda \vec{x}, \quad z \neq 1. \tag{15.144}$$

Here, z is the dynamical exponent which also determines the dispersion relation, $\omega \propto k^z$. The case $z = 1$ would correspond to relativistic scaling. The simplest anisotropic case is given by $z = 2$, for which there is the *Lifshitz field theory*

$$\mathcal{L} = \int \mathrm{d}^2 x \, \mathrm{d} t \, \left((\partial_t \phi)^2 - g(\nabla^2 \phi)^2 \right). \tag{15.145}$$

This theory has a line of fixed points parametrized by g. It describes critical points in strongly correlated electron systems, for instance.

For a gravity dual of a field theory with the scaling properties (15.144), we obtain a phenomenological 'bottom-up' model by writing the metric

$$\mathrm{d}s^2 = L^2 \left(-r^{2z} \mathrm{d}t^2 + r^2 \mathrm{d}\vec{x}^2 + \frac{\mathrm{d}r^2}{r^2} \right) \tag{15.146}$$

with $0 < r < \infty$. $\vec{x}$ stands for the spatial coordinates. This is invariant under the scale transformation

$$t \mapsto \lambda^z t, \qquad \vec{x} \mapsto \lambda \vec{x}, \qquad r \mapsto \lambda^{-1} r. \tag{15.147}$$

For $z = 1$ we recover the usual AdS metric. This metric is non-singular everywhere, although it is not geodesically complete at $r = 0$. Its Ricci scalar is given by

$$R = -2(z^2 + 2z + 3) \frac{1}{L^2}. \tag{15.148}$$

As a toy model we may consider a real scalar field in this background metric. In particular, we can calculate its two-point correlation using the same techniques as in AdS space. However, the result will depend on z and is known in closed form only for $z = 2$.

Dimensional analysis involving the Lifshitz scaling implies that the entropy density scales with the temperature as

$$s \propto T^{2/z}. \tag{15.149}$$

In the limit $z \to \infty$, the space given by (15.146) returns to $AdS_2 \times \mathbb{R}^d$ for d spatial dimensions, and there remains a finite entropy density in the limit $T \to 0$.

15.5.2 Lifshitz and Schrödinger symmetry algebra

Let us consider the symmetries associated with the Lifshitz geometry introduced in the previous section, and compare it to the symmetries of the Schrödinger geometry of section 15.2.5. As displayed in (15.144), in a non-relativistic theory, t and $\vec{x}$ do not necessarily transform under a scale transformation in the same way.

A non-relativistic theory, such as the Lifshitz theory considered in the previous section, has spatial translations P_i, time translations H and rotations M_{ij} as its symmetry generators.

A scale invariant theory will also have the dilatation generator, D. In addition, in a non-relativistic theory, the particle number must be discrete and is associated with a number operator N. The non-zero commutators of these operators are

$$[M^{ij}, M^{kl}] = i\left(\delta^{ik}M^{jl} + \delta^{jl}M^{ik} - \delta^{il}M^{jk} - \delta^{jk}M^{il}\right),$$

$$[M^{ij}, P^k] = i\left(\delta^{ik}P^j - \delta^{jk}P^i\right), \tag{15.150}$$

$$[D, P^i] = -iP^i, \qquad [D, H] = -izH, \qquad [D, N] = i(z - 2)N.$$

This algebra is referred to as the *Lifshitz algebra*. When $z = 1$, this symmetry algebra can be enhanced to the familiar *relativistic* conformal group, with relativistic dispersion relation, $\omega \sim k$.

The case $z = 2$ is also clearly special. When $z = 2$, N becomes a central element of the algebra, commuting with all other elements. Also, in this case the symmetry group can be enhanced to include Galilean boosts K_i which satisfy the additional commutation relations

$$[M^{ij}, K^k] = i\left(\delta^{ik}K^j - \delta^{jk}K^i\right),$$

$$[K^i, P^j] = i\delta^{ij}N, \qquad [H, K^i] = -iP^i \tag{15.151}$$

as well as, for the dilatation operator,

$$[D, K^i] = (z - 1)iK^i. \tag{15.152}$$

Moreover, for $z = 2$ there is a special conformal generator, C, which acts as

$$C: \quad t \mapsto \frac{t}{1 + \lambda t}, \qquad \vec{x} \mapsto \frac{\vec{x}}{1 + \lambda t},$$

with non-zero commutators

$$[D, C] = 2iC, \qquad [H, C] = iD. \tag{15.153}$$

For $z = 2$, the Lifshitz algebra extended by the relations (15.151), (15.152), (15.153) is referred to as the *Schrödinger algebra*.

An important difference from the relativistic case is that now *two* operators can be diagonalised simultaneously, and their eigenvalues can be used to label inequivalent representations of the algebra, namely D and N. In the relativistic case, only the dimension D is used. A further important difference is that non-relativistic conformal symmetry is *not* sufficient to determine the form of two-point functions. Two-point functions are determined only up to an unknown function of $|\vec{x}|^2/t$, which is invariant under the scaling in (15.144) with $z = 2$. Many concepts from relativistic conformal field theories are still valid, however. For example, a primary operator $\mathcal{O}$ obeys $[K^i, \mathcal{O}] = [C, \mathcal{O}] = 0$. The scaling dimension $\Delta_{\mathcal{O}}$ and particle number $N_{\mathcal{O}}$ of a local operator $\mathcal{O}$ are defined by

$$[D, \mathcal{O}] = i\Delta_{\mathcal{O}}\mathcal{O}, \qquad [N, \mathcal{O}] = N_{\mathcal{O}}\mathcal{O}. \tag{15.154}$$

Not all operators will have a well-defined scaling dimension and particle number, however. The Schrödinger algebra in d dimensions may be embedded into the conformal algebra of a space with $d + 1$ spatial directions as follows. Consider a conformal theory in Minkowski space with $d+1$ spatial directions and choose one of the spatial directions, x^{d+1}

to define light-cone coordinates of the form $x^{\pm} = t \pm x^{d+1}$. The Schrödinger algebra is then obtained by retaining only those generators that commute with the light-cone translation generator P^-. The resulting sub-algebra will be the Schrödinger algebra, subject to the identifications

$$N = -\tilde{P}^-, \qquad H = -\tilde{P}^+, \qquad P_i = \tilde{P}_i, \qquad M_{ij} = \tilde{M}_{ij},$$
$$K_i = \tilde{M}_{-i}, \qquad D = \tilde{D} + 2\tilde{M}_{-+}, \qquad C = -\tilde{K}_-. \tag{15.155}$$

Here, we have denoted the generators of the conformal group by a tilde. Notice that we identify the generator of time translations, the Hamiltonian H, with the generator of translations in x^+. In other words, x^+ will play the role of time in the non-relativistic theory.

15.5.3 Backreacting fermions: electron star

A finite density of charged fermions in the bulk does not lead to a condensation process as in the bosonic case, but rather to the formation of a Fermi surface. On the gravity side, this leads to a geometry similar to a neutron star, however since the fermions are charged, it is referred to as an *electron star* [12]. The dual geometry is a renormalisation group flow, similar to those discussed in chapter 9, flowing from AdS_4 in the UV to a Lifshitz space in the IR. While a charged electron star would be unstable in flat space due to the repulsion between equal charges, this is not the case in asymptotically AdS space, which may be viewed as a box.

The starting point for this geometry is the action for a free, charged Dirac fermion added to the Einstein–Maxwell action,

$$\mathcal{L} = \frac{1}{2\kappa^2} \left(R + \frac{6}{L^2} \right) - \frac{1}{4g^2} F_{mn} F^{mn} + \mathcal{L}_{\psi}, \tag{15.156}$$

$$\mathcal{L}_{\psi} = i\bar{\psi} \Gamma^m \left(\partial_m + \frac{1}{4} \omega_{mab} \Gamma^{ab} + iA_m \right) \psi - m \bar{\psi} \psi, \tag{15.157}$$

where Γ^{ab} is an antisymmetrised gamma matrix and ω_{mab} is the spin connection. This leads to the equations of motion

$$R_{mn} - \frac{1}{2} g_{mn} R - \frac{3}{L^2} g_{mn} = \kappa^2 \left(\frac{1}{g^2} \left(F_{mp} F_n{}^p - \frac{1}{4} g_{mn} F_{pq} F^{pq} \right) + T_{mn}^{(f)} \right), \tag{15.158}$$

$$\nabla_n F^{mn} = g^2 J^m, \tag{15.159}$$

where the energy-momentum tensor $T_{mn}^{(f)}$ contains the fermionic degrees of freedom. To study the influence of the fermions on the geometry as given by these equations, we would have to study the backreaction of the fermions. This is generically very involved. There is a complicated interdependence since the energy-momentum tensor depends on the Fermi surface, which in turn again depends on the geometry. We therefore use an approximation leading to a simpler approach. We assume that T_{mn} and J^m in (15.158), (15.159) take the form corresponding to a perfect fluid governed by ideal hydrodynamics. For the energy-momentum tensor and current of a perfect fluid we may write, using the results of chapter 12,

$$T_{mn} = (\epsilon + p)u_m u_n + p g_{mn}, \qquad J_m = \rho u_m, \tag{15.160}$$

with u_m the normalised relativistic four-velocity. This is a general result; in order to obtain information about the fermions considered here, we have to find the relation between ϵ and p, i.e. the equation of state. For this purpose, let us first consider the simpler case of flat space. At zero temperature, for non-interacting fermions of mass m, we just fill the Fermi sea from the lowest energy state $E = m$ up to a chemical potential μ. The relativistic density of states for a given energy reads

$$g(E) = \frac{1}{\pi^2} E \sqrt{E^2 - m^2} \tag{15.161}$$

and thus the total particle number and the energy of the fermionic system are given by

$$\epsilon = \int_m^\mu dE \, E g(E), \qquad \rho = \int_m^\mu dE \, g(E). \tag{15.162}$$

Using this result, we determine the pressure p to be

$$-p = \epsilon - \mu \rho. \tag{15.163}$$

Exercise 15.5.1   Derive the density of states (15.161). Hint: Write the number of states per unit cell in momentum space. Generalise this to general dimensions.

Exercise 15.5.2   Obtain the relation (15.163) from the grand canonical potential $G = U - ST - \mu N$.

Exercise 15.5.3   Calculate ϵ, p and ρ. For this purpose, use (15.161) to calculate the integrals in (15.162) explicitly.

Let us assume that this construction also applies to curved space. This is highly non-trivial and self-consistency has to be checked after performing the calculation. In curved space, the only modification in the approach given above is to replace the chemical potential μ by the local expression

$$\mu_{\text{loc}} = \frac{A_t}{\sqrt{-g_{tt}}}. \tag{15.164}$$

A convenient ansatz for the metric leading to a self-consistent result is the *planar star* ansatz,

$$ds^2 = L^2 \left(-f \, dt^2 + g \, dr^2 + \frac{1}{r^2}(dx^2 + dy^2) \right), \qquad A = \frac{gL}{\kappa} h \, dt \tag{15.165}$$

with f, g, h functions of r. Also ϵ, ρ and thus p depend on r. By solving the resulting equations of motion, we see that the planar star ansatz is self-consistent, provided that $mL \gg 1$. This is evident since in this case, the Compton wavelength of the fermion is smaller than the curvature scale of the planar electron star. mL corresponds to the dimension of the dual operator on the field theory side. Consequently, the approximation as described above corresponds to considering field theory operators of large dimension. Moreover, as a second condition for consistency, the gravitational attractive force has to be of the order of the repulsive electrostatic force and hence we further have $mL \sim gL/\kappa$.

Finally, we note that the planar star solution is indeed a renormalisation group flow to a Lifshitz geometry in the IR, for which the parameter z is determined by the boundary values of the functions g, h. The physical picture which follows from the explicit form of the functions g, h in the bulk is that electric charge is distributed in the bulk outside the black hole horizon. This is in contrast to the case of the AdS Reissner–Nordström black hole, for which the charge is hidden behind the horizon.

Therefore the following picture emerges for holographic finite density systems. The field theory at finite density requires an electric field in the bulk. There are two possibilities for the location of the sources of this field: either they are hidden behind the black hole horizon, or they are located in the bulk outside the horizon. The prime example of the first case is the Reissner–Nordström black hole. For the second case, we have two different scenarios: either charged bosons or charged fermions are present. If the charge carriers are bosonic, at least for low temperatures the global $U(1)$ symmetry is broken spontaneously and we obtain a holographic superfluid as discussed in section 15.3.3. In the fermionic case, we obtain an electron star as described in the present section. Of course, it is also possible to have some of the charges behind the horizon and the remaining charges outside the horizon.

To understand the physical significance of these charge distributions, we have to consider Luttinger's theorem. This theorem states that the volume enclosed by the Fermi surface is equal to the charge density,

$$\frac{q}{(2\pi)^2} V = \tilde{\rho}, \tag{15.166}$$

where the prefactor $1/(2\pi)^2$ arises from the two spatial field theory directions which we consider here. In holographic theories, we have to allow for more than one Fermi surface, and the Luttinger theorem reads

$$\sum_i \frac{q^i}{(2\pi)^2} V^i = \tilde{\rho}. \tag{15.167}$$

In the holographic context, $\tilde{\rho}$ is the charge density outside the horizon. We can relate the charge density $\tilde{\rho}$ to the total charge density ρ by virtue of

$$\tilde{\rho} = \rho - \mathcal{A} \tag{15.168}$$

where $\mathcal{A}$ is the charge density hidden behind the horizon. Thus in the two limiting cases, for the Reissner–Nordström black hole $\rho = 0$, whereas for the electron star, $\rho = \tilde{\rho}$. $\mathcal{A}$ corresponds to the density of *fractionalised* excitations. In QCD language, the fractionalised states correspond to the deconfined degrees of freedom, for example the quarks. On the other hand, the confined states are gauge invariant bound states such as mesons. Consequently, the AdS Reissner–Nordström black hole describes a density of fractionalised or deconfined degrees of freedom, while the electron star is associated with a density of confined mesonic degrees of freedom.

15.5.4 Dilatonic systems and hyperscaling violation

Further physically relevant structures arise when we consider a dilaton in the gravity action in addition to the charged fields considered so far, i.e. the action of *Einstein–Maxwell dilaton theory* [13]. This contains a dilaton field in addition to the metric and the $U(1)$ gauge field,

$$\mathcal{L}_{\text{EMD}} = \frac{1}{2\kappa^2} \left(R - 2\partial^m \Phi \partial_m \Phi - \frac{V(\Phi)}{L^2} \right) - \frac{Z(\Phi)}{4g^2} F^{mn} F_{mn}, \qquad (15.169)$$

with potential $V(\Phi)$ and coupling $Z(\Phi)$ whose precise form will be discussed below. A consistent ansatz for the solution of the equations of motion is given by

$$ds^2 = L^2 \left(-f(r)\mathrm{d}t^2 + g(r)\mathrm{d}r^2 + \frac{1}{r^2}\mathrm{d}\vec{x}^2 \right) \qquad (15.170)$$

for the metric, and

$$A_t = \frac{gL}{\kappa} h(r) \qquad (15.171)$$

for the gauge field.

Exercise 15.5.4 Derive the equations of motion for the action obtained from (15.169) using the ansatz (15.170), (15.171).

Using the equations of motion, it can be shown that the solution is of the hyperscaling violating type (15.173) if the potential and dilaton coupling are chosen to be of the IR asymptotic form

$$V(\Phi) = -V_0 \exp(-\beta\Phi), \qquad Z(\Phi) = Z_0 \exp(\alpha\Phi), \qquad (15.172)$$

for $\Phi \to \infty$ with $\alpha, \beta > 0$.

The equations of motion for Einstein–Maxwell dilaton theory (15.169) with potential and dilaton coupling (15.172) give rise to a *hyperscaling violating geometry*, which is of the form

$$ds^2 = \frac{1}{r^2} \left(-\frac{\mathrm{d}t^2}{r^{2(d-1)(z-1)/(d-1-\theta)}} + r^{2\theta/(d-1-\theta)}\mathrm{d}r^2 + \mathrm{d}\vec{x}^2 \right). \qquad (15.173)$$

For Einstein–Maxwell dilaton theory, the parameters θ and z in (15.173) take the values

$$\theta = \frac{d^2\beta}{\alpha + (d-2)\beta},$$

$$z = 1 + \frac{\theta}{d-1} + \frac{8((d-1)(d-1-\theta)+\theta)^2}{(d-1)^2(d-1-\theta)\alpha^2} \qquad (15.174)$$

in terms of the coefficients α, β, with d the number of dimensions in the boundary theory (i.e. the asymptotically AdS space is of dimension $d+1$).

Let us examine the properties of the metric (15.173) for arbitrary values of the parameters θ, z. Under scale transformations ζ of the form $x_i \to \zeta x_i$, $t \to \zeta^z$, $r \to \zeta^{(d-1-\theta)/(d-1)}r$, the line element transforms as

$$ds \to \zeta^{\theta/(d-1)}ds. \qquad (15.175)$$

This scaling behaviour implies that z is the dynamical critical exponent introduced in (15.2). For non-trivial θ, the line element transforms non-trivially under scale transformations. This implies, in particular, that the volume element scales non-trivially. Generically, the volume element of the bulk theory is related to the thermal entropy density, as discussed in section 11.2.1. In theories with *hyperscaling*, the free energy scales with its engineering dimension, which implies that the entropy scales with the temperature as

$$S \sim T^{(d-1)/z}. \tag{15.176}$$

In the geometry (15.175), however, this relation is violated and the entropy scales as

$$S \sim T^{(d-1-\theta)/z}, \tag{15.177}$$

hence hyperscaling is violated and θ is referred to as the *hyperscaling violation exponent*. Equation (15.177) is obtained by computing the entropy from the area of the black hole horizon in the bulk geometry, which implies that $S \sim r_{\mathrm{h}}^{-d+1}$. Together with $r^{d-1} \sim t^{(d-1-\theta)/z}$, this gives (15.177). Systems which display hyperscaling violation are generically gapless, but not conformal. They are referred to as *compressible* systems, in which the density can be varied freely by varying an external parameter.

Let us comment on further aspects of the physical properties of the field theory dual to the hyperscaling violating geometry. These are related to the *entanglement entropy* of the system, a very important concept which will be introduced in section 15.6 below. Subject to the condition

$$\theta = d - 2, \tag{15.178}$$

the entanglement entropy for the geometry (15.173) has an area law behaviour plus logarithmic corrections. At weak coupling, this logarithmic behaviour in a charged system with unbroken symmetries signals the presence of a Fermi surface. By analogy, it may be possible that this logarithm also signals a Fermi surface in the strongly coupled system considered here. This would be a Fermi surface for the elementary fermionic degrees of freedom of the dual field theory, rather than for composite mesino fermionic operators. This type of Fermi surface is referred to as a *hidden* Fermi surface of fractionalised charges and may provide the required additional contribution to the Luttinger theorem (15.168). For Einstein–Maxwell dilaton theory, the inequality (15.178) amounts to

$$\beta \leq \frac{(d-1)}{(2d-1)}\alpha \tag{15.179}$$

for the parameters of (15.172).

15.6 Entanglement entropy

Entanglement entropy is an important concept of relevance in condensed matter and other areas of physics, which may also be realised in gauge/gravity duality. Entanglement entropy has important properties which are due to its non-local nature. It may help to characterise gapped phases of matter in the absence of both a classical order parameter and spontaneous symmetry breaking. However, the field theory calculations involved in determining the entanglement entropy are very difficult. In contrast, computing the entanglement entropy in the dual gravity approach turns out to be very simple.

15.6.1 Definition and field theory realisation

Consider a quantum mechanical system characterised by a density matrix ρ. This system may be in a pure state $|\Psi\rangle$ in a Hilbert space $\mathcal{H}$. We assume $|\Psi\rangle$ to be normalised to one, $\langle\Psi|\Psi\rangle = 1$. For this pure state, the density matrix reads $\rho = |\Psi\rangle\langle\Psi|$, such that $\rho^2 = \rho$.

On the other hand, for a system in a mixed state we have

$$\rho = \sum_n p_n |\Psi_n\rangle\langle\Psi_n|, \tag{15.180}$$

for an orthonormal basis $|\Psi_n\rangle$ with probabilities satisfying $\sum_n p_n = 1$. For example, for a system in thermal equilibrium we may choose

$$p_n = \frac{1}{Z_{\text{can}}}\exp(-\beta E_n), \tag{15.181}$$

such that ρ becomes ρ_{can} in (11.2).

A quantum mechanical analogue of the thermodynamic entropy is the *von Neumann entropy*, which is given by

$$S_{\text{von Neumann}} = -\text{Tr}\,(\rho \ln \rho). \tag{15.182}$$

This entropy is maximised if the matrix ρ is diagonal, with all entries equal. On the other hand, it vanishes for pure states for which $\rho^2 = \rho$.

The von Neumann entropy or its analogues for non-equilibrium systems are the starting point for defining the *entanglement entropy* which provides a measure for the entanglement of quantum states. Let us consider the zero temperature case and assume that the Hilbert space under consideration has the product structure $\mathcal{H} = \mathcal{H}_A \otimes \mathcal{H}_B$ for two subsystems A and B. Then for the subsystem A we define a reduced density matrix by

$$\rho_A = \text{Tr}_B \rho, \tag{15.183}$$

where ρ is the density matrix of the total system and the trace is taken over states in system B. For system A, the entanglement entropy is defined by

$$S_A = -\text{Tr}_A\,(\rho_A \ln \rho_A). \tag{15.184}$$

This is a measure for entanglement, as may be seen as follows. The reduced density matrix for a pure ensemble can correspond to a mixed ensemble. The entanglement entropy (15.184) for the pure state $|\Psi\rangle$ with density matrix $\rho = |\Psi\rangle\langle\Psi|$ can be non-zero, while the von Neumann entropy for this state vanishes.

The entanglement entropy satisfies a number of relations, which are introduced in the following. If the global system is in a pure state, and A is the complement of B, then

$$S_A = S_B. \tag{15.185}$$

This implies that the entanglement entropy is not extensive. Equation (15.185) does not hold at finite temperature. Moreover, for any three non-intersecting subsystems A, B and C, the inequalities

$$S_{A+B+C} + S_B \leq S_{A+B} + S_{B+C}, \tag{15.186}$$

$$S_A + S_C \leq S_{A+B} + S_{B+C} \tag{15.187}$$

are satisfied. By choosing B to be the empty set (15.186), we obtain

$$S_{A+B} \leq S_A + S_B, \qquad (15.188)$$

which is known as the *subadditivity relation*. A detailed analysis reveals that an even stronger condition, the *strong subadditivity condition*

$$S_A + S_B \geq S_{A \cup B} + S_{A \cap B}, \qquad (15.189)$$

applies for any two subsystems A, B. When A and B do not intersect, this reduces to (15.188).

Within quantum field theory, the entanglement entropy is obtained by considering two complementary regions A, B in space for fixed time $t = t_0$. The two regions are separated by a smooth boundary surface ∂A. The entanglement entropy for region A is obtained by integrating over the degrees of freedom in region B. The result is generically divergent and takes the schematic form

$$S = c_0 (R\Lambda)^{d-2} + c_1 (R\Lambda)^{d-3} + \cdots \qquad (15.190)$$

where R is a scale determining the typical size of region A and Λ is a UV cut-off. d is the spacetime dimension and the c_i are model-dependent coefficients. We note that the leading term in (15.190) scales as R^{d-2}, i.e. it is of the same dimension as the boundary ∂A. Therefore the entanglement satisfies an area law, similar to the black hole entropy. This also suggests how to realise the entanglement entropy holographically.

In addition, for even d there are additional logarithmic terms in (15.190) reminiscent of the conformal anomaly discussed in sections 1.8.2 and 6.3.2,

$$S = \cdots + c_d \ln(R\Lambda) + \cdots . \qquad (15.191)$$

In particular, in two-dimensional conformal field theories the entanglement entropy is given by

$$S = \frac{c}{3} \ln(R\Lambda), \qquad (15.192)$$

where R is the length of the system and c is the central charge of the two-dimensional conformal field theory. For a circle of circumference l, on which the region A is defined as the line segment l_A, the general expression (15.192) becomes

$$S_{\text{circ}} = \frac{c}{3} \cdot \ln\left(\frac{l}{\pi a} \sin\left(\frac{\pi l_A}{l} \right) \right), \qquad (15.193)$$

with a a lattice size cut-off with $a \sim 1/\Lambda$.

15.6.2 Holographic entanglement entropy

For a holographic realisation of the entanglement entropy in a d-dimensional quantum field theory, consider the setup of figure 15.8. It was proposed [14, 15] that within gauge/gravity duality, the entanglement entropy is given by

$$S_A = \frac{\gamma (\Sigma_A)}{4 G_{d+1}}, \qquad (15.197)$$

Box 15.3   **Entanglement entropy in two-dimensional CFT**

For two-dimensional conformal field theories, the entanglement entropy is obtained using the *replica trick* of first evaluating $\mathrm{Tr}_A \rho_A{}^n$, differentiating with respect to n and subsequently taking $n \to 1$. For ρ_A normalised such that $\mathrm{Tr}_A \rho_A = 1$, we have

$$S_A = \lim_{n \to 1} \frac{\mathrm{Tr}_A \rho_A{}^n - 1}{1 - n} = -\frac{\partial}{\partial n} \mathrm{Tr}_A \rho_A{}^n \Big|_{n=1}. \tag{15.194}$$

Let us sketch how to obtain the result (15.193) from (15.194). The n copies of the density matrix lead to a path integral on a Riemann surface with n sheets, labelled by k. In order to obtain the CFT in the flat complex plane, *twisted* boundary conditions have to be imposed on each of the sheets. Equivalently, these boundary conditions correspond to insertions of *twist operators* $\Phi^{+(k)}$, $\Phi^{-(k)}$ in the CFT. This leads to

$$\mathrm{Tr}\rho_A{}^n = \prod_{k=0}^{n-1} \langle \Phi^{+(k)}(u) \Phi^{-(k)}(v) \rangle, \tag{15.195}$$

with u, v coordinates in the complex plane. To obtain the entanglement entropy on a circle, a conformal transformation from the complex plane to the cylinder has to be performed. The appropriate transformation is

$$u = \tan\frac{\pi r}{R}, \qquad v = \tan\frac{\pi s}{R} \tag{15.196}$$

for a cylinder of circumference R. This transformation leads to the trigonometric factor in the result (15.193) for the entanglement entropy.

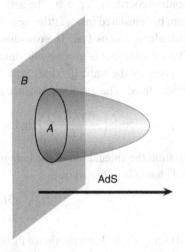

Figure 15.8   Calculation of holographic entanglement entropy.

where Σ_A is the $(d-1)$-dimensional minimal surface in AdS_{d+1} whose boundary is given by ∂A for a fixed time $t = t_0$, and $\gamma(\Sigma_A)$ is the area of this surface. Moreover, G_{d+1} is the $(d+1)$-dimensional Newton constant. As in the holographic calculation of the conformal anomaly, (15.197) is divergent since the surface integral extends all the way to the boundary in the radial direction. The leading divergent term in this integral gives

$$S_A = \frac{L^{d-1}}{G_{d+1}} \frac{\gamma\,(\partial A)}{\epsilon^{d-2}} + \cdots, \tag{15.198}$$

where $\gamma(\partial A)$ is the surface of the $(d-2)$-dimensional boundary ∂A, L is the AdS radius, and ϵ is the UV cut-off in the radial direction, with $\Lambda \propto 1/\epsilon$ in (15.190). The cut-off is the same as in the calculation of the holographic conformal anomaly in section 6.3.2. The leading term in (15.198) is proportional to the area of ∂A and thus gives rise to an area law.

In general, the complete expression for (15.197) takes the form

$$S_A = c_0 \left(\frac{L}{\epsilon}\right)^{d-2} + c_1 \left(\frac{L}{\epsilon}\right)^{d-4} + \cdots + c_{d-2} + \cdots \tag{15.199}$$

in odd dimensions d and

$$S_A = c_0 \left(\frac{L}{\epsilon}\right)^{d-2} + c_1 \left(\frac{L}{\epsilon}\right)^{d-4} + \cdots + c_{d-2}\ln\left(\frac{L}{\epsilon}\right) + \cdots \tag{15.200}$$

in even dimensions d.

To be specific, we consider the case of $d = 2$. In global coordinates, the metric of AdS_3 is given by

$$ds^2 = L^2(-\cosh\rho^2 dt^2 + d\rho^2 + \sinh\rho^2 d\theta^2) \tag{15.201}$$

with dimensionless coordinates t, ρ, θ. The action is regulated by a cut-off ρ_0 with $\rho \le \rho_0$. This cut-off can be translated into a lattice spacing a by $\rho_0 \sim l/a$, with l the circumference of the AdS_3 cylinder of radius $L\rho_0$. The two-dimensional CFT is defined on the space (t, θ) at $\rho = \rho_0$. The subsystem A is given by the interval $0 \le \theta \le 2\pi l_A/l$. The surface $\gamma(\Sigma_A)$ of (15.197) is given by the static geodesic which connects the boundary points $\theta = 0$ and $\theta = 2\pi l_A/l$ with t fixed. The geodesic distance $d(\gamma(\Sigma_A))$ is given by

$$\cosh\left(\frac{d(\gamma(\Sigma_A))}{L}\right) = 1 + 2\sinh^2\rho_0 \, \sin^2 \frac{\pi l_A}{l}. \tag{15.202}$$

We also know from the calculation of the holographic conformal anomaly in section 6.3.2 that the central charge is obtained from

$$c = \frac{3L}{2G_3}. \tag{15.203}$$

Using this and taking $\rho_0 \ll 1$, we obtain for the holographic entanglement entropy (15.197)

$$S = \frac{L}{4G_3}\ln\left(e^{2\rho_0} \sin^2 \frac{\pi l_A}{l}\right) = \frac{c}{3}\ln\left(e^{\rho_0} \sin \frac{\pi l_A}{l}\right). \tag{15.204}$$

Up to an overall constant, which reflects the arbitrariness of choosing a scale in a conformal field theory, i.e. relating ρ_0 and l/a, this coincides with the field theory result (15.193). This is similar to the normalisation issue discussed when testing the AdS/CFT correspondence by calculating the three-point function of one-half BPS operators in section 6.1.

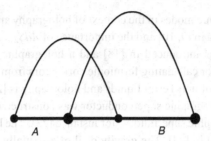

A B

Figure 15.9 Holographic subadditivity.

A similar calculation may also be performed in higher dimensions. For instance, considering $\mathcal{N} = 4$ Super Yang–Mills theory and $AdS_5 \times S^5$, then taking the subsystem A to be a rectangle of size $l_1 \times l_2$ with $l_2 \gg l_1$, the result of the holographic calculation is

$$S = \frac{N^2 l_2^2}{2\pi a^2} - 2\sqrt{\pi} \left(\frac{\Gamma(2/3)}{\Gamma(1/6)} \right)^3 \frac{N l_2^2}{l_1^2}. \tag{15.205}$$

It can be shown that the numerical factors are of the same order of magnitude as obtained in a weak coupling field theory calculation in $\mathcal{N} = 4$ Super Yang–Mills theory.

Moreover, we can check that the holographic entanglement entropy given by (15.197) satisfies the required properties as outlined in section 15.6.1 above, in particular the strong subadditivity relation (15.189). For this purpose, we consider the minimal surfaces Σ_A, Σ_B in the bulk which end on the boundary surfaces ∂A, ∂B. The enclosed volumes are denoted by V_A, V_B, respectively, such that $\partial V_A = A \cup \Sigma_A$, and similarly for B. Now consider the volumes

$$V_{A \cup B} = V_A \cup V_B, \qquad V_{A \cap B} = V_A \cap V_B, \tag{15.206}$$

with surfaces $\partial V_{A \cup B} = (A \cup B) \cup \Sigma_{A \cup B}$, $\partial V_{A \cap B} = (A \cap B) \cup \Sigma_{A \cap B}$. As is visualised in figure 15.9, the surface $A_{A \cup B}$ ends on $\partial(A \cup B)$. $\Sigma_{A \cup B}$ is not necessarily the minimal surface ending on $\partial(A \cup B)$, but its area provides an upper bound on the area, and therefore on the holographic entanglement entropy $S_{A \cup B}$. A similar argument applies to $S_{A \cap B}$. Now since $\Sigma_{A \cup B}$ and $\Sigma_{A \cap B}$ have the same area as Σ_A and Σ_B,

$$\gamma(\Sigma_{A \cup B}) + \gamma(\Sigma_{A \cap B}) = \gamma(\Sigma_A) + \gamma(\Sigma_B), \tag{15.207}$$

we obtain the desired result (15.189). This demonstrates that the proposal (15.197) for the holographic entanglement entropy satisfies strong subadditivity.

15.7 Further reading

An extensive review of quantum phase transitions from the condensed matter point of view is [1]. Standard reviews on holographic methods applied to condensed matter physics are [2, 16, 17]. The holographic quantum critical model with magnetic field for a (3+1)-dimensional field theory as discussed in section 15.2.5 is presented in [5]. Quantum critical transport was first discussed within gauge/gravity duality in [4].

Fermionic modes in the context of holography applied to condensed matter physics were introduced in [9, 10] and the importance of AdS_2 was realised in [3]. Marginal Fermi liquid theory was introduced in [18] and a holographic realisation was proposed in [19]. The methods for calculating fermionic correlators from holography were developed in [20, 21]. A review of non-Fermi liquids and holography is [22].

The holographic superconductor was constructed in [6, 7]. There are earlier approaches to holographic superfluidity, for instance [23]. The holographic p-wave superconductor was introduced in [24]. The gravity dual of a spatially modulated phase of section 15.3.5 was found in [8].

Supergravity embeddings of holographic superconductors include [25, 26]. A top-down approach to holographic p-wave superconductors based on probe branes is described in [27, 28], in which the dual field theory is known explicitly. Fermionic excitations for the D3/D5-brane probe system of section 10.4 were studied in [29]. Moreover, the D3/D5-brane probe system also provides a holographic realisation of Berezinskii–Kosterlitz–Thouless (BKT) phase transitions [30, 31].

Lifshitz spaces are introduced for instance in [11] and the electron star geometry in [12]. Hyperscaling violation was introduced into the holographic context in [32] via a discussion of hidden Fermi surfaces. The dilatonic backgrounds giving rise to hyperscaling violation were introduced in [13].

The entanglement entropy for two-dimensional CFTs was calculated by Calabrese and Cardy. An introduction to their work is found in [33]. The holographic entanglement entropy was proposed by Ryu and Takayanagi in [14, 15]. Arguments towards a proof of this conjecture may be found in [34, 35]. In particular, in [35] it was shown that the Ryu–Takayanagi proposal is obtained by applying the replica trick on the gravity side. The argument for holographic strong subadditivity may be found in [36]. A covariant formulation of the Ryu–Takayanagi proposal describing the time dependence of entanglement entropy was given in [37].

References

[1] Sachdev, S. 2011. *Quantum Phase Transitions*, 2nd edition. Cambridge University Press.

[2] Hartnoll, Sean A. 2009. Lectures on holographic methods for condensed matter physics. *Class. Quantum Grav.*, **26**, 224002.

[3] Faulkner, Thomas, Liu, Hong, McGreevy, John, and Vegh, David. 2011. Emergent quantum criticality, Fermi surfaces, and AdS_2. *Phys. Rev.*, **D83**, 125002.

[4] Herzog, Christopher P., Kovtun, Pavel, Sachdev, Subir, and Son, Dam Thanh. 2007. Quantum critical transport, duality, and M-theory. *Phys. Rev.*, **D75**, 085020.

[5] D'Hoker, Eric, and Kraus, Per. 2012. Quantum criticality via magnetic branes. ArXiv:1208.1925.

[6] Gubser, Steven S. 2008. Breaking an Abelian gauge symmetry near a black hole horizon. *Phys. Rev.*, **D78**, 065034.

[7] Hartnoll, Sean A., Herzog, Christopher P., and Horowitz, Gary T. 2008. Holographic Superconductors. *J. High Energy Phys.*, **0812**, 015.

[8] Nakamura, Shin, Ooguri, Hirosi, and Park, Chang-Soon, 2009. Gravity dual of spatially modulated phase. *Phys. Rev.*, **D81**, 044018.

[9] Liu, Hong, McGreevy, John, and Vegh, David. 2011. Non-Fermi liquids from holography. *Phys. Rev.*, **D83**, 065029.

[10] Cubrovic, Mihailo, Zaanen, Jan, and Schalm, Koenraad. 2009. String theory, quantum phase transitions and the emergent Fermi-liquid. *Science*, **325**, 439–444.

[11] Kachru, Shamit, Liu, Xiao, and Mulligan, Michael. 2008. Gravity duals of Lifshitz-like fixed points. *Phys. Rev.*, **D78**, 106005.

[12] Hartnoll, Sean A., and Tavanfar, Alireza. 2011. Electron stars for holographic metallic criticality. *Phys. Rev.*, **D83**, 046003.

[13] Charmousis, C., Gouteraux, B., Kim, B. S., Kiritsis, E., and Meyer, R. 2010. Effective holographic theories for low-temperature condensed matter systems. *J. High Energy Phys.*, **1011**, 151.

[14] Ryu, Shinsei, and Takayanagi, Tadashi. 2006. Holographic derivation of entanglement entropy from AdS/CFT. *Phys. Rev. Lett.*, **96**, 181602.

[15] Ryu, Shinsei, and Takayanagi, Tadashi. 2006. Aspects of holographic entanglement entropy. *J. High Energy Phys.*, **0608**, 045.

[16] McGreevy, John. 2010. Holographic duality with a view toward many-body physics. *Adv. High Energy Phys.*, **2010**, 723105.

[17] Herzog, Christopher P. 2009. Lectures on holographic superfluidity and superconductivity. *J. Phys.*, **A42**, 343001.

[18] Varma, C. M., Littlewood, P. B., Schmitt-Rink, S., Abrahams, E., and Ruckenstein, A. E. 1989. Phenomenology of the normal state of Cu-O high-temperature superconductors. *Phys. Rev. Lett.*, **63**, 1996–1999.

[19] Faulkner, Thomas, Iqbal, Nabil, Liu, Hong, McGreevy, John, and Vegh, David. 2010. Strange metal transport realized by gauge/gravity duality. *Science*, **329**, 1043–1047.

[20] Iqbal, Nabil, and Liu, Hong. 2009. Universality of the hydrodynamic limit in AdS/CFT and the membrane paradigm. *Phys. Rev.*, **D79**, 025023.

[21] Iqbal, Nabil, and Liu, Hong. 2009. Real-time response in AdS/CFT with application to spinors. *Fortschr. Phys.*, **57**, 367–384.

[22] Iqbal, Nabil, Liu, Hong, and Mezei, Mark. 2011. Lectures on holographic non-Fermi liquids and quantum phase transitions. ArXiv:1110.3814.

[23] Evans, Nick J., and Petrini, Michela. 2001. Superfluidity in the AdS/CFT correspondence. *J. High Energy Phys.*, **0111**, 043.

[24] Gubser, Steven S., and Pufu, Silviu S. 2008. The gravity dual of a p-wave superconductor. *J. High Energy Phys.*, **0811**, 033.

[25] Gubser, Steven S., Herzog, Christopher P., Pufu, Silviu S., and Tesileanu, Tiberiu. 2009. Superconductors from superstrings. *Phys. Rev. Lett.*, **103**, 141601.

[26] Gauntlett, Jerome P., Sonner, Julian, and Wiseman, Toby. 2009. Holographic superconductivity in M-theory. *Phys. Rev. Lett.*, **103**, 151601.

[27] Ammon, Martin, Erdmenger, Johanna, Kaminski, Matthias, and Kerner, Patrick. 2009. Superconductivity from gauge/gravity duality with flavour. *Phys. Lett.*, **B680**, 516–520.

[28] Ammon, Martin, Erdmenger, Johanna, Kaminski, Matthias, and Kerner, Patrick. 2009. Flavour superconductivity from gauge/gravity duality. *J. High Energy Phys.*, **0910**, 067.

[29] Ammon, Martin, Erdmenger, Johanna, Kaminski, Matthias, and O'Bannon, Andy. 2010. Fermionic operator mixing in holographic p-wave superfluids. *J. High Energy Phys.*, **1005**, 053.

[30] Jensen, Kristan, Karch, Andreas, Son, Dam T., and Thompson, Ethan G. 2010. Holographic Berezinskii-Kosterlitz-Thouless transitions. *Phys. Rev. Lett.*, **105**, 041601.

[31] Evans, Nick, Gebauer, Astrid, Kim, Keun-Young, and Magou, Maria. 2011. Phase diagram of the D3/D5 system in a magnetic field and a BKT transition. *Phys. Lett.*, **B698**, 91–95.

[32] Huijse, Liza, Sachdev, Subir, and Swingle, Brian. 2012. Hidden Fermi surfaces in compressible states of gauge-gravity duality. *Phys. Rev.*, **B85**, 035121.

[33] Calabrese, Pasquale, and Cardy, John L. 2006. Entanglement entropy and quantum field theory: a non-technical introduction. *Int. J. Quantum Inf.*, **4**, 429.

[34] Casini, Horacio, Huerta, Marina, and Myers, Robert C. 2011. Towards a derivation of holographic entanglement entropy. *J. High Energy Phys.*, **1105**, 036.

[35] Lewkowycz, Aitor, and Maldacena, Juan. 2013. Generalized gravitational entropy. *J. High Energy Phys.*, **1308**, 090.

[36] Headrick, Matthew, and Takayanagi, Tadashi. 2007. A holographic proof of the strong subadditivity of entanglement entropy. *Phys. Rev.*, **D76**, 106013.

[37] Hubeny, Veronika E., Rangamani, Mukund, and Takayanagi, Tadashi. 2007. A covariant holographic entanglement entropy proposal. *J. High Energy Phys.*, **0707**, 062.

Grassmann numbers

By definition, *Grassmann numbers* $\theta_1, \ldots, \theta_n$ satisfy the anticommutation relations

$$\{\theta_i, \theta_j\} = 0, \tag{A.1}$$

which imply $\theta_k^2 = 0$. Moreover, any product of the form $\theta_{i_1} \theta_{i_2} \ldots \theta_{i_k}$ is antisymmetric under odd permutations of the indices $i_1, i_2, \ldots, i_k$. In particular, all products involving more than n Grassmann numbers have to vanish and

$$\theta_{i_1} \theta_{i_2} \theta_{i_n} = \epsilon_{i_1 i_2 \ldots i_n} \theta_1 \theta_2 \ldots \theta_n. \tag{A.2}$$

The Grassmann numbers θ_i and arbitrary products thereof generate the *Grassmann algebra*. An arbitrary element of this algebra may be written as

$$f = \sum_{k=0}^{n} \sum_{i_1 < \cdots < i_k} f_{i_1 i_2 \cdots i_k} \theta_{i_1} \theta_{i_2} \ldots \theta_{i_k}, \tag{A.3}$$

with $f_{i_1 i_2 \cdots i_k} \in \mathbb{C}$. Therefore the complex dimension of the Grassmann algebra is 2^n. Note that by definition the Grassmann numbers commute with all real and complex numbers.

Functions of Grassmann numbers are defined in terms of the Taylor expansion of the function. This series truncates at a finite order and therefore any function can be expressed in terms of a polynomial of the Grassmann numbers. For example, in the case of just one Grassmann number we have $e^{\theta} = 1 + \theta$. Moreover, we can differentiate with respect to Grassmann numbers applying the following rules

$$\frac{\partial}{\partial \theta_j} \theta_i = \delta_{ij}, \qquad \frac{\partial}{\partial \theta_i} (\theta_j \theta_k) = \delta_{ij} \theta_k - \delta_{ik} \theta_j \tag{A.4}$$

to an arbitrary polynomial of Grassmann numbers, i.e. to a generic function. The differential operator satisfies the anticommutation relations

$$\left\{ \frac{\partial}{\partial \theta_i}, \frac{\partial}{\partial \theta_j} \right\} = 0, \tag{A.5}$$

which imply

$$\frac{\partial^2}{\partial \theta_i^2} = 0. \tag{A.6}$$

Due to (A.4) we also have

$$\left\{ \frac{\partial}{\partial \theta_i}, \theta_j \right\} = \delta_{ij}. \tag{A.7}$$

For Grassmann variables, integration coincides with differentiation. The integral of a function $f(\theta_1, \ldots, \theta_n)$ is given by

$$\int d\theta_1 d\theta_2 \ldots d\theta_n f(\theta_1, \ldots, \theta_n) = \frac{\partial}{\partial \theta_1} \frac{\partial}{\partial \theta_2} \cdots \frac{\partial}{\partial \theta_n} f(\theta_1, \ldots, \theta_n). \tag{A.8}$$

In particular, in the case of a single Grassmann variable θ we have

$$\int d\theta = 0, \qquad \int d\theta \, \theta = 1. \tag{A.9}$$

Note that in the case of more than one integral over Grassmann variables we have to be careful with signs. For example, for two Grassmann variables θ_1 and θ_2 we have

$$\int d\theta_1 \int d\theta_2 \, \theta_1 \theta_2 = -1. \tag{A.10}$$

The integration rule (A.8) is imposed such that the integration measure is invariant under translations,

$$\int d\theta f(\theta + \tilde{\theta}) = \int d\theta f(\theta). \tag{A.11}$$

Moreover, under a linear transformation of the form $\theta_i \mapsto \theta_i' = A_i{}^j \theta_j$ the integration measure transforms as

$$d\theta_1 d\theta_2 \ldots d\theta_n = \det A \, d\theta_1' d\theta_2' \cdots d\theta_n'. \tag{A.12}$$

This is just the opposite of what happens for real or complex numbers, which is due to the fact that for Grassmann numbers, integration is really a differentiation.

The delta distribution of Grassmann variables is defined as in the case of ordinary numbers and integrals. For instance, in the case of one Grassmann variable we have

$$\int d\theta \, \delta(\theta - \tilde{\theta}) f(\theta) = f(\tilde{\theta}). \tag{A.13}$$

Using the integration rules (A.9) we may rewrite $\delta(\theta - \tilde{\theta})$ as

$$\delta(\theta - \tilde{\theta}) = \theta - \tilde{\theta}. \tag{A.14}$$

This result for the delta distribution is easily generalised to n variables,

$$\delta^n(\theta - \tilde{\theta}) = (\theta_1 - \tilde{\theta}_1)(\theta_2 - \tilde{\theta}_2) \cdots (\theta_n - \tilde{\theta}_n). \tag{A.15}$$

There is also an integral representation for the delta distribution. Since

$$\int d\xi \, e^{i\xi\theta} = \int d\xi \, (1 + i\xi\theta) = i\theta \tag{A.16}$$

for ξ and θ both Grassmann variables, we have

$$\delta(\theta) = \theta = -i \int d\xi \, e^{i\xi\theta}. \tag{A.17}$$

Complex conjugation of Grassmann variables is defined by the properties

$$\begin{aligned} (\theta_i)^* = \theta_i^*, \qquad (\theta_i^*)^* = \theta_i, \\ (\theta_i \theta_j)^* = \theta_j^* \theta_i^*. \end{aligned} \tag{A.18}$$

The θ_i^* are also a generating set of Grassmann variables. The second property ensures that the real commuting number $\theta_i\theta_i^*$ satisfies the reality condition $(\theta_i\theta_i^*)^* = \theta_i\theta_i^*$.

The Gaussian integral for a symmetric $n \times n$ matrix M whose entries are commuting numbers, real or complex, is given by

$$\int d\theta_1 d\theta_1^* \cdots d\theta_n d\theta_n^* e^{\theta_i^* M_{ij}\theta_j} = \det M. \tag{A.19}$$

This is shown by performing a variable transformation $\theta_i \mapsto \theta_i' = M_{ij}\theta_j$ and using the new integration measure $d\theta_1' d\theta_1^* \cdots d\theta_n' d\theta_n^* \det M$. Then it suffices to calculate the integral $\int d\theta' d\theta^*(1 + \theta^*\theta') = 1$. Moreover, for the same matrix M and Grassmann numbers η_i we have

$$\int d\theta_1 d\theta_2 \cdots d\theta_n \, e^{\frac{1}{2}\theta^T M\theta + \eta^T\theta} = (\det M)^{1/2} \, e^{\frac{1}{2}\eta^T M^{-1}\eta}, \tag{A.20}$$

as well as, for M a Hermitian matrix,

$$\int d\theta_1 d\theta_1^* \cdots d\theta_n d\theta_n^* \, e^{\theta^\dagger M\theta + \eta^\dagger\theta + \theta^\dagger\eta} = \det M e^{-\eta^\dagger M^{-1}\eta}. \tag{A.21}$$

We may also define a functional derivative with respect to a Grassmann valued function $\psi(t)$, with t a real or complex commutative number. The functional derivative of the functional $F[\psi(t)]$ is given by

$$\frac{\delta F[\psi(t)]}{\delta\psi(s)} = \frac{1}{\epsilon} \left(F[\psi(t) + \epsilon\delta(t - s)] - F[\psi(t)] \right), \tag{A.22}$$

where ϵ is a Grassmann number. Since the Taylor expansion of

$$F[\psi(t) + \epsilon\delta(t - s)] - F[\psi(t)] \tag{A.23}$$

contains only one term, which is linear in ϵ, the symbolic expression $1/\epsilon$ means that only the term linear in ϵ in the following expression should be retained. The division by a Grassmann number is not defined for any other cases. Since there is only one term in the Taylor expansion, which in addition is linear, it is not necessary to take the limit $\epsilon \to 0$. This is reassuring since the Grassmann numbers may not be ordered.

Lie algebras and superalgebras

B.1 Lie groups and Lie algebras

A Lie algebra $\mathfrak{g}$ is a vector space over some field $\mathbb{F}$ with an operation $[\cdot, \cdot] : \mathfrak{g} \times \mathfrak{g} \to \mathfrak{g}$, the *Lie bracket*, which is

- *bilinear,*

$$[\alpha x + \beta y, z] = \alpha[x, z] + \beta[y, z] \qquad \text{for } \alpha, \beta \in \mathbb{F} \text{ and } x, y, z \in \mathfrak{g}, \tag{B.1}$$

- *antisymmetric,*

$$[x, y] = -[y, x] \qquad \text{for } x, y \in \mathfrak{g}, \tag{B.2}$$

- and satisfies the *Jacobi identity*

$$[x, [y, z]] + [y, [z, x]] + [z, [x, y]] = 0 \qquad \text{for } x, y, z \in \mathfrak{g}. \tag{B.3}$$

Choosing a basis of the vector space $\mathfrak{g}$ with basis vectors T_a, $a = 1, \ldots, \dim \mathfrak{g}$, where $\dim \mathfrak{g}$ is the dimension of the vector space $\mathfrak{g}$, any element of $\mathfrak{g}$ may be written as $x = x^a T_a$ where $x^a \in \mathbb{F}$. The Lie bracket for two elements x and y is then specified by the *structure constants* $f_{ab}{}^c$,

$$[T_a, T_b] = i f_{ab}{}^c T_c. \tag{B.4}$$

Because of (B.3) the structure constants have to satisfy

$$f_{ad}{}^e f_{bc}{}^d + f_{cd}{}^e f_{ab}{}^d + f_{bd}{}^e f_{ca}{}^d = 0. \tag{B.5}$$

Lie algebras are intimately connected to Lie groups. A *Lie group* $\mathcal{G}$ is a smooth manifold which also possesses a group structure. In particular, this implies that we can define the product of two elements in $\mathcal{G}$. Moreover, there exists a neutral element which we denote by $\mathbb{1}$, as well as the inverse of group elements. In addition, the group structure and the differentiable structure are compatible in the sense that the product of two group elements, as well as the inverse of a group element, are differentiable maps.

To any Lie group $\mathcal{G}$, we may associate a Lie algebra as follows: the tangent space $T_{\mathbb{1}}(\mathcal{G})$ at the identity element $\mathbb{1}$ of the group forms a Lie algebra which we denote by $\mathfrak{g}$. In other words, the Lie algebra $\mathfrak{g}$ captures the local structure of the Lie group $\mathcal{G}$. We can also reverse the process: starting from a Lie algebra $\mathfrak{g}$ and exponentiating all elements of $\mathfrak{g}$, we obtain a connected Lie group $\mathcal{G}$. Note, however, that the mapping of Lie algebras and Lie groups is not one-to-one. Given a Lie algebra $\mathfrak{g}$ there exist more than one Lie group $\mathcal{G}$ whose Lie algebra is $\mathfrak{g}$.

In physics, the concept of Lie groups is important to describe continuous symmetries, while the generators of an infinitesimal symmetry transformation form a Lie algebra. Most important are *real* or *complex Lie groups* for which the field $\mathbb{F}$ is either $\mathbb{F} = \mathbb{R}$ or $\mathbb{F} = \mathbb{C}$ and which are therefore in particular real or complex finite dimensional manifolds. From now on we will restrict ourselves to real or complex Lie groups and their Lie algebras.

B.1.1 Properties of Lie algebras and Lie groups

A Lie algebra is *Abelian* if all the structure constants vanish, i.e. if all the generators commute with each other. Another interesting class is the *simple* or *semi-simple* Lie algebra. To define these we have to introduce the notion of *invariant subalgebras*. A vector subspace $\mathfrak{h} \subseteq \mathfrak{g}$ is a *subalgebra* if it forms a Lie algebra on its own. In other words, the Lie bracket has to be closed, i.e. for $h_1, h_2 \in \mathfrak{h}$ also $[h_1, h_2] \in \mathfrak{h}$. Moreover, a subalgebra $\mathfrak{h}$ is *invariant* if $[g, h] \in \mathfrak{h}$ for all $g \in \mathfrak{g}$ and all $h \in \mathfrak{h}$. An invariant subalgebra is also called an *ideal*. A $\mathfrak{u}(1)$ *subalgebra* of $\mathfrak{g}$ has just one generator T which commutes with all generators of $\mathfrak{g}$.

A Lie algebra $\mathfrak{g}$ is *simple* if $\mathfrak{g}$ is non-Abelian and if $\{0\}$ and $\mathfrak{g}$ are the only invariant subalgebras. $\mathfrak{g}$ is *semi-simple* if there are no Abelian invariant subalgebras besides $\{0\}$. Any semi-simple Lie algebra $\mathfrak{g}$ may be written as a direct sum of simple Lie algebras. The simple Lie algebras are the basic building blocks. The direct sum of two Lie algebras $\mathfrak{g}_1$ and $\mathfrak{g}_2$ is given by

$$\mathfrak{g}_1 \oplus \mathfrak{g}_2 = \{x_1 + x_2 | x_1 \in \mathfrak{g}_1, x_2 \in \mathfrak{g}_2\} \tag{B.6}$$

with Lie bracket $[x_1 + x_2, y_1 + y_2] = [x_1, y_1] + [x_2, y_2]$ where $x_1, y_1 \in \mathfrak{g}_1$ and $x_2, y_2 \in \mathfrak{g}_2$. Note that $\mathfrak{g}_1$ and $\mathfrak{g}_2$ are subalgebras of $\mathfrak{g}_1 \oplus \mathfrak{g}_2$ and that by construction these two subalgebras commute.

For a real Lie algebra $\mathfrak{g}$ we can also consider its *complex extension* or *complexification*. For a real Lie algebra, we consider only elements $x = x^a T_a$ with $x^a \in \mathbb{R}$. The complexification of a real Lie algebra amounts to allowing $x^a \in \mathbb{C}$. In other words, instead of considering the real vector space $\mathfrak{g}$ we consider $\mathfrak{g} \otimes_{\mathbb{R}} \mathbb{C}$.

There is a complete classification of complex simple Lie algebras due to Cartan in terms of four infinite series A_l, B_l, C_l and D_l, where l is an integer, and the exceptional cases F_2, G_4, E_6, E_7, E_8. For more details and for later reference see table B.1. The corresponding Lie algebras are the complexified versions of $\mathfrak{su}(n), \mathfrak{so}(n), \mathfrak{sp}(n)$ which we discuss in detail below as well as $\mathfrak{g}_2, \mathfrak{f}_4, \mathfrak{e}_6, \mathfrak{e}_7$ and $\mathfrak{e}_8$.

Killing form

Using the structure constants of the Lie algebra, $f_{ab}{}^c$, as defined in (B.4) we may define the Killing form κ_{ab} by

$$\kappa_{ab} = -f_{ac}{}^d f_{bd}{}^c. \tag{B.7}$$

Table B.1 Cartan classification of complex simple Lie algebras				
Cartan	Label	Name	Dimensions	Rank l
A_l	$\mathfrak{su}(n)$	special unitary	$n^2 - 1$	$n - 1$
B_l	$\mathfrak{so}(n)$ (n odd)	orthogonal	$n(n-1)/2$	$(n-1)/2$
C_l	$\mathfrak{sp}(n)$	symplectic	$n(n+1)/2$	$n/2$
D_l	$\mathfrak{so}(n)$ (n even)	orthogonal	$n(n-1)/2$	$n/2$
G_2	$\mathfrak{g}_2$	exceptional	14	2
F_4	$\mathfrak{f}_4$	exceptional	52	4
E_6	$\mathfrak{e}_6$	exceptional	78	6
E_7	$\mathfrak{e}_7$	exceptional	133	7
E_8	$\mathfrak{e}_8$	exceptional	248	8

Note that the Killing form κ is symmetric, $\kappa_{ab} = \kappa_{ba}$. The Killing form may be used to characterise *semi-simple* Lie algebras. A Lie algebra is semi-simple if and only if the determinant of the Killing form does not vanish, i.e. $\det \kappa_{ab} \neq 0$. Note that in this case we can define κ^{ab} to be the inverse matrix, $\kappa^{ab}\kappa_{bc} = \delta^a_c$. Using κ_{ab} and κ^{ab}, we may raise and lower indices of the structure constants, $f_{abc} = f_{ab}{}^d \kappa_{dc}$. Moreover, a real semi-simple Lie algebra is *compact* if its Killing form is negative definite.[1]

Cartan–Weyl form, simple roots and Dynkin diagrams

For a semi-simple Lie algebra $\mathfrak{g}$ there exists a maximal set of linearly independent generators h_i with $i = 1, \ldots, l$ which commute with each other, i.e.

$$[h_i, h_j] = 0 \qquad \text{for } i, j \in \{1, \ldots, l\}. \tag{B.8}$$

The generators h_i form the *Cartan subalgebra*. l is referred to as the rank of the Lie algebra. Note that for the infinite series A_l, B_l, C_l and D_l, and for the exceptional Lie algebras, the rank is given by the index.

It is useful to decompose the generators T_a of the semi-simple Lie algebra $\mathfrak{g}$ into $T_a = (h_i, e_\alpha)$, with h_i the generators of the Cartan subalgebra. In other words, any element of $x \in \mathfrak{g}$ may be written as $x = x^i h_i + x^\alpha e_\alpha$. The commutation relations of h_i with the e_α read

$$[h_i, e_\alpha] = \alpha_i e_\alpha, \tag{B.9}$$

while the commutation relations of e_α and e_β are given by

$$[e_\alpha, e_\beta] = n_{\alpha\beta} e_{\alpha+\beta} \tag{B.10}$$

in the case of $\alpha + \beta \neq 0$ and

$$[e_\alpha, e_{-\alpha}] = \alpha^i h_i \tag{B.11}$$

otherwise. $n_{\alpha\beta}$ are normalisation constants and the l-dimensional vectors α with components α_i are the *roots* of the Lie algebra $\mathfrak{g}$. This form of the semi-simple Lie algebra is referred to as the *Cartan–Weyl form*.

[1] Sometimes compact Lie algebras are defined to be real Lie algebras with negative semi definite Killing form. Any compact Lie algebra of this kind may be decomposed into compact simple Lie algebras and $\mathfrak{u}(1)$ algebras.

The roots α are l-dimensional vectors in the *weight space*. In this weight space we may define a scalar product by $(\alpha, \beta) = \sum_{i=1}^{l} \alpha_i \beta_i$. The roots satisfy the following rules:

- if α is a root, then $-\alpha$ is also a root,
- if α and β are roots, then $2(\alpha, \beta)/(\alpha, \alpha)$ is an integer,
- if α and β are roots, then $\gamma = \beta - 2(\alpha, \beta)/(\alpha, \alpha)\alpha$ is also a root.

The roots are characterised by the following properties. A root α is a *positive* root[2] if $\alpha_1 > 0$ and a null root if $\alpha_1 = 0$. Using just positive and null roots we can reconstruct all roots by using the first rule. This also implies that the number of positive roots is half the number of non-null roots. A *simple* root is a positive root which cannot be decomposed into a sum of two or more positive roots. The set of simple roots α_i is the building block for any positive root β in the sense that β can be written as $\beta = \sum n_i \alpha_i$ where n_i are non-negative integers. It turns out that we can classify all simple root systems. First we realise by using the scalar product $(\cdot, \cdot)$ that we can define an angle φ between two roots α and β by $\cos \varphi = (\alpha, \beta)/\sqrt{(\alpha, \alpha)(\beta, \beta)}$. The rules above imply that $\cos \phi$ and therefore φ can take only particular values,

$$(\cos \varphi)^2 \in \left\{ 0, \frac{1}{4}, \frac{1}{2}, \frac{3}{4}, 1 \right\}. \tag{B.12}$$

For two simple roots α and β, we have in addition $(\alpha, \beta) \le 0$ and therefore φ can only be $\pi/2, 2\pi/3, 3\pi/4$ or $5\pi/6$. We may represent this information in a pictorial way as follows:

- for each simple root α_i, we draw a circle;
- for each pair of simple roots α_i and α_j, we draw a connection depending on the angle φ between them
  - for $\varphi = \pi/2$, the circles are not connected,
  - for $\varphi = 2\pi/3$, we draw a single line between the circles,
  - for $\varphi = 3\pi/4$, we draw a double line between the circles,
  - for $\varphi = 5\pi/6$, we draw a triple line between the circles;
- double and triple lines connecting two roots α_i and α_j are oriented, i.e. we draw an arrow pointing to the shorter root which by definition has the smaller value for (α, α).

These rules give rise to the *Dynkin diagrams*. In general for a given root system, the Dynkin diagram is not connected, but consists of several copies of *connected Dynkin diagrams*. It turns out that we can classify all connected Dynkin diagrams. They are given by

---

[2] Note that the definition of positive or null roots depends on the frame of the weight space which we choose. A different frame leads to different positive or null roots.

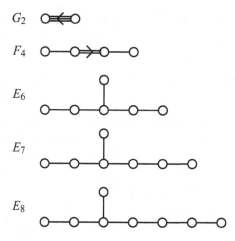

It is not a coincidence that we have used the same labels for the Dynkin diagrams and for the complex simple Lie algebras. In fact, complex simple Lie algebras have a simple root system and their associated Dynkin diagrams are connected and given by the above. The number l of A_l, B_l, C_l and D_l in the Dynkin diagrams is determined by the number of circles.

Enveloping algebra and Casimir operators

Starting with elements $T_a \in \mathfrak{g}$, where $\mathfrak{g}$ is a Lie algebra, we consider products of these generators, $T_{a_1} T_{a_2} \ldots T_{a_p}$ which do not have to be elements of the Lie algebra $\mathfrak{g}$. However, these products $T_{a_1} \ldots T_{a_p}$ as well as sums thereof form an algebra associated with $\mathfrak{g}$, the *enveloping algebra* of $\mathfrak{g}$.

An example of elements of the universal enveloping algebra are *Casimir operators*. By definition a Casimir operator C commutes with all elements $x \in \mathfrak{g}$, i.e. $[C, x] = 0$. A Casimir operator C_p of order p is of the form

$$C_p = \sum_{a_1, \ldots, a_p} f^{a_1 \ldots a_p} T_{a_1} T_{a_2} \ldots T_{a_p} \tag{B.13}$$

where $f^{a_1 \ldots a_p} \in \mathbb{C}$. Using the Killing metric κ_{ab}, it is straightforward to write down the quadratic Casimir operator C_2

$$C_2 = \kappa^{ab} T_a T_b \tag{B.14}$$

for a semi-simple Lie algebra. For a generic semi-simple Lie algebra the quadratic Casimir operator is not the only one. It turns out that a semi-simple Lie algebra $\mathfrak{g}$ of rank l has exactly l *Casimir operators* C.

Compact Lie algebras and groups

For a compact Lie group all the generators T_a of the corresponding Lie algebra $\mathfrak{g}$ may be chosen to be Hermitian and all representations of the compact Lie group are equivalent to unitary representations. Furthermore, for compact Lie groups $\mathcal{G}$ we can define a measure $d\mu(g)$, the so-called *Haar* measure, which is invariant under left and right multiplication with a fixed but arbitrary group element $g_0 \in \mathcal{G}$, i.e. $d\mu(g) = d\mu(gg_0) = d\mu(g_0g)$ and which is finite if integrated over $\mathcal{G}$,

$$\int_{\mathcal{G}} d\mu(g) < \infty. \tag{B.15}$$

B.1.2   Examples of Lie groups and Lie algebras

In the following we consider matrix groups which serve as examples of Lie groups and Lie algebras. The basic building blocks are the *general linear groups* $GL(n, \mathbb{R})$ and $GL(n, \mathbb{C})$ for real and complex numbers, respectively. $GL(n, \mathbb{R})$ is defined to be a set of real $n \times n$ matrices which can be inverted. $SL(n, \mathbb{R})$ is the set of those matrices in $GL(n, \mathbb{R})$ which have unit determinant. Similarly, there are the sets of matrices $GL(n, \mathbb{C})$ and $SL(n, \mathbb{C})$ which are the obvious generalisations to matrices with complex entries. All these examples are real or complex Lie groups. The dimension of the Lie group $GL(n, \mathbb{R})$ is n^2. $SL(n, \mathbb{R})$ has dimension $n^2 - 1$ since we impose that the determinant of the matrix has to be one. The real dimension of $GL(n, \mathbb{C})$ is $2n^2$, while $SL(n, \mathbb{C})$ has dimension $2(n^2 - 1)$.

The corresponding Lie algebras are denoted by $\mathfrak{gl}(n, \mathbb{R})$ and $\mathfrak{sl}(n, \mathbb{R})$ for the real case as well as $\mathfrak{gl}(n, \mathbb{C})$ and $\mathfrak{sl}(n, \mathbb{C})$ for the complex case. It turns out that $\mathfrak{gl}(n, \mathbb{R}) = \mathrm{Mat}(n, \mathbb{R})$, i.e. the Lie algebra of $GL(n, \mathbb{R})$ is given by any real $n \times n$ matrix. Indeed the exponential of A, defined by the power series

$$\exp(A) = \sum_{n=0}^{\infty} \frac{1}{n!} A^n, \tag{B.16}$$

is invertible since $\det(\exp(A)) = e^{\mathrm{Tr}(A)} > 0$. In particular we see that $\mathfrak{sl}(n, \mathbb{R})$ is given by all traceless real $n \times n$ matrices. Using the same arguments, the complex Lie algebras $\mathfrak{gl}(n, \mathbb{C})$ and $\mathfrak{sl}(n, \mathbb{C})$ are given by all complex $n \times n$ matrices, which in the case of $\mathfrak{sl}(n, \mathbb{C})$ also have to be traceless. The results are summarised in table B.2. For later reference, the Lie groups are also listed; these will be discussed in the following sections.

In the following we consider interesting subgroups of $GL(n, \mathbb{R})$ and $GL(n, \mathbb{C})$ which are relevant in the textbook. The complexified Lie algebras of these Lie groups will correspond to the infinite Cartan series A_l, B_l, C_l and D_l.

$O(n)$ and $SO(n)$

The *orthogonal group* $O(n)$ is the group of real *orthogonal* matrices $M \in GL(n, \mathbb{R})$ satisfying $M^t M = \mathbb{1}$. Here, M^t denotes the transposed matrix. Therefore the components $M^i{}_j$ of a real orthogonal matrix M satisfy

$$M^i{}_j \, \delta_{ik} \, M^k{}_l = \delta_{jl}. \tag{B.17}$$

From the definition $M^t M = \mathbb{1}$ it is obvious that the determinant of a real orthogonal matrix M is $\det(\mathrm{M}) = \pm 1$. A subgroup of $O(n)$ is the special orthogonal group $SO(n)$ whose

Table B.2 Examples of classical Lie groups and their Lie algebras

Lie group	Lie algebra	Real dimension	Generators of Lie algebra
$GL(n, \mathbb{R})$	$\mathfrak{gl}(n, \mathbb{R})$	n^2	$T_a \in \text{Mat}(n, \mathbb{R})$
$SL(n, \mathbb{R})$	$\mathfrak{sl}(n, \mathbb{R})$	$n^2 - 1$	$T_a \in \text{Mat}(n, \mathbb{R})$ and $\text{Tr}(T_a) = 0$
$SO(n)$	$\mathfrak{so}(n)$	$\frac{1}{2}n(n-1)$	$T_a \in \text{Mat}(n, \mathbb{R})$ and $(T_a)^{\text{t}} = T_a$, $\text{Tr}(T_a) = 0$
$GL(n, \mathbb{C})$	$\mathfrak{gl}(n, \mathbb{C})$	$2n^2$	$T_a \in \text{Mat}(n, \mathbb{C})$
$SL(n, \mathbb{C})$	$\mathfrak{sl}(n, \mathbb{C})$	$2(n^2 - 1)$	$T_a \in \text{Mat}(n, \mathbb{C})$ and $\text{Tr}(T_a) = 0$
$U(n)$	$\mathfrak{u}(n)$	n^2	$T_a \in \text{Mat}(n, \mathbb{C})$ and $(T_a)^{\dagger} = T_a$
$SU(n)$	$\mathfrak{su}(n)$	$n^2 - 1$	$T_a \in \text{Mat}(n, \mathbb{C})$ and $(T_a)^{\dagger} = T_a$, $\text{Tr}(T_a) = 0$

matrices additionally satisfy $\det(M) = 1$. $O(n)$ and $SO(n)$ are real Lie groups of dimension $\frac{1}{2}n(n-1)$.

Let us construct the associated Lie algebra $\mathfrak{so}(n)$ by considering orthogonal matrices M close to $\mathbb{1}$,

$$M = \mathbb{1} + i\alpha^a T_a + \mathcal{O}(\alpha^2), \tag{B.18}$$

where α is infinitesimal parameterising all orthogonal matrices close to $\mathbb{1}$. To satisfy $M^{\text{t}}M = \mathbb{1}$ to order α, we immediately see that T_a have to be symmetric $n \times n$ matrices. Furthermore, in order to satisfy $\det M = 1$, the matrices T_a have to be traceless. T_a are the generators of the Lie algebra and therefore we conclude that $\mathfrak{so}(n)$ consists of all symmetric traceless $n \times n$ matrices, since there are only $\frac{1}{2}n(n+1) - 1 = \frac{1}{2}n(n-1)$ linear independent symmetric traceless matrices.

In physics, *pseudo-orthogonal groups* $SO(p, q)$ (with $p + q = n$) also play an important role. Instead of $M^{\text{t}}M = \mathbb{1}$, the matrices M in $SO(p, q)$ satisfy

$$M^{\text{t}}\eta M = \eta, \qquad \text{where} \quad \eta = \begin{pmatrix} \mathbb{1}_p & 0 \\ 0 & -\mathbb{1}_q \end{pmatrix}. \tag{B.19}$$

The corresponding Lie algebras are denoted by $\mathfrak{so}(p, q)$. It turns out that $\mathfrak{so}(p, q)$ have (up to signs) the same commutation relations as $\mathfrak{so}(n)$. The signs can be rescaled into the generators if we are allowed to multiply them with complex numbers. In other words, the complexifications of the Lie algebras $\mathfrak{so}(p, q)$ and $\mathfrak{so}(n)$ (provided that $p + q = n$) are identical and thus $\mathfrak{so}(p, q)$ and $\mathfrak{so}(n)$ are just different real forms of the same complex Lie algebra. This complex Lie algebra corresponds to the infinite series B_l and D_l of the Cartan classification, depending on whether n, or equivalently $p + q$, is odd or even , see table B.1. In particular, for n odd, l is related to n by $(n-1)/2$ corresponding to B_l, while for n even, l is given by $l = n/2$ corresponding to D_l.

Note also that the Lie groups $SO(n)$ and $SO(p, q)$ with $p + q = n$ have different properties. In particular, while $SO(n)$ is a compact Lie group, $SO(p, q)$ is for $p, q > 0$ non-compact and therefore, as we will see later, it does not have non-trivial finite dimensional unitary representations.

$U(n)$ and $SU(n)$

The *unitary group* $U(n)$ consists of complex unitary $n \times n$ matrices with $M^\dagger M = \mathbb{1}$. The matrices in the special unitary group $SU(n)$ additionally satisfy $\det M = 1$. $U(n)$ and $SU(n)$ are examples of complex Lie groups. The real dimensions of $U(n)$ and $SU(n)$ are given by n^2 and $n^2 - 1$, respectively.

Linearising $M \in U(n)$ near the identity matrix $\mathbb{1}$ by

$$M(x) = \mathbb{1} + i\alpha^a T_a + \mathcal{O}(\alpha^2), \tag{B.20}$$

where α^a is real and infinitesimal, we see that the condition evaluated to order α implies $(T_a)^\dagger = T_a$, i.e. the Lie algebra $\mathfrak{u}(\mathfrak{n})$ corresponds to Hermitian $n \times n$ matrices. For the Lie algebra $\mathfrak{su}(n)$ we furthermore have to impose the tracelessness condition of T_a. Therefore $\mathfrak{u}(n)$ has n^2 generators while $\mathfrak{su}(n)$ has $n^2 - 1$.

Furthermore, as in the case of real Lie groups we can define the pseudo-unitary groups $SU(n, m)$ by $M^\dagger \eta M = \eta$ for η as in (B.19). The corresponding Lie algebras are denoted by $\mathfrak{su}(p, q)$. Again $\mathfrak{su}(n)$ and $\mathfrak{su}(p, q)$ with $p + q = n$ are two different real forms of the complexified Lie algebra which correspond to the infinite Cartan series A_l with $l = n - 1$.

$Sp(2n, \mathbb{R})$ and $Sp(2n, \mathbb{C})$

The symplectic groups $Sp(2n, \mathbb{R})$ and $Sp(2n, \mathbb{C})$ consist of $2n \times 2n$ real or complex symplectic matrices satisfying $M^\mathsf{T} J + JM = 0$, where J is a $2n \times 2n$ antisymmetric matrix of the form

$$J = \begin{pmatrix} 0 & \mathbb{1}_n \\ -\mathbb{1}_n & 0 \end{pmatrix}. \tag{B.21}$$

while all other entries are zero. Note that both $Sp(2n, \mathbb{R})$ and $Sp(2n, \mathbb{C})$ are non-compact. A compact symplectic group, denoted by $USp(2n)$ or just $Sp(2n)$, consists of matrices belonging to both $Sp(2n, \mathbb{C})$ and to $U(2n)$.

B.1.3   Representations of Lie algebras

Although the definition of the Lie algebra was very formal, all the examples considered so far were realised in terms of matrices. This is not a coincidence. Any finite-dimensional real or complex Lie algebra may be represented in terms of matrices. A *representation* of a Lie algebra $\mathfrak{g}$ is a map, denoted by D, which assigns each element $x \in \mathfrak{g}$ a real or complex $N \times N$ matrix,

$$D : \mathfrak{g} \to \mathrm{Mat}(N, \mathbb{R}) \text{ or } \mathrm{Mat}(N, \mathbb{C}), \tag{B.22}$$

such that the Lie bracket is preserved, i.e.

$$D([x, y]) = [D(x), D(y)] \qquad \text{for any } x, y \in \mathfrak{g}. \tag{B.23}$$

In other words, a representation of the Lie algebra satisfies the same commutation relations (B.4) as the Lie algebra $\mathfrak{g}$. Note that N is the dimension of the representation and does not have to coincide with the dimension dim $\mathfrak{g}$ of the Lie algebra. Sometimes, it is convenient to denote the representation by $\mathbf{N}$ indicating the dimension of the representation. For a *faithful* representation, the mapping D is injective.

Let us discuss a few important examples of such representations.

(1) The trivial or singlet representation assigns each element $x \in \mathfrak{g}$ the (1×1) matrix 0,

$$D : \mathfrak{g} \to \text{Mat}(1, \mathbb{R}) \quad \text{with} \quad D(x) = 0 \quad \text{for all} \quad g \in \mathfrak{g}. \tag{B.24}$$

Thus the singlet representation is one dimensional and not faithful.

(2) If the Lie algebra $\mathfrak{g}$ is given in terms of matrices, and therefore $\mathfrak{g} \subset \text{Mat}(n, \mathbb{R})$ or $\mathfrak{g} \subset \text{Mat}(n, \mathbb{C})$, we can simply take D to be the identity map in (B.22) and $N = n$. This is the fundamental or defining representation which is faithful by definition.

(3) Another important representation is the adjoint representation:

$$D : \mathfrak{g} \to GL(\mathfrak{g}), \tag{B.25}$$

where $GL(\mathfrak{g})$ is the general linear group of the vector space $\mathfrak{g}$ and the mapping D associates every element $x \in \mathfrak{g}$ via a linear mapping, denoted by $D(x)$. For fixed $x \in \mathfrak{g}$, this linear mapping may be defined by $D(x)(y) = [x, y]$ where $y \in \mathfrak{g}$. Using the definition of the structure constants (B.4), the adjoint representation is given in terms of its generators by

$$(T_a^{\text{adj}})^b{}_c = i f_{ac}{}^b. \tag{B.26}$$

Given two representations $D_{1\mathbf{N}}$ and $D_{2\mathbf{M}}$ of dimension N and M respectively, we may consider the *direct sum* of the two representations $\mathbf{N}$ and $\mathbf{M}$, denoted by $\mathbf{N} \oplus \mathbf{M}$

$$x_{\mathbf{N} \oplus \mathbf{M}} = \begin{pmatrix} x_{\mathbf{N}} \\ x_{\mathbf{M}} \end{pmatrix}, \qquad D_{\mathbf{N} \oplus \mathbf{M}} = \begin{pmatrix} D_{1\mathbf{N}} & 0 \\ 0 & D_{2\mathbf{M}} \end{pmatrix}. \tag{B.27}$$

Here, $x_{\mathbf{N}}$ denotes the elements of the vector space $\mathbb{R}^N$ or $\mathbb{C}^N$ associated with the representation and thus $x_{\mathbf{N} \oplus \mathbf{M}}$ is an element of the vector space $\mathbb{R}^{N+M}$ or $\mathbb{C}^{N+M}$ associated with the representation $\mathbf{N} \oplus \mathbf{M}$. To conclude, $D_{\mathbf{N} \oplus \mathbf{M}}$ as defined by (B.27) is a representation of dimension $N + M$.

Another way to build a new representation is to take the tensor product $\mathbf{N} \otimes \mathbf{M}$ given by

$$x_{\mathbf{N} \otimes \mathbf{M} \, \alpha\gamma} = x_{\mathbf{N}\alpha} \cdot x_{\mathbf{M}\gamma}, \qquad D_{\mathbf{N} \otimes \mathbf{M}} = D_{1\,\mathbf{N}} \otimes D_{2\mathbf{M}}. \tag{B.28}$$

Again, $x_{\mathbf{N} \otimes \mathbf{M}}$ are the elements of the $N \cdot M$ dimensional vector space $\mathbb{R}^{N \cdot M}$ or $\mathbb{C}^{N \cdot M}$. Thus $D_{\mathbf{N} \otimes \mathbf{M}}$ is an $N \cdot M$ dimensional representation.

In the following we are only interested in the basic building blocks, i.e. in those representations which cannot be written in a block diagonal form as in (B.27). The representations which can be brought into such a block diagonal form by a suitable transformation of the form $D(g) \mapsto P^{-1}D(g)P$ for fixed P are commonly called *reducible*. All other representations are called *irreducible*.

Let us draw a few important conclusions. For a simple Lie algebra, we know that for each a there has to exist at least a pair of indices b and c such that at least one of the structure constants is non-zero, i.e. $f_{bc}^a \neq 0$. Then tracing (B.4) implies that all the generators are traceless, i.e. $\mathrm{Tr}_{\mathbf{r}}(T_a) = 0$ in any representation $\mathbf{r}$ of the Lie algebra. Moreover, a compact Lie algebra may be represented by finite-dimensional Hermitian matrices. Therefore the generators have to satisfy $(T_a)^\dagger = T_a$.

Let us consider a simple Lie algebra of rank l. By definition the Lie algebra has an l-dimensional Cartan subalgebra and an associated root space. As explained above, the root space furnishes a vector space basis in terms of the simple roots α_i where $1 \leq i \leq l$, and an alternate basis in terms of the reciprocal basis vectors η^j where $1 \leq j \leq l$. The two bases are related by

$$\frac{\langle \eta^j, \alpha_i \rangle}{\langle \alpha_i, \alpha_i \rangle} = \delta_i^j. \tag{B.29}$$

An irreducible representation of the Lie algebra can be characterised via its highest weight vector Λ which can be decomposed into a linear combination of the reciprocal basis vectors with integer coefficients m_j with $1 \leq j \leq l$,

$$\Lambda = m_j \eta^j. \tag{B.30}$$

These coefficients m_j are referred to as *Dynkin labels*. The irreducible representation is therefore fixed by the values of m_j. We denote the representation by $[m_1, \ldots, m_l]$.

In order to determine the dimension of the representation and in particular to determine the decomposition of a product of representations into irreducible representations, it is convenient to associate a Young diagram. A *Young diagram* is a collection of rows of boxes, stacked vertically on top of each other. The rows are aligned to the left. For example, let us consider a Young diagram with l rows. We denote the number of boxes in the ith row by λ_i where $1 \leq i \leq l$. The integers λ_i satisfy

$$\lambda_1 \geq \lambda_2 \geq \lambda_3 \geq \cdots \geq \lambda_l \geq 0, \tag{B.31}$$

such that there are at least as many boxes in row 1 as in row 2, etc.

In particular, the irreducible representation given by the highest weight vector Λ or in other words by $[m_1, \ldots, m_l]$ maps to such a Young diagram in the following way: m_j is the number of columns with length j. For example, the representation $[1, 0, 0, \ldots, 0]$, i.e. with $m_1 = 1$ and $m_j = 0$ for $j \geq 2$ corresponds to the Young diagram with just one column of length one, while the representation $[3, 2, 0, \ldots, 0]$ corresponds to a Young diagram with three columns of length one and two columns of length two.

The dimension of the representation $[m_1, \ldots, m_l]$ and the decompositions of products of representations depend on the Lie algebra and are discussed in the next section for $\mathfrak{su}(N)$.

B.2 Representations of $\mathfrak{su}(N)$ and $\mathfrak{so}(N)$

B.2.1 Representations of $\mathfrak{su}(N)$

Let us recall that the Lie algebra $\mathfrak{su}(N)$ can be realised by Hermitian traceless $N \times N$ complex matrices. These matrices together with the associated vector space on which the matrices act define the fundamental representation which obviously has dimension N. Therefore the fundamental representation is denoted by $\mathbf{N}$. The generators of the fundamental representation are denoted by T_a. Note that the complex conjugated generators $-T_a^*$ satisfy the same commutation relations. These are the generators of the anti-fundamental representation which also has dimension N and thus is denoted by $\overline{\mathbf{N}}$ to distinguish it from the fundamental representation $\mathbf{N}$. From the two representations $\mathbf{N}$ and $\overline{\mathbf{N}}$ we can build other representations by taking the tensor product. For example, we may consider $\mathbf{N} \otimes \mathbf{N}$ and reduce it to its symmetric and antisymmetric parts, which we denote by $\mathbf{N} \otimes_S \mathbf{N}$ and $\mathbf{N} \otimes_A \mathbf{N}$, respectively.

In the same way, we find that the tensor product of the fundamental and the anti-fundamental representations may be decomposed into

$$\mathbf{N} \otimes \overline{\mathbf{N}} = \mathbb{1} \oplus \mathbf{adj}, \tag{B.32}$$

where $\mathbf{1}$ represents the singlet representation while $\mathbf{adj}$ is the adjoint representation.

All irreducible tensor product representations of $\mathfrak{su}(N)$ can be found by considering $\otimes_{k=1}^{f} \mathbf{N}$ subject to certain symmetry constraints, such as considering symmetric or anti-symmetric parts of $\mathbf{N} \otimes \mathbf{N}$ considered above. The information about (anti-)symmetrisation is conveniently stored in a Young diagram. A Young diagram consists of $f \in \mathbb{N}$ boxes, where each box represents a factor $\mathbf{N}$ of the tensor product $\otimes_{k=1}^{f} \mathbf{N}$. All of the boxes are aligned to the left and the length of the rows decreases monotonically. Such a Young tableau stores information about (anti-)symmetrisation among the representations $\mathbf{N}$ since representations $\mathbf{N}$ in rows are symmetrised while representations in columns are antisymmetrised. Therefore the representation $\mathbf{N} \otimes_S \mathbf{N}$ corresponds to the Young diagram with two boxes in a row while $\mathbf{N} \otimes_A \mathbf{N}$ is associated with a Young diagram with one column consisting of two boxes. Of course, there exist representations corresponding to more complicated Young diagrams of the form

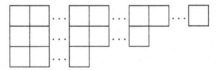

Given the Young diagram we can determine the dimension of the representation which is a ratio of two products of integers – and has to be an integer itself. The numerator is calculated as follows by inserting numbers into the Young diagram. Put N in the uppermost box of the diagram and increase this number by one if you go along a row. If you go along a column, you have to decrease the number by one. The numerator is just the product of all the numbers you obtain in this way. In order to determine the denominator, you have to consider the Hook length of each box of the Young diagram. The Hook length is given by the sum of all boxes to its right (including the box itself) plus all boxes in its own column below it. The denominator is then just the product of all Hook lengths. Alternatively, we may determine the dimension of a representation $[\lambda_1, \ldots, \lambda_{N-1}]$ of $\mathfrak{su}(N)$ from

$$\dim[\lambda_1, \ldots, \lambda_{N-1}] = \prod_{1 \le i < j \le N} \frac{\lambda_i - \lambda_j + j - i}{j - i}, \tag{B.33}$$

where $\lambda_N = 0$ by definition.

In this book, representations of $\mathfrak{su}(4)$ are of particular importance since $SU(4)$ is the R-symmetry group of $\mathcal{N} = 4$ super Yang–Mills theory. The rank of $\mathfrak{su}(4)$ is three and therefore all irreducible representations may be represented by Young diagrams with no more than three rows.

The tensor product representations of $\mathfrak{su}(4)$ are given by the Dynkin labels $[m_1, m_2, m_3]$ which are related to Young diagrams of the form

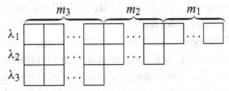

Thus, in the first line of the Young diagram we have $m_1 + m_2 + m_3$ boxes, in the second line $m_2 + m_3$ and in the third line m_3 boxes.

Thus identifying $\lambda_3 = m_3, \lambda_2 = m_2 + m_3$ and $\lambda_1 = m_1 + m_2 + m_3$ and using the formula (B.33), the dimension of the $\mathfrak{su}(4)$ representation with Dynkin labels $[m_1, m_2, m_3]$ is given by

$$\dim[m_1, m_2, m_3] = \frac{1}{12}(m_1 + 1)(m_2 + 1)(m_3 + 1)$$
$$\times (m_1 + m_2 + 2)(m_2 + m_3 + 2)(m_1 + m_2 + m_3 + 3). \quad \text{(B.34)}$$

The trivial representation is given by $[0, 0, 0]$ and has dimension one, hence it is referred to as $\mathbf{1}$. $\mathfrak{su}(4)$ has three fundamental representations, two of them are given by the Dynkin labels $[1, 0, 0]$ and $[0, 0, 1]$. Since $[0, 0, 1]$ is the complex conjugate of $[1, 0, 0]$ we denote the representation $[1, 0, 0]$ by $\mathbf{4}$ while $[0, 0, 1]$ corresponds to $\bar{\mathbf{4}}$. The adjoint representation of $\mathfrak{su}(4)$ – which can easily be constructed by considering the tensor product of $\mathbf{4}$ and $\bar{\mathbf{4}}$ – is given by $[1, 0, 1]$ and is referred to as $\mathbf{15}$. A list of important representations and their Dynkin labels is summarised in table B.3.

Starting from the representations $\mathbf{4}$ and $\bar{\mathbf{4}}$ we can construct other representations by taking appropriate tensor products of these two fundamental representations. For example, as discussed above, we can consider the tensor product $\mathbf{4} \otimes \mathbf{4}$ to obtain the symmetric

Table B.3 Representations of $\mathfrak{su}(4)$

Dynkin label	Representation
$[0, 0, 0]$	$\mathbf{1}$
$[1, 0, 0]$	$\mathbf{4}$
$[0, 0, 1]$	$\bar{\mathbf{4}}$
$[0, 1, 0]$	$\mathbf{6}$
$[2, 0, 0]$	$\mathbf{10}$
$[0, 0, 2]$	$\mathbf{10'}$
$[1, 0, 1]$	$\mathbf{15}$
$[0, 2, 0]$	$\mathbf{20'}$
$[1, 1, 0]$	$\mathbf{20}$

traceless representation $[2, 0, 0]$ of dimension **10** and the antisymmetric representation $[0, 1, 0]$ of $\mathfrak{su}(4)$ referred to as **6** which play an important role.

For generic irreducible representations of $\mathfrak{su}(4)_R$, denoted by $\mathbf{R_1}$ and $\mathbf{R_2}$, to decompose the product $\mathbf{R_1} \otimes \mathbf{R_2}$ into a sum of irreducible representations, there is the following recipe:

(1) Translate the Dynkin variables of $\mathbf{R_1}$ and $\mathbf{R_2}$ into the corresponding Young tableaux as explained above, i.e. calculate the λ_i from the m_i.

(2) Take the Young tableau of $\mathbf{R_1}$ and fill the first row with a's, the second row with b's and the third row with c's.

(3) Add the boxes filled with a to the Young tableau for $\mathbf{R_2}$, such that there is not more than one a in each column and such that the number of boxes in each row decreases. Repeat these steps with the boxes filled with b and c.

(4) If tableaux of the same form appear, and their labelling with a, b, c coincides, remove all tableaux of this form but one.

(5) Count the number n_a of a's to the right and above (also both to the right and above) for each box. Do the same for b and c. Only those Young tableaux are taken into account for which $n_a \geq n_b \geq n_c$ for each box.

(6) If there are four boxes in a column, remove this column. If this leads to a tableau in which there are no columns left, this tableau corresponds to the trivial representation **1**.

(7) Translate all diagrams back into Dynkin labels and the corresponding representations.

For the AdS/CFT correspondence, the representations $[0, k, 0]$ with $k \geq 2$ play an important role since they correspond to the 1/2 BPS operators of dimension $\Delta = k$ introduced in chapter 3. Similarly, the 1/4 BPS operators of dimension $\Delta = k + 2l$ correspond to the representations $[l, k, l]$, $l \geq 1$, and the 1/8 BPS operators of dimension $\Delta = k + 2l + 3m$, $m \geq 1$ correspond to the representations $[l, k, l + 2m]$.

Exercise B.2.1 By generalising the tensor products introduced above, obtain the general result (6.48) in chapter 6.

B.2.2 Representations of $\mathfrak{so}(N)$

For any orthogonal group $SO(p, q)$, and therefore in particular for the Lorentz group $SO(d - 1, 1)$, the metric $\eta_{\mu\nu}$ and the totally antisymmetric tensor are the only two invariant tensors. This implies that any tensor product of fundamental representations can be decomposed into irreducible representations of the metric and the antisymmetric tensor. Examples are discussed in chapter 3, where we also introduced spinorial representations. Here we demonstrate that faithful representations of the Clifford algebra exist by constructing the Dirac matrices explicitly.

Spinor representations of $\mathfrak{so}(m, n)$

Consider flat Minkowski spacetime with $d = m + n$ dimensions, of which n are timelike and m are spacelike. The Clifford algebra is given by

$$\{\gamma_\mu, \gamma_\nu\} = \gamma_\mu \gamma_\nu + \gamma_\nu \gamma_\mu = -2\eta_{\mu\nu} \mathbb{1}, \tag{B.35}$$

where
$$\eta = \mathrm{diag}(\underbrace{-,\ldots,-}_{n \text{ times}},\underbrace{+,\ldots,+}_{m \text{ times}}).$$

Let us first define the gamma matrices for the signature $(d,0)$, i.e. for Euclidean spacetimes. In the following, σ^i are the usual Pauli matrices and $\sigma^0 = -\mathbb{1}$. For Euclidean spacetime with even spacetime dimensions, the gamma matrices are given by

$$\gamma_{2\mu} = i \overset{\mu}{\underset{k=1}{\bigotimes}} \sigma_3 \otimes \sigma_1 \otimes \overset{d/2}{\underset{k=\mu+1}{\bigotimes}} \sigma_0, \qquad \gamma_{2\mu+1} = i \overset{\mu}{\underset{l=1}{\bigotimes}} \sigma_3 \otimes \sigma_2 \otimes \overset{d/2}{\underset{k=\mu+1}{\bigotimes}} \sigma_0, \qquad (\text{B.36})$$

where $\mu = 0,\ldots,\frac{d}{2} - 1$. They satisfy the Clifford algebra (B.35). In particular, (B.36) provides a $2^{d/2}$-dimensional complex representation of the Clifford algebra. For d odd, we define the same gamma matrices as above for

$$\gamma_{2\mu} = i \overset{\mu}{\underset{k=1}{\bigotimes}} \sigma_3 \otimes \sigma_1 \otimes \overset{\frac{d-1}{2}}{\underset{k=\mu+1}{\bigotimes}} \sigma_0, \qquad \gamma_{2\mu+1} = i \overset{\mu}{\underset{k=1}{\bigotimes}} \sigma_3 \otimes \sigma_2 \otimes \overset{\frac{d-1}{2}}{\underset{k=\mu+1}{\bigotimes}} \sigma_0,$$

$$\text{as well as} \quad \gamma_{d-1} = \overset{\frac{d-1}{2}}{\underset{i=1}{\bigotimes}} \sigma_3. \qquad (\text{B.37})$$

These then provide a $2^{(d-1)/2}$-dimensional complex representation of the Clifford algebra, as may be checked by verifying the anticommutation relations.

Let us now consider representations for the gamma matrices for spacetime with the signature (m,n) with $m+n = d$. These are obtained by omitting the factor i in the definition of the first n gamma matrices associated to signature $(d,0)$. By construction, the gamma matrices γ_μ have the following Hermiticity properties

$$\gamma_{\mu<n}^\dagger = \gamma_\mu \qquad \text{as well as} \qquad \gamma_{\mu\geq n}^\dagger = -\gamma_\mu, \qquad (\text{B.38})$$

i.e. $\gamma_\mu^\dagger = \gamma_\mu$ for timelike directions and $\gamma_\mu^\dagger = -\gamma_\mu$ for spacelike directions. Other representations are possible too. For example, equivalent representations are given by $\gamma' = U\gamma U^{-1}$ with U unitary. In even spacetime dimensions, all representations are equivalent in the sense defined in the preceding sentence, while in odd spacetime dimensions, there are two inequivalent representations. In even spacetime dimensions, the antisymmetrised gamma matrices

$$\gamma_{\mu_1\ldots\mu_n} = \gamma_{[\mu_1}\gamma_{\mu_2}\cdots\gamma_{\mu_n]} \qquad (\text{B.39})$$

form a basis of the $2^{d/2} \times 2^{d/2}$ matrices. In odd spacetime dimensions, the space of these matrices is spanned by all $\gamma_{\mu_1\ldots\mu_n}$ in (B.39) with $n \leq \frac{d-1}{2}$.

Spinor representations of $\mathfrak{so}(3,1)$

Here we review the conventions for spinors in four spacetime dimensions used in this book which agree mostly with those of the book of Wess and Bagger, referenced in chapter 3.[3] The algebra $\mathfrak{so}(3, 1)$ is isomorphic to $\mathfrak{su}(2) \oplus \mathfrak{su}(2)$. Left-handed Weyl spinors, transforming in the $(1/2, 0)$ representation of $\mathfrak{su}(2) \oplus \mathfrak{su}(2)$, have undotted indices, i.e. ψ_α with $\alpha = 1, 2$, while right-handed Weyl spinors transforming in $(0, 1/2)$ have dotted indices, $\bar{\psi}_{\dot\alpha}$ with $\dot\alpha = 1, 2$. Spinor indices are raised and lowered by antisymmetric tensors $\epsilon^{\alpha\beta}$ and $\epsilon^{\dot\alpha\dot\beta}$, i.e.

$$\psi^\alpha = \epsilon^{\alpha\beta}\psi_\beta, \qquad \psi_\alpha = \epsilon_{\alpha\beta}\psi^\beta, \tag{B.40}$$

$$\bar{\psi}^{\dot\alpha} = \epsilon^{\dot\alpha\dot\beta}\bar{\psi}_{\dot\beta}, \qquad \bar{\psi}_{\dot\alpha} = \epsilon_{\dot\alpha\dot\beta}\bar{\psi}^{\dot\beta}, \tag{B.41}$$

with

$$\epsilon^{12} = -\epsilon^{21} = -\epsilon_{12} = \epsilon_{21} = 1, \tag{B.42}$$

$$\epsilon^{\dot1\dot2} = -\epsilon^{\dot2\dot1} = -\epsilon_{\dot1\dot2} = \epsilon_{\dot2\dot1} = 1. \tag{B.43}$$

Moreover, we define matrices σ^μ with natural index structure $(\sigma^\mu)_{\alpha\dot\alpha} = \sigma^\mu{}_{\alpha\dot\alpha}$, where σ^μ is given by $\sigma^\mu = (-\mathbb{1}_2, \sigma^i)$ with Pauli matrices

$$\sigma^1 = \begin{pmatrix} 0 & 1 \\ 1 & 0 \end{pmatrix}, \qquad \sigma^2 = \begin{pmatrix} 0 & -i \\ i & 0 \end{pmatrix}, \qquad \sigma^3 = \begin{pmatrix} 1 & 0 \\ 0 & -1 \end{pmatrix}. \tag{B.44}$$

$\bar\sigma^\mu$ with natural index structure $(\bar\sigma^\mu)^{\dot\alpha\alpha} \equiv \sigma^{\mu\dot\alpha\alpha}$ is given by $\bar\sigma^\mu = (-\mathbb{1}_2, -\sigma^i)$ and is related to $\sigma^\mu{}_{\beta\dot\beta}$ by

$$(\bar\sigma^\mu)^{\dot\alpha\alpha} = \sigma^{\mu\dot\alpha\alpha} = \epsilon^{\dot\alpha\dot\beta}\epsilon^{\alpha\beta}\sigma^\mu{}_{\beta\dot\beta}. \tag{B.45}$$

The generators of Lorentz transformations, $\sigma^{\mu\nu}$ and $\bar\sigma^{\mu\nu}$, are defined by

$$(\sigma^{\mu\nu})_\alpha{}^\beta = \frac{i}{4}\left(\sigma^\mu{}_{\alpha\dot\alpha}\bar\sigma^{\nu\dot\alpha\beta} - \sigma^\nu{}_{\alpha\dot\alpha}\bar\sigma^{\mu\dot\alpha\beta}\right),$$

$$(\bar\sigma^{\mu\nu})^{\dot\alpha}{}_{\dot\beta} = \frac{i}{4}\left(\bar\sigma^{\mu\dot\alpha\alpha}\sigma^\nu{}_{\alpha\dot\beta} - \bar\sigma^{\nu\dot\alpha\alpha}\sigma^\mu{}_{\alpha\dot\beta}\right). \tag{B.46}$$

These generators satisfy

$$(\sigma^\mu\bar\sigma^\nu + \sigma^\nu\bar\sigma^\mu)_\alpha{}^\beta = -2\eta^{\mu\nu}\delta_\alpha{}^\beta,$$

$$(\bar\sigma^\mu\sigma^\nu + \bar\sigma^\nu\sigma^\mu)^{\dot\alpha}{}_{\dot\beta} = -2\eta_{\mu\nu}\delta^{\dot\alpha}{}_{\dot\beta}, \tag{B.47}$$

as well as the completeness relations

$$\mathrm{tr}(\sigma^\mu\bar\sigma^\nu) = -2\eta^{\mu\nu}, \qquad \sigma^\mu{}_{\alpha\dot\alpha}\bar\sigma_\mu{}^{\dot\beta\beta} = -2\delta_\alpha{}^\beta\delta_{\dot\alpha}{}^{\dot\beta}. \tag{B.48}$$

Contractions between spinor indices are taken according to the conventions

$$\psi\chi = \psi^\alpha\chi_\alpha = -\psi_\alpha\chi^\alpha = \chi^\alpha\psi_\alpha = \chi\psi,$$

$$\bar\psi\bar\chi = \bar\psi_{\dot\alpha}\bar\chi^{\dot\alpha} = -\bar\psi^{\dot\alpha}\bar\chi_{\dot\alpha} = \bar\chi_{\dot\alpha}\bar\psi^{\dot\alpha} = \bar\chi\bar\psi. \tag{B.49}$$

Since under complex conjugation $(\psi_\alpha)^* = \bar\psi_{\dot\alpha}$, we obtain from (B.49) that

$$(\chi\psi)^* = (\chi^\alpha\psi_\alpha)^* = \bar\psi_{\dot\alpha}\bar\chi^{\dot\alpha} = \bar\chi\bar\psi. \tag{B.50}$$

---

[3] Note however that the definition of γ_5 is different from Wess and Bagger.

In concrete calculations, the Fierz identities

$$\psi^\alpha \psi^\beta = -\frac{1}{2}\epsilon^{\alpha\beta}\psi\psi, \quad \psi_\alpha \psi_\beta = \frac{1}{2}\epsilon_{\alpha\beta}\psi\psi,$$
$$\bar{\psi}^{\dot\alpha}\bar{\psi}^{\dot\beta} = \frac{1}{2}\epsilon^{\dot\alpha\dot\beta}\bar{\psi}\bar{\psi}, \quad \bar{\psi}_{\dot\alpha}\bar{\psi}_{\dot\beta} = -\frac{1}{2}\epsilon_{\dot\alpha\dot\beta}\bar{\psi}\bar{\psi},$$

(B.51)

and the relations

$$\theta\sigma^\mu\bar{\theta}\theta\sigma^\nu\bar{\theta} = -\frac{1}{2}\theta\theta\bar{\theta}\bar{\theta}\eta^{\mu\nu},$$

$$(\theta\psi)(\theta\chi) = -\frac{1}{2}(\theta\theta)(\psi\chi), \quad (\bar{\theta}\bar{\psi})(\bar{\theta}\bar{\chi}) = -\frac{1}{2}(\bar{\theta}\bar{\theta})(\bar{\psi}\bar{\chi})$$

(B.52)

are useful. The Weyl and Dirac notations are related by

$$\Psi = \begin{pmatrix} \psi_\alpha \\ \bar{\psi}^{\dot\alpha} \end{pmatrix}, \quad \gamma^\mu = \begin{pmatrix} 0 & \sigma^\mu \\ \bar{\sigma}^\mu & 0 \end{pmatrix}.$$

(B.53)

In particular, γ^5 is given by

$$\gamma^5 = i\gamma^0\gamma^1\gamma^2\gamma^3 = \begin{pmatrix} \mathbb{1}_2 & 0 \\ 0 & -\mathbb{1}_2 \end{pmatrix}.$$

(B.54)

B.3 Superalgebra

B.3.1 Definition

Besides the usual generators of a Lie algebra – which we will call bosonic generators from now on – a superalgebra also consists of fermionic generators. While bosonic generators have grade 0, the fermionic generators have grade +1. To assign the grade to a product of fields we simply add the grades of the fields. In particular, the product of two fermionic generators is a bosonic generator, since $1 + 1 = 0 \pmod 2$, while the product of a bosonic and a fermionic generator is fermionic.

The (anti-)commutation relations of two generators, denoted by $\mathcal{O}_1$ and $\mathcal{O}_2$ with grades g_1 and g_2, is given by

$$[\mathcal{O}_1, \mathcal{O}_2\} = \mathcal{O}_1\mathcal{O}_2 - (-1^{g_1 g_2})\mathcal{O}_2\mathcal{O}_1.$$

(B.55)

In particular the notation of the bracket $[\cdot, \cdot\}$ suggests that it can be either a commutator or an anticommutator. To be precise, in the case of two fermionic generators the bracket is an anticommutator, while in all other cases it is a commutator.

Here we are interested in those superalgebras which contain at least the Poincaré algebra. For example, in four spacetime dimensions, the algebra should contain $\mathfrak{so}(3,1)$ which is isomorphic to $\mathfrak{su}(2,2)$.

B.3.2 Example: $\mathfrak{su}(2,2|\mathcal{N})$

The superalgebra $\mathfrak{su}(2,2|\mathcal{N})$ contains the bosonic generators $D, J_{\mu\nu}, P_\mu$ and K_μ as well as the fermionic generators $Q_\alpha^a, \bar{Q}_{a\dot\alpha}, S_{a\alpha}, \bar{S}_{\dot\alpha}^a$. Moreover, the internal R-symmetry generators

T and T^j form a $\mathfrak{u}(\mathcal{N})$ subalgebra. T generates the $U(1)$ factor if the R-symmetry group $U(\mathcal{N}) = U(1) \otimes SU(\mathcal{N})$, while the T^i generate $SU(\mathcal{N})$.

For completeness let us state the (anti-)commutation relations of $\mathfrak{su}(2,2|\mathcal{N})$.

$$[J_{\mu\nu}, J_{\rho\sigma}] = -i(\eta_{\mu\rho}J_{\nu\sigma} - \eta_{\mu\sigma}J_{\nu\rho} - \eta_{\nu\rho}J_{\mu\sigma} + \eta_{\nu\sigma}J_{\mu\rho}),$$

$$[J_{\mu\nu}, P_\rho] = i(\eta_{\mu\rho}P_\nu - \eta_{\nu\rho}P_\mu),$$

$$[J_{\mu\nu}, K_\rho] = i(\eta_{\mu\rho}K_\nu - \eta_{\nu\rho}K_\mu), \tag{B.56}$$

$$[J_{\mu\nu}, D] = 0,$$

$$[J_{\mu\nu}, Q^a_\alpha] = -(\sigma_{\mu\nu})_\alpha{}^\beta Q^a_\beta, \qquad [J_{\mu\nu}, \bar{Q}_{a\dot\alpha}] = -\epsilon_{\dot\alpha\dot\beta}(\bar\sigma_{\mu\nu})^{\dot\beta}{}_{\dot\gamma}\bar{Q}^{\dot\gamma}_a,$$

$$[J_{\mu\nu}, S^a_\alpha] = -(\sigma_{\mu\nu})_\alpha{}^\beta S_{a\beta}, \qquad [J_{\mu\nu}, \bar{S}^a_{\dot\alpha}] = -\epsilon_{\dot\alpha\dot\beta}(\bar\sigma_{\mu\nu})^{\dot\beta}{}_{\dot\gamma}\bar{S}^{a\dot\gamma}.$$

These commutation relations fix the representations of the Lorentz group under which the generators of $\mathfrak{su}(2,2|\mathcal{N})$ transform. While D is a scalar transforming in $(0,0)$ of $\mathfrak{su}(2)_L \times \mathfrak{su}(2)_R$, Q^a_α and $S_{a\alpha}$ are in $(1/2,0)$, i.e. they are left-handed Weyl spinors. $\bar{Q}_{a\dot\alpha}$ and $\bar{S}^a_{\dot\alpha}$ transform in $(0,1/2)$, while P_μ, K_μ are in $(1/2,1/2)$. Finally, $J_{\mu\nu}$ is in $(0,1) \oplus (1,0)$. The remaining commutators of the conformal algebra read

$$[P_\mu, P_\nu] = 0, \qquad [K_\mu, K_\nu] = 0$$

$$[K_\mu, P_\nu] = 2i(\eta_{\mu\nu}D - J_{\mu\nu}), \tag{B.57}$$

$$[D, P_\mu] = iP_\mu, \qquad [D, K_\mu] = -iK_\mu.$$

The anticommutation relations of $\mathfrak{su}(2,2|\mathcal{N})$ are given by

$$\{Q^a_\alpha, \bar{Q}_{b\dot\beta}\} = 2\sigma^\mu{}_{\alpha\dot\beta} P_\mu \delta^a{}_b, \qquad \{Q^a_\alpha, Q^b_\beta\} = \epsilon_{\alpha\beta}Z^{ab},$$

$$\{\bar{Q}_{a\dot\alpha}, \bar{Q}_{b\dot\beta}\} = \epsilon_{\dot\alpha\dot\beta}\bar{Z}_{ab}, \qquad\qquad Z^{ab} = (\bar{Z}^\dagger)_{ab},$$

$$\{S_{a\alpha}, \bar{S}^b{}_{\dot\beta}\} = 2\sigma^\mu{}_{\alpha\dot\beta}K_\mu\delta^b{}_a, \qquad \{S_{a\alpha}, S_{b\beta}\} = \{\bar{S}^a_{\dot\alpha}, \bar{S}^b_{\dot\beta}\} = 0,$$

$$\{Q^a_\alpha, S_{b\beta}\} = 2\epsilon_{\alpha\beta}\delta^a{}_b D - i(\sigma^{\mu\nu})_\alpha{}^\gamma\epsilon_{\gamma\beta}J_{\mu\nu}\delta^a{}_b - 4i\epsilon_{\alpha\beta}(\delta^a{}_b T + B^{ia}{}_b T^i), \tag{B.58}$$

$$\{\bar{Q}_{a\dot\alpha}, \bar{S}^b{}_{\dot\beta}\} = 2\epsilon_{\dot\alpha\dot\beta}\delta^b{}_a D - i(\bar\sigma^{\mu\nu})^{\dot\gamma}{}_{\dot\beta}\epsilon_{\dot\alpha\dot\gamma}J_{\mu\nu}\delta^b{}_a + 4i\epsilon_{\dot\alpha\dot\beta}(\delta^b{}_a T + B^i{}_a{}^b T^i),$$

$$\{Q^a_\alpha, \bar{S}^b{}_{\dot\beta}\} = \{\bar{Q}_{a\dot\alpha}, S_{b\beta}\} = 0.$$

Here, the $B^i{}_a{}^b$ are defined by the commutator of the spinor charges and the R-symmetry generators as given in (B.60) below. The commutation relations of Q and S with the conformal generators D, K_μ and P_μ read

$$[Q^a_\alpha, D] = -\frac{i}{2}Q^a_\alpha, \qquad\qquad [\bar{Q}_{a\dot\alpha}, D] = -\frac{i}{2}\bar{Q}_{a\dot\alpha},$$

$$[Q^a_\alpha, P_\mu] = 0, \qquad\qquad [\bar{Q}_{a\dot\alpha}, P_\mu] = 0,$$

$$[Q^a_\alpha, K^\mu] = i\sigma^\mu{}_{\alpha\dot\alpha}\bar{S}^{a\dot\alpha}, \qquad\qquad [\bar{Q}_{a\dot\alpha}, K^\mu] = -i\epsilon_{\dot\alpha\dot\beta}(\bar\sigma^\mu)^{\dot\beta\alpha}S_{a\alpha},$$

$$[D, S_{a\alpha}] = -\frac{i}{2}S_{a\alpha}, \qquad\qquad [D, \bar{S}^a_{\dot\alpha}] = -\frac{i}{2}\bar{S}^a_{\dot\alpha}, \tag{B.59}$$

$$[S_{a\alpha}, P^\mu] = -i\sigma^\mu{}_{\alpha\dot\alpha}\bar{Q}_{a\alpha}{}^{\dot\alpha}, \qquad\qquad [\bar{S}^a_{\dot\alpha}, P^\mu] = i\epsilon_{\dot\alpha\dot\beta}(\bar\sigma^\mu)^{\dot\beta\gamma}Q^a{}_\gamma,$$

$$[S_{a\alpha}, K_\mu] = 0, \qquad\qquad [\bar{S}^a_{\dot\alpha}, K^\mu] = 0.$$

Finally there are the commutation relations with the R-symmetry generators T and T^i. As we will see, these also explain the index structure of Q, S, $\bar{Q}$ and $\bar{S}$ as far as the Latin indices are concerned. We have

$$
\begin{aligned}
[\text{conformal}, T] = [\text{conformal}, T^i] &= 0, && [T, T] = 0, \\
[T^j, T^k] &= if^{jk}{}_l T^l, && [T^j, T] = 0, \\
[Q^a{}_{\dot\alpha}, T^i] &= B^{ia}{}_b Q^b{}_{\dot\alpha}, && [\bar{S}^a{}_{\dot\alpha}, T^i] = B^{ia}{}_b \bar{S}^b{}_{\dot\alpha}, \\
[Q^a{}_\alpha, T] &= \frac{4 - \mathcal{N}}{4\mathcal{N}} Q^a{}_\alpha, && [\bar{Q}_{a\dot\alpha}, T] = -\frac{4 - \mathcal{N}}{4\mathcal{N}} \bar{Q}_{a\dot\alpha}, \\
[S_{a\alpha}, T] &= -\frac{4 - \mathcal{N}}{4\mathcal{N}} S_{a\alpha}, && [\bar{S}^a{}_{\dot\alpha}, T] = \frac{4 - \mathcal{N}}{4\mathcal{N}} \bar{S}^a{}_{\dot\alpha}.
\end{aligned}
\tag{B.60}
$$

We note that $Q^a{}_\alpha$ and $S_{a\alpha}$ transform in conjugated representations of the R-symmetry group. The same applies to $\bar{Q}_{a\dot\alpha}$ and $\bar{S}^a{}_{\dot\alpha}$. This explains the chosen index structure with respect to the Latin indices.

For $\mathcal{N} < 4$ we note that the factors $(4 - \mathcal{N})/4\mathcal{N}$ may be scaled away by rescaling T. However, this is not necessary since for $\mathcal{N} = 4$, we have

$$
[Q^a{}_\alpha, T] = [\bar{Q}_{a\dot\alpha}, T] = [S_{a\alpha}, T] = [\bar{S}^a{}_{\dot\alpha}, T] = 0.
\tag{B.61}
$$

Therefore T, i.e. the $U(1)$ factor of the R-symmetry group, has the same action on all states of the supersymmetry multiplets, and therefore in particular also on states with opposite helicities. Therefore T has to vanish. For this reason, the R-symmetry group for $\mathcal{N} = 4$ is not $U(4)_R$ as expected from general arguments, but rather $SU(4)_R$. Moreover, as argued in chapter 3, we may write

$$
Z^{ab} = A^{ab}{}_j T^j.
\tag{B.62}
$$

The coefficients $B^{ia}{}_b$ and $B^i{}_a{}^b$ are related by

$$
B^{ia}{}_b = (B^i)^\dagger{}_a{}^b
\tag{B.63}
$$

and have to satisfy

$$
B^{ia}{}_b A_j{}^{bc} = -A_j{}^{ab} B^{ic}{}_b.
\tag{B.64}
$$

Conventions

Signature Whenever Lorentzian signature is considered, we use the mostly plus convention for the metric $(-, +, +, \ldots, +)$.

Spinors The Dirac algebra is taken to be

$$\{\gamma^\mu, \gamma^\nu\} = -2\eta^{\mu\nu}\mathbb{1}. \tag{C.1}$$

AdS metric When denoting the radial coordinate by z, the AdS metric is given by

$$ds^2 = \frac{L^2}{z^2}\left(\eta_{\mu\nu}dx^\mu dx^\nu + dz^2\right), \tag{C.2}$$

with the boundary at $z \to 0$. When denoting the radial variable by r, the AdS metric is given by

$$ds^2 = \frac{r^2}{L^2}\eta_{\mu\nu}dx^\mu dx^\nu + \frac{L^2}{r^2}dr^2, \tag{C.3}$$

with the boundary at $r \to \infty$. z and r are related by

$$z = \frac{L^2}{r}. \tag{C.4}$$

In chapter 9, we use the standard formulation

$$ds^2 = e^{A(r)}\eta_{\mu\nu}dx^\mu dx^\nu + dr^2, \tag{C.5}$$

with the boundary at $r \to \infty$ and the interior at $r \to -\infty$.

AdS black brane metric We use the same z and r variables as given above. The range for r is $[r_{\rm h}, \infty]$ and the range for z is $[0, z_{\rm h}]$. In five dimensions, we use a dimensionless variable u with metric

$$ds^2 = \frac{(\pi TL)^2}{u}\left(-f(u)\,dt^2 + d\vec{x}^2\right) + \frac{L^2}{4\,u^2 f(u)}\,du^2, \tag{C.6}$$

where $u = r_{\rm h}^2/r^2$ and $f(u) = 1 - u^2$. The boundary is located at $u = 0$, the horizon at $u = 1$.

For the spacetime indices in different dimensions, we use the notation shown in table C.1.

Table C.1 Index notation	
Ten-dimensional indices	$M, N, \ldots$
$(d + 1)$-dimensional bulk indices	Latin letters $m, n \ldots$
d-dimensional indices at boundary	Greek letters $\mu, \nu, \ldots$
$(d - 1)$-dimensional spatial indices at boundary	i, j, k

Index